Introduction to Stochastic Processes Using R

Sivaprasad Madhira · Shailaja Deshmukh

Introduction to Stochastic Processes Using R

Springer

Sivaprasad Madhira
Savitribai Phule Pune University
Pune, India

Shailaja Deshmukh
Department of Statistics
Savitribai Phule Pune University
Pune, Maharashtra, India

ISBN 978-981-99-5600-5 ISBN 978-981-99-5601-2 (eBook)
https://doi.org/10.1007/978-981-99-5601-2

This Springer imprint is published by the registered company Springer Nature Singapore Pte Ltd.
The registered company address is: 152 Beach Road, #21-01/04 Gateway East, Singapore 189721, Singapore

Paper in this product is recyclable.

Dedicated to Our Respected Teachers and Beloved Students Who Enriched Our Learning

Preface

A stochastic process is a probability model that describes the random evolution of a system over time and hence is useful in modeling a wide variety of phenomena. It quantifies the dynamic relationships of sequences of random events that occur in various disciplines, such as engineering, communication theory, biology, bioinformatics, medicine, computer science, social sciences, economics, demography, epidemiology, meteorology. Being a probability model, it starts with the axioms of probability which lead to a variety of fascinating results. If the model is appropriate, real problems can be solved by analysis within the model. However, models are almost always approximations of reality, hence the precise results within the model become approximations in the real world. In stochastic processes, many models are approximate but good models for the given situation.

In the present book, we introduce some basic stochastic processes, mainly Markov processes. We have adopted a rigorous mathematical approach toward the theoretical concepts. To absorb the notions easily, it is simultaneously supported by numerical computations and simulation studies with R software.

The book begins with a brief introduction to the basic framework of stochastic processes. Chapter 2 is devoted to the study of Markov chains, which is the simplest and the most important class of stochastic processes with discrete time parameter and countable state space. We elaborate in detail the theory of Markov chains such as classification of states, the first passage distributions and the concept of periodicity. In Chap. 3, we discuss the limiting behavior of a Markov chain and study associated stationary and long run distributions. Theory of Markov chains is a foundation to the theories developed in subsequent chapters. We study some typical Markov chains including random walks, gambler's ruin chain and Ehrenfest model in Chap. 4. Chapter 5 covers the detailed treatment of the simplest version of a branching process, known as Bienayme-Galton-Watson branching process. It is a special type of Markov chain and is widely used in many fields such as biological, social and engineering sciences.

Chapter 6 presents the theory of Markov chains when the time parameter is continuous. In such a process, it is observed at any time t, not just at transition times. A continuous time Markov chain is a Markov chain in which the times between transitions are positive-valued random variables and are modeled by a continuous distribution. The Markov property implies that the times spent in any state before making a transition are exponentially distributed.

Chapter 7 is concerned with a special type of a continuous time Markov chain with countable state space, known as Poisson process. It is the most frequently applied stochastic process in a variety of fields. Poisson process is defined in many ways. One approach involves the point process and the other involves the concept of a stochastic process with stationary and independent increments. Three definitions of a Poisson process are given and their equivalence is established. We discuss some operations on Poisson processes, such as superposition of Poisson processes and decomposition or thinning of a Poisson process. We also discuss two generalizations of a Poisson process, namely, non-homogeneous Poisson process and compound Poisson process.

In Chap. 8, we study some other examples of continuous time Markov chains. These include pure birth process, Yule Furry process, pure death process, birth and death process and its variations. These processes play a fundamental role in the theory and applications in queuing and inventory models, population growth, epidemiology and engineering systems. Chapters 2–8 are concerned with the stochastic processes in which time parameter is discrete as well as continuous but the state space is discrete. Chapter 9 is devoted to a stochastic process of historic importance, known as a Brownian motion process, which is a continuous time continuous state space process. We briefly discuss some of its important properties. A few extensions and variations of the Brownian motion process, such as Brownian bridge, geometric Brownian motion process, are covered. These have applications in finance, stock markets, inventory, etc.

The stochastic processes discussed in Chaps. 2–9 are all Markov processes. The last Chap. 10 presents discussion on a non-Markov process, known as a renewal process, which is another extension of a Poisson process.

We have tried to develop the subject with the help of a variety of completely worked out examples. Numerous illustrative examples of different difficulty levels are incorporated throughout each chapter to clarify the concepts. These illustrations and several remarks reveal the depth of the covered theory. For better assimilation of the notions contained in the book, conceptual, computational and multiple choice questions are included at the end of each chapter. Solutions to almost all the conceptual exercises are given in Appendix. We hope that solutions will serve as a motivation to students to solve the exercises and to digest the underlying concepts.

There is a vast literature on stochastic processes in the form of research papers and excellent books. The material discussed in the present book is covered in all these books. However, these books generally do not discuss the computational aspects. An important feature of this book is the augmentation of the theory with `R` software as a tool for simulation and computation. Over the years, we have noted that the concepts from stochastic processes are difficult for the students to grasp due to their abstract nature. To overcome this difficulty, it is essential that the theory and computations must go hand in hand. Hence, in the present book we have adopted an approach to explain theory along with the computations. We are convinced that the development of computational methods, along with the theory, will greatly contribute to better understanding of the theory, which in turn can provide more insight and help students to appreciate the beauty of the subject. We will be deeply rewarded, if the present book helps students to enhance their understanding and to enjoy the subject.

Nowadays R is a leading computational tool for statistics and data analysis. It is free and platform-independent and hence can be used on any operating system. Keeping up with the recent trend of using R software for statistical computations, we too have used it extensively in this book for illustrating the concepts. The R codes are provided in the last section of every chapter which are used to illustrate various concepts and procedures and to solve some examples. These will help the readers to understand the notions with ease, to reveal the hidden aspects of the procedures and to fulfill the need for visual demonstration of the concepts. The codes are deliberately kept simple, so that readers can understand the underlying theory with the minimal effort. In each chapter, computational exercises based on R software are included so as to provide a hands-on experience to students.

Using R, we illustrate how to obtain realizations of various stochastic processes, for a Markov chain how to verify that a state is persistent or transient, how to find the period of a state and how to compute a stationary distribution and a long run distribution. For a branching process, computation of extinction probability, graphically and algebraically with R are discussed. For a continuous time Markov chain with finite state space, various methods of computing the matrix of transition probability functions, given the generator matrix, are illustrated using R. For a renewal process, verification of some limit laws with R is presented.

The book has evolved out of the instructional material prepared for teaching a course on "Elementary Stochastic Processes" for several years at Savitribai Phule Pune University, formerly known as University of Pune and at Modern college, Pune. To some extent, the topics coincide with what we used to cover in the course. The main motive is to provide fairly thorough treatment of basic techniques, theoretically and computationally using R, so that the book will be suitable for self-study, particularly in the present era of online education. The style of the book is purposely kept conversational, so that the reader may feel the vicinity of a teacher.

The mathematical pre-requisites for this book are elementary probability, including conditional distributions and expectations, basic convergence concepts for a sequence of real numbers, some concepts from the linear algebra such as solution of a matrix equation, eigen values and eigen vectors, spectral decomposition of a matrix, etc. Familiarity with the properties of standard discrete and continuous distributions and some background of the probability theory would also be beneficial. These concepts form the mathematical foundation of stochastic processes. In addition, some basic knowledge of R software is desirable. In Chap. 1 for ready reference, we have added a section on R software, which gives a brief introduction to the basic concepts.

The intended target audience of the present book is mainly post-graduate students in a quantitative program, such as Statistics, Data Science, Finance and Mathematics. It will also provide sufficient background material for studying inference in stochastic processes. The book is designed primarily to serve as a textbook for a two-semester introductory course on stochastic processes in any post-graduate statistics program.

We are happy to acknowledge our indebtedness to our teachers Profs. S. R. Adke, B. L. S. Prakasa Rao and M. B. Rajarshi, who have laid the strong foundation of stochastic processes and influenced our understanding, appreciation and taste for the subject. We wish to express our deep gratitude to the R core development team and the authors of contributed packages, who have invested considerable time and effort in creating R as it is today. With the help of such a wonderful computational tool, it is possible to showcase the beauty of the theory of stochastic processes.

We thank Prof. T. V. Ramanathan, Head of the Department of statistics, Savitribai Phule Pune University, for providing the necessary facilities. We take this opportunity to acknowledge Nupoor Singh, editor of the Statistics section of Springer Nature and her team, for providing help from time to time and subsequent processing of the text to its present form.

We are deeply grateful to our family members for their constant support and encouragement. Last but not the least, we owe profound thanks to all our students whom we have taught during the last several years and who have been the driving force to take up this immense task. Their reactions and doubts in the class and our urge to make the theory crystal clear to them, compelled us to pursue this activity and to prepare various illustrations and exercises in this book.

All mistakes and ambiguities in the book are exclusively our responsibility. We would love to know any mistakes that a reader comes across in the book. Feedback in the form of suggestions and comments from colleagues and readers is most welcome.

Pune, India
February 2023

Sivaprasad Madhira
Shailaja Deshmukh

Contents

About the Authors

Sivaprasad Madhira has been Lecturer and Reader in the Department of Statistics, University of Pune, India; Professor of Statistics and Head, Department of Computer Science, Shivaji University, Kolhapur; Professor of Computer Applications, SIBER, Kolhapur; Director (MCA) and Director (IMED), Dean (Faculty of Management Studies) at Bharati Vidyapeeth University, Pune. He retired as Professor, Computer Applications, at IMED and Director-ICT at Bharati Vidyapeeth University, Pune. At present he is a visiting faculty at Savitribai Phule Pune University, formerly known as University of Pune. Prof. Prasad has started new departments/new programs and conducted quality courses in Statistics, Computer Science and Management. He was instrumental in setting up the statistics program and MCA program at Shivaji University; MCA program at SIBER, Kolhapur; and MCA program at Bharati Vidyapeeth University. During his career spanning 49 years of teaching and research, he has published many research papers in national and international journals of repute and supervised students for their PhD and MPhil degrees. He participated in numerous national and international conferences and delivered invited lectures.

Shailaja Deshmukh retired as Professor of Statistics from the Department of Statistics, Savitribai Phule Pune University, formerly known as University of Pune, India, and continues to serve there as a visiting faculty. She has taught around twenty five different theoretical and applied courses. She worked as a visiting professor at the Department of Statistics, University of Michigan, Ann Arbor, Michigan during 2009-2010 academic year. Her areas of interest are inference in stochastic processes, applied probability and analysis of microarray data. She has a number of research publications in various peer-reviewed journals. She has worked as an executive editor and as a chief editor of the Journal of Indian Statistical Association and is an elected member of the International Statistical Institute. She has authored five books-'Microarray Data: Statistical Analysis Using R' (jointly with Dr. Sudha Purohit) in

2007; ‘Statistics Using R’ (jointly with Dr. Sudha Purohit and Prof. Sharad Gore) in 2008; ‘Actuarial Statistics: An Introduction Using R’ in 2009; ‘Multiple Decrement Models in Insurance: An Introduction Using R’ in 2012; and ‘Asymptotic Statistical Inference: A Basic Course Using R’ in 2021; with the last two being published by Springer.

List of Figures

List of Tables

Chapter 1
Basics of Stochastic Processes

1.1 Introduction

Statistics is concerned with the collection of data on the characteristics under study, finding a suitable probability model for the data, its analysis and interpretation of the results. If the data are on a single characteristic or k related characteristics, then we have a variety of families of univariate and multivariate probability distributions to model the data. In many real-life situations, we come across data on a single characteristic or k related characteristics, which are observed over a period of time. For example, we observe various weather parameters of a city for a number of days with the aim to predict the future climatic conditions; in order to decide the premiums for the following year, an insurance company may collect data on a number of portfolios regarding frequency and amount of claims by policy holders for a certain time period; the management committee of a supermarket may collect data on the number of customer arrivals over a period of time and demand for particular items so as to take certain policy decisions. In all such situations which are governed by random mechanisms, we are confronted with countably infinite or uncountable collections of random variables. In order to answer questions regarding predictions and policy decisions, it is essential to identify a suitable probability model for countably infinite or uncountable collections of random variables. Stochastic processes are probability models to deal with such situations.

In the present chapter, we introduce the concept of a stochastic process and the related terms. We discuss the notion of a family of finite dimensional distribution functions associated with a stochastic process and the Kolmogorov extension theorem in Sect. 1.2. A brief introduction is given to some general classes of stochastic processes in Sects. 1.3 and 1.4. Particularly, Sect. 1.3 is devoted to stochastic processes with stationary and independent increments, while Sect. 1.4 is concerned with a brief discussion on stationary processes. R software [25] is used throughout the book to illustrate the various concepts in stochastic processes. Therefore, some preliminary

S. Madhira and S. Deshmukh, *Introduction to Stochastic Processes Using R*,
https://doi.org/10.1007/978-981-99-5601-2_1

information on R software is provided in Sect. 1.5, which helps in understanding the programs in subsequent chapters.

We begin with the definition of a stochastic process.

Definition 1.1.1 Stochastic Process: Suppose that T is a non-empty set and $\{X(t), t \in T\}$ is a family of random variables defined on the same probability space $(\Omega, \mathbb{A}, P)$. Then $\{X(t), t \in T\}$ is said to be a stochastic process.

Note that $X(t)$ can be a random vector; however, in this book, we assume it to be a univariate random variable. A stochastic process is also known as a random process or a random function or a random field. A stochastic process is also denoted by $\{X_t, t \in T\}$. We now define some terms associated with stochastic processes.

Definition 1.1.2 Index Set: Suppose $\{X(t), t \in T\}$ is a stochastic process defined on $(\Omega, \mathbb{A}, P)$. Then the set T is known as the index set of the stochastic process.

Definition 1.1.3 State Space: Suppose $\{X(t), t \in T\}$ is a stochastic process and S_t denotes the set of possible values of $X(t)$ for $t \in T$. Then $S = \bigcup_{t \in T} S_t$ is the set of possible values of all random variables $X(t),\ t \in T$. The set S is known as the state space of the stochastic process.

An element $i \in S$ is known as a state of the process. If for a given $t,\ X(t) = i$, then we say that at time t, the process is in state i. In this book, we study only two types of stochastic processes in which random variables $X(t)$ are either discrete $\forall\ t \in T$ or continuous $\forall\ t \in T$.

An element $t \in T$ is usually referred to as a time parameter, although it may not always indicate time. For example, if $\{W_n, n = 1, 2, \dots\}$ is a stochastic process, where W_n denotes the waiting of the n-th customer in a queue, then n is not time, but the serial number of the customer. If T is a singleton set then the collection reduces to only one random variable, and corresponding probability distribution is sufficient to decide probabilities of various events. If T is a finite set, we have a finite collection of random variables and we can probabilistically study this collection in terms of their joint distribution. Therefore, we assume that T is either a countably infinite set such as W, the set of all non-negative integers, or is uncountable, such as $\mathbb{R}$ or $\mathbb{R}^+$. When the index set T is W, the stochastic process $\{X(t), t \in T\}$ is denoted by $\{X_n, n \geq 0\}$.

A stochastic process $\{X(t), t \in T\}$ is basically a collection of functions $\{X(t, \omega), t \in T, \omega \in \Omega\}$ with two arguments t and ω. For each fixed t, it is a random variable on $(\Omega, \mathbb{A}, P)$. For each fixed $\omega \in \Omega$, $X(t, \omega)$ is a function of t with domain T. Such a function has a special name which is as follows.

Definition 1.1.4 Realization or Sample Path or Trajectory: Suppose $\{X(t), t \in T\}$ is a stochastic process defined on $(\Omega, \mathbb{A}, P)$. For each fixed $\omega \in \Omega$, $X(t, \omega)$, which is a function of $t \in T$, is known as a realization or a sample path or a trajectory of the stochastic process.

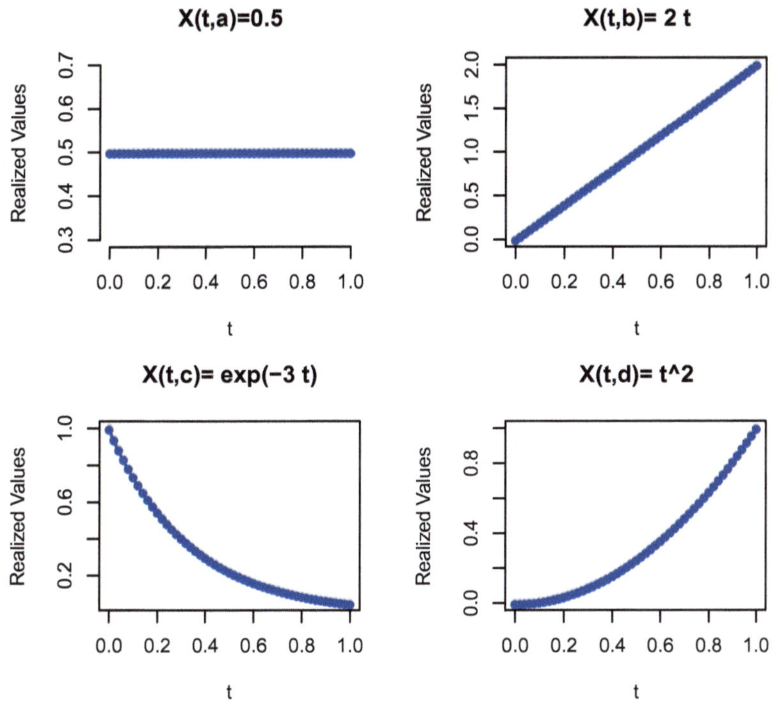

Fig. 1.1 Realization of a Stochastic Process

The following example illustrates the concept of a realization of a stochastic process.

Example 1.1.1 Suppose $\Omega = \{a, b, c, d\}$, $\mathbb{A} = \mathbb{P}(\Omega)$, a power set of Ω. Suppose P is a probability measure on $(\Omega, \mathbb{A})$ such that $P(\{a\}) = 0.2$, $P(\{b\}) = 0.1$, $P(\{c\}) = 0.4$ and $P(\{d\}) = 0.3$. For $0 \le t \le 1$, suppose

$$X(t, a) = 0.5, \quad X(t, b) = 2t, \quad X(t, c) = e^{-3t} \;\;\&\;\; X(t, d) = t^2.$$

Then, the family $\{X(t), 0 \le t \le 1\}$ of random variables is a stochastic process with index set $T = [0, 1]$ and state space $S = [0, 2]$. Since Ω has four elements, there will be four realizations of the process as shown in Fig. 1.1. □

Example 1.1.2 Suppose $\Omega = (0, \pi/2)$ and $\mathbb{A}$ is a Borel field of subsets of Ω. Suppose P is a probability measure on $(\Omega, \mathbb{A})$ such that $P((a, b)) = 2(b - a)/\pi$, $0 < a < b < \pi/2$. If $X(t, \omega) = t \times \tan(\omega)$, for $t \ge 0$, then the family $\{X(t), t \ge 0\}$ is a stochastic process with index set $T = [0, \infty)$ and state space $S = [0, \infty)$. In this example, the realizations are the rays in the first quadrant with initial point at $(0, 0)$. □

Realizations of various processes studied in this book and their R codes are given in the respective chapters.

Some illustrations of stochastic process are given below.

(i) Suppose X_n denotes the state of a system at epoch n. For example, X_n may be the maximum temperature of a city on day n, or rainfall in a city on day n. X_n may be the number of telephone calls on day n, number of transactions in a bank on n-th day, number of emails received on n-th day, number of claims received by an insurance company in n-th month and number of accidents at a certain junction during n-th month. In all these illustrations, $\{X_n, n \geq 0\}$ is a countably infinite collection of random variables. Such a collection $\{X_n, n \geq 0\}$ is known as a discrete time stochastic process, since we observe the system at a set of discrete time points, say at the end of every day, or every hour or every month.

(ii) On the other hand, if we observe the system continuously at all time points, we get a continuous time stochastic process. For example, suppose $X(t)$ denotes the state of a machine at time t; it will be either in a working condition or in a failed condition. $X(t)$ may be number of customers arriving at a mall during the time period $(0, t]$, length of a queue at a bill counter at time t, cash withdrawn from an ATM center during time period $(0, t]$ or $X(t)$ may be the value of a stock at time t. In all these illustrations, $\{X(t), t \geq 0\}$ is a continuous time stochastic process.

As the above examples indicate, an index set T can be either countably infinite or uncountable. Further, state space can be either countable (finite or countably infinite) or uncountable. Depending on the nature of the state space S and the index set T, there are four kinds of stochastic processes, as described below.

(i) If T is countably infinite and S is countable, then the process is said to be a discrete parameter discrete state space stochastic process: For example, status of a machine as either working or failed, at the beginning of day n.
(ii) If T is countably infinite and S is uncountable, then the process is said to be a discrete parameter continuous state space stochastic process: For example, maximum temperature of a city on day n.
(iii) If T is uncountable and S is countable, then the process is said to be a continuous parameter discrete state space stochastic process: For example, length of a queue at a bill counter in a mall at time t.
(iv) If T is uncountable and S is also uncountable, then the process is said to be a continuous parameter continuous state space stochastic process: For example, cash withdrawn from an ATM center in $(0, t]$.

We are interested in studying these stochastic processes to answer a variety of questions. For example, on the basis of climatic conditions for the first few days, we would like to predict the climate condition for the future. Information about the cash withdrawn from an ATM center will help the management about the cash to be kept at the ATM center. Queue length at a bill counter will help to decide whether the number of such counters should be increased or reduced. The data on the status of the machine at various time points will be useful to design maintenance policies.

Information on the number of claims and claim amounts for each claim in the case of vehicle insurance is useful to an insurance company to decide the premium for the next year.

As in the case of a random variable or a random vector, a stochastic process has a probability law or distribution, which is infinite dimensional. However, it is inconvenient to specify distributions over infinite dimensional spaces. The concept of a family of finite dimensional distribution functions associated with a stochastic process and some theorems related to it assure that the specification of distributions over infinite dimensional spaces is not necessary. Many probability computations can be carried out, if we know the family of finite dimensional distribution functions associated with a stochastic process. However, it is important to note that all probability computations cannot be performed based on the family of finite dimensional distributions. In the next section, we discuss the Kolmogorov compatibility conditions related to the family of finite dimensional distribution functions and their role in stochastic processes.

1.2 Kolmogorov Compatibility Conditions

It is known that a probability distribution of a univariate random variable is completely specified by its distribution function. Similarly, a probability distribution of a k-variate random vector is completely specified by its k-dimensional distribution function. We now address the issue of specifying a probability distribution or probability law of a stochastic process. It is specified by the family of finite dimensional distributions. We begin with its definition.

Definition 1.2.1 Family of Finite Dimensional Distribution Functions: Suppose that T is a non-empty set and that for $n \geq 1$ and $t_1 < t_2 < \ldots < t_n \in T$, $F_{t_1,t_2,\ldots,t_n}$ is a n-dimensional distribution function. Then the collection

$$\mathbb{F} = \{F_{t_1,t_2,\ldots,t_n} | t_1, t_2, \ldots, t_n \in T, \ n \geq 1\}$$

is said to be a family of finite dimensional distribution functions on T.

In order for such a family to define the probability law of a stochastic process, the family $\mathbb{F}$ must satisfy some conditions, known as Kolmogorov compatibility conditions. These conditions are as stated below.

Compatibility conditions: Suppose that $\mathbb{F}$ is a family of finite dimensional distribution functions on T.

(i) Consistency condition: The family $\mathbb{F}$ is said to satisfy the consistency condition if, for $n \geq 2$, $1 \leq m < n$ and $1 \leq i_1, i_2, \ldots, i_m \leq n$,

$$F_{t_{i_1},t_{i_2},\ldots,t_{i_m}}(x_{i_1}, x_{i_2}, \ldots, x_{i_m}) = \lim_{\substack{x_j \to \infty \\ j \neq i_1,\ldots,i_m}} F_{t_1,t_2,\ldots,t_{n-1},t_n}(x_1, x_2, \ldots, x_{n-1}, x_n).$$

That is, every m-dimensional marginal distribution obtained from $F_{t_1,t_2,\ldots,t_n}$ in $\mathbb{F}$ is again in $\mathbb{F}$, for all $1 \leq m < n$ and for all n.

(ii) Symmetry condition: The family $\mathbb{F}$ is said to satisfy the symmetry condition if for $n \geq 2$ and for every permutation $i_1, i_2, \ldots, i_n$ of $1, 2, \ldots, n$,

$$F_{t_{i_1},t_{i_2},\ldots,t_{i_n}}(x_{i_1}, x_{i_2}, \ldots, x_{i_n}) = F_{t_1,t_2,\ldots,t_n}(x_1, x_2, \ldots, x_n).$$

These two conditions, the consistency condition and the symmetry condition, together are known as Kolmogorov compatibility conditions.

The properties of a stochastic process are studied in terms of its family of finite dimensional distribution functions which is defined as follows.

Definition 1.2.2 Family of Finite Dimensional Distribution Functions of a Stochastic Process: Suppose $\{X(t), t \in T\}$ is a stochastic process defined on the probability space $(\Omega, \mathbb{A}, P)$. For $n \geq 1$ and $t_1, t_2, \ldots, t_n \in T$, suppose $F_{t_1,t_2,\ldots,t_n}$ is the joint distribution function of the random variables $X(t_1)$, $X(t_2)$, ..., $X(t_n)$ with respect to the probability measure P, defined by

$$F_{t_1,t_2,\ldots,t_n}(x_1, x_2, \ldots, x_n) = P\left[X(t_1) \leq x_1, X(t_2) \leq x_2, ..., X(t_n) \leq x_n\right],$$

$x_i \in \mathbb{R},\ i = 1, 2, \ldots, n$. Then the family $\{F_{t_1,t_2,\ldots,t_n} | t_1, t_2, \ldots, t_n \in T, n \geq 1\}$ is said to be the family of finite dimensional distribution functions of the stochastic process $\{X(t), t \in T\}$.

From this definition, it is clear why it is essential to define all the random variables on the same probability space. We now state an important theorem, which asserts that the family of finite dimensional distribution functions of the stochastic process $\{X(t), t \in T\}$ satisfies the compatibility conditions. Since P is a probability measure, the proof follows from the definition of a n-dimensional distribution function.

Theorem 1.2.1 *The family of finite dimensional distribution functions of a stochastic process $\{X(t), t \in T\}$ satisfies the Kolmogorov compatibility conditions.*

The converse of Theorem 1.2.1 is also true which we state without proof. For proof one may refer to Billingsley [4] or Doob [14]. The theorem is known as Kolmogorov extension theorem or Kolmogorov existence theorem or Kolmogorov consistency theorem. It is also known as the Daniell-Kolmogorov theorem, since it is independently proved by the English mathematician P. J. Daniell and the Russian mathematician A. N. Kolmogorov. The theorem guarantees that a collection of finite dimensional distributions, which satisfies the Kolmogorov compatibility conditions, defines a stochastic process.

Theorem 1.2.2 *Kolmogorov Existence (Extension) Theorem: Suppose T is a non-empty set and*

$$\mathbb{G} = \{G_{t_1,t_2,\ldots,t_n} | t_1 < t_2, \cdots < t_n \in T,\ \ n \geq 1\}$$

is a family of finite dimensional distribution functions. If the family $\mathbb{G}$ satisfies the Kolmogorov compatibility conditions, then there exists a stochastic process $\{X^*(t), t \in T\}$ defined on some probability space $(\Omega^*, \mathbb{A}^*, P^*)$ such that its family of finite dimensional distribution functions is given by $\mathbb{G}$.

Remark 1.2.1 (i) There can be many stochastic processes defined on the same probability space or different probability spaces with $\mathbb{G}$ as their family of finite dimensional distributions. We say that all such stochastic processes are identically distributed. (ii) Theorems 1.2.1 and 1.2.2 together convey that Kolmogorov compatibility conditions are necessary and sufficient, for the existence of a collection of random variables $\{X(t), t \in T\}$ to be a stochastic process.

The following example illustrates that a sequence of independent random variables satisfies the compatibility conditions and hence qualifies to be a stochastic process.

Example 1.2.1 Suppose $\{X_n, n \geq 1\}$ is a sequence of independent random variables with H_n as a distribution function of X_n. In view of independence, the joint distribution function $F_{t_1,t_2,\ldots,t_n}(x_1, x_2, \ldots, x_n),\ x_i \in \mathbb{R},\ t_i \in T = \{1, 2, \ldots, \}$ can be expressed as follows.

$$\begin{aligned}
&F_{t_1,t_2,\ldots,t_n}(x_1, x_2, \ldots, x_n)\\
&\quad= P[X_{t_1} \leq x_1, X_{t_2} \leq x_2, \ldots, X_{t_n} \leq x_n]\\
&\quad= P[X_{t_1} \leq x_1]P[X_{t_2} \leq x_2]\cdots P[X_{t_n} \leq x_n]\\
&\quad= \prod_{i=1}^{n} H_{t_i}(x_i)\,.
\end{aligned}$$

Thus, the family of finite dimensional distribution functions associated with $\{X_n, n \geq 1\}$ is expressible in terms of $H_i,\ i = 1, 2, \ldots, n$. We now examine whether it satisfies Kolmogorov compatibility conditions. Observe that $\forall\ n \geq 2$ and finite, $1 \leq m < n$ and $1 \leq i_1, i_2, \ldots, i_m \leq n$,

$$\begin{aligned}
&\lim_{\substack{x_j \to \infty \\ j \neq i_1,\ldots,i_m}} F_{t_1,t_2,\ldots,t_{n-1},t_n}(x_1, x_2, \ldots, x_{n-1}, x_n)\\
&= \lim_{\substack{x_j \to \infty \\ j \neq i_1,\ldots,i_m}} \prod_{j=1}^{n} H_{t_j}(x_j)\\
&= \prod_{j=i_1,\ldots,i_m} H_{t_j}(x_j) \lim_{\substack{x_j \to \infty \\ j \neq i_1,\ldots,i_m}} \prod_{j=1}^{n} H_{t_j}(x_j)\\
&= \prod_{j=i_1,\ldots,i_m} H_{t_j}(x_j), \quad \text{since the second factor is } 1\\
&= F_{t_{i_1},t_{i_2},\ldots,t_{i_m}}(x_{i_1}, x_{i_2}, \ldots, x_{i_m})\,.
\end{aligned}$$

Thus, the consistency condition is satisfied. Further, suppose $\{i_1, i_2, \ldots, i_n\}$ is a permutation of $\{1, 2, \ldots, n\}$. Then

$$\begin{aligned} F_{t_1,t_2,\ldots,t_n}(x_1, x_2, \ldots, x_n) &= \prod_{j=1}^{n} H_{t_j}(x_j) = \prod_{j=1}^{n} H_{t_{i_j}}(x_{i_j}) \\ &= F_{t_{i_1},t_{i_2},\ldots,t_{i_n}}(x_{i_1}, x_{i_2}, \ldots, x_{i_n}) \,. \end{aligned}$$

Thus, the symmetry condition is also satisfied. □

Remark 1.2.2 (i) From the above example it follows that, if $\{U_n, n \geq 1\}$ is a sequence of independent and identically distributed random variables that are uniformly distributed over (0, 1), then it is a stochastic process on some probability space. It can be shown that such a sequence exists on the probability space $((0, 1), \mathbb{A}, \lambda)$, where $\mathbb{A}$ is a Borel field of subsets of (0, 1) and λ is the Lebesgue measure. (ii) It may also be noted that, if $\{X(t), t \in T\}$ is a stochastic process, then any other family of random variables defined in terms of $X(t)$'s is also a stochastic process. Many stochastic processes are defined in this way. It is illustrated in Example 1.2.3.

We now give two non-trivial examples which illustrate the above two theorems. First, we state a useful theorem, known as the Daniell extension theorem. This theorem is about the consistency condition when the index set $T = \{1, 2, \ldots, \}$ and is implied in the work by Daniell [11].

Theorem 1.2.3 *Daniell Extension Theorem: Suppose* $T = \{1, 2, \ldots\}$ *and* $\mathbb{G}$ *is a family of finite dimensional distribution functions on* T. *Suppose* $\mathbb{G}^* = \{G_{1,2,\ldots,k}, k \geq 1\}$ *is a family of finite dimensional distribution functions. Then, (i) if* $\mathbb{G}$ *is consistent and* $\mathbb{G}^* \subset \mathbb{G}$, *then the family* $\mathbb{G}^*$ *is also consistent; (ii) if* $\mathbb{G}^*$ *is consistent, then there exists a unique* $\mathbb{G}$ *such that it is consistent and* $\mathbb{G}^* \subset \mathbb{G}$.

In view of the above theorem, when $T = \{1, 2, \ldots\}$, it is enough to verify the consistency of the subfamily $\mathbb{G}^*$ for the existence of a stochastic process $\{X(t), t \in T\}$ such that for every n, the joint distribution function of $X_1, \ldots, X_n$ is given by $G_{1,2,\ldots,n} \in \mathbb{G}^*$. The other finite dimensional distributions are derived from G.

Suppose $T = \{1, 2, 3, \ldots\}$ and $\mathbb{F}^* = \{f_{1,2,\ldots,n} | n \geq 1\}$, where for every $n \geq 1$, $f_{1,2,\ldots,n}$ is a n-dimensional probability density function. We observe that (i) If $\mathbb{F}^*$ is consistent, and if $\mathbb{F}$ is the family of all finite dimensional distributions derived from $\mathbb{F}^*$, then the family $\mathbb{F}$ is also consistent. Therefore, there exists a stochastic process having all its finite dimensional distributions derived from $\mathbb{F}^*$. (ii) If $\mathbb{F}^*$ is not consistent, then any family of finite dimensional distributions that contains $\mathbb{F}^*$ is also not consistent. In this case, there is no stochastic process on any probability space, with specified family of finite dimensional distributions, if it were to contain $\mathbb{F}^*$.

We now illustrate the Kolmogorov consistency theorem by considering some examples of $\mathbb{F}^*$.

Example 1.2.2 Suppose $T = \{1, 2, \ldots\}$. We consider the family of finite dimensional distribution functions on T such that for $n \geq 1$, the densities $\{f_{1,2,\ldots,n}; n = 1, 2, ..\}$ are given by

$$f_{1,2,\ldots,n}(x_1, x_2, ...x_n) = \prod_{k=1}^{n} g_k(x_k)$$

where g_k is a probability density function for each k. This family satisfies the compatibility conditions and hence there exists a stochastic process with this family as its family of finite dimensional distributions. □

Example 1.2.3 Suppose $\{X_n, n \geq 1\}$ is a sequence of independent and identically distributed random variables such that X_n follows an exponential distribution with scale parameter λ. Then from Example 1.2.2, $\{X_n, n \geq 1\}$ is a stochastic process. Suppose a sequence $\{S_n, n \geq 1\}$ is defined as $S_n = \sum_{i=1}^{n} X_i, \ \forall \ n$. Then $\{S_n, n \geq 1\}$ is a stochastic process, because each S_n is a function of the stochastic process $\{X_n\}$. In this case, we can verify that the family of finite dimensional distributions of the process $\{S_n\}$ satisfies the Kolmogorov compatibility conditions as follows. We observe that, for $n \geq 1$, the joint probability density function of $S_1, S_2, \ldots, S_n$ is given by

$$f_{1,2,\ldots,n}(s_1, s_2, ..., s_n) = \begin{cases} \lambda^n \exp\{-\lambda s_n\}, \text{ if } 0 \leq s_1 ... \leq s_n < \infty \\ 0, \qquad \text{otherwise,} \end{cases}$$

from which the consistency follows. Hence, the family $\mathbb{F} = \{F_{i_1,i_2,\ldots,i_k}, k \geq 1\}$, which is the family of finite dimensional distributions of the process $\{S_n, n \geq 1\}$ is also consistent. This process arises in the context of a Poisson process which we study in Chap. 7. Another important observation in this case is that the conditional distribution of S_n given that $S_1 = s_1, S_2 = s_2, \ldots, S_n = s_n$ is given by

$$f_{S_n \big| S_1=s_1,\ldots,S_{n-1}=s_{n-1}}(s_n) = \begin{cases} \lambda \exp\{-(s_n - s_{n-1})\}, \text{ if } s_{n-1} < s_n < \infty \\ 0, \qquad \text{otherwise} \end{cases}$$

which depends only on s_{n-1} but not on $s_1, s_2, \ldots, s_{n-2}$. A stochastic process that has this property is known as a "Markov process". □

The following example generalizes the above result.

Example 1.2.4 Suppose $\{X_n, n \geq 1\}$ is a sequence of independent non-negative random variables with distribution function G_n. A sequence $\{S_n, n \geq 1\}$ is defined as $S_n = \sum_{i=1}^{n} X_i, \ \ \forall \ \ n$. Then, $\{S_n, n \geq 1\}$ is a stochastic process, because each S_n is a function of the stochastic process $\{X_n\}$. In this case, we can directly verify that the finite dimensional distributions of the process $\{S_n\}$ satisfy the Kolmogorov compatibility conditions. Suppose that g_n is the probability density function of X_n. Then, the joint probability density function of $S_1, S_2, \ldots, S_n$ is given by

$$f_{1,2,\dots,n}(s_1,\dots,s_n) = \begin{cases} g_1(s_1)\prod_{k=1}^{n} g_k(s_k - s_{k-1}), & \text{if } 0 \le s_1 < \cdots < s_n < \infty \\ 0, & \text{otherwise.} \end{cases}$$

This family is consistent and hence all other finite dimensional distributions can be consistently derived from $\mathbb{F}^*$. Therefore, by the Kolmogorov consistency theorem, there exists a stochastic process on some probability space with $\mathbb{F}^*$, as its finite dimensional distributions and all other finite dimensional distributions can be consistently derived from $\mathbb{F}^*$. Observe that in this case also the conditional density of S_n given $S_1, S_2, \dots, S_{n-1}$ depends only on S_{n-1} and hence this stochastic process is also a "Markov process". □

In the next example, we present a family of finite dimensional distributions $\mathbb{G}$ which violates the Kolmogorov compatibility conditions, and hence there is no stochastic process with $\mathbb{G}$ as its family of finite dimensional distributions.

Example 1.2.5 Suppose $T = \{1, 2, \dots, \}$. We consider the family of finite dimensional distribution functions on T such that for $n \ge 1$, the densities $\{f_{1,2,\dots,n};\ n = 1, 2, ..\}$ are given by

$$f_{1,2,\dots,n}(x_1, x_2, \dots x_n) = \begin{cases} n!\exp\{-\sum_{j=1}^{n} x_j\}, & \text{if } 0 \le x_1 \le x_2 \dots \le x_n < \infty \\ 0, & \text{otherwise.} \end{cases}$$

This family is not consistent, since by integrating out x_n, the resulting density function is not the same as $f_{1,2,\dots,n-1}$ as defined above and hence there is no stochastic process with the above family as its family of finite dimensional distributions. Observe that $f_{1,2,\dots,n}$ is the joint probability density function of the order statistics $X_{(1)}, X_{(2)}, \dots, X_{(n)}$, when X_i has exponential distribution with mean 1, $i = 1, 2, \dots, n$. □

Another example of an inconsistent family of finite dimensional distributions is the following.

Example 1.2.6 Suppose $\mathbb{F}^*$ is given by the specifications

$$f_{1,2,\dots,n}(x_1, x_2, \dots x_n) = \begin{cases} n!, & \text{if } 0 \le x_1 \le x_2 \dots \le x_n < 1 \\ 0, & \text{otherwise.} \end{cases}$$

This family is also not consistent and hence there is no probability space on which a stochastic process can be defined that has above $\mathbb{F}^*$, as its family of finite dimensional distributions. In this case also, $f_{1,2,\dots,n}$ is the joint probability density function of the order statistics $X_{(1)}, X_{(2)}, \dots, X_{(n)}$, when X_i has uniform $U(0, 1)$ distribution, $i = 1, 2, \dots, n$. □

The measure-theoretic approach to stochastic processes starts with a probability space and defines a stochastic process as a family of functions on this probability space. However, in many applications the starting point is the family of finite dimensional distribution functions of the stochastic process. Kolmogorov extension theorem states that if the family of finite dimensional distribution functions satisfies the consistency requirements, one can always identify a probability space on which the process is defined. However, it is rarely specified explicitly.

We can find the probabilities of events of interest related to a stochastic process, if we can find the family of finite dimensional distribution functions of the stochastic process. Thus, the main problem one needs to address is how to specify such joint distributions. In the simplest case, where the sequence $\{X_n, n \geq 1\}$ is a sequence of independent random variables, the joint distribution is the product of marginal distributions, as shown in Example 1.2.1. But in most of the practical situations, we come across collections of dependent random variables. If $X(t), t \in T$ are not independent, we need information about the type of dependence among these random variables and the nature of dependence can be of many types.

A class of stochastic processes, known as Markov processes, specifies a simple yet rich enough dependence structure, with which it is easy to find the family of finite dimensional distribution functions and carry out a variety of probabilistic computations. We give a general definition of a Markov process below. The precise definitions with different types of S and T are given in respective chapters.

Definition 1.2.3 Markov Process: Suppose $\{X(t), t \in T\}$ is a stochastic process. It is said to be a Markov process, if $\forall\, t \in T$, the conditional distribution of $X(t)$ given $X(u)$, $\forall\, u \leq s < t$, is the same as the conditional distribution of $X(t)$ given $X(s)$, or $\forall\, n \geq 1$ and $0 < t_1 < \cdots < t_n \in T$, the conditional distribution of $X(t_n)$ given $X(t_{n-1}), X(t_{n-2}), \ldots, X(t_1)$ is the same as the conditional distribution of $X(t_n)$ given $X(t_{n-1})$.

Thus roughly speaking in a Markov process, the conditional distribution of the future, that is $X(t)$, given the entire past, that is $X(u)$, $\forall\, u \leq s < t$, is the same as the conditional distribution of the future $X(t)$, given immediate past $X(s)$, which is referred to as present. Such a property is known as a Markov property. Hence, for a Markov process, the knowledge of the conditional distributions of $X(t)$ given $X(s)$ for $s < t \in T$ is sufficient to find the family of finite dimensional distribution functions. If the conditional distribution of $X(t)$ given $X(s)$ depends on $t - s$ and not separately on t and s, then the Markov process is known as a time homogeneous process. In the present book, we study some Markov processes with discrete and continuous time parameters, when state space is either countable or uncountable. As already noted, the stochastic processes in Examples 1.2.3 and 1.2.4 are Markov processes.

In Chaps. 2, 3, 4, 5, 6, 7, 8 and 9, we present the theory and applications of various Markov processes. However, discrete time and continuous time auto-regressive processes are not discussed in the present book. Interested readers may refer to Davis [5] and Chatfield and Xing [7].

Another general class of stochastic processes in which the family of finite dimensional distribution functions is completely specified by the distribution of $X(1)$ is the class of stochastic processes with stationary and independent increments. We prove that such a stochastic process is a Markov process. The next section is devoted to a detailed study of stochastic processes with stationary and independent increments.

1.3 Stochastic Processes with Stationary and Independent Increments

Processes with stationary and independent increments are also known as Levy processes. These are continuous time analogues of random walks. Just as random walks provide simple examples of stochastic processes in discrete time, Levy processes provide key examples of stochastic processes in continuous time and provide ingredients for building continuous time stochastic models, particularly in financial mathematics, dam models, etc. Levy processes were first introduced in financial econometrics in 1963, as a model for cotton prices. Since then, a variety of models based on Levy processes have been proposed as models for asset prices and tested on empirical data. Recently, Levy processes and other stochastic processes with jumps have acquired increasing popularity in risk management, option pricing, dam models, etc. General properties of Levy processes are described in Steutel and Harn [29].

Suppose $\{X(t), t \in T\}$ is a stochastic process with state space S. Both the state space S and the index set T can be discrete or continuous. For a stochastic process $\{X(t), t \in T\}$, an increment is the difference in the random variables at two time points, say s and t. For $s < t$, the increment from time s to time t is the difference $X(t) - X(s)$. A process is said to have independent increments if increments over disjoint time intervals are independent random variables. A formal definition is given below.

Definition 1.3.1 Stochastic Process with Independent Increments: A stochastic process $\{X(t), t \in T\}$ is said to be a process with independent increments, if for every $k \geq 2$ and for any $t_0, t_1, t_2, \ldots, t_k \in T$ such that $t_0 < t_1 < t_2 < \cdots < t_k$, the increments $X(t_i) - X(t_{i-1}), i = 1, 2, \ldots, k$ are independent random variables.

Another condition on the distribution of the increment leads to the following definition of a process with stationary increments.

Definition 1.3.2 Stochastic Process with Stationary Increments: A stochastic process $\{X(t), t \in T\}$ is said to be a process with stationary increments, if for every $s < t \in T$, the probability distribution of the increment $X(t) - X(s)$ depends only on $(t - s)$, but not on s and t separately.

Thus, if a process $\{X(t), t \in T\}$ has stationary increments, then the increments $X(t) - X(s)$ and $X(t - s) - X(0)$ are identically distributed random variables. Further, the distribution of $X(t_1) - X(s_1)$ is the same as the distribution of $X(t_2) - X(s_2)$ if $t_1 - s_1 = t_2 - s_2$, where the intervals $(s_1, t_1]$ and $(s_2, t_2]$ may overlap.

Combining these two definitions, we now define a process with stationary and independent increments.

Definition 1.3.3 Stochastic Process with Stationary and Independent Increments: A stochastic process $\{X(t), t \in T\}$ is said to be a process with stationary and independent increments, if

(i) $\forall\ k \geq 2$ and any $t_0, t_1, t_2, \ldots, t_k \in T$ such that $t_0 < t_1 < \ldots < t_k$, the increments $X(t_i) - X(t_{i-1}),\ i = 1, 2, \ldots, k$ are independent random variables and
(ii) $\forall\ s < t$, the probability distribution of the increment $X(t) - X(s)$ depends only on $(t - s)$, but not on s and t separately.

A Poisson process to be discussed in Chap. 7 is a continuous time discrete state space stochastic process with stationary and independent increments, while a Brownian motion process in Chap. 9 is a continuous time continuous state space stochastic process with stationary and independent increments.

We now state and prove some properties of the process with stationary and independent increments, when the index set $T = [0, \infty)$, state space $S = I$, the set of integers, and when $X(0) = 0$. These remain valid with appropriate changes, even if T is countably infinite and S is uncountable.

Suppose $\{X(t), t \geq 0\}$ is a process with independent increments with state space $S = I$ and $X(0) = 0$. Then for $s < t$ and for all $i, j \in S$,

$$\begin{aligned} P[X(s) = i, X(t) = j] &= P[X(s) - X(0) = i, X(t) - X(s) = j - i] \\ &= P[X(s) - X(0) = i]P[X(t) - X(s) = j - i] \\ &= P[X(s) = i]P[X(t) - X(s) = j - i]. \end{aligned}$$

Hence,

$$P_{ij}(s, t) = P[X(t) = j \big| X(s) = i] = P[X(t) - X(s) = j - i].$$

$P_{ij}(s, t)$ is known as the transition probability function of the process $\{X(t), t \geq 0\}$. Thus, the conditional distribution of $X(t)$ given $X(s)$ is given by $P[X(t) - X(s) = j - i]$. As a consequence, a specification of $P[X(t) - X(s) = j - i]$ for all $s, t \geq 0$ and for all $i, j \in S$ and the distribution of $X(s)$ for all s determines the joint distribution of $X(s)$ and $X(t)$, and all the finite dimensional distributions. With the assumption $X(0) = 0$, the distribution of $X(s)$ is the same as the distribution of $X(s) - X(0)$. Hence, it is enough to specify $P[X(t) - X(s) = j - i]$ for all $s, t \geq 0$ and for all $i, j \in S$.

Along with the independence of increments, if the increments are stationary, then we have further simplification for the transition probability function, as shown below.

Suppose $\{X(t), t \geq 0\}$ is a process with stationary and independent increments and $X(0) = 0$. Then for $s < t$ and for all $i, j \in S$,

$$\begin{aligned} P[X(s) = i, X(t) = j] &= P[X(s) - X(0) = i]P[X(t) - X(s) = j - i] \\ &= P[X(s) = i]P[X(t - s) - X(0) = j - i] \\ &= P[X(s) = i]P[X(t - s) = j - i]. \end{aligned}$$

Thus, the transition probability function is given by

$$P_{ij}(s,t) = P[X(t) = j \big| X(s) = i] = P[X(t-s) = j-i].$$

Hence, the conditional distribution of $X(t)$ given $X(s)$ for $s < t$, that is, the transition probability function $P_{ij}(s,t)$, is specified by the marginal distribution of $X(t-s)$. Further, it is to be noted that it is a function of $t-s$ and does not depend on t and s separately which implies that the transition probability function is a stationary transition function. Thus, with $t-s = u$

$$P_{ij}(u) = P[X(s+u) = j \big| X(s) = i] = P[X(u) = j-i] = P_{j-i}(u), \quad \text{say.} \tag{1.3.1}$$

Thus, a specification of $P_i(u) = P[X(u) = i]$, for all $u > 0$ and $i \in S$, determines the transition probability function and the family of finite dimensional distributions of the process with stationary and independent increments. This property is illustrated in the following example.

Example 1.3.1 Suppose $\{X(t), t \geq 0\}$ is a process with stationary and independent increments with $X(0) = 0$. Then for $0 < t_1 < t_2 < t_3$,

$$\begin{aligned}
&P[X(t_1) = i, X(t_2) = j, X(t_3) = k] \\
&= P[X(t_1) = i, X(t_2) - X(t_1) = j-i, X(t_3) - X(t_2) = k-j] \\
&= P[X(t_1) - X(0) = i]P[X(t_2) - X(t_1) = j-i]P[X(t_3) - X(t_2) = k-j] \\
&= P[X(t_1) = i]P[X(t_2 - t_1) = j-i]P[X(t_3 - t_2) = k-j] \\
&= P_i(t_1)P_{j-i}(t_2 - t_1)P_{k-j}(t_3 - t_2).
\end{aligned}$$

Thus, the joint distribution is determined by the one-dimensional distributions. □

From the above example, we note that the conditional probability $P[X(t_3) = k \big| X(t_2) = j, X(t_1) = i]$ is given by $P_{k-j}(t_3 - t_2)$. Thus, the conditional distribution of $X(t_3)$ given $X(t_2)$ and $X(t_1)$ for $0 < t_1 < t_2 < t_3$ is the same as the conditional distribution of $X(t_3)$ given $X(t_2)$, which resembles the Markov property. In fact, it is true that a process with stationary and independent increments is a time homogeneous Markov process. In the following theorem, we prove this assertion when the index set is uncountable and the state space is countable.

Theorem 1.3.1 *A stochastic process $\{X(t), t \geq 0\}$ with stationary and independent increments, with $X(0) = 0$ and with a countable state space S, is a time homogeneous Markov process.*

Proof Suppose $\{x_1, x_2, \ldots, x_n\} \in S$ and $0 < t_1 < \cdots < t_n$ are positive real numbers. Then for any $n \geq 1$,

$$
\begin{aligned}
& P[X(t_n) = x_n | X(t_{n-1}) = x_{n-1}, \dots, X(t_1) = x_1] \\
& = P[X(t_n) = x_n, X(t_{n-1}) = x_{n-1}, \dots, X(t_1) = x_1] \\
& \times \{P[X(t_{n-1} = x_{n-1}, \dots, X(t_1) = x_1]\}^{-1} \\
& = P[X(t_n) - X(t_{n-1}) = x_n - x_{n-1}, \dots, X(t_1) - X(0) = x_1] \\
& \times \{P[X(t_{n-1}) - X(t_{n-2}) = x_{n-1} - x_{n-2}, \dots, X(t_1) - X(0) = x_1]\}^{-1} \\
& = \prod_{r=2}^{n} P[X(t_r) - X(t_{r-1}) = x_r - x_{r-1}] \times P[X(t_1) - X(0) = x_1] \\
& \times \left\{ \prod_{r=2}^{n-1} P[X(t_r) - X(t_{r-1}) = x_r - x_{r-1}] \times P[X(t_1) - X(0) = x_1] \right\}^{-1} \\
& = P[X(t_n) - X(t_{n-1}) = x_n - x_{n-1}].
\end{aligned}
$$

Similarly,

$$
\begin{aligned}
& P[X(t_n) = x_n | X(t_{n-1}) = x_{n-1}] \\
& = P[X(t_n) = x_n, X(t_{n-1}) = x_{n-1}] \times \{P[X(t_{n-1}) = x_{n-1}]\}^{-1} \\
& = P[X(t_n) - X(t_{n-1}) = x_n - x_{n-1}, X(t_{n-1}) - X(0) = x_{n-1}] \\
& \times \{P[X(t_{n-1}) - X(0) = x_{n-1}]\}^{-1} \\
& = P[X(t_n) - X(t_{n-1}) = x_n - x_{n-1}] P[X(t_{n-1}) - X(0) = x_{n-1}] \\
& \times \{P[X(t_{n-1}) - X(0) = x_{n-1}]\}^{-1} \\
& = P[X(t_n) - X(t_{n-1}) = x_n - x_{n-1}].
\end{aligned}
$$

Thus for all $n \geq 1$, for all $\{x_1, x_2, \dots, x_n\} \in S$ and positive real numbers $0 < t_1 < \cdots < t_n$, $P[X(t_n) = x_n | X(t_{n-1}) = x_{n-1}, \dots, X(t_1) = x_1]$ is the same as $P[X(t_n) = x_n | X(t_{n-1}) = x_{n-1}]$. Hence, $\{X(t), t \geq 0\}$ is a Markov process. It is to be noted that to prove the Markov property, we used independence of increments. Using stationarity of increments, we now prove that the Markov process is time homogeneous. Observe that its transition function is given by

$$
\begin{aligned}
P_{x_{n-1}x_n}(t_{n-1}, t_n) &= P[X(t_n) = x_n | X(t_{n-1}) = x_{n-1}] \\
&= P[X(t_n) - X(t_{n-1}) = x_n - x_{n-1}] \\
&= P[X(t_n - t_{n-1}) = x_n - x_{n-1}] \qquad \text{by } (1.3.1)\,.
\end{aligned}
$$

Thus the transition function $P_{x_{n-1}x_n}(t_{n-1}, t_n)$ depends only on the difference $t_n - t_{n-1}$, and hence the Markov process is time homogeneous. □

In the following theorem, we discuss some more properties of a process with stationary and independent increments.

Theorem 1.3.2 *Suppose $\{X(t), t \geq 0\}$ is a stochastic process with stationary and independent increments. Then the stochastic process $\{Y(t), t \geq 0\}$ where $Y(t) = X(t) - X(0)$ is also a process with stationary and independent increments.*

Proof We have $Y(t) = X(t) - X(0)$. Hence, by definition $Y(0) = 0$. For $k \geq 2$ and $0 = t_0 < t_1 < t_2 < \cdots < t_k$, consider the increments $Y(t_i) - Y(t_{i-1})$, $i = 1, 2, \ldots, k$. We have to prove that these k increment random variables are independent. This trivially follows because

$$Y(t_i) - Y(t_{i-1}) = X(t_i) - X(t_{i-1}) \ \forall \ i = 1, 2, \ldots, k$$

and the $\{X(t)\}$ process has independent increments. Similarly, for $s < t$, $Y(t) - Y(s) = X(t) - X(s)$. Since the $\{X(t)\}$ process has stationary increments, the distribution of the increment $X(t) - X(s)$ depends only on $(t - s)$ but not on s and t separately. Hence, the same holds for the $\{Y(t)\}$ process and so it also has stationary increments. □

This theorem conveys that without loss of generality, for a process with stationary and independent increments, we may take $X(0) = 0$. In the next theorem we find expression for the mean function, the variance function and the covariance function of a process with stationary and independent increments.

Theorem 1.3.3 *Suppose $\{X(t), t \geq 0\}$ is a stochastic process with stationary and independent increments with $X(0) = 0$. Then*

(i) If $E(X(t)) < \infty$, then $E(X(t)) = E(X(1))t$.
(ii) If $E(X^2(t)) < \infty$, then $Var(X(t)) = ct$, where $c = Var(X(1))$.
(iii) If $E(X^2(t)) < \infty$, then $Cov(X(s), X(t)) = c\ min\{s, t\}$, where $c = Var(X(1))$.

Proof (i) Suppose $M(t) = E(X(t))$. In view of stationarity of increments for, $s, t \geq 0$

$$\begin{aligned} X(t+s) - X(s) \stackrel{d}{=} X(t) &\Rightarrow E(X(t+s) - X(s)) = E(X(t)) \\ &\Rightarrow M(t+s) - M(s) = M(t) \\ &\Rightarrow M(t+s) = M(t) + M(s)\ . \end{aligned}$$

Thus, the function $M(\cdot)$ satisfies the Cauchy equation $M(t+s) = M(s) + M(t)$, for $s, t \geq 0$. It is known that this equation has a unique solution given by $M(t) = ct$. By setting $t = 1$ in this equation, we get $c = M(1) = E(X(1))$. Thus we have $E(X(t)) = E(X(1))t$. It is to be noted that for this result to hold, we need only stationarity of increments.
(ii) Suppose $v(t) = Var(X(t))$ and consider for $s, t \geq 0$,

$$
\begin{aligned}
v(t+s) &= Var(X(t+s)) = Var\ ((X(t+s) - X(s)) + X(s)) \\
&= Var\ (X(t+s) - X(s)) + Var(X(s)) \\
&+ 2Cov(X(t+s) - X(s), X(s) - X(0)) \\
&= Var\ (X(t+s) - X(s)) + Var(X(s)) + 0 \\
&= Var(X(t)) + Var(X(s)) \text{ because of stationary increments} \\
&= v(t) + v(s).
\end{aligned}
$$

In the third step from below, $Cov(X(t+s) - X(s), X(s) - X(0)) = 0$ in view of independence of increments. Thus, the function $v(\cdot)$ satisfies the Cauchy equation $v(t+s) = v(s) + v(t)$, for $s, t \geq 0$. Hence, its unique solution is given by $v(t) = ct$. With $t = 1$, $c = v(1)$. Thus, $Var(X(t)) = Var(X(1))t$.
(iii) Suppose $s < t$. Then,

$$
\begin{aligned}
Cov(X(t), X(s)) &= Cov\ ((X(t) - X(s)) + X(s), X(s)) \\
&= Cov\ (X(t) - X(s), X(s) - X(0)) + Cov(X(s), X(s)) \\
&= 0 + Var(X(s)), \text{ by independence of increments} \\
&= cs, \text{ where } c = Var(X(1))\ .
\end{aligned}
$$

Thus, $Cov(X(s), X(t)) = cs$ if $t < s$. Similarly, it can be shown that $Cov(X(s), X(t)) = ct$ if $s > t$. Combining both, we get the required result that $Cov(X(s), X(t)) = min\{cs, ct\}$. Observe that $c = Var(X(1)) \geq 0$ implies that $Cov(X(s), X(t)) = c\ min\{s, t\} \geq 0, \ \forall \ s, t \geq 0$. □

The following important corollary follows from the expression of $E(X(t))$.

Corollary 1.3.1 *Suppose $\{X(t), t \in T\}$ is a process with stationary and independent increments. If the index set T is unbounded, then the state space S of the process cannot be finite or a bounded interval.*

Proof We assume the contrary that the state space S is finite or is a bounded interval. Suppose $m \leq X(t) \leq M$ almost surely, where M and m are the maximum/supremum and the minimum/infimum of S respectively. Then for all
$t > 0$,

$$
\begin{aligned}
m \leq X(t) \leq M \ a.s. &\Rightarrow m \leq E(X(t)) \leq M \ \Rightarrow \ m \leq E(X(1))t \leq M \\
&\Rightarrow \frac{m}{E(X(1))} \leq t \leq \frac{M}{E(X(1))}, \ \text{ if } \ E(X(1)) > 0 \\
\& \ E(X(1)) < 0 &\Rightarrow \frac{M}{E(X(1))} \leq t \leq \frac{m}{E(X(1))}.
\end{aligned}
$$

Thus, t is bounded from below and above, which is a contradiction to the fact that T is unbounded. Thus, the assumption that state space S is bounded is wrong. If $E(X(1)) = 0$, then similar arguments hold with $Var(X(t))$. Hence, the state space cannot be finite or a bounded interval. □

The following theorem proves that for a process $\{X(t), t \geq 0\}$ with stationary and independent increments, the distribution of $X(1)$ determines the distribution of $X(t)$ for all t.

Theorem 1.3.4 *Suppose $\{X(t), t \geq 0\}$ is a process with stationary and independent increments, with $X(0) = 0$. Suppose ϕ_t denotes the characteristic function of $X(t)$. Then $\phi_t(u) = \{\phi_1(u)\}^t, \quad \forall \quad u \in \mathbb{R}$.*

Proof For a fixed u

$$\begin{aligned}
\phi_{s+t}(u) &= E(e^{iuX(s+t)}) = E\{e^{iu([X(s+t)-X(s)]+X(s))}\} \\
&= E(e^{iu(X(s+t)-X(s))})\ E(e^{iuX(s)}) \\
&= E(e^{iuX(t)})\ E(e^{iuX(s)}) = \phi_s(u)\ \phi_t(u) \\
\Rightarrow \quad \log(\phi_{s+t}(u)) &= \log(\phi_s(u)) + \log(\phi_t(u)) \\
\Rightarrow \quad \log(\phi_t(u)) &= d(u) \times t,
\end{aligned}$$

as for fixed u, the function $\log(\phi_t(u))$, as a function of t, satisfies the Cauchy equation. Substituting $t = 1$, we get $d(u) = \log(\phi_1(u))$ so that

$$\log(\phi_t(u)) = \log(\phi_1(u)) \times t \quad \Longleftrightarrow \quad \phi_t(u) = \{\phi_1(u)\}^t, \quad u \in \mathbb{R}\,.$$

Thus, if a process has stationary and independent increments, then the distribution of $X(t)$ is determined by the distribution of $X(1)$ $\forall\ t \geq 0$. □

Remark 1.3.1 From Theorem 1.3.4, it follows that the distribution of $X(1)$ is infinitely divisible [29], which in turn implies that the support of $X(1)$ and hence of $X(t)$ for any t cannot be bounded. In Corollary 1.3.1, it is proved that if the index set T is unbounded, then the state space cannot be bounded. From Theorem 1.3.4, it follows that even if the index set T is bounded, $(0, 1)$, say, the state space cannot be bounded. It thus follows that for a stochastic process with stationary and independent increments, the state space cannot be bounded, in particular, it cannot be finite.

Remark 1.3.2 In Theorem 1.3.1, it is proved that a stochastic process with stationary and independent increments satisfies the Markov property. However, the converse of Theorem 1.3.1 is not true. To elaborate on this assertion, suppose $\{X(t), t \in T\}$ is a Markov process with $T = W$ or $T = \mathbb{R}^+$ and finite state space. Suppose it is a process with stationary and independent increments, but then the state space cannot be bounded. It is a contradiction to the assumption that the state space of $\{X(t), t \in T\}$ is finite. Thus, a Markov process with a finite state space cannot be a process with stationary and independent increments.

There are many processes with stationary and independent increments. In the present book, we study only two processes with stationary and independent increments, one is a Poisson process and the other is a Brownian motion process, also known as a Wiener process. In a Poisson process, the marginal distributions are

Poisson while in a Wiener process the marginal distributions are normal. For both the processes, we will observe that the distribution of $X(t)$ is determined by the distribution of $X(1)$ for all $t \geq 0$, as proved in Theorem 1.3.4. A Poisson process is a continuous time discrete state space process with state space $\{0, 1, \ldots, \}$, while a Wiener process is a continuous time continuous state space process with state space $\mathbb{R}$. The sample paths of a Poisson process are step functions, while the sample paths of a Wiener process are continuous.

In the next section, we briefly discuss the concept of stationarity of a stochastic process. In the subsequent chapters, we examine whether a given stochastic process is stationary and under which conditions.

1.4 Stationary Processes

A stationary process is a stochastic process whose probabilistic law remains unchanged under shifts in time. The concept captures a natural notion of a physical system that lacks an inherent time origin. It is an appropriate assumption in a variety of processes in communication theory, astronomy, biology, ecology and economics. Stationarity property of a stochastic process leads to a number of important conclusions. Following is the definition of a stationary process.

Definition 1.4.1 Stationary Process: A stochastic process $\{X(t), t \in T\}$ is said to be a stationary process if $\forall\ n \geq 1, t_1, t_2, \ldots, t_n \in T$ and $h \in \mathbb{R}$ such that $t_1 + h, t_2 + h, \ldots, t_n + h \in T$, the random vectors

$$(X(t_1), X(t_2), \ldots, X(t_n))' \quad \& \quad (X(t_1 + h), X(t_2 + h), \ldots, X(t_n + h))'$$

are identically distributed.

The condition in the definition asserts that the process is in probabilistic equilibrium and the particular times at which we examine the process are of no relevance. In other words, a process is stationary if we choose any fixed point h as the origin, and the process has the same probability law. In particular, for a stationary process all marginal distributions are the same.

Suppose $Var(X(t))$ is finite. Since all marginal distributions are the same, $E(X(t))$ should be the same for all t and $Var(X(t))$ also should be the same for all t, which further implies that these have to be constants, free from t. On similar lines, since all bivariate distributions are identical, $Cov(X(t), X(s))$ is a function of $|t - s|$. Using these results, it is proved in Chap. 5 that a branching process is not stationary and in Chap. 7 that a Poisson process is not stationary. In Chaps. 3 and 6, we prove that a Markov chain in discrete time and in continuous time, respectively, is stationary under certain conditions. It follows easily that a sequence of independent and identically distributed random variables is a stationary stochastic process. However, if the random variables in the sequence are not identically distributed, then it is not a stationary stochastic process.

There are two types of stationary processes: the one defined above is known as a strictly stationary or a strongly stationary process. A strictly stationary process is simply referred to as a stationary process. The condition for a process to be stationary is rather stringent and involves the finite dimensional distributions. We now define the other version which involves only the first two moments. It is known as a weakly stationary or covariance stationary or wide-sense stationary or second-order stationary stochastic process. We first define a second-order stochastic process.

Definition 1.4.2 Second-order Stochastic Process: A stochastic process $\{X(t), t \in T\}$, for which $E(X^2(t)) < \infty$, is known as a second-order stochastic process.

Definition 1.4.3 Covariance Stationary Process: A second-order stochastic process $\{X(t), t \in T\}$ is known as a second-order stationary or a covariance stationary process if $E(X(t)) = c$ and $Cov(X(t), X(s))$ is a function of $|t-s|$ for all $s, t \in T$.

Thus for a covariance stationary process, the first two moments of $X(t)$ are the same for all t and the covariance between $X(s)$ and $X(t)$ depends only on $|t-s|$. These conditions are satisfied if a second-order stochastic process $\{X(t), t \geq 0\}$ is a strictly stationary process. Thus, a stationary process possessing finite first two moments is covariance stationary. In view of this implication, second-order stationary process is labeled as a weakly stationary process, or a strictly stationary process is labeled as a strongly stationary process. The converse is usually not true. It is true only for a Gaussian process. The finite dimensional distributions of a Gaussian process are determined by their means and covariances, hence it follows that a second-order covariance stationary Gaussian process is a stationary process. In Chap. 9, we elaborate on this issue.

Example 1.4.1 Suppose $\{X_n, n \geq 1\}$ is a stochastic process such that $\forall\ n \geq 1$, $(X_1, X_2, \ldots, X_n)'$ follows n-variate normal distribution with mean vector $\underline{0}$ and dispersion matrix $\Sigma = [\sigma_{ij}]$ such that $\sigma_{ij} = 1$ if $i = j$, $\sigma_{ij} = \rho$ if $j = i+1$ or $j = i-1$, and $\sigma_{ij} = 0$ for all other values of i and j. Note that $Cov(X_n, X_m) = \rho$ if $|m-n| = 1$ and 0 otherwise. Hence, the process is covariance stationary. Further, it is a Gaussian process and hence strongly stationary as well.

Following is an example of a covariance stationary process.

Example 1.4.2 Suppose $\{X(t), t \in \mathbb{R}\}$ is a stochastic process where $X(t)$ is defined as $X(t) = U\cos(\alpha t) + V\sin(\alpha t)$, where U and V are uncorrelated random variables with mean 0 and variance 1 for both. It then follows that $E(X(t)) = 0$ and $Var(X(t)) = 1$. Suppose $s < t$. We find $Cov(X(s), X(t))$ as follows:

$$
\begin{aligned}
Cov(X(s), X(t)) &= E(X(s)X(t)) \\
&= E((U\cos(\alpha s) + V\sin(\alpha s))(U\cos(\alpha t) + V\sin(\alpha t))) \\
&= \cos(\alpha s)\cos(\alpha t)E(U^2) + \sin(\alpha s)\sin(\alpha t)E(V^2) \\
&= \cos(\alpha s)\cos(\alpha t) + \sin(\alpha s)\sin(\alpha t) \\
&= \cos(\alpha(s-t)).
\end{aligned}
$$

Similarly, for $s > t$, $Cov(X(s), X(t)) = \cos(\alpha(t - s))$. Thus, $Cov(X(s), X(t))$ is a function of $|s - t|$ and hence $\{X(t), t \in \mathbb{R}\}$ is covariance stationary. In addition, if U and V are independent random variables each following the standard normal distribution, then it is a strictly stationary stochastic process. □

Definition 1.4.4 Evolutionary Process: A stochastic process $\{X(t), t \in T\}$ which is neither strictly stationary nor covariance stationary is known as an evolutionary stochastic process.

In Chap. 5, we note that a branching process is an evolutionary stochastic process and in Chap. 7, we note that a Poisson process is also an evolutionary stochastic process.

For more details on a stationary process and related examples, one may refer to Chap. 9 of Karlin and Taylor [19].

There is a vast literature on stochastic processes, plenty of research papers and number of excellent books. We list a few books such as Adke and Manjunath [1], Bhat [2], Bhat [3], Castañeda et al. [6], Cinlar [8], Doob [14], Feller [15], Feller [16], Hoel, Port and Stone [17], Ibe [18], Karlin and Taylor [19], Karlin and Taylor [20], Kulkarni [21], Medhi [22], Parzen [23], Ross [26], Ross [27] and Taylor and Karlin [28]. The material discussed in the present book is covered in many of these books. A few conceptual exercises are collected from some of these books.

The computational aspect of the stochastic processes has been rarely touched in any of these books. The novelty of the present book is that we have augmented the theory with R software as a tool for simulation and computation. The theory is better understood when the computations and theory go hand in hand. Hence, we have given significant weight to computational features. In every chapter, in each section, the theory and corresponding computations using R code are presented simultaneously. R codes are given in the last section of Chaps. 2, 3, 4, 5, 6, 7, 8, 9 and 10. We adopted a similar approach in all our books, for example, refer to Deshmukh [12] and Deshmukh [13], and it is found to be effective in understanding the subject.

Some readers may be familiar with R software as it has been introduced in the curriculum of many under-graduate and post-graduate statistics programs. In the following section, we give a brief introduction to R, which will be useful to beginners. We have also tried to make the codes given in Chaps. 2, 3, 4, 5, 6, 7, 8, 9 and 10 to be self-explanatory.

1.5 Introduction to R Software and Language

In statistics while analyzing data, a good statistical software is necessary in two phases of analysis. In exploratory data analysis, it is needed to summarize the data, visual presentation of data and some of its important characteristics. In confirmatory data analysis, we need it to carry out certain test procedures of interest. A variety of software is available for the computations such as Excel, Minitab, Matlab and

We use functions `rbinom`, `rnorm` to draw random samples from a binomial distribution and a normal distribution, respectively. Thus, we need to change the family name and add appropriate parameters. The names of all probability distributions can be obtained by giving the path help → manuals (in pdf) → An Introduction to R → Probability distributions, at the R console.

The function `round(d,4)` prints the values of d, rounded to the fourth decimal point. Note that rounding is mainly for printing purpose; original unrounded values of d are stored in the object d and used in computations.

There are some useful built-in functions. Apart from many built-in functions, one can write a suitable function as required in a specific situation. In subsequent chapters you will find many such functions, written to serve the specific purpose.

Code 1.5.3: This code illustrates some commonly used functions with a data set stored in variable x:

```
x=rexp(30,rate=1.5) # random sample of size n=30 from
# exponential distribution with mean 2/3
mean(x); median(x); max(x); min(x); sum(x)
cumsum(x)# cumulative sum
var(x)# divisor is (n-1) and not n
quantile(x,c(.25,.5,.75)) # three quartiles
summary(x) # gives minimum, maximum, three quartiles and mean
```

We can carry out a number of test procedures; the manual from the help menu lists some of these. We illustrate one test procedure after discussing graphical tools of R.

Graphical facility of R software has various features. Each graphical function has many options to bring in flexibility. Graphical device is a graphical window or a file. There are two kinds of graphical functions—the high-level plotting functions, which create a new graph, and low-level plotting functions, which add elements to an already existing graph. The standard high-level plotting functions are `plot()` function for scatter plot, `hist()` function for histogram, `boxplot()` function for box plot, etc. The lower level plotting functions are `lines()` to impose curves on the existing plot, `abline()` to add a line with given intercept and slope, `points()` to add points at appropriate places, etc. These functions take extra arguments that control the graphic. The graphs are produced with respect to graphical parameters, which are defined by default and can be modified with the function "par". If we type `?par` on R console, we get the description on the number of arguments for graphical functions, as documented in R. We explain one among these which is frequently used in the book. It is `par(mfrow=c(2,2))` or `par(mfcol=c(2,2))`. This command divides the graphical window invisibly in 2 rows and 2 columns to accommodate 4 graphs. A function legend() is usually added in plots to specify a list of symbols or colors used in the graphs. The following code illustrates one test procedure and `plot` and `lines` functions.

Suppose we draw a random sample of size 60 from the distribution of Y which is binomial $B(7, 0.49)$. Using Karl Pearson's test procedure, we test whether the observed data are from binomial $B(7, 0.49)$ distribution. Karl Pearson's test statistic T is given by $T = \sum_{i=1}^{k}(o_i - e_i)^2/e_i$, where k denotes the number of classes and o_i and e_i respectively denote the observed frequency and the expected frequency, expected under $B(7, 0.49)$ distribution, of the i-th class. We adopt three approaches of computing T.

In the first approach, according to the standard convention, the frequencies which are less than or equal to 5 are pooled. In the second approach, we do not pool the frequencies. In the third approach, we use the built-in function `chisq.test(ob,p=pr)`. We note that the second and the third approach give the same results. Under the null hypothesis that the observed data are from binomial $B(7, 0.49)$ distribution, T has chi-square distribution. Using it, we find the cut-off point and the p-value to arrive at a decision. We can observe the closeness between the observed and the expected distributions visually from the graph. We thus plot the observed and expected frequencies using `plot` and `lines` functions. The following code incorporates all these features. In the code, we use the function `set.seed` so that every time we run the code, we get the same sample. Another function `data.frame` is a list of vectors of the same length, which are related, in some way. If one of them is shorter, it is "recycled" an appropriate number of times.

Code 1.5.4: This code illustrates Karl Pearson's goodness of fit test procedure for binomial distribution. It also demonstrates some graphical features:

```
set.seed(208); n=60; m=7; p=.49
# argument to set.seed function can be any positive integer
x=rbinom(n,m,p) # random sample from binomial distribution
d=data.frame(table(factor(x, levels = 0:max(x))));d
ob=d$Freq; # observed frequencies
# $ symbol extracts required values from the object, in
# this case frequencies from the object ob
t=sort(unique(x)) # observed values of Y
pr=dbinom(t,m,p) # expected probabilities
e=n*pr # expected frequencies
# pooling the frequencies
o1=c(ob[1]+ob[2]+ob[3],ob[4:5],ob[6]+ob[7]+ob[8]); o1
e1=c(e[1]+e[2]+e[3],e[4:5],e[6]+e[7]+e[8]); e1
t1=sum((o1-e1)^2/e1); t1 # Karl Pearson's test statistic
df=length(o1)-1;df # degrees freedom
b1=qchisq(.95,df);b1 #95% quantile of chi-square distribution
p1=1-pchisq(t1,df); p1 # p-value
# Without pooling the frequencies which are less than 5
d2=length(ob)-1; d2
tn=sum((ob-e)^2/e); tn; b=qchisq(.95,d2); b
pval=1-pchisq(tn,d2); pval
chisq.test(ob,p=pr) # Built-in function
rf=ob/n; a=min(min(rf),min(pr));b=max(max(rf),max(pr))
u=seq(round(a,2),b,.03);u
```

```
plot(t,rf,"h",col="blue",lty=1,lwd=2, xlab="Values of Y",
main="Observed and Expected Distributions",
ylab="Relative Frequency and Expected probability",
ylim=range(c(a,b)),yaxt="n")
axis(2,at=u)
points(t,rf,pch=20)
lines(t,pr,"h",col="red",lty=2,lwd=1)
points(t,pr,pch=20)
legend("topright",legend=c("Observed Distribution",
"Expected Distribution"),cex=.7,col = c("blue","red"),
lty=c(1,2),lwd=c(2,1))
abline(h=0,col="blue") # horizontal line at 0
```

With pooling of the frequencies, $T = 2.5946$. The cut-off and p-value are 7.8147 and 0.4584 respectively, as the null distribution is χ_3^2. Without pooling the frequencies which are less than 5, $T = 10.0204$. The cut-off and p-value are 14.0671 and 0.1874 respectively, since the null distribution is χ_7^2. The built-in function also gives the same cut-off and p-value. Thus, with all the approaches, we arrive at the conclusion that the data may be from binomial $B(7, 0.49)$ distribution. Figure 1.2 also leads to the same conclusion.

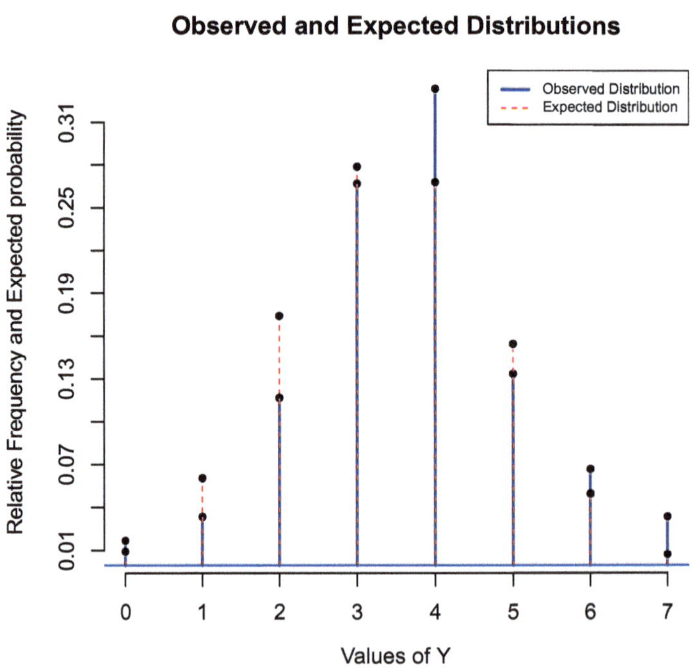

Fig. 1.2 Binomial B(7, 0.49) distribution: observed and expected distributions

Code 1.5.4 illustrates all `d,p,q,r` functions and a variety of graphical functions. We define a and b to set the limits on y-axis using `ylim=range(c(a,b))` argument. In the `plot` function with the argument `yaxt="n"`, we can suppress the default values and set the desired values with a function `axis(2,at=u)`. Observe various arguments in a `legend` function, such as `"topright"` to decide the position of the legend on the graph, `col` to specify colors of vertical lines corresponding to two distributions, `lty` to specify types of lines and `lwd` to specify the width of lines.

All these features are used in a variety of graphs drawn in the subsequent chapters.

There are many books on the introduction to statistics using R, such as Crawley [9], Dalgaard [10], Purohit et al. [24] and Verzani [30]. There is a tremendous amount of information about R on the web at http://cran.r-project.org/ with a variety of R manuals. Following are some links useful for beginners to learn R software.

1. https://www.datacamp.com/courses/free-introduction-to-r
2. http://www.listendata.com/p/r-programming-tutorials.html
3. http://www.r-tutor.com/r-introduction
4. https://www.r-bloggers.com/list-of-free-online-r-tutorials/
5. https://www.tutorialspoint.com/r/
6. https://www.codeschool.com/courses/try-r.

As for any software or programming language, the best way to learn R is to use it for understanding the concepts and solving problems. We hope that this brief introduction will definitely be useful to a reader to be comfortable with R and follow the codes written in subsequent chapters.

A quick recap of the results discussed in this chapter is given below.

Summary

1. Suppose $\{X(t), t \in T\}$ is a collection of random variables defined on the same probability space $(\Omega, \mathbb{A}, P)$, where T is a non-empty set, then $\{X(t), t \in T\}$ is known as a stochastic process.
2. Suppose $\{X(t), t \in T\}$ is a stochastic process and S_t denotes the set of possible values of $X(t)$ for any $t \in T$. Then $S = \bigcup_{t \in T} S_t$ is known as a state space of the stochastic process. The set T is known as the index set.
3. Suppose $\{X(t), t \in T\}$ is a stochastic process defined on $(\Omega, \mathbb{A}, P)$. For each fixed $\omega \in \Omega$, $X(t, \omega)$, which is a function of $t \in T$, is known as a sample path or a realization or a trajectory of the stochastic process.
4. Suppose a family $\mathbb{F}$ of finite dimensional distribution functions on T satisfies the following two conditions.
   (i) For every $n \geq 2$ and finite, $1 \leq m < n$ and $1 \leq i_1, i_2, \ldots, i_m \leq n$,

$$F_{t_{i_1}, t_{i_2}, \ldots, t_{i_m}}(x_{i_1}, x_{i_2}, \ldots, x_{i_m}) = \lim_{\substack{x_j \to \infty \\ j \neq i_1, \ldots, i_m}} F_{t_1, t_2, \ldots, t_{n-1}, t_n}(x_1, x_2, \ldots, x_{n-1}, x_n).$$

Thus, $\forall\ n \geq 1$, $\forall\ m$ such that $1 \leq m \leq n-1$, $F_{t_1,t_2,\ldots,t_n} \in \mathbb{F}$ implies that all the m-dimensional distribution functions of $\{F_{t_1,t_2,\ldots,t_n}\}$ are in $\mathbb{F}$.
(ii) For every permutation $\{i_1, i_2, \ldots, i_n\}$ of $\{1, 2, \ldots, n\}$,

$$F_{t_1,t_2,\ldots,t_n}(x_1, x_2, \ldots, x_n) = F_{t_{i_1},t_{i_2},\ldots,t_{i_n}}(x_{i_1}, x_{i_2}, \ldots, x_{i_n})\,.$$

These two conditions together are known as Kolmogorov compatibility conditions: condition (i) is known as a consistency condition while (ii) is known as a symmetry condition.

5. Suppose $\{X(t), t \in T\}$ is a stochastic process defined on the probability space $(\Omega, \mathbb{A}, P)$. For $n \geq 1$ and $t_1, t_2, \ldots, t_n \in T$, the joint distribution function of the random variables $\{X(t_1), X(t_2), \ldots, X(t_n)\}$, with respect to the probability measure P, is defined by

$$F_{t_1,t_2,\ldots,t_n}(x_1, x_2, \ldots, x_n) = P\left[X(t_1) \leq x_1, X(t_2) \leq x_2, \ldots, X(t_n) \leq x_n\right],$$

$x_i \in \mathbb{R}$, $\quad i = 1, 2, \ldots, n$. Then the family $\{F_{t_1,t_2,\ldots,t_n}, n \geq 1\}$ is known as a family of finite dimensional distribution functions of the stochastic process $\{X(t), t \in T\}$.

6. The family of finite dimensional distribution functions of a stochastic process satisfies the Kolmogorov compatibility conditions.
7. Kolmogorov Existence Theorem: Suppose T is a non-empty set. Suppose

$$\mathbb{G} = \{G_{t_1,t_2,\ldots,t_n} | t_1 < t_2, \cdots < t_n \in T, \ \ n \geq 1\}$$

is a family of finite dimensional distribution functions. If the family $\mathbb{G}$ satisfies the Kolmogorov compatibility conditions, then there exists a stochastic process $\{X^*(t), t \in T\}$ defined on some probability space $(\Omega^*, \mathbb{A}^*, P^*)$ such that its family of finite dimensional distribution functions is given by $\mathbb{G}$.

8. A continuous time stochastic process $\{X(t), t \geq 0\}$ is said to be a process with stationary and independent increments, if (i) for every $k \geq 2$ and $t_1, t_2, \ldots, t_k$ such that $0 = t_0 < t_1 < t_2 < \cdots < t_k$, the increments $X(t_i) - X(t_{i-1})$, $i = 1, 2, \ldots, k$ are independent random variables and (ii) for every $s < t$, the probability distribution of the increment $X(t) - X(s)$ depends only on $(t - s)$, but not on s and t separately.
9. A stochastic process $\{X(t), t \geq 0\}$ with stationary and independent increments with $X(0) = 0$ is a time homogeneous Markov process. However, the converse is not true.
10. For a stochastic process $\{X(t), t \geq 0\}$ with stationary and independent increments with $X(0) = 0$ and $E[X^2(t)] < \infty$, $E(X(t)) = E(X(1))\,t$, $Var(X(t)) = Var(X(1))\,t$ and $Cov(X(s), X(t)) = Var(X(1))\,min\{s, t\}$.
11. The state space S of a process with stationary and independent increments cannot be finite or a bounded interval.
12. For a process $\{X(t), t \geq 0\}$ with stationary and independent increments, the distribution of $X(1)$ determines the distribution of $X(t)$ for all t.

13. A stochastic process $\{X(t), t \in T\}$ is said to be a stationary process if $\forall\, n \geq 1$, $t_1, t_2, \ldots, t_n \in T$, $h \in \mathbb{R}$ such that $t_1 + h, t_2 + h, \ldots, t_n + h \in T$, the random vectors $(X(t_1), X(t_2), \ldots, X(t_n))'$ and $(X(t_1 + h), X(t_2 + h), \ldots, X(t_n + h))'$ are identically distributed.
14. A stochastic process $\{X(t), t \in T\}$ for which $E(X^2(t)) < \infty$ is known as a second-order stochastic process.
15. A second-order stochastic process $\{X(t), t \in T\}$ is known as a second-order stationary or a covariance stationary or weakly stationary process if $E(X(t)) = c$ and $Cov(X(t), X(s))$ is a function of $|t - s|$ for all $s, t \in T$.
16. A stochastic process $\{X(t), t \in T\}$ which is neither strictly stationary nor covariance stationary is known as an evolutionary stochastic process.
17. A stationary process possessing finite first two moments is covariance stationary. The converse is usually not true.

In the next chapter, we discuss in detail a variety of results related to Markov chains.

References

1. Adke, S. R., & Manjunath, S. M. (1984). *An introduction to finite Markov processes*. Wiley Eastern.
2. Bhat, B. R. (2000). *Stochastic models: Analysis and applications*. New Delhi: New Age International.
3. Bhat, U. N. (1984). *Elements of applied stochastic processes* (2nd ed.). New York: Wiley.
4. Billingsley, P. (1986). *Probability and measure* (2nd ed.). New York: Wiley.
5. Brockwell, P. J., & Davis, R. A. (2003). *Introduction to time series analysis*. Berlin: Springer.
6. Castañeda, L. B., Arunachalam, V., & Dharmaraja, D. (2012). *Introduction to probability and stochastic processes with applications*. New York: John Wiley.
7. Chatfield, C., & Xing, H. (2019). *The analysis of time series: An introduction with R*. Chapman and Hall.
8. Cinlar, E. (1975). *Introduction to stochastic processes*. New Jersey: Prentice Hall.
9. Crawley, M. J. (2007). *The R book*. New York: John Wiley.
10. Dalgaard, P. (2008). *Introductory statistics with R* (2nd ed.). New York: Springer.
11. Daniell, P. J. (1919). Functions of limited variation in an infinite number of dimensions. *Annals of Mathematics, 21*, 30–38.
12. Deshmukh, S. R. (2012). *Multiple decrement models in insurance: An introduction using R*. New Delhi: Springer.
13. Deshmukh, S. R., & Kulkarni, M. G. (2021). *Asymptotic statistical inference: A basic course using R*. Singapore: Springer.
14. Doob, J. L. (1953). *Stochastic processes*. New York: John Wiley.
15. Feller, W. (1978). *An introduction to probability theory and its applications* (Vol. I). New York: John Wiley.
16. Feller, W. (2000). *An introduction to probability theory and its applications* (2nd ed., Vol. II). New York: John Wiley.
17. Hoel, P. G., Port, S. C., & Stone, C. J. (1972). *Introduction to stochastic Processes*. Wiley Eastern.
18. Ibe, O. (2005). *Markov processes for stochastic modeling* (2nd ed.). US: Elsevier.

19. Karlin, S., & Taylor, H. M. (1975). *A first course in stochastic processes*. New York: Academic Press.
20. Karlin, S., & Taylor, H. M. (1981). *A second course in stochastic processes*. New York: Academic Press.
21. Kulkarni, V. G. (2011). *Introduction to modeling and analysis of stochastic systems*. New York: Springer.
22. Medhi, J. (1994). *Stochastic processes*. New Delhi: Wiley Eastern.
23. Parzen, E. (1962). *Stochastic processes*. San Francisco, California: Holden-Day.
24. Purohit, S. G., Gore, S. D., & Deshmukh, S. R. (2008). *Statistics using R* (2nd ed.). New Delhi: Narosa Publishing House.
25. R Core Team (2019). *R: a language and environment for statistical computing, R foundation for statistical computing, Vienna, Austria*. https://www.R-project.org/
26. Ross, S. M. (1996). *Stochastic processes* (2nd ed.). New York: John Wiley.
27. Ross, S. M. (2014). *Introduction to Probability Models* (11th ed.). New York: Academic Press.
28. Taylor, H. N., & Karlin, S. (1984). *An introduction to stochastic modeling*. New York: Academic Press.
29. Steutel, F. W., & Harn, K. V. (2004). *Infinite divisibility of probability distributions on the real line*. New York: Marcel Dekker, Inc.
30. Verzani, J. (2005). *Using R for introductory statistics*. New York: Chapman and Hall /CRC Press.

Chapter 2
Markov Chains

2.1 Introduction

Markov chains are the most widely used stochastic processes. The theory of Markov chains is heavily applied in finance, insurance, biotechnology, weather prediction, social sciences and marketing. Its extensions such as hidden Markov chains have applications in bioinformatics. Its origin is however in linguistics.

The year 2013 was celebrated as 100th anniversary of Markov chains. This subject matter is named after A. A. Markov, who laid the foundations of the theory in a series of papers starting in 1907. He was a Russian probabilist in the St. Petersburg School. He was a student of Chebyshev and proved the law of large numbers rigorously in a variety of cases, including for dependent sequences. Markov has given a rigorous solution to the Urn problem, which was first posed by Daniel Bernoulli in 1769 and later analyzed by Laplace in 1812.

Markov was interested in investigating the way the vowels and consonants alternate in Russian literature. He studied a piece of text "Eugene Onegin" by Pushkin and classified 20,000 consecutive characters as vowels or consonants. The aim was to explore patterns in the sequence of consonants and vowels. He found that the sequence of consonants and vowels in the text by Pushkin is well described as a random sequence, where the likely category of a letter depends only on the category of the previous letter only or previous two letters. He also carried out such a study on Aksakov's "The Childhood Years of Bagrov's Grandson". In the analysis of these sequences, Markov invented what are now known as "Markov chains". Similar analysis was applied on the text "Alice's Adventures in Wonderland", by Lewis Carroll. It was found that the Markov model accurately predicts the average distance between vowels. The distributions of distances predicted by the model also agree well with those actually found in the text.

The early work on Markov chains was restricted to the finite state space, and matrix theory played an important role. The infinite state space case was introduced by A. N. Kolmogorov. The present probabilistic approach of studying Markov chains

S. Madhira and S. Deshmukh, *Introduction to Stochastic Processes Using R*,
https://doi.org/10.1007/978-981-99-5601-2_2

was first introduced by Feller. Much of the present terminology originates from his books, Feller [5] and Feller [6].

In the present chapter, we study in detail the theory of Markov chains. We begin with the basic definitions related to Markov chains in this section and illustrate the concept with some examples. Section 2.2 is concerned with the higher step transition probabilities and related results. Section 2.3 is devoted to the notion of a realization of a Markov chain and a brief discussion on the maximum likelihood estimation of the transition probability matrix. In Sects. 2.4 and 2.5, we discuss classification of states of a Markov chain, which is essential to study the limiting behavior of a Markov chain. Section 2.6 elaborates on the first passage distribution. It also plays a key role in the limiting behavior of a Markov chain. In Sect. 2.7, we study the concept of periodicity of a Markov chain. In Sects. 2.2–2.7, various theoretical concepts are illustrated with computations using R software. All the codes used in these illustrations are presented in Sect. 2.8.

We now proceed with the definition of a Markov chain.

Definition 2.1.1 Markov Chain: Suppose $\{X_n, n \geq 0\}$ is a discrete time stochastic process with a countable state space S. It is known as a Markov chain if $\forall\ n \geq 1$ and $\forall\ x_0, x_1, \ldots, x_{n-1}, i, j$ in S,

$$P[X_{n+1} = j|X_n = i, X_{n-1} = x_{n-1}, \ldots X_0 = x_0] = P[X_{n+1} = j|X_n = i], \quad (2.1.1)$$

provided the conditional probabilities are defined.

Remark 2.1.1 In the definition, it is enough to state that the conditional probability on the left-hand side is defined, since it implies that the conditional probability on the right-hand side is also defined.

The property in Eq. (2.1.1) is known as the Markov property, named after Markov. Markov property is usually described as "history independence", because the conditional probability distribution of the state of the system at time $n + 1$, given all the information at previous time points, depends on the state at time n only and does not depend on the states at times prior to n. Suppose the event $[X_{n-1} = x_{n-1}, \ldots X_0 = x_0]$ is referred to as past, the event $[X_n = i]$ is referred to as present and the event $[X_{n+1} = j]$ is referred to as future. Then the Markov property states that, given the past and present, the probabilistic behavior of the future depends only on the present. Hence, the Markov property is also interpreted as a "lack of memory" property. In a Markov chain $\{X_n, n \geq 0\}$, if $X_n = i$, we say that the system is in state i at time n, x_0 being the initial state. A Markov chain with finite state space is referred to as a finite Markov chain.

We now define some basic terms related to Markov chains.

Definition 2.1.2 Initial Distribution: Suppose $\{X_n, n \geq 0\}$ is a Markov chain with state space S and $p_i^{(0)} = P[X_0 = i]$ for $i \in S$. Then $\underline{p}^{(0)} = \{p_i^{(0)},\ i \in S\}$ is the probability mass function of X_0 and is known as the initial distribution of the Markov chain.

Definition 2.1.3 One Step Transition Probability: Suppose $\{X_n, n \geq 0\}$ is a Markov chain with state space S. The conditional probability $P[X_{n+1} = j|X_n = i]$ of transition in one step, from state i at time n to state j at time $n+1$, is known as the one step transition probability. It is denoted by $p_{ij}^{n,n+1}$.

In general, the transition probability $p_{ij}^{n,n+1}$ is a function of i, j and n. Its dependence or non-dependence on n leads to two types of Markov chains as defined below.

Definition 2.1.4 Time Homogeneous Markov Chain: Suppose $\{X_n, n \geq 0\}$ is a Markov chain with one step transition probabilities $p_{ij}^{n,n+1}, i, j \in S, n \geq 0$. If $p_{ij}^{n,n+1}$ depends on n for some $i, j \in S$, then the Markov chain is said to be a non-homogeneous Markov chain. If $p_{ij}^{n,n+1}$ is free from n $\forall\ i, j \in S$, then it is known as a time homogeneous Markov chain or a Markov chain having stationary transition probabilities.

Thus, for a time homogeneous Markov chain, $\forall\ i,\ j \in S$

$$P[X_{n+1} = j|X_n = i] = P[X_{100} = j|X_{99} = i] = P[X_1 = j|X_0 = i].$$

For a homogeneous Markov chain, the one step transition probability is denoted by p_{ij}. These one step transition probabilities are presented in a matrix $P = [p_{ij}]$.

Definition 2.1.5 One Step Transition Probability Matrix of a Homogeneous Markov Chain: Suppose $\{X_n, n \geq 0\}$ is a time homogeneous Markov chain with state space S and one step transition probabilities p_{ij}. A matrix $P = [p_{ij}]$ is known as the one step transition probability matrix, in which rows correspond to a starting state while columns correspond to an ending state after one step transition.

If the state space is countably infinite, then P is an infinite dimensional square matrix. If the state space S consists of M states, then it is of order $M \times M$. It is to be noted that after a unit time, the system may remain in the same state and hence $p_{ii} \geq 0$. Since probabilities are non-negative and since the system must make a transition to some state in the state space in one step, we have

$$p_{ij} \geq 0, \ \ \forall\ i, j \in S \ \ \& \ \sum_{j \in S} p_{ij} = 1, \ \ \forall\ i \in S\ .$$

To be more precise, $\forall\ i \in S$,

$$\sum_{j \in S} p_{ij} = \sum_{j \in S} P[X_{n+1} = j|X_n = i] = \sum_{j \in S} P[X_{n+1} = j, X_n = i]/P[X_n = i] = 1.$$

In view of these conditions on p_{ij}, all elements in the matrix P are non-negative and sum of the elements in each row is 1. Such a matrix is known as a stochastic matrix or a Markov matrix. Thus, the one step transition probability matrix is a stochastic matrix.

Definition 2.1.6 Stochastic Matrix and Doubly Stochastic Matrix: A matrix $A = [a_{ij}]$ is known as a stochastic matrix or a Markov matrix if (i) $a_{ij} \geq 0 \; \forall \; i, j \in S$ and (ii) $\sum_{j \in S} a_{ij} = 1 \; \forall \; i \in S$. If in addition, sum of the elements in each column is also 1, that is, $\sum_{i \in S} a_{ij} = 1 \quad \forall \quad j \in S$, then the matrix is known as a doubly stochastic matrix.

The following examples illustrate the above concepts.

Example 2.1.1 In auto-insurance, the insured drivers are classified into two classes—preferred and standard at the beginning of each year. A Markov model is assumed for transitions of drivers between the two classes, from year to year. At the beginning of each year, 70% of preferred are reclassified as preferred and 30% as standard. Similarly, 80% of standard are reclassified as standard and 20% as preferred. Suppose the preferred state is denoted by 1 and standard by 2. Since the transition probabilities do not depend on the time of transition, we have a homogeneous Markov chain with state space $S = \{1, 2\}$ and the transition probability matrix P, where

$$P = \begin{array}{c} \\ 1 \\ 2 \end{array} \begin{array}{c} \quad 1 \quad\;\; 2 \\ \begin{pmatrix} 0.7 & 0.3 \\ 0.2 & 0.8 \end{pmatrix} \end{array}.$$

It is easy to verify that P is a stochastic matrix. Each row of P represents a probability mass function of a random variable with possible values 1 and 2, with probabilities 0.7 and 0.3 respectively in the first row, and with probabilities 0.2 and 0.8 respectively in the second row. Suppose for $n \geq 1$, X_n denotes the state of a driver at the beginning of the nth year. If we want to find the probability that a driver known to be classified as standard at the beginning of the first year will be reclassified as standard at the beginning of the third year, then we have to find $P[X_3 = 2|X_1 = 2]$. We use the Markov property repeatedly to compute this probability as shown below

$$\begin{aligned}
P[X_3 = 2|X_1 = 2] &= P[X_3 = 2, X_1 = 2]/P[X_1 = 2] \\
&= \left(\sum_{k \in S} P[X_3 = 2, X_2 = k, X_1 = 2]\right) \Big/ P[X_1 = 2] \\
&= \frac{\sum_{k \in S} P[X_3 = 2|X_2 = k, X_1 = 2] P[X_2 = k, X_1 = 2]}{P[X_1 = 2]} \\
&= \sum_{k \in S} P[X_3 = 2|X_2 = k] P[X_2 = k|X_1 = 2] \\
&= p_{12}p_{21} + p_{22}p_{22} = 0.7.
\end{aligned}$$

Thus with 70% chance, a driver known to be classified as standard at the beginning of the first year will be reclassified as standard at the beginning of the third year. □

Example 2.1.2 Operating condition of a machine at any time is classified as follows: State 1: Good; State 2: Deteriorated; State 3: In repair. We observe the condition of the machine at the end of each period in a sequence of periods. Suppose X_n denotes the condition of the machine at the end of nth time period for $n \geq 1$. We assume that the sequence of machine conditions is a Markov chain with transition probability matrix P as given below

$$P = \begin{matrix} & \begin{matrix} 1 & 2 & 3 \end{matrix} \\ \begin{matrix} 1 \\ 2 \\ 3 \end{matrix} & \begin{pmatrix} 0.9 & 0.1 & 0 \\ 0 & 0.9 & 0.1 \\ 1 & 0 & 0 \end{pmatrix} \end{matrix}.$$

Suppose that in the initial time period, the machine is in a good condition. To find the probability that it remains in the good condition till the end of fourth time period, we use the Markov property repeatedly as illustrated below

$$\begin{aligned} & P[X_4 = 1, X_3 = 1, X_2 = 1, X_1 = 1 | X_0 = 1] \\ &= P[X_4 = 1, X_3 = 1, X_2 = 1, X_1 = 1, X_0 = 1]/P[X_0 = 1] \\ &= P[X_4 = 1 | X_3 = 1]P[X_3 = 1 | X_2 = 1]P[X_2 = 1 | X_1 = 1] \\ &\times P[X_1 = 1 | X_0 = 1] = (p_{11})^4 = (0.9)^4 = 0.6561 . \end{aligned}$$

The chance that the machine is in "repair" state for the first time at the end of the third period is given by $P[X_3 = 3, X_2 \neq 3, X_1 \neq 3 | X_0 = 1]$. It is to be noted that $p_{13} = 0$ and hence

$$\begin{aligned} & P[X_3 = 3, X_2 \neq 3, X_1 \neq 3 | X_0 = 1] \\ &= \sum_{i=1}^{2} \sum_{j=1}^{2} P[X_3 = 3, X_2 = j, X_1 = i | X_0 = 1] \\ &= P[X_3 = 3, X_2 = 2, X_1 = 1 | X_0 = 1] \\ &+ P[X_3 = 3, X_2 = 2, X_1 = 2 | X_0 = 1] \\ &= p_{11}p_{12}p_{23} + p_{12}p_{22}p_{23} = 0.018 . \end{aligned}$$

□

Example 2.1.3 An insurer issues an insurance contract to a person when the transitions among four states—1: active; 2: disabled; 3: withdrawn; 4: dead—are governed by a homogeneous Markov model with the following transition probability matrix:

$$P = \begin{matrix} & \begin{matrix} 1 & 2 & 3 & 4 \end{matrix} \\ \begin{matrix} 1 \\ 2 \\ 3 \\ 4 \end{matrix} & \begin{pmatrix} 0.50 & 0.25 & 0.15 & 0.10 \\ 0.40 & 0.40 & 0 & 0.20 \\ 0 & 0 & 1 & 0 \\ 0 & 0 & 0 & 1 \end{pmatrix} \end{matrix}.$$

Suppose we are interested in knowing the chance that after two years, the active person is in a disabled state. Thus, we want to find the probability $P[X_2 = 2|X_0 = 1]$. Observe that the person is in state 1 initially; at time 1, he will be in some state in state space $S = \{1, 2, 3, 4\}$ and then transits to state 2 at time 2. Thus, the probability $P[X_2 = 2|X_0 = 1]$ can be found as follows:

$$\begin{aligned}
P[X_2 = 2|X_0 = 1] &= P[X_2 = 2, X_1 \in S|X_0 = 1] \\
&= \sum_{k=1}^{4} P[X_2 = 2, X_1 = k|X_0 = 1] \\
&= \sum_{k=1}^{4} P[X_2 = 2|X_1 = k, X_0 = 1]P[X_1 = k|X_0 = 1] \\
&= \sum_{k=1}^{4} P[X_2 = 2|X_1 = k]P[X_1 = k|X_0 = 1] \\
&= \sum_{k=1}^{4} p_{1k}p_{k2} = 0.5 \times 0.25 + 0.25 \times 0.40 = 0.225 \,,
\end{aligned}$$

where the second last step follows by the Markov property. Suppose the death benefit is Rs. 10000/-, payable at the end of the year of death and premiums are to be paid at the beginning of each year when the insured is active. Insureds do not pay annual premiums when they are disabled. Using the given transition probability matrix, one can calculate the annual net premium for this insurance, Deshmukh [4]. □

Example 2.1.4 Suppose $\{Y_n, n \geq 0\}$ is a sequence of independent discrete random variables with set S as the set of all possible values of $Y_n, n \geq 0$. Since $Y_n, n \geq 0$ are independent, for any $i, j, \ldots, y_0 \in S$,

$$\begin{aligned}
P[Y_{n+1} = j|Y_n = i, \ldots, Y_0 = y_0] &= P[Y_{n+1} = j] \\
&= P[Y_{n+1} = j|Y_n = i].
\end{aligned}$$

Hence, $\{Y_n, n \geq 0\}$ is a Markov chain. If $\{Y_n, n \geq 0\}$ is a sequence of identically distributed random variables, then it is a time homogeneous Markov chain. Now suppose $X_n = \sum_{i=0}^{n} Y_i$ with S_1 as a set of all possible values of $X_n, n \geq 0$. We examine whether $\{X_n, n \geq 0\}$ is a Markov chain. For $n \geq 1$ and any $x_0, x_1, \ldots, i, j \in S_1$, observe that

$$
\begin{aligned}
&P[X_n = j \big| X_{n-1} = i, \ldots, X_0 = x_0] \\
=&P\Big[\sum_{r=0}^{n} Y_r = j \big| X_{n-1} = i, \ldots, X_0 = x_0\Big] \\
=&P[X_{n-1} + Y_n = j \big| X_{n-1} = i, \ldots, X_0 = x_0] \\
=&P[Y_n = j - i \big| X_{n-1} = i, \ldots, X_0 = x_0] = P[Y_n = j - i].
\end{aligned}
$$

The last equality follows since the conditioning event is a function of $Y_0, \ldots, Y_{n-1}$ only and $\{Y_n, n \geq 0\}$ is a sequence of independent random variables. Similarly,

$$P[X_n = j | X_{n-1} = i] = P[Y_n = j - i].$$

It thus follows that $\{X_n, n \geq 0\}$ is a Markov chain. If $\{Y_n, n \geq 0\}$ is a sequence of identically distributed random variables, then $P[Y_n = j - i] = P[Y_1 = j - i]$ which does not depend on n and hence $\{X_n, n \geq 0\}$ is a time homogeneous Markov chain. □

The following example presents a sequence of random variables which is not a Markov chain.

Example 2.1.5 Suppose $\{Y_n, n \geq 0\}$ is a sequence of independent and identically distributed random variables, each following Bernoulli $B(1, p)$ distribution. For $n \geq 1$, we define random variables X_n as $X_n = Y_n + Y_{n-1}$. Thus, $\{X_n, n \geq 1\}$ is a stochastic process with state space $\{0, 1, 2\}$. We compute $P[X_3 = 1 | X_2 = 1, X_1 = 1]$ and $P[X_3 = 1 | X_2 = 1]$ and examine whether the two conditional probabilities are the same. Observe that

$$
\begin{aligned}
P[X_3 = 1, X_2 = 1, X_1 = 1] &= P[Y_3 + Y_2 = 1, Y_2 + Y_1 = 1, Y_1 + Y_0 = 1] \\
&= P[Y_0 = 1, Y_1 = 0, Y_2 = 1, Y_3 = 0] \\
&+ P[Y_0 = 0, Y_1 = 1, Y_2 = 0, Y_3 = 1] = 2p^2q^2 \\
P[X_2 = 1, X_1 = 1] &= P[Y_2 + Y_1 = 1, Y_1 + Y_0 = 1] \\
&= P[Y_0 = 1, Y_1 = 0, Y_2 = 1] \\
&+ P[Y_0 = 0, Y_1 = 1, Y_2 = 0] = p^2q + pq^2 = pq \\
P[X_3 = 1, X_2 = 1] &= P[Y_3 + Y_2 = 1, Y_2 + Y_1] \\
&= P[Y_1 = 0, Y_2 = 1, Y_3 = 0] \\
&+ P[Y_1 = 1, Y_2 = 0, Y_3 = 1] = p^2q + pq^2 = pq \\
P[X_2 = 1] &= P[Y_2 + Y_1 = 1] \\
&= P[Y_1 = 0, Y_2 = 1] + P[Y_1 = 1, Y_2 = 0] = 2pq.
\end{aligned}
$$

$$\begin{aligned}
\text{Hence,}\quad & P[X_3 = 1|X_2 = 1, X_1 = 1] = 2p^2q^2/pq = 2pq \\
& \& \ P[X_3 = 1|X_2 = 1] = pq/2pq = 1/2 \\
\Rightarrow\quad & P[X_3 = 1|X_2 = 1, X_1 = 1] \neq P[X_3 = 1|X_2 = 1] \ \text{if} \ p \neq 1/2,
\end{aligned}$$

which implies that $\{X_n, n \geq 1\}$ is not a Markov chain if $p \neq 1/2$. It may be noted that the two subsequences $\{X_{2n-1}, n \geq 1\}$ and $\{X_{2n}, n \geq 1\}$ are homogeneous Markov chains. □

Remark 2.1.2 The sequence $\{X_n, n \geq 1\}$ defined in Example 2.1.5 is a particular case of an important stochastic process $\{X_n, n \geq 1\}$, defined by $X_n = \alpha Y_n + Y_{n-1}$. This process is known as a moving average process, which is not a Markov chain.

The following theorem gives one more approach to define a Markov chain. We first prove a lemma used in the proof of the theorem.

Lemma 2.1.1 *Suppose* $(\Omega, \mathbb{A}, P)$ *is a probability space. Suppose* $A, C, D \in \mathbb{A}$ *and* $\{B_1, B_2, \ldots\}$ *is a measurable countable partition of* D. *Then*

$$P(A|C \cap B_i) = p \ \forall \ i \geq 1 \ \Rightarrow \ P(A|C \cap D) = p, \quad p \in (0, 1).$$

Proof Observe that

$$\begin{aligned}
P(A|C \cap D) &= P(A \cap C \cap D)\big/P(C \cap D) \\
&= P(A \cap C \cap (\cup B_i))\big/P(C \cap (\cup B_i)) \\
&= \sum_{i\geq 1} P(A \cap C \cap B_i)\Big/\sum_{i\geq 1} P(C \cap B_i) \\
&= \sum_{i\geq 1} P(A|C \cap B_i)P(C \cap B_i)\Big/\sum_{i\geq 1} P(C \cap B_i) \\
&= p, \quad \text{since} \ P(A|C \cap B_i) = p \ \forall \ i \geq 1.
\end{aligned}$$

□

Theorem 2.1.1 *Suppose* $\{X_n, n \geq 0\}$ *is a discrete time discrete state space stochastic process with state space* S, *such that* $\forall \ n \geq 1$ *and* $\forall \ x_0, x_1, \ldots, x_{n-1}, i, j \in S$,

$$P[X_{n+1} = j|X_n = i, X_{n-1} = x_{n-1}, X_{n-2} = x_{n-2}, \ldots X_0 = x_0] = a(i, j, n, n + 1),$$

where $a(i, j, n, n + 1)$ *does not depend on* $x_0, x_1, \ldots, x_{n-1}$. *Then* $P[X_{n+1} = j|X_n = i] = a(i, j, n, n + 1)$, $\forall \ i, j \in S$, *that is, the stochastic process* $\{X_n, n \geq 0\}$ *satisfies the Markov property and hence is a Markov chain.*

Proof By repeatedly applying Lemma 2.1.1 $(n - 1)$ times, we note that

$$\begin{aligned}
P[X_{n+1} = j|X_n = i] &= P[X_{n+1} = j|X_n = i, X_{n-1} \in S, X_{n-2} \in S, \ldots, X_0 \in S] \\
&= a(i, j, n, n + 1) \ \forall \ i, j \in S.
\end{aligned}$$

Thus, the stochastic process $\{X_n, n \geq 0\}$ satisfies the Markov property and hence is a Markov chain. □

We illustrate the proof of Theorem 2.1.1 for a discrete time discrete state space stochastic process with state space $S = \{0, 1\}$. Suppose

$$P[X_3 = j|X_2 = i, X_1 = x_1, X_0 = x_0] = a(i, j), \;\; x_1, x_0 \in S.$$

Observe that

$$\begin{aligned} P[X_3 = j|X_2 = i] &= P[X_3 = j, X_2 = i, X_1 \in S, X_0 \in S]/P[X_2 = i, X_1 \in S, X_0 \in S] \\ &= P[X_3 = j|X_2 = i, X_1 \in S, X_0 \in S]. \end{aligned}$$

With the notation of Lemma 2.1.1, suppose $A = [X_3 = j]$, $C = [X_2 = i]$ and $D = [X_1 \in S, X_0 \in S]$. Further the event D can be written as

$$\begin{aligned} D &= [X_1 = 0, X_0 = 0] \cup [X_1 = 0, X_0 = 1] \cup [X_1 = 1, X_0 = 0] \cup [X_1 = 1, X_0 = 1] \\ &= B_1 \cup B_2 \cup B_3 \cup B_4. \end{aligned}$$

It is given that $P[X_3 = j|X_2 = i, X_1 = x_1, X_0 = x_0] = P(A|C \cap B_r) = a(i, j)$ and is the same for $r = 1, 2, 3, 4$. Hence,

$$P[X_3 = j|X_2 = i] = P(A|C \cap D) = a(i, j)$$

and the Markov property is satisfied.

Converse of Theorem 2.1.1 is also true and it follows from the Markov property. As a consequence, we have one more definition of a Markov chain as stated below.

Definition 2.1.7 Suppose $\{X_n, n \geq 0\}$ is a discrete time discrete state space stochastic process with state space S, such that

$$P[X_{n+1} = j|X_n = i, X_{n-1} = x_{n-1}, \ldots X_0 = x_0] = a(i, j, n, n + 1),$$

$\forall \; n \geq 1$ and $\forall \; x_0, x_1, \ldots, x_{n-1}, i, j \in S$. Then $\{X_n, n \geq 0\}$ is a Markov chain.

From Definition 2.1.7, we note that to examine whether a given discrete time discrete state space stochastic process is a Markov chain, it is enough to show that the conditional probability $P[X_{n+1} = j|X_n = i, X_{n-1} = x_{n-1}, \ldots X_0 = x_0]$ does not depend on $x_0, x_1, \ldots, x_{n-1}$. We use this definition in many examples.

Remark 2.1.3 The condition in Definition 2.1.7 is a natural extension of that of a sequence $\{X_n, n \geq 0\}$ of independent random variables, in which case

$$P[X_{n+1} = j|X_n = i, X_{n-1} = x_{n-1}, \ldots X_0 = x_0] = b(j, n + 1),$$

$\forall \; n \geq 1$ and $\forall \; x_0, x_1, \ldots, x_{n-1}, i, j \in S$. Thus, the sequence of independent random variables is a Markov chain.

The following example illustrates Theorem 2.1.1.

Example 2.1.6 Suppose $\{Y_n, n \geq 0\}$ is a sequence of independent and identically distributed random variables with possible values $\{0, 1, 2, 3\}$ and with respective probabilities $\{0.1, 0.3, 0.2, 0.4\}$. Suppose $X_n = \max\{Y_0, Y_1, \ldots, Y_n\}$. We examine whether $\{X_n, n \geq 0\}$ is a Markov chain. Observe that by the definition of X_n,

$$\begin{aligned} X_n &= \max\{Y_0, Y_1, \ldots, Y_n\} = \max\{\max\{Y_0, Y_1, \ldots, Y_{n-1}\}, Y_n\} \\ &= \max\{X_{n-1}, Y_n\}. \end{aligned}$$

Now by Theorem 2.1.1, for any $x_0, x_1, \ldots, i, j \in S$ and for any $n \geq 1$,

$$\begin{aligned} & P[X_n = j | X_{n-1} = i, \ldots, X_0 = x_0] \\ =& P[\max\{X_{n-1}, Y_n\} = j | X_{n-1} = i, \ldots, X_0 = x_0] \\ =& P[\max\{i, Y_n\} = j | X_{n-1} = i, \ldots, X_0 = x_0] \\ =& P[\max\{i, Y_n\} = j | X_{n-1} = i] \\ =& P[\max\{i, Y_n\} = j] = a(i, j, n-1, n). \end{aligned}$$

Hence, $\{X_n, n \geq 0\}$ is a Markov chain with state space $\{0, 1, 2, 3\}$. To determine the transition probabilities, note that given X_{n-1}, the possible values of X_n depend on the values of Y_n, as is clear from the following cases. (i) Suppose $X_{n-1} = 0$. If $Y_n = 0$ then $X_n = 0$, if $Y_n = 1$ then $X_n = 1$, if $Y_n = 2$ then $X_n = 2$ and if $Y_n = 3$ then $X_n = 3$. (ii) If $X_{n-1} = 1$, then X_n cannot be 0; X_n is 1, if $Y_n = 0$ or 1; X_n is 2, if $Y_n = 2$ and X_n is 3, if $Y_n = 3$. (iii) If $X_{n-1} = 2$, then X_n cannot be 0, 1; X_n is 2, if $Y_n = 0$, 1 or 2 and X_n is 3 if $Y_n = 3$. (iv) If $X_{n-1} = 3$, then X_n cannot be 0, 1, 2; X_n is 3, if $Y_n = 0, 1, 2$ or 3. Since $\{Y_n, n \geq 0\}$ is a sequence identically distributed random variables, $P[Y_n = j]$ is the same for all n. Hence, $\{X_n, n \geq 0\}$ is a time homogeneous Markov chain. Its one step transition probability matrix P is given by

$$P = \begin{array}{c} \\ 0 \\ 1 \\ 2 \\ 3 \end{array} \begin{array}{c} \begin{array}{cccc} 0 & 1 & 2 & 3 \end{array} \\ \begin{pmatrix} 0.1 & 0.3 & 0.2 & 0.4 \\ 0 & 0.4 & 0.2 & 0.4 \\ 0 & 0 & 0.6 & 0.4 \\ 0 & 0 & 0 & 1 \end{pmatrix} \end{array}.$$

□

The Markov property states that, given the past and present, the probabilistic behavior of the future depends only on the present. If the conditional probability distribution of the state of the system at time $n+1$, given the information at all previous time points, depends on the states of the system for previous r time points, then we have a Markov chain of order r, as defined below.

Definition 2.1.8 Markov Chain of Order r: Suppose $\{X_n, n \geq 0\}$ is a discrete time stochastic process with countable state space S. It is a Markov chain of order r, if $\forall \; n \geq r$ and $\forall \; x_0, x_1, \ldots, x_{n+1} \in S$, the conditional probability $P[X_{n+1} = x_{n+1} | X_n = x_n, \ldots X_0 = x_0]$ is the same as the conditional probability $P[X_{n+1} = x_{n+1} | X_n = x_n, \ldots, X_{n-r+1} = x_{n-r+1}]$, provided these are defined.

A Markov chain defined in Eq. (2.1.1) is the first-order Markov chain and is simply referred to as a Markov chain. A sequence of independent random variables is referred to as a 0th-order Markov chain. It is to be noted that Theorem 2.1.1 is applicable in this setup also. A higher order Markov chain can be reduced to the first-order Markov chain. Example 2.1.7 shows how a second-order Markov chain can be reduced to the first-order Markov chain. As a consequence, all techniques developed for the first-order Markov chain are applicable for the higher order Markov chains also. Hence, we concentrate on the theory of the first-order Markov chains.

The next example illustrates how a Markov chain of order 2 can be converted to the first-order Markov chain.

Example 2.1.7 Suppose weather condition of a certain locality on any day depends on the weather conditions on the previous two days. Suppose X_n denotes the weather condition of the nth day, then $\{X_n, n \geq 0\}$ is assumed to be a second-order Markov chain with state space $S = \{1, 2\}$ where 1 denotes the sunny day and 2 denotes the cloudy day. Thus, $\forall \, n \geq 2$ and $\forall \, x_i \in S$,

$$
\begin{aligned}
&P[X_{n+1} = x_{n+1} | X_n = x_n, X_{n-1} = x_{n-1} \ldots X_0 = x_0] \\
=&P[X_{n+1} = x_{n+1} | X_n = x_n, X_{n-1} = x_{n-1}],
\end{aligned}
$$

provided the conditional probabilities are defined. Suppose that

$$
\begin{aligned}
&P[X_{n+1} = 1 | X_n = 1, X_{n-1} = 1] = 0.8, \;\; P[X_{n+1} = 1 | X_n = 1, X_{n-1} = 2] = 0.6 \\
&P[X_{n+1} = 1 | X_n = 2, X_{n-1} = 1] = 0.4, \;\; P[X_{n+1} = 1 | X_n = 2, X_{n-1} = 2] = 0.1\,.
\end{aligned}
$$

We show that $\{X_n, n \geq 0\}$ can be transformed to a first-order Markov chain as follows. We define a random vector $\underline{Y}_n = (X_n, X_{n-1}), n \geq 1$. The possible values of $\underline{Y}_n$ are as follows. It is $(1, 1)$, if it is sunny on the two consecutive days, it is $(1, 2)$, if it is sunny on nth day and cloudy on $(n-1)$th day, it is $(2, 1)$, if it is cloudy on nth day and sunny on $(n-1)$th day and it is $(2, 2)$, if it is cloudy on the two consecutive days. Thus, the state space of $\underline{Y}_n$ is $S_1 = \{(1, 1), (1, 2), (2, 1), (2, 2)\}$. For any pair in S_1, from the definition of $\underline{Y}_n$, observe that the conditional probability

$$
P[\underline{Y}_{n+1} = (i_{n+1}, j_{n+1}) | \underline{Y}_n = (i_n, j_n), \underline{Y}_{n-1} = (i_{n-1}, j_{n-1}), \ldots, \underline{Y}_1 = (i_1, j_1)]
$$

is defined only if $j_{k+1} = i_k, \; k = 1, 2, \ldots, n+1$ and it is 0 otherwise. Thus, for any $k \geq 1$, the second coordinate of $\underline{Y}_{k+1}$ should be the same as the first coordinate of $\underline{Y}_k$. Under this condition,

$$
\begin{aligned}
&P[\underline{Y}_{n+1} = (i_{n+1}, i_n) | \underline{Y}_n = (i_n, i_{n-1}), \underline{Y}_{n-1} = (i_{n-1}, i_{n-2}), \ldots, \underline{Y}_1 = (i_1, i_0)] \\
&\quad = \frac{P[\underline{Y}_{n+1} = (i_{n+1}, i_n), \underline{Y}_n = (i_n, i_{n-1}), \ldots, \underline{Y}_1 = (i_1, i_0)]}{P[\underline{Y}_n = (i_n, i_{n-1}), \underline{Y}_{n-1} = (i_{n-1}, i_{n-2}), \ldots, \underline{Y}_1 = (i_1, i_0)]} \\
&\quad = \frac{P[X_{n+1} = i_{n+1}, X_n = i_n, X_n = i_n, X_{n-1} = i_{n-1}, \ldots, X_0 = i_0]}{P[X_n = i_n, X_{n-1} = i_{n-1}, \ldots, X_0 = i_0]} \\
&\quad = P[X_{n+1} = i_{n+1} | X_n = i_n, X_{n-1} = i_{n-1}, \ldots, X_0 = i_0] \\
&\quad = P[X_{n+1} = i_{n+1} | X_n = i_n, X_{n-1} = i_{n-1}] \\
&\quad = P[X_{n+1} = i_{n+1}, X_n = i_n | X_n = i_n, X_{n-1} = i_{n-1}] \\
&\quad = P[\underline{Y}_{n+1} = (i_{n+1}, i_n) | \underline{Y}_n = (i_n, i_{n-1})].
\end{aligned}
$$

In the fourth step, we use the assumption that $\{X_n, n \geq 0\}$ is the second-order Markov chain. Thus, it is proved that $\{\underline{Y}_n, n \geq 1\}$ is the first-order Markov chain with state space S_1. From the transition probabilities of the Markov chain $\{X_n, n \geq 0\}$, the transition probability matrix P of $\{\underline{Y}_n, n \geq 1\}$ can be written as follows:

$$
P = \begin{array}{c} (1,1) \\ (1,2) \\ (2,1) \\ (2,2) \end{array}
\overset{\begin{array}{cccc} (1,1) & (1,2) & (2,1) & (2,2) \end{array}}{\begin{pmatrix} 0.8 & 0.2 & 0 & 0 \\ 0 & 0 & 0.4 & 0.6 \\ 0.6 & 0.4 & 0 & 0 \\ 0 & 0 & 0.1 & 0.9 \end{pmatrix}}.
$$

Observe that P is a stochastic matrix. □

It is to be noted that for each n, X_n is a discrete random variable with set S as its support. To compute the probabilities of various events related to a Markov chain, we need to know the joint probability distribution of $\{X_0, X_1, X_2, \ldots, X_n\}$. In the following theorem, using the Markov property, we prove that it can be expressed in terms of the one step transition probabilities and the initial distribution.

Theorem 2.1.2 *Suppose $\{X_n, n \geq 0\}$ is a Markov chain with state space S, one step transition probabilities $p_{ij}^{n,n+1}$, $i, j \in S$ and initial distribution $\underline{p}^{(0)}$. Then the joint probability distribution of $\{X_0, X_1, \ldots, X_n\}$ is completely specified by the one step transition probabilities and the initial distribution.*

Proof For $n \geq 1$ and any $x_0, x_1, \ldots, x_n \in S$, we write the joint probability $P[X_n = x_n, \ldots X_0 = x_0]$ in terms of the conditional probability and use the Markov property repeatedly in the following derivation:

$$\begin{aligned}
& P[X_n = x_n, X_{n-1} = x_{n-1}, \dots, X_0 = x_0] \\
&= P[X_n = x_n | X_{n-1} = x_{n-1}, \dots, X_0 = x_0] \\
&\times P[X_{n-1} = x_{n-1}, \dots, X_0 = x_0] \\
&= P[X_n = x_n | X_{n-1} = x_{n-1}] \text{ by Markov property} \\
&\times P[X_{n-1} = x_{n-1}, X_{n-2} = x_{n-2}, \dots, X_0 = x_0] \\
&= p^{n-1,n}_{x_{n-1}x_n} \times P[X_{n-1} = x_{n-1}, \dots, X_0 = x_0] \\
&= p^{n-1,n}_{x_{n-1}x_n} \times p^{n-2,n-1}_{x_{n-2}x_{n-1}} \times \cdots p^{0,1}_{x_0x_1} \times P[X_0 = x_0]\,.
\end{aligned}$$

Thus, the joint distribution of $\{X_0, X_1, X_2, \dots, X_n\}$ is completely specified in terms of the one step transition probabilities and the initial distribution. □

Remark 2.1.4 Theorem 2.1.2 conveys that if we know $P[X_n = x_n, X_{n-1} = x_{n-1}, \dots, X_0 = x_0]$, for all $n \ge 0$ and for all $x_0, x_1, x_2, \dots, x_n \in S$, then for any $t_0 < t_1 < \cdots < t_n \in T$, we can find $P[X_{t_n} = x_n, X_{t_{n-1}} = x_{t_{n-1}}, \dots, X_{t_0} = x_0]$, by using the method of finding lower dimensional joint distributions from a higher dimensional joint distribution. It is illustrated in the following example. Thus, every element $F_{t_1,t_2,\dots,t_n}$ from the family of finite dimensional distribution functions can be obtained, for all $n \ge 1$, for all $x_0, x_1, x_2, \dots, x_n \in \mathbb{R}$ and for $t_1, t_2, \dots, t_n \in T$. In Chap. 1, we have noted that the probability law of the stochastic process is completely determined by the family of finite dimensional distribution functions. Theorem 2.1.2 conveys that the one step transition probabilities and the initial distribution of a Markov chain determine its family of finite dimensional distribution functions, that is, the probability law of a Markov chain. Hence in subsequent discussions, we specify a time homogeneous Markov chain by the triplet $(S, \underline{p}^{(0)}, P)$ which indicates the state space, the initial distribution and the one step transition probability matrix, respectively. Further, note that the probability structure of a Markov chain is mainly governed by the one step transition probabilities and hence the entire development of the theory of Markov chains is based on the transition probabilities.

In the following example, for a Markov chain $\{X_n, n \ge 0\}$ specified by the triplet $(S, \underline{p}^{(0)}, P)$, we compute the joint distribution of X_k, X_l for some k and l.

Example 2.1.8 Suppose $\{X_n, n \ge 0\}$ is a Markov chain with state space $S = \{1, 2, 3\}$, and the one step transition probability matrix P is given by

$$P = \begin{matrix} \\ 1 \\ 2 \\ 3 \end{matrix} \begin{matrix} \begin{matrix} 1 & 2 & 3 \end{matrix} \\ \begin{pmatrix} 0.4 & 0.4 & 0.2 \\ 0.6 & 0.2 & 0.2 \\ 0.5 & 0.4 & 0.1 \end{pmatrix} \end{matrix}.$$

Suppose the initial distribution is given by $\underline{p}^{(0)} = (0.3, 0.3, 0.4)$. We compute $P[X_2 = 2, X_4 = 3]$ from the joint distribution of $X_0, X_1, \dots, X_4$, using the Markov property as follows:

$$\begin{aligned}
P[X_2 = 2, X_4 = 3] &= \sum_{i,j,l=1}^{3} P[X_4 = 3, X_3 = j, X_2 = 2, X_1 = i, X_0 = l] \\
&= \sum_{i,j,l=1}^{3} p_{li} p_{i2} p_{2j} p_{j3} P[X_0 = l] = \sum_{i,j,l=1}^{3} p_l^{(0)} p_{li} p_{i2} p_{2j} p_{j3} \\
&= \sum_{i,l=1}^{3} p_l^{(0)} p_{li} p_{i2} \times \sum_{j=1}^{3} p_{2j} p_{j3} = (0.332)(0.18) = 0.0598.
\end{aligned}$$

On similar lines, given the initial distribution and the one step transition probability matrix, we can compute $P[X_2 = i, X_4 = j]$ for any $i, j \in S$. □

From the result proved in Theorem 2.1.2, we have the following definition of a Markov chain.

Definition 2.1.9 Markov Chain: Suppose $\{X_n, n \geq 0\}$ is a discrete time stochastic process with countable state space S. It is a Markov chain, if $\forall\ n \geq 1$ and $\forall\ x_0, x_1, \ldots, x_n \in S$,

$$\begin{aligned}
P[X_n = x_n, X_{n-1} = x_{n-1}, \ldots X_0 = x_0] = p_{x_{n-1}x_n}^{n-1,n} \times p_{x_{n-2}x_{n-1}}^{n-2,n-1} \times \cdots \\
\times\ p_{x_0x_1}^{0,1} \times P[X_0 = x_0]\ ,
\end{aligned}$$

where $p_{ij}^{k,k+1} = P[X_{k+1} = j | X_k = i],\ k = 0, 1, \ldots, n-1$.

In the following theorem, we prove that the two definitions of a Markov chain are equivalent.

Theorem 2.1.3 *Definitions 2.1.1 and 2.1.9 of a Markov chain are equivalent.*

Proof From Theorem 2.1.2, it immediately follows that Definition 2.1.1 of a Markov chain implies Definition 2.1.9 of a Markov chain. To prove the other way, we have

$$\begin{aligned}
P[X_n = j | X_{n-1} = i, \ldots X_0 = x_0] &= \frac{P[X_n = j, X_{n-1} = i, \ldots X_0 = x_0]}{P[X_{n-1} = i, \ldots X_0 = x_0]} \\
&= \frac{p_{ij}^{n-1,n} \times p_{x_{n-2}i}^{n-2,n-1} \times \cdots p_{x_0x_1}^{0,1} \times P[X_0 = x_0]}{p_{x_{n-2}i}^{n-2,n-1} \times \cdots p_{x_0x_1}^{0,1} \times P[X_0 = x_0]} \\
&= p_{ij}^{n-1,n} = P[X_n = j | X_{n-1} = i],
\end{aligned}$$

where the last equality follows from Theorem 2.1.1. Hence, Definition 2.1.9 of a Markov chain implies Definition 2.1.1 of a Markov chain. □

Having defined a Markov model, in the next section, we develop the mathematical machinery which is necessary for the analysis of such models.

2.2 Higher Step Transition Probabilities

The analysis of a Markov chain concerns mainly the calculation of the probabilities of the partial realizations of the process. Central to these calculations are the transition probabilities for longer time periods. In Example 2.1.3, we have obtained the probability that after two years, the active person is in a disabled state, that is, $P[X_2 = 2|X_0 = 1]$. The same derivation can be obtained using the following simple result from probability theory. It is used frequently in many theorems. Suppose $(\Omega, \mathbb{A}, P)$ is a probability space. Suppose $A, B \in \mathbb{A}$ and $\{B_1, B_2, \ldots, \}$ is a measurable partition of Ω. Then

$$\begin{aligned}
P(A|B) &= P(A \cap B)/P(B) = P(A \cap \Omega \cap B)/P(B) \\
&= P(A \cap (\cup B_i) \cap B)/P(B) = P(\cup(A \cap B_i \cap B))/P(B) \\
&= \sum_{i \geq 1} P(A \cap B_i \cap B)/P(B) = \sum_{i \geq 1} P(A|B_i \cap B)P(B_i \cap B)/P(B) \\
&= \sum_{i \geq 1} P(A|B_i \cap B)P(B_i|B) \,. \qquad (2.2.1)
\end{aligned}$$

Using this result, $P[X_2 = 2|X_0 = 1]$ in Example 2.1.3 can be computed as follows:

$$\begin{aligned}
P[X_2 = 2|X_0 = 1] &= P[X_2 = 2, X_1 \in S|X_0 = 1] \\
&= \sum_{k=1}^{4} P[X_2 = 2, X_1 = k|X_0 = 1] \\
&= \sum_{k=1}^{4} P[X_2 = 2|X_1 = k, X_0 = 1]P[X_1 = k|X_0 = 1] \\
&= \sum_{k=1}^{4} P[X_2 = 2|X_1 = k]P[X_1 = k|X_0 = 1] = \sum_{k=1}^{4} p_{1k}p_{k2} \\
&= \sum_{k=1}^{4} (1, k)\text{th element of P } \times \ (k, 2)\text{th element of P} \\
&= (1, 2)\text{th element of } \ P^2 \,.
\end{aligned}$$

The fourth step follows by the Markov property. Thus, the probability of transition from 1 to 2 in two steps is given by the $(1, 2)$th element of P^2. It is known as a two step transition probability and is denoted by $p_{12}^{(2)}$. The probability that after 10 years, the active person dies is $P[X_{10} = 4|X_0 = 1]$. Proceeding on similar lines, we can show that $P[X_{10} = 4|X_0 = 1] = p_{14}^{(10)}$, the 10-step transition probability, which is given by the $(1, 4)$th element of P^{10}.

Definition 2.2.1 n-step Transition Probability and n-step Transition Probability Matrix: Suppose $\{X_n, n \geq 0\}$ is a Markov chain with state space S. Then for $i, j \in S$, $p_{ij}^{(n)} = P[X_n = j | X_0 = i]$ is defined as the n-step transition probability from state i to state j. The matrix $P^{(n)} = [p_{ij}^{(n)}]$ is known as the n-step transition probability matrix.

It may be noted that $p_{ij}^{(n)}$ is the probability that starting in state i, the Markov chain is in state j at the nth step, it may or may not be in state j prior to the nth step. To express $P^{(n)}$ in terms of P, we now give one more definition of a Markov chain, which is in terms of higher step transition probabilities.

Definition 2.2.2 Markov Chain: Suppose $\{X_n, n \geq 0\}$ is a sequence of random variables with a countable state space S. The sequence $\{X_n, n \geq 0\}$ is defined as a Markov chain if $\forall\, n \geq 0$, $\forall\, x_0, x_1, \ldots, x_{n+1} \in S$ and $\forall\, 0 = t_0 < t_1 < \cdots < t_{n+1}$,

$$P[X_{t_{n+1}} = x_{n+1} | X_{t_n} = x_n, \ldots, X_0 = x_0] = P[X_{t_{n+1}} = x_{n+1} | X_{t_n} = x_n], \quad (2.2.2)$$

provided the conditional probabilities are defined.

The difference between Definitions 2.1.1 and 2.2.2 is that in the latter the time points need not be consecutive. We prove in the following theorem that the two definitions are equivalent. The proof is mainly based on the definition of matrix multiplication.

Theorem 2.2.1 *Definitions 2.1.1 and 2.2.2 of a Markov chain are equivalent.*

Proof Definition 2.1.1 follows immediately from Definition 2.2.2 by taking $t_1 = 1, \ldots, t_{n+1} = n + 1$. We now prove that Definition 2.2.2 follows from Definition 2.1.1. Thus, we have to prove that $\forall\, n \geq 0$, $\forall\, x_0, x_{t_1}, \ldots, x_{t_{n+1}} \in S$ and $\forall\ 0 \leq t_1 \leq \cdots \leq t_{n+1}$,

$$P[X_{t_{n+1}} = x_{t_{n+1}} | X_{t_n} = x_{t_n}, \ldots X_0 = x_0] = P[X_{t_{n+1}} = x_{t_{n+1}} | X_{t_n} = x_{t_n}].$$

Suppose A denotes a set of all integers strictly between t_j and t_{j-1}, $j = 1, 2, \ldots, n + 1$. Suppose B denotes a set of all integers strictly between t_j and t_{j-1}, $j = 1, 2, \ldots, n$. Observe that

$$
\begin{aligned}
& P[X_{t_{n+1}} = x_{t_{n+1}} | X_{t_n} = x_{t_n}, \dots X_0 = x_0] \\
&= \frac{P[X_{t_{n+1}} = x_{t_{n+1}}, X_{t_n} = x_{t_n}, \dots X_0 = x_0]}{P[X_{t_n} = x_{t_n}, \dots X_0 = x_0]} \\
&= \frac{\sum_{x_j \in S, j \in A} P[X_{t_{n+1}} = x_{n+1}, X_j = x_j, X_0 = x_0]}{\sum_{x_j \in S, j \in B} P[X_{t_n} = x_{t_n}, X_j = x_j, \dots, X_0 = x_0]} \\
&= \frac{\sum_{x_j \in S, j \in A} \prod_{r=0}^{n} p_{x_r, x_{r+1}}}{\sum_{x_j \in S, j \in B} \prod_{r=0}^{n-1} p_{x_r, x_{r+1}}} \quad \text{by Definition 2.1.1} \\
&= p^{(t_{n+1}-t_n)}_{x_{t_n}, x_{t_{n+1}}} = P[X_{t_{n+1}} = x_{t_{n+1}} | X_{t_n} = x_{t_n}],
\end{aligned}
$$

where in the second last step we have used the formula for matrix multiplication. The last step follows from Lemma 2.1.1. Thus, Definition 2.2.2 follows from Definition 2.1.1. □

We illustrate the second part of the proof for particular values of t_r. Observe that

$$
\begin{aligned}
& P[X_5 = x_5 | X_3 = x_3, X_1 = x_1] \\
&= \frac{P[X_5 = x_5, X_3 = x_3, X_1 = x_1]}{P[X_3 = x_3, X_1 = x_1]} \\
&= \frac{\sum_{x_4, x_2, x_0} P[X_5 = x_5, X_4 = x_4, X_3 = x_3, X_2 = x_2, X_1 = x_1, X_0 = x_0]}{\sum_{x_2, x_0} P[X_3 = x_3, X_2 = x_2, X_1 = x_1, X_0 = x_0]} \\
&= \frac{\sum_{x_4, x_2, x_0} p_{x_0,x_1} p_{x_1,x_2} p_{x_2,x_3} p_{x_3,x_4} p_{x_4,x_5} P[X_0 = x_0]}{\sum_{x_2, x_0} p_{x_0,x_1} p_{x_1,x_2} p_{x_2,x_3} P[X_0 = x_0]} \\
&= \frac{(\sum_{x_4} p_{x_3,x_4} p_{x_4,x_5})(\sum_{x_2} p_{x_1,x_2} p_{x_2,x_3})(\sum_{x_0} p_{x_0,x_1} P[X_0 = x_0])}{(\sum_{x_2} p_{x_1,x_2} p_{x_2,x_3})(\sum_{x_0} p_{x_0,x_1} P[X_0 = x_0])} \\
&= \frac{p^{(2)}_{x_3,x_5} p^{(2)}_{x_1,x_3} P[X_1 = x_1]}{p^{(2)}_{x_1,x_3} P[X_1 = x_1]} = p^{(2)}_{x_3,x_5} = P[X_5 = x_5 | X_3 = x_3] \,,
\end{aligned}
$$

where in the second last step we have used the formula for matrix multiplication.

As an application of Theorem 2.2.1, we have the following important theorem.

Theorem 2.2.2 *Suppose $\{X_n, n \geq 0\}$ is a time homogeneous Markov chain with state space S. Then for any $i, j \in S$ and $n, m \geq 0$,*

$$
P[X_{n+m} = j | X_m = i] = P[X_n = j | X_0 = i] = p^{(n)}_{ij} .
$$

Proof Since $\{X_n, n \geq 0\}$ is a time homogeneous Markov chain, by Theorem 2.2.1,

$$P[X_{n+m} = j|X_m = i] = p_{ij}^{(n+m-m)} = p_{ij}^{(n)}$$

$$\text{Similarly } P[X_n = j|X_0 = i] = p_{ij}^{(n)}.$$

□

We thus have four definitions of a Markov chain and all are shown to be equivalent. We use any one of these definitions as required.

In all the following theorems and results, we assume that a Markov chain is time homogeneous. In the following theorem, we derive a general formula for the higher step transition probabilities. The equations derived in the theorem are known as Chapman-Kolmogorov equations. These are heavily used in many derivations. In the proof, we use Definition 2.2.2 of a Markov chain.

Theorem 2.2.3 *Chapman-Kolmogorov Equations: Suppose $\{X_n, n \geq 0\}$ is a time homogeneous Markov chain with state space S and n-step transition probabilities $p_{ij}^{(n)}$, i, j in S, $n \geq 1$. Then $\forall\, n,\ l \geq 1$,*

$$p_{ij}^{(n+l)} = \sum_{k \in S} p_{ik}^{(n)}\, p_{kj}^{(l)} \ \forall\ i, j\ \in S \iff P^{(n+l)} = P^{(n)}P^{(l)}.$$

Proof Observe that

$$\begin{aligned} p_{ij}^{(n+l)} &= P[X_{n+l} = j|X_0 = i] = \sum_{k \in S} P[X_{n+l} = j, X_n = k|X_0 = i] \\ &= \sum_{k \in S} P[X_{n+l} = j|X_n = k, X_0 = i]P[X_n = k|X_0 = i] \\ &= \sum_{k \in S} P[X_{n+l} = j|X_n = k]P[X_n = k|X_0 = i] \\ &= \sum_{k \in S} p_{ik}^{(n)}\, p_{kj}^{(l)} \ \forall\ i, j\ \in S \\ \iff P^{(n+l)} &= P^{(n)}P^{(l)}. \end{aligned}$$

The second step follows from Eq. (2.2.1) and the third follows by Definition 2.2.2.

□

Remark 2.2.1 (i) These equations essentially convey that the event of transition from state i to state j in $n + l$ steps can occur in the mutually exclusive ways of going to some intermediate state k in n transitions, and then going from state k to state j in the remaining l transitions. Hence, the probability of transition from state i to state j in $n + l$ steps is the addition of probabilities of such mutually exclusive events. (ii) In particular, the equations $p_{ij}^{(n+1)} = \sum_{k=1}^{m} p_{ik}^{(n)}\ p_{kj}$ are known as forward Chapman-Kolmogorov equations, and the equations $p_{ij}^{(n+1)} = \sum_{k=1}^{m} p_{ik}\ p_{kj}^{(n)}$ are known as backward Chapman-Kolmogorov equations.

In the following theorem, we prove that the n-step transition probability matrix $P^{(n)}$ can be computed as nth power of P. We have already shown it to be true for $n = 2$. The proof follows from the Chapman-Kolmogorov equations.

Theorem 2.2.4 *Suppose $\{X_n, n \geq 0\}$ is a time homogeneous Markov chain with a one step transition probability matrix P. Then $P^{(n)} = P^n, \;\; n \geq 0$.*

Proof It is to be noted that

$$p_{ij}^{(0)} = P[X_0 = j | X_0 = i] \;\; = \;\; \begin{cases} 1, \text{ if } i = j \\ 0, \text{ if } i \neq j. \end{cases}$$

Thus, we have $P^{(0)} = I$. By convention we write $P^0 = I$. Further, $P^{(1)} = P^1 = P$. Hence, the theorem is true for $n = 0$ and 1. From the Chapman-Kolmogorov equations, $P^{(n+l)} = P^{(n)} P^{(l)}$. With $n = l = 1$, we have $P^{(2)} = P^{(1)} P^{(1)} = P^2$. We assume that $P^{(m-1)} = P^{m-1}$. Then

$$P^{(m)} = P^{(m-1)} P^{(1)} = P^{m-1} P = P^m.$$

Thus, by induction it follows that $P^{(n)} = P^n$ for all $n \geq 0$. □

Remark 2.2.2 The one step transition probability matrix of a Markov chain is a stochastic matrix. It is easy to verify that if A and B are two stochastic matrices of the same order, then the product AB is also a stochastic matrix (see Exercise 2.9). It then follows that P^n is also a stochastic matrix $\forall\, n \geq 1$. If P is a doubly stochastic matrix, then P^n is also a doubly stochastic matrix $\forall\, n \geq 1$. It is proved in the following lemma.

Lemma 2.2.1 *If $P = [p_{ij}]$ is a doubly stochastic matrix, then P^n is also a doubly stochastic matrix $\forall\; n \geq 1$.*

Proof Note that $P = [p_{ij}]$ is a doubly stochastic matrix, that is, it is a stochastic matrix with column sums also equal to 1. Hence, P^n is a stochastic matrix. We now prove that column sums of P^n are also equal to 1. Since P is a doubly stochastic matrix, we have $\sum_{j \in S} p_{ij} = \sum_{i \in S} p_{ij} = 1$. Observe that

$$p_{ij}^{(2)} = \sum_{k \in S} p_{ik} p_{kj} \;\; \Rightarrow \;\; \sum_{i \in S} p_{ij}^{(2)} = \sum_{i \in S} \sum_{k \in S} p_{ik} p_{kj} = \sum_{k \in S} p_{kj} \left(\sum_{i \in S} p_{ik} \right) = 1\,.$$

Assume that $\sum_{i \in S} p_{ij}^{(n-1)} = 1$, which is true for $n = 2, 3$. Now

$$\sum_{i \in S} p_{ij}^{(n)} = \sum_{i \in S} \sum_{k \in S} p_{ik}^{(n-1)} p_{kj} = \sum_{k \in S} p_{kj} \left(\sum_{i \in S} p_{ik}^{(n-1)} \right) = 1\,.$$

Thus by induction it follows that, if P is a doubly stochastic matrix, then P^n is also a doubly stochastic matrix for any $n \geq 1$. □

A Markov chain always satisfies the Chapman-Kolmogorov equations, but the converse is not true. In the following example, we present a finite state space stochastic process whose one step transition probabilities satisfy the Chapman-Kolmogorov equations, but is not a Markov chain; refer to Chan et al. [2].

Example 2.2.1 Suppose $(\Omega, \mathbb{A}, P)$ is a probability space, where $\Omega = \{(0,0,1), (0,1,0), (1,0,0), (1,1,1)\}$, $\mathbb{A}$ is a power set of Ω and for $A \in \mathbb{A}$, $P(A) = n(A)/4$, where $n(A)$ denotes the number of elements in A. Three random variables X_0, X_1, X_2 are defined on $\Omega \to \{0, 1\}$ as follows:

$$X_0(i, j, k) = i, \quad X_1(i, j, k) = j \;\;\&\;\; X_2(i, j, k) = k.$$

These random variables are not mutually independent because X_0, X_1 completely determine X_2. Further, note that

$$P[X_n = i] = 1/2 \;\;\&\;\; P[X_n = i, X_m = j] = 1/4, \;\; i, j = 0, 1; \;\; m \neq n = 0, 1, 2.$$

Thus, the random variables X_0, X_1, X_2 are pairwise independent although not mutually independent. Now we define a stochastic process $\{X_n, n \geq 0\}$ where the first triplet (X_0, X_1, X_2) is as defined above, and each of the subsequent triplets is defined in the same way, but is independent of X_0, X_1, X_2. Thus, if $s \geq 3$, then X_{n+s} and X_n are independent of each other. Now observe that $\forall \;\; n \geq 0$, $P[X_{n+1} = j | X_n = i] = 1/2, \;\; i, j = 0, 1$. Hence, the one step transition probability matrix P of order 2×2 is given by

$$P = \begin{array}{c} \\ 0 \\ 1 \end{array} \begin{array}{c} \begin{array}{cc} 0 & 1 \end{array} \\ \begin{pmatrix} 0.5 & 0.5 \\ 0.5 & 0.5 \end{pmatrix} \end{array}.$$

Further, for all $i, j \in \{0, 1\}$

$$\begin{aligned} p_{ij}^{(2)} &= P[X_2 = j | X_0 = i] = \sum_{k \in S} P[X_2 = j, X_1 = k | X_0 = i] \\ &= \sum_{k \in S} p_{ik} \, p_{kj} = 1/2 \;\; \Rightarrow \;\; P^{(2)} = PP = P^2. \end{aligned}$$

Proceeding on similar lines, it follows that $P^{(n+l)} = P^{(n)} P^{(l)}$, for any $n, l \geq 1$. Thus, the stochastic process $\{X_n, n \geq 0\}$ satisfies the Chapman-Kolmogorov equations. Now observe that

$$P[X_2 = 1 | X_1 = 0, X_0 = 0] = 1 \;\;\&\;\; P[X_2 = 1 | X_1 = 0] = 1/2.$$

Hence, the stochastic process $\{X_n, n \geq 0\}$ does not satisfy the Markov property and hence is not a Markov chain. □

Theorem 2.2.4 is very useful in practice to find the higher order transition probabilities. Method of spectral decomposition of a matrix can be used to find powers of matrices, once we compute its eigenvalues and eigenvectors. We will use this method in Sect. 3.3 to prove a result related to stationary distributions associated with a Markov chain. It is also used in Sect. 6.4 in the computation of a matrix of transition probability functions for a continuous time Markov chain. We briefly describe the method below.

Spectral decomposition of a matrix: Suppose A is a matrix of order $M \times M$ with real-valued elements. Suppose $\underline{x}_i$ and $\underline{y}_i$ denote the right and left eigenvectors respectively corresponding to eigenvalue λ_i of A, $i = 1, 2, \ldots, M$. Hence, we have $A\underline{x}_i = \lambda_i \underline{x}_i$ and $\underline{y}_i' A = \lambda_i \underline{y}_i'$ for $i = 1, 2, \ldots, M$. We assume that the eigenvalues are distinct. Under this assumption $\underline{x}_i$ are linearly independent; similarly $\underline{y}_i$ are linearly independent for $i = 1, 2, \ldots, M$. Suppose D is a diagonal matrix with diagonal elements to be the eigenvalues, R is a matrix of right eigenvectors and L is a matrix of left eigenvectors, where the order of columns in R and L corresponds to eigenvalues λ_i, $i = 1, 2, \ldots, M$. Linear independence of columns of R and that of L implies that both R and L are non-singular matrices. From the two sets of equations $A\underline{x}_i = \lambda_i \underline{x}_i$ and $\underline{y}_i' A = \lambda_i \underline{y}_i'$ for all $i \neq j$, $i, j = 1, 2, \ldots, M$, we have

$$
\begin{aligned}
\underline{y}_i'(A\underline{x}_j) &= \underline{y}_i'(\lambda_j \underline{x}_j) = \lambda_j \underline{y}_i'\underline{x}_j \;\;\&\;\; (\underline{y}_i'A)\underline{x}_j = (\lambda_i \underline{y}_i')\underline{x}_j = \lambda_i \underline{y}_i'\underline{x}_j \\
&\Rightarrow \underline{y}_i'(A\underline{x}_j) - (\underline{y}_i'A)\underline{x}_j = \lambda_j \underline{y}_i'\underline{x}_j - \lambda_i \underline{y}_i'\underline{x}_j = 0 \\
&\Rightarrow (\lambda_j - \lambda_i)\underline{y}_i'\underline{x}_j = 0 \;\Rightarrow\; \underline{y}_i'\underline{x}_j = 0\,,
\end{aligned}
$$

as λ_i, $i = 1, 2, \ldots, M$ are distinct. It is to be noted that $\underline{y}_i'\underline{x}_j$ is the (i, j)th element of the matrix $L'R$. For all $i \neq j$, $\underline{y}_i'\underline{x}_j = 0$ implies that all off-diagonal elements of $L'R$ are 0, thus $L'R$ is a diagonal matrix. Further, R and L are non-singular matrices, implying that $L'R$ is also a non-singular matrix. Hence, all the diagonal elements $\underline{y}_i'\underline{x}_i$ must be non-zero, $i = 1, 2, \ldots, M$. Suppose $\underline{y}_i'\underline{x}_i > 0$ for all $i = 1, 2, \ldots, M$. We define

$$
\underline{v}_i = \underline{x}_i \Big/ \sqrt{\underline{y}_i'\underline{x}_i} \;\;\&\;\; \underline{u}_i = \underline{y}_i \Big/ \sqrt{\underline{y}_i'\underline{x}_i} \;\;\Rightarrow\;\; A\underline{v}_i = \lambda_i \underline{v}_i \;\;\&\;\; \underline{u}_i' A = \lambda_i \underline{u}_i'.
$$

Thus, $\underline{v}_i$ and $\underline{u}_i$ are the right and the left eigenvectors respectively of A, corresponding to the eigenvalue λ_i. Further,

$$
\underline{u}_i'\underline{v}_i = \underline{y}_i'\underline{x}_i / \underline{y}_i'\underline{x}_i = 1 \;\;\&\;\; \underline{u}_i'\underline{v}_j = 0 \;\forall\; i \neq j.
$$

Suppose V is a matrix of the normalized right eigenvectors $\underline{v}_i$ and U is a matrix of the normalized left eigenvectors $\underline{u}_i$, where the order of columns in V and U is corresponding to eigenvalues λ_i, $i = 1, 2, \ldots, M$. Hence,

$$\begin{aligned}
\underline{u}_i'\underline{v}_i &= 1 \ \& \ \underline{u}_i'\underline{v}_j = 0 \ \forall \ i \neq j \\
&\Rightarrow V'U = I = U'V \ \Rightarrow \ U' = V^{-1} \text{ or } V = (U')^{-1} \\
\text{Similarly } A\underline{v}_i = \lambda_i\underline{v}_i &\Rightarrow AV = VD \ \Rightarrow \ A = VDV^{-1} = VDU' \\
\underline{u}_i'A = \lambda_i\underline{u}_i' &\Rightarrow U'A = DU' \ \Rightarrow \ A = (U')^{-1}DU' = VDU' \\
\iff \quad A &= VDU' = \sum_{i=1}^{M} \lambda_i \underline{v}_i \underline{u}_i'.
\end{aligned}$$

The last equation is the spectral decomposition or spectral representation of A. Suppose D^n is a diagonal matrix with diagonal elements λ_i^n, $i = 1, 2, \ldots, M$. In view of the result $U'V = I$, we have

$$A^2 = VDU' \times VDU' = VD^2U', \quad A^3 = VD^2U' \times VDU' = VD^3U'.$$

Continuing in this manner, the nth power of the matrix A is given by

$$A^n = \sum_{i=1}^{M} \lambda_i^n \underline{v}_i \underline{u}_i' \quad \iff \quad A^n = VD^nU' = VD^nV^{-1} = (U')^{-1}D^nU' \ . \quad (2.2.3)$$

Remark 2.2.3 The condition $\underline{y}_i'\underline{x}_i > 0$ for all $i = 1, 2, \ldots, M$ may not be satisfied for some P; see Computational Exercise 2.10. It is known that if $\underline{y}_i'$ is an eigenvector corresponding to eigenvalue λ_i, then $(-1)\underline{y}_i'$ is also an eigenvector corresponding to eigenvalue λ_i. Using this result, in such cases, we multiply $\underline{y}_i'$ or $\underline{x}_i$ by (-1) so that $\underline{y}_i'\underline{x}_i > 0$.

Several matrix-oriented computer packages are now available which give the nth power of a matrix. In R software, nth power of a matrix can be obtained using matrix multiplication. There is no function in the base package, but is available in "matrixcalc" package (Frederick Novomestky [10]) or "expm" package (Vincent et al. [7]). We can write a function to find the nth power of P, if needed for many values of n. In Sect. 2.8, Code 2.8.1 is for the spectral decomposition of P and computation of nth power of P, using spectral decomposition, matrix multiplication, the two packages and a function.

Using Code 2.8.1, in the next example we verify the results related to the spectral decomposition of P and illustrate computation of P^n.

Example 2.2.2 Suppose $\{X_n, n \geq 0\}$ is a Markov chain with state space $S = \{1, 2, 3\}$ and transition probability matrix P as given below

$$P = \begin{matrix} \\ 1 \\ 2 \\ 3 \end{matrix} \begin{matrix} \begin{matrix} 1 & \ \ 2 & \ \ 3 \end{matrix} \\ \begin{pmatrix} 0.5 & 0.4 & 0.1 \\ 0.2 & 0.4 & 0.4 \\ 0.4 & 0.4 & 0.2 \end{pmatrix} \end{matrix}.$$

The matrix D of eigenvalues of P is diag$(1, 0.1, 0)$. R and L are non-singular matrices with determinant -0.3228 and -0.3212, respectively. Further, $RDR^{-1} = L'^{-1}DL = P$ and $L'R = \text{diag}(0.9813, 0.3260, 0.3241)$. We obtain V and U from R and L respectively, by further scaling, as described above and we have $V'U = U'V = I$ and $VDU' = UDV' = P$. Using spectral decomposition, matrix multiplication and the two packages, we have

$$P^5 = \begin{matrix} \\ 1 \\ 2 \\ 3 \end{matrix} \begin{matrix} \begin{matrix} 1 & 2 & 3 \end{matrix} \\ \begin{pmatrix} 0.3556 & 0.4 & 0.2444 \\ 0.3556 & 0.4 & 0.2444 \\ 0.3556 & 0.4 & 0.2444 \end{pmatrix} \end{matrix}.$$

The function to find a power of P^n for $n = 2, 5, 8, 11$ gives

$$P^2 = \begin{matrix} \\ 1 \\ 2 \\ 3 \end{matrix} \begin{matrix} \begin{matrix} 1 & 2 & 3 \end{matrix} \\ \begin{pmatrix} 0.37 & 0.4 & 0.23 \\ 0.34 & 0.4 & 0.26 \\ 0.36 & 0.4 & 0.24 \end{pmatrix} \end{matrix} \;\&\; P^5 = P^8 = P^{11} = \begin{matrix} \\ 1 \\ 2 \\ 3 \end{matrix} \begin{matrix} \begin{matrix} 1 & 2 & 3 \end{matrix} \\ \begin{pmatrix} 0.3556 & 0.4 & 0.2444 \\ 0.3556 & 0.4 & 0.2444 \\ 0.3556 & 0.4 & 0.2444 \end{pmatrix} \end{matrix}.$$

Note that P^n remains the same, up to four decimal places accuracy, for $n \geq 5$ and that all the rows are identical. We will discuss its interpretation in Chap. 3. □

The following examples illustrate the computation of higher step transition probabilities using Code 2.8.1. We begin with a very simple example of a two-state Markov chain.

Example 2.2.3 We consider the Markov chain in Example 2.1.1, with P given by

$$P = \begin{matrix} \\ 1 \\ 2 \end{matrix} \begin{matrix} \begin{matrix} 1 & 2 \end{matrix} \\ \begin{pmatrix} 0.7 & 0.3 \\ 0.2 & 0.8 \end{pmatrix} \end{matrix}.$$

Suppose we wish to find the probability that a driver known to be classified as standard at the start of the year will be classified as preferred at the start of the fourth year, that is $P[X_4 = 1|X_1 = 2]$, then we compute P^3. To find the probability that a driver known to be classified as standard at the start of the year will be classified as standard at the start of the seventh year, that is, to find $P[X_7 = 2|X_1 = 2]$, we compute P^6. The matrices P^3 and P^6 using all the methods are the same and are as follows:

$$P^3 = \begin{matrix} \\ 1 \\ 2 \end{matrix} \begin{matrix} \begin{matrix} 1 & 2 \end{matrix} \\ \begin{pmatrix} 0.475 & 0.525 \\ 0.350 & 0.650 \end{pmatrix} \end{matrix} \;\&\; P^6 = \begin{matrix} \\ 1 \\ 2 \end{matrix} \begin{matrix} \begin{matrix} 1 & 2 \end{matrix} \\ \begin{pmatrix} 0.4094 & 0.5906 \\ 0.3937 & 0.6063 \end{pmatrix} \end{matrix} \quad \text{and} \quad .$$

$$\text{Hence,}\quad P[X_4 = 1|X_1 = 2] = 0.35 \;\&\; P[X_7 = 2|X_1 = 2] = 0.6063.$$

□

The following example is similar to that of Example 2.1.1, but the underlying model is a non-homogeneous Markov chain. Suppose P_n denotes the matrix with (i, j)th element $p_{ij}^{n,n+1}$. For a non-homogeneous Markov chain the k-step transition probability, denoted by $p_{ij}^{n,n+k}$, $k \geq 1$, is defined as $p_{ij}^{n,n+k} = P[X_{n+k} = j | X_n = i]$. Suppose the matrix ${}_kP_n$ denotes the corresponding matrix; prefix k is usually omitted if $k = 1$. Using Markov property repeatedly, we get

$$ {}_kP_n = P_n \times P_{n+1} \times P_{n+2} \cdots \times P_{n+k-1}, \quad k \geq 1 \quad \text{and} \quad {}_0P_n = I, $$

where I is the identity matrix. The following example illustrates such computations.

Example 2.2.4 Suppose the auto-insurer classifies its policyholders according to preferred (1) and standard (2) statuses, starting at time 0 at the start of the first year when they are first insured, with reclassifications occurring at the start of each annual renewal of the policy. We assume that the status of a policyholder in successive years can be modeled by a non-homogeneous Markov chain, with transition probability matrix P_n as given below

$$ P_n = \begin{matrix} & \begin{matrix} 1 & 2 \end{matrix} \\ \begin{matrix} 1 \\ 2 \end{matrix} & \begin{pmatrix} 0.6 & 0.4 \\ 0.3 & 0.7 \end{pmatrix} \end{matrix} + \left(\frac{1}{n+1}\right) \begin{matrix} & \begin{matrix} 1 & 2 \end{matrix} \\ \begin{matrix} 1 \\ 2 \end{matrix} & \begin{pmatrix} 0.15 & -0.15 \\ -0.20 & 0.20 \end{pmatrix} \end{matrix}, \quad n \geq 0. $$

The probability that the insured is in state 1 at the start of the third year, given that the insured is in state 1 at the start of the first year, is $P[X_2 = 1 | X_0 = 1]$, and we need to compute two step transition probability matrix. The probability that the insured who is in state 1 at the start of the first year transits to state 2 at the start of the fourth year is $P[X_3 = 2 | X_0 = 1]$, and we need to compute three step transition probability matrix. The Markov chain is non-homogeneous, hence, ${}_2P_0 = P_0 \times P_1$ and ${}_3P_0 = P_0 \times P_1 \times P_2$. We have

$$ P_0 = \begin{matrix} & \begin{matrix} 1 & 2 \end{matrix} \\ \begin{matrix} 1 \\ 2 \end{matrix} & \begin{pmatrix} 0.75 & 0.25 \\ 0.10 & 0.90 \end{pmatrix} \end{matrix}, \quad P_1 = \begin{matrix} & \begin{matrix} 1 & 2 \end{matrix} \\ \begin{matrix} 1 \\ 2 \end{matrix} & \begin{pmatrix} 0.675 & 0.325 \\ 0.200 & 0.800 \end{pmatrix} \end{matrix} \ \& \ P_2 = \begin{matrix} & \begin{matrix} 1 & 2 \end{matrix} \\ \begin{matrix} 1 \\ 2 \end{matrix} & \begin{pmatrix} 0.650 & 0.350 \\ 0.233 & 0.767 \end{pmatrix} \end{matrix}. $$

Hence using matrix multiplication, we get

$$ {}_2P_0 = \begin{matrix} & \begin{matrix} 1 & 2 \end{matrix} \\ \begin{matrix} 1 \\ 2 \end{matrix} & \begin{pmatrix} 0.5563 & 0.4437 \\ 0.2475 & 0.7525 \end{pmatrix} \end{matrix} \quad \text{and} \quad {}_3P_0 = \begin{matrix} & \begin{matrix} 1 & 2 \end{matrix} \\ \begin{matrix} 1 \\ 2 \end{matrix} & \begin{pmatrix} 0.4650 & 0.5350 \\ 0.3362 & 0.6638 \end{pmatrix} \end{matrix}. $$

Hence, (i) $P[X_2 = 1 | X_0 = 1] = 0.5563$ and (ii)$P[X_3 = 2 | X_0 = 1] = 0.5350$. □

Using Theorem 2.2.4, we can easily find the marginal distribution of X_n for each n as follows. Suppose $\underline{p}^{(0)} = \{p_i^0, \ i \in S\}$ denotes the initial distribution, where $p_i^{(0)} = P[X_0 = i]$. Then

$$
\begin{aligned}
P[X_n = j] &= \sum_{i \in S} P[X_n = j, X_0 = i] \\
&= \sum_{i \in S} P[X_n = j | X_0 = i] P[X_0 = i] = \sum_{i \in S} p_i^{(0)} p_{ij}^{(n)} . \quad (2.2.4)
\end{aligned}
$$

Observe that it is the jth component of the vector $\underline{p}^{(0)}\ P^n$. Suppose $p_i^{(n)} = P[X_n = i]$, then $\underline{p}^{(n)} = \{p_i^{(n)},\ i \in S\}$ denotes the marginal distribution of X_n. Thus, from Eq. 2.2.4 we have

$$\underline{p}^{(n)} = \underline{p}^0\ P^n. \quad (2.2.5)$$

The following example illustrates the computation of the marginal probability distribution. In Sect. 2.8, Code 2.8.2 presents two approaches to compute the marginal distribution of X_n. In the first approach, the code computes the marginal distributions, using Eq. (2.2.5), for consecutive values of n. The second approach uses Eq. (2.2.4), to compute the marginal distribution, for any specified values of n, not necessarily consecutive.

Example 2.2.5 Suppose the transitions among states in an elderly care center are governed by a homogeneous Markov chain with state space $S = \{1, 2, 3\}$, where 1 denotes the healthy state, 2 denotes critically ill state and 3 stands for death. Suppose a transition probability matrix P is as given below, where time unit is taken as a day:

$$
P = \begin{matrix} \\ 1 \\ 2 \\ 3 \end{matrix}
\begin{matrix} \begin{matrix} 1 & \quad 2 & \quad 3 \end{matrix} \\
\begin{pmatrix} 0.92 & 0.05 & 0.03 \\ 0.00 & 0.76 & 0.24 \\ 0 & 0 & 1 \end{pmatrix} \end{matrix} .
$$

Suppose the expenses incurred per day at the care center are Rs. 1200, Rs. 5000 and Rs. 1000 for an individual in states 1, 2, 3, respectively. For budgetary purpose, the management wants to know the expected expense per individual for the next five days. Suppose the initial distribution is $\underline{p}^{(0)} = (1, 0, 0)$. With this initial distribution and given one step transition matrix P, we compute the probabilities of an individual being in each of the states after 1, 2, 3, 4 and 5 days, that is, we find the marginal distribution of X_n using the formula in Eq. (2.2.5). Here X_n denotes the state of the individual at the beginning of the nth day. Using the marginal distribution, we find the expected expenses for each state for the initial day and the next five days and the total expected expenses for these days. In Sect. 2.8, Code 2.8.2 computes the marginal distribution of X_n and expected expenses on nth day for $n = 1, 2, 3, 4, 5, 6$, where day 1 is taken as an initial day. Table 2.1 displays the marginal distributions for the initial and the next 5 days.

Table 2.2 displays the expected cost for each state for the first 6 days. The last column presents the output of a vector "totalcost", which gives the expected total cost per individual for the first 6 days.

Table 2.1 Care center model: marginal distributions of X_1 to X_5

	1	2	3
X_0	1.0000	0.0000	0.0000
X_1	0.9200	0.0500	0.0300
X_2	0.8464	0.0840	0.0696
X_3	0.7787	0.1062	0.1152
X_4	0.7164	0.1196	0.1640
X_5	0.6591	0.1267	0.2142

Table 2.2 Care center model: expected daily expenses

	1	2	3	Total expenses
Day 1	1200.00	0	0	1200.00
Day 2	1104.00	250.00	30.00	1384.00
Day 3	1015.68	420.00	69.60	1505.28
Day 4	934.43	530.80	115.15	1580.38
Day 5	859.67	598.08	163.99	1621.74
Day 6	790.90	633.64	214.19	1638.73

Table 2.3 Care center model: marginal distributions of X_4, X_5, X_9, X_{15}

	1	2	3
X_4	0.7164	0.1196	0.1640
X_5	0.6591	0.1267	0.2142
X_9	0.4722	0.1211	0.4067
X_{15}	0.2863	0.0844	0.6293

It is to be noted that the expenses increase, since proportion of individuals in critically ill state increases. The second part of Code 2.8.2 computes the marginal distribution of X_n, where values of n are not necessarily consecutive. Table 2.3 displays the marginal distributions of X_4, X_5, X_9, X_{15}. Note that as n increases, the probability that the individual is in state 1 decreases while that in state 3 increases. □

Example 2.2.6 Weather conditions on any day in a city are classified as sunny, cloudy and rainy. It is assumed that tomorrow's weather depends only on today's weather, given the climatic conditions for all the previous days. With this assumption, the weather process of the city can be modeled as a Markov chain $\{X_n, n \geq 0\}$, where X_n denotes the weather condition on day n, defined as follows:

$$X_n = \begin{cases} 1, & \text{if nth day is sunny} \\ 2, & \text{if nth day is cloudy} \\ 3, & \text{if nth day is rainy.} \end{cases}$$

Further, a one step transition probability matrix P is

$$P = \begin{matrix} & \begin{matrix} 1 & 2 & 3 \end{matrix} \\ \begin{matrix} 1 \\ 2 \\ 3 \end{matrix} & \begin{pmatrix} 0.5 & 0.3 & 0.2 \\ 0.5 & 0.2 & 0.3 \\ 0.4 & 0.5 & 0.1 \end{pmatrix} \end{matrix}.$$

A group of friends plans on Monday to spend coming weekend at a resort near the city, as Monday was a sunny day. A general insurance company has an insurance product that assures to reimburse the entire package of Rs. 10,000/- if it rains on both Saturday and Sunday and nothing will be reimbursed otherwise. The one-time premium for the insurance is Rs. 1000/-. The group approached their statistician friend to seek advice on whether it is worth to buy the insurance. The statistician finds the probability that it will rain on Saturday and Sunday given that Monday is a sunny day as follows:

$$\begin{aligned} P[X_6 = 3, X_5 = 3 | X_0 = 1] &= P[X_6 = 3 | X_5 = 3] P[X_5 = 3 | X_0 = 1] \\ &= p_{33} p_{13}^{(5)} = 0.1 \times 0.211 = 0.0211, \end{aligned}$$

as five-step transition probability matrix P^5 is given by

$$P^5 = \begin{matrix} & \begin{matrix} 1 & 2 & 3 \end{matrix} \\ \begin{matrix} 1 \\ 2 \\ 3 \end{matrix} & \begin{pmatrix} 0.479 & 0.310 & 0.211 \\ 0.479 & 0.310 & 0.211 \\ 0.479 & 0.310 & 0.211 \end{pmatrix} \end{matrix}.$$

Hence, the expected reimbursement is $E = 10000 \times 0.0211 + 0 \times (1 - 0.0211) = 211$, which is far smaller than the premium of Rs. 1000/-. Thus, naturally the advice of the statistician is not to purchase the insurance. □

From Theorem 2.1.2, we have noted that the family of finite dimensional distributions of a Markov chain is completely determined by the one step transition probability matrix and the initial distribution. We use Code 2.8.3 to find a member from the family of finite dimensional distributions for the given Markov chain. In the following example, we illustrate it by finding the joint distribution of $\{X_3, X_6, X_9\}$ for the weather model of Example 2.2.6.

Example 2.2.7 Suppose $\{X_n, n \geq 0\}$ is a Markov chain as described in Example 2.2.6. Using Code 2.8.3, we find the probability distribution of weather conditions on day 3, day 6 and day 9. Table 2.4 presents the first six and the last six rows of the data frame "jp2", which specifies $P[X_3 = i, X_6 = j, X_9 = k]$ for all triplets (i, j, k).

Table 2.4 Weather model: joint probability distribution

i	j	k	Probability
1	1	1	0.1098
1	1	2	0.0719
1	1	3	0.0477
1	2	1	0.0716
1	2	2	0.0456
1	2	3	0.0323
3	2	1	0.0320
3	2	2	0.0204
3	2	3	0.0145
3	3	1	0.0204
3	3	2	0.0136
3	3	3	0.0086

From Table 2.4, we note that corresponding to the given Markov model, the probability that all the three days are sunny is 0.1098 and the probability that all the three days are rainy is 0.0086. □

From these examples, we note that once we have the initial distribution of a Markov chain and its one step transition probability matrix, we can compute the probabilities of events of interest.

In the next section, we discuss how to obtain a realization of a Markov chain specified by $(S, \underline{p}^{(0)}, P)$. Estimation of transition probability matrix given a realization is also briefly explained.

2.3 Realization of a Markov Chain

A Markov chain in Example 2.1.1 describes the possible transitions between the two types of statuses of a driver. It is of interest to observe the possible transitions of a driver from the given category for the next few years. In Example 2.2.6, the weather condition of a city is modeled as a Markov chain. Using this model, it is of interest to find out the weather conditions for some days. It can be achieved by obtaining a realization of the underlying Markov chain. In Definition 1.1.4, we have defined a sample path or a realization of a stochastic process. In the present section, we discuss how to obtain a realization of a Markov chain. The following theorem proposes a methodology to obtain a realization of a Markov chain specified by $(S, \underline{p}^{(0)}, P)$.

Theorem 2.3.1 *Suppose that the set $S = \{1, 2, \ldots, M\}$ or W, $\underline{p}^{(0)}$ is a probability mass function on S and P is the stochastic matrix of order $M \times M$ or of infinite* *dimensions. Suppose* $\{U_n, n \geq 0\}$ *is a sequence of independent and identically distributed random variables, each having* $U(0, 1)$ *distribution. We define a sequence* $\{X_n, n \geq 0\}$ *of random variables by*

$$X_0 = k, \text{ if } \sum_{j=0}^{k-1} p_j^{(0)} < U_0 \leq \sum_{j=0}^{k} p_j^{(0)}$$

$$\& \text{ for } n \geq 1 \;\; X_n = k \text{ if } \sum_{j=0}^{k-1} p_{X_{n-1},j} < U_n \leq \sum_{j=0}^{k} p_{X_{n-1},j}.$$

Then $\{X_n, n \geq 0\}$ *is a Markov chain with initial distribution* $\underline{p}^{(0)}$ *and one step transition probability matrix* P.

Note that $\{U_n, n \geq 0\}$ is a Markov chain and X_n is a function of U_n and hence $\{X_n, n \geq 0\}$ is a stochastic process, as noted in Chap. 1. Theorem 2.3.1 conveys that given $(S, \underline{p}^{(0)}, P)$, there exists a Markov chain, with S as the state space, $\underline{p}^{(0)}$ as the initial distribution and P as the transition probability matrix. Based on Theorem 2.3.1, we have the following stepwise procedure to obtain a realization of a Markov chain specified by $(S, \underline{p}^{(0)}, P)$.

(i) We first decide the initial state using the initial distribution. In some situations, initial state is specified. Suppose $X_0 = i$.
(ii) The next state X_1 is decided by the probability mass function specified in the ith row. Hence, we draw a random sample of size 1 from the state space S, according to the probability mass function specified in the ith row.
(iii) Suppose the state selected is j, then to find the next state, we draw a random sample of size 1 using the probability mass function specified in the jth row.
(iv) Thus, at each time point, to obtain the next state, we draw a sample of size 1 from the state space, using the probability distribution specified in the row corresponding to the previous state. Continuing in this manner, we obtain a realization from the given Markov chain.

We use Code 2.8.4 to obtain a realization. In the following examples, we illustrate it for the Markov chain in Example 2.2.6 and for the Markov chain in Example 2.1.1.

Example 2.3.1 For the Markov chain as specified in Example 2.2.6, we obtain a realization of length 120. It is assumed that the initial day is sunny. From the output of the function `table(x)`, we note that out of the 120 days, the day is sunny 72 times, cloudy 26 times while rainy 22 times. Figure 2.1 displays the realized weather conditions for 120 days.

From Fig. 2.1, we note that the weather condition is sunny for three days, then it is cloudy, again sunny for two days and then rainy. It is assumed that the transition probabilities remain the same for the period of 120 days. □

Example 2.3.2 Suppose $\{X_n, n \geq 0\}$ is a Markov chain as specified in Example 2.1.1. Using Code 2.8.4, we decide the possible states of the driver for

Table 2.5 Classification of characters in Puskin's "Eugen Onegin"

	Vowel next	Consonant next	Total
Vowel	1106	7532	8638
Consonant	7533	3829	11362
Total	8639	11361	20000

Example 2.3.3 In this example, we compute $\hat{p}_{ijn}$ for the data given in Table 2.5, where 20,000 consecutive characters are classified as vowels or consonants, Guttorp [8]. From the data, it is clear that the nature of the character does depend on the nature of the previous character. Suppose X_n denotes the nth character, which is either a vowel (0) or consonant (1). Suppose the sequence $\{X_n, n \geq 0\}$ is modeled as a Markov chain with state space $S = \{0, 1\}$ and the transition probability matrix $P = [p_{ij}]$. From the given data, we have $n_{00} = 1106,\ n_{01} = 7532,\ n_{10} = 7533$ and $n_{11} = 3829$. Hence, an estimate of P is given by

$$\hat{P}_n = \begin{matrix} & \begin{matrix} 0 & & 1 \end{matrix} \\ \begin{matrix} 0 \\ 1 \end{matrix} & \begin{pmatrix} 1106/8638 & 7532/8638 \\ 7533/11362 & 3839/11362 \end{pmatrix} \end{matrix} = \begin{matrix} & \begin{matrix} 0 & & 1 \end{matrix} \\ \begin{matrix} 0 \\ 1 \end{matrix} & \begin{pmatrix} 0.128 & 0.872 \\ 0.663 & 0.337 \end{pmatrix} \end{matrix}.$$

From $\hat{P}_n$, we note that an estimate of the probability of a consonant followed by a vowel is high. □

Example 2.3.4 The US Weather Service maintains a large number of precipitation records throughout the United States. Data on dry and wet days for 1076 days were available from one station located at the Snoqualmie Falls in the foothills of the Cascade Mountains in western Washington; refer to Guttorp [8]. Data are summarized in Table 2.6, in terms of transition frequencies.

Suppose X_n denotes the state of the nth day as dry or wet, which is assumed to depend on the weather condition on the previous day. If we label dry day as 0 and wet day as 1, then $\{X_n, n \geq 0\}$ can be modeled as a Markov chain with state space $S = \{0, 1\}$ and the transition probability matrix $P = [p_{ij}]$. From the given data, the estimate of P is given by

$$\hat{P}_n = \begin{matrix} & \begin{matrix} 0 & & 1 \end{matrix} \\ \begin{matrix} 0 \\ 1 \end{matrix} & \begin{pmatrix} 186/309 & 123/309 \\ 124/767 & 643/767 \end{pmatrix} \end{matrix} = \begin{matrix} & \begin{matrix} 0 & & 1 \end{matrix} \\ \begin{matrix} 0 \\ 1 \end{matrix} & \begin{pmatrix} 0.6019 & 0.3981 \\ 0.1617 & 0.8383 \end{pmatrix} \end{matrix}.$$

Table 2.6 Data on observed precipitation

	Today weather		
Previous day weather	Dry	Wet	Total
Dry	186	123	309
Wet	124	643	767
Total	310	766	1076

From $\hat{P}_n$, we note that an estimate of the probability of a wet day followed by a wet day is high. □

In the following example, we obtain a realization from the Markov chain in Example 2.2.6, using Code 2.8.4. On the basis of the realization, we illustrate the computation of $\hat{p}_{ijn}$.

Example 2.3.5 In Example 2.2.6, the weather condition in a city is modeled as a Markov chain. We obtain a realization of length $n = 365$ from this model and based on it, we obtain the maximum likelihood estimates of transition probabilities using Code 2.8.5. In the realization, weather conditions are sunny, cloudy and rainy for 187, 102 and 76 days, respectively. The matrix $\hat{P}_n$ (Phat in the code) is the maximum likelihood estimate of P. Both P and $\hat{P}_n$ are given below

$$P = \begin{matrix} & \begin{matrix} 1 & 2 & 3 \end{matrix} \\ \begin{matrix} 1 \\ 2 \\ 3 \end{matrix} & \begin{pmatrix} 0.5 & 0.3 & 0.2 \\ 0.5 & 0.2 & 0.3 \\ 0.4 & 0.5 & 0.1 \end{pmatrix} \end{matrix} \quad \& \quad \hat{P}_n = \begin{matrix} & \begin{matrix} 1 & 2 & 3 \end{matrix} \\ \begin{matrix} 1 \\ 2 \\ 3 \end{matrix} & \begin{pmatrix} 0.5484 & 0.2527 & 0.1989 \\ 0.5000 & 0.1569 & 0.3431 \\ 0.4342 & 0.5132 & 0.0526 \end{pmatrix} \end{matrix}.$$

Observe that (i, j)th elements are close in two matrices for some (i, j). □

From the theorems and examples discussed so far, we note that once we have an initial distribution and a one step transition probability matrix of a Markov chain, we can compute the probabilities of various events of interest.

The next important issue to be addressed is about the limiting behavior of a Markov chain. More precisely, one would like to know whether X_n converge in law as $n \to \infty$. Hence, the issues that we have to sort out are (i) does $\lim_{n\to\infty} P[X_n = j]$ exist? and (ii) how to find it? In the following sections, we explore the technicalities needed to answer these issues.

The next three sections are devoted to the important concept of classification of states. In the classification of states, we examine the nature of a Markov chain, which in turn is determined by the nature of the states of a Markov chain. The concerned results are useful to sort out different possibilities of the limiting behavior of a Markov chain, which we discuss in Chap. 3.

2.4 Classification of States

Suppose $\{X_n, n \geq 0\}$ is a Markov chain specified by $(S, \underline{p}^{(0)}, P)$. Then from Eq. (2.2.4), we have

$$\begin{aligned} & P[X_n = j] = \sum_{i \in S} p_{ij}^{(n)} P[X_0 = i] \\ \Rightarrow \quad & \lim_{n\to\infty} P[X_n = j] = \lim_{n\to\infty} \sum_{i \in S} p_{ij}^{(n)} P[X_0 = i] = \sum_{i \in S} \lim_{n\to\infty} p_{ij}^{(n)} P[X_0 = i]. \end{aligned}$$

If the state space is finite, summation and limit can always be interchanged. If the state space is countably infinite, we use the following theorem to justify the interchange of limit and summation.

Theorem 2.4.1 *Suppose $h_n(x) \to h(x)$ uniformly in x. Then*

$$\lim_{n\to\infty} \sum_{i\geq 1} h_n(x_i) = \sum_{i\geq 1} \lim_{n\to\infty} h_n(x_i) = \sum_{i\geq 1} h(x_i).$$

A sufficient condition for the uniform convergence is that $|h_n(x)| \leq g(x)$ where $\sum_{i\geq 1} g(x_i) < \infty$.

In the derivations and proofs, we verify the sufficient condition to justify the interchange of limit and summation. If the state space is countably infinite, note that $|p_{ij}^{(n)} P[X_0 = i]| \leq P[X_0 = i]$ and $\sum_{i\in S} P[X_0 = i] = 1$. Hence, by Theorem 2.4.1, summation and limit can be interchanged. Thus, to examine whether $\lim_{n\to\infty} P[X_n = j]$ exist, we examine whether $\lim_{n\to\infty} p_{ij}^{(n)}$ exists $\forall\ i,\ j \in S$. We begin the discussion with some illustrations.

(i) For the one step transition probability matrix P of the weather model in Example 2.2.6, P^{10} is given by

$$P^{10} = \begin{matrix} \\ 1 \\ 2 \\ 3 \end{matrix} \begin{matrix} \begin{matrix} 1 & \quad\ 2 & \quad\ 3 \end{matrix} \\ \begin{pmatrix} 0.4790 & 0.3109 & 0.2101 \\ 0.4790 & 0.3109 & 0.2101 \\ 0.4790 & 0.3109 & 0.2101 \end{pmatrix} \end{matrix}.$$

Note that all rows of P^{10} are identical. Hence, all the rows of P^n will also be identical for any $n \geq 10$. Thus, $\lim_{n\to\infty} P^n$ is a matrix with identical rows, which implies that $\lim_{n\to\infty} p_{ij}^{(n)} = a_j,\ \forall\ \ i \in S$, the limit being free from the initial state. It seems reasonable that the influence of the initial state recedes in time. In Example 2.2.6, the interpretation is as follows: whatever may be the weather condition on the initial day, after 10 days, the probability that a day will be sunny is 0.479, it is cloudy is 0.3109 and it is rainy is 0.2101, and these probabilities will remain the same for all days after 10 days. Thus for $j \in S$,

$$\lim_{n\to\infty} P[X_n = j] = \lim_{n\to\infty} \sum_{i\in S} p_{ij}^{(n)} P[X_0 = i] = a_j \sum_{i\in S} P[X_0 = i] = a_j\ .$$

The above method is one of the methods of finding the limit of $P[X_n = j]$.
(ii) Note that for the one step transition probability matrix P of the care center model in Example 2.2.5,

$$P^{64} = \begin{array}{c} \\ 1 \\ 2 \\ 3 \end{array}\begin{array}{c} \begin{array}{ccc} 1 & 2 & 3 \end{array} \\ \begin{pmatrix} 0.0048 & 0.0015 & 0.9937 \\ 0 & 0 & 1 \\ 0 & 0 & 1 \end{pmatrix} \end{array} \quad \& \quad P^n = \begin{array}{c} \\ 1 \\ 2 \\ 3 \end{array}\begin{array}{c} \begin{array}{ccc} 1 & 2 & 3 \end{array} \\ \begin{pmatrix} 0 & 0 & 1 \\ 0 & 0 & 1 \\ 0 & 0 & 1 \end{pmatrix} \end{array}, \ n \geq 100.$$

We thus conclude that after $n \geq 100$ days, according to the given Markov model, all the members are in state 3, whatever may be their initial state. Thus, for this model, $\lim_{n\to\infty} P[X_n = 3] = 1$ and for the other two states 1 and 2, it is 0.
(iii) For the one step transition probability matrix P, given by

$$P = \begin{array}{c} \\ 1 \\ 2 \end{array}\begin{array}{c} \begin{array}{cc} 1 & 2 \end{array} \\ \begin{pmatrix} 1/2 & 1/2 \\ 0 & 1 \end{pmatrix} \end{array}, \quad P^n = \begin{array}{c} \\ 1 \\ 2 \end{array}\begin{array}{c} \begin{array}{cc} 1 & 2 \end{array} \\ \begin{pmatrix} (1/2)^n & 1-(1/2)^n \\ 0 & 1 \end{pmatrix} \end{array} \quad \& \quad P^n \to \begin{array}{c} \\ 1 \\ 2 \end{array}\begin{array}{c} \begin{array}{cc} 1 & 2 \end{array} \\ \begin{pmatrix} 0 & 1 \\ 0 & 1 \end{pmatrix} \end{array}.$$

Thus, for this model, $\lim_{n\to\infty} P[X_n = 2] = 1$ and $\lim_{n\to\infty} P[X_n = 1] = 0$.
(iv) Suppose the one step transition probability matrix is $P = I_3$, an identity matrix of order 3. Then for all $n \geq 1$, $P^n = I_3$, that is, $\lim_{n\to\infty} p_{ij}^{(n)} = 1$ if $i = j$ and 0 otherwise. Thus, $\lim_{n\to\infty} p_{ij}^{(n)}$ depends on i. Observe that

$$\lim_{n\to\infty} P^{(n)} = I_3 \ \Rightarrow \ \lim_{n\to\infty} P[X_n = j] = \lim_{n\to\infty} \sum_{i \in S} p_{ij}^{(n)} P[X_0 = j] = P[X_0 = j] \, .$$

Thus, $\lim_{n\to\infty} P[X_n = j]$, $j \in S$ exists and is the same as the initial distribution.
(v) Suppose the one step transition probability matrix P is given by

$$P = \begin{array}{c} \\ 1 \\ 2 \end{array}\begin{array}{c} \begin{array}{cc} 1 & 2 \end{array} \\ \begin{pmatrix} 0 & 1 \\ 1 & 0 \end{pmatrix} \end{array}.$$

Then for all $n \geq 1$, $P^{2n+1} = P$ and $P^{2n} = I_2$. Thus, powers of P oscillate and hence $\lim_{n\to\infty} p_{ij}^{(n)}, i, \ j \in S$ does not exist. Suppose the one step transition probability matrix P is given by

$$P = \begin{array}{c} \\ 1 \\ 2 \\ 3 \end{array}\begin{array}{c} \begin{array}{ccc} 1 & 2 & 3 \end{array} \\ \begin{pmatrix} 0 & 1 & 0 \\ 0.1 & 0 & 0.9 \\ 0 & 1 & 0 \end{pmatrix} \end{array} \text{ then } P^{2n} = \begin{array}{c} \\ 1 \\ 2 \\ 3 \end{array}\begin{array}{c} \begin{array}{ccc} 1 & 2 & 3 \end{array} \\ \begin{pmatrix} 0.1 & 0 & 0.9 \\ 0 & 1 & 0 \\ 0.1 & 0 & 0.9 \end{pmatrix} \end{array} \quad \& \quad P^{2n-1} = P.$$

Then as in the previous case, $\lim_{n\to\infty} p_{ij}^{(n)}, i, \ j \in S$ does not exist.

Remark 2.4.1 In the last two illustrations, we have noted that $\lim_{n\to\infty} p_{ij}^{(n)}, i, \ j \in S$ does not exist. However, in some cases $\lim_{n\to\infty} P[X_n = j]$, $j \in S$ may exist. For example, suppose $\{X_n, n \geq 1\}$ is a Markov chain with state space $S = \{1, 2\}$, initial distribution $\underline{p}^{(0)} = \{1/2, 1/2\}$ and the one step transition probability matrix P is given by

$$P = \begin{matrix} & \begin{matrix} 1 & 2 \end{matrix} \\ \begin{matrix} 1 \\ 2 \end{matrix} & \begin{pmatrix} 0 & 1 \\ 1 & 0 \end{pmatrix} \end{matrix}.$$

Then for all $n \geq 1$, $P^{2n+1} = P \ and \ P^{2n} = I_2$. From it, we note that for $j = 1, 2$,

$$P[X_n = j] = (1/2)(p_{1j}^{(n)} + p_{2j}^{(n)}) = 1/2 \ \forall \ n \geq 1 \ \Rightarrow \ \lim_{n\to\infty} P[X_n = j] = 1/2.$$

We will discuss this issue in Chap. 3, after introducing the concept of a stationary distribution associated with a Markov chain.

These examples illustrate the distinct types of limiting behavior of a Markov chain. To investigate the limiting behavior, we need some more terminology and mathematical machinery that will be developed in this and the next three sections. We discuss different aspects of a Markov chain, such as reducibility, recurrence and periodicity. These aspects determine the answers to the above questions. Each aspect is quite intuitive: (i) reducibility captures whether or not the transition probability matrix of the given Markov chain can be partitioned in two or more stochastic matrices, (ii) recurrence captures whether or not the chain will return to a given state over and over again and (iii) periodicity captures how often the chain will visit the same state again and again. We now define various properties of the states of a Markov chain and elaborate on the above concepts,

Definition 2.4.1 *i leads to j*: If there exists an integer $n \geq 1$ such that $p_{ij}^{(n)} > 0$, that is, if there is a positive probability that state j can be reached from state i in some finite number of transitions, then i is said to lead to j.

The property i leads to j is denoted by $i \to j$. It is also described as j is accessible or reachable from i.

Remark 2.4.2 In the definition of $i \to j$, some authors (Cinlar [3], Feller [5]) include $n = 0$ in $p_{ij}^{(n)} > 0$. By convention $p_{ij}^{(0)} = \delta_{ij}$, that is, $p_{ii}^{(0)} = 1$ and $p_{ij}^{(0)} = 0$ for all $j \neq i$. Thus, with such a definition one can say that $i \to i$ always.

In the following theorem we prove that "leads to" property is transitive.

Theorem 2.4.2 *If $i \to j$ and $j \to k$, then $i \to k$.*

Proof Observe that

$$\begin{aligned} i \to j &\Rightarrow \exists \ n \geq 1 \ \text{ such that } \ p_{ij}^{(n)} > 0 \\ j \to k &\Rightarrow \exists \ l \geq 1 \ \text{ such that } \ p_{jk}^{(l)} > 0 \\ &\Rightarrow p_{ik}^{(n+l)} = \sum_{u \in S} p_{iu}^{(n)} \, p_{uk}^{(l)} \ \geq \ p_{ij}^{(n)} \, p_{jk}^{(l)} > 0 \\ &\Rightarrow i \to k. \end{aligned}$$

The third step follows from Chapman-Kolmogorov equations. Thus, if $i \to j$ and $j \to k$ then $i \to k$. □

We define below one more property, which is transitive as well as symmetric.

Definition 2.4.2 i communicates with j: A state i is said to communicate with state j, if i leads to j and j leads to i.

The property i communicates with j is denoted by $i \leftrightarrow j$. If two states i and j do not communicate, then either $p_{ij}^{(n)} = 0, \quad \forall \; n \geq 1$ or $p_{ji}^{(n)} = 0, \quad \forall \; n \geq 1$ or both. In Example 2.2.6, all elements of P^5 are positive, which implies that all states communicate with each other. Such a class of states is termed as a communicating class. It is defined below.

Definition 2.4.3 Communicating Class: A set of states such that any two states in the class communicate with each other is said to be a communicating class.

Example 2.4.1 Suppose $\{X_n, n \geq 0\}$ is a Markov chain with state space $S = \{1, 2, 3\}$ and transition probability matrix P is given by

$$P = \begin{array}{c} \\ 1 \\ 2 \\ 3 \end{array} \begin{array}{c} \begin{array}{ccc} 1 & 2 & 3 \end{array} \\ \begin{pmatrix} 1/3 & 2/3 & 0 \\ 1/4 & 3/4 & 0 \\ 1/3 & 1/3 & 1/3 \end{pmatrix} \end{array}.$$

Observe that $1 \to 2$ and $2 \to 1$, thus $1 \leftrightarrow 2$. Further, $3 \to 1$, but from 1 possible transitions are only to 1 and 2. Similarly, from 2 possible transitions are only to 1 and 2. Thus, $1 \nrightarrow 3$. Further, $3 \to 2$ but $2 \nrightarrow 3$. Hence $\{1, 2\}$ is a communicating class. Similarly, $\{3\}$ is a communicating class, since 3 communicates with itself. □

In Example 2.4.1, we have noted that both $\{1, 2\}$ and $\{3\}$ are communicating classes. Further, if the class $\{1, 2\}$ is denoted by C, say, then we note that $\forall \; i \in C, \; \sum_{j \in C} p_{ij} = 1$. Such a property is not valid for a class $\{3\}$. A class of the type $\{1, 2\}$ is known as a closed class. It is defined below.

Definition 2.4.4 Closed Class: A subset C of S is said to be a closed class of states, if $\forall \; i \in C, \; \sum_{j \in C} p_{ij} = 1$, that is, if $[p_{ij}]_{i,j \in C}$ is a stochastic matrix.

In Example 2.4.1, observe that $\{1, 2\}$ is a closed class, however, $\{3\}$ is not a closed class. In the following theorem, we prove that any state in a closed class does not lead to a state outside the closed class; that is the reason for the label "closed class".

Lemma 2.4.1 *A class C is a closed class of states, if and only if*

$$\forall \; i \in C \; \& \; j \notin C, \; p_{ij}^{(n)} = 0 \; \forall \; n > 0.$$

Proof Only if part: Suppose C is a closed class of states. By definition, $\forall\ i \in C$,

$$1 = \sum_{j \in S} p_{ij} = \sum_{j \in C} p_{ij} + \sum_{j \notin C} p_{ij} \Rightarrow\ 1 = 1 - \sum_{j \notin C} p_{ij}$$
$$\Rightarrow\ 0 = \sum_{j \notin C} p_{ij}\ \Rightarrow\ p_{ij} = 0\ \forall\ j \notin C.$$

We assume that $\forall\ i \in C\ and\ j \notin C,\ \ p_{ij}^{(m)} = 0$. It is shown to be true for $m = 1$. Now by the Chapman-Kolmogorov equations, $\forall\ i \in C\ and\ j \notin C$

$$p_{ij}^{(m+1)} = \sum_{k \in S} p_{ik}^{(m)} p_{kj} = \sum_{k \in C} p_{ik}^{(m)} p_{kj} + \sum_{k \notin C} p_{ik}^{(m)} p_{kj}.$$

Observe that the first term on the right-hand side of the above identity is 0, since $p_{kj} = 0$, for $k \in C\ \ and\ \ j \notin C$ and the second term is 0, since $p_{ik}^{(m)} = 0$ by the induction hypothesis. Thus, $p_{ij}^{(m+1)} = 0\ \forall\ i \in C$ and $\forall\ j \notin C$. Hence by induction, for any positive integer n, $\forall\ i \in C$ and $j \notin C,\ \ p_{ij}^{(n)} = 0$. If part: Suppose $p_{ij}^{(n)} = 0$, $\forall\ i \in C$ and $j \notin C$. Note that $\forall\ n > 0$,

$$1 = \sum_{j \in S} p_{ij}^{(n)} = \sum_{j \in C} p_{ij}^{(n)} + \sum_{j \notin C} p_{ij}^{(n)} = \sum_{j \in C} p_{ij}^{(n)}.$$

In particular, with $n = 1$, $\sum_{j \in C} p_{ij} = 1,\ \ \forall\ i \in C$. Hence, C is a closed class. □

Theorem 2.4.3 *If C is a closed class of states, then for all $i \in C$, there exists no $j \notin C$ such that $i \to j$.*

Proof In Lemma 2.4.1, it is proved that if C is a closed class of states, then $\forall\ i \in C\ and\ j \notin C,\ \ p_{ij}^{(n)} = 0\ \forall\ n > 0$, which implies that $i \nrightarrow j$ if $i \in C$ and $j \notin C$. □

Remark 2.4.3 (i) A closed class may not be a communicating class and vice versa. In Example 2.4.1, observe that $\{1, 2\}$ is a closed class as well as a communicating class. However, $\{3\}$ is a communicating class but not a closed class. (ii) Since $\forall\ \ i \in S,\ \ \sum_{j \in S} p_{ij} = 1$, the state space S is a closed class by definition. However, it may not be a communicating class. For example, the state space in Example 2.4.1.

In the following example, we examine which are communicating classes and which are closed classes.

Example 2.4.2 Suppose $\{X_n, n \geq 0\}$ is a Markov chain with state space $S = \{1, 2, 3, 4\}$ and P given by

$$P = \begin{array}{c} \\ 1 \\ 2 \\ 3 \\ 4 \end{array} \begin{array}{c} \begin{array}{cccc} 1 & 2 & 3 & 4 \end{array} \\ \begin{pmatrix} 1/5 & 4/5 & 0 & 0 \\ 5/6 & 1/6 & 0 & 0 \\ 1/3 & 1/3 & 1/3 & 0 \\ 1/4 & 1/4 & 1/4 & 1/4 \end{pmatrix} \end{array}.$$

Observe that $1 \to 2$ and $2 \to 1$, thus $1 \leftrightarrow 2$. Further, $3 \to 1$ but $1 \nrightarrow 3$. Similarly, $3 \to 2$ but $2 \nrightarrow 3$. Further, $3 \to 4$ but $4 \nrightarrow 3$. Thus, $3 \nleftrightarrow 4$. Note that $4 \to 1$ but $1 \nrightarrow 4$; similarly, $4 \to 2 \nrightarrow 4$. Thus, $\{1, 2\}$, $\{3\}$ and $\{4\}$ are three communicating classes, and the state space S can be partitioned into three disjoint classes $\{1, 2\}$, $\{3\}$ and $\{4\}$. Further, observe that $\{1, 2\}$ is a closed class but $\{3\}$ and $\{4\}$ are not closed classes. Thus, $\{1, 2\}$ is a closed communicating class. S is a closed class but not a communicating class. □

A closed class is further labeled as an irreducible or a minimal closed class, if it satisfies a condition as specified in the following definition.

Definition 2.4.5 Minimal Closed Class: A class C is said to be an irreducible or a minimal closed class if (i) C is a closed class and (ii) no proper subset of C is closed.

Definition of a minimal closed class leads to an important classification of a Markov chain as stated in the following definition.

Definition 2.4.6 Irreducible Markov Chain: If the state space of a Markov chain is a minimal closed class, then the Markov chain is said to be an irreducible Markov chain, otherwise it is a reducible Markov chain.

If all states communicate with each other, then the state space is a minimal closed class, which implies that the Markov chain is irreducible. Thus, a minimal closed class is a closed as well as a communicating class. In Example 2.2.6, P^5 has all positive elements, indicating that all states communicate with each other and hence the corresponding Markov chain is irreducible.

Example 2.4.3 Suppose $\{X_n, n \geq 0\}$ is a time homogeneous Markov chain with state space $S = \{1, 2, 3, 4\}$ and P given by

$$P = \begin{array}{c} \\ 1 \\ 2 \\ 3 \\ 4 \end{array} \begin{array}{c} \begin{array}{cccc} 1 & 2 & 3 & 4 \end{array} \\ \begin{pmatrix} 1/2 & 0 & 1/2 & 0 \\ 0 & 1/3 & 0 & 2/3 \\ 0 & 2/3 & 0 & 1/3 \\ 1/4 & 0 & 3/4 & 0 \end{pmatrix} \end{array}.$$

To examine whether $i \leftrightarrow j$, observe that $1 \to 3 \to 4 \to 3 \to 2 \to 1$. Thus, all states communicate with each other and S is a closed class. Suppose $C = \{1, 2, 3\} \subset S$. Observe that $1 \in C$ and $4 \notin C$, but $1 \to 4$. Thus C is not a closed class. It can be verified that no proper subset of S is closed. Hence, S itself is a minimal closed class and the given Markov chain is an irreducible Markov chain. Another approach to decide which states communicate with each other is to find powers of the transition probability matrix. We find P^2 as given below

$$P^2 = \begin{matrix} & \begin{matrix} 1 & 2 & 3 & 4 \end{matrix} \\ \begin{matrix} 1 \\ 2 \\ 3 \\ 4 \end{matrix} & \begin{pmatrix} 0.250 & 0.333 & 0.250 & 0.167 \\ 0.167 & 0.111 & 0.500 & 0.222 \\ 0.083 & 0.222 & 0.250 & 0.444 \\ 0.125 & 0.500 & 0.125 & 0.250 \end{pmatrix} \end{matrix}.$$

We note that all the elements of P^2 are positive, which implies that any state is accessible from any other state. Hence, all states communicate with each other and the given Markov chain is irreducible. □

We now examine whether the Markov chain as specified in Example 2.4.1 is an irreducible Markov chain.

Example 2.4.4 Suppose $\{X_n, n \geq 0\}$ is a Markov chain with state space $S = \{1, 2, 3\}$ and transition probability matrix P given by

$$P = \begin{matrix} & \begin{matrix} 1 & 2 & 3 \end{matrix} \\ \begin{matrix} 1 \\ 2 \\ 3 \end{matrix} & \begin{pmatrix} 1/3 & 2/3 & 0 \\ 1/4 & 3/4 & 0 \\ 1/3 & 1/3 & 1/3 \end{pmatrix} \end{matrix}.$$

In Example 2.4.1, we have noted that $\{1, 2\}$ is a communicating class. Further by the definition of a closed class, it is a closed class. Thus, there exists a proper subset of S which is closed and hence S is not a minimal closed set, implying that the Markov chain is a reducible Markov chain. □

Example 2.4.5 Suppose $\{X_n, n \geq 0\}$ is a time homogeneous Markov chain with state space $S = \{1, 2, 3, 4\}$ and P given by

$$P = \begin{matrix} & \begin{matrix} 1 & 2 & 3 & 4 \end{matrix} \\ \begin{matrix} 1 \\ 2 \\ 3 \\ 4 \end{matrix} & \begin{pmatrix} 1/2 & 0 & 1/2 & 0 \\ 0 & 1/4 & 0 & 3/4 \\ 2/3 & 0 & 1/3 & 0 \\ 0 & 1/4 & 0 & 3/4 \end{pmatrix} \end{matrix}.$$

Observe that $1 \to 3, 3 \to 1 \;\Rightarrow\; 1 \leftrightarrow 3$ and $2 \to 4 \to 2 \;\Rightarrow\; 2 \leftrightarrow 4$. Now it is to be noted that $1 \to 3$ but $3 \to 3$ and $3 \to 1$ only. Same scenario is observed for states 2 and 4. Thus, $1 \leftrightarrow 3$ only and $2 \leftrightarrow 4$ only. Hence $\{1, 3\}$ and $\{2, 4\}$ are two closed classes, and the given Markov chain is a reducible Markov chain. Further, it can be verified that for no n, all elements of P^n are positive. In fact, for $n \geq 8$, P^n is given by

$$P^n = \begin{array}{c} \\ 1 \\ 2 \\ 3 \\ 4 \end{array}\begin{array}{c} \begin{array}{cccc} 1 & 2 & 3 & 4 \end{array} \\ \begin{pmatrix} 0.571 & 0.00 & 0.429 & 0.00 \\ 0.000 & 0.25 & 0.000 & 0.75 \\ 0.571 & 0.00 & 0.429 & 0.00 \\ 0.000 & 0.25 & 0.000 & 0.75 \end{pmatrix} \end{array}.$$

This matrix indicates that states 1 and 3 are not accessible from either 2 or 4 and vice versa. Thus, the state space has two proper subsets $\{1, 3\}$ and $\{2, 4\}$ which are closed and the Markov chain is reducible. Such a Markov chain can be studied by studying two separate Markov chains with state spaces $\{1, 3\}$ and $\{2, 4\}$. It is to be noted that for $n \geq 8$, the first and the third rows of P^n are the same. Similarly, the second and the fourth rows of P^n are the same. We discuss interpretation of such a feature when we study stationary distribution associated with a Markov chain. □

Definition 2.4.7 Absorbing State: If the singleton class $\{i\}$ is closed, then i is said to be an absorbing state.

Absorbing state is such that once the system enters that state, it does not leave that state in any number of transitions. Thus, if i is an absorbing state, then $p_{ii}^{(n)} = 1 \ \forall \ n \geq 0$ and i does not lead to any other state. In view of such a nature of the state, it is termed as the absorbing state. In Example 2.2.5, state 3 is the absorbing state.

The next section is devoted to an elaborate discussion on the concept of recurrence or persistence, which conveys whether or not the Markov chain will return to a given state over and over again.

2.5 Persistent and Transient States

As discussed earlier, the long run behavior of a Markov chain depends on the nature of its states. We now consider the classification of states of the Markov chain as persistent and transient. Toward it, we introduce the concept of the first return to a state. Suppose N_i denotes the number of steps required for the first return to state i by the Markov chain that starts in state i. Then the event $[N_i = n]$ can be expressed as

$$[N_i = 1] = [X_1 = i] \ \& \ [N_i = n] = [X_n = i, X_r \neq i \ \forall \ r = 1, 2, \ldots, n-1], \ n \geq 2.$$

Suppose $f_{ii}^{(n)} = P[N_i = n | X_0 = i], \ n \geq 1$. Then $f_{ii}^{(n)}$ is known as the probability of the first return to the state i at the nth step given that the initial state is i and the distribution of N_i is known as the first return distribution. We define $f_{ii}^{(0)} = p_{ii}^{(0)} = 1$. The difference between $p_{ii}^{(n)}$ and $f_{ii}^{(n)}$ is that in the event corresponding to $p_{ii}^{(n)}$, the Markov chain may visit the state i any number of times before step n. It is to be noted that

$$f_{ii}^{(1)} = p_{ii} \ \& \ [N_i = n] \subset [X_n = i] \ \Rightarrow \ f_{ii}^{(n)} \leq p_{ii}^{(n)} \ \forall \ n \geq 2.$$

Suppose f_{ii} denotes the probability that the Markov chain ever returns to state i. Since $[N_i = n],\ n \geq 1$ are mutually exclusive events, we have

$$f_{ii} = P\Big[\bigcup_{n\geq 1}[N_i = n]|X_0 = i\Big] = \sum_{n=1}^{\infty} P[N_i = n|X_0 = i] = \sum_{n=1}^{\infty} f_{ii}^{(n)} = P[N_i < \infty].$$

We note that $1 - f_{ii}$ is the probability that the Markov chain never returns to state i, that is, $1 - f_{ii} = P[N_i = \infty]$. Thus, N_i is an extended real-valued random variable.

We classify a state as a persistent state or a transient state, depending on the values of f_{ii}. Note that being probability, $0 \leq f_{ii} \leq 1$.

Definition 2.5.1 Persistent and Transient States: A state i is said to be persistent or recurrent if $f_{ii} = 1$ and transient or non-recurrent if $f_{ii} < 1$.

Suppose $\mu_i = E(N_i|X_0 = i)$ denotes the expected number of steps required for the first return to state i or mean recurrence time to state i. If i is transient, then N_i is an extended real-valued random variable and μ_i is infinite. If i is persistent, then $\{f_{ii}^{(n)}, n \geq 1\}$ is a probability mass function and hence μ_i is given by $\mu_i = \sum_{n=1}^{\infty} n f_{ii}^{(n)}$. It may be finite or infinite. For a persistent state i, depending on whether μ_i is finite or infinite, we have a further classification as given below.

Definition 2.5.2 Non-null Persistent and Null Persistent States: A persistent state i is said to be non-null persistent or positive persistent or positive recurrent if $\mu_i < \infty$. It is said to be null persistent or null recurrent if $\mu_i = \infty$.

Observe that $N_i \geq 1$ implies $\mu_i \geq 1$. Another argument for a persistent state i is as follows:

$$\mu_i = \sum_{n=1}^{\infty} n f_{ii}^{(n)} \geq \sum_{n=1}^{\infty} f_{ii}^{(n)} = f_{ii} = 1 \quad \Rightarrow \quad \mu_i \geq 1.$$

Further for a persistent state i,

$$p_{ii}^{(n)} \geq f_{ii}^{(n)} \quad \Rightarrow \quad \sum_{n=1}^{\infty} p_{ii}^{(n)} \geq \sum_{n=1}^{\infty} f_{ii}^{(n)} = 1.$$

Thus for a persistent state i, $\mu_i \geq 1$ and $\sum_{n=1}^{\infty} p_{ii}^{(n)} \geq 1$. In Theorem 2.5.3, we prove that for a persistent state i, $\sum_{n=1}^{\infty} p_{ii}^{(n)}$ is divergent.

A trivial example of a persistent state is an absorbing state. Once the system enters an absorbing state, it remains in that state forever. In the following theorem, we prove this claim.

Theorem 2.5.1 *An absorbing state is a non-null persistent state.*

Proof Suppose i is an absorbing state, that is, $p_{ii} = 1$ and i does not lead to any other state. Then

$$
\begin{aligned}
f_{ii}^{(1)} &= p_{ii} = P[X_1 = i|X_0 = i] = 1 \\
f_{ii}^{(n)} &\le P[X_1 \neq i|X_0 = i] = 0, \ \ \forall \ n \ge 2,
\end{aligned}
$$

which implies that $f_{ii} = \sum_{n=1}^{\infty} f_{ii}^{(n)} = p_{ii} = 1$. Hence, the absorbing state is a persistent state. Further $\mu_i = 1$ and hence it is non-null persistent. □

Example 2.5.1 Suppose $\{X_n, n \ge 0\}$ is a Markov chain with state space $S = \{1, 2\}$ and P given by

$$
P = \begin{matrix} & \begin{matrix} 1 & 2 \end{matrix} \\ \begin{matrix} 1 \\ 2 \end{matrix} & \begin{pmatrix} 1 & 0 \\ \alpha & 1-\alpha \end{pmatrix} \end{matrix}, \quad 0 < \alpha \le 1.
$$

State 1 is an absorbing state and hence is non-null persistent. To find the nature of state 2, observe that

$$
\begin{aligned}
f_{22}^{(1)} &= p_{22} = 1 - \alpha \\
f_{22}^{(2)} &= P[X_2 = 2, X_1 \neq 2|X_0 = 2] = P[X_2 = 2|X_1 = 1]P[X_1 = 1|X_0 = 2] = 0 \\
f_{22}^{(3)} &= P[X_3 = 2, X_2 \neq 2, X_1 \neq 2|X_0 = 2] \\
&= P[X_3 = 2|X_2 = 1]P[X_2 = 1|X_1 = 1]P[X_1 = 1|X_0 = 2] = 0 .
\end{aligned}
$$

Similarly, it follows that $f_{22}^{(n)} = 0, \ \ \forall \ n \ge 4$. Hence $f_{22} = 1 - \alpha$ and hence state 2 is transient if $0 < \alpha \le 1$ and is persistent only if $\alpha = 0$. If $\alpha = 0$, then state 2 is also an absorbing state and hence non-null persistent. □

Remark 2.5.1 In Example 2.5.1, if $\alpha = 1$, $f_{22} = 0$. Note that if $\alpha = 1$, then from 2, the only possible transition is to state 1, and 1 is an absorbing state. Thus, starting in state 2, the chain will never return to state 2.

Example 2.5.2 Suppose $\{X_n, n \ge 0\}$ is a Markov chain with state space $S = \{1, 2, 3, 4\}$ and transition probability matrix

$$
P = \begin{matrix} & \begin{matrix} 1 & 2 & 3 & 4 \end{matrix} \\ \begin{matrix} 1 \\ 2 \\ 3 \\ 4 \end{matrix} & \begin{pmatrix} 1 & 0 & 0 & 0 \\ 1/2 & 0 & 1/2 & 0 \\ 0 & 1/2 & 0 & 1/2 \\ 0 & 0 & 0 & 1 \end{pmatrix} \end{matrix}.
$$

Note that 1 and 4 are absorbing and hence non-null persistent states. To decide the nature of 2 and 3, we find f_{22} and f_{33}. By definition, $f_{22}^{(1)} = p_{22} = 0$. Now

$$
\begin{aligned}
f_{22}^{(2)} &= P[X_2 = 2, X_1 \neq 2|X_0 = 2] \\
&= P[X_2 = 2, X_1 = 1|X_0 = 2] + P[X_2 = 2, X_1 = 3|X_0 = 2] \\
&= p_{12}p_{21} + p_{32}p_{23} = 0 + 1/4 = 1/4.
\end{aligned}
$$

From the transition probability matrix, we observe that from 2 the possible transitions are to 1 and 3 only. Once the system enters 1, it stays in 1 only, it being an absorbing state. If a transition is from 2 to 3, then from 3 possible transitions are to 2 and 4 only. Again 4 is an absorbing state. It thus follows that $f_{22}^{(n)} = 0$ for all $n \geq 3$. Hence $f_{22} = 1/4$ which implies that 2 is a transient state. Using similar arguments, we get $f_{33} = 1/4$ and hence 3 is also a transient state. □

For a two-state Markov chain, we can obtain an explicit expression of the first return distribution and hence the expressions for mean recurrence times, as shown in the following example.

Example 2.5.3 Suppose $\{X_n, n \geq 0\}$ is a Markov chain with state space $S = \{0, 1\}$ and transition probability matrix P given by

$$P = \begin{matrix} & \begin{matrix} 0 & 1 \end{matrix} \\ \begin{matrix} 0 \\ 1 \end{matrix} & \begin{pmatrix} p_{00} & p_{01} \\ p_{10} & p_{11} \end{pmatrix} \end{matrix}.$$

We have $f_{00}^{(1)} = p_{00}$ and for $n \geq 2$,

$$\begin{aligned} f_{00}^{(n)} &= P[X_n = 0, X_r = 1, 0 < r < n | X_0 = 0] \\ &= P[X_n = 0 | X_{n-1} = 1] \prod_{r=2}^{n-1} P[X_r = 1 | X_{r-1} = 1] P[X_1 = 1 | X_0 = 0] \\ &= p_{10} p_{11}^{n-2} p_{01} = (1 - p_{00}) p_{10} (1 - p_{10})^{n-2} . \end{aligned}$$

Thus, the distribution of N_0 is given by

$$f_{00}^{(n)} = \begin{cases} p_{00}, & \text{if } n = 1 \\ (1 - p_{00}) p_{10} (1 - p_{10})^{n-2}, & \text{if } n \geq 2 . \end{cases}$$

Note that it is a mixture of two distributions—one is degenerate at 1 and the second is a shifted geometric distribution with parameter p_{10} on the support $\{2, 3, 4, \ldots, \}$, with mixing probabilities p_{00} and $(1 - p_{00})$, respectively. Thus,

$$\begin{aligned} \mu_0 &= p_{00} + (1 - p_{00})(2 + (1 - p_{10})/p_{10}) = p_{00} + (p_{01}/p_{10})(1 + p_{10}) \\ &= p_{00} + p_{01} + p_{01}/p_{10} = 1 + p_{01}/p_{10} = (p_{10} + p_{01})/p_{10} . \end{aligned}$$

We can also obtain it by the routine calculations, as shown below

$$\begin{aligned} \mu_0 &= \sum_{n=1}^{\infty} n f_{00}^{(n)} = p_{00} + (1 - p_{00}) p_{10} \sum_{n=2}^{\infty} n (1 - p_{10})^{n-2} \\ &= p_{00} + (p_{01}/p_{10})(1 + p_{10}) = 1 + p_{01}/p_{10} = (p_{10} + p_{01})/p_{10} . \end{aligned}$$

On similar lines, the distribution of N_1 is given by

$$f_{11}^{(n)} = \begin{cases} p_{11}, & \text{if } n = 1 \\ (1 - p_{11})p_{01}(1 - p_{01})^{n-2}, & \text{if } n \geq 2\,. \end{cases}$$

Further $\mu_1 = 1 + p_{10}/p_{01} = (p_{01} + p_{10})/p_{01}$. Note that

$$1/\mu_0 + 1/\mu_1 = p_{10}/(p_{10} + p_{01}) + p_{01}/(p_{01} + p_{10}) = 1\,.$$

We will elaborate on the interpretation of this result after discussing the concept of a stationary distribution associated with a Markov chain. For the Markov chain of Example 2.3.4, an estimate of transition probability matrix is given by

$$\hat{P}_n = \begin{matrix} & \begin{matrix} 0 & 1 \end{matrix} \\ \begin{matrix} 0 \\ 1 \end{matrix} & \begin{pmatrix} 0.398 & 0.602 \\ 0.162 & 0.838 \end{pmatrix} \end{matrix}.$$

Hence we get $\hat{\mu}_0 = 1 + 0.602/0.162 = 4.416$ and $\hat{\mu}_1 = 1 + 0.162/0.602 = 1.269$. Thus, on the average between two dry days, the spell of wet days is 4.42 days, while on the average between two wet days, the spell of dry days is 1.27 days. Observe that $1/\hat{\mu}_0 + 1/\hat{\mu}_1 = 0.2120 + 0.7880 = 1$. □

We now briefly discuss one more criterion to examine whether a state is transient. We first define an essential state and then state its relation with a transient state.

Definition 2.5.3 Essential and Inessential States: A state i is essential if i communicates with every state it leads to, that is, i is an essential state if $i \to j \;\Rightarrow\; j \to i$. A state i is inessential if it is not essential, that is, i is an inessential state if there exists a state k, such that $i \to k$ but $k \nrightarrow i$.

Example 2.5.4 Suppose $\{X_n, n \geq 0\}$ is a Markov chain with state space $S = \{1, 2, 3\}$ and transition probability matrix P given by

$$P = \begin{matrix} & \begin{matrix} 1 & 2 & 3 \end{matrix} \\ \begin{matrix} 1 \\ 2 \\ 3 \end{matrix} & \begin{pmatrix} 1/3 & 2/3 & 0 \\ 1/4 & 3/4 & 0 \\ 1/3 & 1/3 & 1/3 \end{pmatrix} \end{matrix}.$$

Observe that $1 \to 2$ and $2 \to 1$, hence 1 is an essential state. Further, $2 \to 1$ and $1 \to 2$, hence 2 is an essential state. However, $3 \to 1$ but $1 \nrightarrow 3$ and hence 3 is an inessential state. □

Remark 2.5.2 In a closed communicating class all states are essential. As a consequence, an absorbing state is essential.

We now state a result about an inessential state in the following theorem. We first define the term "class property". Such a property is useful to verify the nature of the states.

Definition 2.5.4 Class Property: Suppose C is a class of states. We say that "α" is a class property, if one state in C possesses property "α" implying that all other states in C also possess the property "α".

If "α" is a class property, then "not α" is also a class property.

Theorem 2.5.2 *(i) "Being essential" and hence "being inessential" are class properties. (ii) An inessential state is a transient state.*

Proof (i) Suppose i is an essential state, thus it communicates with every state it leads to. Suppose $C(i) = \{j | i \leftrightarrow j\}$ is a class of states. Suppose $j \in C(i)$ and $j \to k$. Then $i \to j \to k$ implies $i \to k$. But i is an essential state, hence $k \to i$. Thus, $k \to i \to j$ implies $k \to j$. Thus, if $j \to k$ then $k \to j$. Hence, j is also an essential state. Thus, if i is essential and $j \leftrightarrow i$, then j is also essential, which implies that "being essential" and hence "being inessential" are class properties. (ii) We defer the proof that an inessential state is a transient state to Sect. 2.6. It is proved in Theorem 2.6.2. □

In Example 2.5.2, $2 \to 1$ but $1 \nrightarrow 2$, thus, 2 is an inessential state and hence transient. We have verified that for this Markov chain, $f_{22} = 1/4$ implying that 2 is a transient state. Similarly, $3 \to 4$ but $4 \nrightarrow 3$, thus, 3 is also an inessential state and hence transient. Further, $2 \leftrightarrow 3$, that is, $\{2, 3\}$ is a communicating class and both the states have the same nature, both being inessential and hence transient states.

In Example 2.5.4, state 3 is inessential and hence is a transient state. We can compute f_{33} and verify whether it is less than 1. By definition $f_{33}^{(1)} = p_{33} = 1/3$. To find $f_{33}^{(2)}$, observe that $3 \to 1$ but $1 \nrightarrow 3$, similarly $3 \to 2$ but $2 \nrightarrow 3$. Hence, it is not possible to go from 3 to 3 in two steps, without visiting 3. Hence, $f_{33}^{(2)} = 0$. Using similar arguments, we get $f_{33}^{(n)} = 0, \ \forall \ n \geq 3$. Thus, $f_{33} = 1/3 < 1$ implying that 3 is a transient state.

Remark 2.5.3 Observe that the second part of Theorem 2.5.2 is equivalent to stating that a persistent state is essential.

In the following example, we verify that a persistent state is essential.

Example 2.5.5 Suppose $\{X_n, n \geq 1\}$ is a Markov chain as given in Example 2.5.4. It is clear that $1 \leftrightarrow 2$. Observe that $f_{11}^{(1)} = p_{11} = 1/3$,

$$f_{11}^{(2)} = p_{12}p_{21} = (2/3)(1/4), \quad f_{11}^{(3)} = p_{12}p_{22}p_{21} = (2/3)(3/4)(1/4).$$

Continuing in this way, we get $f_{11}^{(n)} = (2/3)(3/4)^{n-2}(1/4), \ n \geq 2$. Hence,

$$f_{11} = 1/3 + \sum_{n \geq 2}(2/3)(3/4)^{n-2}(1/4) = 1/3 + 2/3 = 1$$

$$\begin{aligned} \text{Similarly,} \quad f_{22}^{(1)} = 3/4, \ f_{22}^{(n)} &= (1/4)(1/3)^{n-2}(2/3), \ n \geq 2 \\ \Rightarrow \quad f_{22} = 3/4 + \sum_{n \geq 2}(1/4)(1/3)^{n-2}(2/3) &= 1 \, . \end{aligned}$$

Thus, both 1 and 2 are persistent states. Further, it is easy to verify that both communicate with every other state that these lead to and hence these are essential states. Thus, it is verified that these two persistent states are essential. Note that $\{1, 2\}$ is a communicating class and both the states are essential. We further investigate whether these are null persistent or non-null persistent states. To find μ_1 and μ_2 for this chain, observe that

$$\begin{aligned}\mu_1 = \sum_{n=1}^{\infty} n f_{11}^{(n)} &= 1(1/3) + (1/6)\sum_{n\geq 2} n(3/4)^{n-2}\\ &= 1/3 + (1/6)\sum_{r\geq 0}(r+2)(3/4)^r\\ &= 1/3 + (1/6)(3/4)\sum_{r\geq 1} r(3/4)^{r-1} + (2/6)\sum_{r\geq 0}(3/4)^r\\ &= 1/3 + (1/8)(1-3/4)^{-2} + (1/3)(1-3/4)^{-1} = 11/3\,.\end{aligned}$$

Thus, $\mu_1 < \infty$ and hence 1 is a non-null persistent state. On similar lines, we find $\mu_2 = 11/8 < \infty$ and hence 2 is also a non-null persistent state. □

Theorem 2.5.2 states that an inessential state is transient, however, its converse is not true. In an irreducible Markov chain, all states communicate with each other and hence are essential but these may be transient. In Sect. 4.2, we give such an example. In Sect. 4.2, we use the results of Theorem 2.5.2 to decide the nature of states of a random walk with absorbing barriers, when state space is either $S = \{0, 1, \ldots, M\}$ or $S = W$.

The definition of a persistent state given above is in terms of the first return probabilities $f_{ii}^{(n)}$. It can also be defined in terms of n-step transition probabilities $p_{ii}^{(n)}$. We prove in the following theorem that a state i is persistent if and only if $\sum_{n=1}^{\infty} p_{ii}^{(n)}$ is a divergent series. We first prove a lemma which is useful in the proof of the theorem and which also leads to a recurrence relation to compute $f_{ii}^{(n)}$.

Lemma 2.5.1 *Suppose $p_{ii}^{(n)}$ and $f_{ii}^{(n)}$ denote the n-step transition probability and probability of the first return at the nth step, respectively. Then*

$$(i)\ \ p_{ii}^{(n)} = \sum_{r=1}^{n} p_{ii}^{(n-r)} f_{ii}^{(r)} = \sum_{r=0}^{n-1} p_{ii}^{(r)} f_{ii}^{(n-r)}$$

$$\&\ (ii)\ \ f_{ii}^{(n)} = p_{ii}^{(n)} - \sum_{r=1}^{n-1} p_{ii}^{(n-r)} f_{ii}^{(r)}.$$

Proof (i) We have defined the events $[N_i = 1] = [X_1 = i]$ and for $r \geq 2$, $[N_i = r] = [X_r = i, X_s \neq i, s = 1, 2, \ldots, r-1]$, when $X_0 = i$. We now define the events E_r, $r = 1, 2, 3, \ldots, n$ as follows:

$$
\begin{aligned}
E_1 &= [X_0 = i, X_1 = i] \ \text{ and} \\
E_r &= [X_0 = i, N_i = r, X_n = i] \quad \text{for } \ r = 2, 3, \ldots, n.
\end{aligned}
$$

Since the events $[N_i = r]$ for $r \geq 1$ are mutually exclusive, it follows that the n events E_r are also mutually exclusive events and $\bigcup_{r=1}^{n} E_r = [X_n = i, X_0 = i]$. Hence,

$$
\begin{aligned}
p_{ii}^{(n)} &= P[X_n = i | X_0 = i] = P[X_n = i, X_0 = i]/P[X_0 = i] \\
&= P\Big(\bigcup_{r=1}^{n} E_r\Big)\Big/P[X_0 = i] \\
&= \sum_{r=1}^{n} P[X_n = i, N_i = r, X_0 = i]/P[X_0 = i] \\
&= \sum_{r=1}^{n} P[X_n = i | N_i = r, X_0 = i]P[N_r = i | X_0 = i] \\
&= \sum_{r=1}^{n} P[X_n = i | N_i = r]P[N_i = r | X_0 = i] \\
&= \sum_{r=1}^{n} p_{ii}^{(n-r)} f_{ii}^{(r)} = \sum_{r=0}^{n-1} p_{ii}^{(r)} f_{ii}^{(n-r)}.
\end{aligned}
$$

(ii) From part (i), using the fact that $p_{ii}^{(0)} = 1$, we find a recurrence relation to compute $f_{ii}^{(n)}$ as follows:

$$
\begin{aligned}
p_{ii}^{(n)} &= \sum_{r=0}^{n-1} p_{ii}^{(r)} f_{ii}^{(n-r)} = f_{ii}^{(n)} + \sum_{r=1}^{n-1} p_{ii}^{(r)} f_{ii}^{(n-r)} \\
\Rightarrow \quad f_{ii}^{(n)} &= p_{ii}^{(n)} - \sum_{r=1}^{n-1} p_{ii}^{(r)} f_{ii}^{(n-r)} = p_{ii}^{(n)} - \sum_{r=1}^{n-1} p_{ii}^{(n-r)} f_{ii}^{(r)}. \qquad (2.5.1)
\end{aligned}
$$

We know that $f_{ii}^{(1)} = p_{ii}$, hence by Eq. (2.5.1), $f_{ii}^{(2)} = p_{ii}^{(2)} - p_{ii} f_{ii}^{(1)}$. Proceeding in this way, we find the distribution of the first return to state i. □

We state below a lemma needed in the proof of the theorem that follows. It is known as Toeplitz' lemma.

Lemma 2.5.2 *Toeplitz' Lemma: Suppose $\{a_n, n \geq 0\}$ is a sequence of non-negative real numbers such that $a_r \neq 0$ for at least one r. Suppose $\{b_n, n \geq 0\}$ is a sequence of non-negative real numbers such that $b_n \to b$ as $n \to \infty$, and b may be finite or infinite. Then*

$$
\lim_{n\to\infty} a_n \Big/ \sum_{r=0}^{n} a_r = 0 \ \& \ \lim_{n\to\infty} b_n = b \ \Rightarrow \ \lim_{n\to\infty} \Big(\sum_{r=0}^{n} a_r b_{n-r}\Big)\Big/\Big(\sum_{r=0}^{n} a_r\Big) = b.
$$

Remark 2.5.4 The condition $\lim_{n\to\infty} a_n / \sum_{r=0}^{n} a_r = 0$ is satisfied if either

(i) $\sum_{n=0}^{\infty} a_n < \infty$, since for a convergent series, nth term converges to 0 or

(ii) $\sum_{n=0}^{\infty} a_n$ diverges with bounded a_n's.

Theorem 2.5.3 *(i) A state i is persistent if and only if $\sum_{n=1}^{\infty} p_{ii}^{(n)}$ is divergent.*

(ii) A state i is transient if and only if $\sum_{n=1}^{\infty} p_{ii}^{(n)}$ is convergent.

Proof From part (i) of Lemma 2.5.1, we have $p_{ii}^{(n)} = \sum_{r=0}^{n-1} p_{ii}^{(r)} f_{ii}^{(n-r)}$. Hence,

$$\begin{aligned}
\sum_{n=1}^{N} p_{ii}^{(n)} &= \sum_{n=1}^{N}\sum_{r=0}^{n-1} p_{ii}^{(r)} f_{ii}^{(n-r)} \\
&= \sum_{r=0}^{N-1} p_{ii}^{(r)}\Big[\sum_{n=r+1}^{N} f_{ii}^{(n-r)}\Big], \text{ by interchanging sums} \\
&= \sum_{r=0}^{N-1} p_{ii}^{(r)}\Big[\sum_{n=1}^{N-r} f_{ii}^{(n)}\Big] \\
&= \sum_{r=0}^{N} p_{ii}^{(r)}\Big[\sum_{n=1}^{N-r} f_{ii}^{(n)}\Big], \text{ since for } r = N, \sum_{n=1}^{N-r} f_{ii}^{(n)} = 0 \\
&= \sum_{r=0}^{N} a_r b_{N-r}, \text{ with } a_r = p_{ii}^{(r)} \ \&\ b_{N-r} = \sum_{n=1}^{N-r} f_{ii}^{(n)}.
\end{aligned}$$

Note that $a_0 = p_{ii}^{(0)} = 1$, thus at least one $a_r \neq 0$. $b_N = \sum_{n=1}^{N} f_{ii}^{(n)} \to b = f_{ii}$ and $f_{ii} \leq 1$. Further, a sequence $p_{ii}^{(n)} / \sum_{r=0}^{n} p_{ii}^{(r)}$ converges to 0 if $\sum_{r=0}^{\infty} p_{ii}^{(r)}$ is divergent, as the numerator is bounded by 1. If $\sum_{r=0}^{\infty} p_{ii}^{(r)}$ is convergent, then the numerator being the nth term of a convergent series goes to 0 and thus a sequence $p_{ii}^{(n)} / \sum_{r=0}^{n} p_{ii}^{(r)}$ converges to 0. Hence by Lemma 2.5.2, as $N \to \infty$,

$$\frac{\sum_{r=0}^{N} a_r b_{N-r}}{\sum_{r=0}^{N} a_r} \to b = f_{ii} \Rightarrow \frac{\sum_{n=1}^{N} p_{ii}^{(n)}}{\sum_{n=0}^{N} p_{ii}^{(n)}} \to f_{ii} \Rightarrow \frac{\sum_{n=1}^{N} p_{ii}^{(n)}}{1 + \sum_{n=1}^{N} p_{ii}^{(n)}} \to f_{ii},$$

since $p_{ii}^{(0)} = 1$. Suppose $\sum_{n=1}^{N} p_{ii}^{(n)}$ diverges, then

$$\Big(1 + 1/\sum_{n=1}^{N} p_{ii}^{(n)}\Big)^{-1} \to 1 \Rightarrow f_{ii} = 1 \Rightarrow i \text{ is persistent.}$$

Suppose i is persistent, we want to prove that $\sum_{n=1}^{N} p_{ii}^{(n)}$ diverges. We assume the contrary. Suppose $\sum_{n=1}^{\infty} p_{ii}^{(n)} < \infty$, then

$$\sum_{n=1}^{N} p_{ii}^{(n)}/(1+\sum_{n=1}^{N} p_{ii}^{(n)}) \to \sum_{n=1}^{\infty} p_{ii}^{(n)}/(1+\sum_{n=1}^{\infty} p_{ii}^{(n)}) < 1$$
$$\Rightarrow \; f_{ii} < 1 \; \Rightarrow \; i \text{ is transient},$$

which is a contradiction. Hence, a state i is persistent if and only if $\sum_{n=1}^{\infty} p_{ii}^{(n)}$ is divergent and (i) is proved. From (i), it follows that a state i is transient if and only if $\sum_{n=1}^{\infty} p_{ii}^{(n)} < \infty$. □

Example 2.5.6 We have noted at the beginning of Sect. 2.4 that if a one step transition probability matrix is given by

$$P = \begin{matrix} & \begin{matrix} 1 & 2 & 3 \end{matrix} \\ \begin{matrix} 1 \\ 2 \\ 3 \end{matrix} & \begin{pmatrix} 0 & 1 & 0 \\ 0.1 & 0 & 0.9 \\ 0 & 1 & 0 \end{pmatrix} \end{matrix} \text{ then } P^{2n} = \begin{matrix} & \begin{matrix} 1 & 2 & 3 \end{matrix} \\ \begin{matrix} 1 \\ 2 \\ 3 \end{matrix} & \begin{pmatrix} 0.1 & 0 & 0.9 \\ 0 & 1 & 0 \\ 0.1 & 0 & 0.9 \end{pmatrix} \end{matrix} \quad \& \quad P^{2n-1} = P,$$

$\forall \; n \geq 1$. Hence,

$$\begin{aligned} p_{ii}^{(2n-1)} &= 0 \; \forall \; n \geq 1 \; \Rightarrow \; \lim_{n\to\infty} p_{ii}^{(2n-1)} = 0, \quad \text{for } i = 1, 2, 3 \\ p_{11}^{(2n)} &= 0.1 \; \forall \; n \geq 1 \; \Rightarrow \; \lim_{n\to\infty} p_{11}^{(2n)} = 0.1 \\ p_{22}^{(2n)} &= 1 \; \forall \; n \geq 1 \; \Rightarrow \; \lim_{n\to\infty} p_{22}^{(2n)} = 1 \\ \& \; p_{33}^{(2n)} &= 0.9 \; \forall \; n \geq 1 \; \Rightarrow \; \lim_{n\to\infty} p_{33}^{(2n)} = 0.9 \\ \Rightarrow \; \sum_{n=1}^{\infty} p_{ii}^{(n)} &= \sum_{n=1}^{\infty} p_{ii}^{(2n)} = \infty \; \text{ for } i = 1, 2, 3, \end{aligned}$$

terms in the series being constants. Hence, states 1, 2, 3 are persistent states. Note that for this Markov chain, all states communicate with each other. Hence, these are essential states. Observe that $\forall \; i = 1, 2, 3$,

$$\liminf_{n\to\infty} p_{ii}^{(n)} = 0 \neq \limsup_{n\to\infty} p_{ii}^{(n)} \; \Rightarrow \; \lim_{n\to\infty} p_{ii}^{(n)} \text{ does not exist.}$$

□

Intuitively, there is a significant difference between a persistent and a transient state. We clarify it on the basis of the interpretation of the series $\sum_{n=1}^{\infty} p_{ii}^{(n)}$ and its relation with persistent and transient states depending on whether it is a divergent or a convergent series. We also elaborate below on a mathematical link between the series $\sum_{n=1}^{\infty} p_{ii}^{(n)}$ and the expected number of times the process is in state i. Suppose $X_0 = i$ and a random variable $Y_n(i), n \geq 1$ is defined as

$$Y_n(i) = \begin{cases} 1, & \text{if } X_n = i \\ 0, & \text{if } X_n \neq i \, . \end{cases}$$

Then $V_i = \sum_{n=1}^{\infty} Y_n(i)$ denotes the number of times the chain is in state i, excluding the initial state i. Further,

$$\begin{aligned} E(V_i) = E\Big(\sum_{n=1}^{\infty} Y_n(i)|X_0 = i\Big) &= \sum_{n=1}^{\infty} E(Y_n(i)|X_0 = i) \\ &= \sum_{n=1}^{\infty} P[X_n = i|X_0 = i] = \sum_{n=1}^{\infty} p_{ii}^{(n)} \, , \end{aligned}$$

and summation and expectation can be interchanged by the monotone convergence theorem. Thus, $\sum_{n=1}^{\infty} p_{ii}^{(n)} = \infty$ implies that the expected number of times the chain is in state i is ∞. Hence, if a state i is persistent, then the expected number of times the chain returns to state i is infinite, which essentially justifies the word persistent or recurrent. As the term persistent suggests, a state i is persistent or recurrent if after every visit to state i, the chain will eventually return for another visit with probability one. More precisely, when the process starts in state i and i is persistent, with probability 1, the process will eventually re-enter state i. The process will start over again with initial state as i and will again re-enter state i. Thus, state i will eventually be visited again. Continuous repetition of this argument leads to the conclusion that if state i is persistent then, starting in state i, the process will re-enter state i infinitely often. This assertion supports the dictionary meaning of "persistent" as "continue to exist".

On the other hand, if $\sum_{n=1}^{\infty} p_{ii}^{(n)} < \infty$, then the expected number of times the chain is in state i is finite. In other words, a transient state will only be visited a finite number of times and hence it is termed as transient. It is consistent with the dictionary meaning of "transient" as "staying only briefly" or "quickly passing away". More precisely, suppose that state i is transient. Hence, each time the process enters state i, there is a positive probability $1 - f_{ii}$ that it will never again enter that state. Suppose $X_0 = i$ and U_i denotes the number of times the process is in state i, including the initial state i. Thus, starting in state i, the probability that the process will be in state i for exactly n times equals $f_{ii}^{n-1}(1 - f_{ii})$, $n \geq 1$. Hence, $P[U_i = n] = f_{ii}^{n-1}(1 - f_{ii}), \;\; n \geq 1$. Hence, if state i is transient then, starting in state i, the number of times the process will be in state i has a geometric distribution with finite mean $1/(1 - f_{ii})$. Observe that $V_i = U_i - 1$. Hence, for a transient state i,

$$E(V_i) = E(U_i) - 1 = f_{ii}/(1 - f_{ii}) \Rightarrow \sum_{n=1}^{\infty} p_{ii}^{(n)} = f_{ii}/(1 - f_{ii}) \; . \quad (2.5.2)$$

Thus, if we exclude the initial state i, then the expected number of time periods that the process is in a transient state i is given by $f_{ii}/(1 - f_{ii})$, which is the value of the convergent series $\sum_{n=1}^{\infty} p_{ii}^{(n)}$. Note that $\sum_{n=1}^{\infty} p_{ii}^{(n)} < 1$ if $f_{ii} < 1/2$ and $\sum_{n=1}^{\infty} p_{ii}^{(n)} \geq 1$ if $f_{ii} \geq 1/2$.

In essence, a Markov chain spends most of its time in persistent states; however, transient states will, in the long run, be left and never returned to.

The following theorem proves that "being persistent" and "being transient" are class properties, where the class is a communicating class.

Theorem 2.5.4 *(i) If $i \leftrightarrow j$ and i is persistent then j is also persistent.*

(ii) If $i \leftrightarrow j$ and i is transient then j is also transient.

Proof State i is persistent implies that $\sum_{n=1}^{\infty} p_{ii}^{(n)}$ is divergent. To prove that state j is persistent, we examine whether $\sum_{n=1}^{\infty} p_{jj}^{(n)}$ is divergent. Now $i \leftrightarrow j \Rightarrow i \to j$ and $j \to i$ which further implies that $\exists\ s > 0$ such that $p_{ij}^{(s)} > 0$ and $\exists\ r > 0$ such that $p_{ji}^{(r)} > 0$. Observe that if we select only one path of going from j to j in $r + n + s$ steps, we get

$$\begin{aligned} p_{jj}^{(r+n+s)} &= P[X_{r+n+s} = j | X_0 = j] \\ &\geq P[X_{r+n+s} = j, X_{n+r} = i, X_r = i | X_0 = j] \\ &= P[X_{r+n+s} = j | X_{n+r} = i] P[X_{n+r} = i | X_r = i] P[X_r = i | X_0 = j] \\ &= p_{ji}^{(r)} p_{ii}^{(n)} p_{ij}^{(s)} . \end{aligned}$$

$$\begin{aligned} \text{Hence,} \quad \sum_{n=1}^{\infty} p_{jj}^{(r+n+s)} &\geq \sum_{n=1}^{\infty} p_{ij}^{(s)} p_{ii}^{(n)} p_{ji}^{(r)} = p_{ij}^{(s)} p_{ji}^{(r)} \sum_{n=1}^{\infty} p_{ii}^{(n)} \\ &= \infty \quad \text{as } i \text{ is persistent, } \ p_{ij}^{(s)} > 0 \ \&\ p_{ji}^{(r)} > 0. \end{aligned}$$

Observe that $\sum_{n=1}^{\infty} p_{jj}^{(n)} \geq \sum_{n=1}^{\infty} p_{jj}^{(r+n+s)} = \infty$ and hence j is persistent. Since "being persistent" is a class property, "being transient" is also a class property. □

In Example 2.5.2, we have noted that states $2 \leftrightarrow 3$ and both are transient. In Example 2.5.5, states $1 \leftrightarrow 2$ and both are persistent. In Example 2.5.6, all states communicate with each other and all are persistent.

In Theorem 2.5.3, it is proved that i is a transient state, if and only if $\sum_{n=1}^{\infty} p_{ii}^{(n)} < \infty$. Hence, if a state i is transient, then $\lim_{n \to \infty} p_{ii}^{(n)} = 0$, since the nth term of a convergent series converges to 0. Further, i is a persistent state if and only if $\sum_{n=1}^{\infty} p_{ii}^{(n)} = \infty$. In this case, we have following three scenarios about $\lim_{n \to \infty} p_{ii}^{(n)}$.

(i) $p_{ii}^{(n)} \to 0$, for example, if $p_{ii}^{(n)} = 1/n$ for $n \geq 2$.
(ii) $p_{ii}^{(n)} \to p > 0$ as $n \to \infty$, for example, if $p_{ii}^{(n)} = 0.01 + a_n$ for $n \geq 2$, where $\{a_n, n \geq 1\}$ is a sequence of real numbers such that $0 < p_{ii}^{(n)} = 0.01 + a_n \leq 1$ and $\lim_{n \to \infty} a_n = 0$.
(iii) In some cases, the limit may not even exist as shown in the following example.

Example 2.5.7 In a two-state Markov chain with transition probability matrix

$$P = \begin{array}{c} \\ 1 \\ 2 \end{array} \begin{array}{c} \begin{array}{cc} 1 & 2 \end{array} \\ \begin{pmatrix} 0 & 1 \\ 1 & 0 \end{pmatrix} \end{array}$$

for any $n \geq 0$, $P^{2n} = I_2$ and $P^{2n+1} = P$ and hence for $i = 1, 2$,

$$p_{ii}^{(n)} = \begin{cases} 1, \text{ if } n \text{ is even} \\ 0, \text{ if } n \text{ is odd.} \end{cases}$$

Note that for both $i = 1, 2$, $\limsup_{n\to\infty} p_{ii}^{(n)} = 1$ and $\liminf_{n\to\infty} p_{ii}^{(n)} = 0$. Hence, $\lim_{n\to\infty} p_{ii}^{(n)}$ does not exist. But for both $i = 1, 2$, $\sum_{n=1}^{\infty} p_{ii}^{(n)}$ is divergent. In Example 2.5.6 also we have noted that $\lim_{n\to\infty} p_{ii}^{(n)}$ does not exist, but for $i = 1, 2, 3$, $\sum_{n=1}^{\infty} p_{ii}^{(n)}$ is divergent. □

To summarize, if i is a persistent state, then $\lim_{n\to\infty} p_{ii}^{(n)}$ may not exist, but $\limsup_{n\to\infty} p_{ii}^{(n)}$ always exists. We have defined a persistent state as null or non-null persistent in terms of its mean recurrence time. Another definition of a null persistent and a non-null persistent state, in terms of $\limsup_{n\to\infty} p_{ii}^{(n)}$, is given below.

Definition 2.5.5 Null Persistent and Non-null Persistent States: A persistent state i is said to be non-null persistent if $\limsup_{n\to\infty} p_{ii}^{(n)} > 0$ and is said to be null persistent if $\limsup_{n\to\infty} p_{ii}^{(n)} = 0$.

It is to be noted that if $\limsup_{n\to\infty} p_{ii}^{(n)} = 0$, then $\liminf_{n\to\infty} p_{ii}^{(n)} = 0$ and hence for a null persistent state $\lim_{n\to\infty} p_{ii}^{(n)} = 0$.

In Chap. 3, we prove that the two definitions of null and non-null persistent states are equivalent.

Using the definition of a null persistent state in terms of $\limsup_{n\to\infty} p_{ii}^{(n)}$, we now prove that "being null persistent" and being "non-null persistent" are class properties.

Theorem 2.5.5 *(i) If i is null persistent and $i \leftrightarrow j$, then j is also null persistent. (ii) If i is non-null persistent and $i \leftrightarrow j$, then j is also non-null persistent.*

Proof (i) Suppose i is null persistent and $i \leftrightarrow j$. Thus, i is persistent implies that j is also persistent. Now $i \leftrightarrow j \Rightarrow i \to j$ *and* $j \to i$ which further implies that $\exists \; s > 0$ such that $p_{ij}^{(s)} > 0$ and $\exists \; r > 0$ such that $p_{ji}^{(r)} > 0$. Hence, as shown in Theorem 2.5.4

$$p_{ii}^{(s+n+r)} \geq p_{ij}^{(s)} p_{jj}^{(n)} p_{ji}^{(r)}$$

$$\Rightarrow \quad \limsup_{n\to\infty} p_{jj}^{(n)} \leq \frac{1}{p_{ij}^{(s)} p_{ji}^{(r)}} \limsup_{n\to\infty} p_{ii}^{(s+n+r)} = 0.$$

$$f_{ij}^{(1)} = P[N_{ij} = 1] = P[X_1 = j|X_0 = i] = p_{ij} \ \ \& \ \text{ for } n > 1,$$
$$f_{ij}^{(n)} = P[N_{ij} = n] = P[X_n = j, X_r \neq j, r = 1, 2, \ldots, n-1|X_0 = i].$$

The distribution of N_{ij} is known as the first passage distribution. Suppose f_{ij} denotes the probability of the first visit to j from i. As in the case of $f_{ii}^{(n)}$, the events $[X_1 = j]$ and $[X_n = j, X_r \neq j, r = 1, 2, \ldots, n-1]$ for $n > 1$ are mutually exclusive events. Hence, f_{ij} is given by

$$\begin{aligned} f_{ij} &= P\Big[\bigcup_{n=1}^{\infty}[N_{ij} = n]|X_0 = i\Big] \\ &= P\Big[[X_1 = j]\bigcup\Big(\bigcup_{n=2}^{\infty}[X_n = j, X_r \neq j \ \forall \ r = 1, 2, \ldots, n-1]\Big)\Big|X_0 = i\Big] \\ &= \sum_{n=1}^{\infty} f_{ij}^{(n)}. \end{aligned}$$

Being probability, $0 \leq f_{ij} \leq 1$. If $f_{ij} = P[N_{ij} < \infty] = 1$, then N_{ij} is a real-valued random variable. If $f_{ij} < 1$, then it is an extended real-valued random variable and $P[N_{ij} = \infty] = 1 - f_{ij} > 0$. Further, suppose μ_{ij} denotes the mean of N_{ij}; it gives the average number of steps needed for the first visit to j from i. It is infinite if $f_{ij} < 1$. If $f_{ij} = 1$, it may be finite or infinite.

Using a similar approach as in Lemma 2.5.1, it can be shown that

$$p_{ij}^{(n)} = \sum_{r=1}^{n} f_{ij}^{(r)} p_{jj}^{(n-r)}.$$

Thus, we have $p_{ij}^{(2)} = f_{ij}^{(2)} + f_{ij}^{(1)} p_{jj}$, which essentially states that the probability of going from i to j in two steps is the sum of the probabilities of two mutually exclusive events, one corresponding to going from i to j for the first time in two steps and the second corresponding to going from i to j for the first time in the first step and then staying there in the next step. From the identity $p_{ij}^{(n)} = \sum_{r=1}^{n} f_{ij}^{(r)} p_{jj}^{(n-r)}$, it is clear that $p_{ij}^{(n)} \geq f_{ij}^{(n)}$. With $p_{jj}^{(0)} = 1$,

$$p_{ij}^{(n)} = \sum_{r=1}^{n-1} f_{ij}^{(r)} p_{jj}^{(n-r)} + f_{ij}^{(n)} \ \Rightarrow \ f_{ij}^{(n)} = p_{ij}^{(n)} - \sum_{r=1}^{n-1} f_{ij}^{(r)} p_{jj}^{(n-r)},$$

which is a recurrence relation to find $f_{ij}^{(n)}$ with the initial value $f_{ij}^{(1)} = p_{ij}$. Thus, $f_{ij}^{(2)} = p_{ij}^{(2)} - p_{jj}^{(1)} f_{ij}^{(1)} = p_{ij}^{(2)} - p_{jj} f_{ij}^{(1)}$.

The following theorem is concerned with the possible values of f_{ij}, when $i \to j$ and if one of them is a persistent state, also when $i \leftrightarrow j$.

Theorem 2.6.1 *Suppose $i \neq j$. Then (i) $i \to j$ if and only if $f_{ij} > 0$. (ii) $i \leftrightarrow j$ if and only if $f_{ij} f_{ji} > 0$. (iii) If $i \to j$ and i is persistent, then $f_{ji} = 1$ and $j \to i$. Further, j is also persistent. (iv) If $i \leftrightarrow j$ and j is persistent then $f_{ji} = f_{ij} = 1$.*

Proof (i) Observe that the statement "$i \to j$ if and only if $f_{ij} > 0$" is equivalent to the statement "$f_{ij} = 0$ if and only if $i \nrightarrow j$". Hence, we prove the latter statement. Suppose $f_{ij} = 0$. Observe that

$$
\begin{aligned}
f_{ij} = 0 \Rightarrow \sum_{n=1}^{\infty} f_{ij}^{(n)} = 0 \;\; &\Rightarrow \;\; f_{ij}^{(n)} = 0 \;\; \forall \;\; n \geq 1 \\
&\Rightarrow p_{ij}^{(n)} = \sum_{r=1}^{n} p_{jj}^{(n-r)} f_{ij}^{(r)} = 0 \;\; \forall \;\; n \geq 1 \\
&\Rightarrow i \nrightarrow j.
\end{aligned}
$$

Suppose now $i \nrightarrow j$ and hence $p_{ij}^{(n)} = 0 \;\; \forall \;\; n \geq 1$. Observe that for $\forall \;\; n \geq 1$,

$$
f_{ij}^{(n)} \leq p_{ij}^{(n)} \;\; \Rightarrow \;\; f_{ij}^{(n)} = 0 \;\; \forall \;\; n \geq 1 \;\; \Rightarrow \;\; f_{ij} = 0.
$$

Thus, we have proved that $f_{ij} = 0$ if and only if $i \nrightarrow j$, which is equivalent to the claim that $i \to j$ if and only if $f_{ij} > 0$.
(ii) It follows immediately from (i).
(iii) If $i \to j$, then $\exists \;\; n \geq 1$ such that $p_{ij}^{(n)} > 0$. Suppose n_0 is the smallest n such that $p_{ij}^{(n_0)} > 0$. If any of the states in the path are equal to i and j or are the same, then n_0 will not be the smallest integer. Hence, the states in the path from i to j in n_0 steps are distinct and are also distinct from i and j. Thus, j can be reached from i without returning to i in n_0 steps. Suppose α is the probability of this event, that is, $\alpha = p_{ij}^{(n_0)} > 0$. Once j is reached, then the probability of never visiting i is $1 - f_{ji}$. Hence the probability $1 - f_{ii}$ of never returning to i satisfies the relation $1 - f_{ii} \geq \alpha(1 - f_{ji}) \geq 0$. But it is given that i is persistent, implying that $1 - f_{ii} = 0$. Hence $\alpha(1 - f_{ji}) = 0$, but $\alpha > 0$ hence $1 - f_{ji} = 0$ which implies that $f_{ji} = 1$. From (i) $f_{ji} > 0$ implies $j \to i$. Thus, if $i \to j$ and i is persistent, then $i \leftrightarrow j$ and hence by the class property j is also persistent.
(iv) It is given that $i \leftrightarrow j$ and j is persistent. Hence i is also persistent. Now, $i \to j$ and i is persistent and hence by part (iii) $f_{ji} = 1$. Similarly, $j \to i$ and j is persistent and hence again by part (iii) $f_{ij} = 1$. □

Result (iii) of Theorem 2.6.1 conveys that a persistent state leads to a persistent state only; it does not lead to a transient state. We have the following corollary based on this result.

Corollary 2.6.1 *If $i \to j$ but $j \nrightarrow i$, then i is transient.*

Proof Suppose $i \to j$ but $j \nrightarrow i$. If i is persistent, then from part (iii) of Theorem 2.6.1, $j \to i$, which is a contradiction. Hence, i must be transient. □

We have defined an inessential state in Sect. 2.5. From Corollary 2.6.1, we now prove that an inessential state is a transient state. We have already verified this result in many examples.

Theorem 2.6.2 *An inessential state is transient.*

Proof Suppose i is an inessential state, that is, there exists a state j such that $i \to j$ but $j \nrightarrow i$. From Corollary 2.6.1, it then follows that i is transient. □

In Theorem 2.5.3, we have proved that $\left[\sum_{n=1}^{N} p_{ii}^{(n)} \Big/ (1+\sum_{n=1}^{N} p_{ii}^{(n)})\right] \to f_{ii}$ as $N \to \infty$. On similar lines, we have a limit theorem, known as a ratio theorem, with limit f_{ij}. It is proved below using Toeplitz' lemma.

Theorem 2.6.3 *Ratio Theorem: For any two states i and j in S,*

$$\lim_{N\to\infty}\left[\sum_{n=1}^{N} p_{ij}^{(n)} \Big/ (1+\sum_{n=1}^{N} p_{jj}^{(n)})\right] = f_{ij}.$$

Proof If $i \nrightarrow j$, then $p_{ij}^{(n)} = 0 \;\forall\; n \geq 1$ and $f_{ij} = 0$ and the result is trivially true. Suppose now $i \to j$, then $p_{ij}^{(n)} > 0$ for at least one $n \geq 1$ and $f_{ij} > 0$. Using the relation $p_{ij}^{(n)} = \sum_{r=1}^{n} f_{ij}^{(r)} p_{jj}^{(n-r)}$, with $n - r = s$ we have

$$\begin{aligned}
\sum_{n=1}^{N} p_{ij}^{(n)} &= \sum_{n=1}^{N}\sum_{s=0}^{n-1} p_{jj}^{(s)} f_{ij}^{(n-s)} = \sum_{s=0}^{N-1} p_{jj}^{(s)}\Big[\sum_{n=s+1}^{N} f_{ij}^{(n-s)}\Big] \\
&= \sum_{s=0}^{N-1} p_{jj}^{(s)}\Big[\sum_{n=0}^{N-s} f_{ij}^{(n)}\Big] \quad \text{as} \quad f_{ij}^{(0)} = 0 \\
&= \sum_{s=0}^{N} p_{jj}^{(s)}\Big[\sum_{n=0}^{N-s} f_{ij}^{(n)}\Big] \quad \text{as for } s = N, \quad \sum_{n=0}^{N-s} f_{ij}^{(n)} = 0 \\
&= \sum_{s=0}^{N} a_s b_{N-s}, \quad \text{with } a_s = p_{jj}^{(s)} \;\&\; b_{N-s} = \sum_{n=0}^{N-s} f_{ij}^{(n)}.
\end{aligned}$$

Thus, $\sum_{s=0}^{N} a_s b_{N-s} \Big/ \sum_{s=0}^{N} a_s = \sum_{n=1}^{N} p_{ij}^{(n)} \Big/ (1+\sum_{n=1}^{N} p_{jj}^{(n)})$. It is to be noted that $a_0 = p_{jj}^{(0)} = 1$, thus at least one $a_s \neq 0$. Further, as $N \to \infty$, $b_N = \sum_{n=0}^{N} f_{ij}^{(n)}$ converges to $f_{ij} = b$, say. Now, $\sum_{n=1}^{N} a_n = \sum_{n=1}^{N} p_{jj}^{(n)}$ either converges to a finite limit or diverges to ∞ as $N \to \infty$. If it converges to a finite limit, then $p_{jj}^{(n)} \to 0$ and hence $p_{jj}^{(n)} / \sum_{n=0}^{N} p_{jj}^{(n)} \to 0$. If $\sum_{n=1}^{N} p_{jj}^{(n)}$ diverges then also $p_{jj}^{(n)} / \sum_{n=0}^{N} p_{jj}^{(n)} \to 0$, as the numerator is bounded by 1. Hence by Lemma 2.5.2,

$$\lim_{N\to\infty}\left[\sum_{s=0}^{N} a_s b_{N-s} \Big/ \sum_{s=0}^{N} a_s\right] = b \;\Rightarrow\; \lim_{N\to\infty}\left[\sum_{n=1}^{N} p_{ij}^{(n)} \Big/ (1+\sum_{n=1}^{N} p_{jj}^{(n)})\right] = f_{ij}.$$

□

In the following theorem, we discuss the limiting behavior of the series $\sum_{n=1}^{\infty} p_{ij}^{(n)}$. The proof is based on the ratio theorem.

Theorem 2.6.4 *(i) If $i \nrightarrow j$, then $\sum_{n=1}^{\infty} p_{ij}^{(n)} = 0$.*

(ii) If $i \rightarrow j$, then the series $\sum_{n=1}^{\infty} p_{ij}^{(n)}$ converges if and only if j is transient.

(iii) If $i \rightarrow j$, then the series $\sum_{n=1}^{\infty} p_{ij}^{(n)}$ diverges if and only if j is persistent.

Proof (i) If $i \nrightarrow j$, then $p_{ij}^{(n)} = 0, \ \forall \ n \geq 1$ and hence $\sum_{n=1}^{\infty} p_{ij}^{(n)} = 0$.

(ii) Suppose $i \rightarrow j$ and j is a transient state. Then by Theorem 2.5.3, $\sum_{n=1}^{\infty} p_{jj}^{(n)} < \infty$. Thus, the denominator in the ratio $\sum_{n=1}^{\infty} p_{ij}^{(n)} / (1 + \sum_{n=1}^{\infty} p_{jj}^{(n)})$ is finite. By ratio theorem, $\sum_{n=1}^{N} p_{ij}^{(n)} / (1 + \sum_{n=1}^{N} p_{jj}^{(n)}) \rightarrow f_{ij} \leq 1$, hence its numerator must be finite, that is, $\sum_{n=1}^{N} p_{ij}^{(n)}$ converges to a finite limit. Conversely, assume that $\sum_{n=1}^{N} p_{ij}^{(n)}$ converges to a finite limit. In Theorem 2.6.1, it is proved that if $i \rightarrow j$ then $f_{ij} > 0$. Hence, by the ratio theorem, the denominator $\sum_{n=1}^{\infty} p_{jj}^{(n)} < \infty$, otherwise $f_{ij} = 0$. Hence, j is a transient state.

(iii) Suppose $i \rightarrow j$ and j is a persistent state. Then by Theorem 2.5.3, $\sum_{n=1}^{\infty} p_{jj}^{(n)}$ is a divergent series. Thus, the denominator in the ratio $\sum_{n=1}^{N} p_{ij}^{(n)} / (1 + \sum_{n=1}^{N} p_{jj}^{(n)})$ tends to ∞ and hence its numerator also has to tend to ∞, otherwise the ratio would have a limit 0 which is contradictory to the result that $f_{ij} > 0$ when $i \rightarrow j$. Thus, if j is persistent then $\sum_{n=1}^{\infty} p_{ij}^{(n)}$ diverges $\forall \ i$ such that $i \rightarrow j$. Conversely, assume that $\sum_{n=1}^{\infty} p_{ij}^{(n)}$ is divergent and $i \rightarrow j$. Thus, the numerator in the ratio $\sum_{n=1}^{N} p_{ij}^{(n)} / (1 + \sum_{n=1}^{N} p_{jj}^{(n)}) \rightarrow \infty$ and hence its denominator also has to tend to ∞, otherwise $f_{ij} = \infty$, but $f_{ij} \leq 1$. Hence, $\sum_{n=1}^{N} p_{jj}^{(n)} = \infty$ and j is persistent. □

In Sect. 2.5, we have shown that if a state i is transient then the series $\sum_{n=1}^{\infty} p_{ii}^{(n)}$ is convergent and its value is given by $f_{ii}/(1 - f_{ii})$. On similar lines, we find the value of a convergent series $\sum_{n=1}^{\infty} p_{ij}^{(n)}$ when j is transient and $i \rightarrow j$. Suppose $X_0 = i$ and a random variable $Y_n(j)$ is defined as

$$Y_n(j) \ = \ \begin{cases} 1, \text{ if } \ X_n = j \\ 0, \text{ if } \ X_n \neq j \ . \end{cases}$$

Then $W_j = \sum_{n=1}^{\infty} Y_n(j)$ denotes the number of times the chain is in state j given that $X_0 = i$. Hence,

$$
\begin{aligned}
E(W_j|X_0=i) &= E\Big(\sum_{n=1}^{\infty} Y_n(j)|X_0=i\Big) = \sum_{n=1}^{\infty} E(Y_n(j)|X_0=i) \\
&= \sum_{n=1}^{\infty} P[X_n=j|X_0=i] = \sum_{n=1}^{\infty} p_{ij}^{(n)} ,
\end{aligned}
$$

and the summation and expectation can be interchanged by the monotone convergence theorem. Thus, $\sum_{n=1}^{\infty} p_{ij}^{(n)} = \infty$ implies that the expected number of times the chain is in state j is ∞. Hence, if a state j is persistent, then the expected number of times the chain is in state j is infinite. If $\sum_{n=1}^{\infty} p_{ij}^{(n)} < \infty$, then the expected number of times the chain is in state j is finite. By definition, for a transient state j, $f_{jj} < 1$. Observe that

$$
\begin{aligned}
&P[W_j=0|X_0=i] = 1 - f_{ij}, \quad P[W_j=1|X_0=i] = f_{ij}(1-f_{jj}) \\
&P[W_j=2|X_0=i] = f_{ij} f_{jj}(1-f_{jj}).
\end{aligned}
$$

Continuing in this way, we have

$$
\begin{aligned}
& P[W_j=n|X_0=i] = f_{ij} f_{jj}^{n-1}(1-f_{jj}) \ \forall \ n \geq 1 \\
\Rightarrow \ & E(W_j|X_0=i) = f_{ij}/(1-f_{jj}) \\
\Rightarrow \ & \sum_{n=1}^{\infty} p_{ij}^{(n)} = f_{ij}/(1-f_{jj}) \ . \qquad (2.6.1)
\end{aligned}
$$

Theorem 2.6.4 is useful to decide $\lim_{n\to\infty} p_{ij}^{(n)}$, depending on the nature of states i and j. We have noted that (i) if $i \nrightarrow j$, then $p_{ij}^{(n)} = 0 \ \forall \ n \geq 1$. Hence, $\lim_{n\to\infty} p_{ij}^{(n)} = 0$. (ii) If $i \to j$ and j is a transient state, by Theorem 2.6.4, $\sum_{n=1}^{\infty} p_{ij}^{(n)} < \infty$, so that $\lim_{n\to\infty} p_{ij}^{(n)} = 0$. (iii) Suppose now $i \to j$ and j is a persistent state. Then by Theorem 2.6.4, $\sum_{n=1}^{\infty} p_{ij}^{(n)}$ is a divergent series. In this case, as discussed in Sect. 2.5 for the divergent series $\sum_{n=1}^{\infty} p_{ii}^{(n)}$, we have the following three possible scenarios:

(i) $p_{ij}^{(n)} \to 0$, for example, if $p_{ij}^{(n)} = 1/n$ for $n \geq 2$.
(ii) $p_{ij}^{(n)} \to p > 0$ as $n \to \infty$, for example, if $p_{ij}^{(n)} = 0.01 + 1/n$ for $n \geq 2$.
(iii) In some cases, as in the Markov chain of Example 2.5.7, $\lim_{n\to\infty} p_{ij}^{(n)}$ may not exist, but $\sum_{n=1}^{\infty} p_{ij}^{(n)}$ is divergent.

Thus, when j is a persistent state, we need to explore the conditions for the limit of $p_{ij}^{(n)}$ to be zero or positive. We study some of these issues below, using the definition of a null persistent state in terms of $\limsup_{n\to\infty} p_{jj}^{(n)} = 0$.

Theorem 2.6.5 *If $i \to j$, then $\lim_{n\to\infty} p_{ij}^{(n)} = 0$, if and only if j is either transient or null persistent.*

Proof Only if part: Suppose $\lim_{n\to\infty} p_{ij}^{(n)} = 0 \ \forall\ i \in S$ such that $i \to j$. In particular with $i = j$, $\lim_{n\to\infty} p_{jj}^{(n)} = 0$. If $\sum_{n=1}^{\infty} p_{jj}^{(n)} < \infty$, then j is transient. If it is infinite, then j is persistent and $\lim_{n\to\infty} p_{jj}^{(n)} = 0$ implies that it is null persistent.
If part: Suppose j is either transient or null persistent. If j is transient, then $\sum_{n=1}^{\infty} p_{jj}^{(n)} < \infty$. Hence, $\lim_{n\to\infty} p_{jj}^{(n)} = 0$. If j is null persistent, then by definition, $\limsup_{n\to\infty} p_{jj}^{(n)} = 0$ and hence $\lim_{n\to\infty} p_{jj}^{(n)} = 0$. To obtain $\lim_{n\to\infty} p_{ij}^{(n)}$, using the recurrence relation we have

$$\lim_{n\to\infty} p_{ij}^{(n)} = \lim_{n\to\infty} \sum_{r=1}^{n} f_{ij}^{(r)} p_{jj}^{(n-r)} = \lim_{n\to\infty} \sum_{r=1}^{\infty} f_{ij}^{(r)} a(n, r)$$

where

$$a(n, r) = \begin{cases} p_{jj}^{(n-r)}, & \text{if } r \le n \\ 0, & \text{if } r > n. \end{cases}$$

Since $\lim_{n\to\infty} p_{jj}^{(n)} = 0$, $f_{ij}^{(r)} a(n, r) \to 0$ as $n \to \infty$. Further, $|f_{ij}^{(r)} a(n, r)| \le f_{ij}^{(r)}$ and $\sum_{r=1}^{\infty} f_{ij}^{(r)} = f_{ij} \le 1$. Hence by Theorem 2.4.1,

$$\lim_{n\to\infty} p_{ij}^{(n)} = \sum_{r=1}^{\infty} f_{ij}^{(r)} \left(\lim_{n\to\infty} p_{jj}^{(n-r)} \right) = 0 \ \text{ as } \ \lim_{n\to\infty} p_{jj}^{(n-r)} = 0 \ \forall\ r.$$

Thus, if j is either transient or null persistent, then $\lim_{n\to\infty} p_{ij}^{(n)} = 0$. □

When j is non-null persistent, we discuss $\lim_{n\to\infty} p_{ij}^{(n)}$ in Chap. 3.

From part (iii) of Theorem 2.6.1, we note that if $i \to j$ and i is persistent, then $j \to i$ and j is also persistent. Thus, no transient state can be reached from any persistent state. Further, being persistent is a class property. Suppose C denotes the class of all persistent states. Then states in C do not lead to states outside C. Thus, C is a closed class. It may not be a communicating class, since within this set, there may be more than one closed class. The following example illustrates these assertions.

Example 2.6.1 Suppose the transition probability matrix P of a Markov chain is

$$P = \begin{array}{c} \\ 1 \\ 2 \\ 3 \\ 4 \\ 5 \\ 6 \\ 7 \\ 8 \end{array} \begin{array}{c} \begin{array}{cccccccc} 1 & 2 & 3 & 4 & 5 & 6 & 7 & 8 \end{array} \\ \begin{pmatrix} 0.3 & 0.4 & 0.3 & 0 & 0 & 0 & 0 & 0 \\ 0.5 & 0 & 0.5 & 0 & 0 & 0 & 0 & 0 \\ 0 & 0.6 & 0.4 & 0 & 0 & 0 & 0 & 0 \\ 0 & 0 & 0 & 0.4 & 0.6 & 0 & 0 & 0 \\ 0 & 0 & 0 & 0.7 & 0.3 & 0 & 0 & 0 \\ 0.1 & 0.1 & 0.1 & 0.2 & 0.1 & 0.1 & 0.1 & 0.2 \\ 0.2 & 0.2 & 0 & 0.3 & 0.1 & 0 & 0.1 & 0.1 \\ 0.1 & 0 & 0.1 & 0.1 & 0 & 0.3 & 0.2 & 0.2 \end{pmatrix} \end{array}.$$

We note that $C_1 = \{1, 2, 3\}$, $C_2 = \{4, 5\}$ are two closed communicating classes. Further, the states in C_1 and C_2 are essential. Using Code 2.8.6, it can be verified that these are persistent. Hence, $C = \{1, 2, 3, 4, 5\}$ is a class of persistent states, and states in C do not lead to states outside C. Note that C is a closed class but it is not a communicating class. The class $T = \{6, 7, 8\}$ is a class of inessential and hence transient states. □

From Example 2.6.1, we note that for each of the states 1, 2, 3, C_1 is the closed communicating class which includes these three states, while for each of the states 4, 5, C_2 is the closed communicating class which includes these two. Thus, corresponding to each persistent state, there is a closed communicating class which includes it. We prove this observation in the following theorem.

Theorem 2.6.6 *For each persistent state j, there exists a closed communicating class C which includes j.*

Proof Suppose $C = \{l | j \to l\}$, where j is a persistent state. Note that

$$f_{jj} = 1 \;\Rightarrow\; j \to j \;\Rightarrow\; j \in C.$$

Thus, C is a non-empty class. Since j is persistent and $j \to l$, l is also persistent by (iii) of Theorem 2.6.1. It further implies that no transient state can be reached from any persistent state. Thus, the class C is a closed class of all persistent states. We now examine whether it is a communicating class. Thus, for any two states $i, k \in C$, we examine whether $i \leftrightarrow k$. Since j is persistent, note that

$$\begin{aligned} i \in C &\Rightarrow j \to i \;\Rightarrow\; i \to j \\ &\Rightarrow i \leftrightarrow j \\ \text{Similarly } k \in C &\Rightarrow j \to k \;\Rightarrow\; k \to j \\ &\Rightarrow j \leftrightarrow k \\ \text{Further, } \; i \leftrightarrow j \;\&\; j \leftrightarrow k &\Rightarrow i \leftrightarrow k, \end{aligned}$$

and the last step follows due to the transitivity of the "leads to" property. □

We thus note that in a Markov chain, the closed class C of persistent states can be partitioned into closed communicating classes $\{C_1, C_2, \ldots\}$. Each one of these classes defines an irreducible Markov chain with persistent states. Hence in many results, it is sufficient to consider only an irreducible Markov chain with persistent states. In addition to the closed class C of persistent states, in general, the chain contains transient states as well. It is possible for the persistent states to be reached from a transient state, but not vice versa. If the states are relabeled in an appropriate manner, the transition probability matrix can be written in a particular form. For example, if a Markov chain has two closed communicating classes C_1 of r states and C_2 of s states, then P can be expressed as follows, with rearrangement of rows and columns if needed:

$$
P = \begin{array}{c} \\ C_1 \\ C_2 \\ A_1 \end{array} \begin{array}{c} \begin{array}{ccc} C_1 & C_2 & A_1 \end{array} \\ \begin{pmatrix} P_1 & 0 & 0 \\ 0 & P_2 & 0 \\ Q_1 & Q_2 & Q_3 \end{pmatrix} \end{array},
$$

where P_1, P_2 are stochastic matrices and Q_3 is of order $M - r - s \times M - r - s$. All the states in C_1 and C_2 are persistent states, while all the states in A_1 are transient, being inessential states. Since P_1, P_2 are stochastic matrices, we have two Markov chains with these two as transition probability matrices and C_1 and C_2 as state spaces, respectively. In Example 2.6.1, $C_1 = \{1, 2, 3\}$, $C_2 = \{4, 5\}$ are two closed communicating classes and the corresponding matrices $[p_{ij}]_{i,j \in C_1}$ and $[p_{ij}]_{i,j \in C_2}$ are stochastic matrices. Thus, in some cases, one may concentrate on the closed communicating class.

From Theorems 2.6.5 and 2.6.6, we now prove a result which has been noted in many examples.

Theorem 2.6.7 *In a finite Markov chain, (i) all states cannot be transient and (ii) all persistent states are non-null persistent.*

Proof Suppose $S = \{1, 2, \ldots, M\}$. To prove (i), we assume the contrary that all M states are transient. If $i \nrightarrow j$, then $p_{ij}^{(n)} = 0 \ \forall \ n \geq 1$ and hence $\lim_{n\to\infty} p_{ij}^{(n)} = 0$. If $i \to j$ and j is transient, $\lim_{n\to\infty} p_{ij}^{(n)} = 0$, by Theorem 2.6.5. Thus, in both the cases, $\lim_{n\to\infty} p_{ij}^{(n)} = 0$, for all $i, \ j \in S$. In view of the fact that $P^{(n)} = P^n$ is a stochastic matrix, we have $\forall \ i \in S$,

$$
\sum_{j=1}^{M} p_{ij}^{(n)} = 1 \ \Rightarrow \ \lim_{n\to\infty} \sum_{j=1}^{M} p_{ij}^{(n)} = 1 \ \Rightarrow \ \sum_{j=1}^{M} \lim_{n\to\infty} p_{ij}^{(n)} = 1 \ \Rightarrow \ 0 = 1
$$

which contradicts the assumption that all states are transient. Thus, all states cannot be transient.

(ii) Since all states cannot be transient, at least one state is a persistent state. Suppose there are k persistent states and all are null persistent, $0 < k \leq M$. Suppose S_t is the set of all transient states and S_{np} is the set of all null persistent states. If $i \nrightarrow j \in S_{np}$, then $p_{ij}^{(n)} = 0 \ \forall \ n \geq 1$ and hence $\lim_{n\to\infty} p_{ij}^{(n)} = 0$. If $i \to j \in S_{np}$, by Theorem 2.6.5, $\lim_{n\to\infty} p_{ij}^{(n)} = 0$. Note that $\forall \ i \in S$,

$$
1 = \lim_{n\to\infty} \sum_{j=1}^{M} p_{ij}^{(n)} = \sum_{j \in S_t} \lim_{n\to\infty} p_{ij}^{(n)} + \sum_{j \in S_{np}} \lim_{n\to\infty} p_{ij}^{(n)} = 0,
$$

which is a contradiction. Hence, all persistent states cannot be null persistent states. Thus, there exists at least one non-null persistent state. Suppose there are l null persistent states, $0 < l < k$ and $k - l$ non-null persistent states, $0 < k - l \leq k$. Suppose

j is a null persistent state. By Theorem 2.6.6, for each persistent state j, there exists a closed communicating class C which includes j. Further by the class property, $j \in C$ is null persistent implying that all the states in C are null persistent. Suppose $Q = [q_{rs}]$ is a matrix of transition probabilities corresponding to the states in C. Since C is a closed class, Q is a stochastic matrix. Hence Q^n is also a stochastic matrix. Hence, $\forall\; r \in C, \sum_{s \in C} q_{rs}^{(n)} = 1$. Since states in C are null persistent, $\forall\; r, s \in C,\; \lim_{n\to\infty} q_{rs}^{(n)} = 0$. Observe that $\forall\; r \in C$

$$\sum_{s\in C} q_{rs}^{(n)} = 1 \;\Rightarrow\; \lim_{n\to\infty} \sum_{s\in C} q_{rs}^{(n)} = 1 \;\Rightarrow\; \sum_{s\in C} \lim_{n\to\infty} q_{rs}^{(n)} = 1 \;\Rightarrow\; 0 = 1,$$

which contradicts the assumption that state j and hence all the states in C are null persistent. Thus, j cannot be null persistent and must be a non-null persistent state. Using similar arguments for each of the l null persistent states, we arrive at the conclusion that these l states cannot be null persistent. Thus, it follows that in a finite state space Markov chain, all persistent states are non-null persistent. □

Theorem 2.6.7 conveys that in a finite state space Markov chain, null persistent states do not exist. Since being non-null persistent is a class property, we get the following important result for a finite Markov chain.

Theorem 2.6.8 *In an irreducible finite Markov chain, all the states are non-null persistent.*

Proof It is proved in Theorem 2.6.7 that in a finite state space Markov chain, at least one state is non-null persistent. Since the Markov chain is irreducible, all states communicate with each other and hence all states are non-null persistent. □

Remark 2.6.1 It may be noted that Theorem 2.6.8 holds for any finite closed communicating class C, since $[p_{ij}]_{i,j \in C}$ is a stochastic matrix and all the states in C communicate with each other.

The Markov chain for the weather model is finite and irreducible and hence all the three states are non-null persistent. Since Theorem 2.6.8 holds for any finite closed communicating class, all the states in any finite closed communicating class are non-null persistent. For a Markov chain in Example 2.5.4, $C = \{1, 2\}$ is a finite closed communicating class, and we have verified that both the states are non-null persistent. It is a reducible Markov chain. In Examples 2.5.2 and 2.5.8, Markov chains are reducible. For these chains, some states are transient and some are non-null persistent, but no state is null persistent.

From Theorem 2.6.1, we note the following results about the values of f_{ij}:

(i) If i and j are in the same finite closed communicating class, then both are persistent and $f_{ij} = 1$.
(ii) If i is persistent and j is transient then $f_{ij} = 0$, since the class of persistent states is closed.
(iii) If i and j are in different closed communicating classes, then $f_{ij} = 0$.

(iv) If i and j both are transient, then $f_{ij} < 1$. From Eq. (2.6.1), $f_{ij} = (1 - f_{jj}) \sum_{n=1}^{\infty} p_{ij}^{(n)}$.

Thus, the only remaining case we have to investigate is the value of f_{ij} when i is transient and j is persistent. To find the value of f_{ij}, when i is transient and j is in one of the closed communicating classes, we note the result in the following lemma.

Lemma 2.6.1 *Suppose C is a closed communicating class of persistent states. Then for any transient state i and $\forall\ j, k \in C$, $f_{ij} = f_{ik}$.*

Proof By Theorem 2.6.1, $\forall\ j, k \in C$, $f_{jk} = f_{kj} = 1$. Thus, once the chain reaches any one of the states in C, it also visits all the other states and it stays in the same class, it being a closed class. Thus, the Markov chain gets absorbed in C. Hence, for a transient state i and $j, k \in C$, f_{ij}, f_{ik} are the same as the probability that the Markov chain is absorbed in the class C from i. Hence, $f_{ij} = f_{ik}$. □

The above lemma conveys that $f_{ij} = f_{ik}$ is the probability of absorption in a closed communicating class C from a transient state i. We now proceed to discuss how to find the absorption probability and its link with f_{ij}.

Probability of absorption in a class of persistent states from a transient state: Suppose C is a class of persistent states and T is a finite class of $k > 1$ transient states. Then the transition probability matrix $P = [p_{ij}]$ can be expressed as

$$P = \begin{pmatrix} P_C & 0 \\ B & Q \end{pmatrix},$$

where $P_C = [p_{ij}]_{i,j \in C}$, $B = [p_{ij}]_{i \in T, j \in C}$ and $Q = [p_{ij}]_{i,j \in T}$. Suppose the class C is partitioned into closed communicating classes $\{C_1, C_2, \ldots, C_m\}$ and the states are labeled so that the states in C_1 precede those in C_2 and so on; and the transient states are after all the persistent states. Then as discussed previously for the setup of the two closed communicating classes, the transition probability matrix P can be expressed as

$$P = \begin{pmatrix} P_1 & 0 & \cdots & \cdots & 0 \\ 0 & P_2 & \cdots & \cdots & 0 \\ \vdots & \vdots & \vdots & \vdots & \vdots \\ 0 & 0 & \cdots & P_m & 0 \\ B_1 & B_2 & \cdots & B_m & Q \end{pmatrix},$$

where P_r is the stochastic matrix corresponding to the closed communicating class $C_r, r = 1, \ldots, m$. Since we are interested only in absorption in C_j, we lump all the states of C_j together to make one absorbing state. Then P is expressible as $\overline{P}$, where

$$\overline{P} = \begin{pmatrix} 1 & 0 & \cdots & \cdots & 0 \\ 0 & 1 & \cdots & \cdots & 0 \\ \vdots & \vdots & \vdots & \vdots & \vdots \\ 0 & 0 & \cdots & 1 & 0 \\ \underline{b}_1 & \underline{b}_2 & \cdots & \underline{b}_m & Q \end{pmatrix} = \begin{pmatrix} I_m & 0_{m\times k} \\ D_{k\times m} & Q_{k\times k} \end{pmatrix},$$

where $\underline{b}_r$ is a vector with the ith component given by $b_r(i) = \sum_{j\in C_r} p_{ij}$, $r = 1, 2, \ldots m$, and $D_{k\times m}$ is a matrix with columns $\{\underline{b}_1, \underline{b}_2, \ldots, \underline{b}_m\}$. I_m is the identity matrix of order $m \times m$, and $Q_{k\times k}$ is a matrix of the transition probabilities for all $(i, j) \in T$. The following example illustrates the procedure.

Example 2.6.2 We consider the Markov chain in Example 2.6.1, with P given by

$$P = \begin{matrix} & \begin{matrix} 1 & \;\;2 & \;\;3 & \;\;4 & \;\;5 & \;\;6 & \;\;7 & \;\;8 \end{matrix} \\ \begin{matrix} 1 \\ 2 \\ 3 \\ 4 \\ 5 \\ 6 \\ 7 \\ 8 \end{matrix} & \begin{pmatrix} 0.3 & 0.4 & 0.3 & 0 & 0 & 0 & 0 & 0 \\ 0.5 & 0 & 0.5 & 0 & 0 & 0 & 0 & 0 \\ 0 & 0.6 & 0.4 & 0 & 0 & 0 & 0 & 0 \\ 0 & 0 & 0 & 0.4 & 0.6 & 0 & 0 & 0 \\ 0 & 0 & 0 & 0.7 & 0.3 & 0 & 0 & 0 \\ 0.1 & 0.1 & 0.1 & 0.2 & 0.1 & 0.1 & 0.1 & 0.2 \\ 0.2 & 0.2 & 0 & 0.3 & 0.1 & 0 & 0.1 & 0.1 \\ 0.1 & 0 & 0.1 & 0.1 & 0 & 0.3 & 0.2 & 0.2 \end{pmatrix} \end{matrix}.$$

Thus, $C_1 = \{1, 2, 3\}$, $C_2 = \{4, 5\}$ are two closed communicating classes and $T = \{6, 7, 8\}$ is a class of transient states. The last three rows and the first three columns constitute the matrix B_1, the last three rows and the forth and fifth columns constitute the matrix B_2 and the last three rows and the last three columns constitute the matrix Q. We lump the states in C_1 and C_2. Then P is expressible as $\overline{P}$, where

$$\overline{P} = \begin{pmatrix} 1 & 0 & 0 \\ 0 & 1 & 0 \\ \underline{b}_1 & \underline{b}_2 & Q \end{pmatrix} = \begin{pmatrix} I_2 & 0 \\ D & Q \end{pmatrix}, \quad \text{where } \underline{b}_1 = \begin{pmatrix} 0.3 \\ 0.4 \\ 0.2 \end{pmatrix}, \; \underline{b}_2 = \begin{pmatrix} 0.3 \\ 0.4 \\ 0.1 \end{pmatrix}$$

and $D = [\underline{b}_1, \underline{b}_2]$. □

In the following theorem, we find the expression for the absorption probability, for absorption in a closed communicating class from a transient state.

Theorem 2.6.9 *Suppose $\{X_n, n \geq 0\}$ is a Markov chain, with C as a class of persistent states, a finite class T of k transient states and $P = [p_{ij}]$ as*

$$P = \begin{pmatrix} P_C & 0 \\ B & Q \end{pmatrix},$$

where $P_C = [p_{ij}]_{i,j\in C}$ $B = [p_{ij}]_{i\in T, j\in C}$ *and* $Q = [p_{ij}]_{i,j\in T}$. *Suppose the class C is partitioned into closed communicating classes* $\{C_1, C_2, \ldots, C_m\}$ *and after lumping the states of* C_j *together, P is expressed as* $\overline{P}$, *where*

$$\overline{P} = \begin{pmatrix} I_m & 0_{m\times k} \\ D_{k\times m} & Q_{k\times k} \end{pmatrix},$$

where D is a matrix with columns $\{\underline{b}_1, \underline{b}_2, \ldots, \underline{b}_m\}$ *and* $\underline{b}_r$ *is a vector with ith component given by* $b_r(i) = \sum_{j\in C_r} p_{ij}$, $r = 1, 2, \ldots m$. *Suppose* $G_{k\times m} = [g_{ij}]$, *where* g_{ij} *is the probability of absorption in class* C_j, $j = 1, 2, \ldots, m$ *from the transient state* $i \in T$. *Then (i)* $G = (I_k - Q)^{-1}D$ *and (ii)* $G\underline{e}_{m\times 1} = \underline{e}_{k\times 1}$ *where* $\underline{e} = (1, 1, \ldots, 1)'$, *that is, row sums of G are 1.*

Proof Note that the probability of absorption in C_j from the transient state i in the Markov chain with the transition matrix P is the same as the probability of the absorption into class $\{j\}$ from the transient state i by the chain with the transition probability matrix $\overline{P}$. Observe that

$$\overline{P}^2 = \begin{pmatrix} I_m & 0 \\ D & Q \end{pmatrix} \times \begin{pmatrix} I_m & 0 \\ D & Q \end{pmatrix} = \begin{pmatrix} I_m & 0 \\ D_2 & Q^2 \end{pmatrix},$$

where $D_2 = D + QD = (I_k + Q)D$. Thus, by induction for $n \geq 1$ we have

$$\overline{P}^n = \begin{pmatrix} I & 0 \\ D_n & Q^n \end{pmatrix}, \quad \text{where } D_n = (I_k + Q + Q^2 + \cdots + Q^{n-1})D.$$

Note that $D_n(i, j)$, the (i, j)th element in D_n, is the probability that starting from i, the chain enters the class C_j at or before the nth step. Hence, the probability of absorption in class C_j, starting from i, is $\lim_{n\to\infty} D_n(i, j)$, provided the limit exists. To examine whether it exists, note that $D_n = (I_k + A_n)D$, where $A_n = \sum_{l=1}^{n-1} Q^l$. Observe that (i, j)th element of A_n is $\sum_{l=1}^{n-1} p_{ij}^{(l)}$, where $i, j \in T$. Since $j \in T$ is transient, $\sum_{l=1}^{\infty} p_{ij}^{(l)}$ is convergent and hence $\sum_{l=1}^{n-1} p_{ij}^{(l)} \to a_{ij}$, say, as $n \to \infty$. Thus, $A_n \to A = [a_{ij}]$ as $n \to \infty$ and this convergence is uniform in $i, j \in T$, since T is finite. Hence, $\lim_{n\to\infty} D_n$ exists. Thus,

$$G = \lim_{n\to\infty} D_n = \Big(\sum_{n=0}^{\infty} Q^n\Big)D = ED, \quad \text{where } E = \sum_{n=0}^{\infty} Q^n.$$

Now to find E, observe that

$$
\begin{aligned}
E = \sum_{n=0}^{\infty} Q^n &\Rightarrow EQ = QE = Q + Q^2 + \cdots = E - I_k \\
&\Rightarrow E - EQ = E - QE = I_k \;\Rightarrow\; E = (I_k - Q)^{-1} \\
\text{Hence, } G &= ED = (I_k - Q)^{-1} D.
\end{aligned}
$$

Alternatively,

$$
\begin{aligned}
A_n \to A, &\Rightarrow Q^n \to 0, \text{ uniformly in } i, j \in T \\
&\Rightarrow (I_k - Q)D_n = (I_k - Q)(I_k + Q + Q^2 + \cdots + Q^{n-1})D \\
&\Rightarrow (I_k - Q)D_n = (I_k - Q^n)D \to D \;\Rightarrow\; D_n \to (I_k - Q)^{-1} D \\
\text{Hence, } G &= \lim_{n\to\infty} D_n = (I_k - Q)^{-1} D
\end{aligned}
$$

and (i) is proved. To prove (ii), observe that $\overline{P}^n$ being a stochastic matrix, we have $\forall\ n \geq 1$,

$$
\begin{aligned}
(D_n\, Q^n)\underset{\sim}{e} = \underset{\sim}{e} &\Rightarrow \lim_{n\to\infty} (D_n\, Q^n)\underset{\sim}{e} = \underset{\sim}{e} \\
&\Rightarrow G\underset{\sim}{e} = \underset{\sim}{e}, \text{ since } Q^n \to 0\ \&\ D_n \to G.
\end{aligned}
$$

Thus, row sums of matrix G are 1, that is, $\sum_{j=1}^{m} g_{ij} = 1, \forall\ i \in T$. □

Remark 2.6.2 Since T is finite, the Markov chain is in T only for a finite number of steps and ultimately enters into some closed communicating class C_j and stays there forever. Hence, probability of absorption in class C is 1. Thus, a transient state disappears after a while and that is the reason for it to be labeled as a transient state. For the Markov chain in Example 2.6.1, $P^n, \ \forall\ n \geq 11$ is given by

$$
P^n = \begin{array}{c} \\ 1 \\ 2 \\ 3 \\ 4 \\ 5 \\ 6 \\ 7 \\ 8 \end{array}
\begin{array}{c}
\begin{array}{cccccccc} 1 & 2 & 3 & 4 & 5 & 6 & 7 & 8 \end{array} \\
\begin{pmatrix}
0.246 & 0.344 & 0.410 & 0 & 0 & 0 & 0 & 0 \\
0.246 & 0.344 & 0.410 & 0 & 0 & 0 & 0 & 0 \\
0.246 & 0.344 & 0.410 & 0 & 0 & 0 & 0 & 0 \\
0 & 0 & 0 & 0.538 & 0.462 & 0 & 0 & 0 \\
0 & 0 & 0 & 0.538 & 0.462 & 0 & 0 & 0 \\
0.127 & 0.178 & 0.212 & 0.260 & 0.223 & 0 & 0 & 0 \\
0.125 & 0.175 & 0.208 & 0.265 & 0.227 & 0 & 0 & 0 \\
0.140 & 0.196 & 0.234 & 0.231 & 0.198 & 0 & 0 & 0
\end{pmatrix}
\end{array}.
$$

Observe that after $n = 11$, probabilities of transition from transient states to transient states are 0. Sums of probabilities in rows 6, 7, 8 of the first 5 columns of P^n are 1, which indicates that the probability of transition into a class of persistent states is 1.

If T is infinite, then the Markov chain may be in T forever. In Chap. 5, we discuss in detail a particular Markov chain, a Galton-Watson branching process. Its state space is W, 0 is an absorbing state and all other states are transient. Under certain conditions, with positive probability, the branching process is not absorbed in 0 starting from any transient state.

The next example illustrates Theorem 2.6.9.

Example 2.6.3 For the Markov chain in Example 2.6.1, we compute G. Note that

$$Q = \begin{pmatrix} 0.1 & 0.1 & 0.2 \\ 0 & 0.1 & 0.1 \\ 0.3 & 0.2 & 0.2 \end{pmatrix} \quad D = \begin{pmatrix} 0.3 & 0.3 \\ 0.4 & 0.4 \\ 0.2 & 0.1 \end{pmatrix} \Rightarrow G = \begin{pmatrix} 0.5166 & 0.4834 \\ 0.5079 & 0.4921 \\ 0.5707 & 0.4293 \end{pmatrix}.$$

Thus, probabilities of absorption in C_1 and C_2 from the transient state 6 are $0.5166, 0.4834$ respectively, from the transient state 7 are $0.5079, 0.4921$ respectively and from the transient state 8 are $0.5707, 0.4293$, respectively. Observe that the row sums of G are 1. Thus, starting from a transient state, the Markov chain is absorbed in one of the two classes of persistent states. □

Remark 2.6.3 (i) By Theorem 2.6.4, if $i \in C_r, r = 1, 2, \ldots, m$ and $j \in T$, then $i \nrightarrow j$ and hence $\sum_{n=1}^{\infty} p_{ij}^{(n)} = 0$.
(ii) If $i \in C_r$ and $j \in C_s$, $s \neq r = 1, 2, \ldots, m$, then also $i \nrightarrow j$ and hence $\sum_{n=1}^{\infty} p_{ij}^{(n)} = 0$.
(iii) If $i, j \in C_r$ or $i \in T, j \in C_r$, $r = 1, 2, \ldots, m$, then j is persistent and $\sum_{n=1}^{\infty} p_{ij}^{(n)}$ diverges.
(iv) If j is transient, then $\sum_{n=1}^{\infty} p_{ij}^{(n)}$ is convergent. We can find its value from Theorem 2.6.9. In this theorem, we have proved that $E = \sum_{n=0}^{\infty} Q^n = (I - Q)^{-1}$ exists. The elements e_{ij} of the matrix E are values of the convergent series $\sum_{n=0}^{\infty} p_{ij}^{(n)}$ for $i, j \in T$. Thus, for $i = j$, $\sum_{n=1}^{\infty} p_{ii}^{(n)} = e_{ii} - 1$, since $p_{ii}^{(0)} = 1$. For $i \neq j$, $\sum_{n=1}^{\infty} p_{ij}^{(n)} = e_{ij}$, since $p_{ij}^{(0)} = 0$.
(v) For transient states, from Eq. (2.5.2), $f_{ii} = \sum_{n=1}^{\infty} p_{ii}^{(n)} / (1 + \sum_{n=1}^{\infty} p_{ii}^{(n)})$. From Eq. (2.6.1), $f_{ij} = (1 - f_{jj}) \sum_{n=1}^{\infty} p_{ij}^{(n)}$. Thus, for transient states f_{ij} can be computed from $(I - Q)^{-1}$.

In the next example, we compute $\sum_{n=1}^{\infty} p_{ij}^{(n)}$ for $i, j \in T$ and f_{ij} for transient states, for the Markov chain in Example 2.6.1.

Example 2.6.4 For the Markov chain in Example 2.6.1, E and $M = \left[\sum_{n=1}^{\infty} p_{ij}^{(n)}\right]$ are given by

$$E = \begin{matrix} & 6 & 7 & 8 \\ 6 & 1.2216 & 0.2094 & 0.3316 \\ 7 & 0.0524 & 1.1518 & 0.1571 \\ 8 & 0.4712 & 0.3665 & 1.4136 \end{matrix} \quad \& \quad M = \begin{matrix} & 6 & 7 & 8 \\ 6 & 0.2216 & 0.2094 & 0.3316 \\ 7 & 0.0524 & 0.1518 & 0.1571 \\ 8 & 0.4712 & 0.3665 & 0.4136 \end{matrix}.$$

Using Eqs. (2.5.2) and (2.6.1), the matrix $F_T = [f_{ij}]_{i,j \in T}$ is given by

$$F_T = \begin{matrix} & 6 & 7 & 8 \\ 6 & 0.1814 & 0.1818 & 0.2346 \\ 7 & 0.0429 & 0.1318 & 0.1111 \\ 8 & 0.3857 & 0.3182 & 0.2926 \end{matrix}.$$

Observe that for the transient states $f_{ij} < 1$ and $M(i, j) \geq F_T(i, j)$ for all i, j. It is true in general and follows from the following arguments. For any (i, j),

$$p_{ij}^{(n)} \geq f_{ij}^{(n)} \Rightarrow \sum_{n \geq 1} p_{ij}^{(n)} \geq \sum_{n \geq 1} f_{ij}^{(n)} \Rightarrow M(i, j) \geq F_T(i, j).$$

□

The proof of Theorem 2.6.9 simplifies if there is a single closed communicating class C of the state space. Since $G\underline{e} = 1$, we expect that the probability of absorption into C from any transient state must be 1. We prove it in the next theorem.

Theorem 2.6.10 *Suppose* $\{X_n, n \geq 0\}$ *is a Markov chain, with a single closed communicating class* C *of persistent states and a finite class* T *of* k *transient states. Then probability of absorption into* C *from any transient state is* 1.

Proof Since there is a single closed communicating class C of the state space, G, as defined in Theorem 2.6.9, is a column vector of dimension k. In this setup, $\overline{P}$ and $\overline{P}^n$ as defined in Theorem 2.6.9 can be expressed as

$$\overline{P} = \begin{pmatrix} I_1 & 0 \\ \underline{d} & Q \end{pmatrix} \;\&\; \overline{P}^n = \begin{pmatrix} I_1 & 0 \\ \underline{d}_n & Q^n \end{pmatrix} \quad \text{where } \underline{d} = (d_1, d_2, \ldots, d_k)'$$

and $\underline{d}_n = (I_k + Q + Q^2 + \cdots + Q^{n-1})\underline{d}$. Suppose $\underline{e}_{k\times 1} = (1, 1, \ldots, 1)'$. Since $\overline{P}$ is a stochastic matrix, we have

$$\underline{d} + Q\underline{e} = \underline{e} \Rightarrow \underline{d} = (I_k - Q)\underline{e}.$$

Using similar arguments as in Theorem 2.6.9, we have

$$\begin{aligned} \underline{G} = \lim_{n\to\infty} \underline{d}_n &= \Big(\sum_{n=0}^{\infty} Q^n\Big)\underline{d} = (I_k - Q)^{-1}\underline{d} \\ &= (I_k - Q)^{-1}(I_k - Q)\underline{e} = \underline{e}. \end{aligned}$$

Thus, if g_{iC} denotes the probability of absorption from a transient state i to a single closed communicating class C, then $g_{iC} = 1, \;\; \forall\; i = 1, 2, \ldots, k$. □

In the following example, we verify Theorem 2.6.10.

Example 2.6.5 Suppose $\{X_n, n \geq 0\}$ is a Markov chain with P given by

$$P = \begin{array}{c} \\ 1 \\ 2 \\ 3 \\ 4 \end{array}\begin{array}{c} \begin{array}{cccc} 1 & 2 & 3 & 4 \end{array} \\ \begin{pmatrix} 0 & 1 & 0 & 0 \\ 1 & 0 & 0 & 0 \\ 1/8 & 1/2 & 1/8 & 1/4 \\ 1/3 & 1/6 & 1/6 & 1/3 \end{pmatrix} \end{array}.$$

Observe that $C = \{1, 2\}$ is a single closed communicating class and $T = \{3, 4\}$ is a class of transient states. Hence ,

$$\overline{P} = \begin{pmatrix} I_1 & 0 \\ \underline{d} & Q \end{pmatrix}, \quad \text{where} \quad Q = \begin{pmatrix} 1/8 & 1/4 \\ 1/6 & 1/3 \end{pmatrix} \;\; \& \;\; \underline{d} = (5/8, 1/2)'.$$

Thus,

$$\underline{G} = (I_2 - Q)^{-1}\underline{d} = \begin{pmatrix} 1.2308 & 0.4615 \\ 0.3077 & 1.6154 \end{pmatrix} \times (0.625, 0.5)' = (1, 1)',$$

as proved in Theorem 2.6.10. □

In the following theorem, we restate the results of Theorems 2.6.9 and 2.6.10, in terms of f_{ij}, when i is transient and j is in one of the closed communicating classes.

Theorem 2.6.11 *Suppose $\{X_n, n \geq 0\}$ is a Markov chain where the class of transient states is finite.*
(i) Suppose C is a closed communicating class of persistent states. Then for any transient state i which leads to $j, k \in C$, $f_{ij} = f_{ik}$, that is, if $i \notin C$ is such that $i \to j$ and $i \to k$ for $j, k \in C$, then $f_{ij} = f_{ik}$.
(ii) Suppose C is a single closed communicating class. If i is a transient state, that is, $i \notin C$ is such that $i \to j \in C$, then $f_{ij} = 1$.
(iii) Suppose $C_r, r = 1, 2$ are two closed communicating classes. If $i \notin C_1 \cup C_2$ is such that $i \to j \in C_1$ but $i \nrightarrow j \in C_2$, then $f_{ij} = 1$. Similarly, if $i \notin C_1 \cup C_2$ is such that $i \to j \in C_2$ but $i \nrightarrow j \in C_1$, then $f_{ij} = 1$, that is, if $i \notin C_1 \cup C_2$ leads to the states $j \in C_r$ for only one $r = 1, 2$, then $f_{ij} = 1$.

Proof (i) By Theorem 2.6.1, $\forall \; j, k \in C, \; f_{jk} = f_{kj} = 1$, that is, once the chain reaches any one of the states in C, it also visits all the other states and it stays in the same class, it being a closed class. Thus, the Markov chain gets absorbed in C. As noted in Lemma 2.6.1, for a transient state i and $j, k \in C, \; f_{ij}, \; f_{ik}$ are the same as the probability that the Markov chain is absorbed in the class C from i. Hence, $f_{ij} = f_{ik}$.

(ii) If $i \notin C$, then i is a transient state. Since $i \to j \in C$, $f_{ij} > 0$. Further, as in (i) f_{ij} is the same as the probability of absorption g_{iC} for any $j \in C$. Hence by Theorem 2.6.10, $f_{ij} = 1$.

(iii) Since $C_r, r = 1, 2$ are two closed communicating classes, $g_{iC_1} + g_{iC_2} = 1$ for transient state i, as proved in Theorem 2.6.9. If $i \notin C_1 \cup C_2$ is such that $i \to j \in C_1$ but $i \nrightarrow j \in C_2$, then $g_{iC_2} = 0$ and hence $f_{ij} = g_{iC_1} = 1$ in this setup. Similarly, if $i \notin C_1 \cup C_2$ is such that $i \to j \in C_2$ but $i \nrightarrow j \in C_1$, then $g_{iC_1} = 0$ and hence $f_{ij} = g_{iC_2} = 1$. □

Note that in Example 2.6.5, there is a single closed communicating class and hence by part (i) of Theorem 2.6.11, $f_{31} = f_{32} = g_{11} = 1$ and $f_{41} = f_{42} = g_{21} = 1$. Part (iii) of Theorem 2.6.11 conveys that if a transient state i leads to states both in C_1 and C_2, then corresponding f_{ij} may not be 1. We note this feature in Example 2.6.7.

Theorem 2.6.11 is useful in Chap. 3, when we discuss the long run distribution of a Markov chain.

There is one more approach to find the absorption probability. We discuss it in the following theorem.

Theorem 2.6.12 *Suppose $\{X_n, n \geq 0\}$ is a Markov chain with T as a finite set of k transient states and $X_0 = i \in T$. Suppose C is a closed class of persistent states, which is partitioned into closed communicating classes, with R denoting one such class; R may be infinite. Suppose $Q = [p_{ij}]_{i,j \in T}$ and $\underline{g}$ and $\underline{p}$ are vectors of g_{iR} and $p_{iR}, i \in T$ respectively, where g_{iR} is a probability of absorption into the class R and p_{iR} is a probability of absorption into the class R in one step. Then $\underline{g} = (I_k - Q)^{-1}\underline{p}$.*

Proof Observe that, for $i \in T$, g_{iR} can be expressed as

$$g_{iR} = p_{iR} + \sum_{l \in T} p_{il} g_{lR}, \quad \forall \; i \in T \quad \iff \quad (I_k - Q)\underline{g} = \underline{p}. \tag{2.6.2}$$

Thus, $\underline{g}$ is a solution to the system of equations in (2.6.2), provided it exists. It exists, if $(I_k - Q)^{-1}$ exists. We have proved in Theorem 2.6.9 that $(I_k - Q)^{-1}$ exists, when T is finite. Thus, the system of equations has a unique solution given by $\underline{g} = (I_k - Q)^{-1}\underline{p}$. □

The next example illustrates Theorem 2.6.12.

Example 2.6.6 Suppose $\{X_n, n \geq 0\}$ is a Markov chain with state space $S = \{1, 2, 3, 4\}$ and transition probability matrix P given by

$$P = \begin{matrix} & \begin{matrix} 1 & 2 & 3 & 4 \end{matrix} \\ \begin{matrix} 1 \\ 2 \\ 3 \\ 4 \end{matrix} & \begin{pmatrix} 1 & 0 & 0 & 0 \\ 0 & 1 & 0 & 0 \\ 1/4 & 1/4 & 1/4 & 1/4 \\ 1/8 & 2/8 & 4/8 & 1/8 \end{pmatrix} \end{matrix}.$$

We note that $C_1 = \{1\}$, $C_2 = \{2\}$ are two closed communicating classes. $C = \{1, 2\}$ is a closed class of persistent states and $T = \{3, 4\}$ is a class of transient states. We determine the probabilities of absorption into both the classes separately. The system of equations specified in Eq. (2.6.2) is

$$g_{3C_1} = p_{3C_1} + p_{33}g_{3C_1} + p_{34}g_{4C_1} \;\&\; g_{4C_1} = p_{4C_1} + p_{43}g_{3C_1} + p_{44}g_{4C_1},$$

which can be expressed as

$$(1 - p_{33})g_{3C_1} - p_{34}g_{4C_1} = p_{3C_1} \;\&\; - p_{43}g_{3C_1} + (1 - p_{44})g_{4C_1} = p_{4C_1}.$$

These equations in a matrix form are

$$\begin{pmatrix} 1 - p_{33} & -p_{34} \\ -p_{43} & 1 - p_{44} \end{pmatrix} \times \begin{pmatrix} g_{3C_1} \\ g_{4C_1} \end{pmatrix} = \begin{pmatrix} p_{3C_1} \\ p_{4C_1} \end{pmatrix} \iff (I_k - Q)\underline{g} = \underline{p},$$

where

$$Q = \begin{pmatrix} p_{33} & p_{34} \\ p_{43} & p_{44} \end{pmatrix}, \quad g = \begin{pmatrix} g_{3C_1} \\ g_{4C_1} \end{pmatrix} \quad \text{and } p = \begin{pmatrix} p_{3C} \\ p_{4C} \end{pmatrix}.$$

By substituting the relevant values, the equations reduce to

$$\begin{pmatrix} 3/4 & -1/4 \\ -4/8 & 7/8 \end{pmatrix} \times \begin{pmatrix} g_{3C_1} \\ g_{4C_1} \end{pmatrix} = \begin{pmatrix} 1/4 \\ 1/8 \end{pmatrix}.$$

Hence,

$$\begin{pmatrix} g_{3C_1} \\ g_{4C_1} \end{pmatrix} = \begin{pmatrix} 3/4 & -1/4 \\ -4/8 & 7/8 \end{pmatrix}^{-1} \times \begin{pmatrix} 1/4 \\ 1/8 \end{pmatrix} = \begin{pmatrix} 8/17 \\ 7/17 \end{pmatrix} = \begin{pmatrix} 0.4706 \\ 0.4118 \end{pmatrix}.$$

Hence, the probability that starting from the transient state 3, the Markov chain is absorbed in C_1 is $g_{3C_1} = 8/17$ and starting from the transient state 4, it is absorbed in C_1 is $g_{4C_1} = 7/17$. On similar lines, we obtain the probabilities of absorption into the class C_2 as

$$\begin{pmatrix} g_{3C_2} \\ g_{4C_2} \end{pmatrix} = \begin{pmatrix} 3/4 & -1/4 \\ -4/8 & 7/8 \end{pmatrix}^{-1} \times \begin{pmatrix} 1/4 \\ 2/8 \end{pmatrix} = \begin{pmatrix} 9/17 \\ 10/17 \end{pmatrix} = \begin{pmatrix} 0.5294 \\ 0.5882 \end{pmatrix}.$$

Hence, $g_{3C_2} = 9/17$ and $g_{4C_2} = 10/17$. Since $C = C_1 \cup C_2$ and C_1 and C_2 are disjoint, we have $g_{3C} = g_{3C_1} + g_{3C_2} = 1$ and $g_{4C} = g_{4C_1} + g_{4C_2} = 1$. Thus, with probability 1, the Markov chain starting from a transient state gets absorbed into a class of persistent states. □

We now discuss the computation of the first passage distribution and f_{ij} using R and verify some results proved in this section. As in the computation of $f_{ii}^{(n)}$, we adopt two approaches. In the first, we compute $f_{ij}^{(n)}$ and μ_{ij}, if $f_{ij} = 1$, for fixed i and j. In the second approach, we unify the computation of $f_{ij}^{(n)}$ for all i and j, by fixing large n. We adopt the second approach in Code 2.8.7, given in Sect. 2.8.

The following examples illustrate the computation of $f_{ij}^{(n)}$ using the recurrence relation and computation of f_{ij}, using Code 2.8.7. With this code we also get f_{ii} when $i = j$. In the code, we compute means μ_{ij} of the first passage distributions, when corresponding $f_{ij} \approx 1$. Theorem 2.6.1 conveys that if two states communicate with each other and are persistent then $f_{ij} = f_{ji} = 1$. If $i \nrightarrow j$, then $f_{ij} = 0$. These results are verified in the following examples. We also verify Theorem 2.6.11.

Example 2.6.7 Suppose $\{X_n, n \geq 0\}$ is a Markov chain with state space $S = \{1, 2, 3, 4, 5, 6\}$ and transition probability matrix P as given below

$$P = \begin{array}{c} \\ 1 \\ 2 \\ 3 \\ 4 \\ 5 \\ 6 \end{array} \begin{array}{c} \begin{array}{cccccc} 1 & 2 & 3 & 4 & 5 & 6 \end{array} \\ \begin{pmatrix} 1/3 & 0 & 2/3 & 0 & 0 & 0 \\ 0 & 1/2 & 1/4 & 0 & 1/4 & 0 \\ 2/5 & 0 & 3/5 & 0 & 0 & 0 \\ 0 & 1/4 & 1/4 & 1/4 & 0 & 1/4 \\ 0 & 0 & 0 & 0 & 1/2 & 1/2 \\ 0 & 0 & 0 & 0 & 1/4 & 3/4 \end{pmatrix} \end{array}.$$

Observe that $\{1, 3\}, \{2\}, \{4\}, \{5, 6\}$ are communicating classes, out of which $C_1 = \{1, 3\}$ *and* $C_2 = \{5, 6\}$ are closed classes and hence the Markov chain is reducible. The two sub-chains can be analyzed separately. Using Code 2.8.6 and Theorem 2.6.8, it follows that 1 and 3 are non-null persistent states. Similarly, 5 and 6 are also non-null persistent states. It is to be noted that 2 and 4 are inessential and hence transient states. Using Code 2.8.7, we compute f_{ij} $\forall\ i, j$ and verify whether values of f_{ij} support the results derived above. Values of f_{ij} are as displayed in the following matrix $F = [f_{ij}]$:

$$F = \begin{array}{c} \\ 1 \\ 2 \\ 3 \\ 4 \\ 5 \\ 6 \end{array} \begin{array}{c} \begin{array}{cccccc} 1 & 2 & 3 & 4 & 5 & 6 \end{array} \\ \begin{pmatrix} 1 & 0 & 1 & 0 & 0 & 0 \\ 0.5 & 0.5 & 0.5 & 0 & 0.5 & 0.5 \\ 1 & 0 & 1 & 0 & 0 & 0 \\ 0.5 & 0.3333 & 0.5 & 0.25 & 0.5 & 0.5 \\ 0 & 0 & 0 & 0 & 1 & 1 \\ 0 & 0 & 0 & 0 & 1 & 1 \end{pmatrix} \end{array}.$$

We arrive at the following conclusions from the output.

(i) $f_{ii} = 1$ for $i = 1, 3, 5, 6$ and these are persistent states, while $f_{22} = 0.5$ and $f_{44} = 0.25$, supporting the conclusion that 2 and 4 are transient states. From P, it is clear that $f_{22}^{(1)} = p_{22} = 0.5$ and $f_{22}^{(n)} = 0\ \forall\ n \geq 2$. Hence, $f_{22} = 0.5$. Similarly, $f_{44}^{(1)} = p_{44} = 0.25$ and $f_{44}^{(n)} = 0\ \forall\ n \geq 2$. Hence, $f_{44} = 0.25$.

(ii) $f_{ij} = 1$ for $i = 1$ *and* $j = 3$, $i = 3$ *and* $j = 1$, $i = 5$ *and* $j = 6$, $i = 6$ *and* $j = 5$. These communicate with each other and are persistent. The means of the first passage distribution for these states are $\mu_{13} = 1.5$, $\mu_{31} = 2.5$ and $\mu_{56} = 2$, $\mu_{65} = 4$. Further, $\mu_1 = 2.6667$, $\mu_3 = 1.6$ and $\mu_5 = 3$, $\mu_6 = 1.5$.

(iii) $f_{ij} = 0$ for $i = 1, 3$ and $j = 2, 4, 5, 6$ when $i \nrightarrow j$. Similarly, $f_{24} = 0$ and $2 \nrightarrow 4$, $f_{ij} = 0$ for $i = 5, 6$ and $j = 1, 2, 3, 4$ when $i \nrightarrow j$.

(iv) The state 4 leads to all the states and we note that $f_{4j} > 0$ for all $j \in S$.

(v) $f_{42} < 1$, where 2 is a transient state.

(vi) The state $i = 2$ is transient, $2 \to 1 \in C_1$ and $2 \to 5 \in C_2$. We note that $f_{ij} < 1$ for $i = 2$, $j = 1, 5$. Similarly, $i = 4$ is transient, $4 \to 1 \in C_1$ and $4 \to 6 \in C_2$; and $f_{ij} < 1$ for $i = 4$, $j = 1, 6$.

(vii) $2 \to 1 \in C_1, 2 \to 3 \in C_1$ and $f_{21} = f_{23} = 0.5$. Similarly, $4 \to 1 \in C_1, 4 \to 3 \in C_1$ and $f_{41} = f_{43} = 0.5$. Thus, result (i) in Theorem 2.6.11 is verified.

(viii) $2 \to 5 \in C_2, 2 \to 6 \in C_2$ and $f_{25} = f_{26} = 0.5$. Similarly, $4 \to 5 \in C_2, 4 \to 6 \in C_2$ and $f_{45} = f_{46} = 0.5$, again as stated in result (i) of Theorem 2.6.11.

(ix) For $i = 2, 4$, $g_{iC_1} + g_{iC_2} = 0.5 + 0.5 = 1$. □

Example 2.6.8 For the Markov chain in Example 2.6.6, with P given by

$$P = \begin{array}{c} \\ 1 \\ 2 \\ 3 \\ 4 \end{array} \begin{array}{c} \begin{array}{cccc} 1 & 2 & 3 & 4 \end{array} \\ \begin{pmatrix} 1 & 0 & 0 & 0 \\ 0 & 1 & 0 & 0 \\ 1/4 & 1/4 & 1/4 & 1/4 \\ 1/8 & 2/8 & 4/8 & 1/8 \end{pmatrix} \end{array},$$

the matrix $F = [f_{ij}]$ is

$$F = \begin{array}{c} \\ 1 \\ 2 \\ 3 \\ 4 \end{array} \begin{array}{c} \begin{array}{cccc} 1 & 2 & 3 & 4 \end{array} \\ \begin{pmatrix} 1 & 0 & 0 & 0 \\ 0 & 1 & 0 & 0 \\ 0.4706 & 0.5294 & 0.3929 & 0.3333 \\ 0.4118 & 0.5882 & 0.5714 & 0.2917 \end{pmatrix} \end{array}.$$

In Example 2.6.6, we have already noted that $C_1 = \{1\}$, $C_2 = \{2\}$ are two closed communicating classes with $C = \{1, 2\}$ being a closed class of persistent states and $T = \{3, 4\}$ is a class of transient states. We have determined the probabilities of absorption into both the classes as $g_{3C_1} = 0.4706$, $g_{4C_1} = 0.4118$ and $g_{3C_2} = 0.5294$, $g_{4C_2} = 0.5882$. Note that $f_{31} = g_{3C_1} = 0.4706$ and $f_{41} = g_{4C_1} = 0.4118$. Further, $f_{32} = g_{3C_2} = 0.5294$ *and* $f_{42} = g_{4C_2} = 0.5882$. Thus, probability of visit from 3 to 1 is the same as the probability of absorption in class $C_1 = \{1\}$, 1 being an absorbing state. We have similar interpretation for f_{41}, f_{32} and f_{42}. □

In the next example, using Code 2.8.7, we compute f_{ij} and μ_{ij} for the weather model as described in Example 2.2.6.

Example 2.6.9 For the weather model in Example 2.2.6, we find the distribution of the first visit to state j from i, $i, j = 1, 2, 3$ using Code 2.8.7, with appropriate changes in the first part. We have noted that for a weather model, all states communicate with each other and all are persistent. From the output, we note that $f_{ij} = 1$ for all i, j supporting the result proved in Theorem 2.6.1. Values of μ_{ij} are reported in the matrix M:

$$M = \begin{matrix} & 1 & 2 & 3 \\ 1 & 2.0876 & 2.9726 & 4.3987 \\ 2 & 2.1051 & 3.2158 & 3.9988 \\ 3 & 2.2805 & 2.4320 & 4.7589 \end{matrix}.$$

The interpretation of $\mu_{12} = 2.97$ is that starting from the sunny day, on the average after 2.97 days, the day will be cloudy for the first time. μ_{ij} for other i, j can be interpreted on similar lines. Values of μ_{ii} are the same as μ_i obtained in Example 2.5.9. □

Example 2.6.10 Suppose $\{X_n, n \geq 0\}$ is a Markov chain with state space $S = \{1, 2, 3, 4, 5\}$ and transition probability matrix P as given below

$$P = \begin{matrix} & 1 & 2 & 3 & 4 & 5 \\ 1 & 0 & 0.5 & 0 & 0.5 & 0 \\ 2 & 0.3 & 0.2 & 0 & 0.5 & 0 \\ 3 & 0.1 & 0.1 & 0.3 & 0.2 & 0.3 \\ 4 & 0.6 & 0 & 0 & 0.4 & 0 \\ 5 & 0.1 & 0.1 & 0.4 & 0 & 0.4 \end{matrix}.$$

We use Code 2.8.7 to compute $f_{ij}^{(n)}, n \geq 1$ using the recurrence relation, f_{ij} and μ_{ij} if $f_{ij} = 1$. From the output of the function `View(fijn)`, we note that $f_{ij}^{(n)} \approx 0 \; \forall \; n \geq 40$ and $\forall \; i, j$. The matrix $F = [f_{ij}]$ is given by

$$F = \begin{matrix} & 1 & 2 & 3 & 4 & 5 \\ 1 & 1 & 1 & 0 & 1 & 0 \\ 2 & 1 & 1 & 0 & 1 & 0 \\ 3 & 1 & 1 & 0.5 & 1 & 0.4286 \\ 4 & 1 & 1 & 0 & 1 & 0 \\ 5 & 1 & 1 & 0.6667 & 1 & 0.5714 \end{matrix}.$$

We compute μ_{ij} for those (i, j) for which $f_{ij} = 1$. These are presented in the following matrix M:

$$M = \begin{matrix} & 1 & 2 & 4 \\ 1 & 2.9792 & 3.6666 & 2.0000 \\ 2 & 2.2917 & 4.7666 & 2.0000 \\ 3 & 4.3542 & 6.2333 & 4.2000 \\ 4 & 1.6667 & 5.3333 & 2.2000 \\ 5 & 4.9514 & 6.4333 & 5.1333 \end{matrix}.$$

We have already noted in Example 2.5.8 that states 3 and 5 are transient states and $C = \{1, 2, 4\}$ is a single closed communicating class. States in C are persistent states. F matrix supports these conclusions. We have also computed $\mu_1 = 2.9792$, $\mu_2 = 4.7665$ and $\mu_4 = 2.2000$. These are the same as in matrix M. We further note the following results. Observe that

(i) $C = \{1, 2, 4\}$ is a single closed class and $\forall\ i, j \in C,\ \ f_{ij} = 1$, since $i \leftrightarrow j$ and i, j are persistent.
(ii) $\forall\ i \in C, j \notin C,\ \ i \nrightarrow j$ by Lemma 2.4.1 and $f_{ij} = 0$ by Theorem 2.6.1.
(iii) $\forall\ i \notin C, j \in C,\ \ f_{ij} = 1$ by (ii) of Theorem 2.6.11.
(iv) $\forall\ \ i, j \notin C,\ \ f_{ij} < 1$ as j is transient. Observe that $f_{35}^{(1)} = p_{35} = 0.3$ and $f_{35}^{(n)} = (0.3)^{n-1}(0.3)\ \forall\ \ n \geq 2$. Hence $f_{35} = 3/7 = 0.4286$. Similarly, $f_{53}^{(1)} = p_{53} = 0.4$ and $f_{53}^{(n)} = (0.4)^{n-1}(0.4)\ \forall\, n \geq 2$. Hence $f_{53} = 4/6 = 0.6667$.
(v) In Theorem 2.6.10, it is proved that the probability of absorption from any transient state to a single closed communicating class is 1. From F matrix we note that $f_{3j} = f_{5j} = 1\ \ \forall\ \ j \in C$, since once the chain from transient states visits any persistent state in C, it stays in the same class, it being a closed class. □

Example 2.6.11 Suppose $\{X_n, n \geq 0\}$ is a Markov chain with state space $S = \{1, 2, 3, 4, 5, 6\}$ and transition probability matrix P as given below

$$P = \begin{array}{c} \\ 1 \\ 2 \\ 3 \\ 4 \\ 5 \\ 6 \end{array}\begin{array}{c} \begin{array}{cccccc} 1 & 2 & 3 & 4 & 5 & 6 \end{array} \\ \begin{pmatrix} 1/3 & 0 & 2/3 & 0 & 0 & 0 \\ 0 & 1/2 & 1/4 & 1/4 & 0 & 0 \\ 2/5 & 0 & 3/5 & 0 & 0 & 0 \\ 1/4 & 1/4 & 1/4 & 1/4 & 0 & 0 \\ 0 & 0 & 0 & 0 & 1/2 & 1/2 \\ 0 & 0 & 0 & 0 & 1/4 & 3/4 \end{pmatrix} \end{array}.$$

Observe that $C_1 = \{1, 3\}$ $C_2 = \{5, 6\}$ are two closed communicating classes. Thus, $\{1, 3, 5, 6\}$ are persistent states and 2, 4 are transient states. We note that 2, 4 lead to states in C_1 only. Using Code 2.8.7, we compute $F = [f_{ij}]$. It is presented below

$$F = \begin{array}{c} \\ 1 \\ 2 \\ 3 \\ 4 \\ 5 \\ 6 \end{array}\begin{array}{c} \begin{array}{cccccc} 1 & 2 & 3 & 4 & 5 & 6 \end{array} \\ \begin{pmatrix} 1 & 0 & 1 & 0 & 0 & 0 \\ 1 & 0.5833 & 1 & 0.5 & 0 & 0 \\ 1 & 0 & 1 & 0 & 0 & 0 \\ 1 & 0.3333 & 1 & 0.375 & 0 & 0 \\ 0 & 0 & 0 & 0 & 1 & 1 \\ 0 & 0 & 0 & 0 & 1 & 1 \end{pmatrix} \end{array}.$$

We arrive at the following conclusions from the output:

(i) $f_{ii} = 1$ for $i = 1, 3, 5, 6$ and these are persistent states, while $f_{22} = 0.5833$ and $f_{44} = 0.375$, supporting the conclusion that 2 and 4 are transient states.
(ii) $f_{ij} = 1$ for $i,\ j \in C_1$ and C_2.
(iii) $f_{ij} = 0$ for $i = 1, 3$ and $j = 2, 4, 5, 6$ when $i \nrightarrow j$. $f_{ij} = 0$ for $i = 5, 6$ and $j = 1, 2, 3, 4$ when $i \nrightarrow j$.
(iv) $f_{42} < 1$, where 2 is a transient state and $f_{24} < 1$, where 4 is a transient state.
(v) $2 \to j \in C_1$ but $2 \nrightarrow j \in C_2$ and $f_{21} = f_{23} = 1$. Similarly, $4 \to j \in C_1$ but $4 \nrightarrow j \in C_2$ and $f_{41} = f_{43} = 1$. Thus, result (iii) in Theorem 2.6.11 is verified.
(vi) Since the transient states 2, 4 lead to states in C_1 only, the chain gets absorbed in C_1 only. Hence, the probability of absorption from these transient states 2, 4 to C_1 is 1, although there are two closed communicating classes. From F matrix, we note that $f_{2j} = f_{4j} = 1\ \forall\ j \in C_1$, since once the chain from these transient states visits any persistent state in C_1, it stays in the same class, it being a closed class. □

In the next example, we compute $F = [f_{ij}]$ for the Markov chain in Example 2.6.1 and verify the results as obtained in Example 2.6.4.

Example 2.6.12 For the Markov chain in Example 2.6.1, we compute $F = [f_{ij}]$, using Code 2.8.7. It is displayed below

$$F = \begin{array}{c} \\ 1 \\ 2 \\ 3 \\ 4 \\ 5 \\ 6 \\ 7 \\ 8 \end{array}\begin{array}{c} \begin{array}{cccccccc} 1 & 2 & 3 & 4 & 5 & 6 & 7 & 8 \end{array} \\ \begin{pmatrix} 1.0000 & 1.0000 & 1.0000 & 0.0000 & 0.0000 & 0.0000 & 0.0000 & 0.0000 \\ 1.0000 & 1.0000 & 1.0000 & 0.0000 & 0.0000 & 0.0000 & 0.0000 & 0.0000 \\ 1.0000 & 1.0000 & 1.0000 & 0.0000 & 0.0000 & 0.0000 & 0.0000 & 0.0000 \\ 0.0000 & 0.0000 & 0.0000 & 1.0000 & 1.0000 & 0.0000 & 0.0000 & 0.0000 \\ 0.0000 & 0.0000 & 0.0000 & 1.0000 & 1.0000 & 0.0000 & 0.0000 & 0.0000 \\ 0.5166 & 0.5166 & 0.5166 & 0.4834 & 0.4834 & 0.1814 & 0.1818 & 0.2346 \\ 0.5079 & 0.5079 & 0.5079 & 0.4921 & 0.4921 & 0.0429 & 0.1318 & 0.1111 \\ 0.5707 & 0.5707 & 0.5707 & 0.4293 & 0.4293 & 0.3857 & 0.3182 & 0.2926 \end{pmatrix} \end{array}.$$

We have already noted in Example 2.6.1 that $C_1 = \{1, 2, 3\},\ C_2 = \{4, 5\}$ are two closed communicating classes and $T = \{6, 7, 8\}$ is a class of transient states. From F, we have the following conclusions:

(i) $f_{ii} = 1$ for all $i \in C_1, C_2$ and these are persistent states, while $f_{ii} < 1$ for $i \in T$, supporting the conclusion that 6, 7 and 8 are transient states. These are the same as in F_T, obtained in Example 2.6.4.
(ii) $f_{ij} = 1$ for $i,\ j \in C_1$ and $i,\ j \in C_2$, the two being closed communicating classes.
(iii) $f_{ij} = 0$ for $i \in C_1$ and $j \in C_2$ and $j \in T$, when $i \nrightarrow j$. Similarly, $f_{ij} = 0$ for $i \in C_2$ and $j \in C_1$ and $j \in T$, when $i \nrightarrow j$.
(iv) For $i = 6 \in T$, and $j \in C_1$, $f_{ij} = 0.5166 < 1$. It is the same for all j in C_1, verifying result (i) in Theorem 2.6.11. It is also the same as g_{iC_1} as obtained in Example 2.6.3. We have similar results for $i = 7, 8$.

(v) For $i = 6 \in T$, and $j \in C_2$, $f_{ij} = 0.4834 < 1$. It is the same for all j in C_2 and also the same as g_{iC_2} as obtained in Example 2.6.3. We have similar results for $i = 7, 8$.

(vi) Further, $g_{iC_1} + g_{iC_2} = 1, \ \forall \ i \in T$.

(vii) For $i \neq j \in T$, f_{ij} are the same as those obtained in F_T in Example 2.6.4, from the series $\sum_{n \geq 1} p_{ij}^{(n)}$. □

Results about f_{ij} discussed above are useful in the next chapter to study the limiting behavior of a Markov chain. To investigate the limiting behavior of $p_{ij}^{(n)}$, in particular when j is a non-null persistent state, we need the concept of periodicity of a Markov chain. We introduce it in the next section.

2.7 Periodicity

We begin with some examples.

Example 2.7.1 Suppose $\{X_n, n \geq 0\}$ is a Markov chain with state space $S = \{1, 2, 3, 4\}$ and P given by

$$P = \begin{array}{c} \\ 1 \\ 2 \\ 3 \\ 4 \end{array} \begin{array}{c} \begin{array}{cccc} 1 & 2 & 3 & 4 \end{array} \\ \begin{pmatrix} 0 & 1 & 0 & 0 \\ 0 & 0 & 1 & 0 \\ 0 & 0 & 0 & 1 \\ 1 & 0 & 0 & 0 \end{pmatrix} \end{array}.$$

Thus, if the initial state is 1, then the Markov chain moves from $1 \to 2 \to 3 \to 4 \to 1$. If the initial state is 2, then the Markov chain moves from $2 \to 3 \to 4 \to 1 \to 2$. The same scenario is observed for the remaining two states. Thus, if the initial state is i, then it returns to i after 4, 8, 12, ..., steps, that is, the system moves in a cyclic manner. Observe that the length of the cycle for each state is 4. □

Example 2.7.2 Suppose $\{X_n, n \geq 0\}$ is a Markov chain with state space $S = \{1, 2, 3\}$ and P given by

$$P = \begin{array}{c} \\ 1 \\ 2 \\ 3 \end{array} \begin{array}{c} \begin{array}{ccc} 1 & 2 & 3 \end{array} \\ \begin{pmatrix} 0 & 1 & 0 \\ 0.5 & 0 & 0.5 \\ 0 & 1 & 0 \end{pmatrix} \end{array}.$$

Observe the following paths of transitions from state i to state i, $i = 1, 2, 3$:

$$\begin{aligned}
&1 \to 2 \to 1 \to 2 \to 1 \to 2 \to 1 \cdots\\
&1 \to 2 \to 3 \to 2 \to 1 \to 2 \to 1 \cdots\\
&2 \to 1 \to 2 \to 1 \to 2 \to 1 \to 2 \cdots\\
&2 \to 3 \to 2 \to 3 \to 2 \to 3 \to 2 \cdots\\
&2 \to 3 \to 2 \to 1 \to 2 \to 3 \to 2 \cdots\\
&3 \to 2 \to 3 \to 2 \to 3 \to 2 \to 3 \cdots\\
&3 \to 2 \to 1 \to 2 \to 3 \to 2 \to 3 \cdots
\end{aligned}$$

Suppose $D_i = \{n > 0 | p_{ii}^{(n)} > 0\}$. Observe that $D_i = \{2, 4, 6, \ldots\}\ \forall\ i \in S$. Thus, the system moves in a cyclic manner with the length of the cycle for each state being 2. Further note that $p_{ii}^{(n)} = 0$ if n is not a multiple of 2. □

In both the above Markov chains, there is a periodic movement of the chain among its states, leading to the concept of periodicity of the states of a Markov chain. We define below a period of a state.

Definition 2.7.1 Periodic and Aperiodic States: Suppose $\{X_n, n \geq 0\}$ is a Markov chain with state space S and transition probability matrix P. Suppose $D_i = \{n > 0 | p_{ii}^{(n)} > 0\}$ and d_i is the greatest common divisor (g.c.d.) of the set D_i. If $d_i > 1$, then the state i is said to be periodic with period d_i. If $d_i = 1$, then the state i is said to be aperiodic. If $D_i = \emptyset$, then d_i is defined to be 0.

Thus, if a state i is periodic with period d_i and $n = kd_i + b$, where k and b are integers such that $0 < b < d_i$, that is, if n is not a multiple of d_i, then $p_{ii}^{(n)} = 0$. In other words, if the chain is in state i at time n, then it can only return to state i at times of the form $n + kd_i$ for some integer k.

Note that in Example 2.7.1 each state has period 4, and in Example 2.7.2 each state has period 2.

Remark 2.7.1 If $p_{ii}^{(n)} = 0,\ \ \forall\ n \geq 1$, then $D_i = \emptyset$ and the period d_i of state i is defined to be 0, Karlin and Taylor [9]. For example, suppose the transition probability matrices P and Q of two Markov chains are given by

$$P = \begin{array}{c} \\ 1 \\ 2 \\ 3 \end{array}\!\!\begin{array}{c} \begin{array}{ccc} 1 & 2 & 3 \end{array} \\ \begin{pmatrix} 0 & 1/3 & 2/3 \\ 0 & 1 & 0 \\ 0 & 3/4 & 1/4 \end{pmatrix} \end{array} \quad \& \quad Q = \begin{array}{c} \\ 1 \\ 2 \end{array}\!\!\begin{array}{c} \begin{array}{cc} 1 & 2 \end{array} \\ \begin{pmatrix} 1 & 0 \\ 1 & 0 \end{pmatrix} \end{array}.$$

It is easy to check that $p_{11}^{(n)} = q_{22}^{(n)} = 0\ \forall\ n \geq 1$. Hence,

$$D_1 = \{n > 0 | p_{11}^{(n)} > 0\} = \emptyset\ \ \&\ \ D_2 = \{n > 0 | q_{22}^{(n)} > 0\} = \emptyset.$$

As a consequence, g.c.d. of D_1 and D_2 are not defined. In such cases, the period of the state is defined to be 0. In Example 2.5.1, if $\alpha = 1$, then the transition probability matrix P is given by

$$P = \begin{array}{c} \\ 1 \\ 2 \end{array} \begin{array}{c} \begin{array}{cc} 1 & 2 \end{array} \\ \begin{pmatrix} 1 & 0 \\ 1 & 0 \end{pmatrix} \end{array}$$

and we have noted that $f_{22} = 0$. Further,

$$f_{22} = 0 \Rightarrow f_{22}^{(n)} = 0 \, \forall \, n \geq 1 \Rightarrow p_{22}^{(n)} = 0 \, \forall \, n \geq 1.$$

Thus in Example 2.5.1, if $\alpha = 1$, period of the state 2 is 0.

Example 2.7.3 Suppose $\{X_n, n \geq 0\}$ is a Markov chain with state space $S = \{1, 2, 3, 4, 5\}$ and the transition probability matrix P given by

$$P = \begin{array}{c} \\ 1 \\ 2 \\ 3 \\ 4 \\ 5 \end{array} \begin{array}{c} \begin{array}{ccccc} 1 & 2 & 3 & 4 & 5 \end{array} \\ \begin{pmatrix} 1/2 & 1/2 & 0 & 0 & 0 \\ 1/2 & 0 & 1/2 & 0 & 0 \\ 0 & 1/2 & 0 & 1/2 & 0 \\ 0 & 0 & 1/2 & 0 & 1/2 \\ 0 & 0 & 0 & 1 & 0 \end{pmatrix} \end{array}.$$

It is clear that all states communicate with each other. Further, $1 \to 1$ and $1 \to 2 \to 1$ imply that $D_1 = \{1, 2, \ldots\}$ and its g.c.d. is 1. For state 2, we note from matrix P that 2 is not accessible from 2 in one step. Observe that $2 \to 1 \to 2$, $2 \to 1 \to 1 \to 2$, $2 \to 3 \to 2 \to 1 \to 2$. Thus $D_2 = \{2, 3, 4, \ldots, \}$ and its g.c.d. is 1. Similarly, it can be shown that the period of the remaining states is also 1. □

Remark 2.7.2 If the transition probability matrix P of a Markov chain has all positive elements or in particular has positive diagonal elements, then the period of each state of the Markov chain is 1. Thus, an absorbing state is aperiodic.

It is to be noted that in Example 2.7.3, all the states communicate with each other and all have the same period. In the following theorem, we prove that periodicity is a class property, class being a communicating class.

Theorem 2.7.1 *Suppose $i \leftrightarrow j$. Then $d_i = d_j$.*

Proof Observe that

$$\begin{aligned} i \leftrightarrow j &\Rightarrow i \to j \;\&\; j \to i \\ &\Rightarrow \exists \; s > 0 \text{ such that } p_{ij}^{(s)} > 0 \;\&\; \exists \; r > 0 \text{ such that } p_{ji}^{(r)} > 0 \\ &\Rightarrow p_{ii}^{(r+s)} \geq p_{ij}^{(s)} \, p_{ji}^{(r)} > 0 \text{ by Chapman-Kolmogorov equations} \\ &\Rightarrow \exists \; r + s > 0 \text{ such that } p_{ii}^{(r+s)} > 0 \\ &\Rightarrow D_i = \{n > 0 | p_{ii}^{(n)} > 0\} \neq \emptyset. \end{aligned}$$

Hence g.c.d. of D_i is not 0. Suppose g.c.d. of D_i is d_i. For $n \in D_i$, by Chapman-Kolmogorov equations,

$$p_{jj}^{(r+s+n)} = \sum_{l \in S} p_{jl}^{(r)} p_{lj}^{(s+n)} = \sum_{l \in S} p_{jl}^{(r)} \left\{ \sum_{k \in S} p_{lk}^{(n)} p_{kj}^{(s)} \right\} \geq p_{ji}^{(r)} p_{ii}^{(n)} p_{ij}^{(s)} > 0 \, .$$

Thus, $p_{jj}^{(r+s+n)} > 0$. Hence, $D_j = \{n > 0 | p_{jj}^{(n)} > 0\} \neq \emptyset$ and its g.c.d. d_j is also not 0. Note that

$$p_{ii}^{(n)} > 0 \;\Rightarrow\; p_{ii}^{(2n)} \geq p_{ii}^{(n)} p_{ii}^{(n)} > 0 \;\Rightarrow\; p_{jj}^{(r+s+2n)} > 0.$$

Thus, both $r+s+n$ and $r+s+2n$ are multiples of d_j. Suppose $r+s+n = k_1 d_j$ and $r+s+2n = k_2 d_j$, then $n = (k_2 - k_1) d_j$. Hence, if n is such that $p_{ii}^{(n)} > 0$, then n is a multiple of d_j. d_i being a g.c.d. of a set of all such n values, we must have $d_j \leq d_i$. The above arguments are symmetric in i and j and hence using similar arguments, we get $d_i \leq d_j$ which implies $d_i = d_j$. □

We now define a period of a Markov chain.

Definition 2.7.2 Period of a Markov chain: A Markov chain is said to be periodic with period d, if all the states have the same period d. If $d = 1$, then the Markov chain is said to be aperiodic.

Theorem 2.7.1 implies that states in a communicating class are either all aperiodic or all periodic with the same period d. Thus, we have the following theorem for an irreducible Markov chain.

Theorem 2.7.2 *In an irreducible Markov chain, all the states have the same period.*

Proof In an irreducible Markov chain, all the states communicate with each other and hence the proof follows from Theorem 2.7.1. □

Converse of Theorem 2.7.2 is not true. For example, suppose the transition probability matrix P of a Markov chain given by

$$P = \begin{matrix} & \begin{matrix} 1 & 2 \end{matrix} \\ \begin{matrix} 1 \\ 2 \end{matrix} & \begin{pmatrix} 1 & 0 \\ 0 & 1 \end{pmatrix} \end{matrix}.$$

Then the Markov chain is aperiodic but not irreducible.

If all states do not communicate with each other, that is, when a Markov chain is reducible, then some states may be aperiodic and some may be periodic. The following example illustrates this assertion.

Example 2.7.4 Suppose $\{X_n, n \geq 0\}$ is a Markov chain with state space $S = \{1, 2, 3, 4, 5\}$ with transition probability matrix P given by

$$P = \begin{array}{c} \\ 1 \\ 2 \\ 3 \\ 4 \\ 5 \end{array} \begin{array}{c} \begin{array}{ccccc} 1 & 2 & 3 & 4 & 5 \end{array} \\ \begin{pmatrix} 0 & 1 & 0 & 0 & 0 \\ 0 & 0 & 1 & 0 & 0 \\ 1 & 0 & 0 & 0 & 0 \\ 1/10 & 1/5 & 2/5 & 1/10 & 1/5 \\ 1/5 & 1/5 & 1/5 & 1/5 & 1/5 \end{pmatrix} \end{array}.$$

Observe that $\{1, 2, 3\}$ is a closed communicating class. Thus, all three states in this class have the same period. Now $1 \to 2 \to 3 \to 1$, thus $p_{11}^{(3)} > 0$. Proceeding on the same lines, we note that $p_{11}^{(3n)} > 0$ for $n \geq 1$. Hence, the period of state 1 is 3, which further implies that the period of states 2 and 3 is also 3. Now 4 and 5 communicate with each other and hence have the same period. Note that $p_{44} > 0$, $p_{55} > 0$. Hence, both the states 4 and 5 are aperiodic. Thus, states $1, 2, 3$ have period 3, while states 4 and 5 have period 1. □

Aperiodicity is often obvious when we observe the powers of the transition probability matrix. If each element of P^m is positive for some $m \geq 1$, then P^{m+k} has strictly positive elements for all $k > 0$. This follows because each element of P is non-negative. In such a case, all states communicate with each other and have the same period. Further, the set $\{n | p_{ii}^{(n)} > 0\} = \{m, m+1, m+2, \ldots\}$. Hence, period of i and hence of all the states is 1. The following example illustrates this feature.

Example 2.7.5 Suppose $\{X_n, n \geq 0\}$ is a Markov chain with state space $S = \{1, 2, 3, 4, 5\}$ and the transition probability matrix P given by

$$P = \begin{array}{c} \\ 1 \\ 2 \\ 3 \\ 4 \\ 5 \end{array} \begin{array}{c} \begin{array}{ccccc} 1 & 2 & 3 & 4 & 5 \end{array} \\ \begin{pmatrix} 1/2 & 1/2 & 0 & 0 & 0 \\ 1/2 & 0 & 1/2 & 0 & 0 \\ 0 & 1/2 & 0 & 1/2 & 0 \\ 0 & 0 & 1/2 & 0 & 1/2 \\ 0 & 0 & 0 & 1 & 0 \end{pmatrix} \end{array}.$$

It is clear that all states communicate with each other and as noted in Example 2.7.3, all have the same period 1. Hence the Markov chain is aperiodic. The matrix P^8 given below has all non-zero elements, supporting the result that the period of the Markov chain is 1:

$$P^8 = \begin{array}{c} \\ 1 \\ 2 \\ 3 \\ 4 \\ 5 \end{array} \begin{array}{c} \begin{array}{ccccc} 1 & 2 & 3 & 4 & 5 \end{array} \\ \begin{pmatrix} 0.2773 & 0.2227 & 0.2500 & 0.1406 & 0.1094 \\ 0.2227 & 0.3047 & 0.1133 & 0.3281 & 0.0312 \\ 0.2500 & 0.1133 & 0.3828 & 0.0352 & 0.2188 \\ 0.1406 & 0.3281 & 0.0352 & 0.4922 & 0.0039 \\ 0.2188 & 0.0625 & 0.4375 & 0.0078 & 0.2734 \end{pmatrix} \end{array}.$$

□

In the next example, even and odd powers of P have different nature, but lead to the same conclusion that the period of each state is 2.

Example 2.7.6 Suppose $\{X_n, n \geq 0\}$ is a Markov chain with state space $S = \{1, 2, 3, 4\}$ and the transition probability matrix P given by

$$P = \begin{array}{c} \\ 1 \\ 2 \\ 3 \\ 4 \end{array} \begin{array}{c} \begin{array}{cccc} 1 & 2 & 3 & 4 \end{array} \\ \begin{pmatrix} 0 & 1 & 0 & 0 \\ 1/2 & 0 & 1/2 & 0 \\ 0 & 1/2 & 0 & 1/2 \\ 0 & 0 & 1 & 0 \end{pmatrix} \end{array}.$$

Observe that all states communicate with each other and hence all have the same period. Further, $1 \to 2 \to 1$, $1 \to 2 \to 3 \to 2 \to 1$, $1 \to 2 \to 3 \to 4 \to 3 \to 2 \to 1$ which implies that $D_1 = \{2, 4, 6, \ldots\}$ with g.c.d. 2. Hence the Markov chain is periodic with period 2. Powers of the transition matrix P are displayed below

$$P^2 = \begin{array}{c} \\ 1 \\ 2 \\ 3 \\ 4 \end{array} \begin{array}{c} \begin{array}{cccc} 1 & 2 & 3 & 4 \end{array} \\ \begin{pmatrix} 1/2 & 0 & 1/2 & 0 \\ 0 & 3/4 & 0 & 1/4 \\ 1/4 & 0 & 3/4 & 0 \\ 0 & 1/2 & 0 & 1/2 \end{pmatrix} \end{array}, \quad P^4 = \begin{array}{c} \\ 1 \\ 2 \\ 3 \\ 4 \end{array} \begin{array}{c} \begin{array}{cccc} 1 & 2 & 3 & 4 \end{array} \\ \begin{pmatrix} 0.375 & 0 & 0.625 & 0 \\ 0 & 0.688 & 0 & 0.312 \\ 0.312 & 0 & 0.688 & 0 \\ 0 & 0.625 & 0 & 0.375 \end{pmatrix} \end{array}.$$

$$P^3 = \begin{array}{c} \\ 1 \\ 2 \\ 3 \\ 4 \end{array} \begin{array}{c} \begin{array}{cccc} 1 & 2 & 3 & 4 \end{array} \\ \begin{pmatrix} 0 & 0.75 & 0 & 0.25 \\ 0.38 & 0 & 0.62 & 0 \\ 0 & 0.63 & 0 & 0.37 \\ 0.25 & 0 & 0.75 & 0 \end{pmatrix} \end{array}, \quad P^5 = \begin{array}{c} \\ 1 \\ 2 \\ 3 \\ 4 \end{array} \begin{array}{c} \begin{array}{cccc} 1 & 2 & 3 & 4 \end{array} \\ \begin{pmatrix} 0 & 0.688 & 0 & 0.312 \\ 0.344 & 0 & 0.656 & 0 \\ 0 & 0.656 & 0 & 0.344 \\ 0.313 & 0 & 0.687 & 0 \end{pmatrix} \end{array}.$$

It is to be noted that P^2 and P^4 have the same pattern of non-zero elements, and it continues for all even powers of P. Similarly, P^3 and P^5 have the same pattern of non-zero elements and it continues for all odd powers of P, supporting the result that the period of the chain is 2. The diagonal elements of P^2 are positive, which implies that the Markov chain with transition probability matrix P^2 is aperiodic. Further note that the Markov chain with transition probability matrix P^2 is reducible, with two closed classes. Thus, all states in this Markov chain do not communicate but still each state has the same period. □

Remark 2.7.3 (i) Suppose a Markov chain is irreducible and periodic with period d, where d =g.c.d. (A) and $A = \{n | p_{ii}^{(n)} > 0\}$. Suppose a set B is defined as $B = \{n/d | n \in A\}$. Then for each $i \in S$,

$$\begin{aligned} A &= \{d, 2d, 3d, \ldots,\} \ \& \ \text{g.c.d.}(A) = d \\ B &= \{1, 2, 3, \ldots,\} \ \& \ \text{g.c.d.}(B) = 1. \end{aligned}$$

Suppose $Y_n = X_{nd}$, then $\{X_n, n \geq 0\}$ is a Markov chain with period d, implies that $\{Y_n, n \geq 0\}$ is an aperiodic Markov chain with transition probability matrix P^d. In Example 2.7.6, the Markov chain with transition probability matrix as P has period 2, while the Markov chain with transition probability matrix as P^2 has period 1.
(ii) In Theorem 2.7.1, it is proved that for an irreducible Markov chain all the states have the same period. The converse of the theorem is not true; it follows from Example 2.7.6. In this example, the Markov chain is with transition probability matrix as P^2, and each state has the same period, but the chain is reducible.
(iii) It will be shown in Chap. 3 that if P is irreducible with period d, then P^d is reducible with d closed communicating classes. In Example 2.7.6, P^2 is reducible, with two closed communicating classes.

For a periodic Markov chain with period d, if n is not a multiple of d, $p_{ii}^{(n)} = 0$. In the following lemma, we prove that $f_{ii}^{(n)}$ is also 0 if n is not a multiple of d.

Lemma 2.7.1 *g.c.d.* $\{n|f_{ii}^{(n)} > 0\}$ = *g.c.d.* $\{n|p_{ii}^{(n)} > 0\}$.

Proof Suppose $A = \{n|p_{ii}^{(n)} > 0\}$ with g.c.d. $(A) = d$ and $B = \{n|f_{ii}^{(n)} > 0\}$ with g.c.d. $(B) = \delta$. Since $\forall\ n \geq 1,\ \ f_{ii}^{(n)} \leq p_{ii}^{(n)}, n \in B \ \Rightarrow\ n \in A$. It further implies that $\delta \geq d$. For example, if $A = \{1, 2, 3, 4, \ldots, \}$, its g.c.d. is $d = 1$ and if $B = \{2, 4, 6, \ldots\}$, its g.c.d. is $\delta = 2$. Here, $B \subset A$ and $d < \delta$. Now to prove that $\delta \leq d$, we have to prove that δ divides all elements in set A. Assume the contrary and suppose n_0 is the smallest integer such that $p_{ii}^{(n_0)} > 0$ and that δ does not divide n_0. Then $n_0 = m\delta + r$ where $0 < r < \delta$. Using the fact that $f_{ii}^{(k)} = 0$ if k is not a multiple of δ, we have for $t = 1, 2, \ldots, m$,

$$p_{ii}^{(n_0)} = \sum_{k=1}^{n_0} f_{ii}^{(k)} p_{ii}^{(n_0-k)} \iff p_{ii}^{(m\delta+r)} = \sum_{t=1}^{m} f_{ii}^{(t\delta)} p_{ii}^{(m\delta+r-t\delta)} .$$

Now $m\delta + r - t\delta < m\delta + r = n_0 \ \Rightarrow\ p_{ii}^{(m\delta+r-t\delta)} = 0,\ \forall\ t = 1, 2, \ldots, m$, as n_0 is the smallest integer such that $p_{ii}^{(n_0)} > 0$. Hence $p_{ii}^{(n_0)} = \sum_{t=1}^{m} f_{ii}^{(t\delta)} p_{ii}^{(m\delta+r-t\delta)}$ is 0, which is a contradiction. Thus, δ divides all $n \in A$. Now d being the g.c.d. of A, it follows that $\delta \leq d$. Hence, $\delta = d$. □

Lemma 2.7.1 also follows from the recurrence relation for $f_{ii}^{(n)}$, which is given by $f_{ii}^{(n)} = p_{ii}^{(n)} - \sum_{r=1}^{n-1} p_{ii}^{(n-r)} f_{ii}^{(r)}, n \geq 2$ and $f_{ii}^{(1)} = p_{ii}^{(1)}$. We illustrate it for $d = 2$. With $d = 2$, $p_{ii}^{(1)} = 0$ and hence $f_{ii}^{(1)} = 0$:

$$
\begin{aligned}
f_{ii}^{(2)} &= p_{ii}^{(2)} - p_{ii}^{(1)} f_{ii}^{(1)} = p_{ii}^{(2)} \\
f_{ii}^{(3)} &= p_{ii}^{(3)} - \sum_{r=1}^{2} p_{ii}^{(n-r)} f_{ii}^{(r)} = p_{ii}^{(3)} - p_{ii}^{(2)} f_{ii}^{(1)} - p_{ii}^{(1)} f_{ii}^{(2)} = 0 \\
f_{ii}^{(4)} &= p_{ii}^{(4)} - \sum_{r=1}^{3} p_{ii}^{(n-r)} f_{ii}^{(r)} = p_{ii}^{(4)} - p_{ii}^{(3)} f_{ii}^{(1)} - p_{ii}^{(2)} f_{ii}^{(2)} - p_{ii}^{(1)} f_{ii}^{(3)} \\
&= p_{ii}^{(4)} - p_{ii}^{(2)} f_{ii}^{(2)}.
\end{aligned}
$$

Hence, for a periodic state i with period d, the recurrence relation is

$$
f_{ii}^{(nd)} = p_{ii}^{(nd)} - \sum_{r=1}^{n-1} p_{ii}^{(n-r)d} f_{ii}^{(rd)}, \quad n \geq 2 \;\&\; f_{ii}^{(d)} = p_{ii}^{(d)}.
$$

Note that $\sum_{n=1}^{\infty} f_{ii}^{(n)} = \sum_{n=1}^{\infty} f_{ii}^{(nd)}$. Thus, if i is persistent for a Markov chain with transition probability matrix P, it is also persistent for a Markov chain with transition probability matrix P^d. Further, the mean recurrence time for state i is given by

$$
\mu_i = \sum_{n\geq 1} n f_{ii}^{(n)} = \sum_{n\geq 1} nd f_{ii}^{(nd)}.
$$

Note that d is implicitly involved in the definition of μ_i. Further from the expression of μ_i, it follows that if i is non-null persistent or null persistent for a Markov chain with transition probability matrix P, then it has the same nature for a Markov chain with transition probability matrix P^d.

We now proceed to discuss how to compute the period of states of a Markov chain using R. We compute it using the following three methods: (i) By definition, period is g.c.d. of a set of integers. Hence, we write a function to compute the g.c.d. of a set of integers. (ii) In the second method to find g.c.d., we use a built-in function `gcd` in the base package. (iii) In the third approach, we find powers of the transition probability matrix P and decide the period by observing the elements of P^n. These three ways are used to write Code 2.8.8.

The following example illustrates Code 2.8.8 to find the period of the states.

Example 2.7.7 Suppose $\{X_n, n \geq 0\}$ is a Markov chain with state space $S = \{1, 2, 3, 4, 5, 6\}$ and transition probability matrix P as given below

$$
P = \begin{array}{c} \\ 1 \\ 2 \\ 3 \\ 4 \\ 5 \\ 6 \end{array}
\begin{array}{c}
\begin{array}{cccccc} 1 & 2 & 3 & 4 & 5 & 6 \end{array} \\
\begin{pmatrix}
0 & 0 & 1/3 & 2/3 & 0 & 0 \\
0 & 0 & 2/3 & 1/3 & 0 & 0 \\
0 & 0 & 0 & 0 & 0 & 1 \\
0 & 0 & 0 & 0 & 1 & 0 \\
1/6 & 1/6 & 1/6 & 2/6 & 0 & 1/6 \\
1/4 & 0 & 1/8 & 1/8 & 1/2 & 0
\end{pmatrix}
\end{array}.
$$

We compute the period of the states, using three methods in Code 2.8.8. By the Methods 1 and 2, all the states of the Markov chain have the same period as 1. Observe that the Markov chain is irreducible. Thus, all the states are aperiodic. In Method 3, we note that all the elements of P^n are positive for all $n \geq 10$, which again implies that all the states are aperiodic.

The next example is similar to Example 2.7.6. We find the period of the states using Code 2.8.8. □

Example 2.7.8 Suppose $\{X_n, n \geq 0\}$ is a Markov chain with state space $S = \{1, 2, 3, 4\}$ and transition probability matrix P, as given below

$$P = \begin{matrix} & \begin{matrix} 1 & 2 & 3 & 4 \end{matrix} \\ \begin{matrix} 1 \\ 2 \\ 3 \\ 4 \end{matrix} & \begin{pmatrix} 0 & 1 & 0 & 0 \\ 3/4 & 0 & 1/4 & 0 \\ 0 & 2/3 & 0 & 1/3 \\ 0 & 0 & 1 & 0 \end{pmatrix} \end{matrix}.$$

Suppose $\{Y_n, n \geq 0\}$ is a Markov chain with state space $S = \{1, 2, 3, 4\}$ and transition probability matrix P^2. We compute the period of the states of both the Markov chains, using Code 2.8.8. We note that all the states of the Markov chain with transition probability matrix P and P^2 have the same period as 2 and 1, respectively. Thus, the Markov chain with transition probability matrix P is periodic with period 2. The Markov chain with one step transition probability matrix P^2 is aperiodic, and it is a reducible chain. From the output of Method 2 in the code, we get the same results. By Method 3, diagonal elements of only even powers of P are positive, which implies that all states have period 2, as in Example 2.7.6. Further, P^4,P^{10} and P^{16} have the same pattern of non-zero elements, and it continues for all even powers of P. Similarly, P^7, P^{13} and P^{19} have the same pattern of non-zero elements and it continues for all odd powers of P, supporting the result that the period of the chain is 2. Further, $(P^2)^n$ remain the same for $n \geq 7$; it is a reducible chain with period 1 for each state. □

Remark 2.7.4 Example 2.7.8 illustrates some features of periodicity, similar to those in Example 2.7.6. If we observe the transition probability matrices in Example 2.7.8 and in Example 2.7.6, the pattern of non-zero elements is the same in both the matrices. Thus, the value of the non-zero elements is immaterial in the determination of the period.

The next section presents R codes, used in some examples in Sects. 2.2–2.7.

2.8 R Codes

An important feature of the present book is the augmentation of the theory with R software as a tool for simulation and computation. To grasp the abstract concepts from the theory of Markov chains, we have adopted the approach to explain the theory along with the computations. In Sects. 2.2–2.7, we have discussed the spectral decomposition and computation of power of a matrix, computation of marginal and joint distributions, realization of the Markov chain, classification of states and periodicity. In this section, we present codes which are used to illustrate these concepts. Codes are written for a Markov chain with a specified state space and transition probability matrix. All the codes are written in parts, and each part is specific to a particular type of computation. Part I of each code is to input the transition probability matrix. To use these codes for other Markov chains, we need to change the statements in part I related to the state space and the transition probability matrix appropriately.

It is to be noted that in some cases, in view of the accumulation of computational errors, row sums of powers of P may not be exactly equal to 1. In the identity matrix, the diagonal elements may not be exactly equal to 1 but are close to 1; similarly, the off-diagonal elements may not be exactly equal to 0 but are close to 0.

Code 2.8.1 Spectral decomposition and power of a matrix: In this R code, we obtain the spectral decomposition of a transition probability matrix and use it to compute P^n. We compute P^n using matrix multiplication from base package, expm package, matrixcalc package for a specific value of n and a function to find powers of P for more than one value of n. For illustration, we consider a Markov chain with three states:

```
# Part I: Input the tpm
ns=3; states=1:ns
r1=c(.5,.4,.1); r2=c(.2,.4,.4); r3=c(.4,.4,.2)
P=rbind(r1,r2,r3);P # P is the t.p.m.
# Part II Spectral decomposition
# Following functions give eigen values in D and
# right eigen vectors in R
e1=eigen(P)$values;D=diag(e1);D=round(D,4); D
R=eigen(P)$vectors; R # matrix of right eigen vectors
# Following functions give left eigen vectors in L
P1=t(P) # P1: Transpose of P
L=eigen(P1)$vectors; L # matrix of left eigen vectors of P
# The commented lines below gives normalized left and right
# eigen vectors and not necessary for computation of P^n.
# det(R); det(L)
# B1=R%*%D%*%solve(R);solve(R) computes inverse of R
# B1=round(B1,4);B1
# B2=solve(t(L))%*%D%*%t(L);B2=round(B2,4);B2 # B1=B2=P
```

```
# B3=t(L)%*%R;B3=round(B3,4);B3 # diagonal matrix
# B4=diag(B3);B4
# V=cbind(R[,1]/sqrt(B4[1]),R[,2]/sqrt(B4[2]),R[,3]/sqrt(B4[3]))
# U=cbind(L[,1]/sqrt(B4[1]),L[,2]/sqrt(B4[2]),L[,3]/sqrt(B4[3]))
# B5=t(U)%*%V; B5=round(B5,4);B5 # Identity matrix
# B6=V%*%D%*%t(U);B6=round(B6,4);B6
# B7=U%*%D%*%t(V);B7=round(B7,4);B7 # B6=B7=P
# To find P^n (n=5) by using normalized eigen vectors
# P5=V%*%D5%*%solve(V); P5; P5=R%*%D5%*%t(U);P5
# Part III: To find P^n (n=5) by using right eigen vectors
D5=diag(e1^5);D5
P5=R%*%D5%*%solve(R);P5=round(P5,4)
# Part IV: Matrix multiplication to compute P^5
P2=P%*%P; P4=P2%*%P2; P5=P4%*%P; P5 # Matrix multiplication
# Part V:Function from expm library to compute P^5
library(expm)
P5=P%^%5;P5=round(P5,4);P5
# Part VI: Function from matrixcalc library to compute P^5
library(matrixcalc)
P5=matrix.power(P,5);P5=round(P5,4);P5
# Part VII: Function to find P^n for specified n
power=function(n)
{
power=round(P%^%n,4)
return(power)
}
N=seq(2,11,3);N; Pn=list()
for(n in N)
{
 Pn[[n]]=power(n)
}
Pn[N] # Print P^n for n in N
```

□

Code 2.8.2 Marginal distribution: In this code, we present two approaches to compute a marginal distribution of X_n, where $\{X_n, n \geq 0\}$ is a Markov chain specified by $(S, \underline{p}^{(0)}, P)$. (i) In the first approach, the code is based on Eq. 2.2.5, to compute the marginal distribution of X_n for consecutive values of n. (ii) In the second approach, we use Eq. 2.2.4, to write a function to compute marginal distribution of X_n for any specified value of n, not necessarily consecutive. We have illustrated the code for the Markov chain in Example 2.2.5. In the first approach, we computed the marginal distribution of X_n and used it to compute the expected expenses on the nth day for

$n = 1, 2, 3, 4, 5, 6$, where day 1 is taken as the initial day. In the second approach, we have found the marginal distributions of X_4, X_5, X_9, X_{15}:

```
# Part I: Input the t.p.m
ns=3; states=1:ns; r1=c(0.92,0.05,0.03)
r2=c(0,0.76,0.24); r3=c(0,0,1);P=rbind(r1,r2,r3)
# Input initial distribution and cost vector
id = c(1,0,0) # id: initial distribution
cst=c(1200,5000,1000)# Cost vector, if any
# Part II: To obtain marginal distributions of
# X_0,..., X_(upto-1)
upto=6; library(expm)
M=matrix(nrow=upto,ncol=ns); cost=matrix(nrow=upto,ncol=ns)
for(m in 1:upto)
{
M[m,]=id%*%(P%^%(m-1))
cost[m,]= M[m,]*cst
}
row.names(M)=c("X0","X1","X2","X3","X4","X5")
colnames(M)=c("H","C","D"); M=round(M,4); M
row.names(cost)=c("Day1","Day2","Day3","Day4","Day5","Day6")
colnames(cost)=c("H","C","D"); cost=round(cost,2); cost
totalcost=apply(cost,1,sum); totalcost
# Part III: Function mdxm to obtain the marginal distribution
mdxm=function(m,P,a)
{
pj=c() # pj: contains the marginal distribution
for(j in 1:ns)
{
sum=0
for(i in 1:ns)
{
sum=sum+a[i]*(P%^%m)[i,j]
}
pj[j]=sum
}
return(pj)
}
# Using the function mdxm to find marginal
# distributions of X0,X4,X5,X9,X15
u=c(1,4,5,9,15);lnt=length(u); Mrg=matrix(nrow=lnt,ncol=ns)
for(i in 1:lnt)
```

```
{
k=u[i]
Mrg[i,]=mdxm(k,P,id)
}
Mrg=round(Mrg,4)
row.names(Mrg)=c("X1","X4","X5","X9","X15")
colnames(Mrg)=c("H","C","D"); Mrg
```

□

Code 2.8.3 Joint distribution: This code presents a function to find a member from the family of finite dimensional distribution functions for the given Markov chain. We have illustrated it by finding the joint distribution of $\{X_3, X_6, X_9\}$ for the weather model in Example 2.2.6. Thus, we find $P[X_3 = i, X_6 = j, X_9 = k]$, $\forall\ i, j, k \in S$. We write a function to compute $P[X_r = i, X_s = j, X_t = k]$, corresponding to the given initial distribution and the transition probability matrix. The state space of the Markov chain in weather model has 3 states. Hence, the number of possible triplets (i, j, k) is $3^3 = 27$. We present three approaches to display all possible permutations. In the first approach, we create a data frame using `rep` function, and in the second we use a package `gtools` (Gregory et al. [11]) to get permutations. In the third approach, we use a function `expand.grid` from `expm` package:

```
# Part I: Input the tpm
ns=3; states=1:3; r1=c(0.5,0.3,0.2); r2=c(0.5,0.2,0.3)
r3=c(0.4,0.5,0.1); P=rbind(r1,r2,r3) # P: tpm
id=c(.4,0.3,0.3) # id: initial distribution
# Part II: Function to compute P[X_r=i,X_s=j,X_t=k]
jointf=function(P,a,r,s,t,i,j,k)
{
v=(P%^%(t-s))[j,k]*(P%^%(s-r))[i,j]*(a%*%(P%^%r))[i]
return(v)
}
# Part III:Approach I to compute all permutations of i,j,k
d1=rep(1:3,each=9);d2=c(rep(rep(1:3,each=3),3))
d3=rep(1:3,9); d=cbind(d1,d2,d3) # or
d=data.frame(d1,d2,d3)
d # data frame for 27 permutations of i,j,k=1,2,3
library(expm); jp=c(); lnt=ns^3; lnt
for(i in 1:lnt)
{
jp[i]=jointf(P,id,3,6,9,d[i,1],d[i,2],d[i,3])
}
sum(jp); jp1=round(jp,4); jp2=cbind(d,jp1)
colnames(jp2)=c("X3","X6","X9","Probability");jp2
# Part IV: Approach II to find all permutations using
# function permutations() from gtools library
library(gtools)
```

```
perm=permutations(ns,3,v=states,repeats.allowed=TRUE)
for(i in 1:lnt)
{
jp[i]=jointf(P,id, 3,6,9,perm[i,1],perm[i,2],perm[i,3])
}
d =cbind(perm,round(jp,4)); d
# Part V: Approach III to find all permutations using
# function expand.grid() from expm library
val=expand.grid(states,states,states); val=as.matrix(val)
for(i in 1:lnt)
{
jp[i]=jointf(P,id, 3,6,9,val[i,1],val[i,2],val[i,3])
}
d=cbind(val,round(jp,4));d
```

□

Code 2.8.4 Realization of a Markov chain: For a Markov chain specified by $(S, \underline{p}^{(0)}, P)$, we find a realization of length n using the following code. It is illustrated for the Markov chains in Example 2.2.6 (parts I, II and IV of the code) and in Example 2.1.1 (part III of the code). In parts I and II, a realization is drawn when the initial state is given. In part IV of the code, a realization is obtained when the initial distribution is specified and in this case the initial state is drawn randomly. In part III of the code two realizations are obtained corresponding to each of the two states being initial states:

```
# Realization of a weather model Markov chain
# Part I: Input the tpm
ns=3; states=1:ns; r1=c(0.5,0.3,0.2)
r2=c(0.5,0.2,0.3); r3=c(0.4,0.5,0.1); P=rbind(r1,r2,r3)
# Part II: Generate a realization of length n
inistate=1; n=120 # n: length of realization
x=c(); x[1]=inistate; set.seed(11)
# x: realization of length n
for(i in 1:(n-1))
{
x[i+1]=sample(states,1,P[x[i],],replace=T)
}
x; table(x); length(x); u=1:120; par(mfrow=c(1,1))
plot(u,x,"h",main="Weather Conditions",ylab="States",
xlab="Day",yaxt="n",col="blue")
axis(2,at=sort(unique(x)),labels=sort(unique(x)))
points(u,x,pch=20,col="dark blue")
# Part III: Another Example
# Input the tpm
ns=2; states=1:ns
```

```
r1=c(.7,.3); r2=c(.2,.8);P=rbind(r1,r2)
# Generate realization from the above P
n=11 # n: length of realization
x=matrix(nrow=n, ncol=2, byrow=TRUE)
for(j in 1:2)
{
x[1,j]=j
y=set.seed(j+1)
for(i in 1:(n-1))
{
x[i+1,j]=sample(states,1,P[x[i,j],],replace=T)
}
}
x; table(x[,1]);table(x[,2]); u=1:11
name=paste("Initial State ",x[1,j],sep=" ")
par(mfrow=c(2,1))
for(j in 1:2)
{
plot(u,x[,j],"b",main=name[j],ylab="States",xlab="Year",
yaxt="n",col="dark blue",pch=20,lwd=2)
axis(2,at=sort(unique(x[,j])),labels=sort(unique(x[,j])))
}
# Part IV: Realization from weather model when
# intial distribution pi is given
# input the tpm and the initial distribution
ns=3; states=1:ns
r1=c(0.5,0.3,0.2); r2=c(0.5,0.2,0.3); r3=c(0.4,0.5,0.1)
P=rbind(r1,r2,r3); pi=c(0.4,0.5,0.1)# pi:initial distribution
n=15; inistate= sample(states,1, pi,replace=TRUE)
x=c(); x[1]=inistate; set.seed(11)# x: realization of length n
for(i in 1:(n-1))
{
x[i+1]=sample(states,1,P[x[i],],replace=T)
}
x; table(x); u=1:15; par(mfrow= c(1,1))
plot(u,x,"h",main="Weather Conditions",ylab="States",
xlab="Day",yaxt="n",col="blue")
axis(2,at=sort(unique(x)),labels=sort(unique(x)))
points(u,x,pch=20,col="dark blue")
```

□

Code 2.8.5 Computation of maximum likelihood estimate of P: R code for the computation of maximum likelihood estimate of P corresponding to the given realization is given below. Data are obtained by generating a realization from the Markov chain in Example 2.2.6:

```
# Part I: Input the transition probability matrix.
ns=3; states=1:ns; r1=c(0.5,0.3,0.2); r2=c(0.5,0.2,0.3)
r3=c(0.4,0.5,0.1); P=rbind(r1,r2,r3)
# Part II: Generate a realization of length n
n=365; inistate=1; x=c(); # n is in days
x[1]=inistate; set.seed(11)
for(i in 1:(n-1))
{
x[i+1]=sample(states,1,P[x[i],],replace=T)
}
table(x)
# Part III: Obtain the MLE Phat
# Count the number of one step transitions
# between each pair of states
N=matrix(0,nrow=ns,ncol=ns) # matrix of transition counts
for(r in 1:(n-1))
{
for(i in 1:ns)
{
for(j in 1:ns)
{
if(x[r]==states[i] & x[r+1]==states[j])
N[i,j]=N[i,j]+1
}
}
}
N
# Obtain Phat,the MLE of P as proportions of transitions
Phat=matrix(0,nrow=ns,ncol=ns) # Estimate of TPM
rsn=rowSums(N)
for(i in 1:ns)
{
if(rsn[i]!=0) # There is at least 1 transition from state i
Phat[i,]=N[i,]/rsn[i]
}
Phat=round(Phat,4); Phat; P
```

□

Code 2.8.6 Persistent and transient states: In this R code, we adopt the two-step procedure outlined in Sect. 2.5, to examine the nature of the states of a Markov chain. In part II, we compute $f_{ii}^{(n)}, n \geq 1$ (fiin in the code) using the recurrence relation and then f_{ii} (fii in the code). We examine whether it is 1 or less than 1. In part III, we compute mean recurrence times (mu(i) in the code) for those states i for which $f_{ii} = 1$ approximately. We have illustrated the code for the Markov chain in Example 2.5.8:

```
# Part I: Input the tpm
ns=5; states=1:ns
r1=c(0,0.5,0,0.5,0); r2=c(0.3,0.2,0,0.5,0)
r3=c(0.1,0.1,0.3,0.2,0.3); r4=c(0.6,0,0,0.4,0)
r5=c(0.1,0.1,0.4,0,0.4); P=rbind(r1,r2,r3,r4,r5)
# Part II: Computation of fii using the recurrence relation
library(expm)
nmax=60; # maximum value of n for all states
# first return probabilities using recurrence relation
fiin=matrix(nrow=nmax,ncol=ns); fiin[1,]=diag(P)
for(i in 1:ns)
{
for(n in 2:nmax)
{
temp=c()
for(r in 1:(n-1))
{
temp[r]=(P%^%(n-r))[i,i]*fiin[r,i]
}
fiin[n,i]=(P%^%(n))[i,i]-sum(temp)
}
}
#View(fiin)
fii=c(); m=1:ns
fii=apply(fiin,2,sum) # To obtain the sum of fiin
d=data.frame(m,fii);d1=round(d,4);d1
prst=which(fii>.99999);prst # persistent states
# fii=1 approx
trst=states[-prst];trst # transient states
# Part III: Computation of mean recurrence times
# of persistent states
n=1:nmax; mu=c()
for(i in prst)
{
mu[i]=sum(n*fiin[,i])
}
mrt=round(data.frame(prst,mu[prst]),4)
mrt # Mean recurrence times
```

□

Code 2.8.7 Computation of f_{ij}: In this code we compute $f_{ij}^{(n)}, n \geq 1$ using the recurrence relation, f_{ij} and μ_{ij} if $f_{ij} = 1$ approximately. We have illustrated the code for the Markov chain in Example 2.6.7:

```
# Part I: input tpm
ns=6;states = 1:ns; r1=c(1/3,0,2/3,0,0,0)
r2=c(0,1/2,1/4,0,1/4,0); r3=c(2/5,0,3/5,0,0,0)
r4=c(0,1/4,1/4,1/4,0,1/4); r5=c(0,0,0,0,1/2,1/2)
r6=c(0,0,0,0,1/4,3/4); P=rbind(r1,r2,r3,r4,r5,r6)
# Part II: Computation of f_{ij}. In this part fijn is
# a matrix to store the values of f_{ij}. There are
# nmax=60 rows and 6*6=36 columns, arranged as
# f_{11},f_{12},f_{13},f_{14},f_{15},f_{16},
# f_{21}, f_{22},..., f_{61},f_{62},...,
# f_{66}. Since f_{ij}^{(1)} = p_{ij},
# first row is row wise elements of transpose of P
library(expm); nmax=60
fijn=matrix(nrow=nmax,ncol=nrow(P)*ncol(P))
fijn[1,]=t(P); fijn[1,]
k=1
for(i in 1:nrow(P))
{
for(j in 1:nrow(P))
{
for(n in 2:nmax)
{
t=c()
for(r in 1:(n-1))
{
t[r]=(P%^%(n-r))[j,j]*fijn[r,k]
}
fijn[n,k]=(P%^%(n))[i,j]-sum(t)
}
k=k+1
}
}
# tail(fijn); View(fijn)# uncomment to see values of fijn
F1=apply(fijn,2,sum) # f_{ij}
F=matrix(F1,nrow=ns,ncol=ns,byrow=TRUE)
F=round(F,4); F
M=which(F1>.9999);M
# Part III: Computation of mu_{ij} for which f_{ij}=1,
# M contains these pairs.
library(gtools)# has permutations function
mu=c(); # Array of mu_{ij} for f_{ij}=1
per=permutations(ns,2,v=states,repeats.allowed=TRUE)
per1=per[M,]; per1; F2=fijn[,M]
for(i in 1:length(M))
{
```

```
mu[i]=sum(c(1:nmax)*F2[,i])
}
d2=data.frame("i"=per1[,1],"j"=per1[,2],"mean(ij)"=mu)
d2=round(d2,4); d2
```

□

Code 2.8.8 Computation of period: This code presents three methods to compute the period of states of a Markov chain. We have used it for the Markov chain in Example 2.7.7:

```
# Part I: Input tpm
ns=6; states=1:ns
r1=c(0,0,1/3,2/3,0,0); r2=c(0,0,2/3,1/3,0,0)
r3=c(0,0,0,0,0,1); r4=c(0,0,0,0,1,0)
r5=c(1/6,1/6,1/6,2/6,0,1/6); r6=c(1/4,0,1/8,1/8,1/2,0)
P=rbind(r1,r2,r3,r4,r5,r6);P
# Part II: Function to decide gcd of a set of numbers
gcd=function(x)
{g=1; lnt=length(x)
for(j in 2:lnt)
{
a=min(x[1],x[2]); b=max(x[1],x[2])
for(i in 1:a)
{
if((a%%i==0)&&(b%%i==0))
{
g=i
}
}
x[1]=g; x[2]=x[j+1]
}
return(g)
}
x=c(24,60,15,45); gcd(x)# illustration of
# use of gcd function
# Part III: Use the above function to determine the period
period=function(M,ns)
{
count=c(); d=c()
for(i in 1:ns)
{
for(n in 1:20)
{count=append(count,n); count}
d[i]=(gcd(count))
}
```

```
return(d)
}
d1=period(P,5); d1
# Part III: A built-in funtion "gcd" in base package
library(expm)
Pn <- function(n,R,l)
# function to find rii(n) for all i and n
{
M=matrix(nrow=n,ncol=ns)# ns is the no of states
for (i in 1:n)
{
M[i,]=diag(R%^%i)
}
return(M)
}
G1=Pn(10,P,ns)
p1=c(gcd(which(G1[,1]>0)),gcd(which(G1[,2]>0)),
gcd(which(G1[,3]>0)),gcd(which(G1[,4]>0)),
gcd(which(G1[,5]>0))); p1
# Part IV: Computation of p_{ii}^n, for i=1:ns
# To verify the period found above
PowerP=list()
N=seq(1,20,3); k=1
for(r in N)
{
PowerP[[k]]=round(P%^%r,4)
k=k+1
}
list("TPM P"=PowerP)
```

□

The theory of Markov chains developed in this chapter forms a foundation for the study of the limiting behavior of a Markov chain. The next chapter is devoted to this important notion.

A quick recap of the results discussed in the present chapter is given below.

Summary

1 Suppose $\{X_n, n \geq 0\}$ is a sequence of random variables with a countable state space S. The sequence $\{X_n, n \geq 0\}$ is known as a Markov chain if $\forall\ n \geq 1$ and $\forall\ x_0, x_1, \ldots, x_{n+1} \in S,\ \forall\ t_0 = 0 < t_1 < \cdots < t_{n+1}$,

$$P[X_{t_{n+1}} = x_{n+1}|X_{t_n} = x_n, \ldots, X_0 = x_0] = P[X_{t_{n+1}} = x_{n+1}|X_{t_n} = x_n],$$

provided the conditional probabilities are defined.

2 The joint probability distribution of $\{X_0, X_1, X_2, \ldots, X_n\}$ and hence the family of finite dimensional distribution functions is completely specified by one step transition probabilities and the initial distribution.

3 Chapman-Kolmogorov equations: $\forall\ n,\ l \geq 1$,

$$p_{ij}^{(n+l)} = \sum_{k \in S} p_{ik}^{(n)}\ p_{kj}^{(l)},\ \ \forall\ i, j\ \in S\,.$$

4 n-step transition probability matrix $P^{(n)}$ is given by $P^{(n)} = P^n,\ n \geq 1$.

5 Given a realization from a Markov chain with transition probability matrix $P = [p_{ij}]$, the maximum likelihood estimator of p_{ij} is given by $\hat{p}_{ijn} = n_{ij}/n_i$, where n_{ij} is the number of transitions from i to j and n_i is the number of transitions from i in a given realization, $i, j = 1, 2, \ldots, M$.

6 A state i leads to state j, if there exists an integer $n \geq 1$ such that $p_{ij}^{(n)} > 0$.

7 A state i communicates with state j, if i leads to j and j leads to i.

8 A communicating class is a set of states such that any two states in the class communicate with each other.

9 A subset C of S is said to be a closed class of states, if for $i \in C$ and $j \notin C$, $i \nrightarrow j$. A class C is said to be a minimal closed class, if (i) C is a closed class and (ii) no proper subset of C is closed.

10 If the state space of a Markov chain is a minimal closed class then the Markov chain is known as an irreducible Markov chain, otherwise it is known as a reducible Markov chain. In an irreducible Markov chain, all states communicate with each other.

11 A closed communicating class is a minimal closed class.

12 If the class singleton $\{i\}$ is closed, then i is known as an absorbing state. The state i is an absorbing state, if and only if $p_{ii}^{(n)} = 1\ \forall\ n \geq 0$.

13 The probability f_{ii} of ever return to state i starting in i is given by $f_{ii} = \sum_{n=1}^{\infty} f_{ii}^{(n)}$, where $f_{ii}^{(n)}$ is the probability that the Markov chain returns to state i for the first time at the nth step.

14 A state i is said to be persistent or recurrent if $f_{ii} = 1$ and transient or non-recurrent if $f_{ii} < 1$.

15 A persistent state i is non-null persistent or positive recurrent if $\mu_i < \infty$ and is null persistent or null recurrent if $\mu_i = \infty$, where μ_i is the mean recurrence time of state i. Equivalently, a persistent state i is non-null persistent if $\limsup_{n\to\infty} p_{ii}^{(n)} > 0$ and is null persistent if $\limsup_{n\to\infty} p_{ii}^{(n)} = 0 \iff \lim_{n\to\infty} p_{ii}^{(n)} = 0$.

16 If i is a transient state then $\mu_i = \infty$.

17 Suppose $i \leftrightarrow j$. If i is persistent then j is also persistent, if i is transient then j is also transient, if i is null persistent then j is also null persistent and if i is non-null persistent then j is also non-null persistent.

18 An absorbing state is a non-null persistent and aperiodic state.

19 A state i is essential if i communicates with every state it leads to. A state i is inessential if there exists a state k, such that $i \rightarrow k$ but $k \nrightarrow i$.

20 In a closed communicating class, all states are essential.
21 An inessential state is a transient state. A persistent state is essential.
22 A state i is persistent if and only if $\sum_{n=1}^{\infty} p_{ii}^{(n)}$ is divergent.
23 A state i is transient if and only if $\sum_{n=1}^{\infty} p_{ii}^{(n)}$ is convergent.
24 Recurrence relation: $p_{ii}^{(n)} = f_{ii}^{(n)} + \sum_{r=1}^{n-1} p_{ii}^{(n-r)} f_{ii}^{(r)}$. It implies that $f_{ii}^{(n)} = p_{ii}^{(n)} - \sum_{r=1}^{n-1} p_{ii}^{(n-r)} f_{ii}^{(r)}$, with $f_{ii}^{(1)} = p_{ii}$.
25 The probability f_{ij} of the first visit to state j from i is given by $f_{ij} = \sum_{n=1}^{\infty} f_{ij}^{(n)}$, where $f_{ij}^{(n)}$ is the probability that the Markov chain, starting in state i, visits state j for the first time at the nth step.
26 Recurrence relation: $p_{ij}^{(n)} = f_{ij}^{(n)} + \sum_{r=1}^{n-1} p_{jj}^{(n-r)} f_{ij}^{(r)}$. It implies that $f_{ij}^{(n)} = p_{ij}^{(n)} - \sum_{r=1}^{n-1} p_{jj}^{(n-r)} f_{ij}^{(r)}$, with $f_{ij}^{(1)} = p_{ij}$.
27 Suppose $i \neq j$. Then (i) $i \to j$ if and only if $f_{ij} > 0$. (ii) $i \leftrightarrow j$ if and only if $f_{ij} f_{ji} > 0$. (iii) If $i \to j$ and i is persistent, then $f_{ji} = 1$ and $j \to i$. Further, j is also persistent. (iv) If $i \leftrightarrow j$ and j is persistent then $f_{ji} = f_{ij} = 1$.
28 Ratio theorem: For any two states i and j in S, $\lim_{N\to\infty}\left[\sum_{n=1}^{N} p_{ij}^{(n)}/(1 + \sum_{n=1}^{N} p_{jj}^{(n)})\right] = f_{ij}$.
29 (i) If i does not lead to j, then $\sum_{n=1}^{\infty} p_{ij}^{(n)} = 0$.
(ii) The series $\sum_{n=1}^{\infty} p_{ij}^{(n)}$ converges if and only if j is transient and $i \to j$.
(iii) The series $\sum_{n=1}^{\infty} p_{ij}^{(n)}$ diverges if and only if j is persistent and $i \to j$.
30 If $i \to j$, then $\lim_{n\to\infty} p_{ij}^{(n)} = 0$, if and only if j is either transient or null persistent.
31 In a finite state space Markov chain, (i) all states cannot be transient and (ii) all persistent states are non-null persistent.
32 In an irreducible Markov chain with finite state space, all the states are non-null persistent.
33 Suppose C is a closed communicating class of persistent states. Then for any transient state i which leads to $j, k \in C$, $f_{ij} = f_{ik}$. (ii) Suppose C is a single closed communicating class. If $i \notin C$ is such that $i \to j \in C$, then $f_{ij} = 1$. (iii) Suppose $C_r, r = 1, 2$ are two closed communicating classes. If $i \notin C_1 \cup C_2$ is such that $i \to j \in C_1$ but $i \nrightarrow j \in C_2$, then $f_{ij} = 1$. Similarly, if $i \notin C_1 \cup C_2$ is such that $i \to j \in C_2$ but $i \nrightarrow j \in C_1$, then $f_{ij} = 1$.
34 Suppose $\{X_n, n \geq 0\}$ is a Markov chain, with a single closed communicating class C of persistent states and a finite class T of k transient states. Then probability of absorption into C from a transient state is 1.
35 Suppose d_i is the greatest common divisor of the set $\{n | p_{ii}^{(n)} > 0,\ n > 0\}$; the set is assumed to be non-empty. If $d_i > 1$, then the state i is periodic with period d_i. If $d_i = 1$, then the state i is said to be aperiodic.
36 If $i \leftrightarrow j$, then $d_i = d_j$.
37 If each element of P^m is positive for some $m \geq 1$, then the period of all the states is 1.

2.9 Conceptual Exercises

2.9.1 Suppose $\{X_n, n \geq 0\}$ is a sequence of independent and identically distributed random variables with $P[X_n = i] = a_i,\ a_i > 0$ and $\sum_{i \in S} a_i = 1$. Examine whether $\{X_n, n \geq 0\}$ is a Markov chain. Classify the states.

2.9.2 Suppose $\{X_n, n \geq 0\}$ is a sequence of independent and identically distributed random variables with $S = I$ as a set of possible values. Examine whether a sequence $\{Y_n, n \geq 0\}$ defined as $Y_n = X_0 - X_1 + X_2 - \cdots + (-1)^n X_n$ is a Markov chain.

2.9.3 Suppose $\{Y_n, n \geq 1\}$ is a sequence of independent and identically distributed random variables with state space $\{0, 1, 2, 3\}$ and with respective probabilities $\{0.1, 0.3, 0.2, 0.4\}$. (i) Suppose $X_n = \min\{Y_1, Y_2, \ldots, Y_n\}$. Examine whether $\{X_n, n \geq 1\}$ is a Markov chain. If yes, determine its state space and the transition probability matrix. Determine the nature of the states in all respects. (ii) It has been shown in Sect. 2.1 that if $X_0 = 0$ and $X_n = \max\{Y_1, Y_2, \ldots, Y_n\}$, then $\{X_n, n \geq 0\}$ is a Markov chain. Determine the nature of the states in all respects.

2.9.4 Prove or disprove: Product of two stochastic matrices is a stochastic matrix.

2.9.5 Suppose a student takes admission to a course to be completed in four semesters. At the end of the ith semester, depending on the performance during the semester, a student either proceeds to the next semester with probability p_i or quits the course with probability q_i and remains in the same semester with probability $1 - p_i - q_i$, $i = 1, 2, 3, 4$. Assuming that the movement among the semesters can be modeled by a Markov chain, find the one step transition probability matrix.

2.9.6 Suppose the transitions among the states in a care center are governed by a homogeneous Markov chain with state space $S = \{1, 2, 3\}$, where 1 denotes healthy state, 2 denotes critically ill state and 3 stands for death. Suppose the transition probability matrix P is as given below, where time unit is taken as a day.

$$P = \begin{array}{c} \\ 1 \\ 2 \\ 3 \end{array} \begin{array}{c} \begin{array}{ccc} 1 & 2 & 3 \end{array} \\ \begin{pmatrix} 0.92 & 0.05 & 0.03 \\ 0.00 & 0.76 & 0.24 \\ 0 & 0 & 1 \end{pmatrix} \end{array}.$$

Compute the probability that a healthy individual admitted to the center on day 1 (i) remains healthy for the next 6 days, (ii) is critically ill for the first time on day 5, (iii) is critically ill on day 5 and (iv) is critically ill for the first time on day 6 and dies on day 9.

2.9.7 Weather in a city is classified as sunny, cloudy and rainy and the weather condition is modeled as a Markov chain $\{X_n, n \geq 0\}$, where X_n is defined as follows:

$$X_n = \begin{cases} 1, & \text{if nth day is sunny} \\ 2, & \text{if nth day is cloudy} \\ 3, & \text{if nth day is rainy.} \end{cases}$$

Further, the one step transition probability matrix P is given by

$$P = \begin{matrix} & \begin{matrix} 1 & 2 & 3 \end{matrix} \\ \begin{matrix} 1 \\ 2 \\ 3 \end{matrix} & \begin{pmatrix} 0.4 & 0.4 & 0.2 \\ 0.6 & 0.2 & 0.2 \\ 0.5 & 0.4 & 0.1 \end{pmatrix} \end{matrix}.$$

(i) Find the probability that the weather is cloudy for second, third and fourth days, given that the initial day is sunny. (ii) Find the probability that day 2 is sunny, day 3 is cloudy and day 4 is rainy given that day 1 is sunny.

2.9.8 Operating condition of a machine at any time is classified as follows: State 1: Good; State 2: Deteriorated but operating; State 3: In repair. We observe the condition of the machine at 6 : 00 pm every day. Suppose X_n denotes the state of the machine on the nth day for $n = 1, 2, \ldots$. We assume that the sequence of machine conditions is a Markov chain with transition probability matrix P as given below

$$P = \begin{matrix} & \begin{matrix} 1 & 2 & 3 \end{matrix} \\ \begin{matrix} 1 \\ 2 \\ 3 \end{matrix} & \begin{pmatrix} 0.9 & 0.1 & 0 \\ 0 & 0.9 & 0.1 \\ 1 & 0 & 0 \end{pmatrix} \end{matrix}.$$

Find the probability that the machine is in good condition on day 5 given that it is in good condition on day 1.

2.9.9 Suppose $\{X_n, n \geq 0\}$ is a Markov chain as defined in Exercise 2.9.7, with the initial distribution $\underline{p}^{(0)} = (1/3, 1/3, 1/3)'$. Suppose 0.44, 0.81, 0.34, 0.56, 0.18, 0.62 is a random sample of size 6 from uniform $U(0, 1)$ distribution. Using this random sample, in the given order, find X_n for $n = 0, 1, \ldots, 5$.

2.9.10 Suppose $\{X_n, n \geq 0\}$ is a time homogeneous Markov chain. Find $P[X_0 = i | X_n = j]$.

2.9.11 Show that for any two states i and j, $\sum_{n\geq 1} p_{ij}^{(n)} \geq f_{ij}$. Show that for a persistent state i such that $i \leftrightarrow j$, $\sum_{n\geq 1} p_{ij}^{(n)} \geq 1$.

2.9.12 If state j is transient, prove that for any state i, $\sum_{n=1}^{\infty} p_{ij}^{(n)} < 1/(1 - f_{jj})$.

2.9.13 Suppose $\{X_n, n \geq 0\}$ is a Markov chain with state space $S = \{1, 2, 3, 4\}$ and P given by

$$P = \begin{matrix} & \begin{matrix} 1 & 2 & 3 & 4 \end{matrix} \\ \begin{matrix} 1 \\ 2 \\ 3 \\ 3 \end{matrix} & \begin{pmatrix} 1/3 & 2/3 & 0 & 0 \\ 1/4 & 3/4 & 0 & 0 \\ 1/3 & 1/3 & 1/6 & 1/6 \\ 1/4 & 1/4 & 1/4 & 1/4 \end{pmatrix} \end{matrix}.$$

(i) Examine which states communicate with each other. (ii) What are the communicating classes of states? (iii) Which are the closed classes? (iv) Does a minimal closed class exist? If yes, specify. (v) Is the Markov chain reducible or irreducible? Justify. (vi) Decide the nature of the states. (vii) Find f_{11} and μ_1.

2.9.14 Suppose $\{X_n, n \geq 0\}$ is a Markov chain with state space $S = \{1, 2, 3, 4\}$ and P given by

$$P = \begin{matrix} & \begin{matrix} 1 & 2 & 3 & 4 \end{matrix} \\ \begin{matrix} 1 \\ 2 \\ 3 \\ 3 \end{matrix} & \begin{pmatrix} 1/3 & 2/3 & 0 & 0 \\ 1/4 & 3/4 & 0 & 0 \\ 1/3 & 1/3 & 1/3 & 0 \\ 1/4 & 1/4 & 1/4 & 1/4 \end{pmatrix} \end{matrix}.$$

(i) Examine which states communicate with each other. (ii) What are the communicating classes of states? (iii) Which are the closed classes? (iv) Does a minimal closed class exist? If yes, specify. (v) Is the Markov chain reducible or irreducible? Justify. (vi) Decide the nature of the states.

2.9.15 For a Markov chain $\{X_n, n \geq 0\}$ with state space $S = \{1, 2, 3\}$, the transition probability matrix P is given by

$$P = \begin{matrix} & \begin{matrix} 1 & 2 & 3 \end{matrix} \\ \begin{matrix} 1 \\ 2 \\ 3 \end{matrix} & \begin{pmatrix} 0 & 1 & 0 \\ 1 & 0 & 0 \\ 1/3 & 1/3 & 1/3 \end{pmatrix} \end{matrix}.$$

(i) Find the period of each state. (ii) Classify the states. (iii) Find f_{33}. (iv) Find the probability of absorption from a transient state. Verify Theorem 2.6.10.

2.9.16 Suppose $\{X_n, n \geq 0\}$ is a Markov chain with state space $S = \{1, 2, 3, 4\}$ and P given by

$$P = \begin{matrix} & \begin{matrix} 1 & 2 & 3 & 4 \end{matrix} \\ \begin{matrix} 1 \\ 2 \\ 3 \\ 4 \end{matrix} & \begin{pmatrix} 0 & 1 & 0 & 0 \\ 1 & 0 & 0 & 0 \\ 1/8 & 1/2 & 1/8 & 1/4 \\ 1/3 & 1/6 & 1/6 & 1/3 \end{pmatrix} \end{matrix}.$$

(i) Find the period of each state. (ii) Classify the states. (iii) Find f_{33}, f_{44}. (iv) Find the probability of absorption from transient states.

2.9.17 Suppose $\{X_n, n \geq 0\}$ is a Markov chain with state space $S = \{1, 2, 3, 4, 5, 6\}$ and P given by

$$P = \begin{array}{c} \\ 1 \\ 2 \\ 3 \\ 4 \\ 5 \\ 6 \end{array} \begin{array}{c} \begin{array}{cccccc} 1 & 2 & 3 & 4 & 5 & 6 \end{array} \\ \begin{pmatrix} 1/4 & 0 & 3/4 & 0 & 0 & 0 \\ 0 & 1/3 & 1/3 & 0 & 1/3 & 0 \\ 2/7 & 0 & 5/7 & 0 & 0 & 0 \\ 0 & 1/4 & 1/4 & 1/4 & 0 & 1/4 \\ 0 & 0 & 0 & 0 & 4/9 & 5/9 \\ 0 & 0 & 0 & 0 & 1/3 & 2/3 \end{pmatrix} \end{array}.$$

(i) Identify the communicating classes and the closed classes. (ii) Is the Markov chain reducible or irreducible? Justify your answer. (iii) Classify the states as transient, null persistent and non-null persistent. (iv) Find the period of each state.

2.9.18 Suppose the transition probability matrix P of a Markov chain is as given below

$$P = \begin{array}{c} \\ 1 \\ 2 \\ 3 \\ 4 \\ 5 \\ 6 \end{array} \begin{array}{c} \begin{array}{cccccc} 1 & 2 & 3 & 4 & 5 & 6 \end{array} \\ \begin{pmatrix} 2/3 & 0 & 1/3 & 0 & 0 & 0 \\ 1/4 & 1/4 & 1/2 & 0 & 0 & 0 \\ 3/5 & 0 & 2/5 & 0 & 0 & 0 \\ 1/8 & 1/2 & 1/4 & 1/8 & 0 & 0 \\ 0 & 0 & 0 & 0 & 1/3 & 2/3 \\ 0 & 0 & 0 & 0 & 1/2 & 1/2 \end{pmatrix} \end{array}.$$

It is given that $f_{22} = 0.25$, $f_{42} = 0.5714$ and $f_{44} = 0.125$. Identify the remaining elements in $F = [f_{ij}]$ without computing. Use all the relevant results. Justify your answers. Find the probability of absorption from transient states to a class of persistent states.

2.10 Computational Exercises

2.10.1 Suppose $\{X_n, n \geq 0\}$ is a Markov chain with state space $S = \{1, 2, 3, 4\}$ and P given by

$$P = \begin{array}{c} \\ 1 \\ 2 \\ 3 \\ 4 \end{array} \begin{array}{c} \begin{array}{cccc} 1 & 2 & 3 & 4 \end{array} \\ \begin{pmatrix} 0.1 & 0.4 & 0.3 & 0.2 \\ 0.4 & 0.2 & 0.2 & 0.2 \\ 0.4 & 0.4 & 0.1 & 0.1 \\ 0.3 & 0.2 & 0.2 & 0.3 \end{pmatrix} \end{array}.$$

Verify all the results related to the spectral decomposition of P, as in Example 2.2.2.

2.10.2 Suppose $\{X_n, n \geq 0\}$ is a Markov chain with state space $S = \{1, 2, 3, 4\}$ and P given by

$$P = \begin{array}{c} \\ 1 \\ 2 \\ 3 \\ 4 \end{array}\begin{array}{c} \begin{array}{cccc} 1 & 2 & 3 & 4 \end{array} \\ \begin{pmatrix} 0 & 1/5 & 3/5 & 1/5 \\ 1/4 & 1/4 & 1/4 & 1/4 \\ 1 & 0 & 0 & 0 \\ 0 & 1/2 & 1/2 & 0 \end{pmatrix} \end{array}.$$

(i) Taking a suitable initial distribution, find the marginal distribution of X_4, X_9 and X_{12}. (ii) Taking a suitable initial distribution, find the joint distribution of $\{X_3, X_5, X_9\}$.

2.10.3 From purchase to purchase, a particular customer switches brands of tea among products labeled as 1, 2 and 3 according to a Markov chain whose transition probability matrix is

$$P = \begin{array}{c} \\ 1 \\ 2 \\ 3 \end{array}\begin{array}{c} \begin{array}{ccc} 1 & 2 & 3 \end{array} \\ \begin{pmatrix} 0.6 & 0.2 & 0.2 \\ 0.1 & 0.7 & 0.2 \\ 0.1 & 0.1 & 0.8 \end{pmatrix} \end{array}.$$

(i) Taking a suitable initial distribution, find a realization of the Markov chain for 50 purchases. (ii) Draw a plot of the realization. (iii) What is a proportion of purchases of three products in your realization? (iv) On the basis of the realization, find the maximum likelihood estimate of P. Compare the estimate with P.

2.10.4 Suppose $\{X_n, n \geq 0\}$ is a Markov chain with state space $S = \{1, 2, 3, 4\}$ and P as given below

$$P = \begin{array}{c} \\ 1 \\ 2 \\ 3 \\ 4 \\ 5 \end{array}\begin{array}{c} \begin{array}{ccccc} 1 & 2 & 3 & 4 & 5 \end{array} \\ \begin{pmatrix} 0 & 0.4 & 0 & 0.6 & 0 \\ 0.3 & 0.3 & 0 & 0.4 & 0 \\ 0.1 & 0.2 & 0.3 & 0.2 & 0.2 \\ 0.7 & 0 & 0 & 0.3 & 0 \\ 0.1 & 0.2 & 0.4 & 0 & 0.3 \end{pmatrix} \end{array}.$$

(i) Find f_{ii} $\forall\, i \in S$ and hence identify which states are transient and which are persistent.
(ii) Find the mean recurrence time for persistent states and hence classify persistent states as non-null persistent and null persistent.

2.10.5 Suppose $\{X_n, n \geq 0\}$ is a Markov chain with state space $S = \{1, 2, 3, 4\}$ and P given by

$$P = \begin{array}{c} \\ 1 \\ 2 \\ 3 \\ 4 \end{array} \begin{array}{c} \begin{array}{cccc} 1 & 2 & 3 & 4 \end{array} \\ \begin{pmatrix} 0 & 1 & 0 & 0 \\ 2/3 & 0 & 1/3 & 0 \\ 0 & 3/4 & 0 & 1/4 \\ 0 & 0 & 1 & 0 \end{pmatrix} \end{array}.$$

(i) Compute the period of a state using the following three ways: (a) Write a function to compute the g.c.d., (b) use the built-in function `gcd` in the base package to find the g.c.d. and (c) find powers of P and decide the period by observing the elements of P^n. (ii) Suppose $\{Y_n, n \geq 0\}$ is a Markov chain with state space $S = \{1, 2, 3, 4\}$ and transition probability matrix P^2. Find the period of each state. (iii) Comment on the results obtained in (i) and (ii).

2.10.6 Suppose $\{X_n, n \geq 0\}$ is a Markov chain with state space $S = \{1, 2, 3, 4, 5\}$ with P given by

$$P = \begin{array}{c} \\ 1 \\ 2 \\ 3 \\ 4 \\ 5 \end{array} \begin{array}{c} \begin{array}{ccccc} 1 & 2 & 3 & 4 & 5 \end{array} \\ \begin{pmatrix} 0 & 1 & 0 & 0 & 0 \\ 0 & 0 & 1 & 0 & 0 \\ 1 & 0 & 0 & 0 & 0 \\ 1/10 & 1/5 & 2/5 & 1/10 & 1/5 \\ 1/5 & 1/5 & 1/5 & 1/5 & 1/5 \end{pmatrix} \end{array}.$$

Find $F = [f_{ij}]$. Find the probability of absorption into a class of persistent states from the transient states. Comment on the results.

2.10.7 Suppose the transition probability matrix P of a Markov chain is as given below

$$P = \begin{array}{c} \\ 1 \\ 2 \\ 3 \\ 4 \\ 5 \\ 6 \\ 7 \\ 8 \end{array} \begin{array}{c} \begin{array}{cccccccc} 1 & 2 & 3 & 4 & 5 & 6 & 7 & 8 \end{array} \\ \begin{pmatrix} 0.4 & 0.5 & 0.1 & 0 & 0 & 0 & 0 & 0 \\ 0.3 & 0.2 & 0.5 & 0 & 0 & 0 & 0 & 0 \\ 0 & 0.5 & 0.5 & 0 & 0 & 0 & 0 & 0 \\ 0 & 0 & 0 & 0.2 & 0.8 & 0 & 0 & 0 \\ 0 & 0 & 0 & 0.6 & 0.4 & 0 & 0 & 0 \\ 0.1 & 0.1 & 0.2 & 0.1 & 0.1 & 0.1 & 0.2 & 0.1 \\ 0.2 & 0.2 & 0.2 & 0.1 & 0.1 & 0 & 0.1 & 0.1 \\ 0.1 & 0 & 0.1 & 0.1 & 0.1 & 0.2 & 0.2 & 0.2 \end{pmatrix} \end{array}.$$

Find the matrix G of absorption probabilities. Find $F = [f_{ij}]$. Find $F_T = [f_{ij}]_{i,j \in T}$, based on the convergent series, where T is a class of transient states. Compare the values with those in F. Verify all the results in Theorem 2.6.11.

2.11 Multiple Choice Questions

Note: In each of the questions, multiple options may be correct. In each question, $\{X_n, n \geq 0\}$ is a Markov chain with state space S, one step transition probability matrix $P = [p_{ij}]$ and $P^{(n)} = [p_{ij}^{(n)}]$. Some MCQs are similar, with a slight difference. These are included to show that it is possible to generate many MCQs from one MCQ.

2.11.1 Suppose $\{X_n, n \geq 0\}$ is a Markov chain with one step transition probabilities $p_{ij}^{n,n+1} = P[X_{n+1} = j|X_n = i], i, j \in S, n \geq 0$. In a homogeneous Markov chain, which of the following is/are correct?

(a) $p_{ij}^{n,n+1}$ depends on n.
(b) $p_{ij}^{n,n+1}$ does not depend on n.
(c) $P[X_{n+1} = j|X_n = i] = P[X_{n+100} = j|X_{n+99} = i] \ \forall \ n \geq 0$.
(d) $P[X_{100} = x_{100}|X_{99} = x_{99}] = P[X_8 = x_8|X_7 = x_7]$, $x_{100}, x_{99}, x_8, x_7 \in S$.

2.11.2 Which of the following is ALWAYS true?

(a) $\sum_{i \in S} p_{ij} = 1, \ \forall \ j \in S$.
(b) $\sum_{j \in S} p_{ij} = 1, \ \forall \ i \in S$.
(c) $\sum_{j \in S} p_{ij} = 1, \forall i \in S$ and $\sum_{i \in S} p_{ij} = 1, \forall j \in S$.
(d) $\sum_{j \in S} p_{ij} = 1, \forall i \in S$ or $\sum_{i \in S} p_{ij} = 1, \forall j \in S$.

2.11.3 Which of the following is/are correct? A matrix $P = [p_{ij}]$ is known as a stochastic matrix if

(a) $p_{ij} \geq 0, \ \forall \ i, j \in S \ \ and \ \ \sum_{i \in S} p_{ij} = 1, \ \forall \ j \in S$.
(b) $p_{ij} \geq 0, \ \forall \ i, j \in S \ \ and \ \ \sum_{j \in S} p_{ij} = 1, \ \forall \ i \in S$.
(c) $\sum_{i \in S} p_{ij} = 1, \ \forall \ j \in S$.
(d) $\sum_{j \in S} p_{ij} = 1, \ \forall \ i \in S$.

2.11.4 Which of the following is/are correct? A matrix $P = [p_{ij}]$ is known as a doubly stochastic matrix if

(a) $p_{ij} \geq 0, \ \forall \ i, j \in S \ \ and \ \ \sum_{i \in S} p_{ij} = 1, \ \forall \ j \in S$.
(b) $p_{ij} \geq 0, \ \forall \ i, j \in S \ \ and \ \ \sum_{j \in S} p_{ij} = 1, \ \forall \ i \in S$.
(c) $p_{ij} \geq 0, \forall i, j \in S \ and \ \sum_{j \in S} p_{ij} = 1, \forall i \in S \ and \ \sum_{i \in S} p_{ij} = 1, \forall j \in S$.
(d) $p_{ij} \geq 0, \forall i, j \in S \ and \ \sum_{j \in S} p_{ij} = 1, \forall i \in S$ or $\sum_{i \in S} p_{ij} = 1, \forall j \in S$.

2.11.5 Suppose $S = \{1, 2, 3, 4\}$ and P is given by

$$P = \begin{array}{c} \\ 1 \\ 2 \\ 3 \\ 4 \end{array} \begin{array}{c} \begin{array}{cccc} 1 & 2 & 3 & 4 \end{array} \\ \begin{pmatrix} 1/2 & 0 & 1/2 & 0 \\ 0 & 1/3 & 0 & 2/3 \\ 1/4 & 0 & 1/2 & 1/4 \\ 1/4 & 1/4 & 0 & 1/2 \end{pmatrix} \end{array}.$$

Following are two statements: (I) P is a stochastic matrix. (II) P is a doubly stochastic matrix.
Which of the following is correct?

(a) Both (I) and (II) are false
(b) Both (I) and (II) are true
(c) (I) is true but (II) is false
(d) (I) is false but (II) is true.

2.11.6 Suppose $S = \{1, 2, 3\}$ and P is given by

$$P = \begin{matrix} & \begin{matrix} 1 & 2 & 3 \end{matrix} \\ \begin{matrix} 1 \\ 2 \\ 3 \end{matrix} & \begin{pmatrix} 0.3 & \alpha & \beta \\ 0.4 & \gamma & 0.1 \\ \delta & 0.2 & \epsilon \end{pmatrix} \end{matrix}.$$

If P is a doubly stochastic matrix, the value of β

(a) cannot be computed from the given information
(b) is < 0.4
(c) is > 0.4
(d) is 0.4.

2.11.7 For a Markov chain $\{X_n, n \geq 0\}$ with state space $S = \{1, 2, 3\}$, the transition probability matrix P is given by

$$P = \begin{matrix} & \begin{matrix} 1 & 2 & 3 \end{matrix} \\ \begin{matrix} 1 \\ 2 \\ 3 \end{matrix} & \begin{pmatrix} 2/5 & 3/5 & 0 \\ 3/8 & 5/8 & 0 \\ 1/6 & 1/3 & 1/2 \end{pmatrix} \end{matrix}.$$

Suppose $X_0 = 3$. Then realized values of X_1 and X_2 corresponding to random numbers 0.36 and 0.64 from $U(0, 1)$ distribution are

(a) $X_1 = 2$ and $X_2 = 2$
(b) $X_1 = 1$ and $X_2 = 1$
(c) $X_1 = 1$ and $X_2 = 2$
(d) $X_1 = 2$ and $X_2 = 1$.

2.11.8 Following are two statements: For $i, j \in S$, i leads to j, if $p_{ij}^{(n)} > 0$ (I) for some $n \geq 1$. (II) $\forall\ n \geq 1$.
Then which of the following is a correct option?

(a) Both (I) and (II) are false
(b) Both (I) and (II) are true
(c) (I) is true but (II) is false
(d) (I) is false but (II) is true.

2.11.9 Following are two statements: For $i \neq j \in S$, if i leads to j, then (I) $p_{ij}^{(n)} > 0$ for some $n \geq 1$. (II) $f_{ij} > 0$.
Which of the following is correct?

(a) Both (I) and (II) are false
(b) Both (I) and (II) are true
(c) (I) is true but (II) is false
(d) (I) is false but (II) is true.

2.11.10 Which of the following options is/are correct? Two states i and j do not communicate if

(a) $p_{ij}^{(n)} = 0, \ \forall \ n > 0$
(b) $p_{ji}^{(n)} = 0, \ \forall \ n > 0$
(c) $p_{ij}^{(n)} = 0, \ \forall \ n > 0$ or $p_{ji}^{(n)} = 0, \ \forall \ n > 0$
(d) $p_{ij}^{(n)} = 0, \ \forall \ n > 0$ and $p_{ji}^{(n)} = 0, \ \forall \ n > 0$.

2.11.11 Following are three statements. Two states i and j do not communicate with each other if (I) $p_{ij}^{(n)} = 0, \ \forall \ n > 0$, (II) $p_{ji}^{(n)} = 0, \ \forall \ n > 0$,
(III) $p_{ij}^{(n)} = 0, \ \forall \ n > 0$ or $p_{ji}^{(n)} = 0, \ \forall \ n > 0$.
Then which of the following options is correct?

(a) Only (I) is true
(b) Only (II) is true
(c) Only (III) is true
(d) All three are true.

2.11.12 Following are two statements: (I) A closed class may not be a communicating class. (II) A communicating class may not be a closed class. Then which of the following options is correct?

(a) Both (I) and (II) are false
(b) Both (I) and (II) are true
(c) (I) is true but (II) is false
(d) (I) is false but (II) is true.

2.11.13 Following are two statements: (I) A closed class is always a communicating class. (II) A communicating class is always a closed class.
Then which of the following options is correct?

(a) Both (I) and (II) are false
(b) Both (I) and (II) are true
(c) (I) is true but (II) is false
(d) (I) is false but (II) is true.

2.11.14 Following are two statements. In an irreducible Markov chain (I) a closed class is always a communicating class, and (II) a communicating class is always a closed class.
Then which of the following options is correct?

(a) Both (I) and (II) are false
(b) Both (I) and (II) are true
(c) (I) is true but (II) is false
(d) (I) is false but (II) is true.

2.11.15 Suppose f_{ii} denotes the probability of ever return to state i and μ_i denotes the mean recurrence time. Which of the following is/are correct? A state i is a non-null persistent state if

(a) $f_{ii} < 1 \ \& \ \mu_i < \infty$
(b) $f_{ii} = 1 \ \& \ \mu_i = \infty$
(c) $f_{ii} < 1 \ \& \ \mu_i = \infty$
(d) $f_{ii} = 1 \ \& \ \mu_i < \infty$.

2.11.16 Suppose f_{ii} denotes the probability of ever return to state i and μ_i denotes the mean recurrence time. Which of the following is/are correct? A state i is a null persistent state if

(a) $f_{ii} < 1 \ \& \ \mu_i < \infty$
(b) $f_{ii} = 1 \ \& \ \mu_i = \infty$
(c) $f_{ii} < 1 \ \& \ \mu_i = \infty$
(d) $f_{ii} = 1 \ \& \ \mu_i < \infty$.

2.11.17 Which of the following is/are correct? A persistent state i is a null persistent state if as $n \to \infty$

(a) $\limsup p_{ii}^{(n)} = 0$
(b) $\lim p_{ii}^{(n)} = 0$
(c) $\liminf p_{ii}^{(n)} = 0$
(d) $\limsup p_{ii}^{(n)} > 0$.

2.11.18 Following are three statements: A persistent state i is a null persistent state if as $n \to \infty$ (I) $\limsup p_{ii}^{(n)} = 0$. (II) $\lim p_{ii}^{(n)} = 0$.
(III) $\liminf p_{ii}^{(n)} = 0$.
Then which of the following options is correct?

(a) Only (I) is are true
(b) Only (III) is true
(c) Only (II) is true
(d) Either (I) or (II) is true.

2.11.19 Which of the following options is/are correct? A persistent state i is a non-null persistent state if as $n \to \infty$

(a) $\limsup p_{ii}^{(n)} = 0$
(b) $\liminf p_{ii}^{(n)} = 0$
(c) $\liminf p_{ii}^{(n)} > 0$
(d) $\limsup p_{ii}^{(n)} > 0$.

2.11.20 Which of the following options is/are correct? A persistent state i is a non-null persistent state if as $n \to \infty$

(a) $\limsup p_{ii}^{(n)} = 0$
(b) $\liminf p_{ii}^{(n)} = 0$
(c) $\liminf p_{ii}^{(n)} > 0$
(d) $\lim p_{ii}^{(n)} > 0$.

2.11.21 Which of the following options is/are correct? A persistent state i is a non-null persistent state if as $n \to \infty$

(a) $\limsup p_{ii}^{(n)} = 0$
(b) $\liminf p_{ii}^{(n)} = 0$
(c) $\lim p_{ii}^{(n)} = 0$
(d) $\lim p_{ii}^{(n)} > 0$.

2.11.22 Which of the following options is/are correct? A persistent state i is a non-null persistent state if as $n \to \infty$

(a) $\limsup p_{ii}^{(n)} = 0$
(b) $\limsup p_{ii}^{(n)} > 0$
(c) $\liminf p_{ii}^{(n)} = 0$
(d) $\lim p_{ii}^{(n)} = 0$.

2.11.23 Suppose i is a persistent state. If $\liminf p_{ii}^{(n)} > 0$, then which of the following options is/are correct?

(a) i is a non-null persistent state
(b) i is a null persistent state
(c) i is a transient state
(d) The information is not sufficient to decide the nature of state i.

2.11.24 Suppose i is a persistent state. If $\liminf p_{ii}^{(n)} = 0$, then which of the following options is/are correct?

(a) i is a non-null persistent state
(b) i is a null persistent state
(c) i is a transient state
(d) The information is not sufficient to decide the nature of state i.

2.11.25 Suppose f_{ii} denotes the probability of ever return to state i. Which of the following options is/are correct? A state i is a transient state if

(a) $f_{ii} < 1$
(b) $f_{ii} = 1$
(c) $\sum_{n=1}^{\infty} p_{ii}^{(n)} = \infty$
(d) $\sum_{n=1}^{\infty} p_{ii}^{(n)} < \infty$.

2.11.26 Suppose f_{ii} denotes the probability of ever return to state i. Following are four statements. A state i is a transient state if (I) $f_{ii} < 1$.

(II) $f_{ii} = 1$. (III) $\sum_{n=1}^{\infty} p_{ii}^{(n)} = \infty$. (IV) $\sum_{n=1}^{\infty} p_{ii}^{(n)} < \infty$.
Then which of the following options is correct?

(a) Only (I) is true
(b) Both (I) and (III) are true
(c) Both (II) and (III) are true
(d) Both (I) and (IV) are true.

2.11.27 Suppose f_{ii} denotes the probability of ever return to state i. Following are four statements: A state i is a persistent state if (I) $f_{ii} < 1$.
(II) $f_{ii} = 1$. (III) $\sum_{n=1}^{\infty} p_{ii}^{(n)} = \infty$. (IV) $\sum_{n=1}^{\infty} p_{ii}^{(n)} < \infty$.
Then which of the following options is correct?

(a) Only (II) is true
(b) Both (I) and (III) are true
(c) Both (II) and (III) are true
(d) Both (I) and (IV) are true.

2.11.28 Suppose f_{ii} denotes the probability of ever return to state i. Which of the following options is/are correct? A state i is a persistent state if

(a) $f_{ii} < 1$
(b) $f_{ii} = 1$
(c) $\sum_{n=1}^{\infty} p_{ii}^{(n)} = \infty$
(d) $\sum_{n=1}^{\infty} p_{ii}^{(n)} < \infty$.

2.11.29 Suppose $\sum_{n=1}^{\infty} p_{22}^{(n)} = 2.45$. Which of the following options is/are correct?

(a) $f_{22} < 1$
(b) $f_{22} = 1$
(c) $\mu_2 = \infty$
(d) $\lim_{n\to\infty} p_{22}^{(n)} > 0$.

2.11.30 Suppose 3 is a transient state. Following are two statements:
(I) $\sum_{n=0}^{\infty} p_{33}^{(n)} = 2.33$. (II) $\sum_{n=0}^{\infty} p_{33}^{(n)} = 0.33$.
Which of the following options is correct?

(a) Both (I) and (II) are false
(b) Both (I) and (II) are true
(c) (I) is true but (II) is false
(d) (I) is false but (II) is true.

2.11.31 Suppose $\sum\limits_{n=1}^{\infty} p_{33}^{(n)} = 1.33$. Which of the following options is/are correct?

(a) $\limsup p_{33}^{(n)} = 0.54$
(b) $\limsup p_{33}^{(n)} = 0$
(c) $\liminf p_{33}^{(n)} = 0.34$
(d) $\liminf p_{33}^{(n)} = 0$.

2.11.32 Suppose $f_{22} = 2/3$. Following are two statements: (I) $\sum_{n=1}^{\infty} p_{22}^{(n)} = 2$. (II) $\sum_{n=1}^{\infty} p_{22}^{(n)} = 2/5$.
Which of the following options is correct?

(a) Both (I) and (II) are false
(b) Both (I) and (II) are true
(c) (I) is true but (II) is false
(d) (I) is false but (II) is true.

2.11.33 Which of the following options is/are correct? An absorbing state is a

(a) null persistent state
(b) transient state
(c) non-null persistent state
(d) persistent state.

2.11.34 Which of the following options is/are correct? An absorbing state is

(a) null persistent
(b) transient
(c) non-null persistent
(d) periodic state.

2.11.35 Following are four statements about an absorbing state: (I) An absorbing state is a null persistent state. (II) An absorbing state is a transient state. (III) An absorbing state is a non-null persistent state. (IV) An absorbing state is an aperiodic state.
Then which of the following is a correct option?

(a) Both (I) and (IV) are true
(b) Both (II) and (IV) are true
(c) Both (III) and (IV) are true
(d) Only (IV) is true.

2.11.36 Which of the following is/are correct? A state i is essential if

(a) i communicates with every state in S
(b) i leads to every state in S
(c) $i \to j \;\Rightarrow\; j \to i$
(d) i is transient.

2.11.37 Which of the following is/are correct? A state i is essential if

(a) i communicates with every state in S
(b) i leads to every state in S
(c) $i \to j \;\Rightarrow\; j \to i$
(d) i is persistent.

2.11.38 Which of the following is/are correct? A state i is essential if

(a) i is periodic

(b) i leads to every state in S
(c) $i \to j \;\Rightarrow\; j \to i$
(d) i is persistent.

2.11.39 Following are four statements: A state i is essential if (I) i communicates with every state in S. (II) i leads to every state in S. (III) $i \to j \;\Rightarrow\; j \to i$. (IV) i is non-null persistent and aperiodic.
Which of the following is correct?

(a) Only (I) is true
(b) Only (III) is true
(c) Both (I) and (III) are true
(d) Only (IV) is true.

2.11.40 Following are two statements. (I) An essential state is a persistent state. (II) An inessential state is a transient state.
Which of the following is correct?

(a) Both (I) and (II) are false
(b) Both (I) and (II) are true
(c) (I) is true but (II) is false
(d) (I) is false but (II) is true.

2.11.41 Following are two statements: (I) A persistent state is an essential state. (II) An inessential state is a transient state.
Which of the following is correct?

(a) Both (I) and (II) are false
(b) Both (I) and (II) are true
(c) (I) is true but (II) is false
(d) (I) is false but (II) is true.

2.11.42 Suppose $S = \{1, 2, 3, 4\}$ and P is given by

$$P = \begin{array}{c} \\ 1 \\ 2 \\ 3 \\ 4 \end{array} \begin{array}{c} \begin{array}{cccc} 1 & 2 & 3 & 4 \end{array} \\ \begin{pmatrix} 1/2 & 0 & 1/2 & 0 \\ 0 & 1/3 & 0 & 2/3 \\ 2/3 & 0 & 1/3 & 0 \\ 0 & 1/4 & 0 & 3/4 \end{pmatrix} \end{array}.$$

Which of the following options is correct?

(a) States 1 and 3 are essential states
(b) States 1 and 3 are inessential states
(c) States 1 and 3 are transient states
(d) States 1 and 3 are null persistent states.

2.11.43 Suppose $S = \{1, 2, 3, 4\}$ and P is given by

$$P = \begin{array}{c} \\ 1 \\ 2 \\ 3 \\ 4 \end{array} \begin{array}{c} \begin{array}{cccc} 1 & 2 & 3 & 4 \end{array} \\ \begin{pmatrix} 1/2 & 0 & 1/2 & 0 \\ 2/3 & 1/3 & 0 & 0 \\ 1/4 & 1/4 & 1/2 & 0 \\ 1/4 & 1/4 & 1/4 & 1/4 \end{pmatrix} \end{array}.$$

Following are three statements: (I) States 1 and 2 are essential states. (II) State 4 is a transient state. (III) State 3 is a persistent state. Which of the following options is correct?

(a) (I) and (III) are true
(b) (I) is false but (II) is true
(c) All are true
(d) (II) and (III) are true.

2.11.44 Which of the following is/are correct? For $i, \ j \in S$ if

(a) $i \to j$ and i is persistent, then j is also persistent.
(b) $i \to j$ and i is transient, then j is also transient.
(c) $i \leftrightarrow j$ and i is transient then j is also transient.
(d) $i \leftrightarrow j$ and i is persistent then j is also persistent.

2.11.45 Which of the following is/are correct? For $i, \ j \in S$ if

(a) $i \to j$ and i is persistent, then j is also persistent.
(b) $i \to j$ and i is transient, but j may not be transient.
(c) $i \leftrightarrow j$ and i is transient then j is also transient.
(d) $i \leftrightarrow j$ and i is persistent but j is not persistent.

2.11.46 Which of the following is/are correct? For $i, \ j \in S$ if

(a) $i \to j$ and i is non-null persistent, then j is also non-null persistent.
(b) $i \to j$ and i is transient, then j is also transient.
(c) $i \leftrightarrow j$ and i is transient then j is also transient.
(d) $i \leftrightarrow j$ and i is non-null persistent then j is also non-null persistent.

2.11.47 Which of the following is/are correct? For $i, \ j \in S$ if

(a) $i \to j$ and i is null persistent, then j is also null persistent.
(b) $i \to j$ and i is transient, then j is also transient.
(c) $i \leftrightarrow j$ and i is transient but j is not transient.
(d) $i \leftrightarrow j$ and i is null persistent then j is also null persistent.

2.11.48 Suppose f_{ij} denotes the probability of the first visit from i to j. Following are three statements: (I) $i \to j \ \Rightarrow \ f_{ij} > 0$. (II) $i \leftrightarrow j \ \Rightarrow \ f_{ji} > 0$. (III) $i \leftrightarrow j \ \Rightarrow \ f_{ij} f_{ji} > 0$.
Which of the following is correct?

(a) Only (I) is true
(b) Only (II) is true

(c) Only (III) is true
(d) All are true.

2.11.49 Suppose f_{ij} denotes the probability of the first visit from i to j. Which of the following is/are correct?

(a) $i \to j \;\Rightarrow\; f_{ij} > 0$
(b) $i \to j \;\Rightarrow\; f_{ji} > 0$
(c) $i \to j \iff f_{ij} f_{ji} > 0$
(d) $i \to j$ and i is persistent then $f_{ij} = 1$.

2.11.50 Following are three statements: (I) For any state i, $f_{ii}^{(n)} \le p_{ii}^{(n)} \; \forall \, n \ge 1$. (II) For any state i, $\sum_{n=1}^{\infty} p_{ii}^{(n)} \ge 1$. (II) For a persistent state i, $\sum_{n=1}^{\infty} p_{ii}^{(n)} \ge 1$. Which of the following is correct?

(a) All three are true
(b) (I) and (II) both are true
(c) (I) and (III) both are true
(d) Only (I) is true.

2.11.51 Following are three statements. (I) For any state i, $\mu_i \ge f_{ii}$. (II) For a persistent state i, $\mu_i \ge 1$. (III) For a transient state i, $\mu_i < 1$.
Which of the following is correct?

(a) All three are true
(b) (I) and (II) both are true
(c) (I) and (III) both are true
(d) Only (I) is true.

2.11.52 Suppose $i \to j$. Then which of the following is/are correct?

(a) If the series $\sum_{n=1}^{\infty} p_{ij}^{(n)}$ is convergent, then j is transient
(b) The series $\sum_{n=1}^{\infty} p_{ij}^{(n)}$ is convergent if i is transient
(c) j is transient implies that $\sum_{n=1}^{\infty} p_{ij}^{(n)}$ is convergent
(d) i is transient implies that $\sum_{n=1}^{\infty} p_{ij}^{(n)}$ is convergent.

2.11.53 Suppose $i \to j$. Following are four statements: (I) If the series $\sum_{n=1}^{\infty} p_{ij}^{(n)}$ is divergent, then j is persistent. (II) The series $\sum_{n=1}^{\infty} p_{ij}^{(n)}$ is divergent if i is persistent. (III) j is persistent implies that $\sum_{n=1}^{\infty} p_{ij}^{(n)}$ is divergent. (IV) i is persistent implies that $\sum_{n=1}^{\infty} p_{ij}^{(n)}$ is divergent.
Which of the following is correct?

(a) Only (I) is true
(b) Only (III) is true
(c) Both (I) and (III) are true
(d) Both (II) and (IV) are true.

2.11.54 Suppose $i \to j$. Following are three statements. (I) The series $\sum_{n=1}^{\infty} p_{ij}^{(n)}$ is convergent.(II) The series $\sum_{n=1}^{\infty} p_{jj}^{(n)}$ is convergent. (III) $f_{jj} < 1$. If the state j is transient, then which of the following options is correct?

(a) Only (II) is true
(b) Only (III) is true
(c) Both (II) and (III) are true
(d) All three are true.

2.11.55 Suppose $i \to j$. Following are two statements. (I) The series $\sum_{n=1}^{\infty} p_{jj}^{(n)}$ is convergent. (II) The series $\sum_{n=1}^{\infty} p_{ij}^{(n)}$ is convergent. If the state j is transient, then which of the following options is correct?

(a) Both (I) and (II) are false
(b) Both (I) and (II) are true
(c) (I) is true but (II) is false
(d) (I) is false but (II) is true.

2.11.56 Suppose $i \to j$. Following are three statements: (I) The series $\sum_{n=1}^{\infty} p_{ij}^{(n)}$ is divergent. (II) The series $\sum_{n=1}^{\infty} p_{jj}^{(n)}$ is divergent. (III) $f_{jj} = 1$. If the state j is persistent, then which of the following options is correct?

(a) Only (II) is true
(b) Only (III) is true
(c) Both (II) and (III) are true
(d) All three are true.

2.11.57 For a time homogeneous Markov chain, if $i \to j$ and j is a transient state, then which of the following options is/are correct?

(a) $p_{ij}^{(n)} \to 0$
(b) $p_{ij}^{(n)} \to 1$
(c) $p_{ij}^{(n)} \to p \in (0, 1)$
(d) $\lim_{n\to\infty} p_{ij}^{(n)}$ does not exist.

2.11.58 For a time homogeneous Markov chain, if $i \to j$ and j is a persistent state, then which of the following options is/are correct?

(a) $p_{ij}^{(n)} \to 0$
(b) $\limsup_{n\to\infty} p_{ij}^{(n)} = 0$
(c) $p_{ij}^{(n)} \to p \in (0, 1)$
(d) $\lim_{n\to\infty} p_{ij}^{(n)}$ does not exist.

2.11.59 Which of the following options is/are correct? If state j is transient, then for any state i,

(a) $\sum_{n=1}^{\infty} p_{ij}^{(n)} \le \sum_{n=0}^{\infty} p_{jj}^{(n)}$
(b) $\sum_{n=1}^{\infty} p_{ij}^{(n)} \le \sum_{n=1}^{\infty} p_{jj}^{(n)}$

(c) $\sum_{n=1}^{\infty} p_{ij}^{(n)} \geq \sum_{n=0}^{\infty} p_{jj}^{(n)}$
(d) $\sum_{n=1}^{\infty} p_{ij}^{(n)} \geq \sum_{n=1}^{\infty} p_{jj}^{(n)}$.

2.11.60 Which of the following options is/are correct? If $f_{22} = 1/2$ and $f_{12} = 2/3$, then

(a) $\sum_{n=1}^{\infty} p_{12}^{(n)} = 4/3$
(b) $\sum_{n=1}^{\infty} p_{12}^{(n)} = 3/4$
(c) $\sum_{n=1}^{\infty} p_{12}^{(n)} = 4/9$
(d) $\sum_{n=1}^{\infty} p_{12}^{(n)}$ cannot be computed in view of insufficient information.

2.11.61 Suppose $S = \{1, 2, 3, 4\}$ and P is given by

$$P = \begin{array}{c} \\ 1 \\ 2 \\ 3 \\ 4 \end{array} \begin{array}{c} \begin{array}{cccc} 1 & 2 & 3 & 4 \end{array} \\ \begin{pmatrix} 1/2 & 0 & 1/2 & 0 \\ 1/2 & 0 & 1/2 & 0 \\ 0 & 2/3 & 0 & 1/3 \\ 1/4 & 0 & 3/4 & 0 \end{pmatrix} \end{array}.$$

Which of the following options is correct?

(a) All states are null persistent
(b) All states are non-null persistent
(c) All states are transient
(d) Some states are null persistent and some are non-null persistent.

2.11.62 Suppose $S = \{1, 2, 3, 4\}$ and P is given by

$$P = \begin{array}{c} \\ 1 \\ 2 \\ 3 \\ 4 \end{array} \begin{array}{c} \begin{array}{cccc} 1 & 2 & 3 & 4 \end{array} \\ \begin{pmatrix} 1/2 & 0 & 1/2 & 0 \\ 0 & 1/3 & 0 & 2/3 \\ 2/3 & 0 & 1/3 & 0 \\ 0 & 1/4 & 0 & 3/4 \end{pmatrix} \end{array}.$$

Which of the following options is correct?

(a) The Markov chain is irreducible
(b) All states are non-null persistent
(c) All states are transient
(d) States $1, 3$ are transient and $2, 4$ are non-null persistent.

2.11.63 Suppose $S = \{1, 2, 3, 4\}$ and P is given by

$$P = \begin{array}{c} \\ 1 \\ 2 \\ 3 \\ 4 \end{array} \begin{array}{c} \begin{array}{cccc} 1 & 2 & 3 & 4 \end{array} \\ \begin{pmatrix} 1/2 & 0 & 1/2 & 0 \\ 2/3 & 1/3 & 0 & 0 \\ 1/3 & 2/3 & 0 & 0 \\ 1/4 & 1/4 & 1/4 & 1/4 \end{pmatrix} \end{array}.$$

Which of the following options is correct?

(a) All states are transient
(b) All states are non-null persistent
(c) States 4 is null persistent and others are non-null persistent
(d) States 4 is transient and others are non-null persistent.

2.11.64 Following are two statements for a Markov chain with finite state space. (I) If C is a closed communicating class of persistent states, then for any transient state i which leads to $j, k \in C$, $f_{ij} = f_{ik}$. (II) If C is a single closed communicating class and if $i \to j$ where $i \notin C$ and $j \in C$, then $f_{ij} = 1$. Which of the following options is correct?

(a) Both (I) and (II) are false
(b) Both (I) and (II) are true
(c) (I) is true but (II) is false
(d) (I) is false but (II) is true.

2.11.65 Following are two statements. Suppose $C_r, r = 1, 2$ are the only two closed communicating classes. (I) If $i \notin C_1 \cup C_2$ is such that $i \to j \in C_1$ but $i \nrightarrow j \in C_2$, then $f_{ij} = 1$. (II) If $i \notin C_1 \cup C_2$ is such that $i \to j \in C_1$ and $i \to j \in C_2$, then $f_{ij} < 1$. Which of the following options is correct?

(a) Both (I) and (II) are false
(b) Both (I) and (II) are true
(c) (I) is true but (II) is false
(d) (I) is false but (II) is true.

2.11.66 Suppose $S = \{1, 2, 3, 4\}$ and P is given by

$$P = \begin{array}{c} \\ 1 \\ 2 \\ 3 \\ 4 \end{array} \begin{array}{c} \begin{array}{cccc} 1 & 2 & 3 & 4 \end{array} \\ \begin{pmatrix} 1/2 & 0 & 1/2 & 0 \\ 0 & 1/3 & 0 & 2/3 \\ 2/3 & 0 & 1/3 & 0 \\ 0 & 1/4 & 0 & 3/4 \end{pmatrix} \end{array}.$$

Following are four statements: (I) $f_{12} = 0$. (II) $f_{24} = 1$. (III) $f_{13} = 1$. (IV) $f_{33} < 1$.
Which of the following options is correct?

(a) Only (I) and (II) are true
(b) Only (I) and (III) are true

(c) Only (I) and (IV) are true
(d) Only (I), (II) and (III) are true.

2.11.67 Following are two statements: (I) If C is a single closed communicating class of persistent states and set of transient states is finite, then the probability of absorption into C from a transient state is 1. (II) If C is one of the communicating classes of persistent states and set of transient states is finite, then the probability of absorption into C from a transient state is always 1.
Which of the following options is correct?

(a) Both (I) and (II) are false
(b) Both (I) and (II) are true
(c) (I) is true but (II) is false
(d) (I) is false but (II) is true.

2.11.68 Suppose the Markov chain is periodic with period 2. Following are three statements: (I) $f_{ii}^{(2)} = p_{ii}^{(2)} - p_{ii}^{(1)} f_{ii}^{(1)}$. (II) $f_{ii}^{(4)} = p_{ii}^{(4)} - p_{ii}^{(2)} f_{ii}^{(2)}$. (III) $f_{ii}^{(7)} = 0$.
Which of the following options is correct?

(a) Only (III) is true
(b) Only (I) and (III) are true
(c) Only (I) and (II) are true
(d) All are true.

2.11.69 Following are two statements: (I) If $i \to j$, then $d_i = d_j$. (II) If $i \leftrightarrow j$, then $d_i = d_j$.
Then which of the following is a correct option?

(a) Both (I) and (II) are false
(b) Both (I) and (II) are true
(c) (I) is true but (II) is false
(d) (I) is false but (II) is true.

2.11.70 Which of the following options is/are correct? An irreducible time homogeneous Markov chain is aperiodic if

(a) the transition probability matrix P of a Markov chain has all positive elements
(b) the diagonal elements of P are positive
(c) all the elements of the matrix P^n are positive for some $n > 0$
(d) the set D of n for which $p_{ii}^{(n)} > 0$ is $D = \{4, 8, 12, 16, 17, 18, 19, 20, \ldots\}$.

2.11.71 Which of the following is NOT true? An irreducible time homogeneous Markov chain is aperiodic if

(a) the transition probability matrix P of a Markov chain has all positive elements

(b) the diagonal elements of P are positive
(c) all the elements of the matrix P^n are positive for some $n > 0$
(d) the set D of n for which $p_{ii}^{(n)} > 0$ is $D = \{4, 8, 12, 16, \ldots\}$.

2.11.72 Following are two statements: (I) If for each pair $(i, j),\ i, j \in S,\ p_{ij}^{(n)} > 0$ for some n, then the Markov chain is aperiodic. (II) If $i \nleftrightarrow j$ for all $i, j \in S$, then the Markov chain is not aperiodic.
Then which of the following is a correct option?

(a) Both (I) and (II) are false
(b) Both (I) and (II) are true
(c) (I) is true but (II) is false
(d) (I) is false but (II) is true.

2.11.73 Following are two statements: (I) If the Markov chain is irreducible, then each state has the same period. (II) If each state has the same period, then the Markov chain must be irreducible.
Then which of the following is a correct option?

(a) Both (I) and (II) are false
(b) Both (I) and (II) are true
(c) (I) is true but (II) is false
(d) (I) is false but (II) is true.

2.11.74 Following are two statements: (I) In a finite Markov chain, all the states are aperiodic. (II) In a finite Markov chain, some states may be periodic and some states may be aperiodic.
Which of the following is a correct option?

(a) Both (I) and (II) are false
(b) Both (I) and (II) are true
(c) (I) is true but (II) is false
(d) (I) is false but (II) is true.

2.11.75 Following are two statements: (I) In a finite irreducible Markov chain, all the states have the same period. (II) In a finite Markov chain, some states may be periodic and some states may be aperiodic.
Which of the following is a correct option?

(a) Both (I) and (II) are false
(b) Both (I) and (II) are true
(c) (I) is true but (II) is false
(d) (I) is false but (II) is true.

2.11.76 Following are two paths in a Markov chain with state space $S = \{1, 2, 3, 4\}$:

$$1 \to 2 \to 3 \to 2 \ \& \ 1 \to 2 \to 4 \to 1.$$

Following are some statements: (I) State 1 is aperiodic. (II) {2, 3} is a communicating class. (III) {1, 2, 3} is a closed class.
Then which of the following statements is true?

(a) Both (I) and (II) are true
(b) Both (I) and (III) are true
(c) (I) is false but (III) is true
(d) Both (II) and (III) are true.

2.11.77 Suppose $S = \{1, 2, 3\}$ and P is given by

$$P = \begin{matrix} & \begin{matrix} 1 & 2 & 3 \end{matrix} \\ \begin{matrix} 1 \\ 2 \\ 3 \end{matrix} & \begin{pmatrix} 1 & 0 & 0 \\ 1/3 & 1/3 & 1/3 \\ 0 & 0 & 1 \end{pmatrix} \end{matrix}.$$

Suppose $X_0 = 2$. Which of the following options is correct?
The probabilities of absorption into state 1 and state 3 respectively are

(a) 1/3 and 1/3.
(b) 1/3 and 2/3.
(c) 1/2 and 1/2.
(d) 1 and 1.

2.11.78 Suppose $S = \{1, 2, 3\}$ and P is given by

$$P = \begin{matrix} & \begin{matrix} 1 & 2 & 3 \end{matrix} \\ \begin{matrix} 1 \\ 2 \\ 3 \end{matrix} & \begin{pmatrix} 0.3 & \alpha & \beta \\ 0.4 & \gamma & 0.1 \\ \delta & 0.2 & \epsilon \end{pmatrix} \end{matrix}.$$

If P is a doubly stochastic matrix , the values of α, β, γ

(a) cannot be computed from the given information
(b) are $\alpha = 0.5, \beta = 0.2, \gamma = 0.3$
(c) are $\alpha = 0.4, \beta = 0.3, \gamma = 0.4$
(d) are $\alpha = 0.3, \beta = 0.4, \gamma = 0.5$.

References

1. Basawa, I. V., & Rao, B. L. S. P. (1980). *Statistical inference for stochastic processes*. New York: Academic.
2. Chan, K. C., Lenard, C. T., & Mills, T. M. (2013). On Markov Chains. *The Mathematical Gazette, 97*(540), 515–520. https://doi.org/10.2307/3616733. Published online: 23 January 2015

3. Cinlar, E. (1975). *Introduction to stochastic processes*. New Jersey: Prentice Hall.
4. Deshmukh S. R., (2012). *Multiple decrement models in insurance: An introduction using R*. New Delhi: Springer.
5. Feller, W. (1978). *An introduction to probability theory and its applications* (Vol. I). New York: Wiley.
6. Feller, W. (2000). *An introduction to probability theory and its applications,* (2nd ed., Vol. II). Singapore: Wiley Inc.
7. Goulet, V., Dutang, C., Maechler, M., Firth, D., Shapira, M., & Stadelmann, M. (2019). expm: Matrix exponential, log,'etc'. R package version 0.999-4. https://CRAN.R-project.org/package=expm
8. Guttorp, P. (1991). *Statistical inference for branching processes*. New York: Wiley.
9. Karlin, S., & Taylor, H. M. (1975). *A first course in stochastic processes*. New York: Academic.
10. Novomestky., F. (2012). Matrixcalc: Collection of functions for matrix calculations. R package version 1.0-3. https://CRAN.R-project.org/package=matrixcalc
11. Warnes, G. R., Bolker, B., & Lumley, T. (2018). gtools: Various R programming tools. R package version 3.8.1. https://CRAN.R-project.org/package=gtools

Chapter 3
Long Run Behavior of Markov Chains

3.1 Introduction

Suppose $\{X_n, n \geq 0\}$ is a Markov chain specified by $(S, \underline{p}^{(0)}, P)$, where S is the state space, $\underline{p}^{(0)}$ is the initial distribution and P is the one step transition probability matrix. At the beginning of Sect. 2.4, several examples are presented which illustrate the distinct types of limiting behavior of Markov chains. These depend on the nature of the states of the Markov chain determined by recurrence and periodicity. The present chapter elaborates on various aspects of the long run behavior of Markov chains.

In Sect. 3.2, we discuss the long run distribution of a Markov chain, while stationary distributions are studied in Sect. 3.3. Section 3.4 is devoted to the computational aspects of the stationary distributions. Section 3.5 is concerned with a brief introduction to the autocovariance function of a Markov chain. Section 3.6 is concerned with the application of the theory of Markov chains in a Bonus-Malus system. The last section presents R codes used to illustrate the concepts and computation of the long run and the stationary distributions.

We begin with the definitions of a long run distribution and a stationary distribution of a Markov chain.

Definition 3.1.1 Long Run Distribution of a Markov Chain: Suppose $\{X_n, n \geq 0\}$ is a Markov chain specified by $(S, \underline{p}^{(0)}, P)$. Suppose $\lim_{n\to\infty} p_{ij}^{(n)}$ exists for all $j \in S$ and does not depend on i. If $a_j = \lim_{n\to\infty} p_{ij}^{(n)}$, $j \in S$ and $\sum_{j\in S} a_j = 1$, then $\{a_j, j \in S\}$ is known as a long run distribution of the Markov chain.

Remark 3.1.1 (i) Note that the condition $a_j \geq 0 \ \forall \ j \in S$ is always satisfied, since a_j is the limit of a sequence of non-negative numbers. (ii) If a long run distribution exists, then it is unique, being a limit of sequence of real numbers. (iii) If a long run distribution exists, then all rows of $\lim_{n\to\infty} P^n$ are identical and each row is the long run distribution. A long run distribution may not exist, if either $\lim_{n\to\infty} p_{ij}^{(n)}$ does

S. Madhira and S. Deshmukh, *Introduction to Stochastic Processes Using R*,
https://doi.org/10.1007/978-981-99-5601-2_3

not exist or even if it exists, it depends on i. For example, suppose $\{X_n, n \geq 0\}$ is a Markov chain with P given by

$$P = \begin{array}{c} \\ 1 \\ 2 \end{array}\begin{array}{c} \begin{array}{cc} 1 & 2 \end{array} \\ \begin{pmatrix} 0 & 1 \\ 1 & 0 \end{pmatrix} \end{array}.$$

In Sect. 2.4 we have noted that $\forall\ n \geq 1$, $P^{2n+1} = P$ and $P^{2n} = I_2$, an identity matrix of order 2. Hence for $i \neq j$,

$$p_{ij}^{(n)} = \begin{cases} 0, & \text{if } n \text{ is even} \\ 1, & \text{if } n \text{ is odd.} \end{cases}$$

Thus, $\limsup_{n\to\infty} p_{ij}^{(n)} = 1$ and $\liminf_{n\to\infty} p_{ij}^{(n)} = 0$, so that $\lim_{n\to\infty} p_{ij}^{(n)}$ and hence the long run distribution do not exist. On the other hand, suppose $\{X_n, n \geq 0\}$ is a Markov chain with $P = I_3$. Then $\lim_{n\to\infty} P^n = I_3$. Thus, $\lim_{n\to\infty} p_{ij}^{(n)}$ exists, but depends on i and hence the long run distribution does not exist.

We elaborate on this concept in Sect. 3.2.

Definition 3.1.2 Stationary Distribution: Suppose $\{X_n, n \geq 0\}$ is a Markov chain specified by $(S, \underline{p}^{(0)}, P)$. Then $\{\pi_i, i \in S\}$ is said to be a stationary distribution associated with the Markov chain if

$$(i)\ \pi_i \geq 0,\ \ \forall\ i \in S,\ \ (ii) \sum_{i\in S} \pi_i = 1\ \ \&\ \ (iii)\ \pi_j = \sum_{i\in S} \pi_i p_{ij}\ \ \forall\ j \in S.$$

From the definition of a stationary distribution, it follows that if the state space has M states, then there are $M + 1$ equations in M unknowns. However, the M equations in condition (iii) are linearly dependent, since the sum over j of both the sides of the equation $\pi_j = \sum_{i\in S} \pi_i p_{ij}$ is equal to 1. Thus, one of them is always redundant and hence we have a system of M equations in M unknowns. The solution may exist or may not exist. Even if it exists, it may or may not be unique. To elaborate on this issue, suppose $S = \{1, 2, \ldots, M\}$ and $\underline{\pi} = (\pi_1, \pi_2, \ldots, \pi_M)$. Then the third condition in the definition can be expressed as a matrix equation $\underline{\pi} = \underline{\pi}P$. By definition, the stationary distribution is a solution of the equation $\underline{\pi} = \underline{\pi}P$ with $\pi_j \geq 0$ and $\sum_{j\in S} \pi_j = 1$. Thus, we have a system of linear equations

$$P'\underline{\pi}' = \underline{\pi}' \iff (P' - I)\underline{\pi}' = \underline{0} \text{ with } \pi_1 + \pi_2 + \cdots + \pi_M = 1\,.$$

Suppose A denotes a matrix $P' - I$ appended with the last row as $\underline{e} = (1, 1, \ldots, 1)$. Thus, to find $\underline{\pi}$, the system of equations in a matrix form is $A\underline{\pi}' = \underline{b}$ where $\underline{b}' = (0, 0, \ldots, 0, 1)$. It has a solution if $rank(A) = rank(A|b)$. If this condition is satisfied, then the solution is $\underline{\pi}' = A^{-}\underline{b}$, where A^{-} is a generalized inverse of A. In general, the generalized inverse of A is not unique. So multiple solutions may exist and these correspond to more than one stationary distributions. However, Moore-Penrose generalized inverse is unique and it gives one of the stationary distributions, if the stationary distribution is not unique. Note that A is not a square matrix as its order is $(M+1) \times M$, but solving $A\underline{\pi}' = \underline{b}$ is equivalent to solving $A'A\underline{\pi}' = A'\underline{b}$ and $A'A$ is a square matrix. It has a solution if $rank(A'A) = rank(A'A|b)$. Thus, $\underline{\pi}' = (A'A)^{-1}A'\underline{b}$, provided $(A'A)^{-1}$ exists, otherwise $\underline{\pi}' = (A'A)^{-}A'\underline{b}$, involving a generalized inverse of $A'A$. If $(A'A)^{-1}$ exists, then the solution is unique.

The existence of the solution of $A\underline{\pi}' = \underline{b}$ and its uniqueness depend on the nature of the Markov chain. All these aspects are studied in Sect. 3.3.

By repeatedly using the third condition in the definition of the stationary distribution we have

$$\pi_j = \sum_{i \in S} \pi_i p_{ij} = \sum_{i \in S} \left(\sum_{k \in S} \pi_k p_{ki} \right) p_{ij} = \sum_{k \in S} \pi_k p_{kj}^{(2)} \iff \underline{\pi} = \underline{\pi} P^2 .$$

Suppose $\underline{\pi} = \underline{\pi} P^m$ for some $m \geq 1$, which is true for $m = 1$ and is shown to be true for $m = 2$. Then

$$\underline{\pi} P^{m+1} = \underline{\pi} P^m P = \underline{\pi} P = \underline{\pi} \;\Rightarrow\; \underline{\pi} = \underline{\pi} P^n \;\; \forall \; n \geq 1,$$

by induction. Further, for any $j \in S$ and $\forall \; n \geq 1$

$$\underline{\pi} = \underline{\pi} P^n \iff \pi_j = \sum_{i \in S} \pi_i p_{ij}^{(n)} .$$

This identity essentially conveys that if $\underline{\pi}$ is a stationary distribution of a Markov chain with transition probability matrix P, then it is also one of the stationary distributions of a Markov chain with transition probability matrix P^n for any $n \geq 1$.

In the present chapter, our goal is to address various issues regarding the existence of the long run distribution and of the stationary distribution. Another important issue is about the relation between the two distributions and their relation with the limit of marginal distribution of X_n. We begin with the following theorem which addresses one such issue.

Theorem 3.1.1 *Suppose a long run distribution* $a_j = \lim_{n\to\infty} p_{ij}^{(n)}$, $j \in S$ *of a Markov chain exists. Then (i)* $\lim_{n\to\infty} P[X_n = j]$ *exists and is* a_j, $\forall \; j \in S$. *(ii) The vector* $\underline{a} = (a_1, a_2, \ldots)$ *satisfies the equation* $\underline{a} = \underline{a}P$, *that is, the stationary distribution exists, it is unique and is given by* $\underline{a}$.

Proof (i) From Eq. (2.2.4), we have $\forall\ j \in S$,

$$\begin{aligned} P[X_n = j] &= \sum_{i \in S} p_{ij}^{(n)} P[X_0 = i] \\ \Rightarrow \lim_{n\to\infty} P[X_n = j] &= \lim_{n\to\infty} \sum_{i \in S} p_{ij}^{(n)} P[X_0 = i] \\ &= \sum_{i \in S} \lim_{n\to\infty} p_{ij}^{(n)} P[X_0 = i], \text{ by Theorem 2.4.1} \\ &= a_j \sum_{i \in S} P[X_0 = i] = a_j. \end{aligned}$$

(ii) Observe that $\forall\ j \in S$,

$$\begin{aligned} a_j &= \lim_{n\to\infty} P[X_n = j] \iff \underline{a} = \lim_{n\to\infty} \underline{p}^{(0)} P^n \\ \Rightarrow \quad \underline{a} &= \lim_{n\to\infty} \underline{p}^{(0)} P^{n+1} = \left(\lim_{n\to\infty} \underline{p}^{(0)} P^n\right) P = \underline{a}P\,. \end{aligned}$$

Thus, $\underline{a}$ is the solution of the equation $\underline{a} = \underline{a}P$ and hence it is a stationary distribution of the Markov chain. To prove uniqueness of the stationary distribution we assume the contrary. Suppose $\underline{b} = (b_1, b_2, \ldots)$ is another stationary distribution of the Markov chain. Then note that

$$\begin{aligned} b_j &= \sum_{i \in S} b_i p_{ij}^{(n)} \quad \forall\ j \in S \\ \Rightarrow \quad \lim_{n\to\infty} b_j &= \lim_{n\to\infty} \sum_{i \in S} b_i p_{ij}^{(n)} = \sum_{i \in S} b_i \lim_{n\to\infty} p_{ij}^{(n)} \text{ by Theorem 2.4.1} \\ &= a_j \sum_{i \in S} b_i = a_j \quad \forall\ j \in S. \end{aligned}$$

Hence, the stationary distribution, which is the same as the long run distribution, is unique. □

Thus, if the long run distribution exists, then the limit of a marginal distribution of X_n exists and the two are the same. In this case, the stationary distribution also exists, it is unique and is the same as the long run distribution. Converse is however not true, as the following example shows.

Example 3.1.1 Suppose $\{X_n, n \geq 0\}$ is a Markov chain with P given by

$$P = \begin{matrix} & \begin{matrix} 1 & 2 \end{matrix} \\ \begin{matrix} 1 \\ 2 \end{matrix} & \begin{pmatrix} 0 & 1 \\ 1 & 0 \end{pmatrix} \end{matrix}.$$

Solving the matrix equation $\underline{\pi} = \underline{\pi}P$, we get $\pi_1 = \pi_2$ and with $\pi_1 + \pi_2 = 1$, we have $\pi_1 = \pi_2 = 1/2$. Thus, $\underline{\pi} = (1/2, 1/2)$ is the unique stationary distribution of the Markov chain. However, as noted in Remark 3.1.1, for $i \neq j$,

$$p_{ij}^{(n)} = \begin{cases} 0, & \text{if } n \text{ is even} \\ 1, & \text{if } n \text{ is odd.} \end{cases}$$

Thus, $\limsup_{n\to\infty} p_{ij}^{(n)} = 1$ and $\liminf_{n\to\infty} p_{ij}^{(n)} = 0$, so that $\lim_{n\to\infty} p_{ij}^{(n)}$ and hence the long run distribution do not exist. However, the limit of the marginal distribution of X_n exists, if the initial distribution is $\underline{p}^{(0)} = (1/2, 1/2)$. This is because for $j = 1, 2$,

$$P[X_n = j] = (1/2)(p_{1j}^{(n)} + p_{2j}^{(n)}) = (1/2)(1) = 1/2 \ \forall \ n \ \Rightarrow \ P[X_n = j] \to 1/2.$$

□

Note that once we know that the long run distribution $\underline{a}$ exists, then we can obtain the limits of any joint distribution. For example, the limit of the joint distribution of X_{n+1} and X_n is obtained as follows. For $k, \ j \in S$,

$$\begin{aligned} \lim_{n\to\infty} P[X_{n+1} = k, X_n = j] &= \lim_{n\to\infty} \sum_{i\in S} P[X_{n+1} = k, X_n = j, X_0 = i] \\ &= \lim_{n\to\infty} \sum_{i\in S} p_{jk} p_{ij}^{(n)} P[X_0 = i] \\ &= p_{jk} \sum_{i\in S} \lim_{n\to\infty} p_{ij}^{(n)} P[X_0 = i] \\ &= p_{jk} a_j. \end{aligned}$$

Further $\underline{a}$ is the stationary distribution implies that $\sum_{j\in S} p_{jk} a_j = a_k$ and $\sum_{k\in S} a_k = 1$. Thus, $\sum_{k\in S}\sum_{j\in S} \lim_{n\to\infty} P[X_{n+1} = k, X_n = j] = 1$, as expected.

Since the existence of the long run distribution implies the existence of the stationary distribution, we begin with the investigation of the following issues regarding the long run distribution, in Sect. 3.2.

1. Under what conditions on the states, does $\lim_{n\to\infty} p_{ij}^{(n)}$ exist?
2. When does $\lim_{n\to\infty} p_{ij}^{(n)}$ not depend on initial state i?

3.2 Long Run Distribution

We begin the section with the first question and examine how the existence of $\lim_{n\to\infty} p_{ij}^{(n)}$ depends on the nature of the states. We have noted in Chap. 2 that if $i \nrightarrow j$, then $p_{ij}^{(n)} = 0 \ \ \forall \ n \geq 1$. Hence, $\lim_{n\to\infty} p_{ij}^{(n)} = 0$. Further, in Theorem 2.6.5, it is proved that if $i \to j$, then $\lim_{n\to\infty} p_{ij}^{(n)} = 0$, if and only if j is either transient or null persistent. Hence, if the long run distribution $\underline{a}$ exists, then $a_j = 0$ if j is either transient or null persistent. We now consider the remaining case of non-null persistent state. We state below a theorem, known as the basic limit theorem (p. 335, Feller [4]) or Erdos-Feller-Pollard theorem (p. 312, Feller [4]), which is useful to find $\lim_{n\to\infty} p_{ij}^{(n)}$ when j is non-null persistent.

Theorem 3.2.1 *Erdos-Feller-Pollard Theorem: Suppose* $\{f_n, n \geq 0\}$ *is a sequence of non-negative real numbers, such that* $\sum_{n\geq 0} f_n = 1$ *and g.c.d* $\{n|f_n > 0\} = d \geq 1$. *Suppose* $\sum_{n\geq 1} nf_n = \mu$. *If a sequence* $\{u_n, n \geq 0\}$ *is defined as* $u_0 = 1$ *and for* $n \geq 1$, $u_n = f_1u_{n-1} + f_2u_{n-2} + \cdots + f_nu_0$, *then*

$$\lim_{n\to\infty} u_n = \begin{cases} d/\mu, & \text{if } \mu < \infty \\ 0, & \text{if } \mu = \infty. \end{cases}$$

Using the Erdos-Feller-Pollard theorem, we obtain the limiting behavior of $p_{ij}^{(n)}$ in the next theorem, when j is a non-null persistent and aperiodic state.

Theorem 3.2.2 *Suppose j is a non-null persistent aperiodic state. Then*
(i) for any $i \in S$,

$$\lim_{n\to\infty} p_{ij}^{(n)} = f_{ij}/\mu_j \ .$$

(ii) If $i \leftrightarrow j$, *then*

$$\lim_{n\to\infty} p_{ij}^{(n)} = 1/\mu_j \ ,$$

where μ_j *is the mean recurrence time of state* j.

Proof (i) Suppose $i \nrightarrow j$, then $p_{ij}^{(n)} = 0 \ \forall \ n \geq 1$ and $f_{ij} = 0$ and hence the theorem is true. Suppose now $i \to j$. If j is persistent then $\sum_{n\geq 1} f_{jj}^{(n)} = 1$. By Lemma 2.7.1, g.c.d$\{n|f_{jj}^{(n)} > 0\} =$ g.c.d$\{n|p_{jj}^{(n)} > 0\}$. Suppose in the recurrence relation $p_{jj}^{(n)} = \sum_{r=1}^{n} f_{jj}^{(r)} p_{jj}^{(n-r)}$, $p_{jj}^{(n)}$ is replaced by u_n and $f_{jj}^{(n)}$ is replaced by f_n. Then the above equation reduces to $u_n = f_1u_{n-1} + f_2u_{n-2} + \cdots + f_nu_0$. Hence, by the Erdos-Feller-Pollard theorem, if j is aperiodic, then

$$\lim_{n\to\infty} p_{jj}^{(n)} = \begin{cases} 1/\mu_j, & \text{if } j \text{ is non-null persistent} \\ 0, & \text{if } j \text{ is null persistent.} \end{cases}$$

As discussed in Theorem 2.6.5, for any state $i \in S$ which leads to an aperiodic state j,

$$\lim_{n\to\infty} p_{ij}^{(n)} = \lim_{n\to\infty} \sum_{r=1}^{n} f_{ij}^{(r)} p_{jj}^{(n-r)} = \lim_{n\to\infty} \sum_{r=1}^{\infty} f_{ij}^{(r)} a(n, r)$$

where

$$a(n, r) = \begin{cases} p_{jj}^{(n-r)}, & \text{if } r \leq n \\ 0, & \text{if } r > n. \end{cases}$$

Observe that $|f_{ij}^{(r)}a(n,r)| \leq f_{ij}^{(r)}$ and $\sum_{r=1}^{\infty} f_{ij}^{(r)} = f_{ij} \leq 1$. Hence by Theorem 2.4.1,

$$\lim_{n\to\infty} p_{ij}^{(n)} = \sum_{r=1}^{\infty} f_{ij}^{(r)} \left(\lim_{n\to\infty} p_{jj}^{(n-r)}\right) = f_{ij}/\mu_j, \text{ since } \lim_{n\to\infty} p_{jj}^{(n-r)} = 1/\mu_j,$$

for each fixed r.

(ii) If $i \leftrightarrow j$ and j is persistent then by Theorem 2.6.1, $f_{ij} = 1$. Hence, $\lim_{n\to\infty} p_{ij}^{(n)} = 1/\mu_j$. □

We have two definitions of a null persistent state j, one is in terms of the mean recurrence time μ_j being infinite and the other is in terms of $\limsup_{n\to\infty} p_{jj}^{(n)}$ being 0. Using the Erdos-Feller-Pollard theorem, we prove that the two definitions are equivalent. As in Theorem 3.2.2, taking $u_n = p_{jj}^{(n)}$, we have

$$\lim_{n\to\infty} p_{jj}^{(n)} = \lim_{n\to\infty} u_n = \begin{cases} 1/\mu_j, & \text{if } \mu_j < \infty \\ 0, & \text{if } \mu_j = \infty. \end{cases}$$

Thus,

$$\mu_j = \infty \Rightarrow \lim_{n\to\infty} p_{jj}^{(n)} = 0 \Rightarrow \limsup_{n\to\infty} p_{jj}^{(n)} = 0.$$

On the other hand, $\limsup_{n\to\infty} p_{jj}^{(n)} = 0$ implies $\lim_{n\to\infty} p_{jj}^{(n)} = 0$. Suppose $\mu_j < \infty$, then by the Erdos-Feller-Pollard theorem, $\lim_{n\to\infty} p_{jj}^{(n)} = 1/\mu_j \neq 0$. Hence, if $\limsup_{n\to\infty} p_{jj}^{(n)} = 0$, then $\mu_j = \infty$.

The discussion of the limiting behavior of $p_{ij}^{(n)}$ for a non-null persistent and periodic state j is deferred to the end of the section.

We now introduce one more term describing the nature of a state of a Markov chain and state some related results.

Definition 3.2.1 Ergodic State: An aperiodic, non-null persistent state is known as an ergodic state.

Theorem 3.2.3 *Being ergodic is a class property.*

Proof It has been proved that periodicity and being non-null persistent are class properties and hence being ergodic is a class property. □

Remark 3.2.1 If all the states in a Markov chain are ergodic, the Markov chain is known as an ergodic Markov chain. If a finite Markov chain is aperiodic and irreducible, then it is an ergodic chain, since all the states are non-null persistent. On similar lines, if C is a finite closed communicating class of aperiodic states, then all the states in C are ergodic states.

A summary of limiting behavior of $p_{ij}^{(n)}$ depending on the nature of state j is given below.

(i) If $i \nrightarrow j$, then $p_{ij}^{(n)} = 0 \ \forall \ n \geq 1$ and hence $\lim_{n\to\infty} p_{ij}^{(n)} = 0$.
(ii) If $i \to j$, $\lim_{n\to\infty} p_{ij}^{(n)} = 0$, if and only if j is either transient or null persistent.
(iii) If j is a non-null persistent aperiodic state, then for any $i \in S$, $\lim_{n\to\infty} p_{ij}^{(n)} = f_{ij}/\mu_j$ and the limit depends on i.
(iv) If $i \leftrightarrow j$ and if j is an ergodic state, then $\lim_{n\to\infty} p_{ij}^{(n)} = 1/\mu_j$ and the limit does not depend on i.

From the summary, we note that

$$\lim_{n\to\infty} P[X_n = j] = \lim_{n\to\infty} p_{ij}^{(n)} = \begin{cases} 0, & \text{if} \quad j \text{ is transient} \\ & \quad\text{or null persistent} \\ f_{ij}/\mu_j, & \text{if} \quad j \text{ is ergodic.} \end{cases}$$

If $\lim_{n\to\infty} p_{ij}^{(n)}$ exists and depends on i, we can find $\lim_{n\to\infty} P[X_n = j]$, if we know the initial distribution. Suppose $\underline{p}^{(0)}$ and $\underline{p}^{(n)}$ denote the initial distribution and the marginal distribution of X_n, respectively. From Eq. (2.2.5),

$$\underline{p}^{(n)} = \underline{p}^{(0)} P^n \Rightarrow \lim_{n\to\infty} \underline{p}^{(n)} = \underline{p}^{(0)} \lim_{n\to\infty} P^n . \tag{3.2.1}$$

Equation (3.2.1) shows that $\lim_{n\to\infty} P[X_n = j]$ depends on the initial distribution. Example 3.2.1 illustrates such a case. It also illustrates Theorems 2.6.5 and 3.2.2.

Example 3.2.1 For the Markov chain discussed in Example 2.5.8, $S = \{1, 2, 3, 4, 5, 6\}$ and P is as given below.

$$P = \begin{array}{c} \\ 1 \\ 2 \\ 3 \\ 4 \\ 5 \\ 6 \end{array} \begin{array}{c} \begin{array}{cccccc} 1 & 2 & 3 & 4 & 5 & 6 \end{array} \\ \begin{pmatrix} 1/3 & 0 & 2/3 & 0 & 0 & 0 \\ 0 & 1/2 & 1/4 & 0 & 1/4 & 0 \\ 2/5 & 0 & 3/5 & 0 & 0 & 0 \\ 0 & 1/4 & 1/4 & 1/4 & 0 & 1/4 \\ 0 & 0 & 0 & 0 & 1/2 & 1/2 \\ 0 & 0 & 0 & 0 & 1/4 & 3/4 \end{pmatrix} \end{array}.$$

Since all the diagonal elements of P are positive, all the states are aperiodic. The Markov chain is reducible with $C_1 = \{1, 3\}$ and $C_2 = \{5, 6\}$ as two closed communicating classes of ergodic states, while 2 and 4 are transient states. Using R, we find P^n. For all $n \geq 20$, P^n is the same up to four decimal places accuracy and is given by

$$Q = \begin{array}{c} \\ 1 \\ 2 \\ 3 \\ 4 \\ 5 \\ 6 \end{array} \begin{array}{c} \begin{array}{cccccc} 1 & 2 & 3 & 4 & 5 & 6 \end{array} \\ \begin{pmatrix} 0.3750 & 0 & 0.6250 & 0 & 0.0000 & 0.0000 \\ 0.1875 & 0 & 0.3125 & 0 & 0.1667 & 0.3333 \\ 0.3750 & 0 & 0.6250 & 0 & 0.0000 & 0.0000 \\ 0.1875 & 0 & 0.3125 & 0 & 0.1667 & 0.3333 \\ 0.0000 & 0 & 0.0000 & 0 & 0.3333 & 0.6667 \\ 0.0000 & 0 & 0.0000 & 0 & 0.3333 & 0.6667 \end{pmatrix} \end{array}.$$

Thus, $\lim_{n\to\infty} P^n = Q$. We note that, $\lim_{n\to\infty} p_{ij}^{(n)}$ exists for all i and j. Using Code 2.8.7, we have computed $f_{ij}\ \forall\ i, j$ in Example 2.6.7. These are displayed in the following matrix $F = [f_{ij}]$.

$$F = \begin{array}{c} \\ 1 \\ 2 \\ 3 \\ 4 \\ 5 \\ 6 \end{array}\begin{array}{c} \begin{array}{cccccc} 1 & 2 & 3 & 4 & 5 & 6 \end{array} \\ \begin{pmatrix} 1 & 0 & 1 & 0 & 0 & 0 \\ 1/2 & 1/2 & 1/2 & 0 & 1/2 & 1/2 \\ 1 & 0 & 1 & 0 & 0 & 0 \\ 1/2 & 1/3 & 1/2 & 1/4 & 1/2 & 1/2 \\ 0 & 0 & 0 & 0 & 1 & 1 \\ 0 & 0 & 0 & 0 & 1 & 1 \end{pmatrix} \end{array}.$$

The mean recurrence times μ_i for $i = 1, 3, 5, 6$ are obtained in Example 2.5.8 and are given by $\mu_1 = 2.6667$, $\mu_3 = 1.6$, $\mu_5 = 3$, $\mu_6 = 1.5$. We compute f_{ij}/μ_j for $j = 1, 3, 5, 6$ and all $i \in S$. These are presented in the following matrix A.

$$A = \begin{array}{c} \\ 1 \\ 2 \\ 3 \\ 4 \\ 5 \\ 6 \end{array}\begin{array}{c} \begin{array}{cccc} 1 & 3 & 5 & 6 \end{array} \\ \begin{pmatrix} 0.3750 & 0.6250 & 0.0000 & 0.0000 \\ 0.1875 & 0.3125 & 0.1667 & 0.3333 \\ 0.3750 & 0.6250 & 0.0000 & 0.0000 \\ 0.1875 & 0.3125 & 0.1667 & 0.3333 \\ 0.0000 & 0.0000 & 0.3333 & 0.6667 \\ 0.0000 & 0.0000 & 0.3333 & 0.6667 \end{pmatrix} \end{array}.$$

From the matrices A and Q, we note the following results.

(i) For persistent states $j = 1, 3, 5, 6$ and for any $i \in S$, $\lim_{n\to\infty} p_{ij}^{(n)} = f_{ij}/\mu_j$.

(ii) For $i, j \in C_1$ and $i, j \in C_2$, $\lim_{n\to\infty} p_{ij}^{(n)} = 1/\mu_j$.

(iii) For transient states $j = 2, 4$ and for $i \to j$, $\lim_{n\to\infty} p_{ij}^{(n)} = 0$.

(iv) For $i \in C_1$ and $j \in C_2$, $p_{ij}^{(n)} = 0$ as these are closed classes. Hence, $\lim_{n\to\infty} p_{ij}^{(n)} = 0$. Similarly, for $i \in C_2$ and $j \in C_1$, $p_{ij}^{(n)} = 0$, these being closed classes. Hence, $\lim_{n\to\infty} p_{ij}^{(n)} = 0$.

Thus, we have verified the results established in Theorems 2.6.5 and 3.2.2. Observe that $\lim_{n\to\infty} p_{ij}^{(n)}$ exists but depends on i, so the long run distribution does not exist. But we can find $\lim_{n\to\infty} P[X_n = j]$, if we know the initial distribution. We have

$$\begin{aligned} \lim_{n\to\infty} P[X_n = j] &= \sum_{i\in S} \lim_{n\to\infty} p_{ij}^{(n)} p_i^{(0)} = \sum_{i\in C_1} \lim_{n\to\infty} p_{ij}^{(n)} p_i^{(0)} \\ &\quad + \sum_{i\in C_2} \lim_{n\to\infty} p_{ij}^{(n)} p_i^{(0)} + \sum_{i\notin C_1\cup C_2} \lim_{n\to\infty} p_{ij}^{(n)} p_i^{(0)}. \end{aligned}$$

$$\text{For } j \in C_1 \ \lim_{n\to\infty} P[X_n = j] = (1/\mu_j)\sum_{i\in C_1} p_i^{(0)} + \sum_{i\notin C_1\cup C_2} (f_{ij}/\mu_j)p_i^{(0)}$$

$$\text{For } j \in C_2 \ \lim_{n\to\infty} P[X_n = j] = (1/\mu_j)\sum_{i\in C_2} p_i^{(0)} + \sum_{i\notin C_1\cup C_2} (f_{ij}/\mu_j)p_i^{(0)}.$$

Further, if $j \notin C_1 \cup C_2$, then j is transient and hence $\lim_{n\to\infty} p_{ij}^{(n)} = 0$ for any $i \in S$. Hence for $j \notin C_1 \cup C_2$, $\lim_{n\to\infty} P[X_n = j] = 0$. Suppose the initial distribution is $\underline{p}^{(0)} = (0.2, 0.4, 0.1, 0.1, 0.1, 0.1)$. Then using the values of μ_j and f_{ij}, $\lim_{n\to\infty} \underline{p}^{(n)} = (0.2062, 0, 0.3438, 0, 0.15, 0.3)$. Further, using $\lim_{n\to\infty} P^n$ and Eq. (3.2.1),

$$\lim_{n\to\infty} \underline{p}^{(n)} = \underline{p}^{(0)} \lim_{n\to\infty} P^n = (0.2062, 0, 0.3438, 0, 0.15, 0.3).$$

If $\underline{p}^{(0)} = (.2, .1, .1, .2, .2, .2)$, then $\lim_{n\to\infty} \underline{p}^{(n)} = (0.1687, 0, 0.2813, 0, 0.1833, 0.3667)$. Thus, $\lim_{n\to\infty} \underline{p}^{(n)}$ depends on the initial distribution. In Sect. 3.3, we prove that a stationary distribution exists for this Markov chain. In fact, we have an uncountable family of stationary distributions. □

In the following theorems, we investigate under what conditions on the Markov chains, the long run distribution exists.

From Theorem 3.2.2, if j is an ergodic state, then $\lim_{n\to\infty} p_{ij}^{(n)} = 1/\mu_j$ for any state i which communicates with j. In an irreducible Markov chain all states communicate with each other. Hence, in a finite irreducible ergodic Markov chain for all i, $\lim_{n\to\infty} p_{ij}^{(n)} = 1/\mu_j$, which is independent of i. Thus, for such a Markov chain a long run distribution exists, provided $\sum_{j\in S} 1/\mu_j = 1$. We prove this result in the following theorem. We begin with a lemma to prove it.

Lemma 3.2.1 *Suppose $\{X_n, n \geq 0\}$ is a Markov chain with state space S and C is a finite closed communicating class of aperiodic states. Suppose a_j, $j \in S$ is defined as*

$$a_j = \begin{cases} 1/\mu_j, & \text{if } j \in C \\ 0, & \text{if } j \notin C. \end{cases}$$

Then $\{a_j, j \in S\}$ is a probability mass function on S.

Proof Since C is a finite closed communicating class of aperiodic states,

$$\lim_{n\to\infty} p_{ij}^{(n)} = \begin{cases} 1/\mu_j, & \text{if } i \in C, j \in C \\ 0, & \text{if } i \in C, j \notin C, \end{cases}$$

Further, j is an ergodic state that implies $1/\mu_j > 0$. Thus, $a_j \geq 0 \ \forall \ j \in S$. Observe that

$$\forall\ n \geq 1,\ \sum_{j\in C} p_{ij}^{(n)} = 1 \Rightarrow \lim_{n\to\infty}\sum_{j\in C} p_{ij}^{(n)} = 1 \Rightarrow \sum_{j\in C}\lim_{n\to\infty} p_{ij}^{(n)} = 1$$

$$\Rightarrow \sum_{j\in C} 1/\mu_j = 1 \Rightarrow \sum_{j\in S} a_j = 1.$$

Hence, $\{a_j,\ j \in S\}$ is a probability mass function on S. □

The next theorem is about the existence of a long run distribution for a Markov chain with a single closed communicating class.

Theorem 3.2.4 *For a finite Markov chain with a single closed communicating class C of aperiodic states, the long run distribution $\underline{a}$ exists and is given by $a_j = 1/\mu_j$ if $j \in C\ \&\ a_j = 0$ if $j \notin C$.*

Proof Since C is a closed communicating class, all the states in C are non-null persistent. Further, all the states are aperiodic. Hence, from Theorem 3.2.2, $\lim_{n\to\infty} p_{ij}^{(n)} = f_{ij}/\mu_j,\ j \in C$. If $j \notin C$, then j is a transient state and $\lim_{n\to\infty} p_{ij}^{(n)} = 0$. Further,

$$f_{ij} = \begin{cases} 1, \text{ if } & i, j \in C \text{ by Theorem 2.6.1} \\ 0, \text{ if } & i \in C, j \notin C \text{ by Theorem 2.6.1} \\ 1, \text{ if } & i \notin C, i \to j \in C \text{ by Theorem 2.6.11} \\ < 1 \text{ if } & i \notin C, j \notin C \text{ since j is transient} \end{cases}$$

Hence,

$$\lim_{n\to\infty} p_{ij}^{(n)} = \begin{cases} 1/\mu_j, \text{ if } & i, j \in C \\ 0, \text{ if } & i \in C, j \notin C \\ 1/\mu_j, \text{ if } & i \notin C, j \in C \\ 0 \text{ if } & i \notin C, j \notin C, \end{cases}$$

the last equality follows since j is a transient state and hence $\mu_j = \infty$. Thus,

$$a_j = \lim_{n\to\infty} p_{ij}^{(n)} = 1/\mu_j, \text{ if } j \in C\ \&\ a_j = \lim_{n\to\infty} p_{ij}^{(n)} = 0 \text{ if } j \notin C.$$

Hence, $\lim_{n\to\infty} p_{ij}^{(n)}$ exists and does not depend on i. Since $j \in C$ is a non-null persistent state, $1/\mu_j > 0$ and by Lemma 3.2.1 $\sum_{j\in S} a_j = 1$. Hence, the long run distribution exists and is given by $\underline{a}$. □

The next two examples illustrate this theorem.

Example 3.2.2 Suppose $\{X_n, n \geq 0\}$ is a Markov chain with P as given below.

$$P = \begin{array}{c} \\ 1 \\ 2 \\ 3 \\ 4 \\ 5 \end{array}\begin{array}{c} \begin{array}{ccccc} 1 & 2 & 3 & 4 & 5 \end{array} \\ \begin{pmatrix} 0 & 0.5 & 0 & 0.5 & 0 \\ 0.3 & 0.2 & 0 & 0.5 & 0 \\ 0.1 & 0.1 & 0.3 & 0.2 & 0.3 \\ 0.6 & 0 & 0 & 0.4 & 0 \\ 0.1 & 0.1 & 0.4 & 0 & 0.4 \end{pmatrix} \end{array}.$$

Note that $C = \{1, 2, 4\}$ is a single closed communicating class. Thus, the states $1, 2$ and 4 are non-null persistent states and the states 3 and 5 are transient states. All the states are aperiodic. Hence by Theorem 3.2.4, the long run distribution $\underline{a}$ exists and is given by $a_j = 1/\mu_j, \ j \in C \ \ \& \ \ a_j = 0 \ j \notin C$. In Example 2.5.8, using Code 2.8.6, we obtained $\mu_1 = 2.9792$, $\mu_2 = 4.7665$ and $\mu_4 = 2.2$. Thus, $\underline{a} = (0.3357, 0.2098, 0, 0.4545, 0)$. Note that $\sum_{j \in S} a_j = 1$. Using R, it is easy to check that P^n has all identical rows for $n \geq 30$ and each is given by $\underline{a} = (0.3357, 0.2098, 0, 0.4545, 0)$. Further,

$$\begin{aligned} &\lim_{n\to\infty} P[X_n = j] = 0 \ \text{ for } \ j = 3, 5 \\ &\lim_{n\to\infty} P[X_n = j] = a_j = 1/\mu_j \ \text{ for } \ j = 1, 2, 4. \end{aligned}$$

□

Example 3.2.3 For the Markov chain of the care center model in Example 2.2.5, $C = \{3\}$ is the single closed communicating class. 3 being the absorbing state, it is ergodic with $\mu_3 = 1$. The states 1 and 2 are transient states. All the states are aperiodic. Hence by Theorem 3.2.4, the long run distribution $\underline{a}$ exists and is given by $a_j = 1/\mu_j, \ j \in C \ \ \& \ \ a_j = 0 \ j \notin C$. Hence, $\underline{a} = (0, 0, 1)$. Further, for $n \geq 100$, P^n has identical rows given by $(0, 0, 1)$. Thus, for this model,

$$\lim_{n\to\infty} P[X_n = j] = 0, \ \text{ for } \ j = 1, 2 \ \& \ \lim_{n\to\infty} P[X_n = 3] = 1.$$

□

Theorem 3.2.5 *For a finite ergodic Markov chain, the long run distribution $\underline{a}$ exists and is given by $a_j = 1/\mu_j, \ j \in S$.*

Proof For a finite ergodic Markov chain, S is the only closed communicating class of ergodic states. Hence, the proof follows from Theorem 3.2.4, with $C = S$. □

In the following example, we compute the long run distribution for an ergodic Markov chain.

Example 3.2.4 The Markov chain describing the weather model in Example 2.2.6 is an ergodic Markov chain. Hence, $\lim_{n\to\infty} p_{ij}^{(n)} = 1/\mu_j, j \in S$. Using Code 2.8.6, in Example 2.5.9 we have obtained the mean recurrence times as $\mu_1 = 2.0876$, $\mu_2 =$

3.2158, $\mu_3 = 4.7589$. Thus, the long run distribution is $\underline{a} = (0.4790, 0.3109, 0.2101)$. For this Markov chain, all rows of P^n are identical, for all $n \geq 10$. Hence $\lim_{n\to\infty} P^n$ is a matrix with identical rows, each given by $\underline{a}$. Thus, each row represents the long run distribution. □

In the next theorem we examine whether a long run distribution exists, if the state space has two closed communicating classes.

Theorem 3.2.6 *Suppose $\{X_n, n \geq 0\}$ is a finite state space Markov chain, with two closed communicating classes C_1 and C_2 of aperiodic states. Then the long run distribution does not exist.*

Proof Observe that

$$\lim_{n\to\infty} p_{ij}^{(n)} = \begin{cases} 1/\mu_j, & \text{if} \quad i, j \in C_1 \\ 0, & \text{if} \quad i \in C_1, j \notin C_1 \\ 1/\mu_j, & \text{if} \quad i, j \in C_2, \\ 0 & \text{if} \quad i \in C_2, j \notin C_2 \\ f_{ij}/\mu_j & \text{if } i \notin C_1 \cup C_2, j \in S, \end{cases}$$

where f_{ij}/μ_j is defined to be 0 if $\mu_j = \infty$, that is, if j is a transient state. Suppose the rows and columns of P are rearranged so that P is expressible as follows. Then $\lim_{n\to\infty} P^n$ can be presented as shown below.

$$P = \begin{matrix} & C_1 & C_2 & T \\ C_1 & \\ C_2 & \\ T & \end{matrix} \begin{pmatrix} P_1 & 0 & 0 \\ 0 & P_2 & 0 \\ Q_1 & Q_2 & Q_3 \end{pmatrix} \quad \& \quad \lim_{n\to\infty} P^n = \begin{matrix} C_1 \\ C_2 \\ T \end{matrix} \overset{\begin{matrix} C_1 & C_2 & T \end{matrix}}{\begin{pmatrix} A_1 & 0 & 0 \\ 0 & A_2 & 0 \\ A_3 & A_4 & 0 \end{pmatrix}}.$$

Here T is the set of transient states, P_1, P_2 are stochastic matrices corresponding to C_1 and C_2 respectively. The matrices $Q_i, i = 1, 2, 3$ present the transition probabilities from transient states to states in C_1, C_2 and T respectively. In $\lim_{n\to\infty} P^n$, A_i are matrices with identical rows given by $1/\mu_j, j \in C_i, i = 1, 2$. Further, $A_i, i = 3, 4$ are matrices with (i, j)-th element $f_{ij}/\mu_j, i \in T, j \in C_1$ and $i \in T, j \in C_2$ respectively. Thus, $\lim_{n\to\infty} p_{ij}^{(n)}$ exists, but depends on i. Hence, the long run distribution does not exist. □

Example 3.2.5 In Example 3.2.1, we have noted that there are two closed communicating classes $C_1 = \{1, 3\}$ and $C_2 = \{5, 6\}$ of aperiodic states. Hence as proved in Theorem 3.2.6, the long run distribution does not exist. For this example, $A_i, i = 1, 2, 3, 4$ in $\lim_{n\to\infty} P^n$, after suitable rearrangement of rows and columns of P^n, are as shown below

$$A_1 = \begin{array}{c} \\ 1 \\ 3 \end{array}\begin{array}{c} \begin{array}{cc} 1 & 3 \end{array} \\ \begin{pmatrix} 0.3750 & 0.6250 \\ 0.3750 & 0.6250 \end{pmatrix} \end{array}, \quad A_2 = \begin{array}{c} \\ 5 \\ 6 \end{array}\begin{array}{c} \begin{array}{cc} 5 & 6 \end{array} \\ \begin{pmatrix} 3333 & 0.6667 \\ 3333 & 0.6667 \end{pmatrix} \end{array},$$

$$A_3 = \begin{array}{c} \\ 2 \\ 4 \end{array}\begin{array}{c} \begin{array}{cc} 1 & 3 \end{array} \\ \begin{pmatrix} 0.1875 & 0.3125 \\ 0.1875 & 0.3125 \end{pmatrix} \end{array} \ \& \ A_4 = \begin{array}{c} \\ 2 \\ 4 \end{array}\begin{array}{c} \begin{array}{cc} 5 & 6 \end{array} \\ \begin{pmatrix} 0.1667 & 0.3333 \\ 0.1667 & 0.3333 \end{pmatrix} \end{array}.$$

Thus, $\lim_{n\to\infty} P^n$ does not have identical rows and hence the long run distribution does not exist. □

We now extend Lemma 3.2.1, Theorems 3.2.4 and 3.2.5 to the countably infinite S and C.

Lemma 3.2.2 *Suppose $\{X_n, n \geq 0\}$ is a Markov chain with state space S and a countably infinite closed communicating class C of ergodic states. Suppose a_j, $j \in S$ is defined as*

$$a_j = \begin{cases} 1/\mu_j, & \text{if } j \in C \\ 0, & \text{if } j \notin C. \end{cases}$$

Then $\{a_j, j \in S\}$ is a probability mass function on S.

Proof Since C is a countably infinite set, we assume $C = N$, the set of natural numbers. Since all the states in C are ergodic, by Theorem 3.2.2, for any two states $i, j \in C$, $\lim_{n\to\infty} p_{ij}^{(n)} = 1/\mu_j$. Further, states in C which are non-null persistent implies that $1/\mu_i > 0 \ \forall \ i \in C$. Observe that for $i, j \in C$,

$$\begin{aligned} \forall \ n \geq 1, \ \sum_{j\geq 1} p_{ij}^{(n)} = 1 &\Rightarrow \sum_{j=1}^{m} p_{ij}^{(n)} \leq 1, \ \forall \ m \geq 1 \\ &\Rightarrow \lim_{n\to\infty} \sum_{j=1}^{m} p_{ij}^{(n)} = \sum_{j=1}^{m} 1/\mu_j \leq 1, \ \forall \ m \geq 1 \\ &\Rightarrow \lim_{m\to\infty} \sum_{j=1}^{m} 1/\mu_j = \sum_{j=1}^{\infty} 1/\mu_j \leq 1 . \end{aligned}$$

Thus, $\sum_{j=1}^{\infty} 1/\mu_j \leq 1$. Now from the Chapman-Kolmogorov equations, for $i, j \in C$ we have

$$
\begin{aligned}
p_{ij}^{(n+1)} &= \sum_{k\in C} p_{ik}^{(n)} p_{kj} \geq \sum_{k=1}^{m} p_{ik}^{(n)} p_{kj}, \quad \forall\ m \geq 1 \\
\Rightarrow \quad \lim_{n\to\infty} p_{ij}^{(n+1)} &\geq \lim_{n\to\infty} \sum_{k=1}^{m} p_{ik}^{(n)} p_{kj}, \quad \forall\ m \geq 1 \\
\Rightarrow \quad 1/\mu_j &\geq \sum_{k=1}^{m} (1/\mu_k) p_{kj}, \quad \forall\ m \geq 1 \\
\Rightarrow \quad 1/\mu_j &\geq \lim_{m\to\infty} \sum_{k=1}^{m} (1/\mu_k) p_{kj} \ \Rightarrow \ 1/\mu_j \geq \sum_{k=1}^{\infty} (1/\mu_k) p_{kj}\,.
\end{aligned}
$$

Suppose, if possible $1/\mu_j > \sum_{k=1}^{\infty}(1/\mu_k)p_{kj}$ for at least one j. Then summing over $j \geq 1$, we get

$$
\begin{aligned}
\sum_{j=1}^{\infty} 1/\mu_j > \sum_{j=1}^{\infty}\sum_{k=1}^{\infty}(1/\mu_k)p_{kj} &\Rightarrow \sum_{j=1}^{\infty} 1/\mu_j > \sum_{k=1}^{\infty}(1/\mu_k)\left(\sum_{j=1}^{\infty} p_{kj}\right) \\
&\Rightarrow \sum_{j=1}^{\infty} 1/\mu_j > \sum_{k=1}^{\infty}(1/\mu_k)\,,
\end{aligned}
$$

which is impossible since $\sum_{j=1}^{\infty} 1/\mu_j \leq 1$. Hence, $1/\mu_j = \sum_{k=1}^{\infty}(1/\mu_k)p_{kj}$ for all $j \in C$. Thus, we have

$$
1/\mu_j > 0, \qquad \sum_{j=1}^{\infty} 1/\mu_j \leq 1 \quad \& \quad 1/\mu_j = \sum_{k=1}^{\infty}(1/\mu_k)p_{kj} \quad \forall\ j \in C.
$$

Now to examine whether $\sum_{j=1}^{\infty} 1/\mu_j = 1$, observe that

$$
\begin{aligned}
1/\mu_j = \sum_{k=1}^{\infty}(1/\mu_k)p_{kj} &\Rightarrow 1/\mu_j = \sum_{k=1}^{\infty}(1/\mu_k)p_{kj}^{(n)}, \quad \forall\ n \geq 1 \\
&\Rightarrow 1/\mu_j = \lim_{n\to\infty} \sum_{k=1}^{\infty}(1/\mu_k)p_{kj}^{(n)}\,.
\end{aligned}
$$

Note that for all $k \geq 1$, $(1/\mu_k)p_{kj}^{(n)} \to \mu_j/\mu_k$ as $n \to \infty$. Further, $|(1/\mu_k)p_{kj}^{(n)}| \leq 1/\mu_k$ and $\sum_{k=1}^{\infty} 1/\mu_k \leq 1$. Hence, by Theorem 2.4.1,

$$
\begin{aligned}
1/\mu_j = \lim_{n\to\infty}\sum_{k=1}^{\infty}(1/\mu_k)p_{kj}^{(n)} &\Rightarrow 1/\mu_j = \sum_{k=1}^{\infty}(1/\mu_k)\lim_{n\to\infty} p_{kj}^{(n)} \\
&\Rightarrow 1/\mu_j = 1/\mu_j\sum_{k=1}^{\infty}(1/\mu_k) \\
&\Rightarrow \sum_{k=1}^{\infty}(1/\mu_k) = 1\ ,
\end{aligned}
$$

as $1/\mu_j > 0$. We have defined $a_j = 0$ for $j \notin C$. Thus, it follows that

$$a_j \geq 0\ \forall\ j \in S\ \&\ \sum_{j\in S} a_j = 1$$

and $\{a_j,\ j \in S\}$ is a probability mass function on S. □

Theorem 3.2.7 *Suppose $\{X_n, n \geq 0\}$ is a Markov chain with state space S and a countably infinite closed communicating class C of ergodic states. Then the long run distribution $\underline{a}$ exists, where $a_j,\ j \in S$ are given by*

$$a_j = \begin{cases} 1/\mu_j, & \text{if } j \in C \\ 0, & \text{if } j \notin C. \end{cases}$$

Proof The proof is similar to that of Theorem 3.2.4. □

Theorem 3.2.8 *For an ergodic Markov chain with countably infinite state space S, the long run distribution $\underline{a}$ exists, where $a_j = 1/\mu_j,\ j \in S$.*

Proof Since the Markov chain is ergodic, S is the minimal closed class of ergodic states. Hence, the result follows from Theorem 3.2.7 with $C = S$. □

We now examine whether a long run distribution exists for a Markov chain which is either transient or null persistent. It has been proved in Chap. 2, that in a finite state space Markov chain all states cannot be transient and no persistent state is null persistent. Hence, the following theorem is valid for a Markov chain with countably infinite state space.

Theorem 3.2.9 *Suppose all the states of a Markov chain, with countably infinite state space, are either transient or null persistent. Then the long run distribution does not exist.*

Proof We assume the contrary that a long run distribution $\{a_j = \lim_{n\to\infty} p_{ij}^{(n)},\ j \in S\}$ exists, such that $a_j \geq 0,\ \forall\ j \in S$ and $\sum_{j\in S} a_j = 1$. If all the states of a Markov chain are either transient or null persistent, then by Theorem 2.6.5,

$$a_j = \lim_{n\to\infty} p_{ij}^{(n)} = 0, \ \forall \ j \in S \ \Rightarrow \ \sum_{j\in S} a_j = 0\,,$$

which is a contradiction. Hence, the long run distribution does not exist. □

Example 3.2.6 Suppose a Markov chain with state space $\{0, 1, 2, \ldots, \}$ has the transition probability matrix P given by $P = [p_{ij}]$, where $p_{ii} = 1 - p$ and $p_{i,i+1} = p,\ 0 < p < 1,\ \ \forall\ i \in S$. It is easy to verify that all states are inessential and hence transient. Hence, $\lim_{n\to\infty} p_{ij}^{(n)} = a_j = 0$ which does not depend on i, for all $i, j \in S$. However, $\sum_{j\in S} a_j = 0 \neq 1$. Hence, the long run distribution does not exist.

We now investigate whether a long run distribution exists for a non-null persistent periodic Markov chain with period $d > 1$.

Long run distribution for a non-null persistent periodic Markov chain: For the discussion of the limiting behavior of $p_{ij}^{(n)}$, when j is a non-null persistent and periodic state, with period d, we need to study the particular nature of transitions among the states in a periodic Markov chain. Hence we introduce some terminology. Suppose $n = ad + b$, where n, a, d and b are positive integers such that $b = 0, 1, \ldots, d - 1$, then $n - b$ is divisible by d. We write it as $n \equiv b(mod)d$. In other words, when n is divided by d, b is the remainder and hence $b = 0, 1, \ldots, d - 1$. For example, suppose $n = 15, d = 4$, then $n = 15 = 3 \times 4 + 3$, thus $a = 3$ and $b = 3$ and $n \equiv 3(mod)4$. If $n = 17, d = 4$, then $n = 17 = 4 \times 4 + 1$, thus $a = 4$ and $b = 1$ and $n \equiv 1(mod)4$.

Suppose $\{X_n, n \geq 0\}$ is a periodic Markov chain with period d. Thus, all states communicate with each other and all have the same period d. If $i \leftrightarrow j$, then there exists positive integers n & n' such that $p_{ij}^{(n)} > 0$ and $p_{ji}^{(n')} > 0$. Suppose $m \neq n$ is a positive integer such that $p_{ij}^{(m)} > 0$. By the Chapman-Kolmogorov equations,

$$\begin{aligned} p_{ii}^{(n+n')} &= \sum_{k\in S} p_{ik}^{(n)} p_{ki}^{(n')} \geq p_{ij}^{(n)} p_{ji}^{(n')} > 0 \\ \& \ \ p_{ii}^{(m+n')} &= \sum_{k\in S} p_{ik}^{(m)} p_{ki}^{(n')} \geq p_{ij}^{(m)} p_{ji}^{(n')} > 0. \end{aligned}$$

Since the period of state i is d, $p_{ii}^{(n+n')} > 0$ and $p_{ii}^{(m+n')} > 0$ implies that $n + n' \equiv 0(mod)d$, $m + n' \equiv 0(mod)d$ and $m - n \equiv 0(mod)d$. Hence, $n \equiv r(mod)d$ and $m \equiv r(mod)d$, for some integer $r = 0, 1, \ldots, d - 1$. It thus follows that if $p_{ij}^{(n)} > 0$ and $p_{ij}^{(m)} > 0$, then $n \equiv r(mod)d$ and $m \equiv r(mod)d$, for some integer $r = 0, 1, \ldots, d - 1$. Thus, corresponding to every state $j \in S$ and for each fixed $i \in S$, there exists r such that $p_{ij}^{(m)} > 0 \ \ \Rightarrow \ \ m \equiv r(mod)d$, where r is a fixed integer such that $r = 0, 1, \ldots, d - 1$. The integer r is known as the characteristic of state j (Feller [4]).

For fixed i, we define

$$B_r = \{j \mid \text{The characteristic of state } j \text{ is } r\}, \ \ r = 0, 1, \ldots, d - 1.$$

Thus, if $j \in B_r$, then $p_{ij}^{(ad+r)} > 0$ and $p_{ij}^{(n)} = 0$ if $n \neq ad + r$. Since the characteristic of state j is unique, it follows that

$$B_r \cap B_s = \emptyset \ \forall \ r \neq s \ \& \ \bigcup_{r=0}^{d-1} B_r = S,$$

that is, $\{B_r, \ r = 0, 1, \ldots, d-1\}$ is a partition of the state space. If the initial state of a Markov chain is in B_0, then the first transition is to a state in B_1, two step transitions are to a state in B_2, in general, r step transitions are to a state in B_r. After d transitions, the Markov chain returns to the state in B_0. The classes $\{B_0, B_1, \ldots, B_{d-1}\}$ can be ordered cyclically, so that B_r is left to B_{r+1} and B_{d-1} is left to B_0. Hence, $\{B_0, B_1, \ldots, B_{d-1}\}$ are known as cyclically moving classes. With such an ordering, the one step transitions are possible only to a state in the neighboring class to the right and hence a path of d steps leads always to the states of the same class. This implies that in a Markov chain with transition probability matrix as P^d, each class B_r is a closed communicating class of aperiodic states. We summarize these results in the following theorem. For proof one may refer to Sect. 15.9 of Feller [4] or Lemma 6.3.1 and Theorem 6.3.7 of Cinlar [3].

Theorem 3.2.10 *Suppose $\{X_n, n \geq 0\}$ is a non-null persistent and periodic Markov chain with period d. Then all the states in S can be grouped into d disjoint cyclically moving classes $\{B_0, B_1, \ldots, B_{d-1}\}$ such that*

$$p_{ij} = 0 \ \text{unless} \ i \in B_r \ \& \ j \in B_{r+1}, r = 1, 2, \ldots, d-1 \ \text{or} \ i \in B_{d-1} \ \& \ j \in B_0.$$

The classes $\{B_0, B_1, \ldots, B_{d-1}\}$ are closed communicating classes of aperiodic states, for the Markov chain with transition probability matrix P^d. The matrix P^d can be expressed as

$$P^d = \begin{pmatrix} P_0 & 0 & \cdots & 0 \\ 0 & P_1 & \cdots & 0 \\ \vdots & \vdots & \vdots & \vdots \\ 0 & 0 & 0 & P_{d-1} \end{pmatrix},$$

where P_r is the stochastic matrix corresponding to B_r, $r = 0, 1, \ldots, d-1$.

Following example illustrates Theorem 3.2.10. It also shows that long run distribution does not exist for a periodic Markov chain.

Example 3.2.7 The Markov chain in Example 2.7.6 is non-null persistent and periodic with period 2. We now rearrange the rows and columns of P so that rows 2 and 3 are interchanged and columns 2 and 3 are interchanged. Hence, the matrix P with such a rearrangement, labeled as P_a, can be expressed as follows.

$$P = \begin{array}{c} \\ 1 \\ 2 \\ 3 \\ 4 \end{array}\begin{array}{c} \begin{array}{cccc} 1 & 2 & 3 & 4 \end{array} \\ \begin{pmatrix} 0 & 1 & 0 & 0 \\ 1/2 & 0 & 1/2 & 0 \\ 0 & 1/2 & 0 & 1/2 \\ 0 & 0 & 1 & 0 \end{pmatrix} \end{array} \ \& \ P_a = \begin{array}{c} \\ 1 \\ 3 \\ 2 \\ 4 \end{array}\begin{array}{c} \begin{array}{cccc} 1 & 3 & 2 & 4 \end{array} \\ \begin{pmatrix} 0 & 0 & 1 & 0 \\ 0 & 0 & 1/2 & 1/2 \\ 1/2 & 1/2 & 0 & 0 \\ 0 & 1 & 0 & 0 \end{pmatrix} \end{array}.$$

Thus, as specified in Theorem 3.2.10, $B_0 = \{1, 3\}$ and $B_1 = \{2, 4\}$. Observe that $p_{ij} = 0$ for $i, j \in B_0$ and for $i, j \in B_1$. Further observe that

$$P_a^2 = \begin{array}{c} \\ 1 \\ 3 \\ 2 \\ 4 \end{array}\begin{array}{c} \begin{array}{cccc} 1 & 3 & 2 & 4 \end{array} \\ \begin{pmatrix} 1/2 & 1/2 & 0 & 0 \\ 1/4 & 3/4 & 0 & 0 \\ 0 & 0 & 3/4 & 1/4 \\ 0 & 0 & 1/2 & 1/2 \end{pmatrix} \end{array} = \begin{pmatrix} P_0 & 0 \\ 0 & P_1 \end{pmatrix}.$$

Thus, in $d = 2$ steps, transitions are possible from $i \in B_0$ to $j \in B_0$ and from $i \in B_1$ to $j \in B_1$. Note that both P_0 and P_1 are stochastic matrices. Thus, for the Markov chain with transition probability matrix P_a^2, B_0 and B_1 are two closed communicating classes. We further note that

$$P_a^4 = \begin{pmatrix} P_0^2 & 0 \\ 0 & P_1^2 \end{pmatrix}, \ P_a^8 = \begin{pmatrix} P_0^4 & 0 \\ 0 & P_1^4 \end{pmatrix}, \ \text{thus } P_a^{2n} = \begin{pmatrix} P_0^n & 0 \\ 0 & P_1^n \end{pmatrix}, n \geq 1.$$

Hence,

$$\lim_{n\to\infty} P_a^{2n} = \begin{pmatrix} \lim_{n\to\infty} P_0^n & 0 \\ 0 & \lim_{n\to\infty} P_1^n \end{pmatrix}.$$

Thus, $\lim_{n\to\infty} p_{ij}^{(2n)} = 0$, if $i \in B_0$ & $j \in B_1$ or $i \in B_1$ & $j \in B_0$. Since the Markov chains corresponding to P_0 and P_1 are ergodic, $\lim_{n\to\infty} P_0^n$ and $\lim_{n\to\infty} P_1^n$ exist by Theorem 3.2.2. It is easy to check that for $n \geq 20$, P_0^n and P_1^n remain the same. Hence,

$$\lim_{n\to\infty} P_0^n = \begin{array}{c} \\ 1 \\ 3 \end{array}\begin{array}{c} \begin{array}{cc} 1 & 3 \end{array} \\ \begin{pmatrix} 1/3 & 2/3 \\ 1/3 & 2/3 \end{pmatrix} \end{array} \ \& \ \lim_{n\to\infty} P_1^n = \begin{array}{c} \\ 2 \\ 4 \end{array}\begin{array}{c} \begin{array}{cc} 2 & 4 \end{array} \\ \begin{pmatrix} 2/3 & 1/3 \\ 2/3 & 1/3 \end{pmatrix} \end{array}.$$

Thus, $\lim_{n\to\infty} p_{ij}^{(2n)}$ exists but depends on i and hence a long run distribution does not exist for the Markov chain with transition probability matrix P_a^2. The result is consistent with Theorem 3.2.6, as the Markov chain is reducible with two closed communicating classes. We also note that for $n \geq 20$, P^n remains the same. Hence,

$$\lim_{n\to\infty} P^n = \begin{array}{c} \\ 1 \\ 2 \\ 3 \\ 4 \end{array}\begin{array}{c} \begin{array}{cccc} 1 & 2 & 3 & 4 \end{array} \\ \begin{pmatrix} 0.3333 & 0 & 0.6667 & 0 \\ 0 & 0.6667 & 0 & 0.3333 \\ 0.3333 & 0 & 0.6667 & 0 \\ 0 & 0.6667 & 0 & 0.3333 \end{pmatrix} \end{array}.$$

Thus, $\lim_{n\to\infty} p_{ij}^{(n)}$ exists but depends on i and hence a long run distribution does not exist for a periodic Markov chain.

The results about limits of n-step transition probabilities shown in the above example are in general true. We prove these in the following theorems, in which we study $\lim_{n\to\infty} p_{jj}^{(nd)}$ and $\lim_{n\to\infty} p_{ij}^{(nd)}$ for a periodic and non-null persistent state j with period d.

Theorem 3.2.11 *Suppose j is a persistent and periodic state with period d. Then*

$$\lim_{n\to\infty} p_{jj}^{(nd)} = \begin{cases} d/\mu_j, & \text{if } j \text{ is non-null persistent} \\ 0, & \text{if } j \text{ is null persistent.} \end{cases}$$

Proof If j is persistent and periodic state with period d, $p_{jj}^{(n)} = f_{jj}^{(n)} = 0$ if n is not a multiple of d. Hence, $\sum_{n\geq 1} f_{jj}^{(n)} = \sum_{n\geq 1} f_{jj}^{(nd)} = 1$. By Lemma 2.7.1, g.c.d$\{n|f_{jj}^{(nd)} > 0\} =$ g.c.d$\{n|p_{jj}^{(nd)} > 0\} = d$. Suppose in the recurrence relation $p_{jj}^{(nd)} = \sum_{r=1}^{n} f_{jj}^{(rd)} p_{jj}^{(nd-rd)}$, $p_{jj}^{(nd)}$ is replaced by u_n and $f_{jj}^{(nd)}$ is replaced by f_n. We thus have $u_n = f_1 u_{n-1} + f_2 u_{n-2} + \cdots + f_n u_0$. Hence, by the Erdos-Feller-Pollard theorem, $\lim_{n\to\infty} p_{jj}^{(nd)} = 1/\sum_{n\geq 1} n f_{jj}^{(nd)}$. For a non-null persistent and periodic state with period d,

$$\sum_{n\geq 1} n f_{jj}^{(nd)} = (1/d)\sum_{n\geq 1} nd f_{jj}^{(nd)} = \mu_j/d \;\Rightarrow\; \lim_{n\to\infty} p_{jj}^{(nd)} = d/\mu_j.$$

If j is null persistent and periodic state with period d, $p_{jj}^{(n)} = 0$ if n is not a multiple of d. Further, it is proved that if j is null persistent, then $\lim_{n\to\infty} p_{jj}^{(nd)} = 0$. This completes the proof. □

Theorem 3.2.12 *Suppose $\{X_n, n \geq 0\}$ is a non-null persistent and periodic Markov chain with period $d > 1$ and the classes $\{B_0, B_1, \ldots, B_{d-1}\}$ are as defined in Theorem 3.2.10. Then*

$$\lim_{n\to\infty} p_{ij}^{(nd)} = d/\mu_j \;\; \textit{if } i, j \in B_r$$

$$\& \;\; \lim_{n\to\infty} p_{ij}^{(nd)} = 0 \;\; \textit{if } i \in B_r \;\&\; j \in B_s, \;\; s \neq r = 0, 1, \ldots, d-1.$$

Proof As in Theorem 3.2.2, for any state $i, j \in B_r$,

$$\lim_{n\to\infty} p_{ij}^{(nd)} = \lim_{n\to\infty} \sum_{r=1}^{n} f_{ij}^{(rd)} p_{jj}^{(nd-rd)} = \lim_{n\to\infty} \sum_{r=1}^{\infty} f_{ij}^{(rd)} a(n, r)$$

where

$$a(n, r) = \begin{cases} p_{jj}^{(nd-rd)}, & \text{if } r \leq n \\ 0, & \text{if } r > n. \end{cases}$$

Observe that $|f_{ij}^{(rd)} a(n, r)| \leq f_{ij}^{(rd)}$ and $\sum_{r=1}^{\infty} f_{ij}^{(rd)} = f_{ij} \leq 1$. Hence by Theorem 2.4.1,

$$\lim_{n\to\infty} p_{ij}^{(nd)} = \sum_{r=1}^{\infty} f_{ij}^{(rd)} \left(\lim_{n\to\infty} p_{jj}^{(nd-rd)}\right) = d f_{ij}/\mu_j, \text{ since } \lim_{n\to\infty} p_{jj}^{(n-r)d} = d/\mu_j,$$

for each fixed r. If $i, j \in B_r$, then $i \leftrightarrow j$ and $f_{ij} = 1$. Hence, $\lim_{n\to\infty} p_{ij}^{(nd)} = d/\mu_j$ if $i, j \in B_r$. If $i \in B_r$ & $j \in B_s$, $s \neq r$ then $f_{ij} = 0$ and hence the limit is 0. Alternatively, in this case $p_{ij}^{(nd)} = 0$, which also implies that the limit is 0. □

For further discussion on these results, one may refer to Theorem 6.3.10 of Cinlar [3] or Chap. 15 of Feller [4].

Remark 3.2.2 Theorem 3.2.12 conveys that a long run distribution does not exist for a periodic Markov chain.

The following example illustrates Theorem 3.2.12.

Example 3.2.8 Suppose $\{X_n, n \geq 0\}$ is a Markov chain with P given by

$$P = \begin{matrix} & \begin{matrix} 1 & 2 & 3 \end{matrix} \\ \begin{matrix} 1 \\ 2 \\ 3 \end{matrix} & \begin{pmatrix} 0 & 1 & 0 \\ 0.5 & 0 & 0.5 \\ 0 & 1 & 0 \end{pmatrix} \end{matrix}.$$

Since $D_1 = \{n | p_{11}^{(n)} > 0, \ n > 0\} = \{2, 4, 6, \ldots\}, d_1 = 2$. Further, all states communicate with each other and have the same period 2. Thus, it is a non-null persistent, periodic Markov chain with period 2. The odd and even powers of P for $n \geq 1$ are

$$P^{2n} = \begin{matrix} & \begin{matrix} 1 & 2 & 3 \end{matrix} \\ \begin{matrix} 1 \\ 2 \\ 3 \end{matrix} & \begin{pmatrix} 0.5 & 0 & 0.5 \\ 0 & 1 & 0 \\ 0.5 & 0 & 0.5 \end{pmatrix} \end{matrix} \ \& \ P = P^{2n+1} = \begin{matrix} & \begin{matrix} 1 & 2 & 3 \end{matrix} \\ \begin{matrix} 1 \\ 2 \\ 3 \end{matrix} & \begin{pmatrix} 0 & 1 & 0 \\ 0.5 & 0 & 0.5 \\ 0 & 1 & 0 \end{pmatrix} \end{matrix}.$$

From these matrices, we note that for any (i, j), the sequence $\{p_{ij}^{(n)}, n \geq 1\}$ oscillates and hence $\lim_{n\to\infty} p_{ij}^{(n)}$ does not exist. Thus, the long run distribution does not exist for the periodic Markov chain as already noted. Using Code 2.8.6, we obtain the mean recurrence times. These are given by $\mu_1 = 4$, $\mu_2 = 2$ and $\mu_3 = 4$. Here $B_0 = \{1, 3\}$ and $B_1 = \{2\}$. For the Markov chain with P^2 as the transition probability matrix, B_0 and B_1 are closed communicating classes with transition probability matrices P_0 and P_1 respectively, as given below. Further, from powers of P^2, we get $\lim_{n\to\infty} P^{2n}$, as follows.

$$P_0 = \begin{array}{c} \\ 1 \\ 3 \end{array}\begin{array}{c} \begin{array}{cc} 1 & 3 \end{array} \\ \begin{pmatrix} 0.5 & 0.5 \\ 0.5 & 0.5 \end{pmatrix} \end{array}, \quad P_1 = 2 \begin{array}{c} 2 \\ \begin{pmatrix} 1 \end{pmatrix} \end{array} \;\&\; \lim_{n\to\infty} P^{2n} = \begin{array}{c} \\ 1 \\ 2 \\ 3 \end{array}\begin{array}{c} \begin{array}{ccc} 1 & 2 & 3 \end{array} \\ \begin{pmatrix} 0.5 & 0 & 0.5 \\ 0 & 1 & 0 \\ 0.5 & 0 & 0.5 \end{pmatrix} \end{array}.$$

Hence,

$$\lim_{n\to\infty} p_{ij}^{(2n)} = \begin{cases} 0, & \text{if } i = 1, 3 \;\&\; j = 2 \\ 0, & \text{if } i = 2,\ j = 1, 3 \\ d/\mu_1 = 2/4 = 0.5, & \text{if } i = 1, 3,\ j = 1 \\ d/\mu_3 = 2/4 = 0.5, & \text{if } i = 1, 3,\ j = 3 \\ d/\mu_2 = 2/2 = 1, & \text{if } i = 2,\ j = 2. \end{cases}$$

Thus, $\lim_{n\to\infty} p_{ij}^{(2n)} = f_{ij} d/\mu_j,\ j \in S$, where $f_{ij} = 1$ when i, j are in the same class and 0 otherwise. Thus, the long run distribution does not exist. Although the long run distribution does not exist for a periodic Markov chain, the stationary distribution exists. Solving the equation $\underline{\pi} = \underline{\pi} P$, we get $\underline{\pi} = (0.25, 0.5, 0.25)$. Note that $\pi_i = 1/\mu_i,\ i = 1, 2, 3$. □

In Sect. 3.3, we prove that a unique stationary distribution exists for a non-null persistent and periodic Markov chain and is given by $\pi_i = 1/\mu_i,\ i \in S$.

To summarize this section, we note that the long run distribution exists, if the Markov chain is either ergodic or if S has a single closed communicating class of aperiodic states. In all these cases, a stationary distribution and the limit of a marginal distribution also exist and the three distributions are the same. The long run distribution does not exist, if a Markov chain is either transient or null persistent and also for periodic Markov chains.

In the next section, we study some more results about the stationary distributions of a Markov chain.

3.3 Stationary Distribution

In Sect. 3.2, we have noted that whenever a long run distribution exists, then a stationary distribution also exists. We have seen in Examples 3.1.1 and 3.2.8 that even if a long run distribution does not exist, a stationary distribution exists. Thus, it is of interest to find out the other cases for which a stationary distribution exists. Hence, the aim of the present section is to address the following issues related to a stationary distribution.

(i) How the marginal distributions and joint distributions are related to a stationary distribution?
(ii) Under what conditions on a Markov chain, does a stationary distribution exist?
(iii) If a stationary distribution exists, under what conditions is it unique?
(iv) What are the methods to compute a stationary distribution?

In the following two theorems, we address the first issue.

Theorem 3.3.1 *Suppose $\{X_n, n \geq 0\}$ is a Markov chain specified by $(S, \underline{p}^{(0)}, P)$. (i) If the initial distribution of the Markov chain is the same as its stationary distribution, then the distribution of X_n for each n is the same as the stationary distribution and hence also its limit. (ii) Conversely if X_0 and X_1 are identically distributed with common probability distribution $\underline{b}$, then $\underline{b}$ is a stationary distribution of the Markov chain.*

Proof (i) Suppose $\{\pi_i, i \in S\}$ is a stationary distribution and $P[X_0 = i] = \pi_i, i \in S$. Then for any $j \in S$ and $\forall\ n \geq 1$,

$$P[X_n = j] = \sum_{i \in S} P[X_n = j, X_0 = i] = \sum_{i \in S} p_{ij}^{(n)} P[X_0 = i] = \sum_{i \in S} p_{ij}^{(n)} \pi_i = \pi_j \ .$$

Thus X_n for all $n \geq 1$ are identically distributed with the common distribution as the stationary distribution. Hence, $\lim_{n\to\infty} P[X_n = j] = \pi_j,\ j \in S$.

(ii) Conversely, suppose

$$P[X_1 = j] = P[X_0 = j] = b_j,\ \ b_j \geq 0\ \forall\ j \in S\ \&\ \sum_{i \in S} b_j = 1.$$

Observe that $\forall\ j \in S$

$$b_j = P[X_1 = j] = \sum_{i \in S} P[X_1 = j | X_0 = i] P[X_0 = i] = \sum_{i \in S} b_i p_{ij},$$

that is, $\underline{b} = \underline{b}P$. Hence, $\underline{b}$ is the stationary distribution of the Markov chain. By (i), $X_n\ \forall\ n \geq 0$ are identically distributed with common distribution $\underline{b}$. □

We have noted the first result of Theorem 3.3.1 in Example 3.1.1. We now extend Theorem 3.3.1 to the joint distribution of X_n's.

Theorem 3.3.2 *If the initial distribution of a Markov chain is the same as its stationary distribution, then the Markov chain is a stationary process.*

Proof Suppose $P[X_0 = i] = \pi_i, \ i \in S$, where $\{\pi_i, i \in S\}$ is a stationary distribution. We obtain the joint distribution of $\{X_{t_1}, X_{t_2}, \ldots, X_{t_n}\}$ for $t_1 < t_2 < \cdots < t_n \in W$ in terms of higher step transition probabilities using the Markov property repeatedly and the initial distribution. Thus, for any $x_1, x_2, \ldots, x_n \in S$ we have

$$\begin{aligned}
&P[X_{t_n} = x_n, X_{t_{n-1}} = x_{n-1}, \ldots, X_{t_1} = x_1] \\
&\quad = P[X_{t_n} = x_n | X_{t_{n-1}} = x_{n-1}, \ldots, X_{t_1} = x_1] \\
&\quad \times P[X_{t_{n-1}} = x_{n-1}, \ldots, X_{t_1} = x_1] \\
&\quad = P[X_{t_n} = x_n | X_{t_{n-1}} = x_{n-1}] \\
&\quad \times P[X_{t_{n-1}} = x_{n-1}, X_{t_{n-2}} = x_{n-2}, \ldots, X_{t_1} = x_1] \\
&\quad = p_{x_{n-1}x_n}^{(t_n - t_{n-1})} \times P[X_{t_{n-1}} = x_{n-1}, \ldots, X_{t_1} = x_1] \\
&\quad = p_{x_{n-1}x_n}^{(t_n - t_{n-1})} \times p_{x_{n-2}x_{n-1}}^{(t_{n-1} - t_{n-2})} \times \cdots \times p_{x_1x_2}^{(t_2 - t_1)} P[X_{t_1} = x_1] \\
&\quad = \pi_{x_1} \prod_{r=2}^{n} p_{x_{r-1}x_r}^{(t_r - t_{r-1})},
\end{aligned}$$

since $P[X_{t_1} = x_1] = \pi_{x_1}$ as proved in Theorem 3.3.1. Suppose $h > 0$ is any real number such that $t_j + h \in W, \quad j = 1, 2, \ldots, n.$ Then we find $P[X_{t_n+h} = x_n, \ldots, X_{t_1+h} = x_1]$, using the similar steps as in the above derivation with t_j replaced by $t_j + h$. Thus we have

$$\begin{aligned}
P[X_{t_n+h} = x_n, \ldots, X_{t_1+h} = x_1] &= P[X_{t_1+h} = x_1] \prod_{r=2}^{n} p_{x_{r-1}x_r}^{(t_r - t_{r-1})} \\
&= \pi_{x_1} \prod_{r=2}^{n} p_{x_{r-1}x_r}^{(t_r - t_{r-1})},
\end{aligned}$$

since $P[X_{t_1+h} = x_1] = \pi_{x_1}$. Hence, the joint distributions of $\{X_{t_1}, X_{t_2}, \ldots, X_{t_n}\}$ and $\{X_{t_1+h}, X_{t_2+h}, \ldots, X_{t_n+h}\}$ are the same for any $h > 0$. Thus, the family of finite dimensional distributions remain invariant under translations along the time axis. Hence, the Markov chain is a stationary process, if the initial distribution of a Markov chain is the same as its stationary distribution. □

Example 3.3.1 Suppose $\{X_n, n \geq 0\}$ is a Markov chain with $S = \{1, 2, 3\}$ and P given by

$$P = \begin{matrix} & \begin{matrix} 1 & 2 & 3 \end{matrix} \\ \begin{matrix} 1 \\ 2 \\ 3 \end{matrix} & \begin{pmatrix} 0.6 & 0.2 & 0.2 \\ 0.1 & 0.7 & 0.2 \\ 0.1 & 0.1 & 0.8 \end{pmatrix} \end{matrix}.$$

Solving the equation $\underline{\pi} = \underline{\pi}P$, we get the stationary distribution as $\underline{\pi} = (0.2, 0.3, 0.5)$. Suppose the initial distribution $\underline{p}^{(0)} = \underline{\pi}$. Then by Theorem 3.3.2, the finite dimensional distributions remains invariant under translations along the time axis. We verify this result by computing the joint distribution of $\{X_4, X_6, X_9\}$ and $\{X_8, X_{10}, X_{13}\}$, where $h = 4$. From the output of Code 2.8.3, we note that these two joint distributions are the same. A partial output is displayed below.

$$
\begin{aligned}
P[X_4 = 1, X_6 = 1, X_9 = 1] &= P[X_8 = 1, X_{10} = 1, X_{13} = 1] = 0.0240\\
P[X_4 = 1, X_6 = 1, X_9 = 2] &= P[X_8 = 1, X_{10} = 1, X_{13} = 2] = 0.0246\\
P[X_4 = 1, X_6 = 1, X_9 = 3] &= P[X_8 = 1, X_{10} = 1, X_{13} = 3] = 0.0314\\
P[X_4 = 1, X_6 = 2, X_9 = 1] &= P[X_8 = 1, X_{10} = 2, X_{13} = 1] = 0.0098\\
P[X_4 = 1, X_6 = 2, X_9 = 2] &= P[X_8 = 1, X_{10} = 2, X_{13} = 2] = 0.0242\\
P[X_4 = 1, X_6 = 2, X_9 = 3] &= P[X_8 = 1, X_{10} = 2, X_{13} = 3] = 0.0220.
\end{aligned}
$$

□

We now investigate the conditions under which a unique stationary distribution exists for a Markov chain. We begin with the following definition.

Definition 3.3.1 Stationary Distribution Concentrated on a Closed Class: Suppose $\{X_n, n \geq 0\}$ is a Markov chain where C is a closed class. A stationary distribution $\{\pi_i, i \in S\}$ such that $\pi_i = 0$ for $i \notin C$ is known as the stationary distribution concentrated on C.

Theorem 3.3.3 *(i) For a Markov chain with a single closed communicating class C of aperiodic states, a unique stationary distribution $\{\pi_i, i \in S\}$ concentrated on C exists, where $\pi_i = 1/\mu_i$ for $i \in C$ and $\pi_i = 0$ for $i \notin C$.*
(ii) For an ergodic Markov chain with state space S, a unique stationary distribution $\{\pi_i, i \in S\}$ exists, where $\pi_i = 1/\mu_i$ for $i \in S$.

Proof (i) In Sect. 3.2, in Theorems 3.2.4 and 3.2.7, we have proved that a long run distribution exists for a Markov chain with a single closed communicating class C of aperiodic states, when S is finite and S is countably infinite respectively. Hence, by Theorem 3.1.1, a stationary distribution exists and is given by $\{\pi_i, i \in S\}$, where $\pi_i = 1/\mu_i$ for $i \in C$ and $\pi_i = 0$ for $i \notin C$.

We now prove the uniqueness of the stationary distribution. Suppose if possible, $\{a_j, \ j \in S\}$ is another stationary distribution of the Markov chain. Then for all $i \in S, \ a_i \geq 0, \sum_{i \in S} a_i = 1$ and $\forall \ j \in S, a_j = \sum_{i \in S} a_i p_{ij}$. Now for $j \in C$,

$$
a_j = \sum_{i \in S} a_i p_{ij} \ \Rightarrow \ a_j = \sum_{i \in S} a_i p_{ij}^{(n)}, \ \forall \ n \geq 1 \Rightarrow \ a_j = \lim_{n \to \infty} \sum_{i \in S} a_i p_{ij}^{(n)}
$$

$$
\Rightarrow \ a_j = \sum_{i \in S} a_i \lim_{n \to \infty} p_{ij}^{(n)} = \sum_{i \in S} a_i (1/\mu_j) = 1/\mu_j \ .
$$

Thus, $a_j = 1/\mu_j$ for all $j \in C$. Further for $j \notin C$, $\lim_{n\to\infty} p_{ij}^{(n)} = 0$, j being transient. Hence, $a_j = \sum_{i\in S} a_i \lim_{n\to\infty} p_{ij}^{(n)} = 0$. Thus, the uniqueness of the stationary distribution of the Markov chain is proved.
(ii) In Theorems 3.2.5 and 3.2.8, it is proved that a long run distribution exists for an ergodic Markov chain, when S is finite and S is countably infinite respectively. Hence, again by Theorem 3.1.1, a stationary distribution exists and is given by $\{\pi_i, i \in S\}$, where $\pi_i = 1/\mu_i$ for $i \in S$. It also follows from (i) with $C = S$. In this setup uniqueness follows on similar lines as in (i). □

Remark 3.3.1 (i) Theorem 3.3.3 conveys that if the state space S has a single closed communicating class C of aperiodic states, then it is enough to compute the stationary distribution concentrated on C.

We now proceed to the discussion of a stationary distribution for a periodic non-null persistent Markov chain.

Stationary distribution for a non-null persistent periodic Markov chain: We have noted in Sect. 3.2 that a long run distribution does not exist for a periodic Markov chain. However, we have noted in Example 3.2.8, that a stationary distribution exists for such a Markov chain. We now prove this result. We first prove a lemma needed in the proof. In Theorem 3.2.12, it is proved that $\lim_{n\to\infty} p_{ij}^{(nd)} = d/\mu_j$, if $i, j \in B_r$ and 0 otherwise. We now discuss $\lim_{n\to\infty} p_{ij}^{(nd+s)}$.

Lemma 3.3.1 *Suppose $\{X_n, n \geq 0\}$ is a non-null persistent and periodic Markov chain with period $d > 1$. Suppose the classes $\{B_0, B_1, \ldots, B_{d-1}\}$ are as defined in Theorem 3.2.10. Then*

$$\lim_{n\to\infty} p_{ij}^{(nd+s)} = \begin{cases} d/\mu_j, & \text{if} \quad i \in B_r, \ j \in B_{r+s} \\ 0, & \text{otherwise.} \end{cases}$$

If $r + s \geq d$, then B_{r+s} is defined as B_{r+s-d}.

Proof From the Chapman-Kolmogorov equations, $p_{ij}^{(nd+s)} = \sum_{l\in S} p_{il}^{(s)} p_{lj}^{(nd)}$.

$$\begin{aligned} \text{Now, } i \in B_r &\Rightarrow p_{il}^{(s)} = 0 \text{ if } l \notin B_{r+s} \\ \& \ l \in B_{r+s} &\Rightarrow p_{lj}^{(nd)} = 0 \text{ if } j \notin B_{r+s} \\ &\Rightarrow p_{ij}^{(nd+s)} = \sum_{l\in B_{r+s}} p_{il}^{(s)} p_{lj}^{(nd)}, \text{ for } i \in B_r, \ j \in B_{r+s} \end{aligned}$$

for fixed r. Note that $l, j \in B_{r+s}$ and hence $\lim_{n\to\infty} p_{lj}^{(nd)} = d/\mu_j$ by Theorem 3.2.12. Further, $\sum_{l\in B_{r+s}} p_{il}^{(s)} \leq 1$. Hence by Theorem 2.4.1, limit and summation can be interchanged in $\lim_{n\to\infty} \sum_{l\in B_{r+s}} p_{il}^{(s)} p_{lj}^{(nd)}$. Thus for $i \in B_r$ and $j \in B_{r+s}$,

$$
\begin{aligned}
p_{ij}^{(nd+s)} &= \sum_{l \in S} p_{il}^{(s)} p_{lj}^{(nd)} = \sum_{l \in B_{r+s}} p_{il}^{(s)} p_{lj}^{(nd)} \\
\Rightarrow \quad \lim_{n\to\infty} p_{ij}^{(nd+s)} &= \lim_{n\to\infty} \sum_{l \in B_{r+s}} p_{il}^{(s)} p_{lj}^{(nd)} = \sum_{l \in B_{r+s}} p_{il}^{(s)} \lim_{n\to\infty} p_{lj}^{(nd)} \\
&= (d/\mu_j) \sum_{l \in B_{r+s}} p_{il}^{(s)} = (d/\mu_j) \sum_{l \in S} p_{il}^{(s)} = d/\mu_j.
\end{aligned}
$$

If $i \in B_r$ and $j \notin B_{r+s}$, $p_{ij}^{(nd+s)} = 0$ and hence its limit is 0. □

In the following example we verify Lemma 3.3.1.

Example 3.3.2 Suppose $\{X_n, n \geq 0\}$ is a Markov chain with P given by

$$
P = \begin{array}{c} \\ 1 \\ 2 \\ 3 \end{array} \begin{array}{c} \begin{array}{ccc} 1 & 2 & 3 \end{array} \\ \begin{pmatrix} 0 & 1 & 0 \\ 0.5 & 0 & 0.5 \\ 0 & 1 & 0 \end{pmatrix} \end{array}.
$$

In Example 3.2.8, we have noted that it is a non-null persistent, periodic Markov chain with period 2, $B_0 = \{1, 3\}$ and $B_1 = \{2\}$ are cyclically moving classes and the mean recurrence times are given by $\mu_1 = 4$, $\mu_2 = 2$ and $\mu_3 = 4$. Further, for $n \geq 1$

$$
P^{2n+1} = \begin{array}{c} \\ 1 \\ 2 \\ 3 \end{array} \begin{array}{c} \begin{array}{ccc} 1 & 2 & 3 \end{array} \\ \begin{pmatrix} 0 & 1 & 0 \\ 0.5 & 0 & 0.5 \\ 0 & 1 & 0 \end{pmatrix} \end{array} \Rightarrow \lim_{n\to\infty} P^{2n+1} = \begin{array}{c} \\ 1 \\ 2 \\ 3 \end{array} \begin{array}{c} \begin{array}{ccc} 1 & 2 & 3 \end{array} \\ \begin{pmatrix} 0 & 1 & 0 \\ 0.5 & 0 & 0.5 \\ 0 & 1 & 0 \end{pmatrix} \end{array}.
$$

From this matrix we note that

$$
\lim_{n\to\infty} p_{ij}^{(2n+1)} = \begin{cases} 0, & \text{if} \quad i \in B_0 \ \& \ j \in B_0 \\ 0, & \text{if} \quad i \in B_1 \ \& \ j \in B_1 \\ d/\mu_2 = 2/2 = 1, & \text{if} \quad i \in B_0 \ \& \ j \in B_1 = \{2\} \\ d/\mu_1 = 2/4 = 0.5, & \text{if } i \in B_1 = \{2\} \ \& \ j = 1 \in B_0 \\ d/\mu_3 = 2/4 = 0.5, & \text{if } i \in B_1 = \{2\} \ \& \ j = 3 \in B_0. \end{cases}
$$

Thus, $\lim_{n\to\infty} p_{ij}^{(nd+s)} = d/\mu_j$, if $i \in B_r$ and $j \in B_{r+s}$. It is 0 otherwise. □

The next theorem is about the existence of a stationary distribution for a non-null persistent and periodic Markov chain.

Theorem 3.3.4 *For a non-null persistent and periodic Markov chain with period d, a unique stationary distribution* $\underline{\pi} = \{\pi_j, j \in S\}$ *exists and* $\pi_j = 1/\mu_j$.

Proof Since the Markov chain is non-null persistent, $\pi_j = 1/\mu_j > 0 \ \forall \ j \in S$. Note that for each $r = 0, 1, \ldots, d-1$, B_r is a closed communicating class in a Markov chain with transition probability matrix P^d. In Theorem 3.2.12, it is proved that $\lim_{n\to\infty} p_{ij}^{(nd)} = d/\mu_j$, if $i, j \in B_r$. Thus, for the Markov chain with state space as B_r, all states in B_r are aperiodic and hence $\{d/\mu_j,\ j \in B_r\}$ is a long run distribution for the Markov chain with state space as B_r. Hence, by Theorem 3.2.1 or Theorem 3.2.2, $\sum_{j\in B_r} d/\mu_j = 1$. Since $\{B_r,\ r = 0, \ldots, d-1\}$ is a partition of S we have

$$\sum_{j\in S} 1/\mu_j = \sum_{r=0}^{d-1}\sum_{j\in B_r} 1/\mu_j = \sum_{r=0}^{d-1}(1/d)\sum_{j\in B_r} d/\mu_j = \sum_{r=0}^{d-1}(1/d) = 1.$$

To examine whether $\pi_j = 1/\mu_j$ satisfies the third condition in the definition of a stationary distribution, note that by the Chapman-Kolmogorov equations, for $i \in B_r$ and $j \in B_{r+1}$, $p_{ij}^{(nd+1)} = \sum_{l\in S} p_{il}^{(nd)} p_{lj}$. Since $j \in B_{r+1}$, $p_{lj} = 0$ if $l \notin B_r$. Thus,

$$\begin{aligned}
p_{ij}^{(nd+1)} &= \sum_{l\in S} p_{il}^{(nd)} p_{lj} = \sum_{l\in B_r} p_{il}^{(nd)} p_{lj} \\
\Rightarrow \quad \lim_{n\to\infty} p_{ij}^{(nd+1)} &= \lim_{n\to\infty} \sum_{l\in B_r} p_{il}^{(nd)} p_{lj} \\
d/\mu_j &= \sum_{l\in B_r} \lim_{n\to\infty} p_{il}^{(nd)} p_{lj} \quad \text{by Lemma 3.3.1} \\
d/\mu_j &= \sum_{l\in B_r} (d/\mu_l) p_{lj} \quad \text{by Theorem 3.2.12} \\
1/\mu_j &= \sum_{l\in S} (1/\mu_l) p_{lj}\ .
\end{aligned}$$

In the third step, the interchange of limit and summation is justified by Theorem 2.4.1. Thus, $\underline{\pi} = \{\pi_j = 1/\mu_j,\ j \in S\}$ is a stationary distribution of the non-null persistent and periodic Markov chain. Uniqueness follows as in Theorem 3.3.3. □

We now discuss how the stationary distribution $\underline{\pi}$ of a non-null persistent and periodic Markov chain with period d is related to long run distributions of the Markov chains with state space B_r, $r = 0, 1, \ldots, d-1$. As stated in Theorem 3.2.10, for the Markov chain with transition probability matrix P^d, the classes $\{B_0, B_1, \ldots, B_{d-1}\}$ are closed communicating classes of aperiodic states. Hence, as noted in the above proof, for a Markov chain with state space as B_r and having k_r states, a long run distribution $\underline{a}_r = (d/\mu_1, d/\mu_2, \ldots, d/\mu_{k_r})'$ exists $\forall\ r = 0, 1, \ldots, d-1$. Note that $\sum_{r=0}^{d-1} k_r$ is the total number of states in S. Suppose $\underline{a} = (1/d)(\underline{a}_0, \underline{a}_1, \ldots, \underline{a}_{d-1})$. Then

$$\underline{\pi} = \underline{a}, \tag{3.3.1}$$

with appropriate renumbering of the states as in the classes $\{B_0, B_1, \ldots, B_{d-1}\}$. We illustrate such a relation in Example 3.4.10, after discussing the methods of computing the stationary distributions. Following example also illustrates this result.

Example 3.3.3 Suppose $\{X_n, n \geq 0\}$ is a Markov chain with P given by

$$P = \begin{array}{c} \\ 1 \\ 2 \\ 3 \\ 4 \\ 5 \end{array} \begin{array}{c} \begin{array}{ccccc} 1 & 2 & 3 & 4 & 5 \end{array} \\ \begin{pmatrix} 0 & 1 & 0 & 0 & 0 \\ 0 & 0 & 1 & 0 & 0 \\ 1 & 0 & 0 & 0 & 0 \\ 1/10 & 1/5 & 2/5 & 1/10 & 1/5 \\ 1/5 & 1/5 & 1/5 & 1/5 & 1/5 \end{pmatrix} \end{array}.$$

We find its stationary distribution. Observe that $C = \{1, 2, 3\}$ is a closed communicating class and each state in C has period 3, while 4, 5 are aperiodic transient states. Thus, the Markov chain with state space as C is a non-null persistent and periodic chain with period 3. Its stationary distribution $\underline{b}$ exists and it is unique. Hence, to find the stationary distribution $\underline{\pi}$ for the Markov chain with transition probability matrix P, we need to compute only $\underline{b}$, since the states 4, 5 being transient, $\pi_4 = \pi_5 = 0$. It is easy to solve $\underline{b} = \underline{b}P_1$, where P_1 is a stochastic matrix corresponding to state space C. From $\underline{b} = \underline{b}P_1$, we have $b_1 = b_2 = b_3$ and hence $b_i = 1/3 \; \forall \; i = 1, 2, 3$. Thus, $1/\mu_i = 1/3, i \in C$. Hence, the unique stationary distribution is $\underline{\pi} = (1/3, 1/3, 1/3, 0, 0)$. In P^3, we note that there are three closed communicating classes $C_i = \{i\}$ and corresponding $P_i = [1], i = 1, 2, 3$. Hence, the stationary distributions are $\underline{a}_i = (1), i = 1, 2, 3$ and $\underline{a} = (1/d)(\underline{a}_1, \underline{a}_2, \underline{a}_3) = (1/3, 1/3, 1/3)$. Then by Eq. (3.3.1), $\underline{b} = \underline{a} = (1/3, 1/3, 1/3)$ and $\underline{\pi} = (1/3, 1/3, 1/3, 0, 0)$. □

Remark 3.3.2 If a periodic Markov chain with period d and with transition probability matrix P is irreducible, then the Markov chain with transition probability matrix P^d is reducible with d closed communicating classes. We compute in Example 3.4.9, the stationary distribution of the two Markov chains with transition probability matrix P and P^d, after discussing various methods to compute a stationary distribution and note the relation between the two. We will note that the stationary distribution of the Markov chain with transition probability matrix P^d is not unique, but a particular convex combination of the stationary distributions of P^d coincides with the stationary distribution of a Markov chain with transition probability matrix P.

Remark 3.3.3 Theorems 3.3.3 and 3.3.4 state that a unique stationary distribution exists for a non-null persistent Markov chain. Converse of this theorem is also true, that is, if a Markov chain is irreducible and its stationary distribution exists, then the Markov chain is non-null persistent. Thus, a necessary and sufficient condition for determining non-null persistence is simply demonstrating the existence of a stationary distribution. We use this result in Sect. 4.2. Further, these theorems provide a method for computing the mean recurrence times. Thus, we compute the stationary distribution, which is the same as the long run distribution for ergodic Markov chain, and invert the components of interest. This result is also used in Sect. 4.2.

We now proceed to examine the existence of a stationary distribution for a Markov chain which is either null persistent or transient.

Stationary distribution for a null persistent or transient Markov chain: In Theorem 3.2.9, it is proved that a long run distribution does not exist for a Markov chain which is either null persistent or transient. We have a similar result for a stationary distribution. We prove it in Theorem 3.3.6, using the result proved in the following theorem.

Theorem 3.3.5 *Suppose $\underline{\pi} = \{\pi_j, j \in S\}$ is a stationary distribution. If state j is either null persistent or transient, then $\pi_j = 0$.*

Proof Suppose j is either a null persistent state or a transient state. Then for any state $i \to j$, $\lim_{n\to\infty} p_{ij}^{(n)} = 0$ as proved in Theorem 2.6.5. Since $\underline{\pi} = \{\pi_j, j \in S\}$ is a stationary distribution, we have (i) $\pi_i \geq 0, \ \forall \ i \in S$,
(ii) $\sum_{i\in S} \pi_i = 1$ and (iii) $\pi_j = \sum_{i\in S} \pi_i p_{ij}$. Further $\forall \ j \in S$,

$$\begin{aligned}
\pi_j = \sum_{i\in S} \pi_i p_{ij} \quad &\Rightarrow \quad \pi_j = \sum_{i\in S} \pi_i p_{ij}^{(n)}, \quad \forall \ n \geq 1 \\
\Rightarrow \quad \pi_j = \lim_{n\to\infty} \sum_{i\in S} \pi_i p_{ij}^{(n)} &= \sum_{i\in S} \lim_{n\to\infty} \pi_i p_{ij}^{(n)} = 0,
\end{aligned}$$

if j is either null persistent or transient. In the last step, the limit and summation can be interchanged by Theorem 2.4.1, since $\pi_i p_{ij}^{(n)} \to 0$ for all $i \in S$ and $\pi_i p_{ij}^{(n)} \leq \pi_i$ with $\sum_{i\in S} \pi_i \leq 1$. Thus, if j is either null persistent or transient, then $\pi_j = 0$. □

Theorem 3.3.6 *For a null persistent or transient Markov chain with countably infinite state space S, a stationary distribution does not exist.*

Proof To prove that a stationary distribution does not exist, we assume the contrary that a stationary distribution $\{\pi_i, i \in S\}$ exists. Thus, (i) $\pi_i \geq 0$ for all $i \in S$, (ii) $\sum_{i\in S} \pi_i = 1$ and (iii) $\pi_j = \sum_{i\in S} \pi_i p_{ij}$. Since all the states are either null persistent or transient, from Theorem 3.3.5, $\pi_i = 0, \ \forall \ i \in S$, which implies that $\sum_{i\in S} \pi_i = 0$ which is a contradiction. It then follows that for a null persistent or a transient Markov chain, a stationary distribution does not exist. □

Remark 3.3.4 (i) In Theorem 3.3.6 we assume that the state space S is countably infinite because if a state space is finite, then all states cannot be either transient or null persistent, as proved in Theorem 2.6.7.
(ii) If a stationary distribution does not exist, then in view of Theorem 3.1.1, a long run distribution also does not exist. We have already proved this result in Theorem 3.2.9.

In Chap. 4, we show that in an unrestricted random walk with state space as a set of integers, all the states are either transient or null persistent. A stationary distribution does not exist for this Markov chain.

One more illustration of a Markov chain which is either transient or null persistent is a Markov chain with $S = W$, the set of whole numbers, and the transition probability matrix to be a doubly stochastic matrix. In the following Theorem 3.3.7 we prove it.

Theorem 3.3.7 *(i) A Markov chain with state space W and a doubly stochastic transition probability matrix P is either transient or null persistent. (ii) A stationary distribution does not exist for such a Markov chain.*

Proof We prove result (i) by contradiction, hence we assume that the Markov chain is non-null persistent and aperiodic. Hence, $\lim_{n\to\infty} p_{ik}^{(n)} = 1/\mu_k > 0$, since $1 \leq \mu_k < \infty \ \forall \ k \in S$. Since P is a doubly stochastic matrix, by Lemma 2.2.1, P^n is also a doubly stochastic matrix for any $n \geq 1$. Hence for $k \in S$,

$$1 = \sum_{i\in S} p_{ik}^{(n)} \geq \sum_{i=1}^{m} p_{ik}^{(n)}, \ \ \forall \ m \geq 1$$

$$\Rightarrow \ 1 \geq \lim_{n\to\infty} \sum_{i=1}^{m} p_{ik}^{(n)} = \sum_{i=1}^{m} 1/\mu_k = m/\mu_k, \ \ \forall \ m \geq 1 \,.$$

Thus, $m/\mu_k \leq 1 \ \forall \ m \geq 1$ and for all $k \in S$. For this relation to be valid, we must have $1/\mu_k = 0 \iff \mu_k = \infty$, which is a contradiction. If the Markov chain is non-null persistent and periodic, using similar arguments we arrive at the contradiction. Hence, the Markov chain is either transient or null persistent. Result (ii) follows from Theorem 3.3.6. □

We have a different scenario, if the state space is finite and the transition probability matrix is doubly stochastic. It is proved in the following theorem.

Theorem 3.3.8 *For a Markov chain with* $S = \{1, 2, \ldots, M\}$ *and a doubly stochastic transition probability matrix* P*, a stationary distribution* $\underline{\pi}$ *is given by* $\pi_i = 1/M, \ i = 1, 2, \ldots, M$.

Proof To find a stationary distribution, we solve the system of equations $\pi_j = \sum_{i=1}^{M} \pi_i p_{ij}, \ j = 1, 2, \ldots, M$. Suppose $\pi_i = c$ for each $i \in S$. Then $\pi_j = \sum_{i=1}^{M} c p_{ij} = c$ for all $j = 1, 2, \ldots, M$, since $\sum_{i=1}^{M} p_{ij} = 1$. Further, $\sum_{i=1}^{M} \pi_i = 1$ implies that $c = 1/M$. Hence $\{\pi_i = 1/M, \ i = 1, 2, \ldots, M\}$ is a stationary distribution. □

The next example illustrates the Markov chain with a doubly stochastic transition probability matrix.

Example 3.3.4 Suppose Y_n is the sum of outcomes in n independent rolls of a fair die. Then $\{Y_n, n \geq 1\}$ is Markov chain with state space as S as a set of natural numbers. Suppose X_n is the remainder when Y_n is divided by 7. Then $\{X_n, n \geq 1\}$ is a Markov chain on the states $\{0, 1, \ldots, 6\}$ with transition probability matrix P as given below.

$$P = \begin{array}{c} \\ 0 \\ 1 \\ 2 \\ 3 \\ 4 \\ 5 \\ 6 \end{array} \begin{array}{c} \begin{array}{ccccccc} 0 & 1 & 2 & 3 & 4 & 5 & 6 \end{array} \\ \begin{pmatrix} 0 & 1/6 & 1/6 & 1/6 & 1/6 & 1/6 & 1/6 \\ 1/6 & 0 & 1/6 & 1/6 & 1/6 & 1/6 & 1/6 \\ 1/6 & 1/6 & 0 & 1/6 & 1/6 & 1/6 & 1/6 \\ 1/6 & 1/6 & 1/6 & 0 & 1/6 & 1/6 & 1/6 \\ 1/6 & 1/6 & 1/6 & 1/6 & 0 & 1/6 & 1/6 \\ 1/6 & 1/6 & 1/6 & 1/6 & 1/6 & 0 & 1/6 \\ 1/6 & 1/6 & 1/6 & 1/6 & 1/6 & 1/6 & 0 \end{pmatrix} \end{array}.$$

It is to be noted that the matrix P is doubly stochastic and it can be easily verified that $\{X_n, n \geq 1\}$ is an ergodic Markov chain and its unique stationary distribution is $\underline{\pi} = (1/7, 1/7, \ldots, 1/7)$. Note that Y_n is a multiple of 7 if and only if $X_n = 0$. $\pi_0 = 1/7$ is interpreted as, in the long run, the probability that Y_n is a multiple of 7 is $1/7$. We discuss the interpretation of the stationary distribution later in this section. □

We now proceed to discuss one more approach to decide the existence and uniqueness of a stationary distribution and its relation with the long run distribution. It is based on the eigenvalues and the corresponding eigenvectors of the transition probability matrix P.

Eigenvalues and eigenvectors approach for stationary distributions: Suppose the state space is finite. The equation $\underline{\pi} = \underline{\pi}P$ in the third condition of a stationary distribution, is the same as $P'\underline{\pi}' = \underline{\pi}'$. Thus, $\underline{\pi}$ is a left eigenvector of P corresponding to the eigenvalue 1 or $\underline{\pi}'$ is a right eigenvector of P' corresponding to the eigenvalue 1, provided 1 is an eigenvalue of P. It is known that the eigenvalues of P and P' are the same. We prove that 1 is always an eigenvalue of P in the following lemma.

Lemma 3.3.2 *If P is a stochastic matrix of order $M \times M$, then 1 is always an eigenvalue of P and all the eigenvalues are ≤ 1 in absolute value.*

Proof To find the eigenvalues of P, we solve the polynomial equation in λ given by $|P - \lambda I| = 0$. If we add all the columns of the matrix $P - \lambda I$, then all the elements in the first column are $1 - \lambda$, since columns of P add to 1. Thus, $|P - \lambda I| = 0$ implies $(1 - \lambda)|Q| = 0$, where all elements of the first column of the matrix Q are 1 and rest of the columns are the same as those of $P - \lambda I$. Hence, it follows that one of the eigenvalues of P is 1. To show that all the eigenvalues are ≤ 1 in absolute value, suppose λ is an eigenvalue of P and $\underline{x}$ is the corresponding right eigenvector, then we have

$$P\underline{x} = \lambda\underline{x} \iff \sum_{j=1}^{M} p_{ij}x_j = \lambda x_i, \; i = 1, 2, \ldots, M.$$

Suppose $\underline{a} = (a_1, a_2, \ldots, a_M)'$ is a vector of real or complex numbers. We define $|\underline{a}|_{\max} = \max\{|a_1|, |a_2|, \ldots, |a_M|\}$. Observe that

$$
\begin{aligned}
|\lambda \underline{x}|_{\max} &= \max\{|\lambda x_1|, |\lambda x_2|, \ldots, |\lambda x_M|\} \\
&= |\lambda| \max\{|x_1|, |x_2|, \ldots, |x_M|\} = |\lambda||\underline{x}|_{\max}
\end{aligned}
$$

$$
\begin{aligned}
\text{Further, } \quad |P\underline{x}|_{\max} &= \max\left\{\Big|\sum_{j=1}^{M} p_{1j}x_j\Big|, \Big|\sum_{j=1}^{M} p_{2j}x_j\Big|, \ldots, \Big|\sum_{j=1}^{M} p_{Mj}x_j\Big|\right\} \\
&\le \max\left\{\sum_{j=1}^{M} p_{1j}|x_j|, \sum_{j=1}^{M} p_{2j}|x_j|, \ldots, \sum_{j=1}^{M} p_{Mj}|x_j|\right\} \\
&\le \max\{|x_1|, |x_2|, \ldots, |x_M|\} \;=\; |\underline{x}|_{\max} ,
\end{aligned}
$$

since $\sum_{j=1}^{M} p_{ij}|x_j|$ is a weighted average of $\{|x_1|, |x_2|, \ldots, |x_M|\}$, it lies between $\min\{|x_1|, |x_2|, \ldots, |x_M|\}$ and $\max\{|x_1|, |x_2|, \ldots, |x_M|\}$. Thus,

$$
\begin{aligned}
\lambda\underline{x} = P\underline{x} &\Rightarrow |\lambda\underline{x}|_{\max} = |P\underline{x}|_{\max} \;\le\; |\underline{x}|_{\max} \\
&\Rightarrow |\lambda||\underline{x}|_{\max} \le |\underline{x}|_{\max} \;\Rightarrow\; |\lambda| \le 1 .
\end{aligned}
$$

Thus all the eigenvalues of a stochastic matrix are ≤ 1 in absolute value. □

Remark 3.3.5 (i) From Lemma 3.3.2, we note that the maximum eigenvalue of a stochastic matrix P is 1, it is known as the spectral radius of a stochastic matrix. (ii) If the Markov chain is aperiodic with a single closed communicating class, then multiplicity of the eigenvalue 1 of P is 1. Thus, absolute value of all other eigenvalues is less than 1. (iii) If the Markov chain is reducible with $k > 1$ closed communicating classes of aperiodic states, then multiplicity of the eigenvalue 1 is k. (iv) If the Markov chain is irreducible and periodic with period d, then there are d eigenvalues with absolute value 1. We refer to p. 556 of Medhi [7] for these results.

We elaborate below on results (iii) and (iv) in Remark 3.3.5.

Result (iii): In Lemma 3.3.2, while finding the eigenvalues of P, we noted that the polynomial equation $|P - \lambda I| = 0 \iff (1-\lambda)|Q| = 0$. Hence, one of the eigenvalues of P is 1. Suppose P has a single closed communicating class C_1 of r states and all states in C_1 are aperiodic. Then P can be partitioned as follows, with rearrangement of rows and columns if needed,

$$
P = \begin{array}{c} \\ C_1 \\ A_1 \end{array}
\begin{array}{c} \begin{array}{cc} C_1 & A_1 \end{array} \\ \begin{pmatrix} P_1 & 0 \\ Q_1 & Q_2 \end{pmatrix} \end{array},
$$

where P_1 is a stochastic matrix of order $r \times r$ and Q_2 is of order $M - r \times M - r$. It is known that when P is partitioned this way, then $|P - \lambda I| = |P_1 - \lambda I||Q_2 - \lambda I|$ (p. 539, Karlin and Taylor [6]). Hence, the polynomial equation

$$
|P - \lambda I| = 0 \iff |P_1 - \lambda I||Q_2 - \lambda I| = 0 \iff (1-\lambda)|Q_3||Q_2 - \lambda I|,
$$

where Q_3 is similar to Q as in Lemma 3.3.2. It then follows that 1 is the eigenvalue and since all states in C_1 are aperiodic, multiplicity of eigenvalue 1 is 1. If P has two closed communicating classes C_1 of r aperiodic states and C_2 of s aperiodic states, then P can be expressed as follows, with rearrangement of rows and columns if needed,

$$P = \begin{array}{c} \\ C_1 \\ C_2 \\ A_1 \end{array} \begin{array}{c} \begin{array}{ccc} C_1 & C_2 & A_1 \end{array} \\ \begin{pmatrix} P_1 & 0 & B_1 \\ 0 & P_2 & B_2 \\ Q_1 & Q_2 & Q_3 \end{pmatrix} \end{array},$$

where P_1, P_2 are stochastic matrices and Q_3 is of order $M - r - s \times M - r - s$. Then the polynomial equation

$$|P - \lambda I| = 0 \iff |P_1 - \lambda I||P_2 - \lambda I||Q_3 - \lambda I| = 0 \iff (1 - \lambda)^2 |Q_4||Q_5||Q_3 - \lambda I|,$$

where Q_4 and Q_5 are similar to Q as in Lemma 3.3.2. It then follows that 1 is the eigenvalue with multiplicity 2. Using similar arguments, if the Markov chain is reducible with k closed classes of aperiodic states, then multiplicity of the eigenvalue 1 is k.

Result (iv): If the Markov chain with transition probability matrix P is irreducible and periodic with period d, then the Markov chain with transition probability matrix P^d is aperiodic. Hence, P^d has eigenvalue 1 with multiplicity 1. We know that if λ is the eigenvalue of P, then the eigenvalue of P^d is λ^d. Thus, if 1 is the eigenvalue of P^d, then the eigenvalue of P is $1^{1/d}$, thus there are d eigenvalues of P, some of these may be complex numbers, but the absolute value of each is 1. For example, if d is 2, then 1 and -1 are two eigenvalue of P. If d is 3, then cube roots of 1 are the eigenvalues of P. Thus one is 1 and the other two are complex conjugates of each other. As an illustration, suppose P is given by

$$P = \begin{array}{c} \\ 1 \\ 2 \end{array} \begin{array}{c} \begin{array}{cc} 1 & 2 \end{array} \\ \begin{pmatrix} 0 & 1 \\ 1 & 0 \end{pmatrix} \end{array}.$$

Then P is irreducible and periodic with period 2. It is easy to check that $|P - \lambda I| = 0 \Rightarrow (\lambda^2 - 1) = 0$ and hence $\lambda_1 = 1$ and $\lambda_2 = -1$ are two eigenvalue of P.

In Lemma 3.3.2, we have proved that 1 is the eigenvalue of P and hence $\underline{\pi}$ is a left eigenvector of P corresponding to the eigenvalue 1. To label $\underline{\pi}$ as a stationary distribution, we have to examine the first two conditions, which require that all the components in the corresponding left eigenvector are non-negative and add up to 1. It has been proved that for a stochastic matrix for the eigenvalue 1, there corresponds a unique (up to a constant factor) left eigenvector with strictly positive components, refer to Corollary 4.15 on p. 363 of Cinlar [3]. Once we have such a left eigenvector, we normalize it, by dividing by sum of its components, so that $\sum_{j \in S} \pi_j = 1$.

If the right and left eigenvectors corresponding to the eigenvalue 1 satisfy a certain relation, then the sum of the component of the left eigenvector is 1. We prove it by using the spectral decomposition of P. By the spectral decomposition, $P = \sum_{i=1}^{M} \lambda_i \underline{x}_i \underline{y}_i'$ where $\underline{x}_i$ and $\underline{y}_i$ denote the right and left eigenvectors respectively corresponding to the eigenvalue λ_i, with $\underline{x}_i'\underline{y}_i = 1$ and $\underline{x}_i'\underline{y}_j = 0$ for all $i \neq j = 1, 2, \ldots, M$. Suppose $\lambda_1 = 1$ and $\underline{e} = (1, 1, \ldots, 1)'$ is a vector of order $M \times 1$. Then $P\underline{e} = \underline{e} = 1\underline{e}$, thus the right eigenvector of P corresponding to the eigenvalue $\lambda = 1$ is $\underline{e}$. Thus, $x_1 = \underline{e}$. Suppose $\underline{y}_1' = (u_1, u_2, \ldots, u_M)$, then $\underline{x}_1'\underline{y}_1 = 1$ implies that $\sum_{j=1}^{M} u_j = 1$.

Thus to find a stationary distribution $\underline{\pi}$, we find the right eigenvector of P' corresponding to the eigenvalue 1 and normalize it, by dividing by sum of its components, so that $\sum_{j\in S} \pi_j = 1$. As discussed in Remark 3.3.5, multiplicity of eigenvalue 1 depends on the number of closed communicating classes of S. If it is 1, then we get a unique stationary distribution. If it is k, we have k stationary distributions. We use this approach in Sect. 3.4 to compute stationary distributions.

In the following example, we find a stationary distribution for a Markov chain when the transition probability matrix is a doubly stochastic matrix, using the method based on eigenvalues.

Example 3.3.5 In Theorem 3.3.8, we have shown that for a Markov chain with $S = \{1, 2, \ldots, M\}$ and a doubly stochastic transition probability matrix P, the stationary distribution is $\underline{\pi} = (1/M, 1/M, \ldots, 1/M)$. In this example we consider the alternative approach based on eigenvalues and eigenvectors to prove the same result. Since P is doubly stochastic, with $\underline{e}' = (1, 1, \ldots, 1)$ we have

$$\sum_{j=1}^{M} p_{ij} = \sum_{i=1}^{M} p_{ij} = 1 \iff P\underline{e} = \underline{e} \ \& \ \underline{e}'P = \underline{e}' .$$

Thus, right eigenvector and also the left eigenvector of P corresponding to eigenvalue 1 are $\underline{e}$. When we normalize it by dividing by sum of the elements of $\underline{e}$, we get the stationary distribution as $\underline{\pi} = (1/M, 1/M, \ldots, 1/M)$. □

We have noted that whenever a long run distribution exists, it is the unique stationary distribution of the Markov chain. In Theorem 3.2.5, we have proved that for a finite state space ergodic Markov chain, a long run distribution and hence a unique stationary distribution $\{\pi_j, j \in S\}$ exists, where $\lim_{n\to\infty} p_{ij}^{(n)} = 1/\mu_j = \pi_j$. We prove the same result in the following theorem, using the approach of eigenvalues and eigenvectors and Lemma 3.3.2.

Theorem 3.3.9 *For a finite state space ergodic Markov chain, the long run distribution exists and it is the same as the stationary distribution of the Markov chain.*

Proof Suppose $S = \{1, 2, \ldots, M\}$. Since P is a stochastic matrix, by Lemma 3.3.2 all the eigenvalues are ≤ 1 in absolute value. By the spectral decomposition, $P = \sum_{i=1}^{M} \lambda_i \underline{x}_i \underline{y}_i'$ where $\underline{x}_i$ and $\underline{y}_i$ denote the right and left eigenvectors respectively corresponding to the eigenvalue λ_i, with $\underline{x}_i'\underline{y}_i = 1$ and $\underline{x}_i'\underline{y}_j = 0$ for all $i \neq j = 1, 2, \ldots, M$. Since the Markov chain is ergodic, λ_1 is the only eigenvalue which is equal to 1 and all other eigenvalues are < 1 in absolute value. Further, $\underline{e} = (1, 1, \ldots, 1)'$ is the right eigenvector of P corresponding to the eigenvalue 1. Thus, $x_1 = \underline{e}$. Suppose $\underline{y}_1' = (u_1, u_2, \ldots, u_M)$, then $\underline{x}_1'\underline{y}_1 = 1$ implies that $\sum_{j=1}^{M} u_j = 1$. Using Eq. (2.2.3), $\forall\ n \geq 1$,

$$P^n = \sum_{i=1}^{M} \lambda_i^n \underline{x}_i \underline{y}_i' = \underline{x}_1 \underline{y}_1' + \sum_{i=2}^{M} \lambda_i^n \underline{x}_i \underline{y}_i' \;\Rightarrow\; \lim_{n\to\infty} P^n = \underline{x}_1 \underline{y}_1' = \underline{e}\underline{y}_1' = A,$$

where the matrix A has identical rows given by $\underline{y}_1'$. Thus, $\lim_{n\to\infty} p_{ij}^{(n)} = u_j$ for $j = 1, 2, \ldots, M$ and the limit is free i. Hence, the long run distribution of the Markov chain exists and it is $\underline{y}_1'$. Further, a stationary distribution $\underline{\pi}$ is a left eigenvector of P corresponding to the eigenvalue 1. Thus $\underline{\pi} = \underline{y}_1'$. □

Remark 3.3.6 Suppose $\lambda_i = a + ib$ is a complex number with absolute value < 1. It can be shown that $\lambda_i^n \to 0$ as follows. Observe that

$$\begin{aligned} |\lambda_i| < 1 \Rightarrow a^2 + b^2 < 1 \;&\Rightarrow\; |\lambda_i^n| = |\lambda_i|^n = (a^2 + b^2)^{n/2} < 1 \\ &\Rightarrow |\lambda_i^n| \to 0 \;\Rightarrow\; \lambda_i^n \to 0. \end{aligned}$$

Remark 3.3.7 Theorem 3.3.9 gives a method to find a stationary distribution of an ergodic Markov chain or a Markov chain with a single closed communicating class. In this case λ_1 is the only eigenvalue which is equal to 1 and all other eigenvalues are distinct and < 1 in absolute value. Hence, each row of $\lim_{n\to\infty} P^n$ is the same as the normalized left eigenvector corresponding to the eigenvalue 1. It is the long run as well as the stationary distribution of a Markov chain. This theorem also proves that whenever a long run distribution exists, a stationary distribution exists and both are the same.

The next example illustrates the non-existence of a long run distribution for a periodic Markov chain, based on eigenvalues.

Example 3.3.6 Suppose $\{X_n, n \geq 0\}$ is a Markov chain with P given by

$$P = \begin{matrix} & \begin{matrix} 1 & 2 \end{matrix} \\ \begin{matrix} 1 \\ 2 \end{matrix} & \begin{pmatrix} 0 & 1 \\ 1 & 0 \end{pmatrix} \end{matrix}.$$

We have already obtained its stationary distribution, which is $\underline{\pi} = (1/2, 1/2)$. It has been noted for this P, $\lim_{n\to\infty} p_{ij}^{(n)}$ oscillates and thus the limit does not exist. Hence, the long run distribution does not exist. Note that the period of the Markov chain is 2 and the two eigenvalues of P are 1 and -1. Thus, $\lim_{n\to\infty} P^n = \underline{x}_1 \underline{y}_1' +$

$\lim_{n\to\infty}(-1)^n \underline{x}_2\underline{y}_2'$. Hence, limit of P^n does not exist. Thus, non-existence of the long run distribution of this Markov chain is due to the two eigenvalues 1 and -1. □

If a Markov chain is reducible with $k > 1$ closed communicating classes C_i of aperiodic states, then there are k eigenvalues which are equal to 1, corresponding to each C_i. Suppose P_i is a stochastic matrix corresponding to C_i. Then by Theorem 3.3.9, the left eigenvector of P_i, corresponding to the eigenvalue 1, is a long run distribution as well as the stationary distribution of the Markov chain with transition probability matrix P_i, $i = 1, 2, \ldots, k$. Using Theorem 3.3.3, we have a stationary distribution concentrated on C_i, $i = 1, 2, \ldots, k$. Thus, there are k stationary distributions for the Markov chain with transition probability matrix P. We prove this result for $k = 2$ in Theorem 3.3.10 below and illustrate by examples in Sect. 3.4.

In the following lemma, we prove that if there are two stationary distributions of a Markov chain, then there is a uncountable family of stationary distributions.

Lemma 3.3.3 *Suppose $\underline{a}$ and $\underline{b}$ are two stationary distributions of a Markov chain. (i) Then for $\alpha \in [0, 1]$, $\underline{\pi}(\alpha) = \alpha\underline{a} + (1-\alpha)\underline{b}$ is a stationary distribution of the Markov chain. (ii) If $\alpha \neq \beta \in [0, 1]$, then $\underline{\pi}(\alpha) \neq \underline{\pi}(\beta)$.*

Proof Since $\underline{a}$ and $\underline{b}$ are stationary distributions,

$$a_i \geq 0 \;\forall\; i \in S, \;\sum_{i\in S} a_i = 1 \;\&\; \underline{a} = \underline{a}P\,.$$

$$\text{Similarly,}\quad b_i \geq 0 \;\forall\; i \in S, \;\sum_{i\in S} b_i = 1 \;\&\; \underline{b} = \underline{b}P\,.$$

Suppose $\pi_i(\alpha) = \alpha a_i + (1-\alpha)b_i, i \in S, \alpha \in [0, 1]$. Observe that

$$\begin{aligned}
\forall\; i \in S,\; a_i, b_i \geq 0 &\Rightarrow \alpha a_i + (1-\alpha)b_i = \pi_i(\alpha) \geq 0\\
\sum_{i\in S} a_i = 1 \;\&\; \sum_{i\in S} b_i = 1 &\Rightarrow \alpha\sum_{i\in S} a_i + (1-\alpha)\sum_{i\in S} b_i = 1\\
&\Rightarrow \sum_{i\in S}(\alpha a_i + (1-\alpha)b_i) = \sum_{i\in S}\pi_i(\alpha) = 1\\
\text{Further } \underline{a} &= \underline{a}P \;\&\; \underline{b} = \underline{b}P\\
&\Rightarrow \underline{\pi}(\alpha)P = (\alpha\underline{a} + (1-\alpha)\underline{b})P\\
&= \alpha P + (1-\alpha)P = P.
\end{aligned}$$

Thus, the convex combination $\underline{\pi}(\alpha) = \alpha\underline{a} + (1-\alpha)\underline{b}$ is a stationary distribution for any $\alpha \in [0, 1]$.

(ii) Observe that

$$\begin{aligned}
\underline{\pi}(\alpha) = \underline{\pi}(\beta) &\Rightarrow \alpha\underline{a} + (1-\alpha)\underline{b} = \beta\underline{a} + (1-\beta)\underline{b} \\
&\Rightarrow (\alpha-\beta)\underline{a} - (\alpha-\beta)\underline{b} = 0 \\
&\Rightarrow (\alpha-\beta)(\underline{a}-\underline{b}) = 0 \\
&\Rightarrow \text{either } (\alpha-\beta) = 0 \text{ or } (\underline{a}-\underline{b}) = 0.
\end{aligned}$$

Thus each $\alpha \in [0, 1]$ corresponds to a distinct stationary distribution $\underline{\pi}(\alpha)$ and hence if there are two stationary distributions, there is a uncountable family of stationary distributions. □

Lemma 3.3.3 can be extended to a convex combination of more than two stationary distributions.

Theorem 3.3.10 *Suppose $\{X_n, n \geq 0\}$ is a Markov chain with two closed communicating classes C_1 and C_2 of aperiodic states. Then there are two stationary distributions $\underline{a}$ and $\underline{b}$ concentrated on C_1 and C_2 respectively, where*

$$a_i = 1/\mu_i \text{ if } i \in C_1, \;\; a_i = 0 \text{ if } i \notin C_1 \;\; \& \;\; b_i = 1/\mu_i \text{ if } i \in C_2, \;\; b_i = 0 \text{ if } i \notin C_2.$$

Further, $\underline{\pi}(\alpha) = \alpha\underline{a} + (1-\alpha)\underline{b}$, $0 \leq \alpha \leq 1$ is a stationary distribution for $\alpha \in [0, 1]$.

Proof Since C_1 is a closed communicating class with all aperiodic states, by Theorem 3.3.3, a unique stationary distribution concentrated on C_1 exists. Suppose it is $\underline{a}$, where $a_i = 1/\mu_i$ if $i \in C_1$, $a_i = 0$ if $i \notin C_1$. Similarly, since C_2 is a closed communicating class with all aperiodic states, a unique stationary distribution concentrated on C_2 exists. Suppose it is $\underline{b}$, where $b_i = 1/\mu_i$ if $i \in C_2$ and $b_i = 0$ if $i \notin C_2$. By Lemma 3.3.3, the convex combination $\underline{\pi}(\alpha) = \alpha\underline{a} + (1-\alpha)\underline{b}$ is a stationary distribution for any $\alpha \in [0, 1]$. □

The result in Theorem 3.3.10 can be extended if there are more than two closed communicating classes of aperiodic states or periodic states.

Remark 3.3.8 As stated in (iii) of Remark 3.3.5, for a Markov chain with two closed communicating classes C_1 and C_2, the multiplicity of eigenvalue 1 is 2. Further, $\{a_i = 1/\mu_i, i \in C_1\}$ and $\{b_i = 1/\mu_i, i \in C_2\}$ are normalized left eigenvectors corresponding to the eigenvalue 1 of P_1 and P_2, which are stochastic matrices corresponding to C_1 and C_2 respectively.

Remark 3.3.9 If a Markov chain is reducible with 2 closed communicating classes, we have noted that a long run distribution does not exist. However, as proved in Theorem 3.3.10, there are infinitely many stationary distributions.

Interpretation of a stationary distribution: A stationary distribution given by $\{\pi_j = 1/\mu_j = \lim_{n\to\infty} p_{ij}^{(n)}, j \in S\}$ is interpreted as follows. After the process has been in operation for a long duration, the probability that the process in state j is π_j, irrespective of the initial state. π_j is also interpreted as the long run mean fraction of time the system is in state j, whatever may be the initial state. To elaborate on this, suppose $X_0 = i$. For $n \geq 1$ and $j \in S$, a random variable $Y_n(j)$ is defined as follows.

$$Y_n(j) \quad = \quad \begin{cases} 1, \text{ if } X_n = j \\ 0, \text{ if } X_n \neq j \;. \end{cases}$$

Then the fraction of time the system is in state j is a random variable, defined as $N_j(m)/m = \sum_{n=1}^{m} Y_n(j)/m$. Therefore, the mean fraction of time the system is in state j is given by

$$E\left(N_j(m)/m\right) = \sum_{n=1}^{m} P[X_n = j|X_0 = i]/m = \sum_{n=1}^{m} p_{ij}^{(n)}/m \;.$$

We have a result which states that if a sequence $\{a_n, n \geq 1\}$ of real numbers converges to a limit a, then the averages of these numbers $\sum_{n=1}^{m} a_n/m$ also converge to the limit a. Analogously,

$$p_{ij}^{(n)} \to 1/\mu_j \Rightarrow \sum_{n=1}^{m} p_{ij}^{(n)}/m \to 1/\mu_j \quad \Rightarrow \quad E\Big(\sum_{n=1}^{m} Y_n(j)/m\Big) \to 1/\mu_j = \pi_j \;.$$

Hence, π_j is interpreted as a long run mean proportion of time the system is in state j. As a consequence, if each visit to state j incurs a "cost" of c_j units, then the long run average cost per unit time associated with this Markov chain is $\sum_{j\in S} \pi_j c_j$.

Remark 3.3.10 (i) It can be shown that $P\left[\lim_{m\to\infty} N_j(m)/m = \pi_j\right] = 1$, that is, $N_j(m)/m$ converges to π_j almost surely. This result shows that the time averages of a single path of an ergodic Markov chain are equal to the chain's space average. This is incredibly useful result and shows that one way to compute statistics of the stationary distribution is to compute one very long path and average over that path. This result is extensively used in Markov chain Monte Carlo (MCMC) methods.
(ii) Result in (i) also follows for any bounded, continuous function $f : S \to \mathbb{R}$. Thus,

$$P\Big[\lim_{m\to\infty} \sum_{n=0}^{m-1} f(X_n)/m = \sum_{j\in S} f(j)\pi_j\Big] = 1.$$

It again conveys that the time average of a single realization of a function of X_n converges with probability one, to the space averages, obtained by simply taking expectations with respect to the stationary distribution.

We now proceed to address the fourth issue related to the computation of a stationary distribution. In the next section, we discuss some methods to compute the stationary distributions using R, for a Markov chain with state space consisting of M states. A variety of theorems studied in Sects. 3.1–3.3 are illustrated by examples in the next section.

3.4 Computation of Stationary Distributions

We begin with a simple example.

Example 3.4.1 Suppose $\{X_n, n \geq 0\}$ is a Markov chain with $S = \{1, 2, 3\}$ and P given by

$$P = \begin{array}{c} \\ 1 \\ 2 \\ 3 \end{array} \begin{array}{c} \begin{array}{ccc} 1 & 2 & 3 \end{array} \\ \begin{pmatrix} 0.6 & 0.2 & 0.2 \\ 0.1 & 0.7 & 0.2 \\ 0.1 & 0.1 & 0.8 \end{pmatrix} \end{array} .$$

From the transition probability matrix P, it is easy to see that all states communicate with each other and each has period 1. Thus, the Markov chain is ergodic. Hence, by Theorem 3.3.3, a unique stationary distribution exists.
(i) To find the stationary distribution $\underline{\pi}$ of this Markov chain, we solve the matrix equation $\underline{\pi} = \underline{\pi}P$, subject to the condition that $\pi_1 + \pi_2 + \pi_3 = 1$. Thus, we have following four equations in 3 unknowns.

$$\pi_1 = 0.6\pi_1 + 0.1\pi_2 + 0.1\pi_3, \quad \pi_2 = 0.2\pi_1 + 0.7\pi_2 + 0.1\pi_3$$
$$\pi_3 = 0.2\pi_1 + 0.2\pi_2 + 0.8\pi_3, \quad \pi_1 + \pi_2 + \pi_3 = 1 .$$

From the third equation we get $\pi_3 = \pi_1 + \pi_2$, substituting it in last equation we have $\pi_1 + \pi_2 = 1/2$ hence we get $\pi_3 = 1/2$. From the first two equations, we get two equations as $0.4\pi_1 - 0.1\pi_2 = 0.05$ & $0.2\pi_1 - 0.3\pi_2 = -0.05$. Solving these two we get $\pi_2 = 0.3$, hence $\pi_1 = 0.2$. Thus, the stationary distribution $\underline{\pi}$ is given by $\underline{\pi} = (0.2, 0.3, 0.5)$. (ii) Using Code 2.8.6, we compute μ_i and these are $5,\ 3.3333,\ 2$ for $i = 1, 2, 3$ respectively. Thus, $\pi_i = 1/\mu_i$ for $i = 1, 2, 3$. □

From the above example we note that if the number of states is large, finding the stationary distribution using the approach of this example would be tedious. Hence, we now discuss its computation using other methods and using R software.

(i) As discussed in Sect. 3.1, to find the stationary distribution $\underline{\pi}$ we solve a matrix equation $A\underline{\pi}' = \underline{b}$, where A denotes a matrix $P' - I$ appended with the last row as $\underline{e} = (1, 1, \ldots, 1)$ and $\underline{b}' = (0, 0, \ldots, 0, 1)$. It has a solution if $rank(A) = rank(A|b)$. When this condition is satisfied, $\underline{\pi}' = A^-\underline{b}$, where A^- is a generalized inverse of A.
(ii) Alternatively, solving $A\underline{\pi}' = \underline{b}$ is equivalent to solving $A'A\underline{\pi}' = A'\underline{b}$ and $A'A$ is a square matrix. It has a solution if $rank(A'A) = rank(A'A|b)$. Thus, $\underline{\pi}' = (A'A)^{-1}A'\underline{b}$, if $(A'A)^{-1}$ exists, otherwise $\underline{\pi}' = (A'A)^-A'\underline{b}$.

(iii) The third method is based on the eigenvalues and corresponding eigenvectors of P. To find the stationary distribution $\underline{\pi}$, we find the right eigenvector of P' corresponding to the eigenvalue 1 and normalize it, by dividing by sum of its components. For a Markov chain with $k > 1$ closed communicating classes, multiplicity of the eigenvalue 1 is k. The normalized eigenvector corresponding to each eigenvalue 1 is the stationary distribution. Thus, there are k stationary distributions. Further, a convex combination of these is also a stationary distribution.

(iv) If λ_1 is the only eigenvalue which is equal to 1 and all other eigenvalues are distinct and < 1 in absolute value, then all rows of $\lim_{n\to\infty} P^n$ are the same and each row is the long run distribution as well as the stationary distribution of a Markov chain. For a Markov chain with $k > 1$ closed communicating classes $C_i, i = 1, 2, \ldots, k$, the rows of $\lim_{n\to\infty} P^n$, corresponding to the states in C_i are identical. Thus, we get k stationary distributions.

In Sect. 3.7, Codes 3.7.1 and 3.7.2 are based on all these methods. We illustrate the computation of the stationary distribution using these codes in the following examples. We also illustrate the interpretation of a stationary distribution.

Example 3.4.2 Sociologists often assume that the mobility of social classes of successive generations in a family can be modeled as a Markov chain. Thus, the occupation of a son is assumed to depend only on his father's occupation and not on his grandfather's. Suppose there are three classes—lower, middle and upper which we label as 1, 2 and 3 respectively and transitions among these classes are governed by a Markov chain with P given by

$$P = \begin{matrix} \\ 1 \\ 2 \\ 3 \end{matrix} \begin{matrix} \begin{matrix} 1 & \quad 2 & \quad 3 \end{matrix} \\ \begin{pmatrix} 0.40 & 0.50 & 0.10 \\ 0.05 & 0.70 & 0.25 \\ 0.05 & 0.50 & 0.45 \end{pmatrix} \end{matrix}$$

It is of interest to obtain a long run mean proportion of population in classes 1, 2 and 3. To find these, we find the stationary distribution using Code 3.7.1.
(i) From the output, we note that by all the methods, the stationary distribution is $\underline{\pi} = (0.0769, 0.6250, 0.2981)$. (ii) Among the three eigenvalues of P, one is 1 and other two are less than 1. The first column of the matrix "vectors" correspond to the right eigenvector of P' corresponding to eigenvalue 1.
(iii) Further, rows of P^{10} are identical and represent the stationary distribution as proved in Theorem 3.3.9. Thus in the long run, the mean proportion of population in the lower, middle and upper classes is 7.69%, 62.5% and 29.81% respectively.
(iv) It can be verified that a stationary distributions corresponding to P, P^2, P^3 are the same. □

Example 3.4.3 A local train in a mass transit system is operating on a continuous route with intermediate stops. The arrival of the train at a stop is classified into one of three states-1:early arrival, 2:on time arrival and 3:late arrival. Suppose the transitions among successive states are governed by a Markov chain with P given by

$$P = \begin{array}{c} \\ 1 \\ 2 \\ 3 \end{array}\begin{array}{c} \begin{array}{ccc} 1 & 2 & 3 \end{array} \\ \begin{pmatrix} 0.4 & 0.5 & 0.1 \\ 0.2 & 0.5 & 0.3 \\ 0.1 & 0.3 & 0.6 \end{pmatrix} \end{array}$$

Suppose we are interested in knowing the proportion of stops where the train will be late in the long run. We then find the stationary distribution, using Code 3.7.1. It is $\underline{\pi} = (0.2037, 0.4259, 0.3704)$. Thus, over a long period of time, fraction of stops where the train is expected to be late is $\pi_3 = 0.3704$. □

In Example 3.4.2 and in Example 3.4.3, the Markov chains are ergodic and the stationary distributions are unique in both the examples, as proved in Theorem 3.2.5. Using Code 2.8.6, we can compute μ_i and verify that $\pi_i = 1/\mu_i$. On the other hand, knowing π_i, we can find $\mu_i = 1/\pi_i$.

Example 3.4.4 In Example 2.2.5, the transitions among states in a care center model are governed by a Markov chain with P as given below, where the time unit is taken as a day.

$$P = \begin{array}{c} \\ 1 \\ 2 \\ 3 \end{array}\begin{array}{c} \begin{array}{ccc} 1 & 2 & 3 \end{array} \\ \begin{pmatrix} 0.92 & 0.05 & 0.03 \\ 0.00 & 0.76 & 0.24 \\ 0 & 0 & 1 \end{pmatrix} \end{array}.$$

(i) Suppose the expenses incurred per day at the care center are Rs.1200, Rs 5000 and Rs.1000 for an individual in states 1, 2, 3 respectively. In Example 2.2.5, using marginal distributions we have obtained expected expenses per individual on day 2, 3, 4 and 5 to be $Rs.1384$, $Rs.1505.28$, $Rs.1580.38$, $Rs.1621.74$ respectively. The long run cost per period associated with this Markov chain is given by $C = 1200\pi_1 + 5000\pi_2 + 1000\pi_3$. Using Code 3.7.1, we get $\underline{\pi} = (0, 0, 1)$. Hence, $C = 1000$. (ii) We have noted in Sect. 3.2 that P^n for $n > 100$ has all identical rows and these are $(0, 0, 1)$. (iii) It is to be noted that for this Markov model, state 3 is an absorbing state and hence is non-null persistent with $\mu_3 = 1$. Further, both 1 and 2 are inessential states and hence transient with $f_{11} = 0.92$, $f_{22} = 0.76$. (iv) Observe that $p_{11} > 0$, $p_{22} > 0$ and $p_{33} > 0$ implying that period of each state is 1. There is a single closed communicating class $\{3\}$ and the other two states are transient. As proved in Theorem 3.3.3, $\pi_i = 0$ for $i = 1, 2$ and $\pi_3 = 1/\mu_3 = 1$. In fact, when we note that there is one closed class $\{3\}$ and the other two states are transient, we immediately get the stationary distribution as $(0, 0, 1)$, by Theorem 3.3.3. □

Example 3.4.5 Suppose $\{X_n, n \geq 0\}$ is a Markov chain with state space $S = \{1, 2, 3\}$ and P given by

$$P = \begin{array}{c} \\ 1 \\ 2 \\ 3 \end{array} \begin{array}{c} \begin{array}{ccc} 1 & 2 & 3 \end{array} \\ \begin{pmatrix} 1/3 & 2/3 & 0 \\ 1/4 & 3/4 & 0 \\ 1/3 & 1/3 & 1/3 \end{pmatrix} \end{array}.$$

(i) In Example 2.4.4, we have noted that it is a reducible Markov chain, with a single closed communicating class $\{1, 2\}$. As proved in Theorem 3.3.3, the stationary distribution $\underline{\pi}$ is such that $\pi_1 = 1/\mu_1, \pi_2 = 1/\mu_2$ and $\pi_3 = 0$. In Example 2.5.5, we have obtained $\mu_1 = 11/3$ & $\mu_2 = 11/8$. Hence, $\pi_1 = 3/11$ and $\pi_2 = 8/11$. (ii) Using Code 3.7.1, the stationary distribution is given by $\underline{\pi} = (0.2727, 0.7273, 0)'$. (iii) The eigenvalue of P is 1 with multiplicity 1 and the corresponding normalized eigenvector is the same as $\underline{\pi}$. □

Note that the Markov chains in the above two examples have a single closed communicating class. In both the examples, stationary distributions are unique, as proved in Theorem 3.3.3.

In the next two examples, we illustrate Theorem 3.3.10 for reducible Markov chains with two closed communicating classes.

Example 3.4.6 Suppose $\{X_n, n \geq 0\}$ is a Markov chain with P given by

$$P = \begin{array}{c} \\ 1 \\ 2 \\ 3 \\ 4 \\ 5 \end{array} \begin{array}{c} \begin{array}{ccccc} 1 & 2 & 3 & 4 & 5 \end{array} \\ \begin{pmatrix} 1/3 & 2/3 & 0 & 0 & 0 \\ 3/4 & 1/4 & 0 & 0 & 0 \\ 0 & 0 & 1/8 & 1/4 & 5/8 \\ 0 & 0 & 0 & 1/2 & 1/2 \\ 0 & 0 & 1/3 & 0 & 2/3 \end{pmatrix} \end{array}.$$

(i) The Markov chain is reducible with two closed communicating classes $C_1 = \{1, 2\}$ and $C_2 = \{3, 4, 5\}$. Further, period of each state is 1. We use Code 3.7.2 to find the stationary distribution. By the first and the second method, which are the same as in Code 3.7.1, the stationary distribution is

$$\underline{\pi} = (0.2584, 0.2297, 0.1241, 0.0620, 0.3258).$$

It is one of the solutions of the system of equations which involves a generalized inverse. In the computation, Moore-Penrose generalized inverse is used and hence we get only one solution. (ii) In the second method the matrix $A'A$ comes out to be singular, its rank is 4. Hence, to solve $A'Ax = A'b$, we use a generalized inverse. Thus, the solution is $(A'A)^{-}A'\underline{b}$. Note that $(A'A)^{-1}A'\underline{b}$ gives the warning that $A'A$ is a singular matrix. (iii) In the third method, we get two eigenvalues of P, which equal 1 and the corresponding normalized eigenvectors are two stationary distributions. These are

$$\underline{a} = (0.5294, 0.4706, 0, 0, 0) \ \& \ \underline{b} = (0, 0, 0.2424, 0.1212, 0.6364)\ .$$

Any convex combination of $\underline{a}$ and $\underline{b}$ is again a stationary distribution. Thus we have an uncountable family of stationary distributions. (iv) By Theorem 3.3.10, with two closed communicating classes C_1 and C_2, two stationary distributions, concentrated on C_1 and C_2, are given by $\underline{f}$ and $\underline{g}$, respectively where $f_i = 1/\mu_i, i = 1, 2$ & $f_i = 0, i = 3, 4, 5$ and $g_i = 0, i = 1, 2$ & $g_i = 1/\mu_i, i = 3, 4, 5$. Using Code 2.8.6, we compute the mean recurrence times. These are given by a vector $(1.8889, 2.1250, 4.1250, 8.2500, 1.5714)$. Thus we get $\underline{f}$ and $\underline{g}$. Note that $f_i = a_i = 1/\mu_i, i = 1, 2$ and $g_i = b_i = 1/\mu_i, i = 3, 4, 5$. (iv) Further, P^n for all $n \geq 11$ remains the same. Hence,

$$\lim_{n\to\infty} P^n = \begin{array}{c} \\ 1 \\ 2 \\ 3 \\ 4 \\ 5 \end{array} \begin{array}{c} \begin{array}{ccccc} 1 & 2 & 3 & 4 & 5 \end{array} \\ \begin{pmatrix} 0.5294 & 0.4706 & 0.0000 & 0.0000 & 0.0000 \\ 0.5294 & 0.4706 & 0.0000 & 0.0000 & 0.0000 \\ 0.0000 & 0.0000 & 0.2424 & 0.1212 & 0.6364 \\ 0.0000 & 0.0000 & 0.2424 & 0.1212 & 0.6364 \\ 0.0000 & 0.0000 & 0.2424 & 0.1212 & 0.6364 \end{pmatrix} \end{array}.$$

Observe that the first two rows correspond to the stationary distribution $\underline{a}$ and the last three rows correspond to the second stationary distribution $\underline{b}$. Since $\lim_{n\to\infty} p_{ij}^{(n)}$ depends on i, the long run distribution does not exist. (v) The stationary distributions by the first two methods and the last two methods are linked with each other as follows.

$$\underline{\pi} = \alpha\underline{a} + (1-\alpha)\underline{b}, \quad \text{where } \alpha = 0.4881.$$

It is to be noted that the components in $\underline{\pi}$ are not the reciprocals of mean recurrence times. □

Example 3.4.7 Suppose $\{X_n, n \geq 0\}$ is a Markov chain with P given by

$$P = \begin{array}{c} \\ 1 \\ 2 \\ 3 \\ 4 \end{array} \begin{array}{c} \begin{array}{cccc} 1 & 2 & 3 & 4 \end{array} \\ \begin{pmatrix} 1/3 & 2/3 & 0 & 0 \\ 2/3 & 1/3 & 0 & 0 \\ 0 & 0 & 1/4 & 3/4 \\ 0 & 0 & 3/4 & 1/4 \end{pmatrix} \end{array}.$$

(i) The Markov chain is reducible with two closed communicating classes. Further, period of each state is 1. We can study two Markov chains separately, one with state space $C_1 = \{1, 2\}$ and the second with state space $C_2 = \{3, 4\}$. For both the Markov chains, the state space has only two states. Hence, it is easy to solve the matrix equation to get the stationary distribution. For both the Markov chains, these are $(1/2, 1/2)$. Hence, $\underline{a} = (1/2, 1/2, 0, 0)$ and $\underline{b} = (0, 0, 1/2, 1/2)$ are two stationary distributions of the given Markov chain with state space $S = \{1, 2, 3, 4\}$. Any convex combination of these two is again a stationary distribution. (ii) From Code 3.7.2, we note that for the given matrix P, there are two eigenvalues which are equal to 1. The normalized eigenvectors corresponding to $\lambda_1 = 1$ is $\underline{a} = (1/2, 1/2, 0, 0)$

and corresponding to $\lambda_2 = 1$ is $\underline{b} = (0, 0, 1/2, 1/2)$. (iii) Further, P^n for all $n \geq 15$ remains the same. Hence,

$$\lim_{n\to\infty} P^n = \begin{array}{c} \\ 1 \\ 2 \\ 3 \\ 4 \end{array} \begin{array}{c} \begin{array}{cccc} 1 & 2 & 3 & 4 \end{array} \\ \begin{pmatrix} 1/2 & 1/2 & 0 & 0 \\ 1/2 & 1/2 & 0 & 0 \\ 0 & 0 & 1/2 & 1/2 \\ 0 & 0 & 1/2 & 1/2 \end{pmatrix} \end{array}.$$

Thus, the first two rows correspond to the stationary distribution $\underline{a}$ and the last two rows correspond to the second stationary distribution $\underline{b}$. □

In the next example, we consider a Markov chain discussed in Example 3.2.1, in which we have noted that the long run distribution does not exist. However, stationary distributions exist as shown in the next example.

Example 3.4.8 Suppose $\{X_n, n \geq 0\}$ is a Markov chain with state space $S = \{1, 2, 3, 4, 5, 6\}$ and transition probability matrix P as given below.

$$P = \begin{array}{c} \\ 1 \\ 2 \\ 3 \\ 4 \\ 5 \\ 6 \end{array} \begin{array}{c} \begin{array}{cccccc} 1 & 2 & 3 & 4 & 5 & 6 \end{array} \\ \begin{pmatrix} 1/3 & 0 & 2/3 & 0 & 0 & 0 \\ 0 & 1/2 & 1/4 & 0 & 1/4 & 0 \\ 2/5 & 0 & 3/5 & 0 & 0 & 0 \\ 0 & 1/4 & 1/4 & 1/4 & 0 & 1/4 \\ 0 & 0 & 0 & 0 & 1/2 & 1/2 \\ 0 & 0 & 0 & 0 & 1/4 & 3/4 \end{pmatrix} \end{array}.$$

(i) We have already noted in Example 3.2.1 that {1,3}, {5,6} are closed communicating classes and hence the Markov chain is reducible. These two sub chains can be analyzed separately. For the two sub chains with state spaces $\{1, 3\}$, $\{5, 6\}$, solving the matrix equation, the stationary distributions are $(3/8, 5/8)$ and $(1/3, 2/3)$ respectively. (ii) Since there are two closed classes, eigenvalue 1 has multiplicity 2. The stationary distributions corresponding to these two eigenvalues are

$$\underline{a} = (3/8, 0, 5/8, 0, 0, 0) \ \ \& \ \ \underline{b} = (0, 0, 0, 0, 1/3, 2/3).$$

Hence,

$$\underline{\pi}(\alpha) = (\alpha * 3/8,\ 0,\ \alpha * 5/8,\ 0,\ (1-\alpha) * 1/3,\ (1-\alpha) * 2/3)$$

is a stationary distribution $\forall\ \alpha \in [0, 1]$. (iii) By the first method the stationary distribution is $\underline{\pi} = (0.1917, 0, 0.3195, 0, 0.1629, 0.3259)$. If we take $\alpha = 0.5112$ in $\underline{\pi}(\alpha)$, we get $\underline{\pi}$. It is to be noted that from this stationary distribution, we cannot get the mean recurrence times. (iv) In Example 3.2.1, we have obtained $\lim_{n\to\infty} P^n = Q$ and it is given by

$$Q = \begin{matrix} & \begin{matrix} 1 & 2 & 3 & 4 & 5 & 6 \end{matrix} \\ \begin{matrix} 1 \\ 2 \\ 3 \\ 4 \\ 5 \\ 6 \end{matrix} & \begin{pmatrix} 0.3750 & 0 & 0.6250 & 0 & 0.0000 & 0.0000 \\ 0.1875 & 0 & 0.3125 & 0 & 0.1667 & 0.3333 \\ 0.3750 & 0 & 0.6250 & 0 & 0.0000 & 0.0000 \\ 0.1875 & 0 & 0.3125 & 0 & 0.1667 & 0.3333 \\ 0.0000 & 0 & 0.0000 & 0 & 0.3333 & 0.6667 \\ 0.0000 & 0 & 0.0000 & 0 & 0.3333 & 0.6667 \end{pmatrix} \end{matrix}.$$

We note the link between the rows of Q and the stationary distribution. Observe that in $\underline{\pi}(\alpha)$, if $\alpha = 1$, then the first and the third rows of Q are the same as the stationary distribution $\underline{a}$, while if $\alpha = 0$, then the last two rows of Q are the same as the stationary distribution $\underline{b}$. In $\underline{\pi}(\alpha)$, if $\alpha = 1/2$, we get the second and the fourth rows of Q. Thus, all rows of Q represent stationary distributions. □

In the following two examples, we verify Theorem 3.3.4 for a periodic Markov chain.

Example 3.4.9 Suppose we have a Markov chain with P given by

$$P = \begin{matrix} & \begin{matrix} 1 & 2 & 3 \end{matrix} \\ \begin{matrix} 1 \\ 2 \\ 3 \end{matrix} & \begin{pmatrix} 0 & 1 & 0 \\ 0.5 & 0 & 0.5 \\ 0 & 1 & 0 \end{pmatrix} \end{matrix}.$$

It is periodic Markov chain with period 2. (i) The eigenvalues of P are $1, -1$ and 0. Thus, $\lim_{n\to\infty} P^n = \underline{x}_1\underline{y}_1' + \lim_{n\to\infty}(-1)^n\underline{x}_2\underline{y}_2'$. Thus, $\lim_{n\to\infty} P^n$ does not exist and hence the long run distribution does not exist. It is clear from the odd and even powers of P as shown below. For $n \geq 1$,

$$P^2 = P^4 = P^{2n} = \begin{matrix} & \begin{matrix} 1 & 2 & 3 \end{matrix} \\ \begin{matrix} 1 \\ 2 \\ 3 \end{matrix} & \begin{pmatrix} 0.5 & 0 & 0.5 \\ 0 & 1 & 0 \\ 0.5 & 0 & 0.5 \end{pmatrix} \end{matrix} \;\&\; P = P^3 = P^{2n+1} = \begin{matrix} & \begin{matrix} 1 & 2 & 3 \end{matrix} \\ \begin{matrix} 1 \\ 2 \\ 3 \end{matrix} & \begin{pmatrix} 0 & 1 & 0 \\ 0.5 & 0 & 0.5 \\ 0 & 1 & 0 \end{pmatrix} \end{matrix}.$$

(ii) The stationary distribution exists by Theorem 3.3.4 and by solving the equation $\underline{\pi} = \underline{\pi}P$, it is given by $\underline{\pi} = (0.25, 0.50, 0.25)$. Using the four methods listed above we get the same $\underline{\pi}$. We have already noted these results in Example 3.2.8. The mean recurrence time are given by $\mu_1 = 4$, $\mu_2 = 2$, $\mu_3 = 4$. Note that $\pi_1 = 0.25 = 1/\mu_1$, $\pi_2 = 0.5 = 1/\mu_2$ and $\pi_3 = 0.25 = 1/\mu_3$. (iii) The Markov chain with transition probability matrix $P_2 = P^2$ has two closed communicating classes, $C_1 = \{1, 3\}$ and $C_2 = \{2\}$. Hence, the two stationary distributions for the Markov chain with transition probability matrix P_2 are given by $\underline{\pi}_1 = (1/2, 0, 1/2)$ and $\underline{\pi}_2 = (0, 1, 0)$. There are two eigenvalues of P_2 which equal 1. $\underline{\pi}_1$ and $\underline{\pi}_2$ correspond to these two eigenvalues. A convex combination of $\underline{\pi} = \alpha\underline{\pi}_1 + (1-\alpha)\underline{\pi}_2$ is again a stationary distribution for a Markov chain with transition probability matrix P_2. Suppose $\alpha = 1/2$ then $\underline{\pi} = (1/4, 1/2, 1/4)$, which is the same as the stationary distribution

of a Markov chain with transition probability matrix P. Thus, a particular convex combination of the stationary distributions of P^2 coincides with the stationary distribution of a Markov chain with transition probability matrix P.
(iv) With the first method based on a generalized inverse, the stationary distribution for the Markov chain with transition probability matrix $P_2 = P^2$ is $(1/3, 1/3, 1/3)$. Observe that it is $(1/3) * \underline{\pi}_1 + (2/3) * \underline{\pi}_2$. (v) For two closed communicating classes as $\{1, 3\}$ and $\{2\}$, the long run distribution and the stationary distributions are the same and these are given by $\{a_1 = 1/2, a_3 = 1/2\}$ and $\{a_2 = 1\}$ respectively. Observe that $(1/2)(a_1, a_2, a_3) = \underline{\pi}$ as noted in Eq. (3.3.1). □

Example 3.4.10 Suppose $\{X_n, n \geq 0\}$ is a Markov chain with P given by

$$P = \begin{matrix} & 1 & 2 & 3 & 4 \\ 1 \\ 2 \\ 3 \\ 4 \end{matrix}\begin{pmatrix} 0 & 1 & 0 & 0 \\ 0.5 & 0 & 0.5 & 0 \\ 0 & 0.5 & 0 & 0.5 \\ 0 & 0 & 1 & 0 \end{pmatrix}.$$

(i) The Markov chain is irreducible and periodic with period 2. Two eigenvalues of P are 1 and -1, thus P^n oscillates and long run distribution does not exist, as noted in Example 3.2.7. (ii) However, as proved in Theorem 3.3.4, a unique stationary distribution $\underline{\pi}$ exists and is given by $\pi_i = 1/\mu_i, i \in S$. Using Code 2.8.6, we compute the mean recurrence times. These are given by $\mu_1 = 6,\ \mu_2 = 3,\ \mu_3 = 3$ and $\mu_4 = 6$. Hence, $\underline{\pi} = (1/6, 2/6, 2/6, 1/6)$. (iii) By Methods 1, 2 and 3 of Code 3.7.1, we get the same stationary distribution. By Method 2, it is given by $(A'A)^{-1}A'b$, thus it is a unique solution. The generalized inverse in Method 1 is also unique. By the third method, corresponding to eigenvalue 1, which has multiplicity 1, we get the same solution, which again justifies the uniqueness. (iv) We now examine how it is related to the stationary distributions of the Markov chain with transition probability matrix P^2. The matrix P^2 is given by

$$P^2 = \begin{matrix} & 1 & 2 & 3 & 4 \\ 1 \\ 2 \\ 3 \\ 4 \end{matrix}\begin{pmatrix} 0.50 & 0.00 & 0.50 & 0.00 \\ 0.00 & 0.75 & 0.00 & 0.25 \\ 0.25 & 0.00 & 0.75 & 0.00 \\ 0.00 & 0.50 & 0.00 & 0.50 \end{pmatrix}.$$

It is reducible with two closed communicating classes as $B_0 = \{1, 3\}$ and $B_1 = \{2, 4\}$. Thus, there are two stationary distributions for a Markov chain with transition probability matrix P^2. By Code 3.7.2, these are given by $\underline{\pi}_1 = (1/3, 0, 2/3, 0)$ and $\underline{\pi}_2 = (0, 2/3, 0, 1/3)$. There are two eigenvalues of P^2 which equal 1 and $\underline{\pi}_1$ and $\underline{\pi}_2$ correspond to these two eigenvalues. A convex combination $\alpha\underline{\pi}_1 + (1-\alpha)\underline{\pi}_2$, $0 < \alpha < 1$, is again a stationary distribution. With $\alpha = 1/2$, it is the same as the stationary distribution $\underline{\pi}$ of a Markov chain with transition probability matrix P. (iv) It can be easily verified that $(P^2)^n$ for $n \geq 8$ remains the same. Hence,

$$\lim_{n\to\infty}(P^2)^n = \lim_{n\to\infty} P^{2n} = \begin{array}{c} \\ 1 \\ 2 \\ 3 \\ 4 \end{array}\begin{array}{c} \begin{array}{cccc} 1 & 2 & 3 & 4 \end{array} \\ \begin{pmatrix} 1/3 & 0 & 2/3 & 0 \\ 0 & 2/3 & 0 & 1/3 \\ 1/3 & 0 & 2/3 & 0 \\ 0 & 2/3 & 0 & 1/3 \end{pmatrix} \end{array}.$$

From $\lim_{n\to\infty} P^{2n}$, we note that $\lim_{n\to\infty} p_{ij}^{(2n)} = d/\mu_j$ if both $i, j \in B_r, r = 0, 1$ and 0 otherwise, as proved in Theorem 3.2.12. Observe that there is a link between rows of $\lim_{n\to\infty}(P^2)^n$ and the two stationary distributions $\underline{\pi}_1$ and $\underline{\pi}_2$. (v) For two closed communicating classes as $\{1, 3\}$ and $\{2, 4\}$, the long run distribution and the stationary distributions are the same and these are given by $\{a_1 = 1/3, a_3 = 2/3\}$ and $\{a_2 = 2/3, a_4 = 1/3\}$ respectively. Observe that $(1/2)(a_1, a_2, a_3, a_4) = \underline{\pi}$ as noted in Eq. (3.3.1). (vi) for $n \geq 8$, P^{2n+1} remains the same. Hence,

$$\lim_{n\to\infty} P^{2n+1} = \begin{array}{c} \\ 1 \\ 2 \\ 3 \\ 4 \end{array}\begin{array}{c} \begin{array}{cccc} 1 & 2 & 3 & 4 \end{array} \\ \begin{pmatrix} 0 & 2/3 & 0 & 1/3 \\ 1/3 & 0 & 2/3 & 0 \\ 0 & 2/3 & 0 & 1/3 \\ 1/3 & 0 & 2/3 & 0 \end{pmatrix} \end{array}.$$

From this matrix we note that

$$\lim_{n\to\infty} p_{ij}^{(2n+1)} = \begin{cases} 0, & \text{if } \quad i \in B_0 \ \& \ j \in B_0 \\ 0, & \text{if } \quad i \in B_1 \ \& \ j \in B_1 \\ d/\mu_2 = 2/3 = 2/3, & \text{if } i \in B_0 \ \& \ j = 2 \in B_1 = \{2, 4\} \\ d/\mu_4 = 2/6 = 1/3, & \text{if } i \in B_0 \ \& \ j = 4 \in B_1 = \{2, 4\} \\ d/\mu_1 = 2/6 = 1/3, & \text{if } \quad i \in B_1 \ \& \ j = 1 \in B_0 \\ d/\mu_3 = 2/3 = 2/3, & \text{if } \quad i \in B_1 \ \& \ j = 3 \in B_0. \end{cases}$$

Thus, $\lim_{n\to\infty} p_{ij}^{(nd+s)} = d/\mu_j$, if $i \in B_r$ and $j \in B_{r+s}$. It is 0 otherwise. Thus, Lemma 3.3.1 is verified. □

In the next example, we consider a Markov chain with period 3.

Example 3.4.11 Suppose $\{X_n, n \geq 0\}$ is a Markov chain with P given by

$$P = \begin{array}{c} \\ 1 \\ 2 \\ 3 \\ 4 \\ 5 \\ 6 \end{array}\begin{array}{c} \begin{array}{cccccc} 1 & 2 & 3 & 4 & 5 & 6 \end{array} \\ \begin{pmatrix} 0 & 0 & 1/3 & 2/3 & 0 & 0 \\ 0 & 0 & 2/3 & 1/3 & 0 & 0 \\ 0 & 0 & 0 & 0 & 1/4 & 3/4 \\ 0 & 0 & 0 & 0 & 1/3 & 2/3 \\ 1/2 & 1/2 & 0 & 0 & 0 & 0 \\ 1/3 & 2/3 & 0 & 0 & 0 & 0 \end{pmatrix} \end{array}.$$

(i) We observe that all states communicate with each other, thus the Markov chain is irreducible. Further, all states from class $\{1, 2\}$ transit to states in class $\{3, 4\}$ in one step, states in class $\{3, 4\}$ transit to states in class $\{5, 6\}$ in one step and states in class $\{5, 6\}$ transit to states in class $\{1, 2\}$ in one step. Thus, for each state i, transition from i to i is possible in $3, 6, 9, \ldots$ steps, which implies that period of each state i is 3. Using Code 2.8.8 we find the period and it is 3 for each state. Since it is a finite state space irreducible Markov chain, all states are non-null persistent. Using Code 2.8.6 we verify it and find the mean recurrence times. These are given by $\underline{\mu} = (7.8664, 4.8494, 5.5602, 6.5154, 10.4023, 4.2156)$. Thus, the given Markov chain is non-null persistent and periodic Markov chain with period 3. We verify all the results studied in Sects. 3.2 and 3.3 for the periodic Markov chain.
(ii) The matrices P^2 and P^3 are as displayed below.

$$P^2 = \begin{array}{c} \\ 1 \\ 2 \\ 3 \\ 4 \\ 5 \\ 6 \end{array} \begin{array}{c} \begin{array}{cccccc} 1 & 2 & 3 & 4 & 5 & 6 \end{array} \\ \begin{pmatrix} 0 & 0 & 0 & 0 & 0.3055 & 0.6945 \\ 0 & 0 & 0 & 0 & 0.2778 & 0.7222 \\ 0.3750 & 0.6250 & 0 & 0 & 0 & 0 \\ 0.3889 & 0.6111 & 0 & 0 & 0 & 0 \\ 0 & 0 & 0.5000 & 0.5000 & 0 & 0 \\ 0 & 0 & 0.5556 & 0.4444 & 0 & 0 \end{pmatrix} \end{array}$$

$$\& \ P^3 = \begin{array}{c} \\ 1 \\ 2 \\ 3 \\ 4 \\ 5 \\ 6 \end{array} \begin{array}{c} \begin{array}{cccccc} 1 & 2 & 3 & 4 & 5 & 6 \end{array} \\ \begin{pmatrix} 0.3842 & 0.6158 & 0 & 0 & 0 & 0 \\ 0.3796 & 0.6204 & 0 & 0 & 0 & 0 \\ 0 & 0 & 0.5417 & 0.4583 & 0 & 0 \\ 0 & 0 & 0.5371 & 0.4629 & 0 & 0 \\ 0 & 0 & 0 & 0 & 0.2916 & 0.7084 \\ 0 & 0 & 0 & 0 & 0.2870 & 0.7130 \end{pmatrix} \end{array}.$$

From P, P^2 and P^3 we note that $B_0 = \{1, 2\}$, $B_1 = \{3, 4\}$ and $B_2 = \{5, 6\}$ are cyclically moving classes. The path of transitions can be described as $B_0 \to B_1 \to B_2 \to B_0 \to B_1 \to B_2 \to B_0 \ldots$. Thus after 3 steps, states in B_r are visited, $r = 0, 1, 2$. From P^3 we note that the matrices P_0, P_1, P_2 of transition probabilities in 3 steps, corresponding to B_0, B_1, B_2 are stochastic matrices. Thus, the Markov chain with transition probability matrix as P^3 is reducible with three closed communicating classes.
(iii) By finding powers of P we note that for $n \geq 3$, P^{3n}, P^{3n+1}, P^{3n+2} remain the same. Hence,

$$\lim_{n\to\infty} P^{3n} = \begin{array}{c} \\ 1 \\ 2 \\ 3 \\ 4 \\ 5 \\ 6 \end{array}\begin{array}{c} \begin{array}{cccccc} 1 & 2 & 3 & 4 & 5 & 6 \end{array} \\ \begin{pmatrix} 0.3814 & 0.6186 & 0 & 0 & 0 & 0 \\ 0.3814 & 0.6186 & 0 & 0 & 0 & 0 \\ 0 & 0 & 0.5396 & 0.4604 & 0 & 0 \\ 0 & 0 & 0.5396 & 0.4604 & 0 & 0 \\ 0 & 0 & 0 & 0 & 0.2884 & 0.7116 \\ 0 & 0 & 0 & 0 & 0.2884 & 0.7116 \end{pmatrix} \end{array},$$

$$\lim_{n\to\infty} P^{3n+1} = \begin{array}{c} \\ 1 \\ 2 \\ 3 \\ 4 \\ 5 \\ 6 \end{array}\begin{array}{c} \begin{array}{cccccc} 1 & 2 & 3 & 4 & 5 & 6 \end{array} \\ \begin{pmatrix} 0 & 0 & 0.5396 & 0.4604 & 0 & 0 \\ 0 & 0 & 0.5396 & 0.4604 & 0 & 0 \\ 0 & 0 & 0 & 0 & 0.2884 & 0.7116 \\ 0 & 0 & 0 & 0 & 0.2884 & 0.7116 \\ 0.3814 & 0.6186 & 0 & 0 & 0 & 0 \\ 0.3814 & 0.6186 & 0 & 0 & 0 & 0 \end{pmatrix} \end{array}$$

and

$$\lim_{n\to\infty} P^{3n+2} = \begin{array}{c} \\ 1 \\ 2 \\ 3 \\ 4 \\ 5 \\ 6 \end{array}\begin{array}{c} \begin{array}{cccccc} 1 & 2 & 3 & 4 & 5 & 6 \end{array} \\ \begin{pmatrix} 0 & 0 & 0 & 0 & 0.2884 & 0.7116 \\ 0 & 0 & 0 & 0 & 0.2884 & 0.7116 \\ 0.3814 & 0.6186 & 0 & 0 & 0 & 0 \\ 0.3814 & 0.6186 & 0 & 0 & 0 & 0 \\ 0 & 0 & 0.5396 & 0.4604 & 0 & 0 \\ 0 & 0 & 0.5396 & 0.4604 & 0 & 0 \end{pmatrix} \end{array}.$$

From $\underline{\mu}$, we have $\underline{3/\mu} = (0.3814, 0.6186, 0.5396, 0.4604, 0.2884, 0.7116)$. From these three matrices, we note that as proved in Theorem 3.2.12 and Lemma 3.3.1

$$\begin{aligned} \lim_{n\to\infty} p_{ij}^{(3n)} &= 3/\mu_j, \text{ if } i, j \in B_r, r = 0, 1, 2 \;\&\; 0 \text{ otherwise} \\ \lim_{n\to\infty} p_{ij}^{(3n+1)} &= 3/\mu_j, \text{ if } i \in B_r \;\&\; j \in B_{r+1}, r = 0, 1, 2 \;\&\; 0 \text{ otherwise} \\ \lim_{n\to\infty} p_{ij}^{(3n+2)} &= 3/\mu_j, \text{ if } i \in B_r \;\&\; j \in B_{r+2}, r = 0, 1, 2 \;\&\; 0 \text{ otherwise,} \end{aligned}$$

where $B_3 = B_0$, $B_4 = B_1$.
(iv) As proved in Theorem 3.3.4, a unique stationary distribution $\underline{\pi}$ exists and is given by $\pi_i = 1/\mu_i, i \in S$. Hence, $\underline{\pi} = (0.1271, 0.2062, 0.1799, 0.1535, 0.0961, 0.2372)$. By Methods 1, 2 and 3 of Code 3.7.1, we get the same stationary distribution. By Method 2, it is given by $(A'A)^{-1}A'b$, thus it is a unique solution. The generalized inverse in Method 1 is also unique. By the third method, corresponding to eigenvalue 1, which has multiplicity 1, we get the same solution.
(v) We now examine how $\underline{\pi}$ is related to the stationary distributions of the Markov chain with transition probability matrix P^3. From P^3, we note that it is a reducible Markov chain with three closed communicating classes as B_0, B_1, B_2. Thus, there

are three stationary distributions. By Code 3.7.2, these are given by $\underline{\pi}_1 = (0.3814, 0.6186, 0, 0, 0, 0)$, $\underline{\pi}_2 = (0, 0, 0.5396, 0.4604, 0, 0)$ and $\underline{\pi}_3 = (0, 0, 0, 0, 0.2884, 0.7116)$. A convex combination of these three is again a stationary distribution. In particular, $\underline{\pi} = (1/3)(\underline{\pi}_1 + \underline{\pi}_2 + \underline{\pi}_3)$. Suppose $\underline{b}_0 = (0.3814, 0.6186)$, $\underline{b}_1 = (0.5396, 0.4604)$ and $\underline{b}_2 = (0.2884, 0.7116)$ are the stationary distributions corresponding to stochastic matrices P_0, P_1, P_2 respectively. These are the same as the corresponding long run distributions. Note that $(1/3)(\underline{b}_0, \underline{b}_1, \underline{b}_2) = \underline{\pi}$ as noted in Eq. (3.3.1). □

In the next section, we discuss briefly one more feature of a Markov chain and it is one of the measures of the degree of Markov dependence.

3.5 Autocovariance Function

The strength of the dependence in a Markov chain with state space W can be judged by $Cov(X_k, X_{k+n})$, which is known as the autocovariance function of lag n. It is heavily used in auto-regressive time series. It can be computed using the transition probability matrix and the initial distribution. To compute $E(X_k X_{k+n})$, observe that by conditioning on X_k, we have for $n \geq 1$,

$$E(X_k X_{k+n}) = E(X_k(E(X_{k+n}|X_k))) = \sum_{i \in S} \sum_{j \in S} ij p_{ij}^{(n)} p_i^{(k)},$$

where $p_i^{(k)} = P[X_k = i]$. Hence,

$$Cov(X_k, X_{k+n}) = \sum_{i \in S} \sum_{j \in S} ij p_{ij}^{(n)} p_i^{(k)} - \left(\sum_{i \in S} i p_i^{(k)} \right) \left(\sum_{j \in S} j p_j^{(k+n)} \right).$$

Since $\{X_n, n \geq 0\}$ is a Markov chain, the conditional distribution of X_n given the entire past is the same as the conditional distribution of X_n given X_{n-1}. It is thus expected that as n increases $Cov(X_k, X_{k+n})$ will decrease. In the following example, we verify it using Code 3.7.3.

Example 3.5.1 We use Code 3.7.3 to compute $Cov(X_5, X_{5+n})$ for the Markov chain for the care center model discussed in Example 2.2.5. Figure 3.1 displays the nature of the $Cov(X_5, X_{5+n})$ as n varies.

From the graph of the covariance function, we note that as n increases $Cov(X_5, X_{5+n})$ decreases, as expected. □

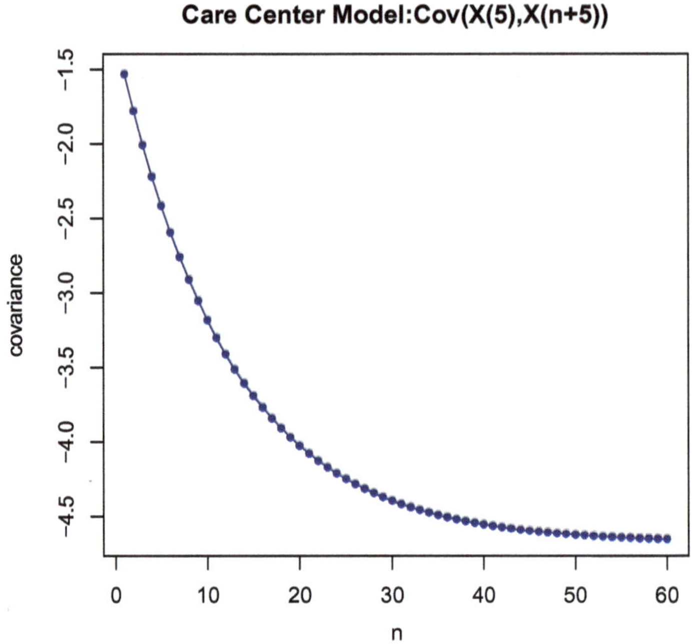

Fig. 3.1 Autocovariance Function: $Cov(X_5, X_{5+n})$

If we assume that the initial distribution of a Markov chain is its stationary distribution, then

$$p_i^{(k)} = P[X_k = i] = P[X_{k+n} = i] = \pi_i \ \forall \ i \in S,$$

where $\{\pi_i, i \in S\}$ is the stationary distribution. Hence,

$$Cov(X_k, X_{k+n}) = \sum_{i \in S} \sum_{j \in S} ijp_{ij}^{(n)} \pi_i - \left(\sum_{i \in S} i\pi_i \right)^2 .$$

Further, note that if $p_{ij}^{(n)} \to \pi_j$ as $n \to \infty$, then $Cov(X_k, X_{k+n}) \to 0$, indicating that the strength of dependence decreases as the lag increases.

If $\{X_n, n \geq 0\}$ is a Markov chain with state space $\{0, 1\}$ then the correlation coefficient between X_k and X_{k+n} is $R_n = Corr(X_k, X_{k+n}) = (p_{11} - p_{01})^n$, (Guttorp [5]). In Example 2.3.4, we have $\hat{p}_{11} = 0.838$ and $\hat{p}_{01} = 0.602$. Hence, $R_n = (0.236)^n$, $n \geq 1$, it decreases as n increases. Table 3.1 displays the values of the correlation coefficient R_n between X_k and X_{k+n} for $n = 1, 2 \ldots, 7$. The values of the correlation coefficient decrease, as expected.

In the following section, we discuss an application of the theory of Markov chains and associated stationary distributions in determining the annual vehicle premium in Bonus-Malus system. For details, refer to Boland [1].

Table 3.1 Values of correlation coefficient

n	1	2	3	4	5	6	7
R_n	0.2360	0.0557	0.0131	0.0031	0.0007	0.00017	0.00004

3.6 Bonus-Malus System

Risk of a motor accident varies from driver to driver. The factors that contribute the most to high risks of claims in motor insurance are driving experience, age, skill of a driver, knowledge and appreciation of the rules of the road travel, care with respect to speed restrictions and driving conditions, good judgment about driving when tired or under the influence of factors such as alcohol, tension and aggressiveness. All these factors are not impossible but difficult to measure effectively. More accessible and the best indicator is individual's past claims history. Bonus-Malus system, also known as No Claim Discount (NCD) system, adjusts premiums on the basis of individual's claims experience. Bonus and malus are the Latin words, bonus means reward and malus means penalty. Thus, policyholders are categorized into relatively homogeneous risk groups who pay premiums relative to their claims experience. Policyholders who have made few claims in recent years are rewarded with discounts on the premium.

As an illustration, suppose there are three discount classes E_0 (no discount), E_1 (20% discount) and E_2 (40% discount). Movement in the system is determined by the following rule. One steps back one discount level (or stays in E_0) with one claim in a year, and returns to a level of no discount if more than one claim is made. A claim-free year results in a step up to a higher discount level (or one remains in class E_2 if already there). Given a set of discount classes and a transition rule for a Bonus-Malus system, it is assumed that (i) movement from discount class E_i in one year to discount class E_j in the next is a random event with probability p_{ij}, which is the same for all insureds in a specific rating group, (ii) movement from any class E_i in one year to another class E_j in the following is independent of how the individual arrived in class E_i to begin with, where p_{ij} is the probability of transition in discount class E_j next year given that the individual is in discount class E_i in this year. Thus, Markov chain is a suitable model for transitions from one class to another. We denote the classes E_0, E_1, E_2 by 0, 1, 2 respectively. Thus state space of the Markov chain is $S = \{0, 1, 2\}$. One step transition probabilities are determined by the number of claims in a year. Suppose probability distribution for number of claims N in a year is as follows.

$$P[N = 0] = 0.7, \quad P[N = 1] = 0.2, \quad P\,[N \geq 2] = 0.1$$

Then matrix P of one step transition probabilities is

$$P = \begin{array}{c} \\ 0 \\ 1 \\ 2 \end{array} \begin{array}{c} \begin{array}{ccc} 0 & 1 & 2 \end{array} \\ \begin{pmatrix} 0.3 & 0.7 & 0 \\ 0.3 & 0 & 0.7 \\ 0.1 & 0.2 & 0.7 \end{pmatrix} \end{array}$$

Suppose the initial distribution is $\underline{p}^{(0)} = (p_0, p_1, p_2)$. If one assumes that a person initially starts in category E_0, then $\underline{p}^{(0)} = (1, 0, 0)$. Now

$$P^2 = \begin{array}{c} \\ 0 \\ 1 \\ 2 \end{array} \begin{array}{c} \begin{array}{ccc} 0 & 1 & 2 \end{array} \\ \begin{pmatrix} 0.30 & 0.21 & 0.49 \\ 0.16 & 0.35 & 0.49 \\ 0.16 & 0.21 & 0.63 \end{pmatrix} \end{array}$$

Thus, if a policyholder is in discount class 1 and paying 80% of the full premium in a given year, then the probability that one is still paying the same premium two years later is 0.35. It is easy to check that

$$P^6 = \begin{array}{c} \\ 0 \\ 1 \\ 2 \end{array} \begin{array}{c} \begin{array}{ccc} 0 & 1 & 2 \end{array} \\ \begin{pmatrix} 0.1883 & 0.2435 & 0.5682 \\ 0.1855 & 0.2463 & 0.5682 \\ 0.1855 & 0.2435 & 0.5710 \end{pmatrix} \end{array}, \quad P^n = \begin{array}{c} \\ 0 \\ 1 \\ 2 \end{array} \begin{array}{c} \begin{array}{ccc} 0 & 1 & 2 \end{array} \\ \begin{pmatrix} 0.1860 & 0.2442 & 0.5698 \\ 0.1860 & 0.2442 & 0.5698 \\ 0.1860 & 0.2442 & 0.5698 \end{pmatrix} \end{array},$$

$\forall \ n \geq 20$. Thus, we conclude that P^n converges to a matrix whose rows are constant, and the row represents the long run distribution of the Markov model. Thus, the long run distribution which is the same as the stationary distribution is $\underline{\pi} = (0.1860, 0.2442, 0.5698)$.

Suppose 2000 policyholders are placed in the discount classes in the initial year according to the probability distribution $\underline{p}^{(0)} = (0.5, 0.3, 0.2)$. Suppose the pure annual premium for the car insurance is Rs.6000. Assuming premiums are paid at the beginning of a year, the expected income from premiums at the beginning of the year would be

$$2000[0.5(6000) + 0.3(4800) + 0.2(3600)] = 1,032,000.00\,.$$

To find the expected income from premiums next year, we find the probabilities that an individual is in states 0, 1 and 2 next year. These are given by $\underline{p}^{(1)} = \underline{p}^{(0)} * P = (0.26, 0.39, 0.35)$ and hence the expected income from premiums at the beginning of the next year is

$$2000[0.26(6000) + 0.39(4800) + 0.35(3600)] = 9,38,400.00\,.$$

Table 3.2 displays probability distributions and the expected income from premiums in rupees for some years.

Table 3.2 Probability Distributions and Expected Income

Year n	p_0	p_1	p_2	Expected income
0	0.500	0.300	0.200	1,032,000.00
1	0.260	0.390	0.350	938,400.00
2	0.230	0.252	0.518	890,880.00
3	0.1964	0.2646	0.539	877,776.00
4	0.1922	0.2453	0.5625	871,123.20
5	0.1875	0.2470	0.5655	869,288.60
6	0.1869	0.2443	0.5688	868,357.20
10	0.1861	0.2442	0.5697	867,915.80
15	0.1860	0.2442	0.5698	867,907.00
30	0.1860	0.2442	0.5698	867,907.00

From Table 3.2, we observe that premium income stabilizes over the years as the proportion of policyholders in three discount groups stabilizes.

Bonus-Malus schemes are popular in motor insurance. Objective of Bonus-Malus systems is higher premiums to those with higher claim rates. Thus, these penalize the poor drivers and reward the good drivers. Introducing discounts for those with a good claims record reduces heterogeneity among policyholders within the various rating classes. It allows the insurer to charge premiums that are more appropriate to the individual risks. Since consequence of making a claim is an increased premium in the following year, so policyholders are naturally discouraged from making small claims which reduces both the number of claims made by the policyholders and the overall management costs incurred by the insurer.

The next section presents `R` codes used for solving examples in Sects. 3.4 and 3.5.

3.7 R Codes

Code 3.7.1 Computation of a stationary distribution for an irreducible Markov chain or a Markov chain with a single closed communicating class: With this `R` code, we obtain the stationary distribution of a Markov chain, using four methods listed in Sect. 3.4. We use a package `Matrix`, (Bates and Maechler [2]), to find the rank of a matrix and a package `MASS` (Venables and Ripley [8]), to find the generalized inverse of a matrix. We illustrate the code for the Markov chain in Example 3.4.2. Since we have assumed that the Markov chain is either irreducible or reducible with only one closed class, multiplicity of eigenvalue 1 is 1. If multiplicity of eigenvalue 1 is more than 1, there are some minor changes in the code, which are discussed in the next code.

```
# Part I: Input tpm and state space
library(Matrix); library(expm); library(MASS)
ns=3;state=1:ns
P=matrix(c(.4,.5,.1,.05,.7,.25,.05,.5,.45), nrow=ns,ncol=ns,
byrow=TRUE); A=t(P)
# Part II: Computation of a stationary distribution by method 1
I=diag(ns); d=rep(1,ns); B=rbind(A-I,d); b=c(rep(0,ns),1)
rank.B=as.numeric(rankMatrix(B));rank.B
rank.Bb=as.numeric(rankMatrix(cbind(B, b)));rank.Bb
a=ginv(B)%*%b; a=round(a,4); a # stationary distribution
# Part III: Computation of a stationary distribution by method 2
D=t(B)%*%B; D; rank.D=as.numeric(rankMatrix(D));rank.D
rank.Db=as.numeric(rankMatrix(cbind(D, t(B)%*%b)));rank.Db
b1=ginv(D)%*%t(B)%*%b; b1=round(b1,4); b1
b2=solve(D,t(B)%*%b);b2; b2=round(b2,4); b2
# b1=b2: stationary distributions
# Part IV: Computation of a stationary distribution by method 3
eigen(A); e1=eigen(A)$values; e1; e2=which(round(e1,4)==1)
v=eigen(A)$vectors; a1=v[,e2]; pi=a1/sum(a1)
pi=round(pi,4);pi # stationary distribution
# Part V: Computation of a stationary distribution by method 4
Power = list(); N=seq(2,14,3); k=1
for(j in N)
{
Power[[k]]=round(P%^%j,4)
k=k+1
}
Power
# Alternative way
pn=list()
for(k in N)
{
pn[[k]]=round(P%^%k,4)
if(pn[[k]][1,]==pn[[k]][2,]&&pn[[k]][2,]==pn[[k]][3,])
break()
}
k; sd=round(P%^%(k),4);sd[1,] # stationary distribution.
```

□

Code 3.7.2 Computation of a stationary distribution for reducible Markov chains: In this code there is a slight variation in the third method of Code 3.7.1 to accommodate the cases where multiplicity of eigenvalue 1 is more than 1. It arises when the Markov chain is reducible with more than one closed classes. We illustrate the code for the Markov chain in Example 3.4.6.

```
# Part I: To input the tpm and state space
ns=5; state=1:ns; r1=c(1/3,2/3,0,0,0)
r2=c(3/4,1/4,0,0,0); r3=c(0,0,1/8,1/4,5/8); r4=c(0,0,0,1/2,1/2)
r5=c(0,0,1/3,0,2/3); P=rbind(r1,r2,r3,r4,r5); A=t(P)
library(Matrix);library(expm); library(MASS)
# Part II: Computation of a stationary distribution by method 3
e1=eigen(A)$values; e1; e2=which(round(e1,4)==1); e2
l=length(e2) # multiplicity of eigen value 1
v=eigen(A)$vectors; a1=v[,e2[1]]; pi1=a1/sum(a1)
p1=round(pi1,4);p1; a2=v[,e2[2]]; pi2=a2/sum(a2)
p2=round(pi2,4);p2
b3=.4881*p1 +(1-.4881)*p2;b3=round(b3,4);b3
# Part III: Computation of a stationary distribution by method 4
Power = list(); N=seq(2,17,3); k=1
for(j in N)
{
Power[[k]]=round(P%^%j,4)
k=k+1
}
Power; n=max(N); Pn=P%^%n;sd1=round(Pn[1,],4);sd1
sd2=round(Pn[3,],4);sd2
mu1=1/pi1[1:2];mu1; mu2=1/pi2[3:5];mu2
mu=round(c(mu1,mu2),4);mu
```

□

Code 3.7.3 Computation of covariance: This code illustrates the computation of $Cov(X_5, X_{5+n})$ for the Markov chain for care center model discussed in Example 2.2.5.

```
# Part I: To input the tpm and state space
ns=3; state=1:ns; r1=c(.92,.05, .03);r2=c(0,.76,.24)
r3=c(0,0,1); P=rbind(r1,r2,r3); a=c(1,0,0)
# Part II: Computation of covariance function
library(matrixcalc)
X=function(P,n) # Marginal distribution of X(n)
{
return(a%*%matrix.power(P,n))
}
EX=function(P,n) # E(X(n))
{
return(state%*%t(X(P,n)))
}
J=function(P,n,i,j) # P[X(n+5)=j, X(5)=i]
{
return((matrix.power(P,n)[i,j])*X(P,5)[i])
```

```
}
JEX=function(P,n) # E(X(n+5)X(5))
{
sum=0
for (i in 1:ns)
{
for (j in 1:ns)
{
sum=sum+i*j*J(P,n,i,j)
}
}
return(sum)
}
Cov=function(P,n)
{
g=JEX(P,n)-EX(P,5)*EX(P,n+5)
return(g)
}
d=c()
for(n in 1:60)
{
d[n]=round(Cov(P,n),3)
}
# Part III: Plot of covariance function
m=1:60
plot(m,d,"o",pch=20,main="Care Center Model:Cov(X(5),X(n+5))",
xlab="n", ylab="covariance",col="darkblue")
```

□

In the next chapter, we discuss some traditional illustrations of Markov chains, such as random walks with state space as a set of all integers or W or a finite set. Ehrenfest chain and Gambler's ruin chain are the particular cases of a random walk with finite state space. In Chap. 5, we study a special type of Markov chain known as Galton Watson branching process.

A quick recap of the results discussed in the present chapter is given below.

Summary

1 If $\lim_{n\to\infty} p_{ij}^{(n)}$ exists $\forall\ j \in S$ and does not depend on i, then $\{a_j = \lim_{n\to\infty} p_{ij}^{(n)},\ j \in S\}$ is known as a long run distribution of the Markov chain, provided $\sum_{j\in S} a_j = 1$.

2 $\{\pi_i, i \in S\}$ is said to be a stationary distribution associated with the Markov chain if (i) $\pi_i \geq 0,\ \forall\ i \in S$, (ii) $\sum_{i\in S} \pi_i = 1$ and (iii) $\pi_j = \sum_{i\in S} \pi_i p_{ij}\ \forall\ j \in S$.

3 If a long run distribution exists, then the stationary distribution and limiting distribution of X_n also exist and all are the same.

4 For a non-null persistent aperiodic state j and $i \in S$, $\lim_{n\to\infty} p_{ij}^{(n)} = f_{ij}/\mu_j$ and if $i \leftrightarrow j$, then $\lim_{n\to\infty} p_{ij}^{(n)} = 1/\mu_j$.
5 An aperiodic non-null persistent state is known as an ergodic state. Being ergodic is a class property.
6 For a Markov chain with a single closed communicating class C of aperiodic states, a unique long run distribution $\underline{a}$ exists and is given by $a_j = 1/\mu_j$ if $j \in C$ & $a_j = 0$ if $j \notin C$.
7 For an ergodic Markov chain, a unique long run distribution $\underline{a}$ exists and is given by $a_j = 1/\mu_j,\ j \in S$.
8 Suppose $\{X_n, n \geq 0\}$ is a finite state space Markov chain, with two closed communicating classes C_1 and C_2 of aperiodic states. Then a long run distribution does not exist.
9 Suppose all the states of a Markov chain are either transient or null persistent. Then the long run distribution does not exist.
10 Suppose $\{X_n, n \geq 0\}$ is a non-null persistent and periodic Markov chain with period d. Then all the states in S can be grouped into d disjoint cyclically moving classes $\{B_0, B_1, \ldots, B_{d-1}\}$ such that

$$p_{ij} = 0 \text{ unless } i \in B_k \ \&\ j \in B_{k+1}, k = 1, 2, \ldots, d-1 \ \&\ i \in B_{d-1} \ \&\ j \in B_0.$$

For the Markov chain with transition matrix P^d, the classes $\{B_0, B_1, \ldots, B_{d-1}\}$ are closed communicating classes of aperiodic states.
11 Suppose j is a persistent and periodic state with period d. Then $\lim_{n\to\infty} p_{jj}^{(nd)} = d/\mu_j$ if j is non-null persistent and 0 if j is null persistent.
12 Suppose $\{X_n, n \geq 0\}$ is a non-null persistent and periodic Markov chain with period $d > 1$. Then (i) $\lim_{n\to\infty} p_{ij}^{(nd)} = d/\mu_j$ if $i, j \in B_r$ and 0 otherwise, and (ii) $\lim_{n\to\infty} p_{ij}^{(nd+s)} = d/\mu_j$ if $i \in B_r,\ j \in B_{r+s}$ and 0 otherwise.
13 If the initial distribution of the Markov chain is the same as its stationary distribution, then the distribution of X_n for each n is the same as the stationary distribution and hence also its limit. (ii) Conversely if X_0 and X_1 are identically distributed with common probability distribution $\underline{b}$, then $\underline{b}$ is the stationary distribution of the Markov chain.
14 If the initial distribution of a Markov chain is the same as its stationary distribution, then the Markov chain is a stationary process.
15 For a Markov chain with a single closed communicating countable class C of aperiodic states, a unique stationary distribution $\{\pi_i, i \in S\}$ exists, where $\pi_i = 1/\mu_i$ for $i \in C$ and $\pi_i = 0$ for $i \notin C$. For an ergodic Markov chain with countable state space S, a unique stationary distribution $\{\pi_i, i \in S\}$ exists, where $\pi_i = 1/\mu_i$ for $i \in S$.
16 For a non-null persistent and periodic Markov chain with period d, a unique stationary distribution $\underline{\pi} = \{\pi_j, j \in S\}$ exists and $\pi_j = 1/\mu_j$.
17 For a null persistent or transient Markov chain with countably infinite state space S, a stationary distribution does not exist.

18 A Markov chain with countably infinite state space W and a doubly stochastic transition probability matrix P is either transient or null persistent. A stationary distribution does not exist for such a Markov chain.

19 For a Markov chain with $S = \{1, 2, \ldots, M\}$ and a doubly stochastic transition probability matrix P, a stationary distribution $\underline{\pi}$ is given by $\pi_i = 1/M, \ i = 1, 2, \ldots, M$.

20 If P is stochastic matrix of order $M \times M$, then 1 is always an eigenvalue of P and all the eigenvalues are ≤ 1 in absolute value.

21 Suppose P is a transition probability matrix of a Markov chain with finite state space with M states. If λ_1 is the only eigenvalue which is equal to 1 and all other eigenvalues are distinct and < 1 in absolute value, then for all $i \in S$, $\lim_{n\to\infty} p_{ij}^{(n)} = \pi_j, \ j = 1, 2, \ldots, M$, where $\{\pi_j, \ j \in S\}$ is the long run as well as the stationary distribution.

22 Suppose $\{X_n, n \geq 0\}$ is a Markov chain with two closed communicating classes C_1 and C_2 of aperiodic states. Then there are two stationary distributions $\underline{a}$ and $\underline{b}$, where

$$a_i = 1/\mu_i \text{ if } i \in C_1, \ a_i = 0 \text{ if } i \notin C_1 \ \& \ b_i = 1/\mu_i \text{ if } i \in C_2, \ b_i = 0 \text{ if } i \notin C_2.$$

Further, $\underline{\pi}(\alpha) = \alpha\underline{a} + (1-\alpha)\underline{b}, \ 0 \leq \alpha \leq 1$ is a stationary distribution for $\alpha \in [0, 1]$.

23 If $\{\pi_i = 1/\mu_i, \ i \in S\}$ is a stationary distribution, then $\pi_i = 1/\mu_i$ is interpreted as a long run mean fraction of time the system is in state i, whatever may be the initial state.

24 A long run distribution may not exist, but a stationary distribution exists.

25 The strength of the dependence in a Markov chain with state space W can be judged by computing the $Cov(X_k, X_{k+n})$. As n increases $Cov(X_k, X_{k+n})$ decreases. If a Markov chain is irreducible and ergodic, then $p_{ij}^{(n)} \to \pi_j$ as $n \to \infty$ and hence,

$$Cov(X_k, X_{k+n}) = \sum_{i\in S}\sum_{j\in S} ijp_{ij}^{(n)}\pi_i - \left(\sum_{i\in S} i\pi_i\right)^2 \to 0 \quad \text{as } n \to \infty.$$

3.8 Conceptual Exercises

3.8.1 Suppose $\{X_n, n \geq 0\}$ is a sequence of independent and identically distributed random variables with $P[X_n = i] = a_i, \ a_i \geq 0$ and $\sum_{i\in S} a_i = 1$. It is known to be a Markov chain. Find the long run and stationary distributions, if exist.

3.8.2 Suppose $\{X_n, n \geq 0\}$ is a Markov chain with state space $S = \{1, 2, 3, 4, 5\}$ and P given by

$$P = \begin{array}{c} \\ 1 \\ 2 \\ 3 \\ 4 \\ 5 \end{array} \begin{array}{c} \begin{array}{ccccc} 1 & 2 & 3 & 4 & 5 \end{array} \\ \begin{pmatrix} 1/3 & 2/3 & 0 & 0 & 0 \\ 3/4 & 1/4 & 0 & 0 & 0 \\ 0 & 0 & 1/8 & 1/4 & 5/8 \\ 0 & 0 & 0 & 1/2 & 1/2 \\ 0 & 0 & 1/3 & 0 & 2/3 \end{pmatrix} \end{array}.$$

(i) Classify the states. (ii) Examine whether the long run distribution exists. (iii) If yes, find it. (iv) Examine whether the stationary distribution exists. (v) If yes, find it. (vi) Comment on the link between the long run distribution and the stationary distribution. (vii) Find the matrix of $F = f_{ij}$. (viii) Verify the relation between f_{ij} and $\lim_{n\to\infty} p_{ij}^{(n)}$.

3.8.3 Suppose $\{X_n, n \geq 0\}$ is a Markov chain with state space $S = \{1, 2, 3, 4, 5\}$ and P given by

$$P = \begin{array}{c} \\ 1 \\ 2 \\ 3 \\ 4 \\ 5 \end{array} \begin{array}{c} \begin{array}{ccccc} 1 & 2 & 3 & 4 & 5 \end{array} \\ \begin{pmatrix} 0.7 & 0.3 & 0 & 0 & 0 \\ 0.3 & 0.7 & 0 & 0 & 0 \\ 0 & 0 & 0.3 & 0.3 & 0.4 \\ 0 & 0 & 0.4 & 0.5 & 0.1 \\ 0 & 0 & 0.3 & 0.2 & 0.5 \end{pmatrix} \end{array}.$$

(i) Obtain the long run distribution if it exists.
(ii) Examine if stationary distributions exist. If yes, find all the stationary distributions.

3.8.4 On a Southern Pacific island, a sunny day is followed by another sunny day with probability 0.9, whereas a rainy day is followed by another rainy day with probability 0.2. Suppose that there are only sunny or rainy days. In the long run what fraction of days is sunny? Find an expression for corresponding $f_{ii}^{(n)}$ and the mean recurrence time. Examine whether the fraction of sunny days is the reciprocal of the corresponding mean recurrence time.

3.8.5 Examine whether a Markov chain with state space $S = \{1, 2, 3\}$ and

$$P = \begin{array}{c} \\ 1 \\ 2 \\ 3 \end{array} \begin{array}{c} \begin{array}{ccc} 1 & 2 & 3 \end{array} \\ \begin{pmatrix} 1/3 & 2/3 & 0 \\ 1/4 & 1/2 & 1/4 \\ 1 & 0 & 0 \end{pmatrix} \end{array}$$

is ergodic. Examine whether the long run distribution and stationary distributions exist. If yes, find the distributions and the mean recurrence times.

3.8.6 For a Markov chain $\{X_n, n \geq 0\}$ with state space $S = \{1, 2, 3\}$, P is given by

$$P = \begin{matrix} & \begin{matrix} 1 & 2 & 3 \end{matrix} \\ \begin{matrix} 1 \\ 2 \\ 3 \end{matrix} & \begin{pmatrix} 0.3 & 0.2 & 0.5 \\ 0.5 & 0.1 & 0.4 \\ 0.5 & 0.2 & 0.3 \end{pmatrix} \end{matrix}.$$

Each visit that the process makes to states 1, 2, 3 incurs cost of Rs.200, 500, 300 respectively. What is the long run cost per visit associated with this Markov chain?

3.8.7 Operating condition of a machine at any time is classified as follows. State 1: Good, State 2: Deteriorated but operating and State 3:In repair. Suppose for $n \geq 1$, X_n denotes the condition of the machine at the end of period n. We assume that the sequence of machine conditions is a Markov chain with the transition probability matrix P as given below.

$$P = \begin{matrix} & \begin{matrix} 1 & 2 & 3 \end{matrix} \\ \begin{matrix} 1 \\ 2 \\ 3 \end{matrix} & \begin{pmatrix} 0.9 & 0.1 & 0 \\ 0 & 0.9 & 0.1 \\ 1 & 0 & 0 \end{pmatrix} \end{matrix}.$$

What is the long run average rate of repairs per unit time?

3.8.8 Suppose there are two groups of drivers: a group of 10,000 relatively good drivers and 10, 000 relatively bad drivers. Suppose discount levels of insurance premiums are 0 (no discount), 1 (20% discount) and 2 (40% discount). The full premium is Rs. 5000. The discount level of a driver changes according to the rule "reduce by one discount level if one claim is made, and move to no discount level if more than one claim is made". Assume that a Markov chain is a suitable model for transitions from one class to another. Probability distributions for number of claims N in a year for the two groups are given below.

$$\text{Good Drivers}: \quad P[N=0]=0.7, \quad P[N=1]=0.2, \quad P\,[N \geq 2]=0.1$$

$$\text{Bad Drivers}: \quad P[N=0]=0.4, \quad P[N=1]=0.4, \quad P\,[N \geq 2]=0.2$$

(i) Obtain the one step transition probability matrices for both the groups. (ii) Assuming that all drivers start in class 0 at the beginning of the year, compute the expected premium income from the two groups for years 1, 2, 3, 4, 8, 16, 32. (iii) Compute the long run expected premium income from the two groups and comment.

3.8.9 Suppose $\{X_n, n \geq 0\}$ is a Markov chain with P given by

$$P = \begin{array}{c} \\ 1 \\ 2 \\ 3 \\ 4 \\ 5 \\ 6 \\ 7 \end{array} \begin{array}{c} \begin{array}{ccccccc} 1 & 2 & 3 & 4 & 5 & 6 & 7 \end{array} \\ \begin{pmatrix} 0 & 0 & 1/2 & 1/4 & 1/4 & 0 & 0 \\ 0 & 0 & 1/3 & 0 & 2/3 & 0 & 0 \\ 0 & 0 & 0 & 0 & 0 & 1/3 & 2/3 \\ 0 & 0 & 0 & 0 & 0 & 1/2 & 1/2 \\ 0 & 0 & 0 & 0 & 0 & 3/4 & 1/4 \\ 1/2 & 1/2 & 0 & 0 & 0 & 0 & 0 \\ 1/4 & 3/4 & 0 & 0 & 0 & 0 & 0 \end{pmatrix} \end{array}.$$

(i) Examine whether the Markov chain is irreducible. (ii) Find the period of each state. (iii) Find the cyclically moving classes.

3.9 Computational Exercises

3.9.1 Suppose $\{X_n, n \geq 0\}$ is a Markov chain with state space $S = \{1, 2, 3, 4, 5, 6\}$ and P given by

$$P = \begin{array}{c} \\ 1 \\ 2 \\ 3 \\ 4 \\ 5 \\ 6 \end{array} \begin{array}{c} \begin{array}{cccccc} 1 & 2 & 3 & 4 & 5 & 6 \end{array} \\ \begin{pmatrix} 1/4 & 0 & 3/4 & 0 & 0 & 0 \\ 0 & 1/3 & 1/3 & 0 & 1/3 & 0 \\ 2/7 & 0 & 5/7 & 0 & 0 & 0 \\ 0 & 1/4 & 1/4 & 1/4 & 0 & 1/4 \\ 0 & 0 & 0 & 0 & 4/9 & 5/9 \\ 0 & 0 & 0 & 0 & 1/3 & 2/3 \end{pmatrix} \end{array}.$$

(i) Find a family of stationary distributions. (ii) Find $\lim_{n\to\infty} p_{ij}^{(n)}$ for all i and j. Comment on your findings. Find f_{ij} for all i and j and examine how these are related to $\lim_{n\to\infty} p_{ij}^{(n)}$.

3.9.2 Suppose $\{X_n, n \geq 0\}$ is a Markov chain with state space $S = \{1, 2, 3, 4, 5\}$ and P given by

$$P = \begin{array}{c} \\ 1 \\ 2 \\ 3 \\ 4 \\ 5 \end{array} \begin{array}{c} \begin{array}{ccccc} 1 & 2 & 3 & 4 & 5 \end{array} \\ \begin{pmatrix} 1/3 & 2/3 & 0 & 0 & 0 \\ 0 & 1/8 & 0 & 1/4 & 5/8 \\ 3/4 & 0 & 1/4 & 0 & 0 \\ 0 & 0 & 0 & 1/2 & 1/2 \\ 0 & 1/3 & 0 & 0 & 2/3 \end{pmatrix} \end{array}.$$

(i) Examine whether the long run distribution exists. (ii) If yes, find it. Comment on the results. (iii) Examine whether the stationary distribution exists. (v) If yes, find it. Use all the methods. (iv) Comment on the link between the long run distribution and the stationary distribution. (v) Find a matrix of f_{ij}. (vi) Examine whether $\lim_{n\to\infty} p_{ij}^{(nd_j)} = f_{ij}d_j/\mu_j$.

3.9.3 Suppose $\{X_n, n \geq 0\}$ is a Markov chain with state space $S = \{1, 2, 3, 4\}$ and P given by

$$P = \begin{matrix} & 1 & 2 & 3 & 4 \\ 1 & 0 & 1/5 & 3/5 & 1/5 \\ 2 & 1/4 & 1/4 & 1/4 & 1/4 \\ 3 & 1 & 0 & 0 & 0 \\ 4 & 0 & 1/2 & 1/2 & 0 \end{matrix}.$$

Classify the states. Examine whether the long run distribution exists. Examine whether the stationary distribution exists. If yes, find these. Use all the methods. Find $\lim_{n\to\infty} P[X_n = 2]$. By taking initial distribution as the stationary distribution, find the joint distribution of $\{X_5, X_9, X_{12}\}$ and $\{X_{10}, X_{14}, X_{17}\}$. Comment on the results.

3.9.4 For the second order Markov chain as defined in Example 2.1.7, find the stationary distribution. What fraction of time in the long run it is sunny?

3.9.5 Suppose a production process changes states according to a Markov chain with P given by

$$P = \begin{matrix} & 0 & 1 & 2 & 3 \\ 0 & 0.3 & 0.5 & 0 & 0.2 \\ 1 & 0.5 & 0.2 & 0.2 & 0.1 \\ 2 & 0.2 & 0.3 & 0.4 & 0.1 \\ 3 & 0.1 & 0.2 & 0.4 & 0.3 \end{matrix}.$$

Suppose states 0 and 1 correspond to "In-Control" status of the production process while states 2 and 3 correspond to "Out-of-Control" status. In the long run, on the average, what fraction of time is the process Out-of-Control?

3.9.6 At the end of a month, a large retail store classifies each receivable account as follows. 0: Current, 1: 30–60 days overdue, 2: 60–90 days overdue and 3: Over 90 days. Each such account moves from state to state according to a Markov chain with transition probability matrix P as given below.

$$P = \begin{matrix} & 0 & 1 & 2 & 3 \\ 0 & 0.95 & 0.05 & 0 & 0 \\ 1 & 0.50 & 0 & 0.50 & 0 \\ 2 & 0.20 & 0 & 0 & 0.80 \\ 3 & 0.10 & 0 & 0 & 0.90 \end{matrix}.$$

In the long run, what mean fraction of accounts are over 90 days overdue?

3.9.7 Suppose $\{X_n, n \geq 0\}$ is a Markov chain with state space $S = \{1, 2, 3, 4\}$ and P given by

$$P = \begin{array}{c} \\ 1 \\ 2 \\ 3 \\ 4 \end{array} \begin{array}{c} \begin{array}{cccc} 1 & 2 & 3 & 4 \end{array} \\ \begin{pmatrix} 0 & 1/5 & 3/5 & 1/5 \\ 1/4 & 1/4 & 1/4 & 1/4 \\ 1 & 0 & 0 & 0 \\ 0 & 1/2 & 1/2 & 0 \end{pmatrix} \end{array}.$$

Taking a suitable initial distribution, find $Cov(X_4, X_{4+n})$. Draw its graph and comment on its nature as n increases.

3.9.8 For the Markov chain in conceptual Exercise 3.8.9, verify Theorem 3.2.12, Lemma 3.3.1, Theorem 3.3.4 and Eq. (3.3.1).

3.10 Multiple Choice Questions

Note: In each of the questions, more than one options may be correct. In each question, $\{X_n, n \geq 0\}$ is a time homogeneous Markov chain with state space S and transition probability matrix P.

3.10.1 Following are four statements about an absorbing state. (I) An absorbing state is a null persistent state. (II) An absorbing state is a transient state. (III) An absorbing state is a non-null persistent state. (IV) An absorbing state is an ergodic state. Then which of the following is a correct option?

(a) Both (I) and (IV) are true
(b) Both (II) and (IV) are true
(c) Only (III) is true
(d) Both (III) and (IV) are true.

3.10.2 Which of the following options is/are always correct? For any state $i \in S$ $\lim_{n\to\infty} p_{ij}^{(n)} = 0$ if

(a) $i \to j$ and j is transient
(b) $i \to j$ and j is null persistent
(c) $i \to j$ and j is non-null persistent
(d) $i \nrightarrow j$.

3.10.3 Which of the following options is/are always correct? For any state $i \in S$, as $n \to \infty$

(a) $\lim p_{ij}^{(n)} = 0$ if $i \to j$ and j is transient
(b) $\lim p_{ij}^{(n)} > 0$ if $i \to j$ and j is null persistent
(c) $\lim p_{ij}^{(n)} = 0$ if $i \to j$ and j is non-null persistent
(d) $\lim p_{ij}^{(n)} = 0$ if $i \nrightarrow j$.

3.10.4 Which of the following is NEVER true? $\lim_{n\to\infty} p_{ij}^{(n)} = 0$ if

(a) $i \to j$ and j is transient
(b) $i \to j$ and j is null persistent
(c) $i \to j$ and j is non-null persistent
(d) $i \nrightarrow j$.

3.10.5 Which of the following options is/are correct? For any state $i \in S$, as $n \to \infty$

(a) $\lim p_{ij}^{(n)} = 0$ if $i \to j$ and j is transient
(b) $\lim p_{ij}^{(n)} = 0$ if $i \to j$ and j is null persistent
(c) $\lim p_{ij}^{(n)} = f_{ij}/\mu_j$ if $i \to j$ and j is ergodic
(d) $\lim p_{ij}^{(n)} = 0$ if $i \nrightarrow j$.

3.10.6 Which of the following is NOT always true? As $n \to \infty$

(a) $\lim p_{ij}^{(n)} = 0$ if $i \to j$ and j is transient
(b) $\lim p_{ij}^{(n)} = 0$ if $i \to j$ and j is null persistent
(c) $\lim p_{ij}^{(n)} = 1/\mu_j$ if $i \to j$ and j non-null persistent
(d) $\lim p_{ij}^{(n)} = 0$ if $i \nrightarrow j$.

3.10.7 Which of the following options is/are correct? In a finite state space irreducible Markov chain all the states are

(a) null persistent
(b) non-null persistent
(c) transient
(d) ergodic.

3.10.8 Which of the following options is/are correct? In a finite state space Markov chain,

(a) some states may be transient and some may be non-null persistent
(b) all the states are non-null persistent
(c) all the states are transient
(d) all the states are ergodic.

3.10.9 Which of the following options is/are correct? For an irreducible Markov chain with countably infinite state space S, a stationary distribution

(a) does not exist if it is null persistent
(b) does not exist if it is transient
(c) does not exist if it is non-null persistent
(d) always exists.

3.10.10 Which of the following options is/are correct? A unique stationary distribution exists if the Markov chain is irreducible and

(a) ergodic
(b) non-null persistent
(c) transient

(d) null persistent.

3.10.11 Following are four statements. A unique stationary distribution exists if the Markov chain is irreducible and (I) ergodic, (II) non-null persistent, (III) transient, (IV) null persistent.
Then which of the following options are correct?

(a) Only (I) is true
(b) Only (II) is true
(c) Both (I) and (II) are true
(d) Either (III) or (IV) are true.

3.10.12 Following are two statements for a Markov chain. (I) If a long run distribution exists, then a stationary distribution exists. (II) If a stationary distribution exists, then a long run distribution exists.
Which of the following is a correct option?

(a) Both (I) and (II) are false
(b) Both (I) and (II) are true
(c) (I) is true but (II) is false
(d) (I) is false but (II) is true.

3.10.13 Suppose $\underline{\pi} = (\pi_1, \pi_2, \ldots,)$ is a stationary distribution of a Markov chain with transition probability matrix P. Following are two statements. (I) $\underline{\pi}$ is a stationary distribution of a Markov chain with transition probability matrix P^4. (II) A stationary distribution of a Markov chain with transition probability matrix P^4 is given by $(\pi_1^4, \pi_2^4, \ldots,)$.
Which of the following is a correct option?

(a) Both (I) and (II) are false
(b) Both (I) and (II) are true
(c) (I) is true but (II) is false
(d) (I) is false but (II) is true.

3.10.14 Which of the following is a correct option? An absorbing state is

(a) a null persistent state
(b) a transient state
(c) an ergodic state
(d) a periodic state.

3.10.15 Suppose $\{X_n, n \geq 0\}$ is a Markov chain with P given by

$$P = \begin{matrix} & \begin{matrix} 1 & 2 & 3 \end{matrix} \\ \begin{matrix} 1 \\ 2 \\ 3 \end{matrix} & \begin{pmatrix} 0 & 1 & 0 \\ 0.4 & 0 & 0.6 \\ 0 & 1 & 0 \end{pmatrix} \end{matrix}.$$

Which of the following is a correct option? As $n \to \infty$

(a) $\lim p_{31}^{(2n+1)} > 0$
(b) $\lim p_{13}^{(2n)} > 0$
(c) $\lim p_{12}^{(2n)} = 0$
(d) $\lim p_{12}^{(2n+1)} > 0$.

3.10.16 Following are the transition probability matrices of the Markov chains with state space $S = \{1, 2\}$.

$$P_1 = \begin{matrix} 1 \\ 2 \end{matrix} \overset{\begin{matrix} 1 & 2 \end{matrix}}{\begin{pmatrix} 1 & 0 \\ 0 & 1 \end{pmatrix}}, \quad P_2 = \begin{matrix} 1 \\ 2 \end{matrix} \overset{\begin{matrix} 1 & 2 \end{matrix}}{\begin{pmatrix} 0 & 1 \\ 1 & 0 \end{pmatrix}}, \quad P_3 = \begin{matrix} 1 \\ 2 \end{matrix} \overset{\begin{matrix} 1 & 2 \end{matrix}}{\begin{pmatrix} 1/2 & 1/2 \\ 2/3 & 1/3 \end{pmatrix}}$$

$$\& \ P_4 = \begin{matrix} 1 \\ 2 \end{matrix} \overset{\begin{matrix} 1 & 2 \end{matrix}}{\begin{pmatrix} 1/2 & 1/2 \\ 0 & 1 \end{pmatrix}}.$$

Which Markov chains have a stationary distribution given by $(1/2, 1/2)$? Markov chains with the transition probability matrices

(a) P_1, P_3
(b) P_1, P_4
(c) P_1, P_2
(d) P_2, P_3.

3.10.17 Suppose a Markov chain with state space $S = \{1, 2, 3, 4\}$ has a unique stationary distribution given by $(1/2, 0, 1/3, 1/6)$. Which of the following options is/are correct?

(a) State 2 is a null persistent state
(b) State 2 is a transient state
(c) The mean recurrence time for state 1 is 2
(d) The mean recurrence time for state 1 is $1/2$.

3.10.18 Suppose a Markov chain with state space $S = \{1, 2, 3\}$ has following three stationary distributions $(1/2, 1/2, 0)$, $(0, 0, 1)$ and $(1/6, 1/6, 2/3)$. Which of the following options is/are correct?

(a) States 1, 2 are null persistent states
(b) State 3 is a transient state
(c) The mean recurrence times for states 1, 2, 3 are 6, 6, 3/2 respectively.
(d) The mean recurrence times for states 1, 2, 3 are 2, 2, 1 respectively.

3.10.19 Following are the transition probability matrices of the Markov chains with state space $S = \{1, 2\}$.

$$P_1 = \begin{matrix} & \begin{matrix} 1 & 2 \end{matrix} \\ \begin{matrix} 1 \\ 2 \end{matrix} & \begin{pmatrix} 1 & 0 \\ 0 & 1 \end{pmatrix} \end{matrix}, \quad P_2 = \begin{matrix} & \begin{matrix} 1 & 2 \end{matrix} \\ \begin{matrix} 1 \\ 2 \end{matrix} & \begin{pmatrix} 0 & 1 \\ 1 & 0 \end{pmatrix} \end{matrix}, \quad P_3 = \begin{matrix} & \begin{matrix} 1 & 2 \end{matrix} \\ \begin{matrix} 1 \\ 2 \end{matrix} & \begin{pmatrix} 1/2 & 1/2 \\ 2/3 & 1/3 \end{pmatrix} \end{matrix}$$

$$\& \ P_4 = \begin{matrix} & \begin{matrix} 1 & 2 \end{matrix} \\ \begin{matrix} 1 \\ 2 \end{matrix} & \begin{pmatrix} 1/2 & 1/2 \\ 0 & 1 \end{pmatrix} \end{matrix}.$$

Which Markov chains have a unique stationary distribution given by $(1/2, 1/2)$? Markov chains with the transition probability matrix.

(a) P_4
(b) P_3
(c) P_2
(d) P_1.

References

1. Boland, P. J. (2007). *Statistical and probabilistic Methods in Actuarial Science*. London: Chapman and Hall.
2. Bates, D., & Maechler, M. (2019). Matrix: Sparse and Dense Matrix Classes and Methods. R package version 1.2-17. https://CRAN.R-project.org/package=Matrix
3. Cinlar, E. (1975). *Introduction to stochastic processes*. New Jersey: Prentice Hall.
4. Feller, W. (1978). *An introduction to probability theory and its applications* (Vol. I). New York: Wiley.
5. Guttorp, P. (1991). *Statistical inference for branching processes*. New York: Wiley.
6. Karlin, S., & Taylor, H. M. (1975). *A first course in stochastic processes*. New York: Academic.
7. Medhi, J. (1994). *Stochastic processes*. New Delhi: Wiley Eastern.
8. Venables, W. N., & Ripley, B. D. (2002). *Modern applied statistics with S* (4th ed.). New York: Springer.

Chapter 4
Random Walks

4.1 Introduction

Random walks are the most extensively applied stochastic processes. These serve as models for discrete approximations to physical processes describing the motion of diffusing particles. If a particle is subjected to collisions and random impulses, then its position fluctuates randomly. The position of the particle at time t is modeled as a continuous time Brownian motion process, which we discuss in Chap. 9. Random walk is a discrete version of a Brownian motion process. Suppose we observe the motion of a particle on the real line at discrete time points. Suppose the position of a particle at time $n = 0$ is denoted by X_0. After a unit interval, it may move one unit to the right or to the left or may stay at the same position. Suppose the random variable Y_i indicates the direction of the movement at the ith step, its value being $+1$ with probability p_i, if the movement is to the right of the current position; -1 with probability q_i, if the movement is to the left of the current position and 0 with probability r_i, if the particle stays at the same position, $p_i, q_i, r_i \geq 0$ and $p_i + q_i + r_i = 1$. Thus, $X_n = \sum_{i=0}^{n} Y_i$ indicates the position of the particle at time n. The stochastic process $\{X_n, n \geq 0\}$ is known as a random walk.

In a game of gambling, a gambler with some initial capital plays a game consisting of a series of bets of an amount of one unit against an adversary. At every trial, he either wins or loses 1 unit. Thus the gambler's capital fluctuates by 1 unit. His capital X_n after n trials is modeled as a random walk. The classical Ehrenfest urn model, which is a discrete version of random diffusion of molecules through a membrane, is a particular type of random walk. Birth-death chain (Hoel et al. [5]) is also modeled as a random walk. It is a discrete version of a birth-death process, which we study in Chap. 8.

In the present chapter, we discuss various versions of random walks, depending on the state space and values of p_i, q_i and r_i. The books Feller [3] and Feller [4] give an exhaustive and interesting discussion on random walks.

The most general version of a random walk is defined as follows (Feller [4]).

S. Madhira and S. Deshmukh, *Introduction to Stochastic Processes Using R*,
https://doi.org/10.1007/978-981-99-5601-2_4

Definition 4.1.1 Random Walk: Suppose $\{Y_i, i \geq 0\}$ is a sequence of independent and identically distributed random variables and X_n is defined as $X_n = \sum_{i=0}^{n} Y_i$. Then the stochastic process $\{X_n, n \geq 0\}$ is known as a random walk.

If Y_i in the above definition are non-negative random variables, then $\{X_n, n \geq 0\}$ is known as a renewal process, which we discuss in Chap. 10. The state space of the random walk depends on the set of possible values of $\{Y_i, i \geq 0\}$. From the definition, it follows that $\{X_n, n \geq 0\}$ is a process with stationary and independent increments and hence a time homogeneous Markov chain, if the state space is countable. In Example 2.1.4, we have discussed this result when $\{Y_i, i \geq 0\}$ are independent and identically distributed discrete random variables. The Markov property of a random walk also follows when we note that

$$X_0 = Y_0 \;\; \& \;\; X_n = X_{n-1} + Y_n \quad \forall \; n \geq 1$$

and hence the conditional distribution of X_n given the entire past is the same as the conditional distribution of X_n given X_{n-1}. Further,

$$p_{ij} = P[X_n = j | X_{n-1} = i] = P[X_{n-1} + Y_n = j | X_{n-1} = i] = P[Y_n = j - i].$$

Thus, p_{ij} depends on i and j only through $j - i$, which is consistent with the result derived in Eq. (1.3.1), for a process with stationary and independent increments. Thus, the common distribution of Y_i determines the transition probabilities of a random walk.

In the present chapter, we consider a particular type of random walk, known as a simple random walk.

Definition 4.1.2 Simple Random Walk: A random walk $\{X_n, n \geq 0\}$ is said to be a simple random walk, if the possible values of Y_i are ± 1, with probabilities p and $1 - p = q$ respectively, $0 < p < 1$.

Definition 4.1.3 Symmetric Simple Random Walk: When $p = 1/2$, a simple random walk is said to be a symmetric simple random walk. It is also known as an unbiased simple random walk.

In the next three sections, we discuss various versions of a simple random walk, which result from variations in the state space and variations in the transition probabilities. We investigate the nature of these random walks by classifying their states and finding the stationary distribution, whenever it exists.

4.2 Random Walk with Countably Infinite State Space

In this section, we study a random walk model when the state space S is either the set $I = \{0, \pm 1, \pm 2, \ldots\}$ of all integers or the set W of whole numbers. We discuss the two cases separately as the nature of the underlying Markov chain differs as the state space changes. We begin with $S = I$. A simple random walk with state space I is known as an unrestricted one-dimensional simple random walk. In terms of a Markov chain, it is defined as follows.

Definition 4.2.1 Unrestricted Simple Random Walk: A Markov chain $\{X_n, n \geq 0\}$ with state space $S = I$ is known as an unrestricted simple random walk, if the transition probabilities are given by

$$p_{i,i+1} = p, \;\; p_{i,i-1} = q = 1 - p \;\; \& \;\; p_{ij} = 0 \;\; \forall \;\; j \neq i+1, \; i-1, \;\; i \in S\,, \quad 0 < p < 1\,.$$

As discussed in Chap. 1, $X_n = \sum_{i=0}^{n} Y_i$ where $\{Y_i, i \geq 0\}$ are independent and all are distributed as Y, where $P[Y = 1] = p$ and $P[Y = -1] = q$.

In the following example, we obtain realizations of the unrestricted random walk.

Example 4.2.1 Suppose $\{X_n, n \geq 0\}$ is an unrestricted simple random walk. We use Code 4.6.1 to obtain a realization of length $n = 25$ of a simple random walk. We assume that the initial state is 0. For comparison we consider 4 values of p, 0.50, 0.33, 0.75 and 0.20. Realizations of the random walk are presented in Fig. 4.1.

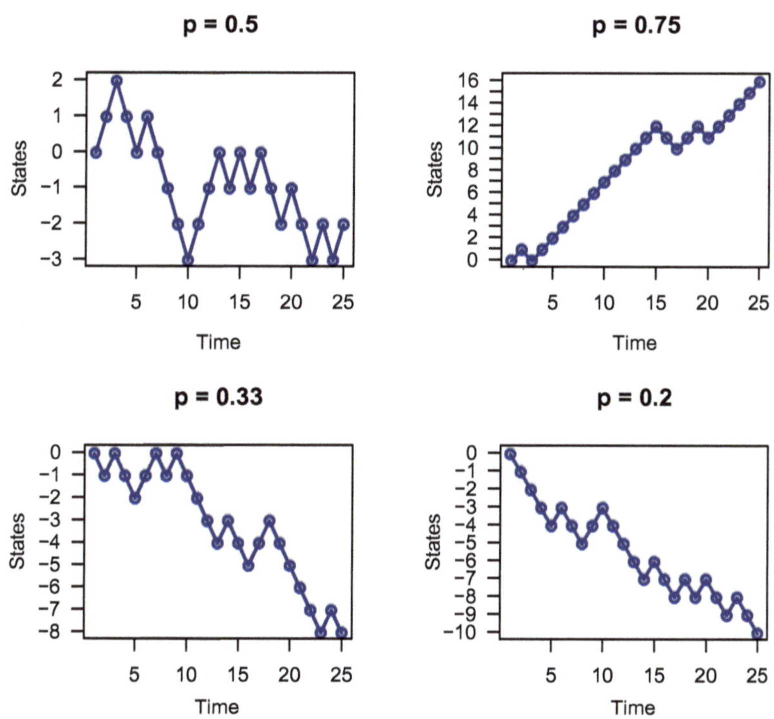

Fig. 4.1 Realization of Unrestricted Simple Random Walk

Note the differences in the realizations for different values of p. In the first graph when $p = 0.5$, values X_n are positive as well as negative integers between -3 and 2. When $p = 0.75 > 0.5$, the shift is toward the positive side, while for $p < 0.5$, the shift is toward the negative side, as expected. □

We now investigate the nature of this model in the following theorem.

Theorem 4.2.1 *Suppose $\{X_n, n \geq 0\}$ is an unrestricted simple random walk. Then (i) the Markov chain is irreducible, (ii) all states are periodic with period 2, (iii) all states are null persistent if $p = 1/2$ and (iv) all are transient if $p \neq 1/2$.*

Proof Since $\{X_n, n \geq 0\}$ is an unrestricted random walk, $S = I$ and the transition probabilities are given by

$$p_{i,i+1} = p, \;\; p_{i,i-1} = q = 1 - p \;\; \& \;\; p_{ij} = 0 \;\; \forall \;\; j \neq i+1, \; i-1, \;\; i \in S \,, \quad 0 < p < 1 \,.$$

(i) Note that for any state

$$i > 0, \;\; p_{0i}^{(i)} = p^i > 0 \;\; \Rightarrow \;\; 0 \to i \;\; \& \;\; p_{i0}^{(i)} = q^i > 0 \;\; \Rightarrow \;\; i \to 0. \text{ Hence } 0 \leftrightarrow i.$$

Similarly for any state,

$$i < 0, \;\; p_{0i}^{(i)} = q^i > 0 \;\; \Rightarrow \;\; 0 \to i \;\; \& \;\; p_{i0}^{(i)} = p^i > 0 \;\; \Rightarrow \;\; i \to 0. \text{ Hence } 0 \leftrightarrow i.$$

Thus, $\forall \; i \in I, 0 \leftrightarrow i$ and hence by transitivity property of communication of states, $i \leftrightarrow j \; \forall \; i, j \in I$. Thus, all the states communicate with each other and the random walk is an irreducible Markov chain. It then follows that all the states are either transient or non-null persistent or null persistent. Further all will have the same period. So we concentrate on state 0 and examine its nature.

(ii) Suppose the chain starts in state 0, it will either transit to 1 or to -1 and may come back to 0 in the second step. Thus, $p_{00}^{(2)} = 2pq > 0$. Thus, the set $\{n | p_{00}^{(n)} > 0\}$ is non-empty. It is clear that if the chain transits from 0, then to return to 0, the number of steps to the right must be the same as the number of steps to the left, that is number n of transitions in this path must be an even number. Thus, it is possible to go from 0 to 0 in $2n$ steps and hence $p_{00}^{(2n)} > 0$. Further, out of these $2n$ transitions, n are to the right and n are to the left, with respective probabilities p and q. Hence,

$$p_{00}^{(2n)} = \binom{2n}{n} p^n q^n, \quad \forall \; n \geq 1 \,.$$

It is impossible to go from 0 to 0 in odd number of steps, hence we have $p_{00}^{(2n-1)} = 0 \; \forall \; n \geq 1$. Thus,

$$\text{g.c.d.}\{n|p_{00}^{(n)} > 0\} = 2 \quad \Rightarrow \quad d_0 = 2 \quad \Rightarrow \quad d_i = 2, \ \ \forall \ i \in I$$

and the Markov chain is periodic with period 2.

(iii) To examine whether 0 is transient or persistent, we examine whether $\sum_{n=1}^{\infty} p_{00}^{(n)}$ converges or diverges. By Stirling's approximation, $n! \approx n^{n+\frac{1}{2}} e^{-n} \sqrt{2\pi}$, where $a_n \approx b_n$ implies $\lim_{n\to\infty} a_n/b_n = 1$. Hence,

$$p_{00}^{(2n)} = \binom{2n}{n} p^n q^n = \frac{(2n)!}{n!n!} p^n q^n \approx \frac{(4pq)^n}{\sqrt{\pi n}} \ \forall \ n \geq 1 \Rightarrow \sum_{n=1}^{\infty} p_{00}^{(n)} \approx \sum_{n=1}^{\infty} \frac{(4pq)^n}{\sqrt{\pi n}},$$

as $p_{00}^{(2n-1)} = 0 \ \forall \ n \geq 1$. Observe that when

$$p = q = 1/2, \quad \sum_{n=1}^{\infty} p_{00}^{(n)} \approx \sum_{n=1}^{\infty} (4pq)^n/\sqrt{\pi n} = \sum_{n=1}^{\infty} 1/\sqrt{\pi n}$$

and the series diverges. Thus if $p = 1/2$, the state 0 is a persistent state. Further, $p_{00}^{(2n)} \approx 1/\sqrt{\pi n} \to 0$ as $n \to \infty$ and $p_{00}^{(2n-1)} = 0 \ \forall \ n \geq 1$. Thus, $\lim_{n\to\infty} p_{00}^{(n)} = 0$ and hence 0 is a null persistent state. As a consequence, when $p = 1/2$, all the states are null persistent.

(iv) Suppose now $p \neq 1/2$. It is known that for $0 < p < 1$, $pq \leq 1/4$ since the unique maximum of pq is attained at $p = q = 1/2$. To examine the convergence or divergence of the series when $p \neq 1/2$, observe that for all $n \geq 1$, $1/\sqrt{n} \leq 1$, hence

$$\sum_{n=1}^{\infty} p_{00}^{(n)} \approx \sum_{n=1}^{\infty} \frac{(4pq)^n}{\sqrt{\pi n}} < \sum_{n=1}^{\infty} \frac{(4pq)^n}{\sqrt{\pi}} = \frac{1}{\sqrt{\pi}}[(1-4pq)^{-1} - 1] < \infty .$$

Thus, if $p \neq 1/2$, the state 0 is a transient state and hence all the states are transient. Intuitively, if $p \neq q$, there is a positive probability that a particle initially at the origin will drift to ∞ if $p > q$ or will drift to $-\infty$ if $p < q$, without ever returning to the origin.

We have thus proved that the unrestricted random walk is an irreducible Markov chain where all the states have period 2, all are transient if $p \neq 1/2$ and all are null persistent if $p = 1/2$. □

From this theorem, we note that the unrestricted random walk is an example of a Markov chain with all states to be either null persistent or transient. The result that all are transient if $p \neq 1/2$ and all are null persistent if $p = 1/2$ can be proved in number of ways. We discuss below some approaches.

(i) If we use the formula, $\sum_{n=0}^{\infty} \binom{2n}{n} \frac{x^{2n}}{2^{2n}} = (1-x^2)^{-1/2}$ when $|x| < 1$, we can find the value of the series $\sum_{n=1}^{\infty} p_{00}^{(n)}$ when $p \neq 1/2$. With $x^2 = 4pq < 1$, observe that

$$\sum_{n=1}^{\infty} p_{00}^{(n)} = \sum_{n=1}^{\infty} \binom{2n}{n} p^n q^n = \sum_{n=1}^{\infty} \binom{2n}{n} \frac{(4pq)^n}{4^n}$$
$$= \sum_{n=1}^{\infty} \binom{2n}{n} \frac{x^{2n}}{2^{2n}} = (1-4pq)^{-1/2} - 1 < \infty .$$

(ii) It is shown in Chap. 3 of Feller [3] that for $p = 1/2$,

$$f_{00}^{(2n+1)} = 0 \ \forall \ n \geq 0 \ \ \& \ \ f_{00}^{(2n)} = p_{00}^{(2n)}/(2n-1) \ \forall \ n \geq 1 .$$

Further, it also satisfies the relation $f_{00}^{(2n)} = p_{00}^{(2n-2)} - p_{00}^{(2n)}$. It then follows that

$$f_{00} = \sum_{n=1}^{\infty} f_{00}^{(2n)} = \sum_{n=1}^{\infty} \left(p_{00}^{(2n-2)} - p_{00}^{(2n)} \right)$$
$$= p_{00}^{(0)} - p_{00}^{(2)} + p_{00}^{(2)} - p_{00}^{(4)} + \cdots = p_{00}^{(0)} = 1$$

supporting the result that for a symmetric random walk, 0 is a persistent state. Further, it is to be noted that

$$\mu_0 = \sum_{n=1}^{\infty} 2n f_{00}^{(2n)} = \sum_{n=1}^{\infty} \frac{2n p_{00}^{(2n)}}{2n-1} \approx \sum_{n=1}^{\infty} \frac{2n}{2n-1} \frac{1}{\sqrt{\pi n}} = \infty.$$

Thus, $\mu_0 = \infty$ supports the result derived earlier that 0 is a null persistent state.

(iii) Using the approach of generating function, it can be shown that $f_{00}^{(2n)} = (-1)^{n-1}\binom{1/2}{n}(4pq)^n$, (Bhat [2]). Hence,

$$f_{00} = \sum_{n=1}^{\infty} f_{00}^{(2n)} = 1 - (1-4pq)^{1/2} = 1 - |2p-1|$$

which is 1 if $p = 1/2$ and is less than 1 if $p \neq 1/2$.

Since an unrestricted random walk is either transient or null persistent, by Theorem 3.3.6, its stationary distribution does not exist. We can verify the same by examining whether $\underline{\pi} = \underline{\pi} P$ has a solution with $\pi_i \geq 0$ and $\sum_{i \in S} \pi_i = 1$. The system of equations $\underline{\pi} = \underline{\pi} P$ can be expressed as $\pi_j = (1/2)(\pi_{j-1} + \pi_{j+1}) \forall \ j \in S$. The only solution to this system of equations is $\pi_j = c > 0 \ \forall \ j \in S$. However, it does not satisfy the requirement that $\sum_{i \in S} \pi_i = 1$.

In the following two examples, we verify some results proved above. These examples illustrate that the approximations are good even for moderate values of n.

Example 4.2.2 Suppose $\{X_n, n \geq 0\}$ is an unrestricted simple random walk. We have shown that

$$p_{00}^{(2n)} = \binom{2n}{n} p^n q^n = \frac{(2n)!}{n!n!} p^n q^n \approx \frac{(4pq)^n}{\sqrt{\pi n}} \ \forall \ n \geq 1.$$

We compute $p_{00}^{(2n)}$ using the exact and approximate formula for $n = 1$ to 20 and for three values of p as 1/3, 1/2, 3/4. We use Code 4.6.2 given in Sect. 4.6. Suppose $p_{00}^{(2n)} = \binom{2n}{n} p^n q^n$ is denoted by $pr(p)$ and $p_{00}^{(2n)} \approx (4pq)^n/\sqrt{\pi n}$ is denoted by $pra(p)$ for the specific value of p. Table 4.1 displays the values of pr and pra for $n = 1$ to 20 and for three values of p as 1/3, 1/2, 3/4. From Table 4.1, we note that the exact and the approximate values of $p_{00}^{(2n)}$ are close for $n \geq 5$, which indicates that one may use the approximate expression of $p_{00}^{(2n)}$ for moderate values of n. Further for $p > 1/2$, $p_{00}^{(2n)}$ approaches 0 for $n \geq 13$, since the probability of transition to the right is high, there is a tendency to shift to the right. Similarly, for $p < 1/2$, $p_{00}^{(2n)}$ approaches 0 for $n > 20$, since the probability of transition to the right is low, there is a tendency to shift to the left. □

Table 4.1 Unrestricted simple random walk: values of $p_{00}^{(2n)}$

n	pr(1/3)	pra(1/3)	pr(1/2)	pra(1/2)	pr(3/4)	pra(3/4)
1	0.44	0.50	0.50	0.56	0.38	0.42
2	0.30	0.32	0.38	0.40	0.21	0.22
3	0.22	0.23	0.31	0.33	0.13	0.14
4	0.17	0.18	0.27	0.28	0.09	0.09
5	0.14	0.14	0.25	0.25	0.06	0.06
6	0.11	0.11	0.23	0.23	0.04	0.04
7	0.09	0.09	0.21	0.21	0.03	0.03
8	0.08	0.08	0.20	0.20	0.02	0.02
9	0.06	0.07	0.19	0.19	0.01	0.01
10	0.05	0.05	0.18	0.18	0.01	0.01
11	0.05	0.05	0.17	0.17	0.01	0.01
12	0.04	0.04	0.16	0.16	0.01	0.01
13	0.03	0.03	0.15	0.16	0.00	0.00
14	0.03	0.03	0.15	0.15	0.00	0.00
15	0.02	0.02	0.14	0.15	0.00	0.00
16	0.02	0.02	0.14	0.14	0.00	0.00
17	0.02	0.02	0.14	0.14	0.00	0.00
18	0.02	0.02	0.13	0.13	0.00	0.00
19	0.01	0.01	0.13	0.13	0.00	0.00
20	0.01	0.01	0.13	0.13	0.00	0.00

Example 4.2.3 Suppose $\{X_n, n \geq 0\}$ is an unrestricted simple random walk. We have noted that

$$\sum_{n=1}^{\infty} p_{00}^{(2n)} = \sum_{n=1}^{\infty} \binom{2n}{n} p^n q^n = (1 - 4pq)^{-1/2} - 1.$$

Thus, for large N, $\sum_{n=1}^{N} p_{00}^{(n)}$ can be approximated by $(1 - 4pq)^{-1/2} - 1$. Hence, we compute $\sum_{n=1}^{N} p_{00}^{(n)}$ for some large value of N and verify if it is equal to $(1 - 4pq)^{-1/2} - 1$ approximately. We take three values of p as $1/3, 2/3, 3/4$. We use Code 4.6.3. From the output, we note that for $p = 1/3, 2/3, 3/4$, $\sum_{n=1}^{70} p_{00}^{(n)}$ is $1.9999, 1.9999, 1$ and $(1 - 4pq)^{-1/2} - 1$ is $2, 2, 1$ respectively. For $n > 70$, $p_{00}^{(2n)} \approx 0$. The values of $\sum_{n=1}^{70} p_{00}^{(n)}$ support the result that when $p \neq 1/2$, all the states are transient. Note that $N = 70$ gives a good approximation. □

We now proceed to study a random walk with state space W. There are three versions of this random walk.

Definition 4.2.2 Random Walk on W with Absorbing Barrier at 0: Suppose $\{X_n, n \geq 0\}$ is a Markov chain with state space W and the transition probabilities as

$$\begin{aligned} p_{00} &= 1, \quad p_{ii} = 0 \ \forall \, i \neq 0 \\ p_{i,i+1} &= p, \quad i = 0, 1, \ldots \ \& \ p_{i,i-1} = q = 1 - p, \ i = 1, 2, \ldots, \end{aligned}$$

$0 < p < 1$. Then $\{X_n, n \geq 0\}$ is a simple random walk on W, with absorbing barrier at 0.

The next example presents realizations of this random walk.

Example 4.2.4 Suppose $\{X_n, n \geq 0\}$ is a simple random walk with absorbing barrier at 0 with state space W. Suppose $p = 0.50, 0.33, 0.75, 0.20$ and $X_0 = 4$. We use Code 4.6.4 to obtain the realizations. These are displayed in Fig. 4.2. The four graphs show the difference in the realizations for different values of p. In the first graph when $p = 0.5$, transitions are to the left and right as well and in 25 transitions state 0 is not visited. When $p = 0.75 > 0.5$, the shift is toward right side of the initial state, as expected. For $p < 0.5$, the random walk gets absorbed at 0. □

We now investigate the nature of this random walk in the next theorem.

Theorem 4.2.2 *Suppose $\{X_n, n \geq 0\}$ is a simple random walk on W, with absorbing barrier at 0. Then (i) states i and j communicate for all $i, j > 0$, (ii) state 0 is aperiodic and non-null persistent state, (iii) all states $i > 0$ have period 2 and (iv) all states $i > 0$ are transient.*

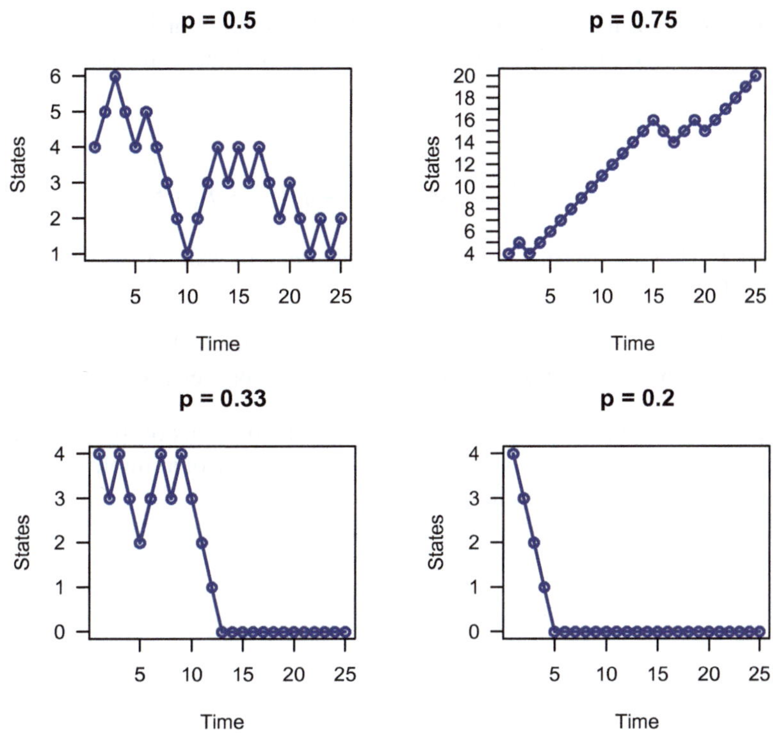

Fig. 4.2 Realization of Simple Random Walk on W with Absorbing Barrier at 0

Proof (i) As shown in Theorem 4.2.1, for any state $i > 0$ and $j = i + r$ with $r > 0$, we have $p_{ij}^{(r)} = p^r$ and $p_{ji}^{(r)} = q^r$. Thus, any state $i > 0$ communicates with any other state $j > 0$. State 0 is an absorbing state and thus does not communicate with any other state.

(ii) Since $p_{00} = 1$, period of 0 is 1. Further, being an absorbing state, it is non-null persistent.

(iii) Any state $i > 0$ cannot return to i in odd number of steps but $p_{ii}^{(2n)} > 0$ for all $n \geq 1$, hence period of any state $i > 0$ is 2. All other states $j > 0$ communicate with $i > 0$ and hence all are periodic with period 2.
(iv) It is to be noted that $1 \to 0$ but $0 \nrightarrow 1$, thus 1 is an inessential state. All other states $i > 0$ communicate with 1 and hence have the same nature, since being inessential is a class property. Thus, by Theorem 2.5.2, all states $i > 0$ are inessential states and hence are transient. □

We now define another version of a random walk on W, where the transition probability from the boundary 0 is different than the previous version.

Definition 4.2.3 Random Walk on W with Reflecting Barrier at 0: Suppose $\{X_n, n \geq 0\}$ is a Markov chain with state space W and the transition probabilities given by

$$p_{01} = 1, \quad p_{ii} = 0 \ \forall \ i \neq 0$$
$$p_{i,i+1} = p, \quad i = 0, 1, \ldots \ \& \ p_{i,i-1} = q = 1 - p, \ i = 1, 2, \ldots,$$

$0 < p < 1$. Then the Markov chain is a simple random walk on W with reflecting barrier at 0.

In this Markov chain, when it visits 0, it gets reflected to 1 at the next transition and hence is labeled as a random walk with reflecting barrier at 0.

Example 4.2.5 Suppose $\{X_n, n \geq 0\}$ is a simple random walk on W, with reflecting barrier at 0. Example 4.2.4 uses Code 4.6.4 for a realization of a simple random walk, with absorbing barrier at 0. In that code if we change

```
else { x[i,j]=0 }
```

to

```
else { x[i,j]=1 }
```

we get a realization of a simple random walk with reflecting barrier at 0. Figure 4.3 displays the realization for $p = 0.50, 0.33, 0.75, 0.20$ and $X_0 = 3$. In the first graph when $p = 0.5$, transitions are to the left and right of X_0 and the random walk gets reflected at 0 only once. When $p = 0.75$, the shift is toward right side of initial state, as expected. For $p < 0.5$, the random walk gets reflected at 0 many times. □

We investigate the nature of this random walk in the next theorem.

Theorem 4.2.3 *Suppose $\{X_n, n \geq 0\}$ is a simple random walk on W, with reflecting barrier at 0. Then (i) the Markov chain is irreducible, (ii) all states have period 2, (iii) all states are non-null persistent if $p < 1/2$ and (iv) all states are either transient or null persistent if $p \geq 1/2$.*

Proof (i) Note that $0 \to 1$ and $1 \to 0$. Further any state $i > 0$ communicates with any other state $j > 0$. Thus all states communicate with each other and hence it is an irreducible Markov chain. As a consequence, nature of all the states is the same.

(ii) To find the period of this chain we consider state 0. State 0 cannot be reached from 0 in odd number of steps, but $p_{00}^{(2n)} > 0 \ \forall \ n \geq 1$, hence period of the state 0 is 2. Hence, all the states are periodic with period 2.

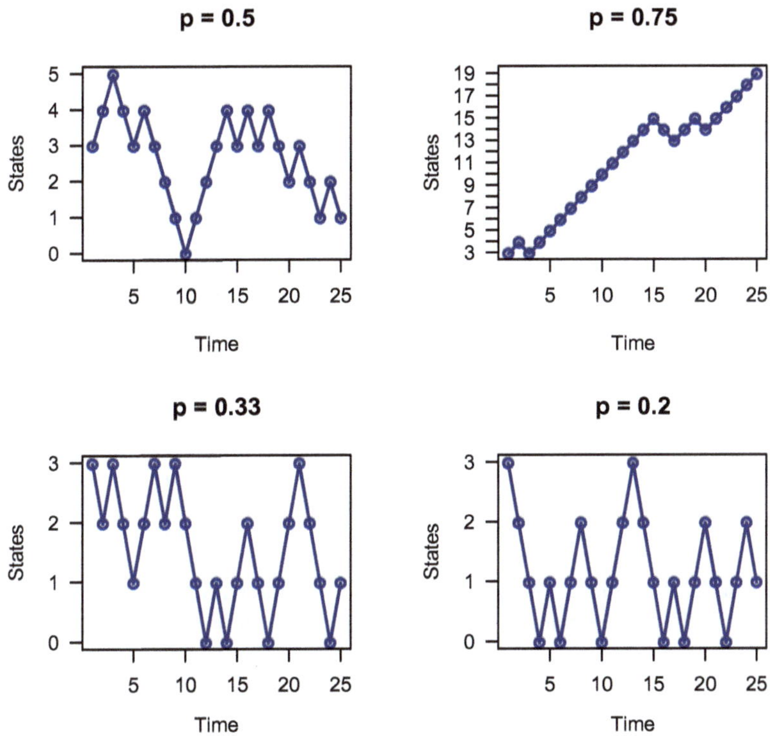

Fig. 4.3 Realization of Simple Random Walk on W with Reflecting Barrier at 0

(iii) To examine whether these are persistent or transient, we adopt an approach, different from that for a random walk with absorbing barrier at 0. We examine when this model has a stationary distribution. If it exists we can claim that all states are non-null persistent by Remark 3.3.3. From the given transition probabilities, the matrix equation $\underline{\pi} = \underline{\pi}P$ is expressible as follows.

$$\pi_0 = q\pi_1, \;\; \pi_1 = \pi_0 + q\pi_2, \;\; \pi_2 = p\pi_1 + q\pi_3 \;\; \& \;\; \pi_j = p\pi_{j-1} + q\pi_{j+1}, \;\; j = 3, 4, \ldots,$$

with the condition that $\pi_j \geq 0$ and $\sum_{j \in S} \pi_j = 1$. We examine whether this system of equations has a solution. From these equations we have

$$\pi_1 = \pi_0/q, \;\; \pi_2 = p\pi_0/q^2, \quad \& \;\; \pi_j = p^{j-1}\pi_0/q^j, \quad j = 3, 4, \ldots.$$

Further for $p/q < 1$, that is, $p < 1/2$,

$$\sum_{j \in S} \pi_j = 1 \Rightarrow \pi_0 + \frac{\pi_0}{q} \sum_{j \geq 1} \left(\frac{p}{q}\right)^{j-1} = 1$$

$$\Rightarrow \pi_0 = \frac{q-p}{q-p+1} = \frac{1-2p}{2-2p} = \frac{1}{2}\left(1 - \frac{p}{q}\right)$$

$$\Rightarrow \pi_j = \frac{1}{2}\left(1 - \frac{p}{q}\right) \frac{p^{j-1}}{q^j} = \frac{1}{2q}\left(1 - \frac{p}{q}\right)\left(\frac{p}{q}\right)^{j-1}, \quad j \geq 1 .$$

Hence, if $p < 1/2$, the stationary distribution exists and hence, the random walk is non-null persistent. The mean recurrence time to state j is given by $\mu_j = 1/\pi_j$.

(iv) For $p \geq 1/2$, $\sum_{j \geq 1} (p/q)^{j-1}$ is divergent and hence the stationary distribution does not exist. In this case, the chain is either null persistent or transient. □

The next version of a random walk on W generalizes the above two random walks.

Definition 4.2.4 Random Walk on W with Elastic Barrier at 0: Suppose $\{X_n, n \geq 0\}$ is a Markov chain with state space W and the transition probabilities as given by

$$p_{i,i+1} = p, \;\; p_{i,i-1} = q = 1 - p \text{ if } i \geq 1, \;\; p_{01} = \delta, \;\; p_{00} = 1 - \delta, \;\; 0 < \delta, p < 1 .$$

It is known as a random walk with elastic barrier at 0 or partially reflecting barrier at 0.

In this model, with some positive probability the process stays in 0 and with some positive probability the transition from 0 to its nearest neighbor 1 takes place. Hence, it is known as a random walk with partially reflecting boundary at 0.

Example 4.2.6 Suppose $\{X_n, n \geq 0\}$ is a simple random walk on W, with elastic barrier at 0. Using Code 4.6.5, we obtain realizations of this random walk. In the realization we take $\delta = p$. For comparison we take four values of p as $p = 0.50, 0.33, 0.75, 0.20$, $n = 25$ and $X_0 = 2$. Figure 4.4 displays the realizations for four values of p. In three graphs, except for $p = 0.75$, we note that when the state 0 is reached, in some transitions the system either remains in 0 or transits to 1. □

We investigate the nature of this random walk in the next theorem.

Theorem 4.2.4 *Suppose $\{X_n, n \geq 0\}$ is a simple random walk on W, with elastic barrier at 0. Then (i) the Markov chain is irreducible, (ii) all states are aperiodic, (iii) all states are non-null persistent if $p < 1/2$ and (iv) all states are either transient or null persistent if $p \geq 1/2$.*

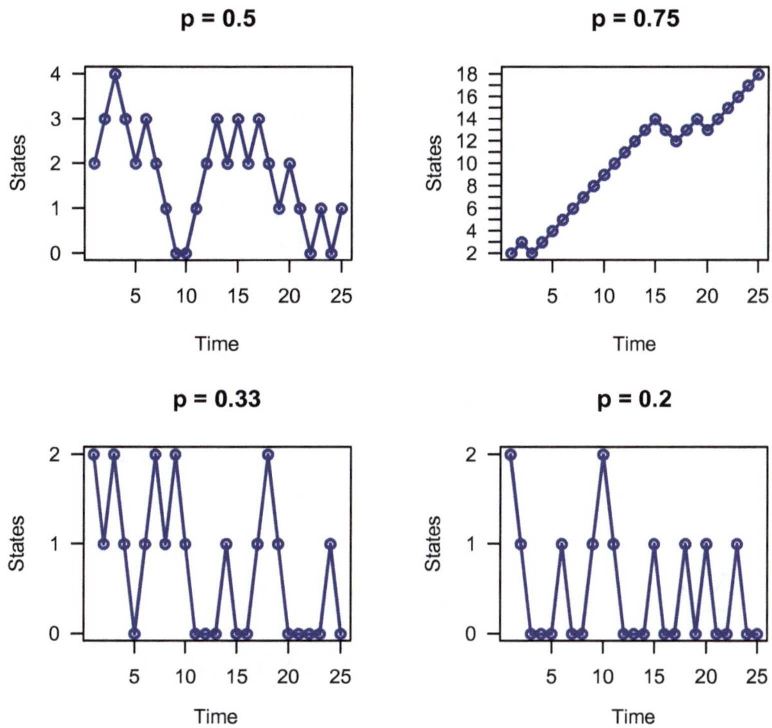

Fig. 4.4 Realization of Simple Random Walk on W with Partially Reflecting Barrier at 0

Proof (i) Note that $0 \leftrightarrow 1$ and any state $i > 0$ communicates with any state $j > 0$ and with 1. Thus, the Markov chain is irreducible and nature of all the states is the same.

(ii) Observe that $p_{00} = 1 - \delta > 0$ and hence period of the state 0 is 1. All states communicate with 0 and hence all states are aperiodic.

(iii) To examine whether these are persistent or transient, we adopt the same approach as that for a random walk with reflecting barrier at 0 and examine when this model has a stationary distribution. A stationary distribution $\underline{\pi}$ for this random walk has to satisfy the following set of equations.

$$\pi_1 q + \pi_0(1 - \delta) = \pi_0, \quad \delta\pi_0 + q\pi_2 = \pi_1$$
$$\& \quad \pi_{j+1}q + \pi_{j-1}p = \pi_j, \quad j > 1,$$

with the condition that $\pi_j \geq 0$ and $\sum_{j \in S} \pi_j = 1$. We examine whether this system of equations has a solution. The general solution to the system of difference equations $\pi_{j+1}q + \pi_{j-1}p = \pi_j$ for $j \geq 2$ is

$$\pi_j = \begin{cases} c_1 + c_2(p/q)^j, & \text{if } p \neq 1/2 \\ c_1 + c_2 j, & \text{if } p = 1/2 . \end{cases}$$

Note that $\pi_0 = q\pi_1/\delta \iff \pi_1 = \delta\pi_0/q$. Similarly, $q\pi_2 = \pi_1 - \delta\pi_0 = p\pi_1$ implies $\pi_2 = (p/q)\pi_1$. However we have

$$\pi_2 = \begin{cases} c_1 + c_2(p/q)^2, & \text{if } p \neq 1/2 \\ c_1 + 2c_2, & \text{if } p = 1/2 . \end{cases}$$

Hence $c_1 = 0$ if $p \neq 1/2$ and $c_2 = 0$ if $p = 1/2$. Therefore, for $j \geq 2$,

$$\pi_j = \begin{cases} c_2(p/q)^j, & \text{if } p \neq 1/2 \\ c_1, & \text{if } p = 1/2 . \end{cases}$$

Since $\pi_2 = c_2(p/q)^2$, $\pi_1 = (q/p)\pi_2 = c_2(p/q)$ and $\pi_0 = (q/\delta)\pi_1 = c_2(p/\delta)$. To find c_2, we use the condition $\sum_{j\in S}\pi_j = 1$. When $p < 1/2$, that is, $p/q < 1$. Observe that

$$\begin{aligned} \sum_{j\in S}\pi_j = 1 &\Rightarrow \pi_0 + \pi_1 + \sum_{j\geq 2} c_2(p/q)^j = 1 \\ &\Rightarrow c_2(p/\delta) + c_2(p/q) + c_2\sum_{j\geq 2}(p/q)^j = 1 \\ &\Rightarrow c_2\Big[(p/\delta) + \sum_{j\geq 1}(p/q)^j\Big] = 1 \\ &\Rightarrow c_2 p\left[\frac{q-p+\delta}{\delta(q-p)}\right] = 1 \;\Rightarrow\; c_2 = \frac{\delta(q-p)}{p(q-p+\delta)}. \end{aligned}$$

Thus if $p < 1/2$, the stationary distribution exists and is given by

$$\pi_0 = \frac{(q-p)}{q-p+\delta} \;\&\; \pi_j = \frac{\delta(q-p)}{p(q-p+\delta)}\left(\frac{p}{q}\right)^j, \quad j \geq 1.$$

If $\delta = p$ then

$$\pi_j = \left(1 - \frac{p}{q}\right)\left(\frac{p}{q}\right)^j, \quad j \geq 0,$$

which is a geometric distribution with parameter $(1 - p/q)$ and support W. Hence when $p < 1/2$, the random walk with elastic barrier at 0 is non-null persistent and the mean recurrence time in state j is given by $\mu_j = 1/\pi_j$.

(iv) If $p > 1/2$, then $p/q > 1$ and the series $\sum_{j\geq 2}\pi_j$ is divergent. If $p = 1/2$, no choice of c_1 could satisfy the condition $\sum_{j\in S}\pi_j = 1$. Thus, a stationary distribution does not exist when $p \geq 1/2$. Hence, for $p \geq 1/2$, the chain is either null persistent or transient. □

The random walk models discussed above may be generalized to allow for the possibility of a particle staying at a given state $i \in S$ during a unit time interval and not moving either to the left or to the right for at least one transition.

In the next section, we discuss random walks with finite state space.

4.3 Random Walk with Finite State Space

Random walks with finite state space have many applications. We discuss two such applications, known as a gambler's chain and an Ehrenfest model of diffusion.

Definition 4.3.1 Random Walk with Finite State Space: It is a Markov chain with state space $S = \{0, 1, \ldots, M\}$ and transition probabilities as specified below.

$$p_{i,i+1} = p, \quad p_{i,i-1} = q = 1 - p \;\; i = 1, 2, \ldots, M - 1, \;\; 0 < p < 1$$
$$p_{00} = \delta_1, \quad p_{01} = 1 - \delta_1 \;\; \& \;\; p_{MM} = \delta_2, \quad p_{M,M-1} = 1 - \delta_2,$$

$0 \leq \delta_1, \delta_2 \leq 1$.

The specific values of transition probabilities δ_1 and δ_2 at the boundaries lead to different types of random walks with finite state space. These are listed below.

(i) Suppose $\delta_1 = \delta_2 = 1$. In this case, $p_{00} = p_{MM} = 1$ and states 0 and M are absorbing states. Thus, we have a random walk with absorption at the barriers.
(ii) Another option for the boundary conditions is to assume that with probability one, the particle transits away from the boundary in the next time unit. If $\delta_1 = \delta_2 = 0$, then $p_{01} = 1 \;\; \& \;\; p_{MM-1} = 1$. We say such a random walk has reflecting barriers.
(iii) The third option for transition probabilities at boundaries is to have a partial reflection. It is achieved if $0 < \delta_1, \delta_2 < 1$. Such a random walk is known as a random walk with elastic boundaries. It is also known as a random walk with partially reflecting barriers.
(iv) We can have any combination of the above conditions at the two boundary points, that is, at 0 there is absorption and at M there is reflection or partial reflection.

One more generalization is to allow for the possibility of a particle staying at a given state $i \in S$ during a unit time interval. Following example illustrates an application of such type of a random walk in cell biology.

Example 4.3.1 One method of transport used in living cells is axonal transport in which certain (motor) proteins carry cargo such as mitochondria, other proteins and other cell parts, on long microtubules. These microtubules can be thought of as the "tracks" of the transportation mechanism, with the motor protein as a particle moving according to a random walk. One natural and simple model for such transport would begin by breaking the microtubule into M equally sized intervals, and then letting X_n be the position of the motor protein on the state space $\{1, 2, \ldots, M\}$. The transition probabilities satisfy

$$p_{i,i+1} = p_i, \;\; p_{i,i-1} = q_i, \;\; p_{ii} = r_i, i = 2, 3, \ldots, M-1, \;\; p_i, q_i, r_i \geq 0$$

and $p_i + q_i + r_i = 1$, with boundary conditions $p_{11} = r_1 + q_1$, $p_{12} = p_1$ and $p_{MM} = 1$. The end of the microtubule associated with state M is the destination of the cargo. In this case, it would be natural to expect $p_i > q_i$. □

We now investigate the nature of these random walks and examine whether stationary distributions exist. We illustrate the results by examples with fixed values of M and p.

Random walk with absorbing barriers at 0 and M: In this random walk 0 and M are absorbing states and hence are ergodic states. Further, $C_1 = \{0\}$ and $C_2 = \{M\}$ are two closed communicating classes. Each state in class $T = \{1, 2, \ldots, M-1\}$ is an inessential state and hence transient. The period of state $i \in T$ is 2. For any stationary distribution $\underline{\pi}$, $\pi_i = 0 \;\forall\; i \in T$. The stationary distribution concentrated on C_1 is $\underline{a}_1 = (1, 0, \ldots, 0)$ and concentrated on C_2 is $\underline{a}_2 = (0, 0, \ldots, 1)$. Hence, any convex combination $\underline{a}_\alpha = \alpha\underline{a}_1 + (1-\alpha)\underline{a}_2 = (\alpha, 0, 0, 1-\alpha)$, $0 < \alpha < 1$ is also a stationary distribution. The matrix P has eigenvalue as 1 with multiplicity 2. The corresponding normalized eigenvectors are $\underline{a}_1$ and $\underline{a}_2$. The probability of absorption into $C = C_1 \cup C_2$ from any state in T is 1, as proved in Theorem 2.6.9. A general formula for probabilities of absorption will be derived in the next section. We observe all these features in the next two examples.

Example 4.3.2 Suppose $\{X_n, n \geq 0\}$ is a symmetric random walk on $S = \{0, 1, 2, 3\}$ with absorbing barriers. Thus, it is a Markov chain with transition probability matrix

$$P = \begin{array}{c} \\ 0 \\ 1 \\ 2 \\ 3 \end{array}\begin{array}{c} \begin{array}{cccc} 0 & 1 & 2 & 3 \end{array} \\ \begin{pmatrix} 1 & 0 & 0 & 0 \\ 1/2 & 0 & 1/2 & 0 \\ 0 & 1/2 & 0 & 1/2 \\ 0 & 0 & 0 & 1 \end{pmatrix} \end{array}.$$

Note that $C_1 = \{0\}$, $C_2 = \{3\}$ are two closed communicating classes. Since 0 and 3 are absorbing states, these are aperiodic and non-null persistent states, with mean recurrence time 1 for each. The class $T = \{1, 2\}$ consists of transient transient states, with period 2. Two stationary distributions are $\underline{a}_1 = (1, 0, 0, 0)$ and $\underline{a}_2 = (0, 0, 0, 1)$. Hence any convex combination $\underline{a}_\alpha = \alpha\underline{a}_1 + (1-\alpha)\underline{a}_2 = (\alpha, 0, 0, 1-\alpha)$ is also a stationary distribution. We can find the probability of absorption into C_1 and C_2 from states in T, using the formula $G = (I - Q)^{-1}D$ derived in Theorem 2.6.9.

With some rearrangement of rows and columns of P, the matrices D, Q and G, as defined in Theorem 2.6.9, are as given below.

$$D = \begin{pmatrix} 1/2 & 0 \\ 0 & 1/2 \end{pmatrix}, \quad Q = \begin{pmatrix} 0 & 1/2 \\ 1/2 & 0 \end{pmatrix} \quad \& \quad G = \begin{pmatrix} 2/3 & 1/3 \\ 1/3 & 2/3 \end{pmatrix}.$$

Thus, from state 1, the probability of absorption into C_1 and C_2 is $2/3$ and $1/3$ respectively. Similarly, from state 2, the probability of absorption into C_1 and C_2 is $1/3$ and $2/3$ respectively. It is to be noted that probability of absorption to the nearest state is higher for both the transient states. □

Following example is similar to the previous example, but the random walk is not symmetric.

Example 4.3.3 Suppose $\{X_n, n \geq 0\}$ is a random walk on $S = \{0, 1, 2, 3\}$ with absorbing barriers and P given by

$$P = \begin{array}{c} \\ 0 \\ 1 \\ 2 \\ 3 \end{array} \begin{array}{c} \begin{array}{cccc} 0 & 1 & 2 & 3 \end{array} \\ \begin{pmatrix} 1 & 0 & 0 & 0 \\ 1/3 & 0 & 2/3 & 0 \\ 0 & 1/3 & 0 & 2/3 \\ 0 & 0 & 0 & 1 \end{pmatrix} \end{array}.$$

Note that $p > 1/2$. The only change in the results from the symmetric random walk in the above example is the matrix G of probabilities of absorption. It is given by

$$G = \begin{pmatrix} 0.4286 & 0.5714 \\ 0.1429 & 0.8571 \end{pmatrix}.$$

Observe that probability of absorption in $\{3\}$ is higher than that in $\{0\}$. It is in view of the fact that the probability p of shift to the right is $> 1/2$. □

Random walk with reflecting barriers at 0 and M: In this random walk all states communicate with each other. Further, for each $i \in S$, $p_{ii}^{(n)} > 0$ if n is a multiple of 2. Thus, it is a non-null persistent periodic Markov chain with period 2. Hence, its long run distribution does not exist, but a unique stationary distribution $\underline{\pi} = (\pi_0, \pi_1, \ldots, \pi_M)$ exists and is given by $\pi_i = 1/\mu_i, i \in S$, where μ_i is the mean recurrence time for state i.

Example 4.3.4 Suppose $\{X_n, n \geq 0\}$ is a symmetric random walk with reflecting barriers and P given by

$$P = \begin{array}{c} \\ 0 \\ 1 \\ 2 \\ 3 \end{array} \begin{array}{c} \begin{array}{cccc} 0 & 1 & 2 & 3 \end{array} \\ \begin{pmatrix} 0 & 1 & 0 & 0 \\ 1/2 & 0 & 1/2 & 0 \\ 0 & 1/2 & 0 & 1/2 \\ 0 & 0 & 1 & 0 \end{pmatrix} \end{array}.$$

The unique stationary distribution is $\underline{\pi} = (1/6, 1/3, 1/3, 1/6)$. Hence, the mean recurrence times are given by $\mu_0 = \mu_3 = 6$ and $\mu_1 = \mu_2 = 3$. □

Random walk with elastic barriers at 0 and M: In this random walk also all states communicate with each other. Further, $p_{00} > 0$ implies that 0 is an aperiodic state and hence all the states are aperiodic. Thus, it is an ergodic Markov chain. Hence, its long run distribution and the unique stationary distribution $\underline{\pi}$ exist, which are the same and are given by $\pi_i = 1/\mu_i, i \in S$.

Random walk with partially reflecting barrier at 0 and reflecting barrier at M: This random walk has the same nature as that of a random walk with partially reflecting barriers at 0 and M. Thus, it is also an ergodic Markov chain. Hence, its long run distribution exists, hence the unique stationary distribution $\underline{\pi}$ exists, the two are the same and are given by $\pi_i = 1/\mu_i, i \in S$. For a random walk with partially reflecting barrier at M and reflecting barrier at 0, we have the same results.

Example 4.3.5 Suppose $\{X_n, n \geq 0\}$ is a Markov chain with state space $S = \{0, 1, 2, 3, 4\}$ and transition probability matrix P given by

$$P = \begin{array}{c} \\ 0 \\ 1 \\ 2 \\ 3 \\ 4 \end{array} \begin{array}{c} \begin{array}{ccccc} 0 & 1 & 2 & 3 & 4 \end{array} \\ \begin{pmatrix} 1/3 & 2/3 & 0 & 0 & 0 \\ 1/4 & 0 & 3/4 & 0 & 0 \\ 0 & 1/4 & 0 & 3/4 & 0 \\ 0 & 0 & 1/4 & 0 & 3/4 \\ 0 & 0 & 0 & 1 & 0 \end{pmatrix} \end{array}.$$

Thus, we have a random walk having reflecting barrier at 4 and partially reflecting barrier at 0. It is clear that all states communicate with each other and all are non-null persistent, state space being finite. Since $p_{00} > 0$, it follows that the period of state 0 is 1 and hence the period of all the states is 1. The stationary distribution exists and is $\underline{\pi} = (0.0186, 0.0497, 0.1491, 0.44720, 0.3354)$. □

A random walk with state space $S = \{0, 1, \ldots, M\}$ with absorbing barriers at 0 and M is the well-known model for gambler's ruin chain with finite total capital. We describe and study some of its properties in the following section.

4.4 Gambler's Ruin Problem

Suppose a gambler with initial capital a plays a game consisting of a series of one unit amount bets against an adversary with capital $M - a$. Suppose the probability that the gambler wins any bet is p and that he loses the bet is $q = 1 - p$. If the gambler's capital ever reaches zero, he is ruined and his capital remains zero thereafter. It is assumed that gambler quits playing when he either loses his entire capital or he earns M units. If X_n denotes the capital of the gambler at the end of nth play, then he quits playing if $X_n = 0$ or $X_n = M$. The gambler's fortune $\{X_n, n \geq 0\}$ is thus a Markov chain having transition probabilities $p_{i,i+1} = p, \quad p_{i,i-1} = q, i = 1, 2, \ldots, M - 1, \; p_{00} = p_{MM} = 1$. It is thus a finite state space random walk with absorbing barriers at states 0 and M. These two are the absorbing and aperiodic states and hence are ergodic. All other states are inessential and hence transient.

In this context, the issues of interest are

(i) what is the probability of the gambler's ruin, that is, what is the probability that his capital ever reaches zero?
(ii) if he is ever ruined, what is the expected number of bets needed, that is, what is the expected duration of the game?
(iii) what is the probability of the gambler's ruin when the total capital is infinite, while the gambler's capital is finite?

These issues are referred to as gambler's ruin problem. We now proceed to their solutions.

As discussed in Sect. 3, $C_1 = \{0\}$ and $C_2 = \{M\}$ are two closed communicating classes of aperiodic states. The class $T = \{1, 2, \ldots, M - 1\}$ is a class transient and periodic states, with period 2. Hence, $\pi_1 = (1, 0, 0, \ldots, 0)$ and $\pi_2 = (0, 0, 0, \ldots, 1)$ are two stationary distributions concentrated on C_1 and C_2 respectively and hence $\underline{\pi}_\alpha = (\alpha, 0, 0, \ldots, 0, 1 - \alpha)$ is a stationary distribution for any $\alpha \in (0, 1)$. The probability of absorption into $C_1 \cup C_2$ from any state in T is one, as proved in Theorem 2.6.9. It thus follows that after some finite amount of time, the gambler will either attain his goal of capital M or will be ruined. Note that the probability of ultimate ruin of the gambler with initial capital a is given by f_{a0}. In the following example we find the probability of absorption into C_i, $i = 1, 2$, when his initial capital is a, $0 < a < M - 1$, by using Theorem 2.6.9.

Example 4.4.1 Suppose $\{X_n, n \geq 0\}$ denotes gambler's fortune with $M = 3$. Suppose a is either 1 or 2. We compute f_{a0} and f_{a3} using Theorem 2.6.9. The transition probability matrix P for this Markov chain is as given below. We rearrange the rows and columns of P to get $\overline{P}$ as defined in Theorem 2.6.9. It is given below.

$$P = \begin{matrix} & \begin{matrix} 0 & 1 & 2 & 3 \end{matrix} \\ \begin{matrix} 0 \\ 1 \\ 2 \\ 3 \end{matrix} & \begin{pmatrix} 1 & 0 & 0 & 0 \\ q & 0 & p & 0 \\ 0 & q & 0 & p \\ 0 & 0 & 0 & 1 \end{pmatrix} \end{matrix} \;\; \& \;\; \overline{P} = \begin{matrix} & \begin{matrix} 0 & 3 & 1 & 2 \end{matrix} \\ \begin{matrix} 0 \\ 3 \\ 1 \\ 2 \end{matrix} & \begin{pmatrix} 1 & 0 & 0 & 0 \\ 0 & 1 & 0 & 0 \\ q & 0 & 0 & p \\ 0 & p & q & 0 \end{pmatrix} \end{matrix} = \begin{pmatrix} I_2 & 0_{2\times 2} \\ D_{2\times 2} & Q_{2\times 2} \end{pmatrix}.$$

Hence, $G = (I_2 - Q)^{-1} D$ is given by

$$G = \begin{pmatrix} f_{10} & f_{13} \\ f_{20} & f_{23} \end{pmatrix} = \begin{pmatrix} q/(1-pq) & p^2/(1-pq) \\ q^2/(1-pq) & p/(1-pq) \end{pmatrix}.$$

It is to be noted that for any $p \in (0, 1),\ pq \leq 1/4$. Further observe that row sums of G matrix are 1. □

We now derive the general expression for probability of ruin of the gambler. It is the probability of absorption into state $C_1 = \{0\}$, when his initial capital is $a,\ 0 < a < M - 1$. We discuss two methods. The arguments in second method are useful to extend to the case when a gambler with finite capital plays against an infinitely rich adversary.

Probability of ruin of the Gambler: Suppose P_a denotes the probability of ultimate ruin of a gambler with initial capital a. Thus, $P_a = f_{a0}$ is the probability of absorption into state 0 from the initial state a. It is clear that $P_0 = 1$ and $P_M = 0$. By conditioning on the outcome of the initial play of the game, with $P_0 = 1$ we obtain

$$P_a = qP_{a-1} + pP_{a+1} \iff pP_{a+1} - P_a + qP_{a-1} = 0, \;\; a = 1, 2, \ldots, M-1 .$$

Method I: The above equations are homogeneous difference equations of the form $(pE^2 - E + q)P_{a-1} = 0$. The auxiliary equation is $pw^2 - w + q = 0$ which has two roots $w_1 = 1$ and $w_2 = q/p$. These are distinct roots when $p \neq q$, that is, $p \neq 1/2$. Hence when $p \neq 1/2$, the general solution is of the form $P_a = c_1 w_1^a + c_2 w_2^a$. To find c_1 and c_2, we use the boundary conditions $P_0 = 1$ and $P_M = 0$. The condition $P_0 = 1$ implies $c_1 + c_2 = 1$ and the condition $P_M = 0$ implies $c_1 + c_2(q/p)^M = 0$. Solving these two equations, for $p \neq 1/2$,

$$c_1 = -\frac{(q/p)^M}{1-(q/p)^M} \;\; \& \;\; c_2 = \frac{1}{1-(q/p)^M}$$

$$\Rightarrow \;\; P_a = c_1 w_1^a + c_2 w_2^a \;\; = \;\; \frac{(q/p)^a - (q/p)^M}{1-(q/p)^M}.$$

When $p = q = 1/2$, $pP_{a+1} - P_a + qP_{a-1} = 0$ is equivalent to $\Delta^2 P_{a-1} = 0$, which implies that $P_a = c_1 + c_2 a$. To find c_1 and c_2 we again use the boundary conditions. Observe that $P_0 = 1$ and $P_M = 0$ implies $c_1 = 1$ and $c_2 = -1/M$. Hence $P_a = 1 - a/M$. Thus, the probability of ruin of the gambler with initial capital a is given by

$$P_a = \begin{cases} \frac{(q/p)^a-(q/p)^M}{1-(q/p)^M}, \text{ if } & p \neq q \\ 1 - a/M, & \text{if } p = q = 1/2 . \end{cases}$$

Method II: From method I, we have for $a = 1, 2, \ldots, M-1$,

$$P_a = qP_{a-1} + pP_{a+1} \iff (p+q)P_a = qP_{a-1} + pP_{a+1}$$
$$\Rightarrow p(P_{a+1} - P_a) = q(P_a - P_{a-1})$$
$$\Rightarrow P_{a+1} - P_a = c(P_a - P_{a-1}), \text{ where } c = q/p$$
$$\Rightarrow P_{a+1} - P_a = c^2(P_{a-1} - P_{a-2}) = c^3(P_{a-2} - P_{a-3}).$$

Continuing in this manner, we have $P_{a+1} - P_a = c^a(P_1 - P_0)$. Now by summing both sides with respect to a from 1 to k, $1 \leq k \leq M-1$, we get

$$P_{k+1} - P_1 = (P_1 - P_0)\sum_{a=1}^{k} c^a \Rightarrow P_{k+1} = P_1 + (P_1 - P_0)c_k, \tag{4.4.1}$$

where c_k is given by

$$c_k = \sum_{a=1}^{k} c^a = \begin{cases} \frac{c^{k+1}-c}{c-1}, \text{ if } c \neq 1 \\ k, \quad \text{if } c = 1 . \end{cases}$$

With $k = M-1$ in Eq. (4.4.1) and using the conditions $P_0 = 1$ and $P_M = 0$, we have

$$0 = P_M = P_1 + (P_1 - 1)c_{M-1} \Rightarrow P_1 = c_{M-1}/(1 + c_{M-1}).$$

Substituting this value of P_1 in P_{k+1} in Eq. (4.4.1), we get for $0 \leq k \leq M$

$$P_k = \begin{cases} \frac{c^M-c^k}{c^M-1}, & \text{if } c \neq 1 \\ 1 - k/M, \text{ if } c = 1 . \end{cases}$$

Substituting for $c = q/p$ with $k = a$, we get

$$P_a = \begin{cases} \frac{(q/p)^a-(q/p)^M}{1-(q/p)^M}, \text{ if } & p \neq q \\ 1 - a/M, & \text{if } p = q = 1/2 . \end{cases}$$

It is the same as obtained by the first method.

Since the probability of absorption into $\{0\} \cup \{M\}$ is 1, the probability that the gambler earns the capital M is $1 - P_a$. Suppose Q_a denotes the probability of ruin of the adversary with initial capital a. It is to be noted that, it is the same as the probability that the gambler wins the capital M. It is given by one minus the probability of the ruin of the gambler with initial capital $M - a$. Hence,

$$Q_a = 1 - P_{M-a} = \begin{cases} 1 - \frac{(q/p)^{M-a} - (q/p)^M}{1-(q/p)^M} = \frac{1-(q/p)^{M-a}}{1-(q/p)^M}, & \text{if } p \neq q \\ 1 - (1 - (M-a)/M) = 1 - a/M, & \text{if } p = q. \end{cases}$$

In the following example we compute the probability of ruin.

Example 4.4.2 Suppose Anil and Sunil play the game of gambling. Anil has probability 0.44 of winning at each trial. Suppose Anil and Sunil both start with 1000 rupees as initial capital. It is decided that the one who wins at the trial has to give 100 rupees to the other. We take 100 rupees as one unit. To compute the probability that Sunil will win the game, we compute the probability $P_{10}^{(A)}$ that Anil, with initial capital of 1000 rupees is ruined. Thus, we compute $P_{10}^{(A)}$ when $p = 0.44$ and $M = 20$ units. It is given by

$$P_{10}^{(A)} = \frac{(56/44)^{10} - (56/44)^{20}}{1 - (56/44)^{20}} = 0.9177\,.$$

Thus, Sunil will win the game with probability 0.9177. Hence, the probability that Anil will win the game is $1 - 0.9177 = 0.0823$. We can also compute it on similar lines, as we have computed the probability that Sunil will win the game. Thus, Anil will win the game is obtained as follows.

$$P_{10}^{(S)} = \frac{(44/56)^{10} - (44/56)^{20}}{1 - (44/56)^{20}} = 0.0823.$$

□

Using the expression of P_a, we now obtain the expected gain $E(G(a))$ of the gambler with initial capital a units. If the gambler wins the game, his gain is $M - a$ with probability $1 - P_a$. If he loses the game, then his gain is $-a$ which happens with probability P_a. Hence, the total expected gain of the gambler by the time the game ends is given by

$$E(G(a)) = \begin{cases} (M-a)(1-P_a) + (-a)P_a = M - MP_a - a, & \text{if } p \neq 1/2 \\ (M-a)(a/M) + (-a)(1-a/M) = 0, & \text{if } p = 1/2\,. \end{cases}$$

Thus, if $p = q = 1/2$, then the expected gain is 0. Conversely,

$$E(G(a)) = 0 \;\Rightarrow\; M - MP_a - a = 0 \;\Rightarrow\; P_a = 1 - a/M \;\Rightarrow\; p = q = 1/2\,.$$

This implies that the game is fair if and only if $p = q = 1/2$, that is, if and only if expected gain at any trial of the game is also 0.

Example 4.4.3 For the game of gambling in Example 4.4.2, the expected gain $E(G(a))$ of Anil with initial capital $a = 10$ units is -8.354 units, that is, -835.40 rupees. Thus, on the average Anil will have to suffer a loss of 835.40 rupees. Note that for Anil the probability of ruin is 0.9177, which is quite high, so the game will result in a loss. On similar lines, the expected gain $E(G(a))$ of Sunil is 8.354 units, that is, 835.40 rupees. □

In gambler's ruin problem, if the gambler's capital is a but the total capital of two players is infinite, that is, when a gambler with finite capital plays against an infinitely rich adversary, then the corresponding random walk has state space as $S = W$. In this case, as already noted, state 0 is absorbing and hence non-null persistent and aperiodic. All other states are transient and periodic with period 2. The probability $P_a(\infty)$ of ruin of the gambler when the total capital of two players is infinite is the same as the probability f_{a0} of absorption into state 0. We derive it on similar lines as in the case of derivation in the second method of P_a. In Method II we have obtained $P_{k+1} = P_1 + (P_1 - P_0)c_k$, where

$$c_k = \begin{cases} \frac{c^{k+1}-c}{c-1}, & \text{if } c \neq 1 \\ k, & \text{if } c = 1\,. \end{cases}$$

Note that

$$\lim_{k\to\infty} c_k = \begin{cases} c/(1-c), & \text{if } c < 1 \\ \infty, & \text{if } c \geq 1\,. \end{cases}$$

Recall that in Method II we have noted that

$$P_{k+1} - P_k = c^k(P_1 - P_0) = c^k(P_1 - 1) \leq 0.$$

Hence $\{P_k, k \geq 0\}$ is a monotonically decreasing sequence bounded below by 0. Hence, $b = \lim_{k\to\infty} P_k$ exists. By taking limits as $k \to \infty$ on both sides of $P_{k+1} = P_1 + (P_1 - 1)c_k$, we get

$$b = \begin{cases} P_1 + (P_1 - 1)\frac{c}{1-c}, & \text{if } c < 1 \\ \infty, & \text{if } c \geq 1\,. \end{cases}$$

However $b \in (0, 1)$ implies that for $c \geq 1$, P_1 must be 1. Thus for $b \in (0, 1)$,

$$P_1 = \begin{cases} (1-c)b + c, & \text{if } c < 1 \\ 1, & \text{if } c \geq 1\,. \end{cases}$$

Substituting the values of P_1 in $P_{k+1} = P_1 + (P_1 - 1)c_k$, then replacing $k+1$ by a and c by q/p, we get

$$P_a(\infty) = \begin{cases} b(1-(q/p)^a) + (q/p)^a, & \text{if } q < p \\ 1, & \text{if } q \geq p\,. \end{cases}$$

Usually b is taken to be 0, however no formal proof is found.

We now proceed to compute the expected duration of the game, when the capital of both the gambler and the adversary is finite. As in the derivation of P_a, here also we use two methods.

Expected duration of the game: Suppose D_a denotes the expected duration of the game when the gambler has initial capital of a units, $a = 1, \ldots, M-1$. As in the derivation of P_a, we condition on the outcome of the initial trial of the game. Note that if the initial trial results in success with probability p, then at the next trial his capital is $a+1$ and then expected duration of the game will be $D_{a+1}+1$. Similarly, if the initial trial results in failure with probability of q, then at the next trial his capital is $a-1$ and then expected duration of the game will be $D_{a-1}+1$. Similar arguments lead to expressions of D_1 and D_{M-1}. Thus we have the following system of equations.

$$\begin{aligned} D_a &= p(D_{a+1}+1) + q(D_{a-1}+1),\ a = 2, 3, \ldots, M-2 \\ D_1 &= p(D_2+1) + q \\ D_{M-1} &= p + q(D_{M-2}+1)\,. \end{aligned}$$

With boundary conditions $D_0 = D_M = 0$, the expressions for D_a, D_1 and D_{M-1} can be combined as

$$pD_{a+1} - D_a + qD_{a-1} = -1,\ \ a = 1, 2, \ldots, M-1\,. \tag{4.4.2}$$

Method I: Note that Eq. (4.4.2) is a system of non-homogeneous difference equations. We solve it to get the expression for D_a. The solution depends on whether $p = q$ or $p \neq q$. So we make two cases.

Case(i) $p \neq q$: A particular solution of the difference equation $pD_{a+1} - D_a + qD_{a-1} = -1$ is given by $D_a = a/(q-p)$ and the most general solution of the corresponding homogeneous equation $pD_{a+1} - D_a + qD_{a-1} = 0$, which is same as that for P_a, is given by $c_1 + c_2(q/p)^a$. Hence, the most general solution of the non-homogeneous equation $pD_{a+1} - D_a + qD_{a-1} = -1$ is given by

$$D_a = a/(q-p) + c_1 + c_2(q/p)^a,\ \ a = 1, 2, \ldots, M-1\,.$$

We find the constants c_1 and c_2 from the boundary conditions $D_0 = D_M = 0$. Thus,

$$D_0 = 0 \;\Rightarrow\; c_1 = -c_2 \;\&\; D_M = 0 \;\Rightarrow\; c_2 = \frac{M}{q-p}\frac{1}{(1-(q/p)^M)}$$

$$\Rightarrow\; D_a = \frac{a}{q-p} - \frac{M}{q-p}\frac{(1-(q/p)^a)}{(1-(q/p)^M)} \text{ when } p \neq q\,.$$

Case(ii) $p = q = 1/2$: In this case the difference equation $pD_{a+1} - D_a + qD_{a-1} = -1$ reduces to $D_{a+1} - 2D_a + qD_{a-1} = -2$. A particular solution of this difference equation is $D_a = -a^2$ and the most general solution of the corresponding homogeneous equation $D_{a+1} - 2D_a + qD_{a-1} = 0$ is given by $c_1 + c_2 a$. Hence, the most general solution of the non-homogeneous equation $D_{a+1} - 2D_a + qD_{a-1} = -2$ is given by $D_a = -a^2 + c_1 + c_2 a$. The constants c_1 and c_2 are obtained from the boundary conditions $D_0 = D_M = 0$. Thus,

$$\begin{aligned} D_0 = 0 \Rightarrow c_1 = 0 \;\&\; D_M = 0 \;\Rightarrow\; c_2 = a \\ \Rightarrow D_a = Ma - a^2 = a(M-a) \text{ when } p = q = 1/2\,. \end{aligned}$$

Thus, the expected duration of the game when the gambler has initial capital of a units is given by

$$D_a = \begin{cases} \frac{a}{q-p} - \frac{M}{q-p}\frac{(1-(q/p)^a)}{(1-(q/p)^M)}, & \text{if } p \neq q \\ a(M-a), & \text{if } p = q = 1/2\,. \end{cases}$$

Method II: In this method we rewrite Eq. (4.4.2) as follows. Suppose q/p is denoted by c.

$$\begin{aligned} D_{a+1} - D_a &= c(D_a - D_{a-1}) - \frac{1}{p}, \quad a = 1, 2, \ldots, M-1 \\ &= c\left(c(D_{a-1} - D_{a-2}) - \frac{1}{p}\right) - \frac{1}{p} \\ &= c^2(D_{a-1} - D_{a-2}) - \frac{c}{p} - \frac{1}{p}, \quad \text{continuing in this way} \\ &= c^a(D_1 - D_0) - \frac{1}{p}(1 + c + c^2 + \ldots + c^{a-1}). \end{aligned}$$

Thus, with $D_0 = 0$

$$D_{a+1} - D_a = \begin{cases} c^a D_1 - \frac{1-c^a}{p(1-c)}, & \text{if } c \neq 1 \\ D_1 - 2a, & \text{if } c = 1\,. \end{cases}$$

The above equation is true $\forall\; a = 1, 2, \ldots M-1$. By summing both sides from $a = 1$ to k, we get for $k = 1, 2, \ldots M-1$,

$$D_{k+1} - D_1 = \begin{cases} D_1 \frac{(c-c^{k+1})}{1-c} - \frac{k}{p(1-c)} + \frac{c-c^{k+1}}{p(1-c)^2}, & \text{if } c \neq 1 \\ D_1 k - k(k+1), & \text{if } c = 1 . \end{cases}$$

Hence, for $k = 1, 2, \ldots, M-1$,

$$D_{k+1} = \begin{cases} \frac{D_1(1-c^{k+1})}{1-c} - \frac{k+1}{p(1-c)} + \left[\frac{c-c^{k+1}}{p(1-c)^2} + \frac{1}{p(1-c)}\right], & \text{if } c \neq 1 \\ (k+1)(D_1 - k), & \text{if } c = 1 . \end{cases}$$

Now with $k = M-1$, we get

$$0 = \begin{cases} \frac{D_1(1-c^M)}{1-c} - \frac{M}{p(1-c)} + \frac{1-c^M}{p(1-c)^2}, & \text{if } c \neq 1 \\ M(D_1 - M + 1), & \text{if } c = 1 . \end{cases}$$

Hence,

$$D_1 = \begin{cases} \left[\frac{M}{p(1-c)} - \frac{1-c^M}{p(1-c)^2}\right] \frac{1-c}{1-c^M} = \frac{M}{p(1-c^M)} - \frac{1}{p(1-c)}, & \text{if } c \neq 1 \\ M-1, & \text{if } c = 1 . \end{cases}$$

Substituting the expression for D_1 in the expression for D_{k+1} we get

$$D_{k+1} = \begin{cases} \left[\frac{M}{p(1-c^M)} - \frac{1}{p(1-c)}\right] \frac{1-c^{k+1}}{1-c} - \frac{k+1}{p(1-c)} + \frac{1-c^{k+1}}{p(1-c)^2}, & \text{if } c \neq 1 \\ (k+1)(M-1-k), & \text{if } c = 1 . \end{cases}$$

This expression simplifies to

$$D_{k+1} = \begin{cases} \frac{M}{p(1-c^M)} \frac{1-c^{k+1}}{1-c} - \frac{k+1}{p(1-c)}, & \text{if } c \neq 1 \\ (k+1)(M-1-k), & \text{if } c = 1 . \end{cases}$$

This equation is true for $k = 1, 2, \ldots, M-1$. Hence with $c = q/p$ and $k+1 = a$ we have

$$D_a = \begin{cases} \frac{a}{q-p} - \frac{M}{q-p} \frac{(1-(q/p)^a)}{(1-(q/p)^M)}, & \text{if } p \neq q \\ a(M-a), & \text{if } p = q = 1/2 . \end{cases}$$

Example 4.4.4 For the game of gambling in Example 4.4.2, with $p = 0.44$, that is, treating Anil as a gambler, $D_{10} = 69.62$, thus, on the average, the game will be played for 70 trials. It is to be noted that the expected duration of the game should be the same, whether any one of the two is treated as a gambler. Observe that, with $p = 0.44$,

$$D_{10} = \frac{10}{0.56 - 0.44} - \frac{20}{0.56 - 0.44}\frac{(1 - (56/44)^{10})}{(1 - (56/44)^{20})} = 69.62.$$

With $p = 0.56$, that is, treating Sunil as a gambler,

$$D_{10} = \frac{10}{0.44 - 0.56} - \frac{20}{0.44 - 0.56}\frac{(1 - (44/56)^{10})}{(1 - (44/56)^{20})} = 69.62.$$

□

It is shown that $P_a(\infty) < 1$ if $q < p$, that is, $p > 1/2$. Thus, in this case the game may go on forever and expected duration may be infinite. It can be shown that (Ash [1]) the expected duration $D_a(\infty)$ of the game, when the capital of the gambler is a but the capital of the adversary is infinite is

$$D_a(\infty) = \begin{cases} a/(q-p), & \text{if } p < 1/2 \\ \infty, & \text{if } p \geq 1/2 . \end{cases}$$

Various versions of random walk including gambler's ruin chain can be unified as follows. Suppose for

$$n \geq 1, \;\; U_n \sim B(1, p), \;\; V_n \sim B(1, \delta) \;\; \& \;\; W_n \sim B(1, \eta)$$

and all these are independent and identically distributed random variables. Suppose a sequence $\{X_n, n \geq 0\}$ of random variables is defined as follows.

1. $X_n = X_{n-1} + (2U_n - 1)$
2. $X_n = X_{n-1} + (2U_n - 1)I[X_{n-1} \neq 0]$
3. $X_n = X_{n-1} + (2U_n - 1)I[X_{n-1} \neq 0, M]$
4. $X_n = X_{n-1} + (2U_n - 1)I[X_{n-1} \neq 0, M] + I[X_{n-1} = 0] - I[X_{n-1} = M]$
5. $X_n = X_{n-1} + (2U_n - 1)I[X_{n-1} \neq 0, M] + V_n I[X_{n-1} = 0]$
   $- W_n I[X_{n-1} = M]$

All the above models can be written in the form $X_n = g(X_{n-1}, \Delta_n)$, where $\Delta_n, n \geq 1$ are independent and identically distributed random variables/vectors, which proves that all are Markov chains.

(i) Markov chain in (1) is an unrestricted random walk on I.
(ii) Markov chain in (2) is an unrestricted random walk on W with an absorbing barrier at 0. This is also the Markov chain for gambler's ruin chain when total capital is infinite.

(iii) Markov chain in (3) is a random walk on finite state space $\{0, 1, \ldots, M\}$ with absorbing barriers at 0 and M. This is also the Markov chain for gambler's ruin chain with finite total capital.
(iv) Markov chain in (4) is a random walk on $\{0, 1, \ldots, M\}$ with reflecting barriers at 0 and M.
(v) Markov chain in (5) is a random walk on $\{0, 1, \ldots, M\}$ with elastic barriers at 0 and M.

In all the above types of random walks, success probability in U_n, V_n and W_n does not depend on the current state X_n. In the next section, we consider two particular Markov chains, which are random walks, in which the probability of transition from X_n depends on X_n.

4.5 Ehrenfest Chain and Birth-Death Chain

We briefly discuss below a particular type of Markov chain with finite state space $\{0, 1, \ldots, M\}$ and with reflecting barriers at 0 and M. It is a classical Ehrenfest urn model.

Ehrenfest chain: A molecule diffuses at random through the membrane. A classical mathematical description of its diffusion through a membrane is the famous Ehrenfest urn model. It is also a model for exchange of heat or gas molecules between two isolated bodies. The physical interpretation of this model is as follows. Suppose we have two urns A and B containing a total of M balls (molecules). A ball is selected at random from M balls, and it is transferred to the other. Suppose X_n denotes the number of balls in urn A at the end of nth drawing. Then

$$X_{n+1} = \begin{cases} X_n \pm 1, & \text{if } X_n = 1, 2, \ldots, M-1 \\ X_n + 1, & \text{if } \quad X_n = 0 \\ X_n - 1, & \text{if } \quad X_n = M \,. \end{cases}$$

Thus, X_{n+1} is expressible in terms of X_n only and hence $\{X_n, n \geq 0\}$ is a Markov chain with finite state space $\{0, 1, \ldots, M\}$. This is an example of state dependent time homogeneous Markov chain. Further, the one step transition probabilities p_{ij} are given by

$$p_{ij} = \begin{cases} 1, & \text{if} \quad i = 0 \;\&\; j = 1 \\ 1, & \text{if} \quad i = M \;\&\; j = M-1 \\ i/M = 1 - p_i, & \text{if} \quad j = i-1, i = 1, 2, \ldots, M-1 \\ 1 - i/M = p_i, & \text{if} \quad j = i+1, i = 1, 2, \ldots, M-1 \,, \end{cases}$$

where $p_i = 1 - i/M$ for $i = 1, 2, \ldots, M-1$. Thus, the possible transitions are to the right and left with a jump of one unit, however the probabilities of transition from i depend on i. Note that for $i = 1, 2, \ldots, M-1$,

$$p_{ij} = i/M \text{ if } j - i = -1 \ \& \ 1 - i/M \text{ if } j - i = 1.$$

Thus, p_{ij} is not just a function of $j - i$, but also depends on i. Thus, the Ehrenfest model is a process where increments are independent but are not stationary. Further, there are reflecting barriers at boundaries 0 and M. Thus, the Ehrenfest chain is a Markov chain with finite state space $\{0, 1, \dots, M\}$ and with reflecting barriers at 0 and M, with probabilities of transition from i depending on i. It is a non-null persistent and periodic Markov chain with period 2 and hence it has a unique stationary distribution. We find it by solving the system of equations $\pi_j = \sum_{i=0}^{M} \pi_i p_{ij}$, $j = 0, 1, 2, \dots, M$. From the given p_{ij}'s we have

$$\begin{aligned}
\pi_0 &= \pi_1 p_{10} = \pi_1/M, \\
\pi_j &= \pi_{j-1} p_{j-1,j} + \pi_{j+1} p_{j+1,j} = \pi_{j-1}\frac{M - j + 1}{M} + \pi_{j+1}\frac{j+1}{M} \\
\pi_M &= \pi_{M-1}/M \ .
\end{aligned}$$

From these equations we have

$$\pi_1 = M\pi_0 = \binom{M}{1}\pi_0, \quad \pi_1 = \pi_0 + \frac{2}{M}\pi_2 \ \Rightarrow \ \pi_2 = \frac{M(M-1)}{2}\pi_0 = \binom{M}{2}\pi_0 \ .$$

We assume $\pi_r = \binom{M}{r}\pi_0, \ \ r \le j$. It is true for $j = 1, 2$. From the equation $\pi_j = \pi_{j-1}(M - j + 1)/M + \pi_{j+1}(j + 1)/M$, we obtain

$$\begin{aligned}
\pi_{j+1} &= \frac{M}{j+1}\left[\binom{M}{j} - \frac{M - j + 1}{M}\binom{M}{j-1}\right]\pi_0 \\
&= \frac{M}{j+1}\left[\frac{M!}{j!(M-j)!} - \frac{M - j + 1}{M}\frac{M!}{(j-1)!(M-j+1)!}\right]\pi_0 \\
&= \frac{M}{j+1}\left[\frac{M!}{j!(M-j)!} - \frac{(M-1)!}{(j-1)!(M-j)!}\right]\pi_0 \\
&= \frac{M!}{(j+1)!(M-j-1)!}\left[\frac{M}{M-j} - \frac{j}{M-j}\right]\pi_0 \\
&= \binom{M}{j+1}\pi_0 \ .
\end{aligned}$$

Hence by induction, $\pi_j = \binom{M}{j}\pi_0$ for all $j = 0, 1, 2, \dots, M$. The condition

$$\sum_{j=1}^{M} \pi_j = 1 \ \Rightarrow \ \pi_0 = 1/2^M \ \Rightarrow \ \pi_j = \binom{M}{j}\Big/2^M, \ j = 0, 1, 2, \dots, M.$$

Thus, the unique stationary distribution associated with the Ehrenfest chain is binomial $B(M, 1/2)$.

Another Markov chain in which the probabilities of transition from i depend on i is the birth-death chain. We discuss its nature below.

Birth-death chain: Suppose $\{X_n, n \geq 0\}$ is a Markov chain with state space $S = W$ or $S = \{0, 1, \ldots, M\}$. Further, the one step transition probabilities p_{ij} are given by

$$p_{ij} = \begin{cases} q_i, & \text{if} \quad j = i - 1 \\ r_i, & \text{if} \quad i = j \\ p_i, & \text{if} \quad j = i + 1 \\ 0, & \text{otherwise,} \end{cases}$$

where $q_0 = 0,\ p_i, q_i, r_i \geq 0$ and $p_i + q_i + r_i = 1$. Then the Markov chain is known as a birth-death chain. The possible transitions are to the right and left with a jump of one unit; or the chain may stay in the same state after one unit. Thus, transitions are similar as in a random walk, but the probabilities of transition from i depend on i. As in the Ehrenfest model, p_{ij} is not just a function of $j - i$, but also depends on i. Thus, it is a process where increments are independent but are not stationary.

It is labeled as a birth-death chain in view of its application where the state of the chain is the size of the population of living organisms. The transition from i to $i + 1$ corresponds to the occurrence of a birth while the transition from i to $i - 1$ corresponds to the occurrence of a death. No event may take place in one unit time and the population will be in the same state. In this setup $p_0 = q_0 = 0$. If S is finite then $p_M = 0$. Chapter 8 is devoted to the detailed study of the continuous time birth-death process. The continuous time birth-death process is the more realistic model in applications. We will note that the birth-death chain is a Markov chain embedded in the continuous time birth-death process.

If $S = W,\ p_0, r_0 > 0\ \&\ q_0 = 0$, then the birth-death chain is labeled as a queuing chain as it has applications in queuing theory. In this case, X_n indicates the number of customers in the queue at time point n. In one unit time queue size increases by one unit due to arrival of a new customer, decreases by one unit if the customer leaves the queue on completion of the service. If queue size is 0 at some time point, it increases to 1 when a new customer arrives. Hence, $p_0 > 0$. This is the main difference in the queuing chain and the birth-death chain. If the capacity of the waiting room is limited then the state space of the queuing chain is finite. We discuss these chains in Chapter 8 in a continuous parameter setup.

The parameters p_i, q_i, r_i are defined suitably at the boundary points. The particular values p_i, q_i, r_i lead to different models as listed below.

(i) For $S = W$, if $r_0 = 1, r_i = 0,\ p_i = p$ and $q_i = 1 - p,\ \forall\, i \geq 1$, then the birth-death chain is the same as the simple random walk with absorbing barrier at 0.

(ii) For $S = W$, if $p_0 = 1, r_i = 0, \forall\, i \geq 0,\ p_i = p$ and $q_i = 1 - p,\ \forall\, i \geq 1$, then the birth-death chain reduces to the simple random walk with reflecting barrier at 0.

(iii) For $S = W$, if $p_0, r_0 > 0, r_i = 0, p_i = p$ and $q_i = 1 - p, \ \forall \ i \geq 1$, then the birth-death chain is the simple random walk with partially reflecting barrier at 0.
(iv) For $S = \{0, 1, \ldots, M\}$, if $r_0 = r_M = 1, r_i = 0, p_i = p$ and $q_i = 1 - p, \forall \ i = 1, \ldots, M - 1$, then the birth-death chain is the simple random walk with finite state space and absorbing barrier at 0 and M. It is also the gambler's ruin chain.
(v) For $S = \{0, 1, \ldots, M\}$, if $p_0 = q_M = 1, r_i = 0, \forall \ i \in S, p_i = p$ and $q_i = 1 - p, \forall \ i = 1, \ldots, M - 1$, then the birth-death chain is the simple random walk with reflecting barrier at 0 and M.
(vi) For $S = \{0, 1, \ldots, M\}$, if $p_0, r_0 > 0, q_M, r_M > 0, q_M + r_M = 1, r_i = 0, p_i = p$ and $q_i = 1 - p, \ \forall \ i = 1, \ldots, M - 1$, then the birth-death chain is the same as the simple random walk with partially reflecting barrier at 0 and M.
(vii) If $S = \{0, 1, \ldots, M\}, r_i = 0, \ \ \forall \ i \in S$ and $q_1 = q_M = 1$, then the birth-death chain is the same as the Ehrenfest chain.

We now discuss the nature of the birth-death chain.

(i) When the state space is finite, it is the same as the simple random walk on finite state space, although p_i, q_i and r_i depend on i.
(ii) Suppose $S = W$. If $r_0 = 1$, then 0 is an absorbing state and hence ergodic. If $p_i, q_i > 0 \ \forall \ i \geq 1$, then all states $i > 0$ communicate with each other. These states are inessential and hence transient. If $r_i > 0 \ \forall \ i \geq 1$, then the chain is aperiodic. If $r_i = 0 \ \forall \ i$, then as in the random walk on W, the chain can return to its initial state in even number of transitions. For example, $p_{00}^{(2)} = p_0 q_1 > 0$. Thus the chain is periodic with period 2.
(iii) Suppose $S = W, q_0 = 0, p_i, q_i, r_i > 0, \ \forall \ i$. Then the birth-death chain, which is a queuing chain, is irreducible and aperiodic. We examine whether its stationary distribution exists in the following theorem. If it exists, then then the chain is non-null persistent.

Theorem 4.5.1 *A birth-death chain with $S = W$ and $q_0 = 0, p_i, r_i, q_i > 0, \forall \ i$ has a stationary distribution if $\sum_{i \geq 1} \rho_i < \infty$, where $\rho_i = (p_0 p_1 \cdots p_{i-1})/(q_1 q_2 \cdots q_i)$.*

Proof To examine whether a stationary distribution $\underline{\pi}$ exists, we examine whether $\underline{\pi} = \underline{\pi} P$ has a solution and under which conditions. For the given birth-death chain, the first row of P is $(r_0, p_0, 0, 0, \ldots)$, the second row is $(q_1, r_1, p_1, 0, 0, \ldots)$, the third row is $(0, q_2, r_2, p_2, 0, \ldots)$ and so on. Hence, the equation $\underline{\pi} = \underline{\pi} P$ leads to the following system of equations. Using the condition $p_i + r_i + q_i = 1 \ \forall \ i$, these can be expressed in terms p_i and q_i as follows.

$$
\begin{aligned}
\pi_0 &= r_0\pi_0 + q_1\pi_1 \iff p_0\pi_0 = q_1\pi_1 \\
\pi_1 &= p_0\pi_0 + r_1\pi_1 + q_2\pi_2 \iff p_0\pi_0 + q_2\pi_2 = (p_1 + q_1)\pi_1 \\
\pi_i &= p_{i-1}\pi_{i-1} + r_i\pi_i + q_{i+1}\pi_{i+1}, \ \ i \geq 2 \\
\iff (p_i + q_i)\pi_i &= p_{i-1}\pi_{i-1} + q_{i+1}\pi_{i+1}, \ \ i \geq 2.
\end{aligned}
$$

From these equations we have

$$\begin{aligned}&\pi_1 = (p_0/q_1)\pi_0, \quad q_2\pi_2 = (p_1 + q_1)(p_0/q_1)\pi_0 - p_0\pi_0\\ \Rightarrow\ &\pi_2 = (p_0 p_1/q_1 q_2)\pi_0, \text{ thus, } \pi_i = \rho_i \pi_0 > 0, \ i \geq 1\\ &\sum_{i \in S} \pi_i = 1 \ \Rightarrow \ \pi_0\Big(1 + \sum_{i \geq 1} \rho_i\Big) = 1 \ \Rightarrow \ \pi_0 = \Big(1 + \sum_{i \geq 1} \rho_i\Big)^{-1},\end{aligned}$$

provided $\sum_{i\geq 1} \rho_i < \infty$. Thus, under this condition, the stationary distribution exists and the birth-death chain is non-null persistent. If the condition is not satisfied, then the chain is either null persistent or transient. □

With $p_i = p$ and $r_i = 0$, the condition of convergence of the series reduces to $p < 1/2$ as proved for the simple random walk with partially reflecting barriers.

The next section presents `R` codes used in Sects. 2 and 3 to illustrate various concepts related to random walks.

4.6 R Codes

Code 4.6.1 Realization of an unrestricted random walk: With this code we obtain a realization of an unrestricted random walk, for specified values of p and specified value of n, the length of the realization. By definition, $X_n = \sum_{i=1}^{n} Y_i$, where $\{Y_1, Y_2, \ldots, Y_n\}$ are independent and all are distributed as Y, with $P[Y = 1] = p$ and $P[Y = -1] = q = 1 - p$. Hence at each stage, we generate a random observation from $\{-1, 1\}$, with probability assigned to 1 being p and obtain X_n as $X_n = \sum_{i=1}^{n} Y_i$. In the code we have taken four values of p, initial state as 0 and $n = 25$. The code is illustrated for Example 4.2.1.

```
# Part I: Input four sets of values of p
S=c(-1,1); p1=c(1/2,1/2); p2=c(2/3,1/3); p3=c(1/4,3/4)
p4=c(4/5,1/5); p=c(p1[2],p2[2],p3[2],p4[2])
P=rbind(p1,p2,p3,p4); P
# Part II: Generate realizations
n=25 # length of the relization
x=matrix(nrow=n,ncol=4); x[1,]=c(0,0,0,0)
for(j in 1:4)
{
set.seed(j)
for(i in 2:n)
{
x[i,j]=x[i-1,j]+sample(S,1,P[j,],replace=T)
}
}
```

```
x;
# Part III: Graphs of realizations
par(mfcol=c(2,2));p=round(p,2);p
prob=paste("p =",p,sep=" ")
for(j in 1:4)
{
plot(1:n,x[,j],"o",main=prob[j],ylab="States",xlab="Time",
col="dark blue",yaxt="n",lwd=2)
axis(2,at=sort(unique(x[,j])),labels=sort(unique(x[,j])),las=2)
}
```

□

Code 4.6.2 Computation of exact and approximate values of $p_{00}^{(2n)}$ in an unrestricted random walk: In an unrestricted simple random walk,

$$p_{00}^{(2n)} = \binom{2n}{n} p^n q^n = \frac{(2n)!}{n!n!} p^n q^n \approx \frac{(4pq)^n}{\sqrt{\pi n}} \quad \forall\ n \geq 1.$$

This code computes $p_{00}^{(2n)}$ using the exact and approximate formulae for $n = 1$ to 20 and for three values of p as 1/3, 1/2, 3/4, to examine the degree of approximation. It is illustrated for Example 4.2.2.

```
# Part I: Input values of p
p=c(1/3,1/2,3/4)
# Part II: Computation of probabilities
n=1:20; pi=3.141593
pr=pra=matrix(0,nrow=length(n),ncol=length(p))
for(j in 1:length(p))
{
for(i in 1:length(n))
{
x=(factorial(2*n[i]))*p[j]^n[i]*(1-p[j])^n[i]
y=(factorial(n[i]))^2
pr[i,j]=x/y
pra[i,j]=((4*p[j]*(1-p[j]))^n[i])/((pi*n[i])^(.5))
}
}
d=data.frame(n,pr[,1],pra[,1],pr[,2],pra[,2],pr[,3],pra[,3])
d1=round(d,2); d1
```

□

Code 4.6.3 Approximation of $\sum_{n=1}^{\infty} p_{00}^{(2n)}$ by $(1 - 4pq)^{-1/2} - 1$ in an unrestricted random walk: In an unrestricted random walk, we have noted that

$$\sum_{n=1}^{\infty} p_{00}^{(2n)} = \sum_{n=1}^{\infty} \binom{2n}{n} p^n q^n = (1-4pq)^{-1/2} - 1.$$

The code computes $\sum_{n=1}^{N} p_{00}^{(n)}$ for some large value of N to verify whether it is equal to $(1-4pq)^{-1/2} - 1$ approximately. We take three values of p as $1/3, 2/3, 3/4$. It is illustrated for Example 4.2.3.

```
# Part I: Input values of p
p=c(1/3,2/3,3/4)
# Part II: Computation of probabilities
n=1:70
pr=matrix(0,nrow=length(n),ncol=length(p))
sum=a=c()
for(j in 1:length(p))
{
a[j]=(1-4*p[j]*(1-p[j]))^(-.5)-1
for(i in 1:length(n))
{
x=(factorial(2*n[i]))*p[j]^n[i]*(1-p[j])^n[i]
y=(factorial(n[i]))^2
pr[i,j]=x/y
}
sum[j]=sum(pr[,j])
}
d=round(data.frame(p,sum,a),4);d
```

□

Code 4.6.4 Realization of a random walk on W with absorbing barrier at 0: Suppose $\{X_n, n \geq 0\}$ is a simple random walk on W, with absorbing barrier at 0. The following R code obtains a realization of the random walk. For comparison we take 4 values of p as $p = 0.50, 0.33, 0.75, 0.20$. For each p, we take $X_0 = 4$ and length of realization as $n = 25$. It is illustrated for Example 4.2.4.

```
# Part I: Input values of p
S=c(-1,1);p1=c(1/2,1/2);p2=c(2/3,1/3); p3=c(1/4,3/4);p4=c(4/5,1/5)
p=c(p1[2],p2[2],p3[2],p4[2])
P=rbind(p1,p2,p3,p4); P
# Part II: Generate realizations
n=25;x=matrix(nrow=n,ncol=length(p)); x[1,]=c(4,4,4,4)
for(j in 1:length(p))
{
set.seed(j)
for(i in 2:n)
{
if(x[i-1,j]>0)
```

```
{
x[i,j]=x[i-1,j]+sample(S,1,P[j,],replace=T)
}
else
{
x[i,j]=0
}
}
}
x; p=round(p,2)
# Part III: Graphs of realizations
par(mfcol=c(2,2))
prob=paste("p =",p,sep=" ")
for(j in 1:4)
{
plot(1:n,x[,j],"o",main=prob[j],ylab="States",xlab="Time",
col="dark blue",yaxt="n",lwd=2)
axis(2,at=sort(unique(x[,j])),labels=sort(unique(x[,j])),las=2)
}
```

□

Code 4.6.5 Realization of a random walk on W with partially reflecting barrier at 0: Suppose $\{X_n, n \geq 0\}$ is a simple random walk on W, with partially reflecting barrier at 0. The following R code gives a realization of the random walk. For comparison we take 4 values of p as $p = 0.50, 0.33, 0.75, 0.20$. For each p, we take $X_0 = 2$ and length of realization as $n = 25$. It is illustrated for Example 4.2.6.

```
# Part I: Input values of p
S=c(-1,1);p1=c(1/2,1/2);p2=c(2/3,1/3); p3=c(1/4,3/4)
p4=c(4/5,1/5); p=c(p1[2],p2[2],p3[2],p4[2])
P=rbind(p1,p2,p3,p4); P;
# Part II: Generate realizations
n=25; x=matrix(nrow=n,ncol=length(p)); x[1,]=c(2,2,2,2)
S1=c(0,1)
for(j in 1:length(p))
{
set.seed(j)
for(i in 2:n)
{
if(x[i-1,j]>0)
{
x[i,j]=x[i-1,j]+sample(S,1,P[j,],replace=T)
}
else
{
```

```
x[i,j]=sample(S1,1,P[j,],replace=T)
}
}
}
x;
# Part III: Graphs of realizations
p=round(p,2); par(mfcol=c(2,2))
prob=paste("p =",p,sep=" ")
for(j in 1:length(p))
{
plot(1:n,x[,j],"o",main=prob[j],ylab="States",xlab="Time",
col="dark blue",yaxt="n",lwd=2)
axis(2,at=sort(unique(x[,j])),labels=sort(unique(x[,j])),las=2)
}
```

□

A quick recap of the results discussed in the present chapter is given below.

Summary

1 A stochastic process $\{X_n, n \geq 0\}$ in which X_n is defined as $X_n = \sum_{i=0}^{n} Y_i$ where $\{Y_i, i \geq 0\}$ are independent and identically distributed random variables is known as a random walk. It is said to be a simple random walk if $Y_i = \pm 1$, with $P[Y_i = 1] = p$ and $P[Y_i = -1] = 1 - p = q, 0 < p < 1$. When $p = 1/2$, the simple random walk is known as a symmetric simple random walk.

2 A Markov chain $\{X_n, n \geq 0\}$ whose state space $S = I$ is known as an unrestricted one-dimensional random walk, if the transition probabilities are given by

$$p_{i,i+1} = p, \quad p_{i,i-1} = q = 1 - p \;\; \& \;\; p_{ij} = 0 \; \forall \; j \neq i+1, \; i-1, \;\; i \in S \,, \quad 0 < p < 1 \,.$$

3 An unrestricted random walk is an irreducible Markov chain where all states have period 2, all are transient if $p \neq 1/2$ and all are null persistent if $p = 1/2$.

4 Suppose $\{X_n, n \geq 0\}$ is a Markov chain with state space $S = W$ and with the transition probabilities given by

$$p_{i,i+1} = p, \quad p_{i,i-1} = q = 1 - p \;\; \& \;\; p_{ij} = 0 \, \forall \, j \neq i+1, \; i-1, \; i = 1, 2, \ldots, \;\; p_{00} = 1,$$

$0 < p < 1$. Then it is known as a simple random walk on W, with absorbing barrier at 0. For this random walk, (i) all states $i > 0$ communicate with states $j > 0$, (ii) state 0 is aperiodic and non-null persistent state, (iii) all states $i > 0$ have period 2 and (iv) all states $i > 0$ are transient.

5 Suppose $\{X_n, n \geq 0\}$ is a Markov chain with state space $S = W$ and with the transition probabilities given by

$$p_{i,i+1} = p, \;\; p_{i,i-1} = q = 1 - p \;\; \& \;\; p_{ij} = 0 \,\forall\, j \neq i+1, \; i-1, \;\; i = 1, 2, \ldots, \;\; p_{01} = 1,$$

$0 < p < 1$. This Markov chain is known as a simple random walk on W, with reflecting barrier at 0. For this random walk, all states (i) communicate with all other states, (ii) have period 2, (iii) are non-null persistent if $p < 1/2$ and (iv) are either transient or null persistent if $p \geq 1/2$.

6 Suppose $\{X_n, n \geq 0\}$ is a Markov chain with state space $S = W$ where the transition probabilities are given by

$$p_{i,i+1} = p, \;\; p_{i,i-1} = q = 1 - p \text{ if } i \geq 1, \;\; p_{01} = p, \;\; p_{00} = 1 - p, \;\; 0 < p < 1\,.$$

It is known as a random walk with elastic barriers or with partially reflecting boundary or barrier at 0. For this random walk, all states (i) communicate with all other states, (ii) are aperiodic, (iii) are non-null persistent if $p < 1/2$ and (iv) are either transient or null persistent if $p \geq 1/2$.

7 Suppose $\{X_n, n \geq 0\}$ is a Markov chain with state space $S = \{0, 1, \ldots, M\}$ and transition probabilities as specified below.

$$\begin{aligned} p_{i,i+1} &= p, \;\; p_{i,i-1} = q = 1 - p \;\; i = 1, 2, \ldots, M-1, \;\; 0 < p < 1 \\ p_{00} &= \delta_1, \;\; p_{01} = 1 - \delta_1 \;\; \& \;\; p_{MM} = \delta_2, \;\; p_{M,M-1} = 1 - \delta_2, \end{aligned}$$

$0 \leq \delta_1, \delta_2 \leq 1$. Then it is a random walk with finite state space. Depending on the values of δ_1 and δ_2, we get a random walk with absorbing barriers at one or both the boundaries, or a reflecting random walk, or a partially reflecting random walk at one or both the boundaries.

8 Gambler's ruin chain is a random walk with state space $S = \{0, 1, \ldots, M\}$ with absorbing barriers at 0 and M.

9 The probability of ruin of a gambler with initial capital a is given by

$$P_a = \begin{cases} \frac{(q/p)^a - (q/p)^M}{1-(q/p)^M}, & \text{if } \quad p \neq q \\ 1 - a/M, & \text{if } p = q = 1/2\,. \end{cases}$$

10 The expected duration of the game when the gambler has initial capital of a units is given by

$$D_a = \begin{cases} \frac{a}{q-p} - \frac{M}{q-p}\frac{1-(q/p)^a}{1-(q/p)^M}, & \text{if } \quad p \neq q \\ a(M-a), & \text{if } p = q = 1/2\,. \end{cases}$$

11 Ehrenfest urn model describing random diffusion of molecules through a membrane, is a random walk with finite state space $\{0, 1, \ldots, M\}$ and with reflecting barriers at 0 and M, where transition probabilities from state i depend on i. The

unique stationary distribution associated with the Ehrenfest model is binomial $B(M, 1/2)$.

12 Suppose $\{X_n, n \geq 0\}$ is a Markov chain with state space $S = W$ or $S = \{0, 1, \ldots, M\}$ and with the one step transition probabilities p_{ij} given by

$$p_{ij} = \begin{cases} q_i, & \text{if} \quad j = i - 1 \\ r_i, & \text{if} \quad i = j \\ p_i, & \text{if} \quad j = i + 1 \\ 0, & \text{otherwise,} \end{cases}$$

where $q_0 = 0$, $p_i, q_i, r_i \geq 0$ and $p_i + q_i + r_i = 1$. Then the Markov chain is known as a birth-death chain.

13 A birth-death chain with $S = W$ and $q_0 = 0$, $p_i, r_i, q_i > 0, \forall\ i$ has a stationary distribution if $\sum_{i \geq 1}(p_0 p_1 \cdots p_{i-1})/(q_1 q_2 \cdots q_i) < \infty$.

4.7 Conceptual Exercises

4.7.1 Suppose $\{X_n, n \geq 0\}$ is an unrestricted symmetric random walk, with state space I. Find (i) $\liminf_{n\to\infty} p_{66}^{(n)}$, (ii) $\limsup_{n\to\infty} p_{88}^{(n)}$, (iii) $p_{00}^{(8)}$ and $f_{00}^{(8)}$ (iv) $p_{00}^{(9)}$, $f_{00}^{(9)}$ and (v) approximate value of $p_{00}^{(40)}$ and $p_{00}^{(49)}$.

4.7.2 Suppose $\{X_n, n \geq 0\}$ is an unrestricted random walk, with state space I and $p = 1/3$. Find (i) $\liminf_{n\to\infty} p_{66}^{(n)}$, (ii) $\limsup_{n\to\infty} p_{88}^{(n)}$, (iii) $p_{00}^{(8)}$ and $f_{00}^{(8)}$, (iv) $p_{00}^{(9)}$, $f_{00}^{(9)}$ and (v) approximate value of $p_{00}^{(40)}$ and $p_{00}^{(49)}$.

4.7.3 Suppose $\{X_n, n \geq 0\}$ is a random walk with state space W and absorbing barrier at 0. Find as $n \to \infty$ (i) $\liminf p_{55}^{(n)}$, (ii) $\limsup p_{66}^{(n)}$, (iii) $\limsup p_{00}^{(n)}$ and (iv) $\limsup f_{00}^{(n)}$ and $\liminf f_{00}^{(n)}$.

4.7.4 Suppose $\{X_n, n \geq 0\}$ is a random walk with state space W and reflecting barrier at 0. If $p = 2/3$, find as $n \to \infty$ (i) $\lim p_{00}^{(n)}$ and (ii) $\limsup p_{77}^{(n)}$.

4.7.5 Suppose $\{X_n, n \geq 0\}$ is a random walk with state space W and reflecting barrier at 0. If $p = 1/3$, find the stationary distribution.

4.7.6 Suppose $\{X_n, n \geq 0\}$ is a random walk with state space W and partially reflecting barrier at 0. If $p = 2/3$, find as $n \to \infty$ (i) $\lim p_{00}^{(n)}$ and (ii) $\limsup p_{77}^{(n)}$.

4.7.7 Suppose $\{X_n, n \geq 0\}$ is a random walk with state space W and partially reflecting barrier at 0. If $p = \delta = 1/3$, find the stationary distribution.

4.7.8 Suppose $\{X_n, n \geq 0\}$ is a random walk with state space $S = \{0, 1, 2, 3\}$ and absorbing barriers at 0 and 3. The transition probability matrix P is as given below.

$$P = \begin{array}{c} \\ 0 \\ 1 \\ 2 \\ 3 \end{array} \begin{array}{c} \begin{array}{cccc} 0 & 1 & 2 & 3 \end{array} \\ \begin{pmatrix} 1 & 0 & 0 & 0 \\ 1/6 & 0 & 5/6 & 0 \\ 0 & 2/7 & 0 & 5/7 \\ 0 & 0 & 0 & 1 \end{pmatrix} \end{array}.$$

(i) Decide the nature of the states. (ii) Find the period of each state. (iii) Find the probability of absorption into $\{0\}$ and $\{3\}$ from states 1 and 2.

4.7.9 Suppose $\{X_n, n \geq 0\}$ is a random walk with state space is $\{0, 1, 2, 3, 4\}$ and partially reflecting barrier at 0 and 4. If $p = 1/4$, what is the long run proportion of time the random walk is in state 0 and state 4?

4.7.10 Suppose $\{X_n, n \geq 0\}$ is a random walk with partially reflecting barriers at 0 and M, that is, it is an irreducible Markov chain with finite state space $S = \{0, 1, 2, \ldots, M\}$ and a transition probability matrix $P = [p_{ij}]$ given by

$$p_{00} = 1 - p = q, \;\; p_{01} = p, \, p_{ij} = q \text{ if } j = i - 1 \;\&\; p_{ij} = p \text{ if } j = i + 1,$$

$i = 1, 2, 3, \ldots, M - 1$ and $p_{Mj} = q$ if $j = M - 1$, $p_{MM} = p$. Find a stationary distribution associated with this Markov chain.

4.7.11 Suppose Ajay and Vijay play a game of gambling. Ajay has probability 0.4 of winning at each flip of the coin. Suppose Ajay's initial capital is 2000 rupees while Vijay's initial capital is 3000 rupees. It is decided that the one who wins at the trial has to give 200 rupees to the other. (i) Compute the probability that Ajay will win the game. Find his expected gain. (ii) Compute the probability that Vijay will win the game. Find his expected gain. (iii) Find the expected duration of the game.

4.8 Computational Exercises

4.8.1 Suppose $\{X_n, n \geq 0\}$ is a random walk with the state space I. Obtain a realization of a random walk for a fixed number of transitions. Take $p = 1/2$, any $p < 1/2$ and any $p > 1/2$. Comment on your findings.

4.8.2 Suppose $\{X_n, n \geq 0\}$ is a random walk with the state space W and absorbing barrier at 0. Obtain a realization for a fixed number of transitions. Take $p = 1/2$, any $p < 1/2$ and any $p > 1/2$. Comment on your findings.

4.8.3 Suppose $\{X_n, n \geq 0\}$ is a random walk with the state space W and reflecting barrier at 0. Obtain a realization for a fixed number of transitions. Take $p = 1/2$, any $p < 1/2$ and any $p > 1/2$. Comment on your findings.

4.8.4 Suppose $\{X_n, n \geq 0\}$ is a random walk with the state space W and partially reflecting barrier at 0. Obtain a realization for a fixed number of transitions. Take $p = 1/2$, any $p < 1/2$ and any $p > 1/2$. Comment on your findings.

4.8.5 Obtain a realization of the Ehrenfest urn model with $M = 5$. Obtain a stationary distribution associated with the Ehrenfest model with $M = 5$ and verify whether it is the same as derived in Sect. 4.5.

4.8.6 Obtain a realization of a gambler's ruin model with total capital M units, initial capital of a gambler a units and bet of one unit at each trial. Take $p = 1/2$, any $p < 1/2$ and any $p > 1/2$. Comment on your findings.

4.9 Multiple Choice Questions

Note: In each of the questions, more than one option may be correct.

4.9.1 Which of the following options is/are correct? A simple random walk is a

(a) process with stationary and independent increments
(b) process where increments are always stationary
(c) process with independent increments
(d) Markov chain

Note: In questions 2 to 18, $\{X_n, n \geq 0\}$ is a simple random walk, with state space I and $p_{i,i+1} = p,\ 0 < p < 1$.

4.9.2 Which of the following options is/are correct?

(a) All states communicate with each other
(b) All states have period 2
(c) All states are null persistent if $p = 1/2$
(d) All states are transient if $p \neq 1/2$.

4.9.3 Which of the following options is/are correct?

(a) All states communicate with each other
(b) All states are aperiodic
(c) All states are null persistent if $p = 1/2$
(d) All are transient if $p \neq 1/2$.

4.9.4 Which of the following is NOT correct?

(a) All states communicate with each other
(b) All states are aperiodic
(c) All states are null persistent if $p = 1/2$
(d) All are transient if $p \neq 1/2$.

4.9.5 Which of the following options is/are correct?

(a) All states communicate with each other
(b) All states have period 2
(c) All states are non-null persistent if $p = 1/2$
(d) All are transient if $p \neq 1/2$.

4.9.6 Which of the following is NOT correct?

(a) All states communicate with each other
(b) All states have period 2
(c) All states are non-null persistent if $p = 1/2$
(d) All are transient if $p \neq 1/2$.

4.9.7 Which of the following options is/are correct?

(a) All states communicate with each other
(b) All states have period 2
(c) All states are transient if $p = 1/2$
(d) All states are null persistent if $p \neq 1/2$.

4.9.8 Which of the following options is/are correct?

(a) $p_{00}^{(2n)} = \binom{2n}{n} p^n q^n \ \forall \ n \geq 1$.
(b) $p_{00}^{(n)} = \binom{2n}{n} p^n q^n \ \forall \ n \geq 1$.
(c) $p_{00}^{(2n-1)} = \binom{2n-1}{n} p^n q^n \ \forall \ n \geq 1$.
(d) $p_{ii}^{(2n)} = \binom{2n}{n} p^n q^n \ \forall \ n \geq 1$, where $i \neq 0$ is any integer.

4.9.9 Which of the following options is/are correct?

(a) $p_{00}^{(2n)} \approx (4pq)^n/\pi n$
(b) $p_{00}^{(2n)} \approx (4pq)^n/\sqrt{n}$
(c) $p_{00}^{(2n)} \approx (4pq)^n/\sqrt{\pi n}$
(d) $p_{00}^{(2n)} \approx (pq)^n/\sqrt{\pi n}$

4.9.10 Which of the following options is/are NOT correct? As $n \to \infty$,

(a) $\lim p_{00}^{(n)} = 0$
(b) $\limsup p_{00}^{(n)} = 0$
(c) $\liminf p_{00}^{(n)} = 0$
(d) $\lim p_{00}^{(n)}$ exists but is not 0.

4.9.11 Following are two statements. (I) In a symmetric random walk all states are null persistent. (II) In a symmetric random walk all states are periodic. Then which of the following is a correct option?

(a) Both (I) and (II) are false
(b) Both (I) and (II) are true
(c) (I) is true but (II) is false
(d) (I) is false but (II) is true

4.9.12 Following are two statements. (I) In a symmetric random walk all states are non-null persistent. (II) In a symmetric random walk all states are periodic. Then which of the following is a correct option?

(a) Both (I) and (II) are false
(b) Both (I) and (II) are true

(c) (I) is true but (II) is false
(d) (I) is false but (II) is true

4.9.13 Following are two statements. (I) In a symmetric random walk all states are transient. (II) In a symmetric random walk all states are periodic.
Then which of the following is a correct option?

(a) Both (I) and (II) are false
(b) Both (I) and (II) are true
(c) (I) is true but (II) is false
(d) (I) is false but (II) is true

4.9.14 Suppose the random walk is not symmetric. Following are two statements. (I) All states are transient. (II) All states are periodic.
Then which of the following is a correct option?

(a) Both (I) and (II) are false
(b) Both (I) and (II) are true
(c) (I) is true but (II) is false
(d) (I) is false but (II) is true

4.9.15 Suppose the random walk is not symmetric. Following are two statements. (I) All states are non-null persistent. (II) All states are aperiodic.
Which of the following options is/are correct?

(a) Both (I) and (II) are false
(b) Both (I) and (II) are true
(c) (I) is true but (II) is false
(d) (I) is false but (II) is true

4.9.16 Suppose the random walk is not symmetric. Following are two statements. (I) All states are null persistent. (II) All states are aperiodic.
Then which of the following is a correct option?

(a) Both (I) and (II) are false
(b) Both (I) and (II) are true
(c) (I) is true but (II) is false
(d) (I) is false but (II) is true

4.9.17 Suppose the random walk is not symmetric. Following are two statements. (I) All states are null persistent. (II) All states are periodic.
Then which of the following is a correct option?

(a) Both (I) and (II) are false
(b) Both (I) and (II) are true
(c) (I) is true but (II) is false
(d) (I) is false but (II) is true

4.9.18 Following are two statements. (I) A unique stationary distribution of the unrestricted random walk exists. (II) A limiting distribution of the unrestricted random walk does not exist.
Then which of the following is a correct option?

(a) Both (I) and (II) are false
(b) Both (I) and (II) are true
(c) (I) is true but (II) is false
(d) (I) is false but (II) is true

Note: In questions 19 **to** 29, $\{X_n, n \geq 0\}$ **is a simple random walk, with state space** W **and** $p_{i,i+1} = p,\ 0 < p < 1$.

4.9.19 Suppose the random walk has absorbing barrier at 0. Which of the following options is/are correct?

(a) All states $i > 0$ communicate with states $j > 0$
(b) State 0 is aperiodic and non-null persistent state
(c) All states $i > 0$ have period 2
(d) All states $i > 0$ are transient.

4.9.20 Suppose the random walk has absorbing barrier at 0. Which of the following options is/are correct?

(a) All states $i > 0$ communicate with states $j > 0$
(b) State 0 is aperiodic and non-null persistent state
(c) All states $i > 0$ have period 2
(d) All states $i > 0$ are null persistent.

4.9.21 Suppose the random walk has absorbing barrier at 0. Then which of the following is NOT true?

(a) All states $i > 0$ communicate with states $j > 0$
(b) State 0 is aperiodic and non-null persistent state
(c) All states $i > 0$ have period 2
(d) All states $i > 0$ are null persistent.

4.9.22 Suppose the random walk has absorbing barrier at 0. Which of the following options is/are correct?

(a) All states $i > 0$ communicate with states $j > 0$
(b) State 0 is a non-null persistent state
(c) All states are aperiodic
(d) All states $i > 0$ are null persistent.

4.9.23 Suppose the random walk has absorbing barrier at 0. Which of the following options is/are correct?

(a) All states $i > 0$ communicate with states $j > 0$
(b) State 0 is aperiodic and non-null persistent state

(c) All states $i > 0$ have period 2
(d) All states $i > 0$ are inessential states.

4.9.24 Suppose the random walk has reflecting barrier at 0. Which of the following options is/are correct?

(a) All states communicate with all other states
(b) All states have period 2
(c) All states are non-null persistent if $p < 1/2$
(d) All states are either transient or null persistent if $p \geq 1/2$.

4.9.25 Suppose the random walk has reflecting barrier at 0. Which of the following options is/are correct?

(a) All states communicate with all other states
(b) All states have period 2
(c) All states are non-null persistent if $p \geq 1/2$
(d) All states are either transient or null persistent if $p < 1/2$.

4.9.26 Suppose the random walk has reflecting barrier at 0. Which of the following options is/are correct?

(a) All states communicate with all other states
(b) All states are aperiodic
(c) All states are non-null persistent if $p < 1/2$
(d) All states are either transient or null persistent if $p \geq 1/2$.

4.9.27 Suppose the random walk has elastic barrier at 0. Which of the following options is/are correct?

(a) All states communicate with all other states
(b) All states are aperiodic
(c) All states are non-null persistent if $p < 1/2$
(d) All states are either transient or null persistent if $p \geq 1/2$.

4.9.28 Suppose the random walk has elastic barrier at 0. Which of the following options is/are correct?

(a) All states communicate with all other states
(b) All states are aperiodic
(c) All states are non-null persistent if $p \geq 1/2$
(d) All states are either transient or null persistent if $p < 1/2$.

4.9.29 Suppose the random walk has elastic barrier at 0. Which of the following options is/are correct?

(a) All states communicate with all other states
(b) All states have period 2
(c) All states are non-null persistent if $p < 1/2$

(d) All states are either transient or null persistent if $p \geq 1/2$
Note: In questions 30 to 39, $\{X_n, n \geq 0\}$ is a Gambler's ruin chain with state space $\{0, 1, \ldots, M\}$.

4.9.30 Which of the following options is/are correct?

(a) All states communicate with all other states
(b) All states are non-null persistent
(c) State 0 and M are non-null persistent and other states are transient
(d) State 0 and M are non-null persistent and other states are null persistent.

4.9.31 Which of the following options is/are correct?

(a) All states are periodic
(b) All states are aperiodic
(c) State 0 and M are periodic and other states are aperiodic
(d) State 0 and M are aperiodic and other states are periodic.

4.9.32 Following are three statements: (I) States $\{1, 2, \ldots, M-1\}$ have period 2. (II) State 0 is aperiodic. (III) State M is aperiodic.
Which of the following options is correct?

(a) Only (I) is true
(b) Only (II) is true
(c) Only (III) is true
(d) (I), (II) and (III) are true.

4.9.33 Following are three statements : (I) $\underline{\pi}_1 = (0, 0, 0, \ldots, 0, 1)$ is a stationary distribution. (II) $\underline{\pi}_2 = (1, 0, 0, \ldots, 0, 0)$ is a stationary distribution.
(III) $\underline{\pi}_3 = (\alpha, 0, 0, \ldots, 0, 1-\alpha)$, where $0 < \alpha < 1$, is a stationary distribution.
Which of the following is a correct option?

(a) Only (I) is true
(b) Only (II) is true
(c) Only (III) is true
(d) (I), (II) and (III) are true.

4.9.34 Suppose P_a denotes the probability of ultimate ruin of a gambler with initial capital a. Which of the following options is/are correct?

(a) $P_a = \frac{(q/p)^a-(q/p)^M}{1-(q/p)^M}$ if $p \neq q$
(b) $P_a = \frac{1-(q/p)^a}{1-(q/p)^M}$ if $p \neq q$
(c) $P_a = \frac{(q/p)^M-(q/p)^a}{1-(q/p)^M}$ if $p \neq q$
(d) $P_a = \frac{1-(q/p)^M}{1-(q/p)^a}$ if $p \neq q$.

4.9.35 Which of the following options is/are correct? The expected gain of a gambler with initial capital a units and $p = 1/2$

(a) $M - a$
(b) M
(c) a
(d) 0.

4.9.36 Suppose probability that gambler wins the bet is p. Following are two statements. (I) The expected gain of a gambler with initial capital a units is 0, if $p = 1/2$. (II) If the expected gain of a gambler with initial capital a units is 0, then $p = 1/2$. Then which of the following is a correct option?

(a) Both (I) and (II) are false
(b) Both (I) and (II) are true
(c) (I) is true but (II) is false
(d) (I) is false but (II) is true.

4.9.37 Which of the following options is/are correct? The expected duration of the game when the gambler has initial capital of a units is

(a) $a(M - a)$ if $p > q$
(b) $a(M - a)$ if $p = q$
(c) $a(M - a)$ if $p < q$
(d) $M(M - a)$ if $p = q$.

4.9.38 Which of the following options is/are correct? The expected duration of the game when the gambler's initial capital is a and total capital of two players is infinite is given by

(a) $a/(q - p)$ if $p > q$
(b) $a/(q - p)$ if $p < q$
(c) ∞ if $p = q$
(d) ∞ if $p > q$.

4.9.39 Which of the following options is/are correct? The Ehrenfest chain on $\{0, 1, 2, \ldots, M\}$ is

(a) irreducible
(b) aperiodic
(c) non-null persistent
(d) has infinitely many stationary distributions.

4.9.40 Which of the following options is/are correct? The Ehrenfest chain on $\{0, 1, 2\}$ has

(a) $p_{11} = 1/2$
(b) a unique stationary distribution is given by $(1/4, 1/2, 1/4)'$
(c) a unique stationary distribution is given by $(1/2, 0, 1/2)'$.
(d) $f_{12} = 1/2$.

References

1. Ash, R. B. (2008). *Basic probability theory*. New York: Dover Publications.
2. Bhat, B. R. (2000). *Stochastic models: Analysis and applications*. New Delhi: New Age International.
3. Feller, W. (1978). *An introduction to probability theory and its applications* (Vol. I). New York: Wiley.
4. Feller, W. (2000). *An introduction to probability theory and its applications* (2nd ed., Vol. II). Singapore: Wiley.
5. Hoel, P. G., Port, S. C., & Stone, C. J. (1972). *Introduction to stochastic processes*. Wiley Eastern.
6. Karlin, S., & Taylor, H. M. (1975). *A first course in stochastic processes*. New York: Academic.

Chapter 5
Bienayme Galton Watson Branching Process

5.1 Introduction

Branching processes are special types of Markov chains that are widely used as models in disciplines such as biological, social and engineering sciences. These have been applied to problems of genetics, nuclear fission, queuing theory and demography. Epidemiological studies also use branching processes to estimate the spread of an epidemic and to decide a proportion of population to be vaccinated. More recently, branching processes have been successfully used to illuminate problems in the areas of molecular biology, cell biology, developmental biology, immunology, evolution, ecology, medicine and others. These processes are studied extensively in literature and have been generalized in many ways. The present chapter is devoted to Bienayme Galton Watson (BGW) branching process, which is the simplest version of the branching process.

Branching processes model the population growth. Suppose we have a population consisting of particles, which may be individuals, cells, molecules, neutrons or electrons. These particles live for a random duration, and at some point of time, during their lifetime or at the moment of death, they produce a random number of offspring of the same kind. Processes allowing production of new individuals during a parent's lifetime, for example, populations of higher organisms, like vertebrates and plants, are called general branching processes. Processes assuming that the offsprings are produced at the terminal point of the parent entity's lifetime, for example, populations of biological cells, genes or bio-molecules, are called the classical branching processes. The nature and the analysis of the process depend on the distribution of particle's lifetime T. In a classical BGW branching process, T is assumed to be degenerate at 1. If the distribution of T is exponential, then the resulting process is called a Markov branching process. If T is an arbitrary non-negative random variable, then the resulting process is called an "age-dependent" or Bellman-Harris process.

The study of branching processes started with the problem of finding the chance of extinction or survival of family names. It was first studied by Bienayme [5], in order

S. Madhira and S. Deshmukh, *Introduction to Stochastic Processes Using R*,
https://doi.org/10.1007/978-981-99-5601-2_5

to find a mathematical, rather than social or genetic, explanation for the fact that a large proportion of the family names seemed to be dying out, when viewed over a long period of time. This problem was further studied by Galton and Watson [6]. Francis Galton in 1873 proposed the following problem. Suppose a male member of a family, labeled as an ancestor, produces j male offspring according to a certain probability distribution. These male offspring carry the surname of the ancestor. It is assumed that these male offspring produce according to the same probability distribution. The problem of interest was to find the chance of survival of family names or family tree. In honor of these three scientists, a branching process, as defined below, is referred to as BGW branching process. We study various aspects of this process in the present chapter.

We now define a BGW branching process and discuss some of its applications.

Definition 5.1.1 BGW Branching Process: Suppose $\{Y_{n,i}, n = 0, 1, 2, \ldots; i = 1, 2, \ldots\}$ is a double array of independent and identically distributed, non-negative integer valued random variables, such that for $n = 0, 1, 2, \ldots$ and $i = 1, 2, \ldots,$

$$P[Y_{n,i} = j] = p_j, \ j = 0, 1, \ldots, \quad p_j \ge 0 \ \ \& \ \sum_{j \ge 0} p_j = 1 .$$

Suppose $Z_0 = k \ge 1$ and Z_n for $n \ge 1$ is defined by,

$$Z_n = \begin{cases} \sum_{i=1}^{Z_{n-1}} Y_{n-1,i}, & \text{if } Z_{n-1} > 0 \\ 0, & \text{if } Z_{n-1} = 0 . \end{cases}$$

Then the stochastic process $\{Z_n, n \ge 0\}$ is known as a BGW branching process. If $Z_0 = 1$, it is referred to as a simple branching process.

Thus, a BGW branching process models the evolution of a population, which develops according to the following scheme, across various generations:

(i) Suppose there exists k individuals, known as ancestors, who constitute the 0th generation, with size $Z_0 = k$.
(ii) An ancestor, independently of others, produces j offspring with probability p_j, $j = 0, 1, 2, \ldots,$ by the end of its lifetime, $p_j \ge 0$ and $\sum_{j \ge 0} p_j = 1$.
(iii) All offspring produced by the ancestors at zero-th generation constitute the first generation and its size is denoted by Z_1. The individuals of the first generation produce independently of each other and with the same probability distribution $\{p_j, j = 0, 1, 2, \ldots,\}$.
(iv) In general, individuals of the nth generation produce independently of each other and with the same probability distribution $\{p_j, j = 0, 1, 2, \ldots,\}$ and the direct descendents of the nth generation individuals constitute the $(n + 1)$th generation.
(v) If at any time n, $Z_n = 0$ then $Z_m = 0$ for all $m > n$.

Then $\{Z_n, n \geq 0\}$ is a BGW branching process, where Z_n denotes the size of the population at the nth generation.

The probability mass function of Z_n for each $n \geq 1$ can be obtained in terms of the probability distribution $\{p_j, j \geq 0\}$. Hence, it has a special nomenclature, which is defined below.

Definition 5.1.2 Offspring Distribution: The probability distribution $\underline{p} = \{p_j, j = 0, 1, 2, \ldots, \}$ is known as the offspring distribution of the branching process.

Following figure shows a realization of a branching process when $Z_0 = 2$.

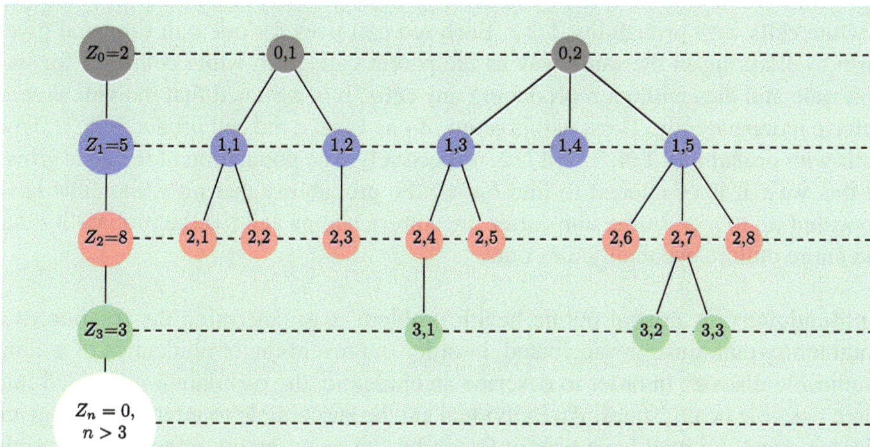

Figure-1: A realization of the BGW process with $Z_0 = 2$

In the above figure (i, j) enclosed in a circle indicate jth individual in ith generation. The first individual in 0th generation produces 2 offspring, while the second produces 3 offspring, hence the size of the first generation is 5. Three individuals in the third generation do not produce any offspring, hence Z_4 and all further generation sizes are 0.

From the above figure, it is clear that the branching process with k ancestors can be viewed as the sum of k independent branching processes, each with one ancestor and the same offspring distribution. More precisely, we have the following lemma.

Lemma 5.1.1 *Suppose $\{Z_n, n \geq 0\}$ is a BGW branching process with $Z_0 = k$ and offspring distribution $\underline{p}$. For $r = 1, 2, \ldots, k$, suppose $Z_n^{(r)}$ denotes the size of the nth generation, corresponding to the rth individual in the 0th generation. Then,*

(i) $\{Z_n^{(r)}, n \geq 0\}$ *is a BGW branching process with* $Z_0^{(r)} = 1$ *and with offspring distribution* $\underline{p}$, *for* $r = 1, 2, \ldots, k$.

(ii) $\{Z_n^{(r)}, n \geq 0\}$ *are independent processes, for* $r = 1, 2, \ldots, k$.

(iii) $Z_n = \sum_{r=1}^{k} Z_n^{(r)}$, *for* $n = 0, 1, \ldots$.

(iv) If $Z_m = 0$ *for some m, then* $Z_n = 0 \;\forall\; n > m$.
(v) $[Z_m = 0] = [Z_m^{(r)} = 0, \;\forall\; r = 1, 2, \ldots, k.]$

In view of this lemma, all the results in subsequent sections are proved when $Z_0 = 1$, from which results for $Z_0 = k > 1$ follow.

There are numerous examples of branching processes that arise naturally in various scientific disciplines. We elaborate below on some of the more prominent cases.

Cell biology: Suppose at time 0, a blood culture starts with one red cell. At the end of one unit time, the red cell dies and is replaced by one of the combinations—(i) 2 red cells with probability $1/4$, (ii) 1 red and 1 white cells with probability $2/3$ and (iii) 2 white cells with probability $1/12$. Each red cell lives for one unit time and gives birth to offspring in the same way as the parent cell. Each white cell lives for one unit time and dies without reproducing any cells. It is assumed that individual cells behave independently. Thus, in this setup, $Z_0 = 1$ and a red cell produces $2, 1, 0$ red cells with probability $1/4, 2/3, 1/12$, respectively. The population of red cells grows in this way. It is of interest to find out (i) the probability that no white cells have appeared up to $n + 1$ time units after the culture begins and (ii) the probability that the entire culture eventually dies out.

Epidemiology: A crucial public health problem is to determine the fraction of a community that must be vaccinated, in order to prevent major epidemics of a communicable disease. In order to describe an epidemic, the population is divided into three possible health states. An individual can be susceptible to infection by a given disease agent, he may have been infected by the agent and is infectious (possibly after a latent period), or he is removed from the epidemic by death, by isolation, or by immunity or other natural loss of infectiousness. Initially, all members of the population are susceptible to infection. The epidemic starts when one or many infectious individuals enter the population and come into contact with its members. A susceptible person is infected if he has adequate contact with an infectious individual. Individuals infected by an infectious individual are termed as offspring. Branching processes have conventionally been used to model the spread of the infectious diseases, refer to the book Bailey [3]. This modeling proves to be useful for estimating the growth rate of the epidemic. The severity of the epidemic can be decided on the basis of the growth rate. Further, estimation of the extinction probability enables us to decide, whether the disease will be completely wiped out from the given closed community. If the estimate of extinction probability turns out to be less than one, we can use the estimate of offspring mean to determine the proportion of population that needs to be vaccinated, so as to make the extinction certain. It has been discussed by Becker [4].

A second order branching process, that is, a second order Markov chain satisfying certain property, known as the branching property, has been used to model the Swine flu data, Kashikar and Deshmukh [11]. The data consist of number of cases tested positive on each day in Pune and La-Gloria, Mexico. For Pune, the data were recorded from July 15 to August 4, 2009. In case of La-Gloria, Mexico, the data were recorded from March 15 to April 30, 2009. Both the data sets were modeled by a second order branching process. Some parameters of interest were estimated on the basis of available data. Growth rates estimated on the basis of given data turn out to be larger than 1 for both the data sets, indicating that the epidemic may not die out unless some efforts are taken to curb its spread. In fact in Pune, at the later stage, the disease spread rapidly and the number of cases reported in a single day reached 97, on August 10, 2009. In Pune, as well as in Mexico, respective Governments adopted some ways such as closing down the educational institutes, theaters and malls etc. to bring the epidemic in control, which turned out to be fruitful. The proportion of population required to be vaccinated, to guarantee the elimination of the disease was also obtained. For Pune data it was 6.16%, while for Mexico data it was 1.59%.

Social science: Branching processes are also used to model spread of a rumor or innovation, the offspring being the recipient of the rumor originally started by an individual, termed as an ancestor. In the early 1950, the Washington Public Opinion Laboratory in Seattle carried out a project to study the diffusion of information. A small experiment took place in a village with 210 housewives. Forty-two of these were told a slogan about a particular brand of coffee. Thus, $Z_0 = 42$. Each housewife was asked to pass on the information. As an incentive, participants were told that they would get a free pound of coffee if they knew the slogan when an interviewer called 48 hours later. It was possible to trace the route by which each hearer had obtained the slogan, so that they could be classified by generations. The data are given in Table 5.1.

Queuing theory: Suppose an individual arriving at a service counter is lucky in the sense that there is no queue and hence he immediately gets service. During his service period, suppose j individuals arrive according to a certain probability distribution and join the queue. We label the lucky customer as an ancestor and individuals arriving during his service period as his offspring. Such a model is useful to obtain the distribution of a busy period of a server.

Physics: Branching processes have applications in physics to model nuclear chain reaction. This application became familiar in connection with atomic bomb. A nucleus is split by a chance collision with a neutron. The resulting fission yields a random number of new neutrons. Each of these secondary neutrons may hit some other nucleus, producing a random number of additional neutrons and so forth. In this case, the initial number of neutrons is $Z_0 = 1$. The first generation of neutrons comprises all those produced from the fission caused by the initial neutron. The size of the first generation is a random variable Z_1. In general, the population Z_n at the nth generation is produced by the chance hits of the individual neutrons of the $(n-1)$th

Table 5.1 Spread of a slogan

Generation	1	2	3	4	5
Size	69	53	14	2	4

generation. Suppose p denotes the probability that a neutron is hit by any other particle and produces m particles, these m particles are considered to be the offspring or descendents of the original neutron. With probability $1-p$, the particle remains inactive and does not have descendents. It may happen that the first particle remains inactive and the process never starts, on the other hand, there will be m particles in the first generation, m^2 in the second generation and so on. If p is close to 1, then the number of particles increases very rapidly.

Another application in physics is in the study of electron multipliers. An electron multiplier is a device that amplifies a weak current of electrons. A series of plates are set up in the path of electrons emitted by a source. Each electron, as it strikes the first plate, generates a random number of new electrons, which in turn strike the next plate and produce more electrons and so forth. Suppose Z_0 denotes the number of electrons initially emitted, Z_1 denotes the number of electrons produced on the first plate by the impact due to the Z_0 initial electrons. In general, suppose Z_n is the number of electrons emitted from the nth plate due to electrons emanating from the $(n-1)$th plate. The sequence of random variables $\{Z_n, n \geq 0\}$ constitutes a branching process.

The book by Harris [7] gives a complete exposure to the developments in the field of branching processes, up to that time, whereas the book by Athreya and Ney [2] presents a unified treatment of the theory of branching processes and various related limit theorems. The book by Jagers [8] discusses BGW process and its many variations while the book by Asmussen and Hering [1] also discusses many applications of the branching processes.

Branching processes described above assume that all the individuals are statistically identical and reproduce according to the same probability distribution. A natural generalization of the BGW process is to allow for a number of distinguishable particles/individuals, having different probabilistic behavior. Such processes, named as multi-type branching process are discussed in detail by Mode [13]. Both the BGW and multi-type branching processes assume that the offspring distributions are time homogeneous. However, there are situations where the offspring distribution depends on some external factors like environment, which may change with time. Branching processes in random environments, discussed by Smith and Wilkinson [15], allow for different offspring distributions for different generations. Controlled branching process, as another generalization of BGW process, was introduced by Sevast'yanov and Zubkov [14].

Branching processes are mainly used for modeling population growth. Therefore, the state space is naturally the set of non-negative integers. Later on, the theory was extended to include the continuous state space. A monograph by Kallenberg [9] focuses on branching processes with continuous state space.

In the present chapter, we study in detail the BGW branching process. In Sect. 5.2, it is shown that $\{Z_n, n \geq 0\}$ is a Markov chain with set of non-negative integers as its state space. We prove that it is a non-ergodic Markov chain, where 0 is an absorbing state and under some reasonable conditions, all other states are transient. A BGW branching process is one of the simplest non-ergodic stochastic models. Further, it is a Markov chain with a typical property, known as the branching property. We discuss it in Sect. 5.3. It is of special interest to study the probability of extinction, that is, probability of ultimate absorption into the state 0, since it has wide variety of applications. Section 5.4 is devoted to the concept of extinction probability. Due to this feature of extinction of a population, although a branching process is a Markov chain, it is studied separately. Section 5.5 is concerned with a realization of a branching process and computation of extinction probability graphically and algebraically. Section 5.6 presents R codes used in solving examples in previous sections.

5.2 Markov Property

In the present section, we show that a BGW process is a Markov chain with state space W.

Theorem 5.2.1 *Suppose $\{Z_n, n \geq 0\}$ is a BGW branching process with offspring distribution $\underline{p}$. Then it is a time homogeneous Markov chain with state space $S = W$.*

Proof From Definition 5.1.1, observe that

(i) If $Z_n = 0$ then $Z_{n+1} = 0$ with probability 1.
(ii) If $Z_n > 0$ then $Z_{n+1} \stackrel{d}{=} \sum_{k=1}^{Z_n} Y_{n,k}$. Hence, $\forall\ z_1, z_2, \ldots, z_{n-1}, i, j \in S$ and $\forall\ n \geq 1$,

$$P[Z_{n+1} = j|Z_n = i, Z_{n-1} = z_{n-1}, \ldots, Z_0 = 1]$$

$$= P[Y_{n,1} + \cdots + Y_{n,i} = j|Z_n = i, Z_{n-1} = z_{n-1}, \ldots, Z_0 = 1] \quad (5.2.1)$$
$$= P[Y_{n,1} + \cdots + Y_{n,i} = j] \quad (5.2.2)$$
$$= P[Z_{n+1} = j|Z_n = i].$$

Note that in Eq. (5.2.1), in the conditional probability $P(A|B)$, event A is defined in terms of random variables $Y_{n,k}, k = 1, 2, \ldots, Z_n$ and event B is defined in terms of random variables $Y_{r,k}, r = 0, 1, \ldots, n-1; k = 1, 2, \ldots$. Since all $Y_{r,k}$'s are independent random variables, events A and B are independent. Hence, we get Eq. (5.2.2). Thus, $\{Z_n, n \geq 0\}$ is a Markov chain with state space W. Suppose p_{ij} denotes the one step transition probability of transition from state i to state j. Then by definition

$$p_{00} = P[Z_{n+1} = 0|Z_n = 0] = 1$$
$$\& \quad p_{ij} = P[Z_{n+1} = j|Z_n = i] = P[Y_{n,1} + Y_{n,2} + \cdots + Y_{n,i} = j], \quad i > 0.$$

In view of the assumption that all individuals in any generation produce according to the same offspring distribution, the distribution of $Y_{n,i}$ does not depend on $n \geq 0$. Thus, p_{ij} does not depend on n and hence the Markov chain is time homogeneous. In this setup we write p_{ij} as $p_{ij} = P[Y_1 + Y_2 + \cdots + Y_i = j]$, where $\{Y_n, n \geq 1\}$ is a sequence of independent and identically distributed random variables with the probability mass function $\underline{p}$. □

Suppose P denotes the transition probability matrix of a BGW branching process, which is a time homogeneous Markov chain. From $p_{00} = 1$, it follows that the first row of P is $(1, 0, 0, \ldots)$. The second row is the same as the offspring distribution $\underline{p}$. For $k \geq 2$, the kth row is the k-fold convolution of the offspring distribution. It is to be noted that the knowledge of the offspring distribution is sufficient to obtain the transition probabilities. Thus, the role of the initial distribution and the transition probability matrix in a general Markov chain, is played by the offspring distribution of a BGW branching process.

Although p_{ij} can be obtained from the offspring distribution, it is usually not easy to find the i-fold convolution. Another approach is to find the transition probabilities from the probability generating function of $Y_1 + Y_2 + \cdots + Y_i$. Suppose $P(s) = \sum_{j \geq 0} p_j s^j$, $|s| \leq 1$, denotes the probability generating function of the offspring distribution. If $Z_0 = 1$, then the probability distribution of Z_1 is the same as the offspring distribution $\underline{p}$. Thus, $P(s)$ is also the probability generating function of Z_1 when $Z_0 = 1$. The generating function of $\{p_{ij}, j \geq 0\}$ for each fixed $i \geq 1$ is given by

$$\begin{aligned}\sum_{j\geq 0} p_{ij}s^j &= \sum_{j\geq 0} P[Z_{n+1} = j | Z_n = i] s^j \\ &= \sum_{j\geq 0} P[Y_1 + Y_2 + \cdots + Y_i = j] s^j = (P(s))^i\,, \qquad (5.2.3)\end{aligned}$$

since $\{Y_i, i \geq 1\}$ is a sequence of independent and identically distributed random variables each having the same probability generating function $P(s)$. Hence, p_{ij} is the coefficient of s^j in the expansion of $(P(s))^i$. In particular,

$$\begin{aligned}p_{i0} &= \text{coefficient of } s^0 \text{ in the expansion of } (P(s))^i \\ &= \text{constant term in the expansion of } (p_0 + p_1 s + p_2 s^2 + \cdots)^i = p_0^i\,.\end{aligned}$$

Following examples illustrate that once we know the initial generation size and the offspring distribution, we can compute the probabilities of events of interest.

Example 5.2.1 Suppose the offspring distribution has the probability generating function $P(s) = 0.2 + 0.3s + 0.3s^2 + 0.2s^3$. Then p_{ij} is the coefficient of s^j in the expansion of $(0.2 + 0.3s + 0.3s^2 + 0.2s^3)^i$, which can be obtained using the multinomial theorem. Thus,

$$\begin{aligned}&(0.2 + 0.3\,s + 0.3\,s^2 + 0.2\,s^3)^i \\ &= \sum_{i_1,i_2,i_3,i_4} \frac{i!}{i_1!i_2!i_3!i_4!}(0.2)^{i_1}(0.3\,s)^{i_2}(0.3\,s^2)^{i_3}(0.2\,s^3)^{i_4} \\ &= \sum_{i_1,i_2,i_3,i_4} \frac{i!}{i_1!i_2!i_3!i_4!}(0.2)^{i_1}(0.3)^{i_2}(0.3)^{i_3}(0.2)^{i_4}s^{i_2+2i_3+3i_4}\,,\end{aligned}$$

where i_1, i_2, i_3, i_4 are non-negative integers such that $i_1 + i_2 + i_3 + i_4 = i$. Hence, to compute p_{ij}, we set $j = i_2 + 2i_3 + 3i_4$ and $i_1 = i - (i_2 + i_3 + i_4)$. Thus, the values of i_1, i_2, i_3, i_4 are decided by the values of i and j. For example, to compute $P[Z_2 = 0|Z_1 = 3] = p_{30}$, $i = 3$ and $j = 0$, so that $i_1 = 3$ and $i_2 = i_3 = i_4 = 0$. Thus, $p_{30} = (0.2)^3 = 0.008$. Observe that if none of the three individuals in the first generation produces an offspring, the second generation size is 0. Since individuals in any generation produce independently of each other and according to the same offspring distribution, $p_{30} = (0.2)^3 = 0.008$. Once the values of i_1, i_2, i_3, i_4 are decided, we can use `dmultinom` function from `R` to compute the probabilities. Code 5.6.1 computes such probabilities. With this code we have $p_{30} = p_{39} = 0.008$ and $p_{31} = 0.036$. Note that $p_{3j} = 0$ for all $j \geq 10$. □

In the next example, we express Z_n in terms of Y_i to compute the required probabilities.

Example 5.2.2 Suppose in a BGW process, $Z_0 = 1$ and the offspring distribution is given by, $p_0 = 1/4$, $p_1 = 2/4$ and $p_2 = 1/4$. Then

$$\begin{aligned} P[Z_2 = 3|Z_1 = 2] &= P[Y_1 + Y_2 = 3] \\ &= P[Y_1 = 1, Y_2 = 2] + P[Y_1 = 2, Y_2 = 1] \\ &= P[Y_1 = 1]P[Y_2 = 2] + P[Y_1 = 2]P[Y_2 = 1] = 1/4. \end{aligned}$$

$$\begin{aligned} \text{Similarly,} \quad P[Z_3 = 2|Z_2 = 3] &= P[Y_1 + Y_2 + Y_3 = 2] \\ &= p_0p_0p_2 + p_0p_2p_0 + p_2p_0p_0 + p_1p_1p_0 \\ &\quad + p_1p_0p_1 + p_0p_1p_1 = 15/64. \end{aligned}$$

$$\begin{aligned} P[Z_2 = 2] &= P[Z_2 = 2|Z_1 = 1]P[Z_1 = 1] + P[Z_2 = 2|Z_1 = 2]P[Z_1 = 2] \\ &= P[Y_1 = 2]P[Z_1 = 1] + P[Y_1 + Y_2 = 2]P[Z_1 = 2] \\ &= p_2p_1 + (p_0p_2 + p_1p_1 + p_2p_0)p_2 = 7/32. \qquad \square \end{aligned}$$

Suppose $\{Z_n, n \geq 0\}$ is a BGW branching process with offspring distribution $\underline{p}$. We classify the states of the branching process as persistent or transient, under the assumption that $0 < p_0 < 1$. This condition seem to be reasonable because (i) if $p_0 = 0$, that is, if probability that the individual gives birth to no offspring is 0, then the population will explode and extinction will never happen and (ii) if $p_0 = 1$, then population will not develop at all and extinction will occur at the first generation itself. Thus, to model various real life situations, we assume $0 < p_0 < 1$.

Theorem 5.2.2 *Suppose $\{Z_n, n \geq 0\}$ is a branching process with offspring distribution $\underline{p}$, where $0 < p_0 < 1$. Then the state 0 is a non-null persistent, aperiodic state and all other states are transient.*

Proof In a branching process, if $Z_n = 0$, then $Z_{n+1} = 0$ with probability 1, that is, 0 is an absorbing state and hence is non-null persistent. Since $p_{00} = 1$, it is aperiodic. Note that since $p_0 < 1$, states $i > 0$ are feasible. Further, for any state $i > 0, \quad p_{i0} = p_0^i > 0$ since $p_0 > 0$. Now $p_{i0} > 0 \quad \Rightarrow \quad i \to 0$. However, $0 \nrightarrow i$. Thus, each state $i > 0$ is an inessential state and hence is a transient state. Moreover, if $i > 0$ and $j > 0$, then $i \leftrightarrow j$ and all states are of the same nature. Thus, all states $i > 0$ are transient states. We can also compute a bound on f_{ii} for $i > 0$, leading to the same conclusion. For a state $i > 0$, using the fact that $f_{ii}^{(1)} = p_{ii}$, we have

$$\begin{aligned}
f_{ii} &= \sum_{n=1}^{\infty} f_{ii}^{(n)} = p_{ii} + \sum_{n=2}^{\infty} f_{ii}^{(n)} = p_{ii} + \sum_{n=2}^{\infty} \left\{ \sum_{k=1,k\neq i}^{\infty} p_{ik} f_{ki}^{(n-1)} \right\} \\
&= p_{ii} + \sum_{n=2}^{\infty} \left\{ \sum_{k=1,k\neq 0,i}^{\infty} p_{ik} f_{ki}^{(n-1)} \right\}, \text{ as } f_{0i}^{(n-1)} = 0 \ \forall \ i > 0 \\
&= p_{ii} + \sum_{k=1,k\neq 0,i}^{\infty} p_{ik} \left\{ \sum_{n=2}^{\infty} f_{ki}^{(n-1)} \right\} = p_{ii} + \sum_{k=1,k\neq 0,i}^{\infty} p_{ik} \left\{ \sum_{n=1}^{\infty} f_{ki}^{(n)} \right\} \\
&\leq p_{ii} + \sum_{k=1,k\neq 0,i}^{\infty} p_{ik}, \text{ as } \sum_{n=1}^{\infty} f_{ki}^{(n)} \leq 1 \\
&= \sum_{k=1}^{\infty} p_{ik} = 1 - p_{i0} = 1 - p_0^i < 1, \text{ since } p_0 > 0 \\
\Rightarrow \quad f_{ii} &< 1.
\end{aligned}$$

In the above derivation, interchange of series is permissible as it is a series of non-negative numbers and is convergent with value at most 1. As $f_{ii} < 1$, state $i > 0$ is a transient state. Thus, the state 0 is an ergodic state and all other states are transient. □

We now examine whether a long-run distribution exists for a BGW branching process. We have noted that 0 is an absorbing state and hence it is a non-null persistent aperiodic state with $\mu_0 = 1$ and all states $j > 0$ are transient states. Observe that, by (i) of Theorem 3.2.2,

$$\begin{aligned}
\text{For } i > 0, \ i \to 0 &\Rightarrow \lim_{n\to\infty} p_{i0}^{(n)} = f_{i0}/\mu_0 = f_{i0} \\
0 \text{ is an absorbing state} &\Rightarrow \lim_{n\to\infty} p_{00}^{(n)} = 1 \\
\text{For } i, \ j > 0, \ i \to j &\Rightarrow \lim_{n\to\infty} p_{ij}^{(n)} = 0, \text{ since } j \text{is transient} \\
\text{For } j > 0, \ 0 \nrightarrow j &\Rightarrow \lim_{n\to\infty} p_{0j}^{(n)} = 0 \\
\text{Thus, } \lim_{n\to\infty} p_{ij}^{(n)} &= 0 \ \forall \ i \geq 0, \ j > 0 \ \& \ \lim_{n\to\infty} p_{i0}^{(n)} = f_{i0} \ \forall \, i \geq 0.
\end{aligned}$$

Thus, $\lim_{n\to\infty} p_{ij}^{(n)}$ exists and but depends on i, since f_{i0} may not be 1 $\forall\ i$. Hence, the long-run distribution does not exist.

We have noted in Chap. 3 that even if a long-run distribution does not exist, a stationary distribution may exist. To find it, observe that $\{0\}$ is the only closed communicating class and all states $i > 0$ are transient states. Hence, by Theorem 3.3.3, $\underline{\pi}$ where $\pi_0 = 1$ and $\pi_i = 0\ \forall\ i > 0$ is the unique stationary distribution. Further, note that $p_{00} = 1$ implies that the first row of the transition probability matrix P is $(1, 0, 0, \ldots,)$. It is easy to verify that $\underline{\pi}P = \underline{\pi}$.

The next section is devoted to a derivation of an important property of a branching process, known as branching property.

5.3 Branching Property

In the previous section, we have noted that a branching process is a Markov chain with state space W. The one step transition probabilities are expressed in terms of convolution of the offspring distribution. It is a typical property of a branching process and we investigate it in more detail in the present section. In the development of the theory of a branching process, the probability generating function plays a key role. We have already used it to obtain p_{ij}. Suppose $P_n(s)$ denotes the probability generating function of Z_n conditional on $Z_0 = 1$, then by definition,

$$P_n(s) = \sum_{i=0}^{\infty} P[Z_n = i|Z_0 = 1]s^i = \sum_{i=0}^{\infty} p_{1i}^{(n)} s^i \ .$$

The following recurrence relation is useful to find the higher step transition probabilities and also in the proof of a theorem related to the extinction probability.

Theorem 5.3.1 *Suppose $\{Z_n, n \geq 0\}$ is a branching process with $P(\cdot)$ as the probability generating function of the offspring distribution and $Z_0 = 1$. Then $\forall\ n \geq 0$, $(i) P_{n+1}(s) = P_n(P(s))$ and $(ii) P_{n+1}(s) = P(P_n(s))$.*

Proof (i) From the definition of $P_n(s)$, it follows that $P_0(s) = s$. Observe that

$$P_1(s) = \sum_{i=0}^{\infty} P[Z_1 = i|Z_0 = 1]s^i = \sum_{i=0}^{\infty} p_i s^i = P(s) \;\Rightarrow\; P_1(s) = P(s) = P_0(P(s)).$$

Thus, the relation (i) is true for $n = 0$. By the Chapman-Kolmogorov equations,

$$
\begin{aligned}
P_{n+1}(s) &= \sum_{i=0}^{\infty} P[Z_{n+1} = i|Z_0 = 1]s^i = \sum_{i=0}^{\infty} p_{1i}^{(n+1)} s^i \\
&= \sum_{i=0}^{\infty}\Big[\sum_{k=0}^{\infty} p_{1k}^{(n)} p_{ki}\Big]s^i = \sum_{k=0}^{\infty} p_{1k}^{(n)}\Big[\sum_{i=0}^{\infty} p_{ki}s^i\Big] \\
&= \sum_{k=0}^{\infty} p_{1k}^{(n)}(P(s))^k \qquad \text{by (5.2.3)} \\
&= P_n(P(s))\,.
\end{aligned}
$$

Thus, $P_{n+1}(s) = P_n(P(s)), \;\; \forall \; n \geq 0$.
(ii) We prove $P_{n+1}(s) = P(P_n(s))$ by induction. As shown in (i) $P_1(s) = P(s)$ and hence $P_1(s) = P(P_0(s))$. Further, $P_2(s) = P_1(P(s)) = P(P_1(s))$. We assume that $P_n(s) = P(P_{n-1}(s))$ for some n, which is true for $n = 1, 2$. Now,

$$
\begin{aligned}
P_{n+1}(s) &= P_n(P(s)) = P(P_{n-1}(P(s))) \;\; \text{by induction hypothesis} \\
&= P(P_n(s)) \;\; \text{by (i).}
\end{aligned}
$$

□

Remark 5.3.1 The functional iterative property of the generating function of the offspring distribution established in Theorem 5.3.1 is known as the branching property and it is a sort of defining property of a branching process. A Markov chain with state space W, satisfying the branching property is a branching process. This property essentially follows from the fact that if $Z_n > 0$, then $Z_{n+1} \stackrel{d}{=} \sum_{i=1}^{Z_n} Y_i$ and $P_n(s)$ and $P(s)$ are probability generating functions of Z_n and Z_1, respectively. Using the functional iteration $P_{n+1}(s) = P_n(P(s))$, we can find the probability generating function of Z_n for each n, when we know the probability generating function of the offspring distribution. Thus, again we note that the offspring distribution of a Branching process plays the role of one step transition probability matrix of the Markov chain.

Surprisingly, the second result of Theorem 5.3.1 is not valid if there are more than one ancestors. While $P_{n+1}(s) = P_n(P(s))$ remains valid, $P_{n+1}(s) = P(P_n(s))$ is not valid when $Z_0 = k \neq 1$. We demonstrate this as follows.

Suppose $P_n^{(i)}(s) = \sum_{j=0}^{\infty} P[Z_n = j|Z_0 = i]s^j$ denotes the probability generating function of Z_n when $Z_0 = i > 1$. When there are i ancestors, we have i independent and identical versions of the branching process with 1 ancestor. Thus, the population size of the nth generation is the total of the offspring in the nth generation of each of the i ancestors. Hence,

$$
P_{n+1}^{(i)}(s) = (P_{n+1}(s))^i = (P_n(P(s)))^i = P_n^{(i)}(P(s))\,.
$$

But,

$$
P_2^{(i)}(s) = (P_1(P(s)))^i = (P(P_1(s)))^i \neq P(P_1^{(i)}(s))\,.
$$

Thus, the relation $P_{n+1}(s) = P(P_n(s))$ does not extend to the case when there are more than one ancestors.

In the next example, we find $P_n(s)$ using the recurrence relation derived in Theorem 5.3.1, when the offspring distribution is geometric with success probability 1/2.

Example 5.3.1 Suppose $\{Z_n, n \geq 0\}$ is a branching process when $Z_0 = 1$ and the offspring distribution is geometric with probability mass function $P[Y = y] = (1/2)^{y+1},\ y = 0, 1, 2, \ldots$. Then,

$$P(s) = P_1(s) = \sum_{i\geq 0}(1/2)^{i+1}s^i = (1/2)(1 - s/2)^{-1} = 1/(2 - s).$$

We show by induction that $P_m(s) = (m - (m - 1)s)/((m + 1) - ms)$. It is true for $m = 1$. Now

$$\begin{aligned} P_{m+1}(s) &= P(P_m(s)) = 1/(2 - P_m(s)) \\ &= \frac{(m + 1 - ms)}{m + 2 - (m + 1)s} = \frac{m + 1 - (m + 1 - 1)s}{m + 1 + 1 - (m + 1)s}. \end{aligned}$$

Hence, by induction, we conclude that $P_n(s) = (n - (n - 1)s)/((n + 1) - ns)$, $n \geq 1$. We now obtain the probability distribution of Z_n from its probability generating function $P_n(s)$. Observe that,

$$\begin{aligned} P_n(s) &= \frac{n - (n - 1)s}{(n + 1) - ns} = \frac{n - (n - 1)s}{n + 1}(1 - ns/(n + 1))^{-1} \\ &= \frac{n - (n - 1)s}{n + 1}\sum_{r=0}^{\infty} s^r(n/(n + 1))^r \\ &= \frac{n}{n + 1}\sum_{r=0}^{\infty}\left(\frac{n}{n + 1}\right)^r s^r - \frac{n - 1}{n + 1}\sum_{r=0}^{\infty}\left(\frac{n}{n + 1}\right)^r s^{r+1}. \end{aligned}$$

From $P_n(s)$, it follows that

$$\begin{aligned} P[Z_n = 0] &= \frac{n}{n + 1} \\ \& \ P[Z_n = i] &= \left(\frac{n}{n + 1}\right)^{i+1} - \frac{n - 1}{n + 1}\left(\frac{n}{n + 1}\right)^{i-1}, \quad i \geq 1 \\ &= \frac{n^{i-1}}{(n + 1)^{i+1}} = \frac{1}{n + 1}\frac{1}{n + 1}\frac{n^{i-1}}{(n + 1)^{i-1}}, \quad i \geq 1 \\ &= \left(1 - \frac{n}{n + 1}\right)\frac{1}{n + 1}\left(1 - \frac{1}{n + 1}\right)^{i-1}, \quad i \geq 1 \end{aligned}$$

Thus, the probability distribution of Z_n is a mixture of two distributions, one is degenerate at 1 and the second is geometric with support $\{1, 2, \ldots\}$ and parameter $1/(n+1)$; the mixing proportions are $n/(n+1)$ and $1/(n+1)$, respectively. It is to be noted that

$$\frac{n}{n+1} + \sum_{i=1}^{\infty} \frac{n^{i-1}}{(n+1)^{i+1}} = \frac{n}{n+1} + \frac{1}{n+1} = 1.$$

We observe that for $n = 10, 30, 50, 70, 90, 110$, $P[Z_n = 0] = n/(n+1)$ is $0.9091, 0.9677, 0.9804, 0.9859, 0.9890, 0.9910$, respectively. Thus, if the offspring distribution is geometric with success probability $1/2$, then with high chance $Z_n = 0$ for $n \geq 90$. It is to be noted that $\lim_{n\to\infty} P[Z_n = 0|Z_0 = 1] = 1$. We now obtain $P[Z_n = i]$ for $n = 1, 4, 7, 10$ and $i = 0$ to 10, using Code 5.6.2 given in Sect. 5.6. The probabilities are displayed in Table 5.2.

From Table 5.2, we note that the first column displays the first 11 probabilities in the offspring distribution, for $i > 10$, the probabilities are almost 0. Further, as n increases, $P[Z_n = 0]$ increases for all n, as expected. For $i \geq 1$, probabilities decrease as n increases for $n = 4, 7, 10$. For all n, probabilities decrease as i increases and for $i > 10$, these are close to 0. □

In general, it is difficult to find $P_n(s)$ and hence the distribution of Z_n, even if we have the recurrence relations $P_{n+1}(s) = P_n(P(s))$ and $P_{n+1}(s) = P(P_n(s))$.

We now proceed to find the first two moments of Z_n given $Z_0 = 1$, which will help further to investigate the nature of a BGW branching process. We have noted above that the offspring distribution determines the distribution of Z_n for any n, and hence it is expected that moments of Z_n will be in terms of the moments of the offspring distribution. We prove below that this is indeed true. Suppose $\mu = E(Y_1)$ and $\sigma^2 = Var(Y_1)$ denote the mean and variance of the offspring distribution, respectively. These are known as the offspring mean and the offspring variance.

Table 5.2 $P[Z_n = i]$, $n = 1, 4, 7, 10$ and $i = 0$ to 10

i	$n = 1$	$n = 4$	$n = 7$	$n = 10$
0	0.5000	0.8000	0.8750	0.9091
1	0.2500	0.0400	0.0156	0.0083
2	0.1250	0.0320	0.0137	0.0075
3	0.0625	0.0256	0.0120	0.0068
4	0.0312	0.0205	0.0105	0.0062
5	0.0156	0.0164	0.0092	0.0056
6	0.0078	0.0131	0.0080	0.0051
7	0.0039	0.0105	0.0070	0.0047
8	0.0020	0.0084	0.0061	0.0042
9	0.0010	0.0067	0.0054	0.0039
10	0.0005	0.0054	0.0047	0.0035

Theorem 5.3.2 *Suppose $\{Z_n, n \geq 0\}$ is a branching process with $Z_0 = 1$. Suppose μ and σ^2 are the mean and the variance of the offspring distribution respectively. Then $\forall\, n \geq 1,\;\; E(Z_n) = \mu^n$ and $Var(Z_n)$ is given by*

$$Var(Z_n) \;\; = \;\; \begin{cases} \sigma^2\mu^{n-1}(1-\mu^n)/(1-\mu), & \text{if } \mu \neq 1 \\ n\sigma^2, & \text{if } \mu = 1\,. \end{cases}$$

Proof With $Z_0 = 1$, $Z_1 \stackrel{d}{=} Y_1$. Thus,

$$\mu = E(Y_1) = E(Z_1) \;\; \& \;\; \sigma^2 = Var(Y_1) = Var(Z_1).$$

Since $Z_n \stackrel{d}{=} \sum_{i=1}^{Z_{n-1}} Y_i$ and Y_i's are independent and identically distributed random variables, each with mean μ and variance σ^2, we have

$$E(Z_n|Z_{n-1}) = \mu Z_{n-1} \quad \& \quad Var(Z_n|Z_{n-1}) = \sigma^2 Z_{n-1}\,.$$

Hence, for all $n \geq 2$,

$$E(Z_n) = E(E(Z_n|Z_{n-1})) = \mu E(Z_{n-1}) = \mu^2 E(Z_{n-2}) = \mu^3 E(Z_{n-3})\,.$$

Continuing in this manner, we have $E(Z_n) = \mu^n, \;\; \forall\; n \geq 1$. To find $Var(Z_n)$, we use the formula

$$Var(Z_n) = E(Var(Z_n|Z_{n-1})) + Var(E(Z_n|Z_{n-1}))\,.$$

Hence,

$$\begin{aligned} Var(Z_n) &= E(Var(Z_n|Z_{n-1})) + Var(E(Z_n|Z_{n-1})) \\ &= \sigma^2 E(Z_{n-1}) + Var(\mu Z_{n-1}) \\ &= \sigma^2\mu^{n-1} + \mu^2\left(E(Var(Z_{n-1}|Z_{n-2})) + Var(E(Z_{n-1}|Z_{n-2}))\right) \\ &= \sigma^2\mu^{n-1} + \mu^2\left(\sigma^2\mu^{n-2} + \mu^2 Var(Z_{n-2})\right) \\ &\;\;\vdots \\ &= \sigma^2(\mu^{n-1} + \mu^n + \cdots + \mu^{2n-3}) + \mu^{2n-2} Var(Z_1) \\ &= \sigma^2(\mu^{n-1} + \mu^n + \cdots + \mu^{2n-2}) \\ &= \sigma^2\mu^{n-1}(1 + \mu + \cdots + \mu^{n-1})\,. \end{aligned}$$

Hence,

$$Var(Z_n) \;\; = \;\; \begin{cases} \sigma^2\mu^{n-1}(1-\mu^n)/(1-\mu), & \text{if } \mu \neq 1 \\ n\sigma^2, & \text{if } \mu = 1\,. \end{cases}$$

$E(Z_n)$ and $Var(Z_n)$ can also be obtained from the probability generating function of Z_n, by taking repeated derivatives and substituting $s = 1$. □

Remark 5.3.2 If a stochastic process $\{X_n, n \geq 0\}$ is a stationary process, then as discussed in Chap. 1, $E(X_n)$ is the same for all n. Since for a BGW branching process, $E(Z_n)$ depends on n, the BGW branching process is not a stationary process; it is an evolutionary stochastic process.

In the next section, we discuss the concept of the probability of extinction and derive some related results. It is the most distinguishing feature of a branching process.

5.4 Extinction Probability

The problem of determining the probability of extinction was first raised in connection with the extinction of family surnames by Galton in 1889. Extinction occurs when the generation size is zero for the first time at some generation. The random time of extinction is thus the first time point n for which $Z_n = 0$, and then, by definition, $Z_k = 0 \ \forall \ k > n$. In Markov chain terminology, 0 is an absorbing state, thus extinction occurs when the chain transits to state 0. We have noted that all states $i > 0$ are transient states. Thus, the set T of transient states is not finite. As a consequence, it is possible that the chain keeps moving among the states in T. We investigate below these two possibilities. It is of immense interest to find the probability of ultimate extinction or explosion of the process as it has many applications.

Suppose q denotes the probability of ultimate extinction, that is, the probability that the population will eventually die out, under the assumption that $Z_0 = 1$. Thus, q is the probability of the first visit to 0 from 1. It is defined as

$$q = P[Z_n = 0 \text{ for some } n \geq 1 | Z_0 = 1] = f_{10} = P[N_{10} < \infty],$$

where $[N_{10} = k] = [Z_1 \neq 0, Z_2 \neq 0, \ldots, Z_{k-1} \neq 0, Z_k = 0]$, $k \geq 2$ and $[N_{10} = 1] = [Z_1 = 0]$. If there are k ancestors, then the population will eventually die out if and only if each of the k families started by each ancestor eventually dies out. Its probability is q^k, in view of the fact that individuals produce independently. Thus,

$$q^k = f_{k0} = f_{10}^k, \quad k \geq 1.$$

We derive another expression for q in terms of $P_n(s)$ in the following lemma.

Lemma 5.4.1 *Suppose $P_n(s)$ denotes the probability generating function of Z_n conditional on $Z_0 = 1$. Then $\lim_{n\to\infty} P_n(0)$ exists and is equal to q.*

Proof By definition, $P_n(s) = \sum_{i=0}^{\infty} P[Z_n = i | Z_0 = 1]s^i$. Suppose events A_n for $n \geq 1$ are defined as $A_n = [Z_n = 0]$. Now,

$$\begin{aligned}
Z_n = 0 \Rightarrow Z_{n+1} = 0 \quad &\Longleftrightarrow \quad A_n \subset A_{n+1} \ \forall \ n \geq 1 \\
&\Rightarrow \{A_n, n \geq 1\} \text{ is a non-decreasing sequence of events} \\
&\Rightarrow \lim_{n\to\infty} A_n = \bigcup_{n=1}^{\infty} A_n .
\end{aligned}$$

By definition, q is given by,

$$\begin{aligned}
q &= P[Z_n = 0 \text{ for some } n \geq 1 | Z_0 = 1] \\
&= P\left[\bigcup_{n=1}^{\infty}[Z_n = 0] \big| Z_0 = 1\right] = P\left[\bigcup_{n=1}^{\infty} A_n \big| Z_0 = 1\right] \\
&= P\left[\lim_{n\to\infty} A_n \big| Z_0 = 1\right] \\
&= \lim_{n\to\infty} P[A_n \big| Z_0 = 1] \text{ since probability measure is continuous} \\
&= \lim_{n\to\infty} P[Z_n = 0 \big| Z_0 = 1] = \lim_{n\to\infty} P_n(0) ,
\end{aligned}$$

provided the limit exists. To examine whether $\lim_{n\to\infty} P_n(0)$ exists, note that $P(s)$ is a power series with non-negative coefficients. Further, $p_0 < 1$ and hence $P(s)$ is a strictly increasing function of $s \in [0, 1]$. Now

$$\begin{aligned}
P_0(s) = s \ \Rightarrow \ P_0(0) = 0 \ \& \ P_1(s) = P(s) \ &\Rightarrow \ P_1(0) = P(0) = p_0 \in (0, 1) \\
\Rightarrow P_1(0) = p_0 > 0 = P_0(0) \quad &\Longleftrightarrow \quad P_1(0) > P_0(0) \\
\Rightarrow P(P_1(0)) > P(P_0(0)) \quad &\text{as } P(s) \text{ is an increasing function} \\
\Rightarrow P_2(0) > P_1(0) \ .
\end{aligned}$$

We assume $P_n(0) > P_{n-1}(0)$, which is true for $n = 1, 2$. Now

$$P_{n+1}(0) = P(P_n(0)) > P(P_{n-1}(0)) = P_n(0) \text{ by branching property.}$$

Hence, by induction we conclude that $P_n(0) > P_{n-1}(0)$ for all $n \geq 1$, that is, $\{P_n(0), n \geq 1\}$ is an increasing sequence of non-negative real numbers and is bounded above by 1 and hence $\lim_{n\to\infty} P_n(0)$ exists and it is q. Alternatively, $q = \lim_{n\to\infty} P[Z_n = 0\big|Z_0 = 1] = \lim_{n\to\infty} p_{10}^{(n)}$. Since 0 is an ergodic state, the limit exists and is $f_{10}/\mu_0 = f_{10}$. □

While proving Theorem 5.2.2, we have imposed the condition $0 < p_0 < 1$ on the offspring distribution $\underline{p}$, to classify the states as persistent or transient. Now we further assume that $0 < p_0 + p_1 < 1$. This condition is also reasonable. If $p_0 + p_1 = 1$, an individual produces either no offspring or only one offspring. If no offspring is produced, then the next generation size will be 0, and if 1 offspring is produced, then the size of the population will remain the same, as the individual will get replaced by its offspring, that is Z_1 is either 0 or 1, Z_2 is either 0 or 1 and so on. If $p_0 + p_1 = 0$, then both p_0 and p_1 are 0 and the population will explode. Under the conditions $0 < p_0 < 1$ and $0 < p_0 + p_1 < 1$, we prove that the probability of ultimate extinction depends on the offspring mean. It is to be noted that $0 < p_0 < 1$ and $0 < p_0 + p_1 < 1$ together imply that $0 < p_0, p_1 < 1$ and $p_i > 0$ for at least one $i \geq 2$.

Theorem 5.4.1 *Suppose* $\{Z_n, n \geq 0\}$ *is a branching process with* $Z_0 = 1$ *and* $P(s) = \sum_{j=0}^{\infty} p_j s^j$ *as the probability generating function of the offspring distribution. Then (i) the probability* q *of ultimate extinction is a solution to the equation* $P(s) = s$*. (ii) Under the assumptions* $0 < p_0 < 1$ *and* $0 < p_0 + p_1 < 1$*, if* $\mu \leq 1$ *then* $q = 1$ *and if* $\mu > 1$*, then* q *is the smallest positive root of the equation* $P(s) = s$*.*

Proof (i) From the definition of q,

$$\begin{aligned} q &= P[Z_n = 0 \text{ for some } n \geq 1 | Z_0 = 1] \\ &= \sum_{i=0}^{\infty} P[Z_n = 0 \text{ for some } n \geq 1, Z_1 = i | Z_0 = 1] \\ &= \sum_{i=0}^{\infty} P[Z_n = 0 \text{ for some } n \geq 1 | Z_1 = i] P[Z_1 = i | Z_0 = 1] \\ &= \sum_{i=0}^{\infty} q^i p_i = P(q) \,, \end{aligned}$$

since the probability of extinction is q^i, when $Z_1 = i$. Thus, q is a solution of the equation $P(s) = s$. Alternatively, by Lemma 5.4.1, $q = \lim_{n\to\infty} P_n(0)$. Further, $P(s)$ being a power series, is differentiable any number of times and differentiation and summation in $P(s) = \sum_{j\geq 0} p_j s^j$ can be interchanged. Being differentiable, it is continuous on [0, 1]. Observe that

$$\begin{aligned} P_{n+1}(0) = P(P_n(0)) \;\Rightarrow\; & \lim_{n\to\infty} P_{n+1}(0) = \lim_{n\to\infty} P(P_n(0)) = P(\lim_{n\to\infty} P_n(0)) \\ \Rightarrow\; & q = P(q), \end{aligned}$$

the second last step follows due to continuity of $P(\cdot)$.

(ii) Observe that the equation $P(s) = s$ may have many solutions and $P(1) = 1$ implies that 1 is always a solution of the equation $P(s) = s$. To investigate the other solutions and to identify which one of them is q, we define the function f as $f(s) = P(s) - s, s \in [0, 1]$. It is differentiable any number of times and is continuous. Note that, the condition $0 < p_0 + p_1 < 1$ implies that $p_i > 0$ for at least one $i \geq 2$. Hence,

$$f'(s) = P'(s) - 1 \;\;\&\;\; f''(s) = P''(s) = \sum_{j=2}^{\infty} j(j-1)p_j s^{j-2} > 0,$$

which implies that f is a convex function and $f'(s) = P'(s) - 1$ is an increasing function on [0, 1]. Further,

$$f'(1) = P'(1) - 1 = \mu - 1 \;\Rightarrow\; f'(1) \le 0 \text{ if } \mu \le 1 \;\&\; f'(1) > 0 \text{ if } \mu > 1.$$

Hence, to examine the solutions of $f(s) = 0$ for $s \in (0, 1)$, we consider two cases depending on values of μ as $\mu \le 1$ and $\mu > 1$.
Case 1 - $\mu \le 1$: Since $\mu \le 1$, $f'(1) \le 0$, further $f'(s)$ is a strictly increasing function on [0, 1]. Hence, $f'(s) = P'(s) - 1 < 0, \; \forall \; s \in [0, 1)$. By the mean value theorem,

$$\forall \; s \in [0, 1], \;\; \exists \, c \in (0, 1) \;\text{ such that }\; P(s) - P(1) = (s - 1)P'(c)\,.$$

Hence, $\forall \; s \in (0, 1)$,

$$\begin{aligned} P(s) - s &= P(s) - P(1) + P(1) - s \\ &= (1 - s)(1 - P'(c)) > 0, \;\; \forall \; c \in (0, 1) \;\Rightarrow\; P(s) > s\,. \end{aligned}$$

Further, $P(s) > s$ at $s = 0$ also since $P(0) = p_0 > 0$. Thus, $P(s) > s$ for all $s \in [0, 1)$, see Fig. 5.3 on p. 301 and Fig. 5.5 on p. 303. Hence, the equation $P(s) = s$ has no solution in [0, 1), and $s = 1$ is the only solution of $P(s) = s$ when $\mu \le 1$. Thus, $q = 1$ when $\mu \le 1$.

Case 2 - $\mu > 1$: If $\mu > 1$, then $f'(1) > 0$. Further f' is a continuous function. Hence, $f'(1) > 0$ implies that in a neighborhood of 1, $f'(s) > 0$ which further implies that f is an increasing function in that neighborhood. Thus, there exists a point $s_0 > 0$ such that f is increasing in $[s_0, 1]$. Observe that

$$f(1) = P(1) - 1 = 0 \;\Rightarrow\; \forall \; s \in [s_0, 1), \; f(s) = P(s) - s < 0.$$

Now $f(0) = p_0 > 0$. Thus,

$$f(0) > 0, \;\; f(1) = 0 \;\&\; f(s) < 0 \;\forall\; s \in [s_0, 1) \;\Rightarrow\; f(s) = 0 \text{ for some } s \in (0, s_0),$$

which further implies that $f(s) = 0$ has at least one solution in (0, 1). Thus, the curve of f crosses the x axis at least once. If it crosses more than once, then $f(s)$ is negative and positive in many intervals in $(0, s_0)$, which is not possible since f is a convex function. Thus, the graph of f against s cannot cross the x axis more than once, see Figs. 5.3 and 5.5. We now prove this assertion algebraically. Suppose $s_1 < s_2$ are two solutions of $f(s) = 0$ in (0, 1), and further $f(1) = 0$, hence by Rolle's theorem,

$$\exists \; s \in (s_1, s_2) \;\text{ such that }\; f'(s) = 0 \;\;\&\;\; \exists \; y \in (s_2, 1) \;\text{ such that }\; f'(y) = 0.$$

Thus, $f'(s) = 0$ and $f'(y) = 0$, and hence again by Rolle's theorem,

$$\exists \; u \in (s, y) \text{ such that } f''(u) = 0\,.$$

It is a contradiction to the result that $f''(s) > 0, \;\; \forall \; s \in (0, 1)$. Thus, there cannot be two solutions $s_1 < s_2$ of $f(s) = 0$ in $(0, 1)$. As a consequence, the equation $f(s) = 0$ has one solution in $(0, 1)$ and one is at $s = 1$. To decide which solution is q, we proceed as follows. Suppose $q = 1$. Then by Lemma 5.4.1,

$$1 = q = \lim_{n\to\infty} P_n(0) \;\Rightarrow\; \text{given } \epsilon > 0, \; \exists \; n_0(\epsilon) \text{ such that } 1 - P_n(0) < \epsilon, \; \forall \; n \geq n_0\,.$$

Suppose $\epsilon = 1 - s_0$ where $s_0 \in (0, 1)$ is such that $P(s) - s < 0$ for all $s \in (s_0, 1)$. With $\epsilon = 1 - s_0$,

$$\exists \; n_0(s_0) \text{ such that } 1 - P_n(0) < 1 - s_0 \;\; \Leftrightarrow \;\; P_n(0) > s_0 \;\; \forall \; n \geq n_0(s_0),$$

that is $P_n(0) \in (s_0, 1)$. Hence,

$$\begin{aligned} P_{n+1}(0) - P_n(0) &= P(P_n(0)) - P_n(0) < 0, \;\; \forall \; n \geq n_0(s_0) \\ &\Rightarrow P_{n+1}(0) < P_n(0) \;\; \forall \; n \geq n_0(s_0)\,. \end{aligned}$$

It is a contradiction to the result that $P_n(0) \leq P_{n+1}(0), \;\; \forall \; n \geq 1$. Thus, q cannot be 1 and hence q is the smallest positive root of $P(s) = s$. □

Figure 5.3 on p. 301 and Fig. 5.5 on p. 303 display the graphs of some probability generating function $P(\cdot)$ and the line $y = s$ imposed on it to determine the value of q graphically. We note that when $\mu \leq 1$, the graph of $P(\cdot)$ is always above the line $y = s$, while when $\mu > 1$, the graph of $P(s)$ crosses the line $y = s$ once at some s less than 1 and at $s = 1$.

Remark 5.4.1 Note that we have used the conditions $0 < p_0 < 1$ and $0 < p_0 + p_1 < 1$ to show that f is a convex and f' strictly increasing function. These properties of f are used in both the cases: $\mu \leq 1$ and $\mu > 1$.

(i) Observe that $\mu = p_1 + 2p_2 + 3p_3 + \cdots$. Hence, both the conditions $0 < p_0 < 1$ and $0 < p_0 + p_1 < 1$ are always satisfied when $\mu > 1$.
(ii) In Example 5.3.1, when the offspring distribution is geometric with parameter $1/2$, both the conditions $0 < p_0 < 1$ and $0 < p_0 + p_1 < 1$ are satisfied. Further, for this offspring distribution, $\mu = 1$. We have already noted that

$$q = \lim_{n\to\infty} P[Z_n = 0|Z_0 = 1] = \lim_{n\to\infty} P_n(0) = \lim_{n\to\infty} n/(n+1) = 1.$$

In the same example, we have obtained $P(s) = 1/(2 - s)$. Observe that $P(s) = s \iff (s-1)^2 = 0$. Thus, both the roots are 1 and hence $q = 1$.
(iii) We now show that the condition $0 < p_0 + p_1 < 1$ is sufficient but not necessary. Suppose the offspring distribution is Bernoulli, then $0 < p_0 < 1$ but $p_0 + p_1 = 1$. In this case, $P(s) = p_0 + (1 - p_0)s$ and $\mu = p_1 = 1 - p_0 < 1$. Hence, $q = 1$. Further,

the solution of $P(s) = s \iff p_0(1 - s) = 0$ is $s = 1$ since $0 < p_0 < 1$, which again implies that $q = 1$. There is one more approach to prove that $q = 1$ when $\mu < 1$, where we do not need these two conditions. We discuss it in the next theorem.
(iv) If $\mu = 1$, then the condition $p_1 < 1$ is necessary to claim that $f''(s) > 0$ and f is convex.

Theorem 5.4.2 *Suppose $\{Z_n, n \geq 0\}$ is a branching process with $Z_0 = 1$ and the offspring mean $\mu < 1$. Then q is 1.*

Proof Suppose a random variable Z is defined as $Z = \sum_{k=0}^{\infty} Z_k$. It represents the total number of individuals that ever existed in the population. Since $Z_k \geq 0$, $P[Z < \infty] + P[Z = \infty] = 1$. Note that,

$$Z_n = 0 \text{ for some n } \Rightarrow Z_k = 0 \ \forall \ k > n \ \Rightarrow \ Z < \infty.$$

Conversely,

$$Z < \infty \Rightarrow Z_k \to 0 \ a.s. \ as \ k \to \infty \Rightarrow \ Z_k = 0 \ \forall \ k > n \text{ for some } n,$$

that is, $Z < \infty$ if and only if the population becomes extinct. Suppose $\mu < 1$. Since $E(Z_k) = \mu^k$, we have

$$\begin{aligned} E(Z) = E\Big(\sum_{k=0}^{\infty} Z_k\Big) &= \sum_{k=0}^{\infty} \mu^k = 1/(1-\mu) < \infty \\ \Rightarrow \ P[Z < \infty] = 1 \ \Rightarrow \ q = 1. \end{aligned}$$

□

Remark 5.4.2 (i) Results proved in Theorem 5.4.1 are intuitively appealing. If $\mu < 1$, then on an average each individual gives birth to less than one individual. Hence, the population dies out eventually. On the other hand, if $\mu > 1$, then on the average, each individual gives birth to more than one individual. Hence, the probability of population growing rapidly is positive. The case $\mu = 1$ is a borderline.
(ii) Note that

$$\mu^n = E(Z_n) = \sum_{j=0}^{\infty} jP[Z_n = j] \geq \sum_{j=1}^{\infty} P[Z_n = j] = P[Z_n \geq 1] \,.$$

If $\mu < 1$, $\mu^n \to 0$. Hence, $P[Z_n \geq 1] \to 0$ and hence $P[Z_n = 0] \to 1$. Thus, if $\mu < 1$ then $q = 1$ as proved in Theorem 5.4.1.

Theorem 5.4.1 implies that the long-run behavior of the generation sizes depends on the offspring mean. An important classification of the BGW process is based on the offspring mean. We have proved that $E(Z_n) = \mu^n$. Therefore, in the expected value sense, the process grows geometrically if $\mu > 1$, stays constant if $\mu = 1$ and decays geometrically if $\mu < 1$. According to these three cases, BGW process is labeled as super-critical, critical and sub-critical, respectively. In Theorem 5.4.1, it is proved that the extinction probability q is 1 in critical and sub-critical cases; while in the super-critical case $q < 1$. The super-critical and sub-critical processes behave as expected from the expression for the mean. The behavior of the critical branching process is counter intuitive. Although the mean stays constant and equals 1, the probability that the process becomes extinct is 1. In epidemiology, extinction of the branching process is interpreted as a minor epidemic and non-extinction as a major epidemic.

Following examples illustrate the computation of the extinction probability.

Example 5.4.1 Suppose $\{Z_n, n \geq 0\}$ is a branching process with $Z_0 = 1$ and the offspring distribution as a geometric distribution with parameter p. Thus, its probability mass function given by $p_j = p(1-p)^j, \quad j = 0, 1, 2, \ldots, 0 < p < 1$. Its mean μ and probability generating function $P(s)$ are

$$\mu = (1-p)/p \ \ \& \ \ P(s) = \sum_{j \geq 0} p(1-p)^j s^j = p(1-(1-p)s)^{-1}, \ s \in [0, 1] \, .$$

Observe that $\mu > 1$ if $p < 1/2$ and $\mu \leq 1$ if $p \geq 1/2$. Thus, the probability of extinction $q = 1$ if $p \geq 1/2$ and it is the smallest positive root of the equation $P(s) = s$, if $p < 1/2$. Now $P(s) = s$ leads to a quadratic equation $(1-p)s^2 - s + p = 0$ with two roots $(1 \pm \sqrt{1-4p(1-p)})/2(1-p)$. Note that for any $p \in (0, 1/2)$, $p(1-p) < 1/4$. Hence, both the roots are real. Observe that

$$(1 \pm \sqrt{1-4p(1-p)})/2(1-p) = (1 \pm (1-2p))/2(1-p) = 1 \ \text{ or } \ p/(1-p).$$

When $p < 1/2$, $p/(1-p) < 1$ and it is the smallest root. Thus, if $p < 1/2$, the probability of ultimate extinction $q = p/(1-p) = 1/\mu$. □

Example 5.4.2 Suppose $\{Z_n, n \geq 0\}$ is a branching process with $Z_0 = 1$ and the offspring distribution with probability mass function given by $p_0 = p_3 = 1/2$ and $p_j = 0, \ \forall \ j \neq 0, 3$. Its mean is $\mu = 3/2 > 1$ and its probability generating function is $P(s) = (1+s^3)/2, \ s \in [0, 1]$. Since $\mu > 1$, the probability of extinction q is the smallest positive root of the equation $P(s) = s$. Now,

$$\begin{aligned} P(s) = s \Rightarrow s^3 - 2s + 1 = 0 \ &\Rightarrow \ (s-1)(s^2 + s - 1) = 0 \\ &\Rightarrow s = 1 \ \text{ or } \ s = (-1 \pm \sqrt{5})/2. \end{aligned}$$

Note that the smallest root in $(0, 1)$ is $(-1+\sqrt{5})/2$. Hence, the probability of extinction $q = (-1+\sqrt{5})/2 = 0.6180$. □

Example 5.4.3 Suppose $\{Z_n, n \geq 0\}$ is a branching process with $Z_0 = 1$ and the offspring distribution to be Poisson with mean $\lambda > 1$. To find the probability of ultimate extinction when $\lambda > 1$, we solve the equation

$$P(s) = s \quad \Longleftrightarrow \quad e^{-\lambda(1-s)} = s \quad \Longleftrightarrow \quad \lambda(1-s) + \log s = 0,$$

by the Newton-Raphson method, using Code 5.6.3. For $\lambda = 1.5$, the extinction probability is 0.4172 and for $\lambda = 2$, it is 0.2032. Note that for higher λ, the probability is smaller, as expected. □

In the next example, using the recurrence relation $P_{n+1}(0) = P(P_n(0))$, we find $P[Z_n = 0|Z_0 = 1]$ for various values of n.

Example 5.4.4 Suppose $\{Z_n, n \geq 0\}$ is a branching process with $Z_0 = 1$. We find $P[Z_n = 0]$ for $n = 25, 50, 75, 100, 125, 150$ and for three offspring distributions specified as follows, along with their offspring means:

(i) $Q_1 = (p_0 = 0.5, p_1 = 0.1, p_2 = 0.4), \quad \mu = 0.9 < 1.$
(ii) $Q_2 = (p_0 = 0.25, p_1 = 0.50, p_2 = 0.25), \quad \mu = 1$
(iii) $Q_3 = (p_0 = 0.25, p_1 = 0.40, p_2 = 0.35), \quad \mu = 1.1 > 1$

From Theorem 5.4.1, it is clear that corresponding to offspring distributions Q_1 and Q_2, $q = 1$ and corresponding to offspring distribution Q_3, q is the smallest positive root of the equation $P(s) = s$, which comes out to be 0.7143, by solving the quadratic equation $P(s) = s \iff 0.25 + 0.4s + 0.35s^2 = s$. In Lemma 5.4.1, it is also proved that the sequence $\{P_n(0), n \geq 1\}$ converges to q. We compute $P_n(0)$ for $n =$ 25, 50, 75, 100, 125, 150, using the recurrence relation $P_{n+1}(0) = P(P_n(0))$, for the three offspring distributions, where $P(s) = p_0 + p_1 s + p_2 s^2$. We use Code 5.6.4. The output is organized in Table 5.3.

From Table 5.3, we note that when $\mu = 0.9 < 1$, $P[Z_n = 0]$ increases to 1 as n increases and stabilizes at $q = 1$ for $n = 100$. For $\mu = 1$, $P[Z_n = 0]$ increases to 1 as n increases and is close to $q = 1$. Similarly, when $\mu = 1.1 > 1$, $P[Z_n = 0]$ increases as n increases and stabilizes at $q = 0.7143$ for $n = 100$. □

We now briefly discuss the concept of extinction time. Suppose $Z_0 = 1$. It is known that if $Z_n = 0$, then $Z_{n+k} = 0$ for all $k \geq 1$. Thus, time to extinction is the first n for which $Z_n = 0$. Suppose T denotes the time to extinction, then $T = \min\{n|Z_n = 0\}$. Thus, for $n \geq 1$

$$T = n \iff Z_{n-1} \neq 0 \;\&\; Z_n = 0 \;\Rightarrow\; P[T = n] = P[Z_n = 0, Z_{n-1} \neq 0|Z_0 = 1].$$

We obtain the probability distribution of T in terms of $P_n(0)$. Observe that

Table 5.3 $P[Z_n = 0|Z_0 = 1]$ for $n = 25, 50, 75, 100, 125, 150$

μ	$n = 25$	$n = 50$	$n = 75$	$n = 100$	$n = 125$	$n = 150$
0.9	0.9877	0.9992	0.9999	1.0000	1.0000	1.0000
1.0	0.8718	0.9296	0.9513	0.9628	0.9698	0.9747
1.1	0.7010	0.7134	0.7142	0.7143	0.7143	0.7143

$$\begin{aligned} P_n(0) &= P[Z_n = 0|Z_0 = 1] = P[Z_n = 0, Z_{n-1} = 0|Z_0 = 1] \\ &+ P[Z_n = 0, Z_{n-1} \neq 0|Z_0 = 1] \\ &= P[Z_n = 0|Z_{n-1} = 0]P[Z_{n-1} = 0|Z_0 = 1] \\ &+ P[Z_n = 0, Z_{n-1} \neq 0|Z_0 = 1] \\ &= 1 \times P_{n-1}(0) + P[T = n] \\ \Rightarrow \quad P[T = n] &= P_n(0) - P_{n-1}(0), \quad n \geq 1. \end{aligned}$$

Remark 5.4.3 In the super-critical case, the probability that the population explodes is positive. In this case, T is an extended real valued random variable. On the other hand, when $\mu \leq 1$, T is a real random variable.

The next two examples illustrate the computation of the distribution of T.

Example 5.4.5 In Example 5.3.1, we have obtained $P_n(s)$ for a branching process with $Z_0 = 1$ and the offspring distribution to be geometric with probability mass function $P[Y = y] = (1/2)^{y+1}, \ y = 0, 1, 2, \ldots$. Note that the offspring mean $\mu = 1$ and hence $q = 1$ implying that $P[T < \infty] = 1$. From $P_n(s)$, we have $P_n(0) = P[Z_n = 0|Z_0 = 1] = n/(n + 1)$. Hence, for $n \geq 1$,

$$\begin{aligned} P[T = n] &= P_n(0) - P_{n-1}(0) = n/(n + 1) - (n - 1)/n = 1/n(n + 1) \\ \Rightarrow \quad \sum_{n\geq 1} P[T = n] &= \sum_{n\geq 1} 1/n(n + 1) < \infty \\ \& \quad \sum_{n\geq 1} 1/n(n + 1) &= \sum_{n\geq 1}(1/n - 1/(n + 1)) = 1, \end{aligned}$$

supporting the result that T is a proper random variable. □

As shown in the next example, it is not always possible to find the form of $P_n(s)$ and hence $P_n(0)$ explicitly. We may use the functional recurrence relation to find $P_n(0)$ and hence the distribution of extinction time.

Example 5.4.6 Suppose $\{Z_n, n \geq 0\}$ is a branching process with $Z_0 = 1$ and offspring distribution as $\{p_0 = 0.48, \ p_1 = 0.1, \ p_2 = 0.42\}$. The offspring mean is 0.94, hence the probability of ultimate extinction is 1. We find the distribution of extinction time T using Code 5.6.5. From the function `head(pr)`, we note that $P[T = n] = 0.4800, 0.1448, 0.0816, 0.0538, 0.0385, 0.0291$ for $n = 1, 2, 3, 4, 5, 6$, respectively. From `tail(pr)`, we observe that $P[T = n] = 0.0001$ for $n = 63$ to 79. For all $n > 79$, $P[T = n]$ is 0 up to four decimal places accuracy. Further, $\sum_{n=1}^{106}[T = n] = 0.9994$. Figure 5.1 displays the probability distribution of T. It is a positively skewed distribution with high probability at 1 and decreasing probabilities as n increases. Further, from the probability distribution, we have $E(T) = 4.4216$. □

The next section is devoted to the discussion on a realization of a branching process and computation of extinction probability graphically and algebraically using R. We also discuss how to obtain the estimate of $P[Z_n = 0]$ and how to estimate $E(T)$ using simulation.

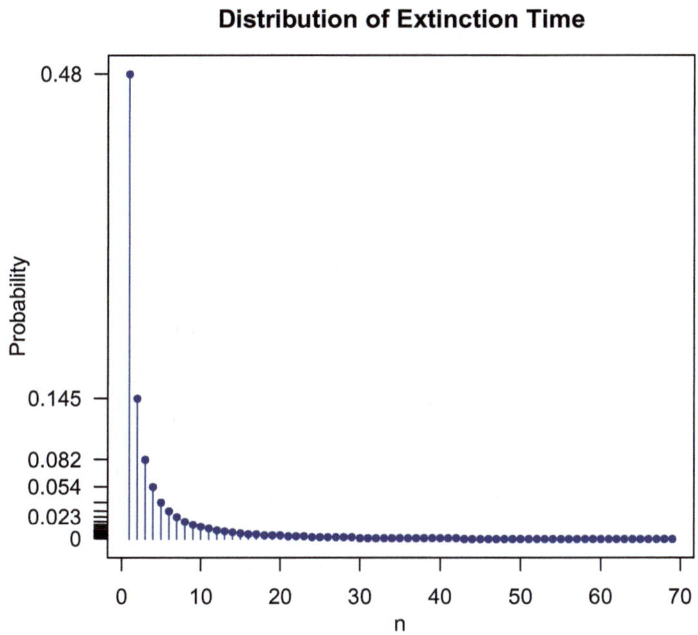

Fig. 5.1 Probability mass function of extinction time

5.5 Realization of a Process and Computation of Extinction Probability

We have noted that a BGW branching process is completely specified by the number of ancestors and the offspring distribution. Further it is a Markov chain, however its transition probabilities cannot be specified explicitly. Hence, to obtain a realization from a BGW branching process, we use its definition and obtain a random sample from the offspring distribution depending on the size of the previous generation. We use Code 5.6.6 to obtain a realization of the branching process corresponding to a given offspring distribution, when the number of ancestors is $k \geq 1$. Further, the evolution of the branching process is different in the super-critical, critical and sub-critical cases. Hence, we obtain the realizations in these three cases separately. Code 5.6.6 has three components: (i) to obtain a realization of the branching process when $Z_0 = 1$ and $Z_0 > 1$; in sub-critical, critical and super-critical cases, (ii) for graphical representation of the realizations and (iii) to obtain multiple realizations of a branching process when $Z_0 = 1$ and $\mu > 1$ and to estimate certain probabilities.

Example 5.5.1 Suppose $\{Z_n, n \geq 0\}$ is a BGW branching process where supports of the following three offspring distributions are $\{0, 1, \ldots, 6\}$:

Table 5.4 Realization of a branching process

y	0	1	2	3	4	5	6			
$P[Y=y]$	0.4	0.2	0.11	0.11	0.05	0.09	0.04			
Mean = 1.64										
	Z_0	Z_1	Z_2	Z_3	Z_4	Z_5	Z_6	Z_7	Z_8	Z_9
$Z_0=1$	1	3	3	4	3	3	8	10	9	12
$Z_0=2$	2	4	3	6	3	8	13	12	24	37
$Z_0=4$	4	6	7	11	14	17	28	44	69	111
y	0	1	2	3	4	5	6			
$P[Y=y]$	0.7	0.05	0.05	0.07	0.05	0.04	0.04			
Mean = 1										
	Z_0	Z_1	Z_2	Z_3	Z_4	Z_5	Z_6	Z_7	Z_8	Z_9
$Z_0=1$	1	1	3	5	5	9	2	0	0	0
$Z_0=2$	2	4	10	9	2	1	5	4	0	0
$Z_0=4$	4	4	10	9	3	5	4	1	0	0
y	0	1	2	3	4	5	6			
$P[Y=y]$	0.7	0.05	0.09	0.04	0.04	0.04	0.04			
Mean = 0.95										
	Z_0	Z_1	Z_2	Z_3	Z_4	Z_5	Z_6	Z_7	Z_8	Z_9
$Z_0=1$	1	2	0	0	0	0	0	0	0	0
$Z_0=2$	2	2	0	0	0	0	0	0	0	0
$Z_0=4$	4	2	0	0	0	0	0	0	0	0

(i) $\underline{p}_1 = (0.4, 0.2, 0.11, 0.11, 0.05, 0.09, 0.04)$ with $\mu = 1.64$,
(ii) $\underline{p}_2 = (0.7, 0.05, 0.05, 0.07, 0.05, 0.04, 0.04)$ with $\mu = 1$ and
(iii) $\underline{p}_3 = (0.7, 0.05, 0.09, 0.04, 0.04, 0.04, 0.04)$ with $\mu = 0.95$.

For each case, we obtain a realization of length $n = 9$. Table 5.4 presents the realizations corresponding to these offspring distributions, when the number of ancestors is $1, 2$ and 4.

From Table 5.4, we observe that for the three values of Z_0, when the offspring mean is greater than 1, then generation sizes increase. If the offspring mean is equal to 1, sizes increase initially but then decrease and are 0. When the offspring mean is less than 1, then generation sizes are positive for the first few generations and then are 0 for all further generations. Figure 5.2 displays the realization of the branching process for the three offspring distributions, when $Z_0 = 1$. It shows the same features as noted in Table 5.4.

From Table 5.4 and Fig. 5.2, it seems that for $\mu > 1$, the generation sizes always increase. However, even if $\mu > 1$, in some cases, the generation sizes may be 0 as well. We have proved in Theorem 5.4.1 that when $\mu > 1$, probability q of extinction is less than 1, but it is positive. Table 5.5 presents multiple realizations of the branching process with $Z_0 = 1$, when $\mu > 1$. From Table 5.5, we note that in 6 realizations out

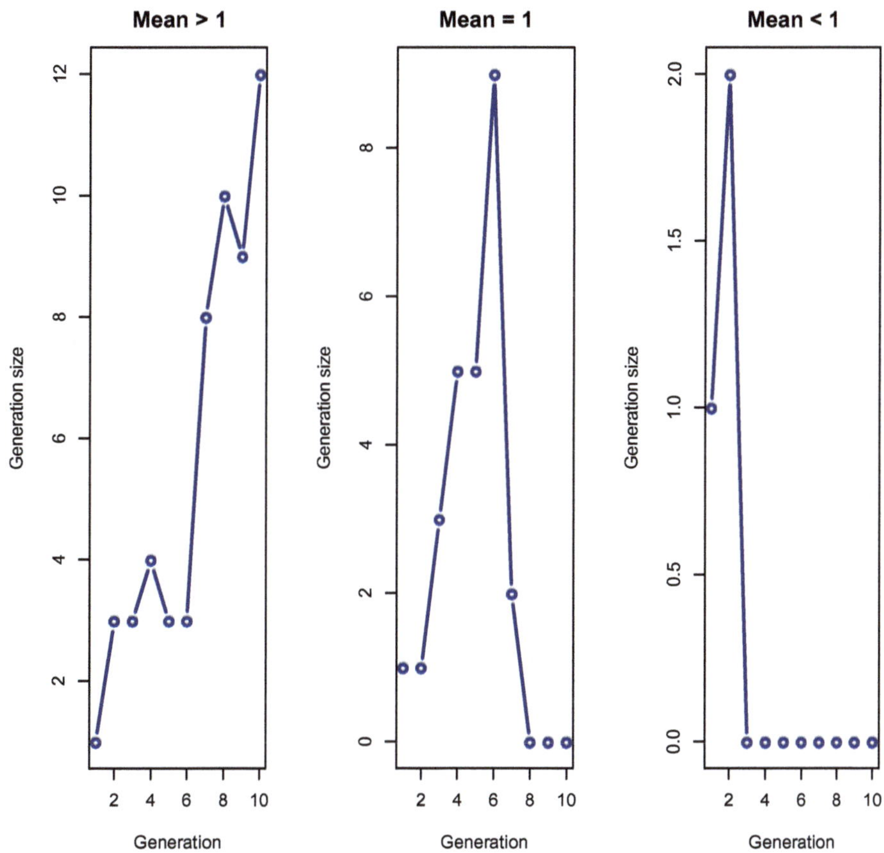

Fig. 5.2 Realization of a branching process

of 10, the population becomes extinct at the 9th generation. In Example 5.5.2, we compute q for this setup and it is 0.597.

From 10 realizations in Table 5.5, the estimate of $P[Z_9 = 0|Z_0 = 1]$ is 0.6, close to q. Based on $m = 200$ realizations, the estimate of $P[Z_{25} = 0|Z_0 = 1]$ is 0.645. □

In Theorem 5.4.1, we have proved that the extinction probability is a solution of the equation $P(s) = s$. In the next example, using Code 5.6.7, we compute it graphically and algebraically. We draw the graph of $P(\cdot)$ against s and impose on it the line $y = s$. The intersection point of the curve and the line is the estimate of extinction probability. We use the following two algebraic methods:

(i) We use the Newton-Raphson procedure to solve $P(s) = s$. We have used it in the previous section, as given in Code 5.6.3, to solve the equation $P(s) = s$ when the offspring distribution is Poisson.

Table 5.5 Multiple realizations with $Z_0 = 1$

y	0	1	2	3	4	5	6			
$P[Y = y]$	0.4	0.2	0.11	0.11	0.05	0.09	0.04			
$\mu = 1.64$										
m	Z_0	Z_1	Z_2	Z_3	Z_4	Z_5	Z_6	Z_7	Z_8	Z_9
1	1	0	0	0	0	0	0	0	0	0
2	1	0	0	0	0	0	0	0	0	0
3	1	2	0	0	0	0	0	0	0	0
4	1	0	0	0	0	0	0	0	0	0
5	1	3	9	20	37	74	133	240	406	690
6	1	3	1	5	10	15	41	70	96	168
7	1	0	0	0	0	0	0	0	0	0
8	1	5	9	12	16	24	30	53	64	126
9	1	0	0	0	0	0	0	0	0	0
10	1	5	15	10	15	23	31	49	110	203

(ii) When $P(s) - s = 0$ is a polynomial equation, there is a built-in function "`polyroot`" in `R`, to obtain the solution of a polynomial equation. Suppose $P(s) = p_0 + p_1 s + p_2 s^2 + p_3 s^3$, then the equation

$$P(s) = s \iff p_0 + (p_1 - 1)s + p_2 s^2 + p_3 s^3 = 0.$$

The function `polyroot(e)` gives the roots, where $e = (p_0, p_1 - 1, p_2, p_3)'$. Some roots are real and some are complex. Depending on the value of the offspring mean, we select the appropriate root as the value of q.

Example 5.5.2 Suppose $\{Z_n, n \geq 0\}$ is a BGW branching process as in Example 5.5.1. Thus, the supports of the three offspring distributions $\underline{p}_1, \underline{p}_2$ and $\underline{p}_3$ are $\{0, 1, \ldots, 6\}$, with respective means 1.64, 1 and 0, 95. For each distribution, we obtain the graphs of $P(s)$ against s and estimate of extinction probability q graphically. We also find q, algebraically, using the Newton-Raphson procedure and the built-in function in `R`. Figure 5.3 displays the graphs of $P(s)$ for the super-critical, critical and sub-critical cases, from left to right, respectively.

From the first graph, we note that the solution of $P(s) = s$ is 0.597 when $\mu > 1$; 0.998, that is, almost 1, when $\mu = 1$ in the second graph and 1 when $\mu < 1$ in the third graph. Algebraically, with the Newton-Raphson procedure, we get the solution as 0.5970, 0.9999 and 1 in the super-critical, critical and sub-critical cases, respectively. With the built-in function `polyroot`, some roots are real and some are complex. When $\mu > 1$, there are two real roots given by 0.597 and 1. Hence $q = 0.597$. When $\mu = 1$, there are two real roots and both are 1. When $\mu < 1$, there are two real roots 1 and 1.033. Being the probability, $q \leq 1$. Hence, $q = 1$. □

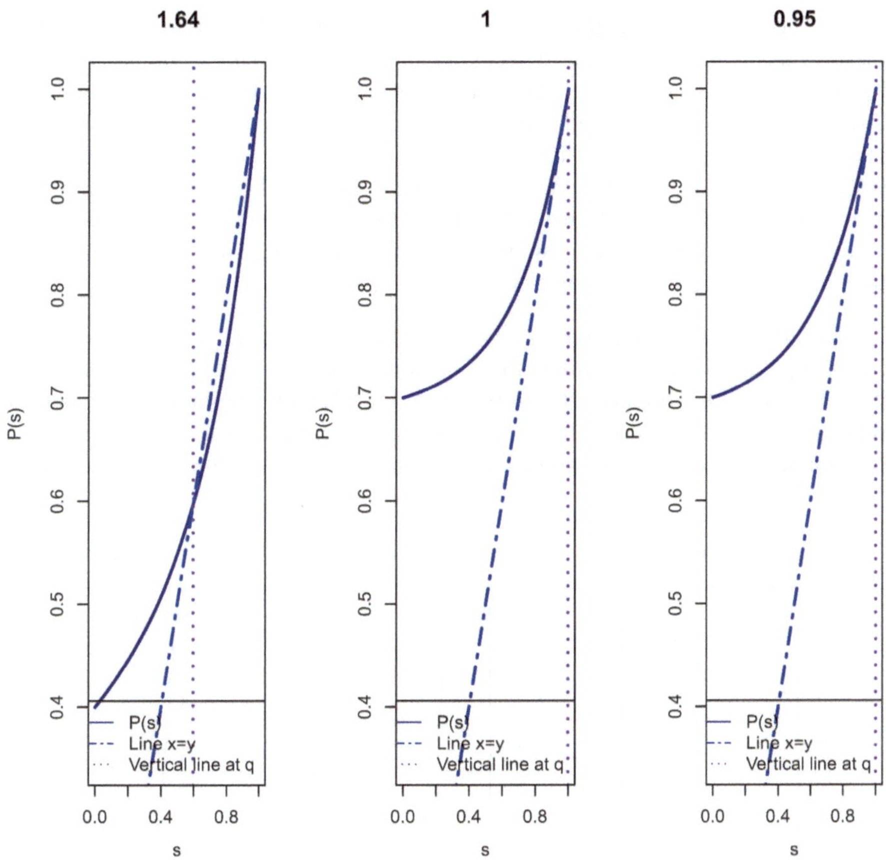

Fig. 5.3 Extinction probability

For a branching process with $\mu \leq 1$, the probability of ultimate extinction is 1. In the following example, using Code 5.6.8, we estimate the expected extinction time $E(T)$ when $\mu < 1$ and $\mu = 1$.

Example 5.5.3 Suppose $\{Z_n, n \geq 0\}$ is a BGW branching process where the offspring distribution is $\{p_0 = 0.7, p_1 = 0.05, p_2 = 0.09, p_3 = 0.04, p_4 = 0.04, p_5 = 0.04, p_6 = 0.04\}$, that is, $\underline{p}_3$ as in Example 5.5.1. The offspring mean μ is 0.95. We obtain $m = 200$ realizations of the process, each for 50 generations. For ith realization, we observe the first n for which $Z_n = 0$, which is the realized value T_i of T. In some realizations, Z_n may not be 0 up to $n = 50$. We omit such $r < m$ realizations. We then estimate $E(T)$ as $\sum_{i=1}^{m-r} T_i/(m-r)$. Code 5.6.8 obtains the multiple realizations, realized values T_i and estimate of $E(T)$. From vector a in the output, we note that out of 200 realizations, in one realization, numbered 135, $Z_n \neq 0$ for $n = 1$ to 51. Hence we omit this realization. Vector b specifies the first n for which $Z_n = 0$. However, $n = 1$ corresponds to Z_0, hence we obtain b1 from b by subtracting 1. The vector `b1` specifies the realized values of T. The mean `e` $= 2.1608$ of these values is the estimate of expected extinction time. Using Code 5.6.5, we can find the distribution of extinction time and hence its mean. It comes out to be 2.442. Thus, the estimate of expected extinction time by simulation is close to $E(T)$. If we take the offspring distribution as $\{p_0 = 0.7,\ p_1 = 0.05,\ p_2 = 0.05,\ p_3 = 0.07,\ p_4 = 0.05,\ p_5 = 0.04,\ p_6 = 0.04\}$, that is, $\underline{p}_2$ as in Example 5.5.1, then the offspring mean is $\mu = 1$. Using Code 5.6.8, we obtain the estimate of expected extinction time. It comes out to be 2.3182. Again using Code 5.6.5, we find the distribution of extinction time and hence its mean. It comes out to be 2.7395. □

Since 0 is an absorbing state, $Z_n = 0$ for some n implies $Z_{n+k} = 0$ for all $k \geq 1$. In Example 5.5.1, we have noted that once the generation size is 0, it continues to be 0. Further, if the offspring mean is greater than 1, then generation sizes increase rapidly in some realizations. In the next example, we use `break` function so that if the generation size is 0 for two consecutive generations then the next 0 generation

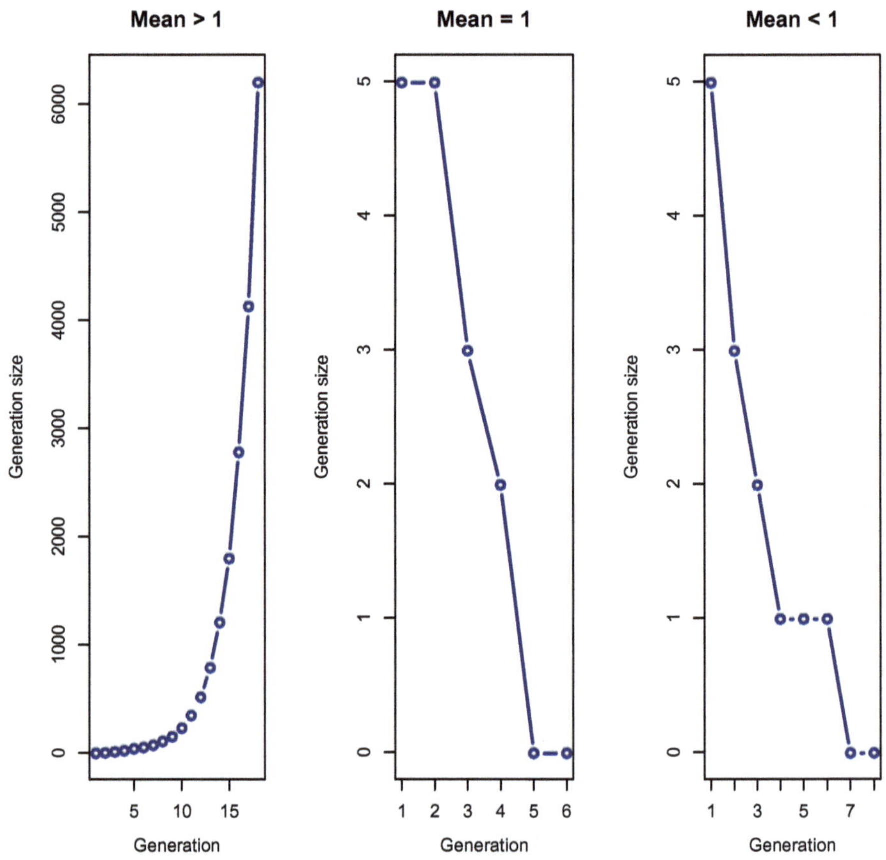

Fig. 5.4 Realization of BGW process: binomial offspring distribution

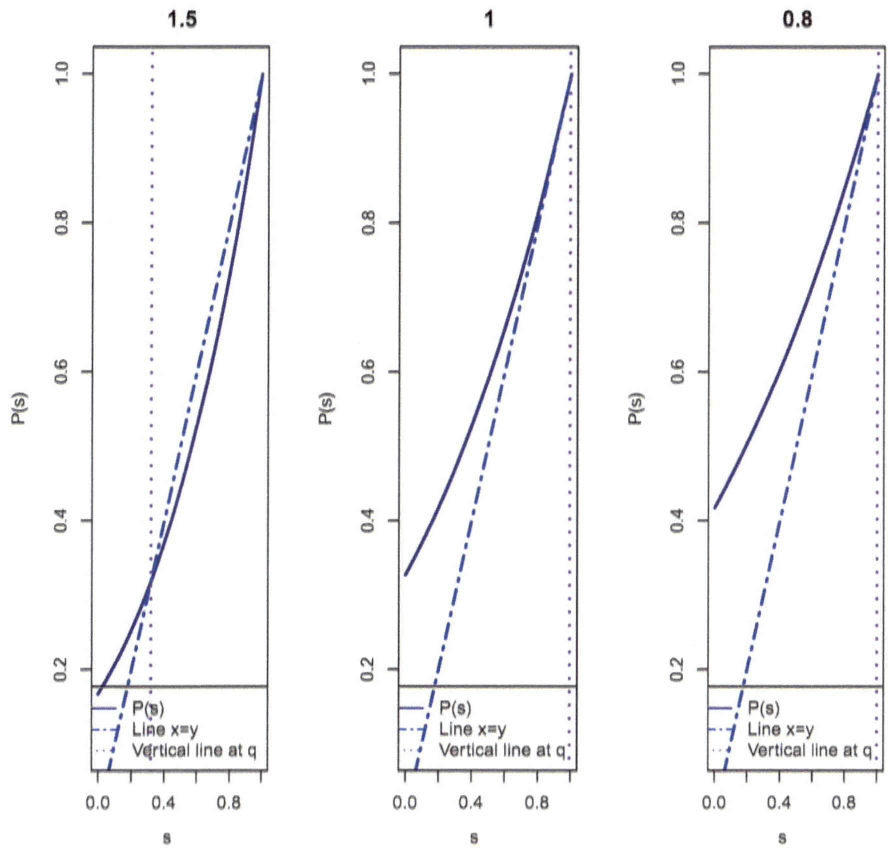

Fig. 5.5 Extinction probability: binomial offspring distribution

sizes are omitted. Similarly, when the offspring mean is greater than 1, we fix a large number, beyond which, except the next generation, the generation sizes are ignored. This is helpful to have a proper scale on the y axis. In the next example, we illustrate this feature when the offspring distribution is binomial $B(5, p)$ and $Z_0 = 5$. We take three values of p to cover super-critical, critical and sub-critical cases. The extinction probability is also calculated graphically and algebraically.

Example 5.5.4 Suppose $\{Z_n, n \geq 0\}$ is a BGW branching process when the offspring distribution is binomial $B(5, p)$ and $Z_0 = 5$. We take $p = 0.3, 0.2$ and 0.16, so that the offspring mean is 1.5, 1 and 0.8 respectively. The extinction probability q when $Z_0 = 1$ is a solution of the equation $P(s) = s$ where $P(s) = (1 - p + ps)^5$. Using Code 5.6.9, we obtain the realization for 20 generations and find q graphically and algebraically using Newton-Raphson method. From the output, we note that when $\mu = 1.5$, the realization is

$$5, 10, 19, 30, 48, 59, 80, 116, 159, 238, 354, 522, 797, 1214, 1803, 2787, 4135, 6207.$$

In this case, the `break` value is given as 5000. In the realization, we observe that only one generation size 6207 larger than 5000 is given and the next two are skipped. When $\mu = 1$, the realization is $5, 5, 3, 2, 0, 0$ and when $\mu = 0.8$, the realization is $5, 3, 2, 1, 1, 1, 0, 0$. When the offspring mean is less than or equal to 1, then after two consecutive 0 generation sizes, the next are omitted. The same is revealed in Fig. 5.4, which displays the realizations for three values of p.

The extinction probabilities obtained, graphically and algebraically, are $0.3189, 1, 1$ when $\mu > 1$, $\mu = 1$, $\mu < 1$ respectively. Figure 5.5 displays the extinction probability graphically. □

The next sections presents R codes used in the examples in this and previous sections.

5.6 R Codes

The following code computes the probabilities using the multinomial theorem. It is illustrated in Example 5.2.1.

Code 5.6.1 Computation of probabilities using the multinomial theorem:

```
# Part I: Input offspring distribution
prob=c(.2,.3,.3,.2) # offspring distribution
# Part II: Computation of probabilities
x1=c(3,0,0,0) # vector corresponding to i_1=3 and i_2=i_3=i_4=0
p30=dmultinom(x1,size=3,prob) # size=sum of components of x
p30
x2=c(2,1,0,0) # vector corresponding to i_1=2,i_2=1,i_3=i_4=0
p31=dmultinom(x2,size=3,prob); p31
x3=c(0,0,0,3) # vector corresponding to i_4 3 and i_1=i_2=i_3=0
p39=dmultinom(x3,size=3,prob); p39
```

□

With the following R code, we compute $P[Z_n = i|Z_0 = 1]$ for $n = 1, 4, 7, 10$ and $i = 0$ to 10, when the offspring distribution is geometric with parameter $1/2$, as specified in Example 5.3.1.

Code 5.6.2 Computation of $P[Z_n = i|Z_0 = 1]$ for the geometric offspring distribution: Suppose $\{Z_n, n \geq 0\}$ is a branching process with $Z_0 = 1$ and the offspring distribution to be geometric with probability mass function $P[Y = y] = (1/2)^{y+1}$, $y = 0, 1, 2, \ldots$. We compute $P[Z_n = i|Z_0 = 1]$ for $n = 1, 4, 7, 10$ and $i = 0$ to 10 with the following code:

```
n=c(1,4,7,10)
pr0=round(n/(n+1),4);pr0; pr=matrix(0,ncol=length(n),nrow=10)
for(j in 1:length(n))
{
for(i in 1:10)
{
pr[i,j]=n[j]^(i-1)/(n[j]+1)^(i+1)
}
}
pr=round(pr,4); P=rbind(pr0,pr);P
```

□

Code 5.6.3 Newton-Raphson procedure: We have proved in Theorem 5.4.1 that when $\mu > 1$, then the probability of ultimate extinction is a solution of the equation $P(s) = s$. In some cases, as in Example 5.4.3, we need iterative procedures to solve this equation. In this code, we illustrate the Newton-Raphson procedure to solve the equation, when the offspring distribution is Poisson with mean $\lambda > 1$. Now,

$$P(s) = s \iff e^{-\lambda(1-s)} = s \iff \lambda(1-s) + \log s = 0.$$

We solve it by the Newton-Raphson procedure, using the following code.

```
#Part I: Function for Newton-Raphson procedure
ep=function(m)
{
g1=function(s)
{
y=m*(1-s)+log(s); return(y)
}
dg=function(s)
{
y1 = -m+1/s; return(y1)
}
u=c(); u[1]=.3;i=1;diff=1;
while(diff>10^(-5))
{
u[i+1]=u[i]- g1(u[i])/dg(u[i])
diff=abs(u[1+i]-u[i]);i=i+1
}
sol=u[i];return(sol)
}
# PartII: Solution for two values of mean
m1=1.5; m2=2; ep1=round(c(ep(m1),ep(m2)),4);ep1
```

□

The next code illustrates computation of $P[Z_n = 0|Z_0 = 1]$ for various values of n, using the recurrence relation $P_{n+1}(0) = P(P_n(0))$, for the BGW process in Example 5.4.4.

Code 5.6.4 Computation of $P[Z_n = 0|Z_0 = 1]$: Suppose $\{Z_n, n \geq 0\}$ is a branching process with $Z_0 = 1$. We find $P[Z_n = 0]$ for $n = 25, 50, 75, 100, 125, 150$ and for three offspring distributions specified below, along with their offspring means:

1. $Q_1 = (p_0 = 0.5, p_1 = 0.1, p_2 = 0.4), \quad \mu = 0.9 < 1.$
2. $Q_2 = (p_0 = 0.25, p_1 = 0.50, p_2 = 0.25), \quad \mu = 1$
3. $Q_3 = (p_0 = 0.25, p_1 = 0.40, p_2 = 0.35), \quad \mu = 1.1 > 1$

We use the recurrence relation $P_{n+1}(0) = P(P_n(0))$ for the three offspring distributions. For these offspring distributions, $P(s)$ is a quadratic function given by $P(s) = p_0 + p_1 s + p_2 s^2$.

```
# Part I: Input offspring distributions
r1=c(.5,.1,.4);r2=c(.25,.5,.25);r3=c(.25,.4,.35)
P=matrix(c(r1,r2,r3),nrow=3,byrow=TRUE);mean=c(0.9,1,1.1)
# Part II: Function to compute probabilities
ext=function(n)
{
a=matrix(0,nrow=n,ncol=3)
a[1,]=P[,1]
for(i in 1:3)
{
for(j in 2:n)
{
a[j,i]=P[i,1]+P[i,2]*a[j-1,i]+P[i,3]*a[j-1,i]^2
}
}
a=round(a,4)
return(a[n,])
}
# Part III: Calculation for specific values of n
d=data.frame(mean,ext(25),ext(50),ext(75),ext(100),ext(125),
ext(150)); d
```

□

Code 5.6.5 Distribution of extinction time: This code illustrates the computation of the distribution of extinction time T for the branching process in Example 5.4.6.

```
# Part I: Input offspring distribution
p=c(.48,.10,.42)
# Part II: Compute distribution of extinction time
a=c();a[1]=p[1]; eps=0.00001; d=1; n=2
while(d > eps)
```

```
{
a[n]=p[1]+p[2]*a[n-1]+p[3]*a[n-1]^2
d=abs(a[n]-a[n-1]); n=n+1
}
l=length(a); l; b=c(0,a[-l]); pr=round(a-b,4);pr; head(pr)
tail(pr); sum(pr); n=1:l; mean=sum(n*pr); mean; pr1=round(pr,3)
# Part III: Graph of distribution
plot(n,pr1,"h",main="Distribution of Extinction Time",
ylab="Probability",xlab="n",yaxt="n",col="blue")
axis(2,at=sort(unique(pr1)),labels=sort(unique(pr1)),las=2)
points(n,pr1,pch=20,col="dark blue")
```

□

The next code is for obtaining the realization of a branching process and its graphical presentation.

Code 5.6.6 Realization of a branching process and its graphical presentation: This code consists of three parts: (i) to obtain a realization of a branching process when $Z_0 = 1$ and $Z_0 > 1$, in sub-critical, critical and super-critical cases, (ii) for graphical presentation of the realizations and (iii) for multiple realizations of a branching process when $Z_0 = 1$ and $\mu > 1$, and estimation of certain probabilities. We illustrate the code for the BGW branching process in Example 5.5.1, where the supports of the offspring distributions are $\{0, 1, \dots, 6\}$ for each of the three sets of probability mass functions, so that offspring mean $\mu > 1$, $\mu = 1$, $\mu < 1$.

```
# Part I: function for realization
x=c(0,1,2,3,4,5,6) # support of offspring distribution.
f=function(z0,pr,n) # function f for realization
{
z=c(); z[1]=z0
for(i in 2:n)
{
if(z[i-1]>0)
{
z[i]=sum(sample(x,z[i-1],pr,replace=T))
}
else
{
z[i]=0
}
}
return(z)
}
# Part II: Realizations
# Offspring distribution, mean > 1
p1=c(0.4, 0.2, 0.11, 0.11, 0.05, 0.09, 0.04)
```

```
mu1=sum(p1*x); n=10; set.seed(50); r11=f(1,p1,n)
set.seed(50); r12=f(2,p1,n); set.seed(50); r14=f(4,p1,n)
# Offspring distribution, mean = 1
p2=c(0.7, 0.05, 0.05, 0.07, 0.05, 0.04, 0.04)
mu2=sum(p2*x); set.seed(20); r21=f(1,p2,n)
set.seed(20); r22=f(2,p2,n);set.seed(20); r24=f(4,p2,n)
# Offspring distribution, mean < 1
p3=c(0.7, 0.05, 0.09, 0.04, 0.04, 0.04 ,0.04)
mu3=sum(p3*x); set.seed(60); r31=f(1,p3,n)
set.seed(60);r32=f(2,p3,n);set.seed(60);r34=f(4,p3,n)
mu=c(mu1,mu2,mu3);mu
Realization=rbind(r11,r12,r14,r21,r22,r24,r31,r32,r34);Realization
# Part III: Graphs of realizations
par(mfrow=c(1,3))
plot(r11,type="b",main="Mean > 1",xlab="Generation",
ylab="Generation size",col="dark blue",lwd=2)
plot(r21,type="b",main="Mean=1",xlab="Generation",
ylab="Generation size",col="dark blue",lwd=2)
plot(r31,type="b",main="Mean < 1",xlab="Generation",
ylab="Generation size",col="dark blue",lwd=2)
# Part IV: Multiple realizations with Z0=1 and
# offspring distribution p1
m=10; n=10;R=matrix(0,nrow=m,ncol=n)
for(i in 1:m)
{
set.seed(i+10); R[i,]=f(1,p1,n)
}
R
# Part V: To find estimate of P[Z_25 = 0] based on m realizations
m=200; n=26; R1=matrix(0,nrow=m,ncol=n); a=c()
for( i in 1:m)
{
set.seed(i)
R1[i,]=f(1,p1,n)
a[i]=R1[i,n]
}
b=length(which(a==0))/m
b # estimate of P[Z_25 = 0]
```

□

In Theorem 5.4.1, we have proved that the extinction probability q is a solution of the equation $P(s) = s$. We use the following code to compute q, graphically and algebraically.

Code 5.6.7 Computation of extinction probability: This code computes q, graphically and algebraically. We illustrate the code for the BGW branching process in Example 5.5.1, where the supports of the offspring distributions are $\{0, 1, \ldots, 6\}$, for each of the three sets of probability mass functions, so that offspring mean $\mu > 1, \mu = 1, \mu < 1$.

```
# Part I: Input offspring distributions
x=c(0,1,2,3,4,5,6)
# Offspring distribution, mean > 1
p1=c(0.4, 0.2, 0.11, 0.11, 0.05, 0.09, 0.04); mu1=sum(p1*x)
# Offspring distribution, mean = 1
p2=c(0.7, 0.05, 0.05, 0.07, 0.05, 0.04, 0.04); mu2=sum(p2*x)
# Offspring distribution, mean < 1
p3=c(0.7, 0.05, 0.09, 0.04, 0.04, 0.04 ,0.04); mu3=sum(p3*x)
mu=c(mu1,mu2,mu3);mu
# Part II: Function to obtain q graphically
g=function(pr)
{
s=seq(0,1,0.0001); ls=length(s); pgf=c()
for(i in 1:ls)
{
pgf[i]=sum(s[i]^x*pr)
}
pgf
epsilon=0.000005
k=min(which(abs(s-pgf)<=epsilon)); extprob=s[k]
plot(s,pgf,type="l", xlab="s", ylab="P(s)",xlim=c(0,1),
ylim=c(.35,1),col="dark blue",lwd=2,main=sum(pr*x))
lines(x=s,y=s,lty=6,col="dark blue",lwd=2)
abline(v=extprob,lty=3,lwd=2,col="purple")
legend("bottomright",legend=c("P(s)","Line x=y",
"Vertical line at q"),lty=c(1,6,3),
col=c("dark blue","blue","purple"))
return(extprob)
}
par(mfrow=c(1,3)); g(p1);g(p2);g(p3)
# Part III: To find q algebraically by solving P(s)=s
# Newton-Raphson method
ep=function(pr) # function for offspring distribution pr
{
g1=function(x)
{
a=c(1,x,x^2,x^3,x^4,x^5,x^6); y=sum(a*pr)-x; return(y)
}
dg=function(x)
```

```
{
p=pr[-1]; b=c(1,2*x,3*x^2,4*x^3,5*x^4,6*x^5);y1=sum(b*p)-1
return(y1)
}
u=c(); u[1]=.4;i=1;diff=1;
while(diff>10^(-4))
{
u[i+1]=u[i]- g1(u[i])/dg(u[i])
diff=abs(u[1+i]-u[i]);i=i+1
}
sol=u[i];return(sol)
}
ep1=round(c(ep(p1),ep(p2),ep(p3)),4);ep1
# Part IV: To find q using built-in function
extp=function(pr)
{
e=c(0,1,0,0,0,0,0); e1=pr-e; r=polyroot(e1); return(r)
}
round(extp(p1),3); round(extp(p2),3); round(extp(p3),3)
```

□

For a branching process with offspring mean ≤ 1, the probability of ultimate extinction is 1. The following code is to estimate the expected extinction time $E(T)$.

Code 5.6.8 Estimation of expected extinction time: In this code, we obtain the multiple realizations and based on it estimate $E(T)$ as follows. We simulate the process for 50 generations for $m = 200$ times. For ith realization, we note the first n for which $Z_n = 0$, which is the realized value of T_i of T. In some realizations, Z_n may not be 0 up to $n = 50$. We omit such realizations, say r. We then estimate $E(T)$ as $\sum_{i=1}^{m-r} T_i/(m-r)$. We illustrate the code for Example 5.5.3. Thus, $\{Z_n, n \geq 0\}$ is the BGW branching process when the offspring distribution is $\{p_0 = 0.7, p_1 = 0.05, p_2 = 0.09, p_3 = 0.04, p_4 = 0.04, p_5 = 0.04, p_6 = 0.04\}$ with μ is 0.95.

```
# Part I: Function for realization
x=c(0,1,2,3,4,5,6) # support of offspring distribution.
f=function(z0,pr,n) # function f for realization
{
z=c(); z[1]=z0
for(i in 2:n)
{
if(z[i-1]>0)
{
z[i]=sum(sample(x,z[i-1],pr,replace=T)
)
}
```

```
else
{
z[i]=0
}
}
return(z)
}
# Part II: Realization
p3=c(0.7,0.05,0.09,0.04,0.04,0.04,0.04) # offspring distribution
mu3=sum(p3*x);mu3 ;m=200; n=51
R3=matrix(0,nrow=m,ncol=n);b=c()
for( i in 1:m)
{
set.seed(i+10); R3[i,]=f(1,p3,n)
b[i]=min(which(R3[i,]==0))
}
a=which(b==Inf);a # Realization no for which z50 is not 0
b1=b[-a];b1 # To remove n corresponding to z0
e=sum(b1-1)/(m-length(a));e
```

□

Code 5.6.9 Realization of a branching process and computation of q: In this code, we obtain the realization for 20 generations and find q graphically and algebraically using Newton-Raphson method. At the beginning, a general code to generate a random sample from three discrete distributions, Poisson, binomial and geometric, is given. One can take any distribution as the offspring distribution. The code is illustrated for the branching process in Example 5.5.4. Thus, the offspring distribution is binomial $B(5, p)$ and $Z_0 = 5$. We take $p = 0.3, 0.2$ and 0.16, so that the offspring mean is $1.5, 1$ and 0.8, respectively. The extinction probability q when $Z_0 = 1$ is a solution of the equation $P(s) = s$ where $P(s) = (1 - p + ps)^5$.

```
# Part I: Function to obtain random sample from specified family
# rans(sample size,family, par1,par2);2nd parameter par2 may be 0.
rans=function(a,s,b,c)
{ if(s=="pois"){return(rpois(a,b))}
if(s=="binom"){return(rbinom(a,b,c))}
if(s=="geom"){return(rgeom(a,b))}}
# Part II: Realization of branching process
# For a realization of branching process we define a function f1 as
# f1(initial population Zo, No.of generations, offspring dist,
# parameters)
f1=function(zo,n,s,b,c)
{
z=c(); z[1]=zo
for(i in 2:n)
```

```
{
if(z[i-1]!=0)
{
z[i]=sum(rans(z[i-1],s,b,c))
if(z[i]>5000){break}
}
else
{
z[i]=0
break
}
}
return(z)
}
# For Z0=5, and binomial distribution B(n,p)
set.seed(110); r1=f1(5,20,"binom",5,0.3) # mean > 1
set.seed(111); r2=f1(5,20,"binom",5,0.2) # mean = 1
set.seed(130); r3=f1(5,20,"binom",5,0.16) # mean < 1
r1; r2; r3
# Part III: Graphs of realizations
par(mfrow=c(1,3))
plot(r1,type="b",main="Mean > 1",xlab="Generation",
ylab="Generation size", col="dark blue",lwd=2)
plot(r2,type="b",main="Mean=1",xlab="Generation",
ylab="Generation size",col="dark blue",lwd=2)
plot(r3,type="b",main="Mean < 1",xlab="Generation",
ylab="Generation size",col="dark blue",lwd=2)
g=function(p) ## to find q graphically
{
s=seq(0,1,0.00001); ls=length(s); pgf=c()
for(i in 1:ls)
{
pgf[i]=(1-p+p*s[i])^n
}
epsilon=0.00001; k=min(which(abs(s-pgf)<=epsilon)); extprob=s[k]
plot(s,pgf,type="l", xlab="s", ylab="P(s)",xlim=c(0,1),
ylim=c(.1,1),col="dark blue",main=n*p,lwd=2)
lines(x=s,y=s,lty=6,col="dark blue",lwd=2)
abline(v=extprob,lty=3,lwd=2, col="purple")
legend("bottomright",legend=c("P(s)","Line x=y",
"Vertical line at q"),lty=c(1,6,3),
col=c("dark blue","blue","purple"))
return(extprob)
}
n=5; p1=0.3;p2=0.2;p3=.16
```

```
par(mfrow=c(1,3))
g(p1);g(p2);g(p3)
# Part IV: To find q algebraically, using Newton Raphson method
ep=function(p)
{g1=function(s){y=(1-p+p*s)^n -s; return(y)}
dg=function(s){y1=n*p*((1-p+p*s)^(n-1))-1; return(y1)}
u=c(); u[1]=.3;i=1;diff=1;
while(diff>10^(-5))
{
u[i+1]=u[i]- g1(u[i])/dg(u[i])
diff=abs(u[1+i]-u[i]);i=i+1
}
sol=u[i];return(sol)
}
ep1=round(c(ep(p1),ep(p2),ep(p3)),4);ep1
```

□

A quick recap of the results discussed in the present chapter is given below.

Summary

1. Suppose $\{Y_{n,i}, n = 0, 1, 2, \ldots; i = 1, 2, \ldots\}$ is a double array of independent and identically distributed, non-negative integer valued random variables, such that for $n = 0, 1, 2, \ldots$ and $i = 1, 2, \ldots$,

$$P[Y_{n,i} = j] = p_j, \; j = 0, 1, \ldots \;\; p_j \geq 0 \;\; \& \; \sum_{j \geq 0} p_j = 1 \, .$$

Suppose $Z_0 = k \geq 1$ and Z_n for $n \geq 1$, is defined as follows:

$$Z_n \; = \; \begin{cases} \sum_{i=1}^{Z_{n-1}} Y_{n-1,i}, & \text{if } Z_{n-1} > 0 \\ 0, & \text{if } Z_{n-1} = 0 \, . \end{cases}$$

Then a stochastic process $\{Z_n, n \geq 0\}$ is known as a BGW branching process. The probability distribution $\underline{p} = \{p_j, j = 0, 1, 2, \ldots, \}$ is known as the offspring distribution.
2. A BGW branching process is a time homogeneous Markov chain with state space W.
3. For a BGW branching process state 0 is a non-null persistent, aperiodic state and under the assumption $0 < p_0 < 1$, all other states are transient.
4. If $P(s)$ is a probability generating function of the offspring distribution, then $\forall \; n \geq 0$, (i) $P_{n+1}(s) = P_n(P(s))$ and (ii) $P_{n+1}(s) = P(P_n(s))$. This property is known as a branching property.
5. For a branching process with offspring mean μ and offspring variance σ^2, $E(Z_n) = \mu^n$ and $Var(Z_n)$ is given by,

$$Var(Z_n) = \begin{cases} \sigma^2\mu^{n-1}\left(\frac{1-\mu^n}{1-\mu}\right), & \text{if } \mu \neq 1 \\ n\sigma^2, & \text{if } \mu = 1 . \end{cases}$$

6. Suppose $\{Z_n, n \geq 0\}$ is a branching process with $P(s) = \sum_{j=0}^{\infty} p_j s^j$ as a probability generating function of an offspring distribution and with $Z_0 = 1$. Suppose $0 < p_0 < 1$ and $0 < p_0 + p_1 < 1$. (i) Then the probability q of ultimate extinction is a solution of the equation $P(s) = s$, (ii) if $\mu \leq 1$ then $q = 1$ and (iii) if $\mu > 1$, then q is the smallest positive root of the equation $P(s) = s$.
7. BGW process is labeled as super-critical, critical and sub-critical, if $\mu > 1$, $\mu = 1$ and $\mu < 1$, respectively.

5.7 Conceptual Exercises

5.7.1 Suppose $\{Z_n, n \geq 0\}$ is a branching process with $Z_0 = 1$ and offspring distribution given by $p_0 = 0.5$, $p_1 = 0.1$, $p_3 = 0.4$. What is the probability that the population becomes extinct in the second generation, given that it is not extinct in the first generation?

5.7.2 Suppose $\{Z_n, n \geq 0\}$ is a branching process with $Z_0 = 1$. (i) Show that the probability generating function of the conditional distribution of Z_n given that $Z_n > 0$ is given by $(P_n(s) - P_n(0))/(1 - P_n(0))$, $|s| \leq 1$, where $P_n(s)$ is the probability generating function of Z_n given $Z_0 = 1$. Find (ii) $P[Z_n = i|Z_n > 0]$, $i \geq 1$ and (iii) $E(Z_n|Z_n > 0)$, when the offspring distribution is geometric with parameter $1/2$. Medhi [12].

5.7.3 In every generation of a population, each individual in the population dies with probability $1/2$ or doubles with probability $1/2$. Suppose Z_n denotes the number of individuals in the population in the nth generation. Find the mean and variance of Z_n.

5.7.4 The number of offspring of an individual in a population is 0, 1 or 2 with respective probabilities $a > 0$, $b > 0$ and $c > 0$, where $a + b + c = 1$. Express the mean and the variance of the offspring distribution in terms of b and c. Find the mean and variance of Z_5 given that $Z_0 = 1$.

5.7.5 Suppose a parent has no offspring with probability $1/2$ and has two offspring with probability $1/2$. If a population of such individuals begins with a single parent and evolves as a branching process, find the probability that the population is extinct by the nth generation, for $n = 1, 2, 3, 4, 5$.

5.7.6 At each stage of an electron multiplier, each electron, upon striking the plate, generates a random number Y of electrons for the next stage, where Y follows Poisson distribution with mean λ. Determine the mean and the variance of the number of electrons at the nth stage.

5.7.7 At time 0, a blood culture starts with one red cell. At the end of one minute, the red cell dies and is replaced by one of the following combinations—2 red cells with probability $1/4$, 1 red and 1 white cells with probability $2/3$ and

2 white cells with probability $1/12$. Each red cell lives for one minute and gives birth to offspring in the same way as the parent cell. Each white cell lives for one minute and dies without reproducing. Assume that individual cells behave independently. (i) At time $n+1$ minutes after the culture begins, what is the probability that no white cells have yet appeared? (ii) What is the probability that the entire culture eventually dies out? Karlin and Taylor [10].

5.7.8 Suppose $\{Z_n, n \geq 1\}$ is a BGW branching process with $P(s) = as^2 + bs + c$, where a, b, c are positive and $P(1) = 1$. Assume that the probability of extinction is $q \in (0, 1)$. Prove that (i) $c < a$ and (ii) $q = c/a$. Karlin and Taylor [10].

5.7.9 The offspring distributions of some branching processes with $Z_0 = 1$, are as follows:

(a) $p_0 = 0.3,\ p_1 = 0.6,\ p_2 = 0.05,\ p_3 = 0.05$
(b) $p_0 = 0.2,\ p_1 = 0.2,\ p_2 = 0.3,\ p_3 = 0.3$
(c) $p_0 = 0.25,\ p_1 = 0.50,\ p_2 = 0.25$
(d) $p_0 = 0.25,\ p_1 = 0.40,\ p_2 = 0.35$
(e) $p_0 = 0.5,\ p_1 = 0.1,\ p_2 = 0.4$.

Find the offspring mean and determine the probability of eventual extinction in each case. Comment on the relation between the offspring mean and the extinction probability.

5.7.10 One-fourth of the married couples in a society have no children. The other three-fourths of families continue to have children until they have a girl and then cease childbearing. Assume that each child is equally likely to be a boy or a girl. (i) What is the probability that a particular husband will have k male offspring, $k = 0, 1, 2, \ldots,$? (ii) What is the probability that the husband's male line will cease to exist by the 5th generation? Taylor and Karlin [16].

5.8 Computational Exercises

5.8.1 Suppose $\{Z_n, n \geq 0\}$ is a branching process. The support of the offspring distribution is $\{0, 1, 2, 3, 4\}$ with the following three sets of respective probabilities: (a) $\{0.53, 0.17, 0.15, 0.1, 0.05\}$, (b) $\{0.45, 0.25, 0.2, 0.05, 0.05\}$ and (c) $\{0.38, 0.25, 0.22, 0.1, 0.05\}$.

(i) For each of the offspring distributions, find the offspring mean and a realization of the branching process till 20 generations when $Z_0 = 1, 3, 5$. Comment on the findings. Draw the graphs of realization for $Z_0 = 1$.
(ii) Obtain multiple realizations with $Z_0 = 1$ and based on these realizations estimate $P[Z_{20} = 0]$ for the three offspring distributions.
(iii) Using the recurrence relation $P_{n+1}(0) = P(P_n(0))$, find $P[Z_n = 0]$ for $n = 30, 60, 90, 120, 150$, for the three offspring distributions.

(iv) Find the probability of ultimate extinction graphically, algebraically using Newton-Raphson method and `polyroot` function.
(v) Comment on the results obtained in (iii) and (iv).
(vi) When the offspring mean is ≤ 1, obtain the estimate of time to extinction.

5.8.2 Suppose $\{Z_n, n \geq 0\}$ is a branching process with the offspring distribution as Poisson $P(\lambda)$.

(i) For three values of $\lambda < 1, = 1, > 1$, obtain a realization of the branching process till 25 generations when $Z_0 = 1, 4, 6$. Use `break` function. Comment on the findings. Draw the graphs of realization for $Z_0 = 1$.
(ii) Obtain multiple realizations with $Z_0 = 1$ and based on these realizations estimate $P[Z_{25} = 0]$ for the three offspring distributions corresponding to three values of $\lambda < 1, = 1, > 1$.
(iii) Using the recurrence relation $P_{n+1}(0) = P(P_n(0))$, find $P[Z_n = 0]$ for $n = 30, 60, 90, 120, 150$, for the three offspring distributions.
(iv) Find the probability of ultimate extinction graphically and algebraically using Newton-Raphson method.
(v) Comment on the results obtained in (iii) and (iv).
(vi) When the offspring mean is ≤ 1, obtain the estimate of time to extinction.
(vii) When the offspring mean is ≤ 1, obtain the distribution of time to extinction.

5.8.3 Suppose $\{Z_n, n \geq 0\}$ is a branching process with the offspring distribution as $\{p_0 = p, p_5 = 1 - p\}$. Find the probability of ultimate extinction using Newton-Raphson method and built-in-function `polyroot` for $p = 0.5, 0.7, 0.3$.
5.8.4 Suppose $\{Z_n, n \geq 0\}$ is a branching process with $Z_0 = 1$ and offspring distribution as binomial $B(6, p)$. Find the distribution of extinction time corresponding to 2 values p, such that offspring mean is less than 1 and is 1.
5.8.5 Suppose $\{Z_n, n \geq 0\}$ is a branching process with $Z_0 = 1$ and the offspring distribution to be geometric with parameter $1/2$. Find the probability distribution of Z_3, Z_6 and Z_9. Comment on the results.
5.8.6 In Conceptual Exercise 5.7, you have obtained $P[Z_n = i | Z_n > 0]$, when the offspring distribution is geometric with parameter $1/2$ and $Z_0 = 1$. Compute the probability distribution for $n = 5$, 10 and 15 and its expectation. Examine whether the expectation is close to $n + 1$. Comment on your findings.

5.9 Multiple Choice Questions

Note: In each of the questions, more than one options may be correct:

5.9.1 Which of the following options is/are always correct? Suppose $\{Z_n, n \geq 0\}$ is a BGW branching process with offspring distribution $\{p_j, j \geq 0\}$. Under the condition $0 < p_0 < 1$

(a) state 0 is a non-null persistent state
(b) states $i > 0$ are transient
(c) states $i > 0$ are null persistent
(d) all states $i > 0$ are aperiodic.

5.9.2 Which of the following is/are NOT correct? Suppose $\{Z_n, n \geq 0\}$ is a BGW branching process with offspring distribution $\{p_j, j \geq 0\}$. Under the assumption $0 < p_0 < 1$,

(a) state 0 is an absorbing state
(b) states $i > 0$ are transient
(c) states $i > 0$ are null persistent
(d) all states are non-null persistent.

5.9.3 Suppose $\{Z_n, n \geq 0\}$ is a BGW branching process with offspring distribution $\{p_j, j \geq 0\}$ and f_{ii} denotes the probability of ever return to state $i > 0$. Which of the following options is/are correct?

(a) $f_{ii} < 1 - p_0^i$
(b) $f_{ii} = 1 - p_0^i$
(c) $f_{ii} = (1 - p_0)^i$
(d) $f_{ii} < (1 - p_0)^i$.

5.9.4 Suppose $\{Z_n, n \geq 0\}$ is a BGW branching process with the offspring mean 0.98 and offspring variance 1.32. Which of the following options is/are correct? The probability of ultimate extinction

(a) is 1
(b) is .98
(c) is 0
(d) cannot be computed in view of insufficient information.

5.9.5 Following are two statements: (I) Long-run distribution of a BGW branching process exists. (II) Stationary distribution of a BGW branching process is $(1, 0, 0, \ldots)'$.
Which of the following is a correct option?

(a) Both (I) and (II) are false
(b) Both (I) and (II) are true
(c) (I) is true but (II) is false
(d) (I) is false but (II) is true.

5.9.6 Following are two statements. (I) Long-run distribution of a BGW branching process does not exist. (II) Long-run distribution of a BGW branching process is $(1, 0, 0, \ldots)'$.
Which of the following is a correct option?

(a) Both (I) and (II) are false
(b) Both (I) and (II) are true
(c) (I) is true but (II) is false

(d) (I) is false but (II) is true.

5.9.7 Suppose $\{Z_n, n \geq 0\}$ is a BGW branching process with $P(s)$ as a probability generating function of the offspring distribution with $Z_0 = 1$ and $P_n(s)$ as a probability generating function of Z_n. Following are two statements. $\forall\ n \geq 0$, (I) $P_{n+1}(s) = P_n(P(s))$ (II) $P_{n+1}(s) = P(P_n(s))$.
Which of the following is a correct option?

(a) Both (I) and (II) are false
(b) Both (I) and (II) are true
(c) (I) is true but (II) is false
(d) (I) is false but (II) is true.

5.9.8 Suppose $\{Z_n, n \geq 0\}$ is a BGW branching process with $P(s)$ as a probability generating function of the offspring distribution with $Z_0 = 5$ and $P_n(s)$ as a probability generating function of Z_n. Following are two statements. $\forall\ n \geq 0$, (I) $P_{n+1}(s) = P_n(P(s))$ (II) $P_{n+1}(s) = P(P_n(s))$.
Which of the following is a correct option?

(a) Both (I) and (II) are false
(b) Both (I) and (II) are true
(c) (I) is true but (II) is false
(d) (I) is false but (II) is true.

5.9.9 Suppose $\{Z_n, n \geq 0\}$ is a BGW branching process with $Z_0 = 1$, $\mu = E(Z_1)$ and $\sigma^2 = Var(Z_1)$. Then which of the following is a correct option?

(a) $E(Z_n) = \mu^{n-1}$ & $Var(Z_n) = \sigma^2\mu^{n-1}(1 + \mu + \cdots + \mu^{n-1})$
(b) $E(Z_n) = \mu^n$ & $Var(Z_n) = \sigma^2\mu^n(1 + \mu + \cdots + \mu^n)$
(c) $E(Z_n) = \mu^n$ & $Var(Z_n) = \sigma^2\mu^{n-1}(1 + \mu + \cdots + \mu^n)$
(d) $E(Z_n) = \mu^n$ & $Var(Z_n) = \sigma^2\mu^{n-1}(1 + \mu + \cdots + \mu^{n-1})$.

5.9.10 Suppose $\{Z_n, n \geq 0\}$ is a BGW branching process with $Z_0 = 1$ and q denotes the probability of ultimate extinction. Suppose $0 < p_0 < 1$ and $0 < p_0 + p_1 < 1$. Which of the following options is/are correct?

(a) q is a solution of the equation $P(s) = s$
(b) $q = 1$ if $\mu < 1$
(c) $q = 1$ if $\mu = 1$
(d) $q < 1$ if $\mu > 1$.

5.9.11 Suppose $\{Z_n, n \geq 0\}$ is a BGW branching process with $Z_0 = 1$ and q denotes the probability of ultimate extinction. Suppose $0 < p_0 < 1$ and $0 < p_0 + p_1 < 1$. Which of the following options is/are correct?

(a) q is a solution of the equation $P(s) = 0$.
(b) $q = 1$ if $\mu < 1$
(c) $q = 1$ if $\mu = 1$
(d) $q < 1$ if $\mu > 1$.

5.9.12 Suppose $\{Z_n, n \geq 0\}$ is a BGW branching process with $Z_0 = 1$ and q denotes the probability of ultimate extinction. Suppose $0 < p_0 < 1$ and $0 < p_0 + p_1 < 1$. Then which of the following is NOT correct?

(a) q is a solution of the equation $P(s) = 0$.
(b) $q = 1$ if $\mu < 1$
(c) $q = 1$ if $\mu = 1$
(d) $q < 1$ if $\mu > 1$.

5.9.13 Suppose $\{Z_n, n \geq 0\}$ is a BGW branching process with $Z_0 = 1$ and q denotes the probability of ultimate extinction. Suppose $0 < p_0 < 1$ and $0 < p_0 + p_1 < 1$. Which of the following options is/are correct?

(a) q is a solution of the equation $P(s) = s$.
(b) $q = 1$ if $\mu > 1$
(c) $q = 1$ if $\mu = 1$
(d) $q < 1$ if $\mu < 1$.

5.9.14 Suppose $\{Z_n, n \geq 0\}$ is a BGW branching process with $Z_0 = 5$. Suppose q denotes the solution of the equation $P(s) = s$. Then the probability of ultimate extinction

(a) $q/5$
(b) q
(c) q^5
(d) $q^{1/5}$.

5.9.15 Suppose $\{Z_n, n \geq 0\}$ is a BGW branching process with $Z_0 = 1$ and $\mu = E(Z_1)$. Which of the following options is/are correct?

(a) The branching process is super-critical if $\mu > 1$
(b) The branching process is critical if $\mu = 1$
(c) The branching process is sub-critical if $\mu < 1$
(d) The probability of extinction is 0 if $\mu > 1$.

5.9.16 Suppose $\{Z_n, n \geq 0\}$ is a BGW branching process with $Z_0 = 1$ and $\mu = E(Z_1)$. Following are three statements. The branching process is (I) super-critical if $\mu < 1$, (II) critical if $\mu = 1$ and (III) sub-critical if $\mu > 1$. Then which of the following statements is true?

(a) Both (I) and (II) are true
(b) Both (I) and (III) are true
(c) Both (II) and (III) are true
(d) Only (II) is true.

5.9.17 Suppose $\{Z_n, n \geq 0\}$ is a BGW branching process with $Z_0 = 1$ and $\mu = E(Z_1)$. Following are three statements. In the expected value sense, the process grows (I) geometrically if $\mu > 1$, (II) stays constant if $\mu = 1$ and (III) decays geometrically if $\mu < 1$.
Which of the following options is/are correct?

(a) Only (I) and (II) are true
(b) Only (I) and (III) are true
(c) Only (II) and (III) are true
(d) All three are true.

5.9.18 Suppose the offspring distribution of a BGW branching process is given by $P(s) = 0.25 + 0.5s + 0.25s^2$. Which of the following options is/are correct?

(a) The branching process is sub-critical
(b) The branching process is super-critical
(c) The probability of ultimate extinction is 0
(d) The probability of ultimate extinction is 1.

5.9.19 Suppose the offspring distribution of a BGW branching process $\{Z_n, n \geq 0\}$ is Bernoulli $B(1, p),\ 0 < p < 1$ and $Z_0 = 5$. Which of the following options is/are correct?

(a) The probability of ultimate extinction is 1
(b) The probability of ultimate extinction is p^5
(c) The probability of ultimate extinction is $(1 - p)^5$
(d) The probability of ultimate extinction is 0.

References

1. Asmussen, S., & Hering, S. (1983). *Branching processes*. Boston: Birkhausser.
2. Athreya, K. B., & Ney, P. (1972). *Branching processes*. Berlin: Springer.
3. Bailey, N. T. J. (1975). *The mathematical theory of infectious diseases and its applications*. London: Griffin.
4. Becker, N. (1976). Estimation for an epidemic model. *Biometrics, 32*(4), 769–777.
5. Bienaymme, I. J. (1845). De la loi de Multiplication et de la Duree des Familles. *Société Philomathique de Paris, 5*, 37–39.
6. Galton, F., & Watson, H. W. (1875). On the probability of extinction of the families. *Journal of the Anthropological Society of London, 4*, 138–144.
7. Harris, T. E. (1963). *The Theory of branching processes*. Berlin: Springer.
8. Jagers, P. (1975). *Branching processes with biological applications*. New York: Wiley.
9. Kallenberg, P. J. M. (1979). Branching processes with continuous state space. Mathematical Centre Tracts No. 117, Amsterdam.
10. Karlin, S., & Taylor, H. M. (1975). *A first course in stochastic processes*. New York: Academic Press.
11. Kashikar, A. S., & Deshmukh, S. R. (2014). Estimation in second order branching processes with application to swine flu data. *Communications in Statistics—Theory and Methods*.
12. Medhi, J. (1994). *Stochastic processes*. New Delhi: Wiley Eastern.
13. Mode, C. J. (1971). *Multitype branching processes*. New York: Elsevier.
14. Sevast'yanov, B. A., & Zubkov, A. M. (1974). Controlled branching process. *Thoery of Probability and Its Applications, XIX*(1), 14–24.
15. Smith, W. L., & Wilkinson, W. E. (1969). On branching processes in random environments. *Annals of Mathematical Statistics, 40*, 814–827.
16. Taylor, H. N., & Karlin, S. (1984). *An introduction to stochastic Modeling*. New York: Academic Press.

Chapter 6
Continuous Time Markov Chains

6.1 Introduction

Markov chains investigated in Chaps. 2–5 are stochastic processes, discrete in both time and state space. In Markov chains, we concentrated on the probabilities of transition from one state to the another in a fixed number of steps. In applications, when considering processes which are evolving in "real time", it means concentrating on the transitions but ignoring the actual times spent in states in between the transitions. For example, in the queuing chain introduced in Chap. 4, we concentrate on the probabilities of queue size increasing by one unit or decreasing by one unit in one unit time. The one unit time is actually a random variable, governed by two random mechanisms—(i) the random duration between arrivals of customers and (ii) the random service time of a customer.

The processes we introduce in the present chapter take into account not only transitions from a state, but also the actual times spent in a state before transition. Thus, a natural generalization of the Markov chain in discrete time is to allow the time between successive transitions to be a continuous random variable.

As an illustration, suppose a machine is in a working state for a random amount of time. When it fails, it gets repaired. The repair time is also a random variable with a certain probability distribution. Suppose a workshop has two such machines which work independently of each other and with the same failure and repair time distributions. Suppose $X(t)$ denotes the number of machines working at time t. Then the state space is $S = \{0, 1, 2\}$. The process is observed continuously, that is, at each time point we observe the state of the process. Thus, $\{X(t), t \geq 0\}$ is a continuous time discrete state space stochastic process. Suppose initially both the machines are in working condition. Thus, $X(0) = 2$. Figure 6.1 displays possible scenarios when the process is observed over a period of 20 h.

At $t = 0$, both the machines are in working condition. After time 4.55 h, one of the two machines fails and undergoes repair. It is repaired and after 20 min, both the machines are in working condition for certain duration. One of the two fails after some time and is under repair. After certain duration, the one which is working also

S. Madhira and S. Deshmukh, *Introduction to Stochastic Processes Using R*,
https://doi.org/10.1007/978-981-99-5601-2_6

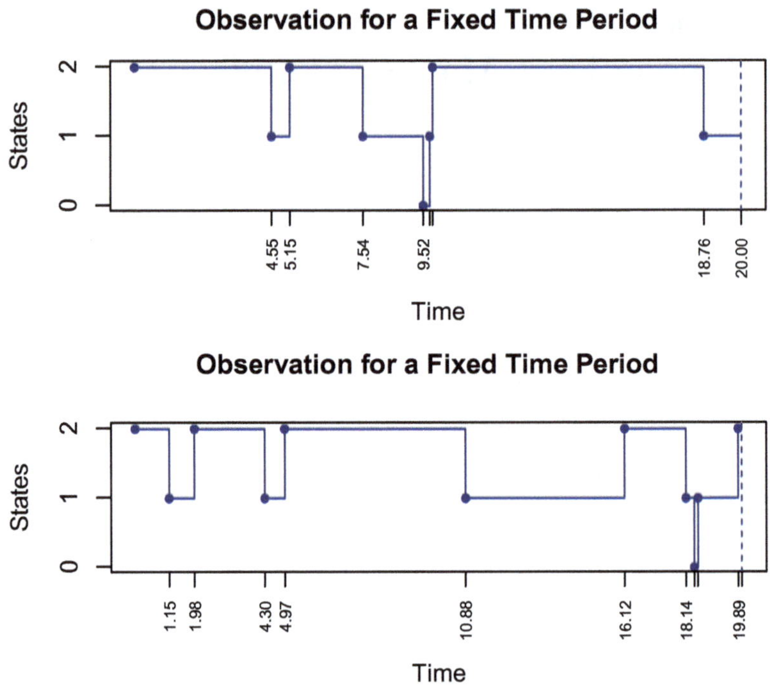

Fig. 6.1 Two realizations of a process for a fixed time period

fails, thus, the state of the system is 0. One of the two gets repaired and $X(t) = 1$. After some time, the second one is also in a working condition. Thus, the evolution of the process $\{X(t), t \geq 0\}$ for a period [0, 20] can be presented as follows.

$$X(t) = \begin{cases} 2, \text{ if } \; 0 \leq t < 4.55 \\ 1, \text{ if } 4.55 \leq t < 5.15 \\ 2, \text{ if } 5.15 \leq t < 7.54 \\ \quad \vdots \\ 1, \text{ if } 18.76 \leq t \leq 20. \end{cases}$$

The realization from the second graph can also be presented as above. Note that in the second graph, the state of the system, the failure times and repair times are different from the first for the period of 20 h. Thus, from these two graphs, we observe that the state of the system is governed by two random mechanisms, one corresponds to changes of states and the other corresponds to the duration for which the system remains in a particular state, such a duration for this illustration is dictated by the probability distributions of failure times and the repair times of the machines.

In the stochastic process $\{X(t), t \geq 0\}$ of the above illustration, we note that the process has jumps at random time points and it is constant between the jumps. The sample paths of the process are right continuous for all t. Such processes are known as jump processes. We define its one version below.

Definition 6.1.1 Pure Jump Process: Suppose $\{S_n, n \geq 0\}$ is a sequence of non-negative random variables such that (i) $S_0 = 0$, (ii) $S_n < S_{n+1}\ \forall\ n$ and (iii) $S_n \to \infty$ as $n \to \infty$. Suppose $\{X_n, n \geq 0\}$ is a sequence of random variables with countable state space S. The stochastic process $\{X(t), t \geq 0\}$ defined as

$$X(t) = X_n, \text{ if } S_n \leq t < S_{n+1}, \forall\ n \geq 0$$

is said to be a pure jump process.

The third condition $S_n \to \infty$ as $n \to \infty$ implies that there can be only a finite number of jumps in any finite interval. Hence, it is labeled as the pure jump process. It is also known as a non-explosive jump process, refer to Cinlar [1] and Hoel, Port and Stone [2]. If the state space is finite, then the jump process is necessarily non-explosive. A random variable $H_n = T_{X_{n-1}} = S_n - S_{n-1}$ is known as the holding time or sojourn time in state $X_{n-1}, n \geq 1$, where $T_i,\ i \in S$ is the sojourn time in state i. Note that S_n is the epoch of jump and $X(S_n) = X_n$. From the definition we note that the process has jumps at S_n, it is constant and is equal to X_n till the next jump, the magnitude of the jump at S_n being $X_n - X_{n-1}, n \geq 1$. Therefore, the sample paths of the pure jump process are right continuous for all t.

It is of interest to model the random duration H_n by an appropriate probability distribution. Different distributions lead to different stochastic processes. In the next section, we define a Markov pure jump process, in which the distribution of H_n is exponential. It then satisfies the Markov property and is known as a Markov process. Another approach to define a Markov process is to define it as a continuous time discrete state space stochastic process which satisfies the Markov property. From the Markov property, it then follows that the distribution of the holding time random variable T_i in state i is exponential. A Markov process is completely specified by two components: one is the sojourn time random variables T_i and the other component is the sequence $\{X_n, n \geq 0\}$ of states visited at transitions. It is proved that $\{X_n, n \geq 0\}$ is a Markov chain which is known as an embedded Markov chain associated with the Markov process $\{X(t), t \geq 0\}$. Hence, $\{X(t), t \geq 0\}$ is also known as a continuous time Markov chain. We discuss in detail the probabilistic structure of a Markov process in Sect. 6.2.

In Sect. 6.3, we investigate some properties of transition probability function. Section 6.4 is devoted to an important concept known as the infinitesimal generator of the continuous time Markov chain, which determines the evolution of the process, that is, the distribution of the holding times and transition probabilities of the embedded Markov chain. Section 6.5 is concerned with the solution of Kolmogorov's forward and backward differential equations and some methods to compute transition probability function numerically. In Sect. 6.6, we discuss the long run behavior of the Markov process in terms of the associated stationary distributions and the long run distribution. Section 6.7 presents the `R` codes.

We need some properties of an exponential distribution to establish the Markov property. Exponential distribution also plays a major role in Poisson processes which we study in Chap. 7. For ready reference, some results about the exponential distribution are listed below.

Properties of exponential distribution: Suppose a random variable X has exponential distribution with location parameter 0 and scale parameter, also known as a rate parameter, λ where $\lambda > 0$.

(i) The probability density function f and the distribution function F of X are given by

$$f(x) = \begin{cases} \lambda e^{-\lambda x}, & \text{if } x \geq 0 \\ 0, & \text{if } x < 0 . \end{cases}$$

$$F(x) = \begin{cases} 0, & \text{if } x < 0 \\ 1 - e^{-\lambda x}, & \text{if } x \geq 0 . \end{cases}$$

(ii) $E(X) = 1/\lambda, \ Var(X) = 1/\lambda^2, \ \mu_r' = r!/\lambda^r, \ r \geq 1$, coefficient of skewness given by $\mu_3/\mu_2^{3/2}$ is 2 and coefficient of kurtosis given by $\mu_4/\mu_2^2 - 3$ is 6, for any value of λ.

(iii) The unique feature of an exponential distribution is that it is the only continuous distribution with the lack of memory property, which can be stated as

$$P[X > t + s | X > s] = P[X > t] \ \forall \ s, t > 0 .$$

(iv) Suppose $X_1, X_2, \ldots, X_n$ are independent and identically distributed random variables each having exponential distribution with scale parameter λ. Then the distribution of $S_n = X_1 + X_2 + \cdots + X_n$ is gamma $G(\lambda, n)$ with scale parameter λ and shape parameter n. When shape parameter is an integer, the gamma distribution is known as an Erlang distribution. In Poisson process, the epochs of occurrence of events, that is, times at which changes in states occur, have an Erlang distribution.

(v) Suppose $X_1, X_2, \ldots, X_n$ are independent random variables and X_i has exponential distribution with scale parameter $\lambda_i, \ i = 1, 2 \ldots, n$. Then the distribution of $Z_n = \min\{X_1, X_2, \ldots, X_n\}$ is exponential with scale parameter $\theta = \sum_{i=1}^n \lambda_i$ and the probability that $Z_n = X_i$ is λ_i / θ.

(vi) The conditional distribution of residual life $X - t$ given that $X > t$ is again exponential with scale parameter λ.

(vii) The hazard rate $r(t) = f(t)/(1 - F(t)) = \lambda$ is a constant. Exponential distribution is the only distribution with constant hazard rate function.

6.2 Definition and Properties

We begin with the definition of a Markov pure jump process. It is also referred to as a Markov process with discrete state space, or as a continuous time Markov chain. Throughout the book, we use any one of these three terms.

Definition 6.2.1 Markov Pure Jump Process: Suppose $\{X(t), t \geq 0\}$ is a pure jump process, where $T_i,\ i \in S$ is the sojourn time in state i. It is said to be a Markov pure jump process, if the following three conditions are satisfied.

(i) $\{X_n, n \geq 0\}$ is a homogeneous Markov chain with countable state space S, initial distribution $\underline{p}^{(0)}$ and transition probability matrix $P = [p_{ij}], i, j \in S$ such that, for each $i \in S$, $p_{ii} = 0$ or 1.

(ii) $\{H_n, n \geq 1\}$ is a sequence of independent random variables and $H_n \equiv T_{X_{n-1}}$ has exponential distribution with parameter $\lambda_{X_{n-1}}$ for $n = 1, 2, \ldots$.

(iii) $\hat{\lambda} = \sup_{i \in S} \lambda_i < \infty$ and $\lambda_i = 0$ if and only if $p_{ii} = 1$.

The process $\{X_n, n \geq 0\}$ is known as the embedded Markov chain of the process $\{X(t), t \geq 0\}$.

The definition states that $\{X_n, n \geq 0\}$ is a Markov chain. In the following theorem, we prove that the process $\{X(t), t \geq 0\}$ also satisfies the Markov property.

Theorem 6.2.1 *Suppose $\{X(t), t \geq 0\}$ is a Markov pure jump process with countable state space S. Then $\forall\ s < t$ and $j \in S$,*

$$P[X(t) = j | X(u), 0 \leq u \leq s] = P[X(t) = j | X(s)], \quad \textit{almost surely},$$

that is, $\{X(t), t \geq 0\}$ satisfies the Markov property.

Proof Suppose $X(s) = i$ and R_i is the remaining holding time in state i. Since total holding time in state i has exponential distribution with rate parameter λ_i, by the memoryless property of the exponential distribution, R_i has the same distribution, which does not depend on what happened before time s. Now, at time $s + R_i$, the process jumps to some state $k \neq i$ with probability p_{ik}. This transition to state k does not depend on what happened prior to time s. Thus, the remaining holding time in state i, as well as transition to next state k, depends only on i, the state at time s, but not on what happened prior to time s. Also, after moving to state k, further evolution does not depend even on state i and time s. Hence, the past before time s has no effect on future after time s, given the current state at time s. Hence the result. □

The equation $P[X(t) = j | X(u), 0 \leq u \leq s] = P[X(t) = j | X(s)]$ is the Markov property in continuous time. It is analogous to the Markov property for a Markov chain in discrete time.

Markov pure jump process is also defined as a pure jump process which satisfies the Markov property, refer to Hoel, Port and Stone [2]. The following derivation of an expression for $P[X(t) = j | X(s) = i]$ is adapted from Hoel, Port and Stone [2].

Theorem 6.2.2 *Suppose $P_{ij}^{(n)}(s,t)$ denotes the conditional probability that the process makes n transitions during $(s,t]$ and it will be in state j at time t, given that it is in state i at time s. Suppose $N(t)$ denotes the number of state changes, that is, jumps in $(0,t]$, so that $N(t)-N(s)$ represents the number of state changes in $(s,t]$. Then,*

$$P[X(t)=j|X(s)=i]=\sum_{n=0}^{\infty}P_{ij}^{(n)}(s,t), \quad where \quad P_{ij}^{(0)}(s,t)=\delta_{ij}\ e^{-\lambda_i(t-s)}$$

$$\&\ P_{ij}^{(n)}(s,t)=\int_0^{t-s}\lambda_i\ e^{-\lambda_i\ u}\sum_{k\in S-\{i\}}p_{ik}P_{kj}^{(n-1)}(s+u,t)\ du, \quad \forall\ n\geq 1,$$

where δ_{ij} is the Kronecker delta function where $\delta_{ij}=1$, if $i=j$ and 0 if $i\neq j$.

Proof Observe that

$$\begin{aligned}P[X(t)=j|X(s)=i]&=\sum_{n=0}^{\infty}P[X(t)=j,N(t)-N(s)=n|X(s)=i]\\&=\sum_{n=0}^{\infty}P_{ij}^{(n)}(s,t).\end{aligned}$$

Note that $P_{ij}^{(0)}(s,t)$ is the conditional probability that the process will be in state j at time t without changing the state during $(s,t]$, given that it is in state i at time s. This probability is 0, if $j\neq i$. If $j=i$, this probability is the same as the probability that the remaining holding time in state i is greater than $t-s$. By the memoryless property of exponential distribution, it is equal to $e^{-\lambda_i(t-s)}$. Hence, $P_{ij}^{(0)}(s,t)=\delta_{ij}\ e^{-\lambda_i(t-s)}$.

Suppose now $n\geq 1$. Thus, there is at least one state change in $(s,t]$. If $N(t)-N(s)=n$, the first change occurs at some time $s+u\in(s,t]$, the process jumps to some state $k\neq i$ at time $s+u$; and there will be $(n-1)$ state changes in $(s+u,t]$. Suppose R_i denotes the remaining holding time in state i, then we get $P_{ij}^{(n)}(s,t)$ as shown below.

$$P_{ij}^{(n)}(s,t)=P[X(t)=j,N(t)-N(s)=n|X(s)=i]$$

$$=\int_0^{t-s}P[X(t)=j,N(t)-N(s)=n,R_i=u|X(s)=i]f_{R_i}(u)\ du$$

$$=\int_0^{t-s}P[X(t)=j,X(s+u)\neq i,N(t)-N(s+u)=n-1,R_i=u|X(s)=i]f_{R_i}(u)\ du$$

$$=\int_0^{t-s}\Big(\sum_{k\in S-\{i\}}P[X(t)=j,X(s+u)=k,N(t)-N(s+u)=n-1,R_i=u|X(s)=i]\Big)$$

$$\times f_{R_i}(u)\ du$$

$$=\int_0^{t-s}\Big(\sum_{k\in S-\{i\}}P[X(t)=j,N(t)-N(s+u)=n-1|X(s+u)=k,R_i=u,X(s)=i]$$

$$\times P[X(s+u)=k,R_i=u|X(s)=i]\Big)f_{R_i}(u)\ du$$

$$
\begin{aligned}
&= \int_0^{t-s} f_{R_i}(u) \sum_{k \in S-\{i\}} P[X(t)=j, N(t)-N(s+u)=n-1 | X(s+u)=k] p_{ik}\, du \\
&= \int_0^{t-s} f_{R_i}(u) \sum_{k \in S-\{i\}} p_{ik} P_{kj}^{(n-1)}(s+u,t)\, du \\
&= \int_0^{t-s} \lambda_i\, e^{-\lambda_i u} \sum_{k \in S-\{i\}} p_{ik} P_{kj}^{(n-1)}(s+u,t)\, du.
\end{aligned}
$$

□

Corollary 6.2.1 *The Markov pure jump process is a homogeneous Markov process.*

Proof Suppose $h = t - s$ in the last step of the proof of Theorem 6.2.2. Then, we get

$$P_{ij}^{(n)}(s, s+h) = \int_0^h \lambda_i\, e^{-\lambda_i u} \sum_{k \in S-\{i\}} p_{ik} P_{kj}^{(n-1)}(s+u, s+h)\, du.$$

Since, $P_{ij}^{(0)}(s, s+h) = \delta_{ij} e^{-\lambda_i h}$ depends only on h, the length of the interval, but not on the beginning of the interval, by induction on n, it follows that for all n, $P_{ij}^{(n)}(s, s+h)$ also depends only on h, but not on s. Hence,

$$
\begin{aligned}
P_{ij}^{(n)}(s, s+h) &= P_{ij}^{(n)}(0, h), \ \ \forall\ n \\
\Rightarrow\ P[X(t)=j | X(s)=i] &= P[X(t-s)=j | X(0)=i].
\end{aligned}
$$

Thus, the Markov pure jump process is a homogeneous Markov process. □

We denote $P[X(t)=j|X(0)=i] = P[X(s+t)=j|X(s)=i]$ by $P_{ij}(t)$. $P_{ij}(t)$ is known as a transition probability function and is similar to n-step transition probabilities of a Markov chain in discrete time. Observe the difference in the notations $P_{ij}(t)$ and p_{ij}. The first with capital P is the transition probability function of the $\{X(t), t \geq 0\}$ process and the second is the transition probability of the embedded Markov chain $\{X_n, n \geq 0\}$.

For a homogeneous Markov process, the results in Theorem 6.2.2 can be expressed as follows.

Corollary 6.2.2 *Suppose $\{X(t), t \geq 0\}$ is a homogeneous Markov process, then $P[X(t)=j|X(0)=i] = \sum_{n=0}^{\infty} P_{ij}^{(n)}(t)$, where $P_{ij}^{(0)}(t) = \delta_{ij}\, e^{-\lambda_i t}$ and $\forall\ n \geq 1$,*

$$P_{ij}^{(n)}(t) = \int_0^t \lambda_i\, e^{-\lambda_i u} \sum_{k \neq i} p_{ik} P_{kj}^{(n-1)}(t-u)\, du.$$

In Definition 6.2.1, continuous time Markov chain is defined by specifying the probabilistic structure of the embedded chain and the sojourn time random variable. In Theorem 6.2.1, it is proved that the process satisfies the Markov property. We now give an alternative definition of a continuous time Markov chain, as a stochastic process which satisfies the Markov property (Karlin and Taylor [3]) and derive the probabilistic structure of the embedded chain and sojourn times.

Definition 6.2.2 Continuous Time Markov Chain: A continuous time discrete state space stochastic process $\{X(t), t \geq 0\}$ is said to be a continuous time Markov chain if for $t > s$ and $j \in S$,

$$P[X(t) = j|X(u), 0 \leq u \leq s] = P[X(t) = j|X(s)] \text{ almost surely}$$

or for $n \geq 1$,

$$P[X(t) = j|X(s) = i, X(t_{n-1}) = i_{n-1}, \ldots, X(t_0) = i_0] = P[X(t) = j|X(s) = i],$$

where $0 \leq t_0 < t_1 < \cdots t_{n-1} < s < t$ and $i_0, i_1, \ldots, i_{n-1}, i, j \in S$.

The conditional probability $P[X(t) = j|X(s) = i]$ may, in general, depend on both t and s and of course on i and j. If it depends on t and s via $t - s$, then a Markov process is time homogeneous. In this chapter we assume that the process is time homogeneous.

Remark 6.2.1 In Chap. 1, it is proved that a stochastic process with stationary and independent increments satisfies the Markov property. However, in view of the Corollary 1.3.1, the state space of a stochastic process with stationary and independent increments cannot be finite or a bounded interval. Hence, if the state space of a continuous time Markov chain is finite, it cannot be a process with stationary and independent increments. Even if the state space is countably infinite, it may not be a process with stationary and independent increments, in view of Remark 1.3.1.

By Theorem 6.2.1 and Corollary 6.2.1, it follows that Definition 6.2.1 of a Markov process implies Definition 6.2.2. To examine whether Definition 6.2.2 implies Definition 6.2.1, we note the following results.

A continuous time Markov chain evolves as shown in Fig. 6.1. Initially, at time $S_0 = T_0 = 0$, the process is in some state $X_0 \in S$. It remains in that state for some random time $T_{X_0} = S_1$ and then jumps to a new state X_1, independent of how long the system is in state X_0. The process remains in the new state for some random time T_{X_1} and jumps at $S_2 = T_{X_0} + T_{X_1}$ to another state X_2, independent of how long the system is in state X_1 and so on. Thus, $\{X_n, n \geq 0\}$ is a sequence of states of the process at successive transitions, where X_n is the state visited at S_n, the epoch of nth transition. With $H_n = T_{X_{n-1}} = S_n - S_{n-1}$, $\{H_n, n \geq 1\}$ is the sequence of holding times or sojourn times, between nth and $(n-1)$th transitions. Using Definition 6.2.2, we now investigate the probability structure of random variables H_n and also of states X_n, which need a couple of technical results, stated below.

(i) For a continuous time stochastic process $\{X(t), t \geq 0\}$, a random variable U with values in $[0, \infty)$ is a stopping time if, for each $t \in [0, \infty)$, the event $[U \leq t]$ depends

only on $[X(s), s \leq t]$.
(ii) The epoch S_n of jump is a stopping time of $\{X(t), t \geq 0\}$, $\forall\ n \geq 1$. Norris [6]
(iii) If $\{X(t), t \geq 0\}$ is a continuous time Markov chain and U is a stopping time such that $P[U < \infty] = 1$, then $\{X(U + t), t \geq 0\}$ is also a continuous time Markov chain such that

$$P[X(U+t) = j|X(U) = i, X(s) = x_s, 0 \leq s \leq U] = P[X(t) = j|X(0) = i].$$

For details refer to Norris [6]. This result is known as "strong Markov property". We use these results to obtain the distribution of sojourn times, for details refer to Miranda [5].

Theorem 6.2.3 *Suppose $\{X(t), t \geq 0\}$ is a time homogeneous continuous time Markov chain with transition probability function $P_{ij}(t)$, according to Definition 6.2.2. If $X(S_{n-1}) = i$, then the distribution of the sojourn time in state i, given by $H_n = T_{X_{n-1}} = S_n - S_{n-1}$ is exponential with scale parameter λ_i.*

Proof It is enough to show that the distribution of H_n satisfies the lack of memory property. For $s \geq 0$ the event $[H_n > s]$ is equivalent to the event that no transition has taken place up to s, which is equivalent to the event $[X(S_{n-1} + u) = i$ for $0 \leq u \leq s]$. Hence,

$$\begin{aligned}
&P[H_n > s + t|H_n > s, X(S_{n-1} + u) = i] \\
&= P[H_n > s + t|X(S_{n-1} + u) = i \text{ for } 0 \leq u \leq s] \\
&= P[H_n > s + t|X(S_{n-1} + s) = i] \text{ by strong Markov property} \\
&= P[H_n > t|X(S_{n-1}) = i] \text{ by time homogeneity.} \qquad (6.2.1)
\end{aligned}$$

Thus, the distribution of H_n satisfies the lack of memory property, given that the process starts in state i. Hence, the distribution of the sojourn time in state i is exponential with some parameter λ_i. □

Remark 6.2.2 A state i is said to an absorbing state if $\lambda_i = 0$, stable or non-absorbing state if $0 < \lambda_i < \infty$ and instantaneous state if $\lambda_i = \infty$, Cinlar [1]. If the state i is an absorbing state, then the system stays in i forever once it enters i. In this case the sojourn time random variable T_i is an extended valued random variable which equals ∞ with probability 1. If state i is stable, then the process stays in the state i for a positive but finite time, that is, $P[0 < T_i < \infty] = 1$. If i is instantaneous, $\lambda_i = \infty$ and the process jumps out of an instantaneous state as soon as it enters it. In this case $P[T_i = 0] = 1$. If the state space is finite then there are no instantaneous states, Cinlar [1]. The third condition $\hat{\lambda} = \sup_{i \in S} \lambda_i < \infty$ in Definition 6.2.1 assures that there are no instantaneous states. In this case the process is known as a regular process, Cinlar [1]. In this chapter, we restrict to regular continuous time Markov chains.

To investigate the nature of random variables X_n, we begin with a simple example of a continuous time Markov chain with state space $\{1, 2, 3\}$. Suppose initial state is 1. Then the system stays in state 1 for random time T_1 and the distribution of T_1 is exponential with scale parameter λ_1, which is the rate of transition from state 1. After leaving state 1, we need further information to figure out whether transition is to state 2 or to state 3. Therefore, we need to know the states visited at every transition, with some associated probabilities. Suppose X_n denotes the state visited at the nth transition and p_{ij} denotes the probability of transition from state i to state j at any transition epoch, that is, it does not depend on the time point of transition. From the Markov property of $\{X(t), t \geq 0\}$, it follows that $\{X_n, n \geq 0\}$ is a discrete time Markov chain with state space S and transition probabilities $\{p_{ij}, i, \ j \in S\}$. It is to be noted that $p_{ii} = 0$, if the state i is a non-absorbing state and $p_{ii} = 1$ and $p_{ij} = 0$ for all $j \neq i \in S$ if the state i is an absorbing state. Further, $\sum_{j \in S} p_{ij} = 1 \ \forall \ i \in S$.

The following theorem makes the above discussion precise. It specifies the probabilistic structure of a continuous time Markov chain. The proof involves the concept of strong Markov property, hence it is omitted. For proof we refer to Cinlar [1].

Theorem 6.2.4 *For* $n \geq 1, i, \ j \in S$ *and* $u > 0$,

$$P[X_n = j, H_n > u | X_0, X_1, \ldots, X_{n-1} = i; H_1, H_2, \ldots, H_{n-1}] = p_{ij} e^{-\lambda_i u} .$$

Remark 6.2.3 Implications of Theorem 6.2.4 are as follows. (i) If $u \downarrow 0$, then

$$P[X_n = j | X_0, X_1, \ldots, X_{n-1} = i; H_1, H_2, \ldots, H_{n-1}] = p_{ij}.$$

Thus, the sequence $\{X_n, n \geq 0\}$ of successive states visited is a Markov chain with the transition probability matrix $P = [p_{ij}]$. The conditional distribution of X_n given the entire past depends only on X_{n-1} and not on $H_1, H_2, \ldots, H_{n-1}$.
(ii) For each $i \in S$, $\sum_{j \in S} p_{ij} e^{-\lambda_i u} = e^{-\lambda_i u}$. Thus, the sojourn time random variable $H_n = T_i$ in state $X_{n-1} = i$ has an exponential distribution with parameter λ_i. Further, the distribution of H_n depends only on X_{n-1} and not on $H_1, H_2, \ldots, H_{n-1}$.
(iii) From the right-hand side of the identity in the above theorem, we have

$$\begin{aligned} P[H_n > u | X_n = j, X_{n-1} = i] &= \frac{P[H_n > u, X_n = j, X_{n-1} = i]}{P[X_n = j, X_{n-1} = i]} \\ &= \frac{P[H_n > u, X_n = j | X_{n-1} = i]}{P[X_n = j | X_{n-1} = i]} \\ &= \frac{p_{ij} e^{-\lambda_i u}}{p_{ij}} = e^{-\lambda_i u}, \quad u > 0, \end{aligned}$$

which is independent of j. Thus, the distribution of H_n is exponential with parameter depending on the state from which the transition occurs and not on the next state j.
(iv) By similar arguments, it can be proved that for $u_1, u_2, \ldots, u_n \in [0, \infty)$,

$$P[H_1 > u_1, \ldots, H_n > u_n | X_0 = i_0, \ldots, X_{n-1} = i_{n-1}] = e^{-\lambda_{i_0} u_1} \cdots e^{-\lambda_{i_{n-1}} u_n} .$$

Thus, the sojourn time random variables are conditionally independent of each other, given the successive states visited and each such sojourn time has an exponential distribution with the parameter depending on the state from which the transition occurs.

We now investigate one more property of sojourn time random variables. Since the sojourn time T_i in state i follows exponential distribution, for $h > 0$

$$P[T_i > h|X(0) = i] = e^{-\lambda_i h} = 1 - \lambda_i h + \frac{(\lambda_i h)^2}{2!} - \frac{(\lambda_i h)^3}{3!} + \cdots = 1 - \lambda_i h + o(h)$$

where $o(h)/h \to 0$ as $h \to 0$. Observe that $P[T_i > h|X(0) = i]$ is the probability of no transition in an interval of length h. Using this result, in the next theorem we prove that the probability of two transitions in a small interval of length h is negligible.

Theorem 6.2.5 *Suppose $X_{n-1} = i$ and $X_n = j$. If T_i and T_j denote the sojourn time in state i and j respectively, then*

$$P[T_i + T_j \leq h|X_{n-1} = i, X_n = j] = o(h) \;\; \forall \;\; n \geq 1.$$

Proof Since T_i and T_j are the sojourn times in state i and j respectively, T_i and T_j have exponential distribution with rate λ_i and λ_j respectively. Further, T_i and T_j are conditionally independent random variables, given X_{n-1}, X_n. Note that $[T_i + T_j \leq h] \subset [T_i \leq h, T_j \leq h]$. Hence

$$\begin{aligned} P[T_i + T_j \leq h|X_{n-1} = i, X_n = j] &\leq P[T_i \leq h|X_{n-1} = i]P[T_j \leq h|X_n = j] \\ &= (\lambda_i h + o(h)) \times (\lambda_j h + o(h)) \\ &= \lambda_i \, \lambda_j h^2 + o(h) = o(h), \;\; \forall \;\; n \geq 1 \end{aligned}$$

and the theorem is proved. □

Remark 6.2.4 It follows that for $k > 2$,
$P[T_{i_1} + T_{i_2} + \cdots T_{i_k} \leq h|X_0 = i_1, X_1 = i_2, \ldots, X_{k-1} = i_k]$
$\leq P[T_{i_1} + T_{i_2} \leq h|X_0 = i_1, X_1 = i_2] = o(h)$.
Thus, the probability of two or more transitions in a small interval of length h is $o(h)$. Further, probability of no transition in $(0, h)$ is $1 - \lambda_i h + o(h)$. It then follows that probability of exactly one transition in $(0, h)$ is $\lambda_i h - 2o(h)$. Hence if $X(0) = i$, then

(i) probability of no transition in $(0, h) = 1 - \lambda_i h + o(h)$,
(ii) probability of exactly one transition in $(0, h) = \lambda_i h + o(h)$,
(iii) probability of more than 1 transitions in $(0, h) = o(h)$.

To sum up, a continuous time Markov chain is completely specified by two components. One is the sojourn time random variables T_i, having exponential distribution with rate parameter λ_i for all $i \in S$. The second component is the sequence

$\{X_n, n \geq 0\}$ of states visited at transitions, which is a Markov chain with transition probabilities $\{p_{ij},\ i,\ j \in S\}$. Thus, a continuous time Markov chain is a stochastic process that moves from state to state in accordance with a Markov chain with the property that the amount of time it spends in each state, before proceeding to the next state, has an exponential distribution.

In the next section, we investigate some properties of transition probability function.

6.3 Transition Probability Function

Suppose $\{X(t), t \geq 0\}$ is a time homogeneous continuous time Markov chain with transition probability function $P_{ij}(t) = P[X(t) = j|X(0) = i]$. Suppose $P(t) = [P_{ij}(t)],\ t \geq 0$ denotes the matrix of transition probability functions. As in the case of a Markov chain in discrete time,

$$\forall\ t \geq 0,\ \ P_{ij}(t) \geq 0\ \ \&\ \sum_{j \in S} P_{ij}(t) = 1,\ \ \forall\ i \in S.$$

In addition, we postulate that

$$\lim_{t \to 0+} P_{ij}(t) = \begin{cases} 1, & \text{if } i = j \\ 0, & \text{if } i \neq j. \end{cases}$$

The above limit in a matrix form is

$$\lim_{t \to 0+} P(t) = I. \tag{6.3.1}$$

If the transition probability function satisfies Eq. (6.3.1), then it is known as the standard transition function, Cinlar [1].

Suppose $\{P_i^{(0)} = P[X(0) = i],\ i \in S\}$ denotes the initial distribution. As in the case of a Markov chain in discrete time, the initial distribution and the transition probability function $\{P_{ij}(t), t > 0\}$ determine the distribution of $X(t),\ \forall\ t > 0$, as follows.

$$\begin{aligned} P_j^{(t)} = P[X(t) = j] &= \sum_{i \in S} P[X(t) = j|X(0) = i]P[X(0) = i] \\ &= \sum_{i \in S} P_{ij}(t)P_i^{(0)},\ \ \forall\ j \in S. \end{aligned}$$

Suppose the initial distribution and the marginal distribution of $X(t)$ are denoted by row vectors $\underline{P}^{(0)} = \{P_i^{(0)}, i \in S\}$ and $\underline{P}^{(t)} = \{P_i^{(t)}, i \in S\}$ respectively. Then $\underline{P}^{(t)}$ can be expressed as

$$\underline{P}^{(t)} = \underline{P}^{(0)} P(t).$$

On similar lines, given the initial distribution and the transition probability function, we can find any member from the family of finite dimensional distribution functions. As an illustration, we find the joint distribution of $X(t_1)$, $X(t_2)$, $X(t_3)$ for $t_1 < t_2 < t_3$. Suppose $i_1, i_2, i_3 \in S$. Then using the Markov property repeatedly we get

$$\begin{aligned}
& P[X(t_3) = i_3, X(t_2) = i_2, X(t_1) = i_1] \\
& = \sum_{i \in S} P[X(t_3) = i_3, X(t_2) = i_2, X(t_1) = i_1, X(0) = i] \\
& = \sum_{i \in S} P[X(t_3) = i_3 | X(t_2) = i_2, X(t_1) = i_1, X(0) = i] \\
& \times P[X(t_2) = i_2, X(t_1) = i_1, X(0) = i] \\
& = P_{i_2 i_3}(t_3 - t_2) P_{i_1 i_2}(t_2 - t_1) \sum_{i \in S} P_{i i_1}(t_1) P[X(0) = i] \\
& = P_{i_2 i_3}(t_3 - t_2) P_{i_1 i_2}(t_2 - t_1) P[X(t_1) = i_1] \\
& = P_{i_2 i_3}(t_3 - t_2) P_{i_1 i_2}(t_2 - t_1) P_{i_1}^{(t_1)} .
\end{aligned}$$

Thus, to find the marginal distribution or to find the joint distribution, we need to know $P_{ij}(t)$ for all t. But to have knowledge of $P_{ij}(t)\ \forall\ t$ is too demanding. We need to have some simpler specification of the probability structure of the continuous time Markov chain $\{X(t), t \geq 0\}$. In the present section, we establish a simpler way of specifying a continuous time Markov chain.

In Chap. 2, it is proved that transition probabilities of a Markov chain satisfy the Chapman-Kolmogorov equations. We have a similar result for continuous time Markov chains as shown below.

Lemma 6.3.1 *Chapman-Kolmogorov Equations: Suppose* $\{X(t), t \geq 0\}$ *is a continuous time Markov chain with state space* S *and transition probability functions* $P_{ij}(t),\ i, j \in S$. *Then for any* $i, j \in S$ *and for any* $s, t \geq 0$,

$$P_{ij}(t+s) = \sum_{k \in S} P_{ik}(t) P_{kj}(s) \iff P(t+s) = P(t)P(s).$$

Proof By definition $P_{ij}(t+s)$ is a probability of transition from state i at time 0 to state j at time $t+s$. By assuming that $X(t) = k$ for some $k \in S$, we have

$$
\begin{aligned}
P_{ij}(t+s) &= P[X(t+s) = j|X(0) = i] \\
&= \sum_{k\in S} P[X(t+s) = j, X(t) = k|X(0) = i] \\
&= \sum_{k\in S} P[(X(t+s) = j|X(t) = k, X(0) = i]P[X(t) = k|X(0) = i] \\
&= \sum_{k\in S} P[X(t+s) = j|X(t) = k]P[X(t) = k|X(0) = i] \\
&= \sum_{k\in S} P[X(s) = j|X(0) = k]P[X(t) = k|X(0) = i] \\
&= \sum_{k\in S} P_{ik}(t)P_{kj}(s) \;\; \forall \; i, j \in S.
\end{aligned}
$$

The fourth step follows from the Markov property and the fifth follows by time homogeneity. In a matrix form, the Chapman-Kolmogorov equations can be written as $P(t+s) = P(t)P(s)$. □

From the Chapman-Kolmogorov equations, we note that with $t = s = 0$,

$$
\begin{aligned}
P(t+s) = P(t)P(s) \Rightarrow P(0) = (P(0))^2 \;\; &\Rightarrow \;\; P(0) = I \\
&\Rightarrow \lim_{t\to 0+} P(t) = P(0) = I \;\text{ by Equation (6.3.1)} \\
&\Rightarrow P(t) \;\text{ is continuous at }\; t = 0.
\end{aligned}
$$

We now examine whether $P(t)$ is continuous $\forall\, t > 0$.

Theorem 6.3.1 *The probability transition function $P(t)$ is continuous $\forall\, t > 0$.*

Proof From the Chapman-Kolmogorov equations observe that $\forall\; t \geq 0$,

$$
\begin{aligned}
P(t+h) = P(t)P(h) &\Rightarrow \lim_{h\to 0+} P(t+h) = P(t) \lim_{h\to 0+} P(h) \\
&\Rightarrow \lim_{h\to 0+} P(t+h) = P(t)I = P(t) \; . \qquad (6.3.2)
\end{aligned}
$$

Since $\lim_{h\to 0+} P(h) = I$, for $0 < h < t$, we note that

$$
\begin{aligned}
P(t) &= P(t-h)P(h) \\
\Rightarrow \quad \lim_{h\to 0+} P(t) &= \lim_{h\to 0+} P(t-h) \lim_{h\to 0+} P(h) \\
\Rightarrow \quad P(t) &= \lim_{h\to 0+} P(t-h) \; . \qquad (6.3.3)
\end{aligned}
$$

From Eqs. (6.3.2) and (6.3.3), it follows that $P(t)$ is continuous for all $t > 0$. □

In discrete time Markov chains, one can obtain the n-step transition probabilities from transition probability matrix P as $P^{(n)} = P^n$. We now examine whether we have a similar result for continuous time Markov chains. We begin by finding an integral equation satisfied by $P_{ij}(t)$ in the following theorem.

Theorem 6.3.2 *Suppose $\{X(t), t \geq 0\}$ is a continuous time Markov chain with transition probability function $P_{ij}(t)$. Then $\forall\ i, j \in S\ \&\ t \geq 0$,*

$$P_{ij}(t) = e^{-\lambda_i t}\delta_{ij} + \int_0^t \lambda_i e^{-\lambda_i u} \sum_{k \neq i} p_{ik} P_{kj}(t-u)\, du$$

$$= e^{-\lambda_i t}\delta_{ij} + \lambda_i e^{-\lambda_i t} \int_0^t e^{\lambda_i v} \sum_{k \neq i} p_{ik} P_{kj}(v)\, dv.$$

Proof We have,

$$P_{ij}(t) = P[X(t) = j|X(0) = i]$$
$$= P[X(t) = j, T_i > t|X(0) = i] + P[X(t) = j, T_i \leq t|X(0) = i].$$

The event in the first term on the right-hand side of the above equation conveys that the first transition has not occurred till t. Thus, $X(t) = j$ if and only if $j = i$ with probability 1. Thus,

$$P[X(t) = j, T_i > t|X(0) = i] = P[X(t) = j|T_i > t, X(0) = i] \times P[T_i > t|X(0) = i] = \delta_{ij} e^{-\lambda_i t}.$$

To find the second term $P[X(t) = j, T_i \leq t|X(0) = i]$, note that some transitions take place in $[0, t]$. Suppose the first transition is at some time u in $(0, t]$ and to some state $k \in S$. Then process moves to the destination state j during the remaining time $t - u$ from state k. The probability of this event is $P_{kj}(t-u)$. Hence,

$$P[X(t) = j, T_i \leq t|X(0) = i] = \int_0^t \lambda_i e^{-\lambda_i u} \sum_{k \neq i} p_{ik} P_{kj}(t-u) du.$$

Since the first transition is at any time u in $(0, t]$, we take the integral over $[0, t]$ and since the first transition is to some state $k \neq i$, we take the sum over $k \neq i$. Combining the expressions for the two terms we have,

$$P_{ij}(t) = e^{-\lambda_i t}\delta_{ij} + \int_0^t \lambda_i e^{-\lambda_i u} \sum_{k \neq i} p_{ik} P_{kj}(t-u)\, du.$$

In this expression, substituting $t - u = v$, we get the second expression. □

Remark 6.3.1 In Corollary 6.2.2, we have proved that for a homogeneous Markov process, $P_{ij}(t) = P[X(t) = j|X(0) = i] = \sum_{n=0}^{\infty} P_{ij}^{(n)}(t)$, where $P_{ij}^{(0)}(t) = \delta_{ij}\ e^{-\lambda_i t}$ and $\forall\ n \geq 1$,

$$P_{ij}^{(n)}(t) = \int_0^t \lambda_i \, e^{-\lambda_i \, u} \sum_{k \neq i} p_{ik} P_{kj}^{(n-1)}(t-u) \, du.$$

Hence,

$$\begin{aligned} P_{ij}(t) = \sum_{n=0}^{\infty} P_{ij}^{(n)}(t) &= \delta_{ij} \, e^{-\lambda_i t} + \sum_{n=1}^{\infty} \int_0^t \lambda_i \, e^{-\lambda_i u} \sum_{k \in S - \{i\}} p_{ik} P_{kj}^{(n-1)}(t-u) \, du \\ &= \delta_{ij} e^{-\lambda_i t} + \int_0^t \lambda_i e^{-\lambda_i u} \sum_{k \neq i} p_{ik} \sum_{n=1}^{\infty} P_{kj}^{(n-1)}(t-u) \, du \\ &= \delta_{ij} e^{-\lambda_i t} + \int_0^t \lambda_i e^{-\lambda_i u} \sum_{k \neq i} p_{ik} P_{kj}(t-u) \, du, \end{aligned}$$

provided the sums on S and $n \geq 1$ can be interchanged. The last expression is then the same as in Theorem 6.3.2.

From the expression of $P_{ij}(t)$ derived in Theorem 6.3.2, we obtain its derivative provided it exists. We state below one theorem which conveys the differentiability of $P_{ij}(t)$. For details refer to Cinlar [1].

Theorem 6.3.3 *If the continuous time Markov chain is a regular process, then $P_{ij}(t)$ for $t \geq 0$ is differentiable and the derivative is continuous.*

We have assumed that the process is regular and hence the derivative of $P_{ij}(t)$ exists. In the following theorem, we obtain the derivative of $P_{ij}(t)$ and in particular at $t = 0$. At $t = 0$, these are right hand derivatives as $t \geq 0$. In the proof, we use the following formula known as Leibnitz' rule.

Leibnitz' rule: Suppose $f(x, y)$ is a totally differentiable function, g_1 and g_2 are differentiable functions, then $h(y) = \int_{g_1(y)}^{g_2(y)} f(x, y) \, dx$ is also a differentiable function and $h'(y)$ is given by

$$h'(y) = g_2'(y) f(g_2(y), y) - g_1'(y) f(g_1(y), y) + \int_{g_1(y)}^{g_2(y)} \frac{\partial}{\partial y} f(x, y) \, dx .$$

Theorem 6.3.4 *Suppose $\{X(t), t \geq 0\}$ is a continuous time Markov chain with transition function $P_{ij}(t)$. Then*
(i) $P_{ij}^{'}(t) = -\lambda_i P_{ij}(t) + \lambda_i \sum_{k \neq i} p_{ik} P_{kj}(t)$.
(ii) $P_{ii}^{'}(0) = -\lambda_i$ and $P_{ij}^{'}(0) = \lambda_i p_{ij}, \;\; i \neq j \in S$.
(iii) For $h > 0$, $P_{ii}(h) = 1 - \lambda_i h + o(h)$ and $P_{ij}(h) = h\lambda_i p_{ij} + o(h)$.

Proof (i) We use the second expression of $P_{ij}(t)$ in Theorem 6.3.2 and obtain its derivative using Leibnitz' rule in the following derivation.

$$
\begin{aligned}
P_{ij}(t) &= e^{-\lambda_i t}\delta_{ij} + \lambda_i e^{-\lambda_i t}\int_0^t e^{\lambda_i v}\sum_{k\neq i} p_{ik}P_{kj}(v)\,dv \\
\Rightarrow\ P_{ij}^{'}(t) &= -\lambda_i e^{-\lambda_i t}\delta_{ij} - \lambda_i^2 e^{-\lambda_i t}\int_0^t e^{\lambda_i v}\sum_{k\neq i} p_{ik}P_{kj}(v)\,dv \\
&\quad +\lambda_i e^{-\lambda_i t}e^{\lambda_i t}\sum_{k\neq i} p_{ik}P_{kj}(t) \\
&= -\lambda_i\Big(\delta_{ij}e^{-\lambda_i t} + \lambda_i e^{-\lambda_i t}\int_0^t e^{\lambda_i v}\sum_{k\neq i} p_{ik}P_{kj}(v)\,dv\Big) \\
&\quad +\lambda_i\sum_{k\neq i} p_{ik}P_{kj}(t) \\
&= -\lambda_i P_{ij}(t) + \lambda_i\sum_{k\neq i} p_{ik}P_{kj}(t).
\end{aligned}
$$

Note that we need to use only the first term from Leibnitz' rule.
(ii) In (i), suppose $i = j$ & $t = 0$. Then

$$
P_{ii}^{'}(0) = -\lambda_i P_{ii}(0) + \lambda_i\sum_{k\neq i} p_{ik}P_{ki}(0) = -\lambda_i,
$$

since $P_{ii}(0) = 1$ and $P_{ki}(0) = 0,\ \ \forall\ k \neq i$. Alternative approach to find $P_{ii}^{'}(0)$ is as follows. Observe that,

$$
\begin{aligned}
P_{ii}(h) &= P[X(h) = i|X(0) = i] \\
&= P[T_i > h|X(0) = i] = 1 - \lambda_i h + o(h) \\
\Rightarrow\ \lim_{h\to 0}\left(\frac{1 - P_{ii}(h)}{h} + \frac{o(h)}{h}\right) &= \lambda_i \\
\Rightarrow\ \lim_{h\to 0}\frac{P_{ii}(h) - P_{ii}(0)}{h} &= -\lambda_i \ \text{ since } \ P_{ii}(0) = 1 \\
\Rightarrow\ P_{ii}^{'}(0) &= -\lambda_i.
\end{aligned}
$$

To find $P_{ij}^{'}(0)$, in (i) suppose $i \neq j$ & $t = 0$. Then

$$
P_{ij}^{'}(0) = -\lambda_i P_{ij}(0) + \lambda_i\sum_{k\neq i} p_{ik}P_{kj}(0) = \lambda_i p_{ij},
$$

since $P_{ij}(0) = 0$, in the sum in the second term $P_{kj}(0) = 1$ when $k = j$ and $P_{kj}(0) = 0$ when $k \neq j$.

(iii) From the definition of a derivative, note that

$$\begin{aligned}\lambda_i = -P'_{ii}(0) &= -\lim_{h\to 0}\frac{(P_{ii}(h) - P_{ii}(0))}{h}\\ &= \lim_{h\to 0}(1 - P_{ii}(h))/h \;\text{ since }\; P_{ii}(0) = 1\\ \Rightarrow\; P_{ii}(h) &= 1 - h\lambda_i + o(h),\end{aligned}$$

for $h > 0$. For $i \neq j$,

$$\begin{aligned}\lambda_i p_{ij} = P'_{ij}(0) &= \lim_{h\to 0}\frac{P_{ij}(h) - P_{ij}(0)}{h}\\ &= \lim_{h\to 0} P_{ij}(h)/h \;\text{ since }\; P_{ij}(0) = 0\\ \Rightarrow\; P_{ij}(h) &= h\lambda_i p_{ij} + o(h), \;\text{ for }\; h > 0.\end{aligned}$$

□

Remark 6.3.2 The probabilities $P_{ii}(h) = 1 - \lambda_i h + o(h)$ and $P_{ij}(h) = h\lambda_i p_{ij} + o(h)$ are known as the infinitesimal transition probabilities.

We have noted that a continuous time Markov chain has two sets of parameters: (i) $\{\lambda_i, i \in S\}$ and (ii) the matrix $P = [p_{ij}]$ of transition probabilities of the embedded Markov chain. We now define a set of parameters $\{q_{ij}, i, j \in S\}$ which combines these two sets of parameters. These are defined as follows.

$$\begin{aligned}&\text{If } \lambda_i \neq 0,\; q_{ij} = \lambda_i p_{ij},\; i \neq j \;\&\; q_{ii} = -\sum_{j\neq i} q_{ij}.\\ &\text{If } \lambda_i = 0,\; q_{ij} = 0 \;\forall\; j \in S.\end{aligned}$$

Observe that when $\lambda_i \neq 0 \;\&\; i \neq j$,

$$\begin{aligned}q_{ij} = \lambda_i p_{ij} \;&\Rightarrow\; \sum_{j\neq i} q_{ij} = \sum_{j\neq i}\lambda_i p_{ij} = \lambda_i\\ \Rightarrow\; \lambda_i = \sum_{j\neq i} q_{ij} &= -q_{ii} \;\;\&\;\; p_{ij} = q_{ij}/(-q_{ii}).\end{aligned}$$

The transition probability p_{ij} is interpreted as the conditional probability of transition from state i to state j, given that a transition from state i has occurred.

If $\lambda_i = 0$, that is, if the state i is an absorbing state, then the system stays in i forever, once it visits i. Hence in this case, we define $q_{ij} = 0 \;\forall\; j \in S$. Thus,

$$q_{ij} = 0 \;\forall\; j \in S \;\Rightarrow\; \lambda_i = 0,\;\; p_{ii} = 1 \;\&\; p_{ij} = 0 \;\forall\; j \neq i \in S.$$

Thus, q_{ij}'s combine the two sets of parameters. Note that $\sum_{j\in S} q_{ij} = 0$ in contrast to $\sum_{j\in S} p_{ij} = 1 \ \forall \ i \in S$. The parameters q_{ij} are known as intensity rates. The term "rate" is clear from Theorem 6.3.4. From this theorem we have,

$$P'_{ii}(0) = -\lambda_i = q_{ii} \ \ \& \ \ P'_{ij}(0) = \lambda_i \ p_{ij} = q_{ij}, \ \ i \neq j \in S.$$

Thus, q_{ij}'s are the derivatives and hence termed as rates. The expression $\lambda_i = -P'_{ii}(0)$ is interpreted as an instantaneous or infinitesimal transition rate at which the process makes a transition from the state i. Hence, the scale parameter λ_i is also referred to as the rate parameter, representing the rate of transition out of state i. We have $E(T_i) = 1/\lambda_i$, which implies that the higher the rate λ_i, the smaller the expected time for the transition to occur, which is intuitively appealing. Further, $q_{ij} = P'_{ij}(0)$ is interpreted as an instantaneous or infinitesimal transition rate at which the process makes transition from state i to state j. For time homogeneous continuous time Markov chain, these are constants. However, these may depend on t if the process is not time homogeneous. In the next section, we discuss some more aspects of intensity rates.

In Theorem 6.3.4 we obtained the derivative of $P_{ij}(t)$. It can be expressed in terms of intensity rates as follows.

$$P'_{ij}(t) = -\lambda_i P_{ij}(t) + \lambda_i \sum_{k\neq i} p_{ik} P_{kj}(t) = q_{ii} P_{ij}(t) + \sum_{k\neq i} q_{ik} P_{kj}(t) = \sum_{k\in S} q_{ik} P_{kj}(t).$$

In the following theorem, we derive the same identity using the Chapman-Kolmogorov equations. We also obtain one more expression for $P'_{ij}(t)$.

Theorem 6.3.5 *Suppose $\{X(t), t \geq 0\}$ is a continuous time Markov chain with transition function $P_{ij}(t)$. Then*

$$(i) \ P'_{ij}(t) = \sum_{k\in S} P_{ik}(t) \ q_{kj} \quad \& \quad (ii) \ P'_{ij}(t) = \sum_{k\in S} q_{ik} P_{kj}(t).$$

Proof (i) From the Chapman-Kolmogorov equations, we have

$$\begin{aligned} P_{ij}(t+h) &= \sum_{k\in S} P_{ik}(t) \ P_{kj}(h) = \sum_{k\in S-\{j\}} P_{ik}(t) P_{kj}(h) + P_{ij}(t) P_{jj}(h) \\ &= \sum_{k\in S-\{j\}} P_{ik}(t)(h\lambda_k p_{kj} + o(h)) + P_{ij}(t)(1 - h\lambda_j + o(h)). \end{aligned}$$

Hence,

$$
\begin{aligned}
\lim_{h \to 0} \frac{P_{ij}(t+h) - P_{ij}(t)}{h} &= \lim_{h \to 0} \sum_{k \in S - \{j\}} P_{ik}(t)(\lambda_k p_{kj} + o(h)/h) \\
&\quad + \lim_{h \to 0} (-\lambda_j + o(h)/h)\ P_{ij}(t) \\
\Rightarrow\ P'_{ij}(t) &= \sum_{k \in S - \{j\}} P_{ik}(t) q_{kj} + q_{jj}\ P_{ij}(t) \qquad (6.3.4) \\
&= \sum_{k \in S} P_{ik}(t) q_{kj}.
\end{aligned}
$$

In Eq. (6.3.4), the limit as $h \to 0$ is taken inside the summation sign. It is valid if S is finite. If S is infinite, some additional conditions are required. We verify these in examples we consider.
(ii) To derive the second system of equations, we again use the Chapman-Kolmogorov equations. Thus,

$$
\begin{aligned}
P_{ij}(t+h) &= \sum_{k \in S} P_{ik}(h) P_{kj}(t) = \sum_{k \in S - \{i\}} P_{ik}(h) P_{kj}(t) + P_{ii}(h) P_{ij}(t) \\
&= \sum_{k \in S - \{i\}} (h\lambda_i p_{ik} + o(h)) P_{kj}(t) + (1 - h\lambda_i + o(h)) P_{ij}(t).
\end{aligned}
$$

Hence,

$$
\begin{aligned}
\lim_{h \to 0} \frac{P_{ij}(t+h) - P_{ij}(t)}{h} &= \lim_{h \to 0} \sum_{k \in S - \{i\}} (\lambda_i p_{ik} + o(h)/h) P_{kj}(t) \\
&\quad + \lim_{h \to 0} (-\lambda_i + o(h)/h)\ P_{ij}(t) \\
\Rightarrow\ P'_{ij}(t) &= \sum_{k \in S - \{i\}} \lambda_i p_{ik} P_{kj}(t) + q_{ii} P_{ij}(t) \qquad (6.3.5) \\
&= \sum_{k \in S} q_{ik} P_{kj}(t).
\end{aligned}
$$

In Equation (6.3.5), the limit as $h \to 0$ can be taken inside the sum when S is finite. If S is infinite, then observe that $(\lambda_i p_{ik} + o(h)/h) P_{kj}(t) \le \lambda_i p_{ik}$ and $\sum_{k \in S - \{i\}} \lambda_i p_{ik} \le \lambda_i < \infty$. Further, $(\lambda_i p_{ik} + o(h)/h) P_{kj}(t) \to \lambda_i p_{ik} P_{kj}(t)$. Hence, by Theorem 2.4.1, sum and limit can be interchanged. □

The equations $P'_{ij}(t) = \sum_{k \in S} P_{ik}(t) q_{kj},\ i, j \in S$ are called Kolmogorov's forward differential equations and the equations $P'_{ij}(t) = \sum_{k \in S} q_{ik} P_{kj}(t),\ i, j \in S$ are called Kolmogorov's backward differential equations. We note that Kolmogorov's backward differential equations are always true.

Remark 6.3.3 We have noted that Kolmogorov's forward differential equations may not hold always. It can be proved that these are valid if the process is non-explosive. As stated in Sect. 6.1, the process is non-explosive or pure if finitely many events occur in a finite interval. Further, if $\lambda_i \leq c \ \forall \ i \in S$ for some $c > 0$ or if the state space S is finite, then continuous time Markov chain is non-explosive, refer to Theorem 2.7.1 of Norris [6]. It is proved that a continuous time Markov chain is non-explosive if and only if $\sum_{n=0}^{\infty} 1/\lambda_{X_{n-1}} = \infty$ almost surely, refer to Theorem 2.3.2 of Norris [6]. We show in Chap. 8 that for a linear birth and death process $\sum_{n=0}^{\infty} 1/\lambda_{X_{n-1}} = \sum_{n=0}^{\infty} 1/n(\lambda + \mu) = \infty$ and hence Kolmogorov's forward equations are satisfied for a birth-death process and as a particular case, for a Yule Furry process and for a linear death process.

These equations can be derived using one more approach, which makes it clear why these are termed as backward and forward differential equations. In this approach, the first step is again via Chapman-Kolmogorov equations. Further we use the continuity of $P_{ij}(t)$ as a function of t, that is, we use the result that $\lim_{s \downarrow 0} P_{ij}(t-s) = P_{ij}(t)$. Observe that for $0 < s < t$,

$$\begin{aligned}
P_{ij}(t) = P[X(t) = j|X(0) = i] &= \sum_{k \in S} P[X(t) = j, X(s) = k|X(0) = i] \\
&= \sum_{k \in S} P_{ik}(s)P_{kj}(t-s) = P_{ii}(s)P_{ij}(t-s) + \sum_{k \neq i} P_{ik}(s)P_{kj}(t-s).
\end{aligned}$$

$$\begin{aligned}
\text{Hence,} \quad P_{ij}(t) - P_{ij}(t-s) &= \sum_{k \neq i} P_{ik}(s)P_{kj}(t-s) - (1 - P_{ii}(s))P_{ij}(t-s) \\
\Rightarrow \quad \lim_{s \downarrow 0} \frac{P_{ij}(t) - P_{ij}(t-s)}{s} &= \lim_{s \downarrow 0} \sum_{k \neq i} \frac{P_{ik}(s)}{s} P_{kj}(t-s) \\
&\quad - \lim_{s \downarrow 0} \frac{1 - P_{ii}(s)}{s} P_{ij}(t-s) \\
\Rightarrow \quad P'_{ij}(t) &= \sum_{k \neq i} q_{ik}P_{kj}(t) - \lambda_i P_{ij}(t) \\
&= \sum_{k \in S} q_{ik}P_{kj}(t) \,, \qquad (6.3.6)
\end{aligned}$$

provided limit and summation can be interchanged. Note that Eq. (6.3.6) is Kolmogorov's backward differential equations. In this derivation, the limit is obtained as s tends to 0 from above, hence the term backward equations seems appropriate for Eq. (6.3.6). For Kolmogorov's forward differential equations, we allow s to tend to t from below, as shown in the following derivation.

$$\begin{aligned}
P_{ij}(t) &= P[X(t)=j|X(0)=i] = \sum_{k\in S} P[X(t)=j, X(s)=k|X(0)=i] \\
&= \sum_{k\in S} P_{ik}(s)P_{kj}(t-s) = P_{ij}(s)P_{jj}(t-s) + \sum_{k\neq j} P_{ik}(s)P_{kj}(t-s) \\
&\Rightarrow P_{ij}(t) - P_{ij}(s) = \sum_{k\neq j} P_{ik}(s)P_{kj}(t-s) - (1-P_{jj}(t-s))P_{ij}(s) \\
&\Rightarrow \lim_{s\uparrow t} \frac{P_{ij}(t)-P_{ij}(s)}{t-s} = \lim_{s\uparrow t} \sum_{k\neq j} \frac{P_{kj}(t-s)}{t-s} P_{ik}(s) \\
&- \lim_{s\uparrow t} \frac{1-P_{jj}(t-s)}{t-s} P_{ij}(s) \\
&\Rightarrow P'_{ij}(t) = \sum_{k\neq j} P_{ik}(t)q_{kj} - \lambda_j P_{ij}(t) = \sum_{k\in S} P_{ik}(t)q_{kj}, \qquad (6.3.7)
\end{aligned}$$

provided limit and summation can be interchanged. In this derivation s increases to t and hence the term forward in Eq. (6.3.7) is justified.

Observe that Kolmogorov's backward and forward equations are expressible in a matrix form as

$$P'(t) = QP(t) \quad \& \quad P'(t) = P(t)Q$$

respectively, where $Q = [q_{ij}]$. It is to be noted that the difference between the two is the order of the multiplication of $P(t)$ and Q. These matrix equations can also be derived from the matrix expression for Chapman-Kolmogorov equations. For all $i, j \in S$, the relations

$$\begin{aligned}
-\lambda_i &= P'_{ii}(0) = q_{ii} = \lim_{h\to 0} \frac{P_{ii}(h)-1}{h} \quad \& \quad q_{ij} = P'_{ij}(0) = \lim_{h\to 0} \frac{P_{ij}(h)-0}{h} \\
&\Longleftrightarrow \lim_{h\to 0} \frac{P(h)-I}{h} = Q
\end{aligned}$$

in a matrix form. With this relation,

$$\begin{aligned}
P(t+h) &= P(h)P(t) \text{ by Chapman-Kolmogorov equations} \\
\Rightarrow \quad P(t+h) - P(t) &= P(h)P(t) - P(t) = (P(h)-I)P(t) \\
\Rightarrow \quad \lim_{h\to 0} \frac{P(t+h)-P(t)}{h} &= \lim_{h\to 0} \frac{(P(h)-I)}{h} P(t) \\
\Rightarrow \quad P'(t) &= QP(t)\,,
\end{aligned}$$

which are Kolmogorov's backward differential equations in a matrix form. Similarly, to derive Kolmogorov's forward differential equations in a matrix form, observe that

$$
\begin{aligned}
P(t+h) &= P(t)P(h) \text{ by Chapman-Kolmogorov equations} \\
\Rightarrow \quad P(t+h) - P(t) &= P(t)P(h) - P(t) = P(t)(P(h) - I) \\
\Rightarrow \quad \lim_{h \to 0} \frac{P(t+h) - P(t)}{h} &= P(t) \lim_{h \to 0} \frac{(P(h) - I)}{h} \\
\Rightarrow \quad P'(t) &= P(t)Q\,.
\end{aligned}
$$

Thus, we have $P'(t) = P(t)Q = QP(t)$.

In Sect. 6.5, we discuss how to solve $P'(t) = P(t)Q = QP(t)$ to find $P(t)$.

In many stochastic models, based on physical phenomena, infinitesimal probabilities are prescribed relating to the process. From these we can derive an explicit expression for $P^{'}_{ij}(t)$ and hence for $P_{ij}(t)$. We adopt such an approach for Poisson process in the next chapter and Yule-Furry process in Chap. 8.

The next section is concerned with some more aspects of intensity rates introduced in this section.

6.4 Infinitesimal Generator

Suppose $\{X(t), t \geq 0\}$ is continuous time Markov chain with the intensity rates q_{ij}, as defined in Sect. 6.3. We note that

(i) $q_{ii} \leq 0 \ \forall \ i \in S$ and $q_{ij} \geq 0 \ \forall \ j \neq i \in S$.
(ii) $\sum_{j \in S} q_{ij} = 0$, so that $-q_{ii} = \sum_{j \in S - \{i\}} q_{ij}$.

If state i is an absorbing state, then $q_{ij} = 0$ for all $j \in S$. In such a case for the corresponding embedded chain $p_{ii} = 1$. The rates q_{ij} are usually organized in a matrix $Q = [q_{ij}]$, such that the diagonal elements are $q_{ii} = -\lambda_i$ and off-diagonal elements are the transition rates q_{ij}. Thus for the absorbing state i, the ith row is a $\underline{0}$ vector.

Conversely, a square matrix $Q = [q_{ij}]$, $i, j \in S$, as defined below, determines the continuous time Markov chain.

Definition 6.4.1 Suppose S is a countable set. A square matrix $Q = [q_{ij}]$, $i, j \in S$ is said to be an infinitesimal generator if it satisfies the following three conditions.

$$
(i)\ q_{ii} \leq 0 \ \forall \ i \in S, \quad (ii)\ q_{ij} \geq 0 \ \forall \ j \neq i \in S \ \& \ (iii) \sum_{j \in S} q_{ij} = 0.
$$

Theorem 6.4.1 *The infinitesimal generator Q determines the Markov pure jump process, that is, the distribution of the holding times and transition probabilities p_{ij} can be determined from the Q matrix.*

Proof Suppose Q is an infinitesimal generator whose elements satisfy the three properties listed in Definition 6.4.1. We define a stochastic matrix $[p_{ij}]$, $i, j \in S$ as follows. If $q_{ii} = 0$, that is, if all elements of the ith row of Q are zero, set $p_{ii} = 1$. If

$q_{ii} < 0$ then, set $p_{ii} = 0$ and $p_{ij} = q_{ij}/(-q_{ii})$ for $i, j \in S$. This defines the embedded Markov chain $\{X_n, n \geq 0\}$. Suppose the holding time in state i has exponential distribution with rate parameter $-q_{ii}$. The above construction defines the Markov pure jump process in terms of the Q matrix. □

In view of Theorem 6.4.1, in most of the models, continuous time Markov chain $\{X(t), t \geq 0\}$ is specified by the intensity rates, which are determined by the infinitesimal transition probabilities. Following are some illustrations.

Poisson process: Suppose $X(0) = 0$,

$$P_{ii}(h) = 1 - \lambda h + o(h), \quad P_{i,i+1}(h) = \lambda h + o(h) \;\; \& \;\; P_{ij}(h) = o(h) \;\; \forall \;\; j \neq i + 1$$

or equivalently in terms of intensity rates,

$$q_{ii} = -\lambda, \quad q_{i,i+1} = \lambda \;\; \& \;\; q_{ij} = 0 \;\; \forall \;\; j \neq i + 1.$$

Then $\{X(t), t \geq 0\}$ is a Poisson process with intensity rate λ.

Birth-death process: Suppose $X(0) = a$ denotes a population of individuals, which increases due to births and decreases due to deaths according to the following infinitesimal probabilities.

$$\begin{aligned} P_{i,i+1}(h) &= P[X(t+h) = i+1|X(t) = i] = \lambda_i h + o(h) \\ P_{i,i-1}(h) &= P[X(t+h) = i-1|X(t) = i] = \mu_i h + o(h) \\ P_{ii}(h) &= P[X(t+h) = i|X(t) = i] = 1 - (\lambda_i + \mu_i)h + o(h) \\ P_{ij}(h) &= o(h) \;\; \forall \;\; j \neq i+1, i-1 \,. \end{aligned}$$

Then $\{X(t), t \geq 0\}$ is known as a birth-death process. The intensity rates for this process are given by

$$q_{i,i+1} = \lambda_i, \quad q_{i,i-1} = \mu_i, \quad q_{ii} = -(\lambda_i + \mu_i) \quad \& \quad q_{ij} = 0 \;\; \forall \;\; j \neq i+1, i-1 \,.$$

The queuing system with one service counter can be modeled as a birth death process. The state $X(t)$ of the queuing system at any time is represented by the number of people in the system at that time, that is, number of people in the queue and the one getting service. Suppose that whenever there are i people in the system, then (i) new arrivals enter the system according to an exponential distribution with rate λ_i and (ii) people leave the system according to an exponential distribution with rate μ_i. Then infinitesimal transition probabilities $P_{ij}(h)$ are the same as given above. If $\mu_i = 0$ and $\lambda_i = \lambda$ for all i, then a birth-death process reduces to a Poisson process.

Linear birth process: In a birth-death process, if $\mu_i = 0$ and $\lambda_i = i\lambda$ for all i, then it reduces to a linear birth process. It is known as Yule Furry process after G. Yule and Furry, who used it in mathematical theory of evolution.

Linear death process: In a birth-death process, if $\mu_i = i\mu$ and $\lambda_i = 0$ for all i, then it reduces to a linear death process.

A linear growth model with immigration: A birth-death process in which $\mu_i = i\mu$ and $\lambda_i = i\lambda + \theta$ is called a linear growth process with immigration. Such processes occur naturally in the study of biological reproduction and population growth. Each individual in the population is assumed to give birth with rate λ, in addition, there is an increase of the population due to an external source such as immigration, with rate θ. Hence, the total birth rate when there are i persons in the system is $i\lambda + \theta$. Deaths are assumed to occur at a rate μ for each member of the population, so $\mu_i = i\mu$.

Poisson process and birth-death processes play a fundamental role in theory and applications in queuing and inventory models, population growth models, engineering systems, etc. We study Poisson process in detail in the next chapter. Birth-death processes and their versions are discussed in Chap. 8.

Following example illustrates how to obtain a transition probability matrix of the corresponding embedded Markov chain and the rate parameters of the sojourn time random variables when the intensity matrix of a continuous time Markov chain is given.

Example 6.4.1 A machine in a workshop can be in one of the three states, 1 indicating good working condition, 2 indicating deteriorated working condition and 3 indicating failed condition. The intensity rates are $q_{12} = \mu_2, q_{23} = \mu_1, q_{32} = \lambda_2$ and $q_{21} = \lambda_1$. Thus the Q matrix is given by

$$Q = \begin{matrix} & \begin{matrix} 1 & 2 & 3 \end{matrix} \\ \begin{matrix} 1 \\ 2 \\ 3 \end{matrix} & \begin{pmatrix} -\mu_2 & \mu_2 & 0 \\ \lambda_1 & -(\lambda_1 + \mu_1) & \mu_1 \\ 0 & \lambda_2 & -\lambda_2 \end{pmatrix} \end{matrix}.$$

Here no state is absorbing and hence the matrix P of corresponding embedded Markov chain is given by

$$P = \begin{matrix} & \begin{matrix} 1 & 2 & 3 \end{matrix} \\ \begin{matrix} 1 \\ 2 \\ 3 \end{matrix} & \begin{pmatrix} 0 & 1 & 0 \\ \lambda_1/(\lambda_1 + \mu_1) & 0 & \mu_1/(\lambda_1 + \mu_1) \\ 0 & 1 & 0 \end{pmatrix} \end{matrix}.$$

The rates parameters of the sojourn time random variables are μ_2, $\lambda_1 + \mu_1$ and λ_2 for states $1, 2, 3$ respectively. □

Example 6.4.2 Suppose the lifetime of a high-altitude satellite has exponential distribution with parameter λ. Once it fails, it remains in the same state as repair is not possible. Suppose $X(t)$ denotes the state of the satellite at time t with $X(t) = 1$ if it is operational at time t and $X(t) = 0$ if in failed state. We first examine that $\{X(t), t \geq 0\}$ is a continuous time Markov chain. Observe that, $P[X(t+s) = 1|X(s) = 1, X(u), 0 \leq u < s]$ is the probability of the event that the system operational up to s, remains operational at $s+t$. Thus, it is the probability that the remaining lifetime in state 1 is larger than t and hence it is $e^{-\lambda t}$. Similarly, $P[X(s+t) = 1|X(s) = 1] = e^{-\lambda t}$. Hence,

$$\begin{aligned} P[X(t+s) = 1|X(s) = 1, X(u), 0 \leq u < s] &= P[X(s+t) = 1|X(s) = 1] \\ &= e^{-\lambda t} = P[X(t) = 1|X(0) = 1] \,. \end{aligned}$$

Now $P[X(t+s) = 0|X(s) = 1, X(u), 0 \leq u < s]$ is the probability of the event that the system operational up to s, fails before $s+t$. Thus, it is the probability that the remaining lifetime in state 1 is less than t and hence it is $1 - e^{-\lambda t}$. Similarly, $P[X(s+t) = 0|X(s) = 1] = 1 - e^{-\lambda t}$. Hence,

$$\begin{aligned} P[X(t+s) = 0|X(s) = 1, X(u), 0 \leq u < s] &= P[X(s+t) = 0|X(s) = 1] \\ &= P[X(t) = 0|X(0) = 1] \,. \end{aligned}$$

It is given that once it fails, it remains in the same state as repair is not possible, thus 0 is an absorbing state. Hence

$$P[X(t+s) = 0|X(s) = 0, X(u), 0 \leq u < s] = P[X(s+t) = 0|X(s) = 0] = 1$$

and

$$P[X(t+s) = 1|X(s) = 0, X(u), 0 \leq u < s] = P[X(s+t) = 1|X(s) = 0] = 0.$$

Thus, $\{X(t), t \geq 0\}$ is a continuous time Markov chain. The matrix $P(t)$ of transition probability functions, matrix Q of intensity rates and the transition probability matrix P of the embedded Markov chain are given by

$$P(t) = \begin{matrix} & \begin{matrix} 0 & 1 \end{matrix} \\ \begin{matrix} 0 \\ 1 \end{matrix} & \begin{pmatrix} 1 & 0 \\ 1 - e^{-\lambda t} & e^{-\lambda t} \end{pmatrix} \end{matrix} \quad Q = \begin{matrix} & \begin{matrix} 0 & 1 \end{matrix} \\ \begin{matrix} 0 \\ 1 \end{matrix} & \begin{pmatrix} 0 & 0 \\ \lambda & -\lambda \end{pmatrix} \end{matrix} \quad P = \begin{matrix} & \begin{matrix} 0 & 1 \end{matrix} \\ \begin{matrix} 0 \\ 1 \end{matrix} & \begin{pmatrix} 1 & 0 \\ 1 & 0 \end{pmatrix} \end{matrix}.$$

Since 0 is an absorbing state, we have $q_{00} = 0$ and hence $p_{00} = 1$, $p_{01} = 0$. It is to be noted that as t increases, matrix $P(t)$ converges to a matrix with identical rows $(1, 0)$, which implies that in the long run satellite will be in failed condition. It seems reasonable as transition function from 1 to 0 is positive and once the system enters 0 it remains in state 0. □

Example 6.4.3 Suppose a machine is in a working state for a random amount of time having an exponential distribution with parameter μ. When it fails, it gets repaired. The repair time has an exponential distribution with parameter λ and is independent of the past. The machine is as good as new after the repair is complete. Suppose a workshop has two such machines which work independently of each other and with the same failure and repair time distributions. Suppose $X(t)$ denotes the number of machines working at time t. Then $\{X(t), t \geq 0\}$ can be modeled as a Markov process with state space $S = \{0, 1, 2\}$. When $X(t) = 0$, both machines are down and hence under repair. Due to memoryless property, the remaining repair time of the two machines are independent and identically distributed random variables, each having exponential distribution with parameter λ. The system moves from state 0 to state 1 as soon as one of the two machines is repaired. Here we assume that both the machines can be repaired simultaneously. Hence, the sojourn time in state 0 is minimum of two repair times, and it has exponential distribution with parameter 2λ. Thus, $\lambda_0 = -q_{00} = 2\lambda, q_{01} = 2\lambda$ and $p_{01} = 1$. When $X(t) = 1$, one machine is down and one is working. Suppose W_1 denotes remaining repair time of the down machine and W_2 denotes remaining lifetime of the working machine. Again due to memoryless property, W_1 has exponential distribution with parameter λ and W_2 has exponential distribution with parameter μ and these are independent. Hence, the sojourn time in state 1 is $\min\{W_1, W_2\}$ and its distribution is exponential with parameter $\lambda + \mu$. Thus, $\lambda_1 = -q_{11} = \lambda + \mu$. The system moves to state 2, if $W_1 < W_2$ and probability of this event is $P[W_1 < W_2] = \lambda/(\lambda + \mu)$. It moves to state 0, if $W_2 < W_1$ and probability of this event is $\mu/(\lambda + \mu)$. It is to be noted that next state is always independent of sojourn time. Hence,

$$p_{12} = \frac{\lambda}{\lambda + \mu} \quad \& \quad p_{10} = \frac{\mu}{\lambda + \mu}, \quad q_{12} = \lambda \quad \& \quad q_{10} = \mu \, .$$

A similar analysis of state 2, when both the machines are in working condition, gives $\lambda_2 = 2\mu$, $p_{21} = 1$ and $q_{21} = 2\mu$. Thus, the generator matrix Q and the transition probability matrix P of the associated embedded Markov chain are given by

$$Q = \begin{array}{c} \\ 0 \\ 1 \\ 2 \end{array} \begin{array}{c} \begin{array}{ccc} 0 & 1 & 2 \end{array} \\ \begin{pmatrix} -2\lambda & 2\lambda & 0 \\ \mu & -(\lambda + \mu) & \lambda \\ 0 & 2\mu & -2\mu \end{pmatrix} \end{array} \quad \& \quad P = \begin{array}{c} \\ 0 \\ 1 \\ 2 \end{array} \begin{array}{c} \begin{array}{ccc} 0 & 1 & 2 \end{array} \\ \begin{pmatrix} 0 & 1 & 0 \\ \mu/(\lambda + \mu) & 0 & \lambda/(\lambda + \mu) \\ 0 & 1 & 0 \end{pmatrix} \end{array}.$$

Suppose on the average each of the two machines is in a working condition for 10 h and on the average repair time for each of the two machines is 2 h, that is, $\lambda = 0.5$ and $\mu = 0.1$. The generator matrix Q for this Markov process is then given by

$$Q = \begin{array}{c} \\ 0 \\ 1 \\ 2 \end{array} \begin{array}{c} \begin{array}{ccc} 0 & 1 & 2 \end{array} \\ \begin{pmatrix} -1 & 1 & 0 \\ 0.1 & -0.6 & 0.5 \\ 0 & 0.2 & -0.2 \end{pmatrix} \end{array}.$$

□

We now discuss how to obtain a realization of a continuous time Markov chain given the generator matrix Q. We have noted that a continuous time Markov chain involves two building blocks. The first is the embedded Markov chain, $\{X_n, n \geq 0\}$ with transition probabilities p_{ij} where $p_{ii} = 0$ for a non-absorbing state i. It determines the sequence of successive states. The second is a sequence of exponential random variables with parameters governed by the sequence of states $\{X_n, n \geq 0\}$ to decide the holding times. A stepwise procedure to obtain a realization is as follows.

(i) Initial state $X(0) = X_0$ is selected according to the initial distribution $\underline{p}^{(0)}$. In many cases it is degenerate.
(ii) Suppose $S_0 = 0$ and H_1 is generated from the exponential distribution with parameter λ_{X_0}, it is the sojourn time in state $X(0) = X_0$.
(iii) Suppose $S_1 = H_1$, then $X(t) = X_0$ for all $t \in [S_0, S_1)$.
(iv) Suppose X_1 is generated according to the transition matrix P and H_2 is generated from the exponential distribution with parameter λ_{X_1}. H_2 is the sojourn time in state X_1.
(v) Suppose $S_2 = S_1 + H_2$, then $X(t) = X_1$ for all $t \in [S_1, S_2)$.
(vi) The process is continued.

Note that two sets of random variables are needed at each iteration of algorithm, one to compute the holding time, and one to compute the next state of the embedded Markov chain. The first step in obtaining the realization is to find a transition probability matrix P of the embedded Markov chain, from the intensity matrix Q. Two methods are given in Code 6.7.1 and one in Code 6.7.2 to compute P from Q. These are illustrated in the following examples.

Example 6.4.4 Suppose the intensity matrix Q is as given below.

$$Q = \begin{array}{c} \\ 0 \\ 1 \\ 2 \\ 3 \\ 4 \end{array} \begin{array}{c} \begin{array}{ccccc} 0 & 1 & 2 & 3 & 4 \end{array} \\ \begin{pmatrix} -7.6 & 2.1 & 1.1 & 1.2 & 3.2 \\ 1.1 & -7.7 & 3.2 & 2.1 & 1.3 \\ 2.3 & 1.1 & -5.7 & 1.2 & 1.1 \\ 1.3 & 1.2 & 1.1 & -6.8 & 3.2 \\ 2.3 & 1.1 & 2.1 & 3.2 & -8.7 \end{pmatrix} \end{array}.$$

From both the methods in Code 6.7.1, we get the matrix P as shown below.

$$P = \begin{array}{c} \\ 0 \\ 1 \\ 2 \\ 3 \\ 4 \end{array} \begin{array}{c} \begin{array}{ccccc} 0 & 1 & 2 & 3 & 4 \end{array} \\ \begin{pmatrix} 0.0000 & 0.2763 & 0.1447 & 0.1579 & 0.4211 \\ 0.1429 & 0.0000 & 0.4156 & 0.2727 & 0.1688 \\ 0.4035 & 0.1930 & 0.0000 & 0.2105 & 0.1930 \\ 0.1912 & 0.1765 & 0.1618 & 0.0000 & 0.4706 \\ 0.2644 & 0.1264 & 0.2414 & 0.3678 & 0.0000 \end{pmatrix} \end{array}.$$

In Code 6.7.1, it is implicitly assumed that diagonal elements of Q are not 0. However, for an absorbing state i, $q_{ii} = 0$. Code 6.7.2 incorporates this feature while computing P from Q. It is illustrated below.

$$Q = \begin{array}{c} \\ 0 \\ 1 \\ 2 \\ 3 \end{array} \begin{array}{c} \begin{array}{cccc} 0 & 1 & 2 & 3 \end{array} \\ \begin{pmatrix} -5.2 & 2.3 & 1.4 & 1.5 \\ 0 & 0 & 0 & 0 \\ 2.4 & 1.8 & -7.4 & 3.2 \\ 1.9 & 1.5 & 2.6 & -6.0 \end{pmatrix} \end{array} \;\&\; P = \begin{array}{c} \\ 0 \\ 1 \\ 2 \\ 3 \end{array} \begin{array}{c} \begin{array}{cccc} 0 & 1 & 2 & 3 \end{array} \\ \begin{pmatrix} 0.0000 & 0.4423 & 0.2692 & 0.2885 \\ 0.0000 & 1.0000 & 0.0000 & 0.0000 \\ 0.3243 & 0.2432 & 0.0000 & 0.4324 \\ 0.3167 & 0.2500 & 0.4333 & 0.0000 \end{pmatrix} \end{array}.$$

□

The procedure of obtaining a realization using Code 6.7.3 of a continuous time Markov chain given Q matrix is illustrated in the following example.

Example 6.4.5 Suppose $\{X(t), t \geq 0\}$ is continuous time Markov chain as in Example 6.4.3 with the intensity matrix Q given by

$$Q = \begin{array}{c} \\ 0 \\ 1 \\ 2 \end{array} \begin{array}{c} \begin{array}{ccc} 0 & 1 & 2 \end{array} \\ \begin{pmatrix} -1 & 1 & 0 \\ 0.1 & -0.6 & 0.5 \\ 0 & 0.2 & -0.2 \end{pmatrix} \end{array}.$$

We obtain a realization of the process till 10 transitions occur. We assume that initially both the machines are working. The states visited, the sojourn times and the epochs of transition are presented in Table 6.1. The two machines working initially remain in the same state for 0.93 h, then one of the two fails and repair time required is 1.96 h and then both are again working for 6.02 h and so on. Among 10 transitions, only 1 time both the machines are down, while 5 times one is down and 4 times both are in working condition. Figure 6.2 shows the realization of the process till 10 transitions. Realization of the process $\{X(t), t \geq 0\}$ is presented as follows.

$$X(t) = \begin{cases} 2, \text{ if } & t < 0.93 \\ 1, \text{ if } & 0.93 \leq t < 2.89 \\ 2, \text{ if } & 2.89 \leq t < 8.91 \\ & \vdots \\ 1, \text{ if } & 35.87 \leq t < 37.36. \end{cases}$$

At time epoch 37.36, process transits to state 2. When the number of transitions is fixed at n, the epoch of nth transition is a random variable, S_n. Its distribution is a sum of n independent random variables having exponential distributions with different scale parameters. In the above realization value of random variable S_{10} is 37.36 h. □

In the next example, we find realization of the same process when observed for a fixed time period, using Code 6.7.4.

Table 6.1 Realization of a Markov process for a fixed number of transitions

n	X_n	T_{X_n}	S_{n+1}
0	2	0.93	0.93
1	1	1.96	2.89
2	2	6.02	8.91
3	1	7.54	16.45
4	2	3.79	20.24
5	1	3.59	23.83
6	2	8.18	32.00
7	1	2.93	34.93
8	0	0.94	35.87
9	1	1.49	37.36
10	2		

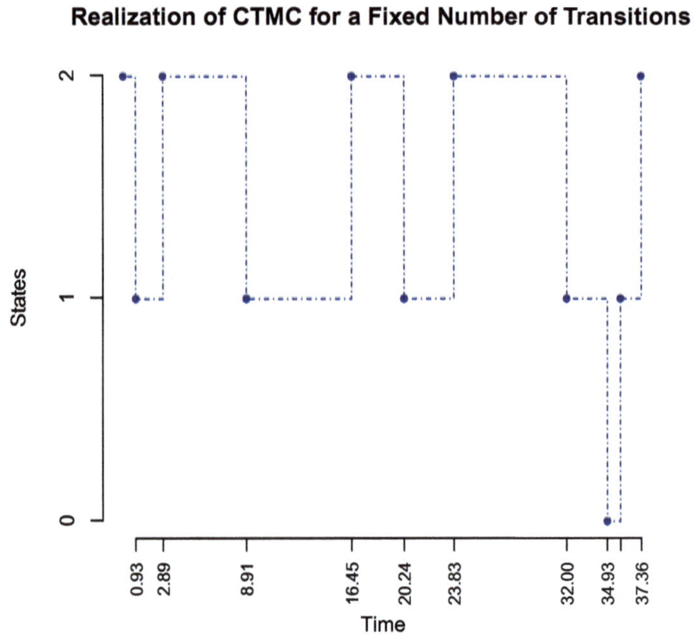

Fig. 6.2 Realization of CTMC: sojourn times and states visited

Example 6.4.6 Suppose $\{X(t), t \geq 0\}$ is continuous time Markov chain as in Example 6.4.3. Suppose we observe the process for a fixed time period $[0, T]$, where $T = 40$ h. In this case, the number of transitions is a random variable. The realization of the process $\{X(t)\}$ is presented as follows.

$$X(t) = \begin{cases} 2, \text{ if } \quad t < 1.15 \\ 1, \text{ if } 1.15 \leq t < 1.98 \\ 2, \text{ if } 1.98 \leq t < 4.30 \\ \quad \vdots \\ 1, \text{ if } 38.71 \leq t \leq 40. \end{cases}$$

Figure 6.3 displays the realization. From the output and the figure, we note that with initial state 2, the sequence of subsequent states visited till $T = 40$ is

$$1, 2, 1, 2, 1, 2, 1, 0, 1, 2, 1, 2, 1, 2, 1, 2, 1.$$

Among these 18 transitions, only one time both the machines are down, only one machine is working 9 times and both the machines are in working condition 8 times. The epochs of successive transitions are 1.15, 1.98, 4.30, 4.97, 10.88, 16.12, 18.14, 18.42, 18.54, 19.89, 26.04, 26.76, 27.46, 28.41, 28.78, 32.94, 38.71. The last transition occurs at time point 38.71 h to state 1 and the process remains in state 1 till $T = 40$ h. After time epoch 40, it may remain in the same state 1 for random time, which is the remaining sojourn time in state 1 and it again has exponential distribution with rate λ_1. □

Remark 6.4.1 From Figs. 6.2 and 6.3, we note that the sample paths of the process are right continuous. If there are no instantaneous states, then the sample paths are always right continuous and have finite left hand limits, Cinlar [1].

In the next section, we discuss the solution of Kolmogorov's forward or backward differential equations and some methods to compute transition probability function numerically.

6.5 Computation of Transition Probability Function

We use Kolmogorov's forward or backward differential equations to get the expression for $P(t)$ under the initial condition $P(0) = I$. Though the backward and forward equations are two different sets of differential equations, these have the same solution. It is illustrated in the following examples.

Example 6.5.1 In telecommunications some source alternates between off-and on-states. Suppose 0 indicates "off" state and 1 indicates "on" state. Holding time random variables in states 0 and 1 are independent, having exponential distributions with

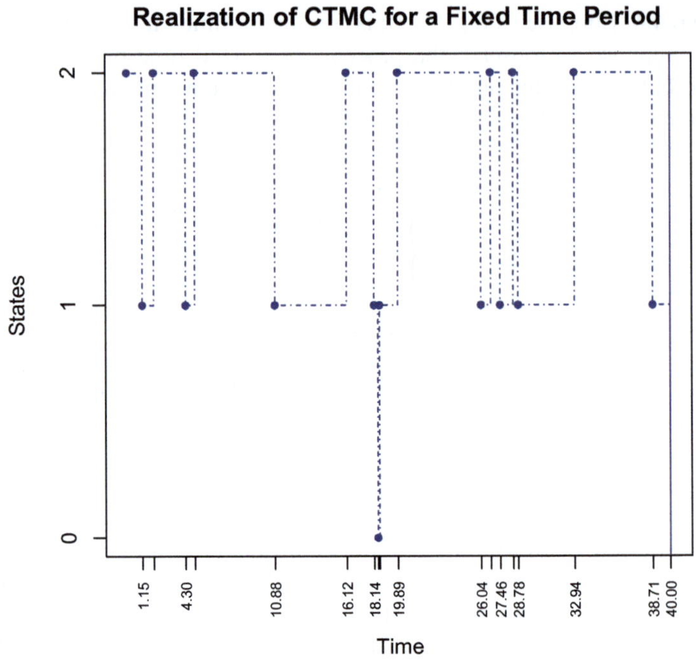

Fig. 6.3 Realization of CTMC For a fixed time period

parameters λ and μ respectively. Suppose $X(t)$ denotes the state of the source at time t, then $\{X(t), t \geq 0\}$ can be modeled as a Markov process with state space $S = \{0, 1\}$. Further $\lambda_0 = \lambda = -q_{00}$. Since row sums of Q are 0, we have $q_{01} = \lambda$. Similarly, $\lambda_1 = \mu = -q_{11}$ and $q_{10} = \mu$. Thus, the generator matrix Q is given by

$$Q = \begin{array}{c} \\ 0 \\ 1 \end{array} \begin{array}{c} \begin{array}{cc} 0 & 1 \end{array} \\ \begin{pmatrix} -\lambda & \lambda \\ \mu & -\mu \end{pmatrix} \end{array}.$$

Now we solve $P'(t) = P(t)Q$, Kolmogorov's forward differential equation, with initial condition $P(0) = I$. Since the state space is finite, forward differential equations exist. Thus,

$$\begin{aligned} P'_{00}(t) &= -\lambda P_{00}(t) + \mu P_{01}(t) = -\lambda P_{00}(t) + \mu(1 - P_{00}(t)) \\ &= \mu - (\lambda + \mu)P_{00}(t) \, . \end{aligned} \qquad (6.5.1)$$

Suppose $h(t) = P_{00}(t) - \mu/(\lambda + \mu)$. Then

$$
\begin{aligned}
h'(t) &= P'_{00}(t) = \mu - (\lambda + \mu)\,(h(t) + \mu/(\lambda + \mu)) \\
&= -(\lambda + \mu)h(t) \\
\Rightarrow \quad \log h(t) &= -(\lambda + \mu)t + c \;\; \Rightarrow \;\; h(t) = ke^{-(\lambda+\mu)t} \\
\Rightarrow \quad P_{00}(t) &= ke^{-(\lambda+\mu)t} + \mu/(\lambda + \mu) \\
P_{00}(0) = 1 \;\; \Rightarrow \;\; k &= \lambda/(\lambda + \mu) \\
\Rightarrow \quad P_{00}(t) &= \frac{\mu}{\lambda + \mu} + \frac{\lambda}{\lambda + \mu} e^{-(\lambda+\mu)t} \\
\& \;\; P_{01}(t) = 1 - P_{00}(t) &= \frac{\lambda}{\lambda + \mu} - \frac{\lambda}{\lambda + \mu} e^{-(\lambda+\mu)t} .
\end{aligned}
$$

Similarly from the equation $P'(t) = P(t)Q$, we have

$$
P'_{10}(t) = -\lambda P_{10}(t) + \mu P_{11}(t) = -\lambda P_{10}(t) + \mu(1 - P_{10}(t)) = \mu - (\lambda + \mu)P_{10}(t) .
$$

Suppose $h(t) = P_{10}(t) - \mu/(\lambda + \mu)$. Then

$$
\begin{aligned}
h'(t) &= P'_{10}(t) = \mu - (\lambda + \mu)\,(h(t) + \mu/(\lambda + \mu)) \\
&= -(\lambda + \mu)h(t) \\
\Rightarrow \quad \log h(t) &= -(\lambda + \mu)t + c \;\; \Rightarrow \;\; h(t) = ke^{-(\lambda+\mu)t} \\
\Rightarrow \quad P_{10}(t) &= ke^{-(\lambda+\mu)t} + \mu/(\lambda + \mu) \\
P_{10}(0) = 0 \;\; \Rightarrow \;\; k &= -\mu/(\lambda + \mu) \\
\Rightarrow \quad P_{10}(t) &= \frac{\mu}{\lambda + \mu} - \frac{\mu}{\lambda + \mu} e^{-(\lambda+\mu)t} \\
\& \;\; P_{11}(t) = 1 - P_{10}(t) &= \frac{\lambda}{\lambda + \mu} + \frac{\mu}{\lambda + \mu} e^{-(\lambda+\mu)t} .
\end{aligned}
$$

We now discuss how to solve Kolmogorov's backward differential equation $P'(t) = QP(t)$, with initial condition $P(0) = I$. From this equation we have,

$$
\begin{aligned}
P'_{00}(t) &= -\lambda P_{00}(t) + \lambda P_{10}(t) \;\; \& \;\; P'_{10}(t) = \mu P_{00}(t) - \mu P_{10}(t) \\
&\Rightarrow \mu P'_{00}(t) + \lambda P'_{10}(t) = 0 \\
&\Rightarrow \mu P_{00}(t) + \lambda P_{10}(t) = c, \quad \text{a constant free from t.}
\end{aligned}
$$

Observe that at

$$
t = 0, \;\; P_{00}(t) = 1 \;\&\; P_{10}(t) = 0 \;\; \Rightarrow \;\; c = \mu \;\; \Rightarrow \;\; \mu P_{00}(t) + \lambda P_{10}(t) = \mu.
$$

Using this relation we get

$$
P'_{00}(t) = -\lambda P_{00}(t) + \lambda P_{10}(t) = -\lambda P_{00}(t) + \mu - \mu P_{00}(t) = \mu - (\mu + \lambda)P_{00}(t) ,
$$

which is the same as Eq. (6.5.1). Hence we get

$$P_{00}(t) = \frac{\mu}{\lambda+\mu} + \frac{\lambda}{\lambda+\mu}e^{-(\lambda+\mu)t}$$
$$\& \ P_{01}(t) = 1 - P_{00}(t) = \frac{\lambda}{\lambda+\mu} - \frac{\lambda}{\lambda+\mu}e^{-(\lambda+\mu)t} .$$

Further,

$$\begin{aligned} P_{10}(t) &= \frac{\mu}{\lambda} - \frac{\mu}{\lambda}P_{00}(t) = \frac{\mu}{\lambda} - \frac{\mu}{\lambda}\left(\frac{\mu}{\lambda+\mu} + \frac{\lambda}{\lambda+\mu}e^{-(\lambda+\mu)t}\right) \\ &= \frac{\mu}{\lambda+\mu} - \frac{\mu}{\lambda+\mu}e^{-(\lambda+\mu)t} . \end{aligned}$$

From $P_{10}(t)$, we get

$$P_{11}(t) = 1 - P_{10}(t) = \frac{\lambda}{\lambda+\mu} + \frac{\mu}{\lambda+\mu}e^{-(\lambda+\mu)t} .$$

Thus, both the forward and the backward differential equations lead to the same solution. Hence, the matrix $P(t)$ of transition probability functions is given by

$$P(t) = \begin{array}{c} \\ 0 \\ 1 \end{array} \begin{array}{c} \begin{array}{cc} \quad 0 & \qquad\qquad 1 \end{array} \\ \begin{pmatrix} \frac{\mu}{\lambda+\mu} + \frac{\lambda}{\lambda+\mu}e^{-(\lambda+\mu)t} & \frac{\lambda}{\lambda+\mu} - \frac{\lambda}{\lambda+\mu}e^{-(\lambda+\mu)t} \\ \frac{\mu}{\lambda+\mu} - \frac{\mu}{\lambda+\mu}e^{-(\lambda+\mu)t} & \frac{\lambda}{\lambda+\mu} + \frac{\mu}{\lambda+\mu}e^{-(\lambda+\mu)t} \end{pmatrix} \end{array} .$$

It is to be noted that it is simpler to solve the forward differential equations than the backward differential equations. □

Remark 6.5.1 The method adopted in Example 6.5.1 to find $P(t)$ from the given generator matrix Q is a standard procedure and can be found in many books, for example see Ross [7].

In Example 6.5.1 we are able to find $P_{ij}(t)$ explicitly. Using the expression of transition probability function we can study the behavior of the process over a finite time interval $[0, T]$. The expected length of time the process spends in a given state during a given interval of time is known as an occupancy time of the given state, Kulkarni [4]. Following theorem gives a method to compute occupancy time when $P_{ij}(t)$'s are known explicitly and when state space is finite. Suppose $\mu_{ij}(T)$ denotes the occupancy time in state j during a given interval $[0, T]$ when $X(0) = i$.

Theorem 6.5.1 *Suppose $\{X(t), t \geq 0\}$ is a continuous time Markov chain with M states and probability transition function $P_{ij}(t)$. Then*

$$\mu_{ij}(T) = \int_0^T P_{ij}(t)\,dt, \quad 1 \leq i, j \leq M.$$

Proof Suppose $X(0) = i$. For fixed i and j, we define a random variable $Y_j(t)$ as $Y_j(t) = 1$ if $X(t) = j$ and 0 otherwise. Then the total amount of time spent in state j by the process during $[0, T]$ is given by $\int_0^T Y_j(t)\,dt$. Hence,

$$\begin{aligned}
\mu_{ij}(T) &= E\Big(\int_0^T Y_j(t)\,dt \big| X(0) = i\Big) = \int_0^T E(Y_j(t)\big|X(0) = i)\,dt \\
&= \int_0^T P[Y_j(t) = 1\big|X(0) = i]\,dt = \int_0^T P[X(t) = j\big|X(0) = i]\,dt \\
&= \int_0^T P_{ij}(t)\,dt\ .
\end{aligned}$$

Note that in the second step, the interchange of expectation and integral is valid. □

We illustrate this theorem by computing occupancy time in the following example.

Example 6.5.2 Suppose a machine is in a working state for random amount of time having an exponential distribution with parameter μ. When it fails, it gets repaired. The repair time has an exponential distribution with parameter λ and is independent of the past. The machine is as good as new after the repair is complete. Suppose $X(t)$ denotes the state of the machine at time t, $X(t) = 0$, if it is down and $X(t) = 1$ if it is working. Then $\{X(t), t \geq 0\}$ can be modeled as a Markov process with state space $S = \{0, 1\}$. Further $\lambda_0 = \lambda = -q_{00}$ and $q_{01} = \lambda$. Similarly, $\lambda_1 = \mu = -q_{11}$ and $q_{10} = \mu$. Thus, the generator matrix Q is given by

$$Q = \begin{array}{c} \\ 0 \\ 1 \end{array}\!\!\begin{array}{c} \begin{array}{cc} 0 & \ \ 1 \end{array} \\ \begin{pmatrix} -\lambda & \lambda \\ \mu & -\mu \end{pmatrix} \end{array}.$$

It is the same as in Example 6.5.1, so $P_{ij}(t)$ for all i and j are the same as in Example 6.5.1. Suppose the expected time until failure of the machine is 10 days, while the expected repair time is 1 day, that is, $\lambda = 1$ and $\mu = 0.1$. Suppose the machine is working at the beginning of January. It is of interest to compute the expected total uptime of the machine in the month of January, that is, we want to compute $\mu_{11}(31)$ in time units of days. From Theorem 6.5.1, $\mu_{11}(T) = \int_0^T P_{11}(t)\,dt$. Now

$$\begin{aligned}
P_{11}(t) &= \frac{\lambda}{\lambda+\mu} + \frac{\mu}{\lambda+\mu}e^{-(\lambda+\mu)t} = \frac{10}{11} + \frac{1}{11}e^{-1.1t} \\
\Rightarrow \quad \mu_{11}(31) &= \int_0^{31} P_{11}(t)\,dt = \int_0^{31}\left(\frac{10}{11} + \frac{1}{11}e^{-1.1t}\right)dt \\
&= \frac{31 \times 10}{11} + \frac{1}{11}\frac{1}{1.1}(1 - e^{-34.1}) = 28.26 \text{ days}\ .
\end{aligned}$$

Thus, the expected time machine is in a working state during January is 28.26 days. Hence, the expected downtime is $31 - 28.26 = 2.74$ days. □

We now proceed to find a general form for the solution of the differential equation $P'(t) = P(t)Q = QP(t)$, when the state space is finite. It is known that the solution of the scalar differential equation $f'(t) = cf(t)$ is $f(t) = f(0)e^{ct}$. On similar lines, it can be shown that

$$P'(t) = P(t)Q = QP(t) \;\Rightarrow\; P(t) = P(0)e^{Qt} = e^{Qt} \text{ since } P(0) = I$$

$$\Rightarrow\; P(t) = e^{tQ} \equiv \sum_{n=0}^{\infty} (tQ)^n/n! \tag{6.5.2}$$

$$\&\; P(t) = e^{Qt} = (P(1))^t, \;\; \forall\; t \geq 0 \text{ since } P(1) = e^{Q}\,. \tag{6.5.3}$$

The identity in Eq. (6.5.3) conveys that to find $P(t)$, it is enough to know $P(1)$, that is Q. This result is analogous to the result $P^{(n)} = P^n$, for Markov chains with discrete time parameter. We now discuss various methods to compute $P(t) = e^{tQ}$.

Method based on the spectral decomposition of Q: Suppose D denotes the diagonal matrix with diagonal elements as the eigenvalues α_i of Q and V denotes the matrix of corresponding normalized right eigenvectors. Then as derived in Eq. (2.2.3), the spectral decomposition of Q is $Q = VDV^{-1}$. Since α_i's are eigenvalues of Q, the eigenvalues of e^{tQ} are $e^{\alpha_i t}$, $i = 1, 2, \ldots$. If D_t denotes the diagonal matrix with diagonal elements as $e^{\alpha_i t}$, then $P(t)$ can be obtained as

$$P(t) = e^{Qt} = VD_tV^{-1}\,.$$

Computation of $P(t)$ using Eq. (6.5.2) and the method based on spectral decomposition are illustrated in the following examples. In these examples, Q is of dimension 2×2 and hence $P(t)$ can be computed explicitly and we can verify the results given by these two methods.

Example 6.5.3 Suppose the lifetime of a high-altitude satellite has exponential distribution with parameter λ. Once it fails, it remains in the same state as repair is not possible. Suppose $X(t)$ denotes the state of the satellite at time t with $X(t) = 1$ if it is operational at time t and $X(t) = 0$ if in failed state. In Example 6.4.2 we have examined that $\{X(t), t \geq 0\}$ is a continuous time Markov chain with matrix $P(t)$ of probability transition functions and matrix Q of intensity rates as given below.

$$P(t) = \begin{array}{c} \\ 0 \\ 1 \end{array}\!\!\begin{array}{c} \begin{array}{cc} 0 & \quad 1 \end{array} \\ \begin{pmatrix} 1 & 0 \\ 1 - e^{-\lambda t} & e^{-\lambda t} \end{pmatrix} \end{array} \quad Q = \begin{array}{c} \\ 0 \\ 1 \end{array}\!\!\begin{array}{c} \begin{array}{cc} 0 & \;1 \end{array} \\ \begin{pmatrix} 0 & 0 \\ \lambda & -\lambda \end{pmatrix} \end{array}.$$

We now examine whether $P(t) = e^{Qt}$. Since the row sums of Q matrix are 0, it follows that 0 is always the eigenvalue of Q. Since sum of the eigenvalues is the trace of Q, the other eigenvalue is $-\lambda$. To find the corresponding eigenvectors, we solve the equations $Q\underline{x} = 0\underline{x}$ and $Q\underline{x} = -\lambda\underline{x}$, where $\underline{x} = (x_1, x_2)'$. Equation $Q\underline{x} = 0\underline{x}$ implies $x_1 = x_2 = c$. Thus, the normalized right eigenvector corresponding to eigenvalue 0 is $(1/\sqrt{2}, 1/\sqrt{2})'$. Equation $Q\underline{x} = -\lambda\underline{x}$ implies $x_1 = 0, x_2 = c$. Thus, the normalized right eigenvector corresponding to eigenvalue $-\lambda$ is $(0, 1)'$. Hence, the matrix V of right eigenvectors and its inverse are given by

$$V = \begin{pmatrix} 0 & 1/\sqrt{2} \\ 1 & 1/\sqrt{2} \end{pmatrix} \quad V^{-1} = \begin{pmatrix} -1 & 1 \\ \sqrt{2} & 0 \end{pmatrix}.$$

The eigenvalues of e^{Qt} are given by $e^{-\lambda t}$ and $e^{0t} = 1$. Thus,

$$\begin{pmatrix} 0 & 1/\sqrt{2} \\ 1 & 1/\sqrt{2} \end{pmatrix} \times \begin{pmatrix} e^{-\lambda t} & 0 \\ 0 & 1 \end{pmatrix} \times \begin{pmatrix} -1 & 1 \\ \sqrt{2} & 0 \end{pmatrix} = \begin{pmatrix} 1 & 0 \\ 1 - e^{-\lambda t} & e^{-\lambda t} \end{pmatrix} = P(t).$$

□

Example 6.5.4 A molecule transits between states 0 and 1. The intensity rates are $q_{01} = 3$ and $q_{10} = 1$. Hence, the generator matrix is

$$Q = \begin{matrix} & \begin{matrix} 0 & 1 \end{matrix} \\ \begin{matrix} 0 \\ 1 \end{matrix} & \begin{pmatrix} -3 & 3 \\ 1 & -1 \end{pmatrix} \end{matrix}.$$

Proceeding exactly on the same lines as in Example 6.5.1 we get

$$P_{00}(t) = \frac{1}{4} + \frac{3}{4}e^{-4t} \quad P_{01}(t) = \frac{3}{4} - \frac{3}{4}e^{-4t}$$
$$P_{10}(t) = \frac{1}{4} - \frac{1}{4}e^{-4t} \quad P_{11}(t) = \frac{3}{4} + \frac{1}{4}e^{-4t}.$$

Thus, the matrix $P(t)$ is given by

$$P(t) = \begin{matrix} & \begin{matrix} 0 & 1 \end{matrix} \\ \begin{matrix} 0 \\ 1 \end{matrix} & \begin{pmatrix} \frac{1}{4} + \frac{3}{4}e^{-4t} & \frac{3}{4} - \frac{3}{4}e^{-4t} \\ \frac{1}{4} - \frac{1}{4}e^{-4t} & \frac{3}{4} + \frac{1}{4}e^{-4t} \end{pmatrix} \end{matrix} = \begin{matrix} & \begin{matrix} 0 & 1 \end{matrix} \\ \begin{matrix} 0 \\ 1 \end{matrix} & \begin{pmatrix} \frac{1}{4} & \frac{3}{4} \\ \frac{1}{4} & \frac{3}{4} \end{pmatrix} \end{matrix} - \frac{e^{-4t}}{4}Q = I + \frac{1 - e^{-4t}}{4}Q.$$

We now use the method based on the spectral decomposition of Q. The eigenvalues of Q are $0, -4$ and corresponding right eigenvectors are $(1, 1)'$ and $(-3, 1)'$. Thus,

$$V = \begin{pmatrix} -3 & 1 \\ 1 & 1 \end{pmatrix}, \quad V^{-1} = \begin{pmatrix} -1/4 & 1/4 \\ 1/4 & 3/4 \end{pmatrix} \quad \& \quad D_t = \begin{pmatrix} e^{-4t} & 0 \\ 0 & 1 \end{pmatrix}.$$

Hence, $P(t) = e^{Qt} = VD_tV^{-1}$ is given by

$$P(t) = \begin{pmatrix} \frac{1}{4} + \frac{3}{4}e^{-4t} & \frac{3}{4} - \frac{3}{4}e^{-4t} \\ \frac{1}{4} - \frac{1}{4}e^{-4t} & \frac{3}{4} + \frac{1}{4}e^{-4t} \end{pmatrix}.$$

It is the same as $P(t)$ obtained by solving Kolmogorov's forward equations. We now examine whether $P(t)$ can be expressed as $P(t) = e^{tQ} \equiv \sum_{n=0}^{\infty}(tQ)^n/n!$. It is easy to verify that $Q^2 = -4Q$, $Q^3 = 4^2Q$ and $Q^4 = -4^3Q$ and so on. Thus,

$$\begin{aligned} P(t) = e^{tQ} = \sum_{n=0}^{\infty} \frac{(tQ)^n}{n!} &= I + t\,Q - 4\frac{t^2}{2!}Q + 4^2\frac{t^3}{3!}Q - 4^3\frac{t^4}{4!}Q + \cdots \\ &= I - Q\frac{1}{4}\left(-4t + \frac{4^2t^2}{2!} - \frac{4^3t^3}{3!} + \frac{4^4t^4}{4!} - \cdots\right) \\ &= I + \left(\frac{1 - e^{-4t}}{4}\right) Q\,. \end{aligned}$$

It is exactly the same expression obtained above. □

Remark 6.5.2 If the intensity matrix is of the form

$$Q = \begin{array}{c c} & \begin{array}{cc} 0 & 1 \end{array} \\ \begin{array}{c} 0 \\ 1 \end{array} & \begin{pmatrix} -\lambda & \lambda \\ \mu & -\mu \end{pmatrix} \end{array}$$

then we always have $Q^2 = -(\lambda + \mu)Q$, $Q^3 = (\lambda + \mu)^2Q$ and $Q^4 = -(\lambda + \mu)^3Q$ and so on. Thus, we have $P(t) = I + \left((1 - e^{-(\lambda+\mu)t})/(\lambda + \mu)\right)\,Q$.

In both the above examples, it is possible to verify that $P(t) = e^{Qt}$, as the matrices involved are of order 2×2. However, it may not be possible always and we need to have some method to compute $P(t)$, at least approximately. Following are some methods to compute $P(t)$ approximately. These approximation methods are routine methods and are also suggested by other authors, for example refer to Kulkarni [4] and Ross [7]. In this book we illustrate how to use these methods using R software.

Approximation method 1: This method is based on Eq. (6.5.2), which states that $P(t) = e^{tQ} \equiv \sum_{n=0}^{\infty}(tQ)^n/n!$. The expression in terms of infinite series is obtained by Taylor's series expansion for matrices. We use first few terms of this series to compute the approximate expression for $P(t)$. Thus, for some large N, approximate expression for $P(t)$ is

$$P(t) = e^{tQ} \approx \sum_{n=0}^{N} \frac{(tQ)^n}{n!}.$$

In R we can compute such an approximation of e^{tQ} using the function `expm(Q*t)` from the library `expm`.

The direct use of this approximation to compute $P(t)$ numerically turns out to be very inefficient for two reasons. The first reason is that there is a problem of computer round-off error when we compute the powers of Q, since the matrix Q contains both positive and negative elements. Secondly, we often have to compute many terms in the infinite sum specified in Eq. (6.5.2) to arrive at a good approximation. There are certain indirect ways that we can utilize to efficiently approximate the matrix $P(t)$. We present three such approximation methods below.

Approximation method 2: In a scalar case we have $e^x = \lim_{n\to\infty}(1 + x/n)^n$. It is known that this convergence is slower than Taylor's series. Its matrix equivalent is given by

$$e^{Qt} = \lim_{n\to\infty}\left(I + Q\frac{t}{n}\right)^n \approx \left(I + Q\frac{t}{n}\right)^n \text{ for large n.}$$

The diagonal elements of Q are negative and the diagonal elements of the identity matrix I are equal to 1. Hence by choosing n large enough corresponding to given t, we can guarantee that the matrix $I + Qt/n$ has all non-negative elements. Thus, we approximate $P(t)$ by $(I + Qt/n)^n$ for sufficiently large n, corresponding to a specified value of t.

Approximation method 3: We also have $e^{-x} = \lim_{n\to\infty}(1 - x/n)^n$. In this approach, we use its matrix version, which gives the following identity

$$e^{-Qt} = \lim_{n\to\infty}\left(I - Q\frac{t}{n}\right)^n \approx \left(I - Q\frac{t}{n}\right)^n \text{ for large n.}$$

It can be shown that the matrix $(I - Qt/n)^{-1}$ has all non-negative elements and hence $P(t)$ can be approximated as

$$P(t) = e^{Qt} = \left(I - Q\frac{t}{n}\right)^{-n} = \left[\left(I - Q\frac{t}{n}\right)^{-1}\right]^n \text{ for large n .}$$

Approximation method 4: Suppose $\lambda = \max_{i\in S}\lambda_i$. Since the state space is finite, λ is finite. Suppose probabilities p^*_{ij} are defined as follows.

$$p^*_{ij} = \begin{cases} 1 - \lambda_i/\lambda, & \text{if } i = j \\ q_{ij}/\lambda, & \text{if } i \neq j\,. \end{cases}$$

Suppose $P^* = [p^*_{ij}]$. Then $P(t)$ can be obtained as

$$P(t) = \sum_{n=0}^{\infty} e^{-\lambda t}\frac{(\lambda t)^n}{n!}(P^*)^n \approx \sum_{n=0}^{N} e^{-\lambda t}\frac{(\lambda t)^n}{n!}(P^*)^n,$$

where N is sufficiently large. A good rule of thumb is to choose $N = \max\{\lambda t + 5 * \sqrt{\lambda t},\ 20\}$. This technique of approximating $P(t)$ is known as uniformization technique. For more details and proof, we refer to Kulkarni [4] and Ross [7].

In the following example, we illustrate the computation of $P(t)$ using Code 6.7.5, based on the following five formulae.

(i) $P(t) = V D_t V^{-1}$,
(ii) $P(t) = e^{tQ} = \sum_{n=0}^{N} (tQ)^n/n!$ for large N,
(iii) $P(t) = (I + Qt/n)^n$ for large n,
(iv) $P(t) = \left[(I - Qt/n)^{-1}\right]^n$ for large n and
(v) $P(t) = \sum_{n=0}^{N} e^{-\lambda t}(\lambda t)^n (P^*)^n/n!$.

Example 6.5.5 Suppose $\{X(t), t \geq 0\}$ is a continuous time Markov chain with state space $S = \{1, 2, 3, 4\}$ and intensity matrix Q as given below.

$$Q = \begin{array}{c} \\ 1 \\ 2 \\ 3 \\ 4 \end{array} \begin{array}{c} \begin{array}{cccc} 1 & 2 & 3 & 4 \end{array} \\ \begin{pmatrix} -5 & 3 & 1 & 1 \\ 1 & -1 & 0 & 0 \\ 2 & 1 & -4 & 1 \\ 0 & 2 & 2 & -4 \end{pmatrix} \end{array}.$$

We compute $P(t)$ for $t = 1, 2, 5$ using the formulae listed above. From the output, we note that $P(t)$ for $t = 1, 2$ and 5 by all the five methods are the same up to three decimal places accuracy. These are presented below.

$$P(1) = \begin{array}{c} \\ 1 \\ 2 \\ 3 \\ 4 \end{array} \begin{array}{c} \begin{array}{cccc} 1 & 2 & 3 & 4 \end{array} \\ \begin{pmatrix} 0.171 & 0.669 & 0.087 & 0.072 \\ 0.168 & 0.718 & 0.062 & 0.053 \\ 0.174 & 0.631 & 0.111 & 0.085 \\ 0.168 & 0.637 & 0.107 & 0.088 \end{pmatrix} \end{array},$$

$$P(2) = \begin{array}{c} \\ 1 \\ 2 \\ 3 \\ 4 \end{array} \begin{array}{c} \begin{array}{cccc} 1 & 2 & 3 & 4 \end{array} \\ \begin{pmatrix} 0.169 & 0.696 & 0.074 & 0.061 \\ 0.169 & 0.700 & 0.071 & 0.060 \\ 0.169 & 0.693 & 0.075 & 0.062 \\ 0.169 & 0.694 & 0.075 & 0.062 \end{pmatrix} \end{array}$$

$$\& \ P(5) = \begin{array}{c} \\ 1 \\ 2 \\ 3 \\ 4 \end{array} \begin{array}{c} \begin{array}{cccc} 1 & 2 & 3 & 4 \end{array} \\ \begin{pmatrix} 0.169 & 0.699 & 0.072 & 0.06 \\ 0.169 & 0.699 & 0.072 & 0.06 \\ 0.169 & 0.699 & 0.072 & 0.06 \\ 0.169 & 0.699 & 0.072 & 0.06 \end{pmatrix} \end{array}.$$

We observe that rows of the matrix $P(5)$ are approximately identical. It can be verified that for $t > 5$, we arrive at the same conclusion. It implies that for large t,

$P_{ij}(t)$ is free from initial state i and $P_{ij}(t)$ converges to P_j, say, a constant free from t. Here $P_1 = 0.169$, $P_2 = 0.699$, $P_3 = 0.072$ and $P_4 = 0.06$. In the next section, we discuss in detail the limiting behavior of $P(t)$. □

Following example gives an interesting application of computation of $P(t)$, Kulkarni [4].

Example 6.5.6 A commercial jet airplane has four engines, two on each wing. Each engine works for a random amount of time which has an exponential distribution with parameter λ. If the failure takes place in flight, there can be no repair. The airplane needs at least one engine on each wing to function properly in order to fly safely. It is of interest to predict the probability of a trouble-free flight. Hence, we need a suitable model to predict this probability. Suppose $X_L(t)$ denotes the number of functioning engines on the left wing at time t and $X_R(t)$ denotes the number of functioning engines on the right wing at time t. The state of the system at time t is given by $\underline{X}(t) = (X_L(t), X_R(t))$. We assume that the engine failures are independent of each other. Then exponential distribution for the lifetime of the engine implies that $\underline{X}(t)$ can be modeled as a continuous time Markov chain with state space S given by

$$S = \{(0,0), (0,1), (0,2), (1,0), (1,1), (1,2), (2,0), (2,1), (2,2)\}.$$

Suppose we number the states as $1, 2, \ldots, 9$ in the order listed in S. The Q matrix is obtained using the similar arguments as in Example 6.4.3. For example, suppose the state of the process is (2, 2), that is, both the engines on the two wings are in a working condition. If the engine on the left wing fails, the next state will be (1, 2) and the rate of such a transition is 2λ as any one of the two engines fails. The rate of transition from (2, 2) to (2, 1) is the same. Thus, the intensity matrix Q is as follows.

$$Q = \begin{array}{c} \\ 1 \\ 2 \\ 3 \\ 4 \\ 5 \\ 6 \\ 7 \\ 8 \\ 9 \end{array} \begin{array}{c} \begin{array}{ccccccccc} 1 & 2 & 3 & 4 & 5 & 6 & 7 & 8 & 9 \end{array} \\ \begin{pmatrix} 0 & 0 & 0 & 0 & 0 & 0 & 0 & 0 & 0 \\ \lambda & -\lambda & 0 & 0 & 0 & 0 & 0 & 0 & 0 \\ 0 & 2\lambda & -2\lambda & 0 & 0 & 0 & 0 & 0 & 0 \\ \lambda & 0 & 0 & -\lambda & 0 & 0 & 0 & 0 & 0 \\ 0 & \lambda & 0 & \lambda & -2\lambda & 0 & 0 & 0 & 0 \\ 0 & 0 & \lambda & 0 & 2\lambda & -3\lambda & 0 & 0 & 0 \\ 0 & 0 & 0 & 2\lambda & 0 & 0 & -2\lambda & 0 & 0 \\ 0 & 0 & 0 & 0 & 2\lambda & 0 & \lambda & -3\lambda & 0 \\ 0 & 0 & 0 & 0 & 0 & 2\lambda & 0 & 2\lambda & -4\lambda \end{pmatrix} \end{array}.$$

We assume that all the four engines are in a working condition at the beginning of the flight, that is $\underline{X}(0) = (2, 2)$. The flight is safe if $X_L(t) \geq 1$ and $X_R(t) \geq 1$, that is, if the state of the system is in the set $S_1 = \{(1,1), (1,2), (2,1), (2,2)\}$. Hence, the probability of safe flight is $P[\underline{X}(t) \in S_1 | \underline{X}(0) = (2,2)]$. It is then obtained by the addition of probabilities in the last row of $P(t)$, since $X(0) = (2,2) \equiv 9$, corresponding to columns 5, 6, 8, 9, since these columns correspond to states in S_1.

Table 6.2 Probability of safe flight

	Mean in hours		
Hours	100	200	400
4	0.9973	0.9993	0.9998
9	0.9886	0.9966	0.9991
15	0.9755	0.9916	0.9976

Code 6.7.6 computes the probability for $t = 4, 9, 15$ h and $\lambda = 0.01, 0.005, 0.0025$, that is, when on the average engine works for 100, 200 and 400 h respectively. The output is displayed in Table 6.2. We note that the probability of safe flight in 4 h is 0.9973, in 9 h is 0.9886 and in 15 h is 0.9755, corresponding to $\lambda = 0.01$. If on the average the engine works for 200 h, that is if $\lambda = 0.005$, then the probability of safe flight is computed with $Q_1 = 0.5Q$. In this case, the probability of safe flight in 4 h is 0.9993, in 9 h is 0.9966 and in 15 h is 0.0.9916. Thus, the probability of crash for a 15 hour flight with $\lambda = 0.005$ is 0.0084, which is large for any commercial airline. It implies that efficiency of the machines needs to be increased so that average life of the machine is larger than 200 h. If it is 400 h, that is, $\lambda = 0.0025$, then the respective probabilities are 0.9998, 0.9991 and 0.9976. □

The next section is concerned with the long run behavior of the continuous time Markov chains.

6.6 Long Run Behavior

Suppose $\{X(t), t \geq 0\}$ is a time homogeneous continuous time Markov chain with transition probability function $P_{ij}(t)$. We study the long run behavior of the process in terms of the associated stationary distributions and the long run distribution. Theoretical aspects are not discussed in detail. Interested reader may refer to Cinlar [1], Hoel, Port and Stone [2] and Norris [6].

In the previous section, we have discussed how to compute, at least approximately, $P_{ij}(t)$ for fixed t. Our aim is now to investigate its behavior as t increases. In Example 6.5.5, we have noted that rows of the matrix $P(t), t \geq 5$ are approximately identical. We can find the limit of the $P_{ij}(t)$, when we can obtain it explicitly as in Example 6.5.1. However, it may not be always possible.

In Chap. 3, we have noted that the existence of the long run and stationary distributions depends on the nature of the corresponding Markov chain. The same is true in the setup of a continuous time Markov chain. Hence, as in the case of Markov chains in discrete time, to examine whether these distributions exist, our starting point is classification of states of a continuous time Markov chain. Viewing a continuous time Markov chain as an embedded discrete time Markov chain with exponential holding times, the classification of states is analogous to the discrete time setting as discussed in Chap. 2.

The properties such as accessibility, irreducibility, transience and recurrence are all defined similarly as for Markov chains in discrete time, via the embedded Markov chain with transition probabilities p_{ij}. The results are summarized below.

(i) The communicating classes of a continuous time Markov chain are the same as the communicating classes of the embedded Markov chain. If there is only one communicating closed class, we say the chain is irreducible, otherwise it is said to be reducible. In particular, $\{X(t), t \geq 0\}$ is irreducible if and only if $P_{ij}(t) > 0$ for some $t > 0$ and for all $i, j \in S$.

(ii) The concept of periodicity no longer plays a role, or even makes sense to define, as time is no longer discrete.

(iii) It is to be noted that $X(t)$ returns to a state i infinitely often if and only if the embedded discrete time chain does. Hence, a state $i \in S$ is persistent in $\{X(t), t \geq 0\}$, if i is persistent in the embedded discrete time chain $\{X_n, n \geq 0\}$. A transient state is defined on similar lines. However, non-null or null persistence cannot be defined in terms of embedded Markov chain. It is defined as follows, Hoel, Port and Stone [2].

Definition 6.6.1 Suppose $X(0) = i$ and T_i is the sojourn time in state i. A random variable τ_i is defined as $\tau_i = \inf\{t \geq T_i \text{ such that } X(t) = i\}$. The state i is non-null persistent if $E(\tau_i) < \infty$ and null persistent if $E(\tau_i) = \infty$.

Thus, at time point 0 the system is in state i, after staying in state i for random time T_i, it transits to any other state and after visiting some more states, returns to state i. A random variable τ_i denotes the first return time to state i after the process leaves i. If τ_i has finite mean, then i is non-null persistent, otherwise it is null persistent. An absorbing state is considered to be non-null persistent. It can be shown that i may be non-null persistent for embedded Markov chain, but not for $\{X(t)\}$ and vice-versa, Cinlar [1].

(iv) As in the discrete time setting, persistence, transience and non-null persistence are class properties.

We now define a stationary distribution associated with a continuous time Markov chain. It is analogous to that in the discrete time parameter setup.

Definition 6.6.2 Stationary Distribution: Suppose $\{X(t), t \geq 0\}$ is a continuous time Markov chain with state space S and transition probability function $P(t)$. A row vector $\underline{\eta} = \{\eta_i, \ i \in S\}$ is said to be a stationary distribution associated with $\{X(t), t \geq 0\}$ if (i) $\eta_i \geq 0 \ \forall \ i \in S$, (ii) $\sum_{i \in S} \eta_i = 1$ and
(iii) $\underline{\eta} = \underline{\eta} P(t) \iff \eta_j = \sum_{i \in S} \eta_i P_{ij}(t) \ \forall \ j \in S \ \& \ \forall \ t \geq 0$.

As in the setup of discrete time Markov chains, if we set the initial distribution, that is, distribution of $X(0)$ to be $\underline{\eta}$, then the distribution of $X(t)$ is also $\underline{\eta}$ for all $t > 0$. It is shown below.

$$P[X(t) = j] = \sum_{i \in S} P_{ij}(t) P[X(0) = i] = \sum_{i \in S} P_{ij}(t)\eta_i = \eta_j \;\; \forall \;\; j \in S \; .$$

We state below some results about the stationary distribution, Hoel, Port and Stone [2].

(i) A stationary distribution is concentrated on a set of non-null persistent states. Hence, a process that is transient or null persistent does not have a stationary distribution. It is similar to the result in discrete time Markov chains.
(ii) An irreducible non-null persistent process has a unique stationary distribution $\underline{\eta}$. Further, $\eta_i = 1/(\lambda_i E(\tau_i)), i \in S$.
(iii) η_i is interpreted as the long run expected proportion of time spent in state i.

In general it is difficult to find $E(\tau_i)$ and decide whether the process is non-null persistent. Following theorem states some equivalent conditions (Norris [6]), which are useful to decide whether the process is non-null persistent.

Theorem 6.6.1 *Suppose Q is a generator matrix of an irreducible continuous time Markov chain. Then the following statements are equivalent.*
(i) Every state is non-null persistent and (ii) continuous time Markov chain is non-explosive and has a stationary distribution.

The condition $S_n \to \infty$ as $n \to \infty$ in the definition of pure jump process guarantees that the process is non-explosive. Thus, if a stationary distribution exists, then every state is non-null persistent. We have a similar result for discrete time Markov chains. If a stationary distribution exists, we can compute $E(\tau_i)$ from $\eta_i = 1/(\lambda_i E(\tau_i))$. We illustrate it in Example 6.6.1.

Definition of a stationary distribution requires that $\underline{\eta} = \underline{\eta} P(t) \; \forall t \geq 0$. Thus to find $\underline{\eta}$, we need to know $P(t) \; \forall \; t > 0$, which is in general difficult. The following theorem conveys a simple way to find a stationary distribution, under certain conditions. The proof is given only for the finite state space.

Theorem 6.6.2 *Suppose $\{X(t), t \geq 0\}$ is an irreducible and persistent continuous time Markov chain with generator matrix Q. Then*

$$\underline{\eta} = \underline{\eta} P(t) \;\; \forall \;\; t \geq 0 \quad \Longleftrightarrow \quad \underline{\eta} Q = \underline{0}.$$

Proof (i) We first assume $\underline{\eta} Q = \underline{0}$. From Kolmogorov's backward differential equations, we have

$$P'(t) = QP(t) \Rightarrow \underline{\eta} P'(t) = \underline{\eta} QP(t) \;\; \Rightarrow \;\; \underline{\eta} P'(t) = \underline{0} \;\; \Rightarrow \;\; \underline{\eta} P(t) = \underline{\alpha},$$

is one of the solutions of $\underline{\eta} P'(t) = \underline{0}$, where $\underline{\alpha}$ is a vector with components free from t. Since $P(0) = I$, at $t = 0$

$$\underline{\alpha} = \underline{\eta} P(0) = \underline{\eta} \;\; \Rightarrow \;\; \underline{\eta} P(t) = \underline{\eta} \;\; \forall \;\; t \geq 0.$$

Thus, $\underline{\eta}Q = \underline{0} \;\Rightarrow\; \underline{\eta} = \underline{\eta}P(t) \;\; \forall \; t \geq 0$.
(ii) Now we assume that $\underline{\eta} = \underline{\eta}P(t) \;\; \forall \; t \geq 0$. Thus with $t = h$,

$$\begin{aligned} \underline{\eta} = \underline{\eta}P(h) &\Rightarrow \underline{\eta}(P(h) - I) = 0 \\ &\Rightarrow \lim_{h \to 0-} \underline{\eta}(P(h) - I)/h = \underline{\eta}P'(0) = 0 \;\Rightarrow\; \underline{\eta}Q = 0, \end{aligned}$$

since $P'(0) = Q$. □

Thus, a stationary distribution can be obtained as a solution of $\underline{\eta}Q = \underline{0}$.

Remark 6.6.1 The interchange of differentiation and summation in the last step of the above derivation, cannot in general be justified if the state space is infinite and a different proof is needed. We refer to Norris [6] for the proof in case of a countable state space.

There is a link between the stationary distribution of a continuous time Markov chain and the stationary distribution of the corresponding embedded Markov chain, provided both exist. It is proved in the following theorem.

Theorem 6.6.3 *Suppose $\underline{\eta}$ and $\underline{\pi}$ are stationary distributions of $\{X(t), t \geq 0\}$ and of the corresponding embedded Markov chain $\{X_n, n \geq 0\}$ respectively.*

$$\textit{If } \lambda_i > 0 \;\forall\; i \in S, \;\; \textit{then } \pi_i = c\eta_i\lambda_i \;\;\forall\; i \in S, \;\; \textit{where } c = 1\Big/\sum_{i \in S} \eta_i\lambda_i.$$

$$\& \textit{ if } \lambda_i = 0 \textit{ for one } i, \;\; \textit{then } \underline{\pi} = \underline{\eta} = (0, 0, \ldots, 1, 0, \ldots, 0),$$

where ith component is 1.

Proof If the stationary distribution $\underline{\eta}$ exists, then by Theorem 6.6.2, $\underline{\eta}Q = \underline{0}$.
Case (i): Suppose $\lambda_i > 0 \;\forall\; i \in S$.

$$\begin{aligned} \underline{\eta}Q = \underline{0} &\iff \sum_{i \in S} \eta_i q_{ij} = 0 \iff \sum_{i \neq j} \eta_i q_{ij} = -\eta_j q_{jj}, \forall \; j \in S \\ &\iff \sum_{i \neq j} \eta_i \lambda_i p_{ij} = \eta_j \lambda_j \iff \sum_{i \in S} \eta_i \lambda_i p_{ij} = \eta_j \lambda_j \;\text{ since }\; p_{jj} = 0 \\ &\iff \sum_{i \in S} \pi_i p_{ij} = \pi_j \;\text{ where }\; \pi_i = c\eta_i\lambda_i \end{aligned}$$

and $c = 1/\sum_{i \in S} \eta_i\lambda_i$. Thus, $\underline{\pi}$ is the stationary distribution of the embedded chain and it is related to $\underline{\eta}$ such that $\pi_i = c\eta_i\lambda_i, i \in S$.

Case (ii): As a particular case, suppose $\lambda_1 = 0$. Then the state 1 is absorbing and the first row of Q is 0. State 1 is also the absorbing state in the embedded Markov chain and all other states being inessential are transient. Hence, in this case $\underline{\pi}$, where $\pi_1 = 1$ and $\pi_i = 0 \;\forall\; i \neq 1$ is the unique stationary distribution of the embedded chain. Thus, $\sum_{i\in S}\pi_i p_{ij} = \pi_j = 0 \ \forall \ j \neq 1$. Further, from $q_{1j} = 0 \ \forall \ j \in S$ we have,

$$\begin{aligned}
\underline{\eta}Q = \underline{0} &\iff \sum_{i\in S}\eta_i q_{ij} = 0 \iff \eta_1 q_{1j} + \sum_{i\neq 1}\eta_i q_{ij} = 0 \ \forall \ j \in S \\
&\iff \sum_{i\neq 1,j}\eta_i q_{ij} = -\eta_j q_{jj} \iff \sum_{i\neq 1,j}\eta_i\lambda_i p_{ij} = \eta_j\lambda_j \ \forall \ j \in S \\
&\iff \sum_{i\in S}\eta_i\lambda_i p_{ij} = \eta_j\lambda_j \ \text{ since } \lambda_1 = 0 \ \& \ p_{jj} = 0 \\
&\iff \sum_{i\in S} a_i p_{ij} = a_j \ \text{ where } a_i = \eta_i\lambda_i \Big/ \sum_{i\in S}\eta_i\lambda_i \ \forall \ j \in S.
\end{aligned}$$

However, as noted above $a_1 = 1$ and $a_i = 0 \ \forall \ i \neq 1$ is the only solution to this system of equations. Thus, $\eta_i = 0 \ \forall \ i \neq 1$ and since $\sum_{i\in S}\eta_i = 1$, $\eta_1 = 1$. Thus, if $\lambda_i = 0$ for any i, then $\pi_i = \eta_i = 1$ and $\pi_j = \eta_j = 0 \ \forall \ j \neq i$. □

We illustrate case (ii) in Example 6.6.2.

Remark 6.6.2 The equation $\underline{\eta}Q = \underline{0}$ conveys that $\underline{\eta}$ is the right eigenvector of Q' corresponding to the eigenvalue 0. Since the row sums of Q are 0, it follows that 0 is always the eigenvalue of Q. Thus, to find $\underline{\eta}$, we find the right eigenvector of Q' corresponding to the eigenvalue 0 and normalize it. If the ith row of Q is 0, then a vector $\underline{\alpha}$ with all components, except the ith component, as 0, is the right eigenvector of Q' corresponding to the eigenvalue 0. After normalization we get $\underline{\eta}$, with the ith component to be 1 and all other components as 0. Thus, as for the discrete time Markov chains, the stationary distribution can be obtained using eigenvalues and eigenvectors.

We use the following methods to find a stationary distribution in the next two examples.

(i) Solve the matrix equation $\underline{\eta}Q = \underline{0} \iff Q'\underline{\eta} = 0$, subject to the condition that sum of the components of $\underline{\eta}$ is 1.
(ii) Find the right eigenvector of Q' corresponding to the eigenvalue 0 and normalize it to get $\underline{\eta}$.
(iii) Find $\underline{\pi}$, the stationary distribution of the embedded Markov chain and hence $\underline{\eta}$, where $\eta_j = c\pi_j/\lambda_j$ and $c = 1/\sum_{j\in S}(\pi_j/\lambda_j)$, provided $\lambda_j > 0$ for all $j \in S$.

Code 6.7.7 incorporates the above methods to compute the stationary distribution. Following example illustrates the code.

Example 6.6.1 Suppose $\{X(t), t \geq 0\}$ is a continuous time Markov chain with state space $S = \{1, 2, 3, 4\}$ and Q as given below.

Table 6.3 Stationary distribution of a continuous time Markov Chain

State j	1	2	3	4
η_j	0.1687	0.6988	0.0723	0.0602
π_j	0.4070	0.3372	0.1395	0.1163
α_j	0.1687	0.6988	0.0723	0.0603
$E(\tau_j)$	1.1855	1.4310	3.4578	4.1528
μ_j	2.4571	2.9655	7.1667	8.6000

$$Q = \begin{array}{c} \\ 1 \\ 2 \\ 3 \\ 4 \end{array}\begin{array}{c} \begin{array}{cccc} 1 & 2 & 3 & 4 \end{array} \\ \begin{pmatrix} -5 & 3 & 1 & 1 \\ 1 & -1 & 0 & 0 \\ 2 & 1 & -4 & 1 \\ 0 & 2 & 2 & -4 \end{pmatrix} \end{array}.$$

Using the first method, the solution of $\underline{\eta} Q = 0$ is $\underline{\eta} = (0.1687, 0.6988, 0.0723, 0.0602)$. With the second method, we find the right eigenvector of Q' corresponding to eigenvalue 0 and divide it by sum of its elements. It is the same as $\underline{\eta}$. In the third method, we find the stationary distribution $\underline{\pi}$ of the embedded Markov chain. Using it, we get the same $\underline{\eta}$. From the relation $\eta_j = 1/(\lambda_j E(\tau_j))$, we find $E(\tau_j)$, the mean recurrence time in state j in the continuous time setup. From π_j, we find μ_j, the mean recurrence time in state j for the embedded Markov chain. Suppose $\alpha_j = c\pi_j/\lambda_j$ where $c^{-1} = \sum_{j \in S} \pi_j/\lambda_j$. The output is presented in Table 6.3. From Table 6.3, we note that $\eta_j = \alpha_j$ for all j and $E(\tau_j)$ and μ_j are different. □

Example 6.6.2 Suppose $\{X(t), t \geq 0\}$ is a continuous time Markov chain with state space $S = \{1, 2, 3, 4\}$, Q and P as given below. Since state 2 is absorbing, the transition matrix P is obtained from Q using Code 6.7.2

$$Q = \begin{array}{c} \\ 0 \\ 1 \\ 2 \\ 3 \end{array}\begin{array}{c} \begin{array}{cccc} 0 & 1 & 2 & 3 \end{array} \\ \begin{pmatrix} -5 & 2 & 1 & 2 \\ 0 & 0 & 0 & 0 \\ 2 & 2 & -7 & 3 \\ 1 & 1 & 2 & -4 \end{pmatrix} \end{array} \;\&\; P = \begin{array}{c} \\ 0 \\ 1 \\ 2 \\ 3 \end{array}\begin{array}{c} \begin{array}{cccc} 0 & 1 & 2 & 3 \end{array} \\ \begin{pmatrix} 0.0000 & 0.4000 & 0.2 & 0.4000 \\ 0.0000 & 1.0000 & 0.0 & 0.0000 \\ 0.2857 & 0.2857 & 0.0 & 0.4286 \\ 0.2500 & 0.2500 & 0.5 & 0.0000 \end{pmatrix} \end{array}.$$

Using the first two methods of Code 6.7.7, we obtain $\underline{\eta} = (0, 1, 0, 0)$. From P we obtain the corresponding stationary distribution as $\underline{\pi} = (0, 1, 0, 0)$, which is the same as $\underline{\eta}$. □

We now discuss the long run distribution of a continuous time Markov chain. Following theorem states the conditions under which the long run distribution exists for a continuous time Markov chain, Norris [6].

Theorem 6.6.4 *If a continuous time Markov chain is irreducible and non-null persistent, then $P_j = \lim_{t\to\infty} P_{ij}(t)$ exists and independent of i with $\sum_{j \in S} P_j = 1$.*

The limit P_j is interpreted as a long run mean proportion of time spent in state j, $j \in S$.

Example 6.6.3 We examine whether $P_j = \lim_{t\to\infty} P_{ij}(t)$ exists for the transition probability function derived in Example 6.5.1. We have

$$P_{00}(t) = \frac{\mu}{\lambda+\mu} + \frac{\lambda}{\lambda+\mu}e^{-(\lambda+\mu)t}, \quad P_{01}(t) = \frac{\lambda}{\lambda+\mu} - \frac{\lambda}{\lambda+\mu}e^{-(\lambda+\mu)t} .$$
$$P_{10}(t) = \frac{\mu}{\lambda+\mu} - \frac{\mu}{\lambda+\mu}e^{-(\lambda+\mu)t}, \quad P_{11}(t) = \frac{\lambda}{\lambda+\mu} + \frac{\mu}{\lambda+\mu}e^{-(\lambda+\mu)t} .$$

It is clear that $P_j = \lim_{t\to\infty} P_{ij}(t)$ exists for $i, j \in \{0, 1\}$ and is independent of i. Thus,

$$P_0 = \lim_{t\to\infty} P_{00}(t) = \lim_{t\to\infty} P_{10}(t) = \frac{\mu}{\lambda+\mu}$$
$$P_1 = \lim_{t\to\infty} P_{01}(t) = \lim_{t\to\infty} P_{11}(t) = \frac{\lambda}{\lambda+\mu} . \quad \square$$

From the above example, it is clear that if we know the form of $P_{ij}(t)$, then one can find a long run distribution. However, it is rare to know the form of $P_{ij}(t)$. To deal with such cases, we derive a method to compute $\underline{P}$ in the following theorem, provided it exists. It is based on Kolmogorov's forward and backward differential equations.

Theorem 6.6.5 *Suppose $\{X(t), t \geq 0\}$ is a continuous time Markov chain with intensity matrix Q and matrix $P(t) = [P_{ij}(t)]$ of transition probability functions. Suppose $P_j = \lim_{t\to\infty} P_{ij}(t)$ exists and $\underline{P} = \{P_j, j \in S\}$. Then $\underline{P}\,Q = \underline{0}$.*

Proof Kolmogorov's backward differential equations are given by $P'_{ij}(t) = \sum_{k\in S} q_{ik}P_{kj}(t)$. Observe that $P_{kj}(t) \to P_j$, $|q_{ik}P_{kj}(t)| \leq q_{ik}$ and $\sum_{k\in S} q_{ik} = 0$. Hence, in $\lim_{t\to\infty}\sum_{k\in S} q_{ik}P_{kj}(t)$, the sum and limit can be interchanged. We allow t to tend to ∞ in Kolmogorov's backward differential equation in the following derivation.

$$\begin{aligned} P'_{ij}(t) = \sum_{k\in S} q_{ik}P_{kj}(t) &\Rightarrow \lim_{t\to\infty} P'_{ij}(t) = \lim_{t\to\infty}\sum_{k\in S} q_{ik}P_{kj}(t) \\ &\Rightarrow \lim_{t\to\infty} P'_{ij}(t) = \sum_{k\in S} q_{ik}\lim_{t\to\infty} P_{kj}(t) \\ &\Rightarrow \lim_{t\to\infty} P'_{ij}(t) = \sum_{k\in S} q_{ik}P_j = P_j\sum_{k\in S} q_{ik} \\ &\Rightarrow \lim_{t\to\infty} P'_{ij}(t) = 0. \end{aligned} \quad (6.6.1)$$

Now from Kolmogorov's forward differential equations, assuming that these exist, we have,

$$\begin{aligned}
P'_{ij}(t) = \sum_{k\in S} P_{ik}(t)q_{kj} &\Rightarrow \lim_{t\to\infty} P'_{ij}(t) = \lim_{t\to\infty} \sum_{k\in S} P_{ik}(t)q_{kj} \\
&\Rightarrow \lim_{t\to\infty} P'_{ij}(t) = \sum_{k\in S} \lim_{t\to\infty} P_{ik}(t)q_{kj} \\
&\Rightarrow \lim_{t\to\infty} P'_{ij}(t) = \sum_{k\in S} P_k q_{kj} \\
&\Rightarrow 0 = \sum_{k\in S} P_k q_{kj}, \text{ by } (6.6.1) \\
&\Rightarrow \underline{P}Q = \underline{0},
\end{aligned}$$

assuming that sum and limit can be interchanged in the second step, which is valid if S is finite.
Alternatively for any $i, j \in S$ we have,

$$\begin{aligned}
P'_{ij}(t) = \lim_{h\to 0} \frac{P_{ij}(t+h) - P_{ij}(t)}{h} &\Rightarrow \lim_{t\to\infty} P'_{ij}(t) = \lim_{t\to\infty} \lim_{h\to 0} \frac{P_{ij}(t+h) - P_{ij}(t)}{h} \\
&\Rightarrow \lim_{t\to\infty} P'_{ij}(t) = \lim_{h\to 0} \lim_{t\to\infty} \frac{P_{ij}(t+h) - P_{ij}(t)}{h} \\
&\Rightarrow \lim_{t\to\infty} P'_{ij}(t) = \lim_{h\to 0} \frac{P_j - P_j}{h} \\
&\Rightarrow \lim_{t\to\infty} P'_{ij}(t) = 0\,,
\end{aligned}$$

where in the third step we have assumed that the two limits can be interchanged. Thus, $P'(t)$ converges to a null matrix. Note that $P_{ij}(t) \to P_j$ implies the matrix $P(t)$ converges to a matrix P^* with all identical rows, each given by $\underline{P}$. Now, using Kolmogorov's forward differential equation in matrix form we have

$$P'(t) = P(t)Q \;\Rightarrow\; \lim_{t\to\infty} P'(t) = \lim_{t\to\infty} P(t)Q \;\Rightarrow\; 0 = P^*Q\,,$$

where 0 denotes a null matrix. Since P^* has all identical rows $\underline{P}$, say, $P^*Q = 0$ implies $\underline{P}Q = \underline{0}$. □

Thus, given the generator matrix Q, we can find $\underline{P}$ as a solution of the matrix equation $\underline{P}Q = 0$ under the condition that sum of the components of $\underline{P}$ is 1. Further, it can also be obtained using eigenvalues and eigenvectors of Q. The equation $\underline{P}Q = \underline{0}$ indicates that $\underline{P}$ is a left eigenvector corresponding to eigenvalue 0 of Q. Note that the stationary distribution $\underline{\eta}$, whenever exists, is given by $\underline{\eta}Q = 0$. Thus, the two distributions are the same.

In Sect. 6.3, we have noted that

$$P_j^{(t)} = P[X(t) = j] = \sum_{i \in S} P[X(t) = j | X(0) = i] P[X(0) = i]$$
$$= \sum_{i \in S} P_{ij}(t) P_i^{(0)}, \quad \forall \ j \in S.$$

Suppose $\lim_{t \to \infty} P_{ij}(t)$ exists and is given by P_j. Then $\forall \ j \in S$,

$$\lim_{t \to \infty} P_j^{(t)} = \lim_{t \to \infty} \sum_{i \in S} P_{ij}(t) P_i^{(0)}$$
$$= \sum_{i \in S} \lim_{t \to \infty} P_{ij}(t) P_i^{(0)} = \sum_{i \in S} P_j P_i^{(0)} = P_j. \tag{6.6.2}$$

Note that $P_{ij}(t) P_i^{(0)} \leq P_i^{(0)}$ and $\sum_{i \in S} P_i^{(0)} = 1$. Hence, sum and limit can be interchanged in $\lim_{t \to \infty} \sum_{i \in S} P_{ij}(t) P_i^{(0)}$. Thus, whenever the long run distribution exists, $\lim_{t \to \infty} P[X(t) = j]$ exists and is P_j, independent of the initial state. These results are analogous to those for discrete time Markov chains.

The following example illustrates the computation of $\underline{P}$.

Example 6.6.4 We consider the continuous time Markov chain discussed in Example 6.5.4 in which a molecule transits between states 0 and 1 with intensity rates as given in the generator matrix Q, where

$$Q = \begin{array}{c} \\ 0 \\ 1 \end{array} \begin{array}{c} \begin{matrix} 0 & \ 1 \end{matrix} \\ \begin{pmatrix} -3 & 3 \\ 1 & -1 \end{pmatrix} \end{array}.$$

In Example 6.5.4 we have obtained $P_{ij}(t)$ and it is easy to check that $\lim_{t \to \infty} P_{ij}(t)$ exists and is given by

$$P^* = \begin{array}{c} \\ 0 \\ 1 \end{array} \begin{array}{c} \begin{matrix} 0 & \quad 1 \end{matrix} \\ \begin{pmatrix} 1/4 & 3/4 \\ 1/4 & 3/4 \end{pmatrix} \end{array}.$$

Thus, $\underline{P} = (1/4, 3/4)$. Now we verify whether solving $\underline{P} Q = \underline{0}$, we get the same vector $\underline{P}$. The equation $\underline{P} Q = \underline{0}$ gives $3P_1 = P_2$. Using the condition $P_1 + P_2 = 1$ we get $\underline{P} = (1/4, 3/4)$. Further, two eigenvalues of Q are -4 and 0 and the normalized left eigenvector corresponding to eigenvalue 0 is $(1/4, 3/4)$. □

The system of equations $\underline{P} Q = \underline{0}$ has the following nice interpretation. These equations can be expressed as follows. The jth equation in $\underline{P} Q = \underline{0}$ is given by

$$-\lambda_j P_j + \sum_{i \neq j} q_{ij} P_i = 0 \iff \lambda_j P_j = \sum_{i \neq j} q_{ij} P_i \, .$$

On the left hand side, P_j is the long run proportion of time the process is in state j, while λ_j is the rate of leaving state j when the process is in state j. Thus, the product $\lambda_j P_j$ is interpreted as the long run rate of leaving state j. On the right hand side, q_{ij} is the rate of going to state j when the process leaves the state i, so the product $P_i q_{ij}$ is interpreted as the long run rate of going from state i to state j. Summing over all $i \neq j$ then gives the long run rate of going to state j. Hence,

$$\lambda_j P_j = \sum_{i \neq j} q_{ij} P_i \iff \text{long run rate out of state } j = \text{ long run rate into state } j.$$

The equations $\underline{P}Q = \underline{0}$ balance the rates and hence these are known as balance equations. Given the Q matrix we can write down such balance equations and solve these to get the long run distribution. We illustrate this procedure in the following example.

Example 6.6.5 A machine in a workshop can be in one of the three states, 1 indicating good working condition, 2 indicating deteriorated working condition and 3 indicating failed condition. The generator matrix Q is given by

$$Q = \begin{array}{c} \\ 1 \\ 2 \\ 3 \end{array} \begin{array}{c} \begin{array}{ccc} 1 & \quad 2 & \quad 3 \end{array} \\ \begin{pmatrix} -\mu_2 & \mu_2 & 0 \\ \lambda_1 & -(\lambda_1 + \mu_1) & \mu_1 \\ 0 & \lambda_2 & -\lambda_2 \end{pmatrix} \end{array}.$$

To obtain the long run distribution $\underline{P}$, the equations $\underline{P}Q = \underline{0}$ are as follows.

$$-\mu_2 P_1 + \lambda_1 P_2 = 0, \;\; \mu_2 P_1 - (\lambda_1 + \mu_1) P_2 + \lambda_2 P_3 = 0 \;\; \& \;\; \mu_1 P_2 - \lambda_2 P_3 = 0\,.$$

The first and the third equations imply that $P_2 = \frac{\mu_2}{\lambda_1} P_1 \;\; \& \;\; P_3 = \frac{\mu_1}{\lambda_2} P_2 = \frac{\mu_1 \mu_2}{\lambda_1 \lambda_2} P_1$. Substituting in $P_1 + P_2 + P_3 = 1$ gives $P_1 = \left(1 + \frac{\mu_2}{\lambda_1} + \frac{\mu_1 \mu_2}{\lambda_1 \lambda_2}\right)^{-1}$ and hence P_2 and P_3 are obtained. Thus, we get the long run distribution as

$$P_1 = \left(1 + \frac{\mu_2}{\lambda_1} + \frac{\mu_1 \mu_2}{\lambda_1 \lambda_2}\right)^{-1}, \;\; P_2 = \frac{\mu_2}{\lambda_1}\left(1 + \frac{\mu_2}{\lambda_1} + \frac{\mu_1 \mu_2}{\lambda_1 \lambda_2}\right)^{-1}$$

$$\& \;\; P_3 = \frac{\mu_1 \mu_2}{\lambda_1 \lambda_2}\left(1 + \frac{\mu_2}{\lambda_1} + \frac{\mu_1 \mu_2}{\lambda_1 \lambda_2}\right)^{-1}.$$

Now we write down the balance equations by noting the rate going out of the state and the rate of going into the state. Observe that

$$\begin{aligned} \text{Rate of going out of the state } &= \text{Rate of going into the state} \\ \text{State 1 :} \quad P_1\mu_2 &= P_2q_{21} + P_3q_{31} = P_2\lambda_1 \\ \text{State 2 :} \quad P_2(\lambda_1 + \mu_1) &= P_1q_{12} + P_3q_{32} = P_1\mu_2 + P_3\lambda_2 \\ \text{State 3 :} \quad P_3\lambda_2 &= P_1q_{13} + P_2q_{23} = P_2\mu_2 \,. \end{aligned}$$

Solving these we again get the same solution, as expected. □

In summary, whenever the long run distribution $\underline{P}$ of a continuous time Markov chain with generator matrix Q exists, we can find it using following methods.

(i) We find $P(t)$ for large t and if the rows are identical, then any row is the long run distribution $\underline{P}$ of a continuous time Markov chain. We can verify whether $\underline{P}Q = 0$ and whether $\underline{P}$ satisfies the balance equations.
(ii) We solve the matrix equation $\underline{P}Q = \underline{0} \iff Q'\underline{P} = 0$, subject to the condition that sum of the components of $\underline{P}$ is 1.
(iii) We find the right eigenvector of Q' corresponding to the eigenvalue 0 and normalize it to get $\underline{P}$.

In the following example, we illustrate all these methods using Code 6.7.8.

Example 6.6.6 Suppose $\{X(t), t \geq 0\}$ is a continuous time Markov chain with state space $S = \{1, 2, 3, 4\}$ and intensity matrix Q as given below.

$$Q = \begin{matrix} & \begin{matrix} 1 & 2 & 3 & 4 \end{matrix} \\ \begin{matrix} 1 \\ 2 \\ 3 \\ 4 \end{matrix} & \begin{pmatrix} -5 & 3 & 1 & 1 \\ 1 & -1 & 0 & 0 \\ 2 & 1 & -4 & 1 \\ 0 & 2 & 2 & -4 \end{pmatrix} \end{matrix}.$$

From the output we note that $\forall \ t \geq 4, \ \ P(t)$ has identical rows up to four decimal places accuracy. These are given by $\underline{P} = (0.1687, 0.6988, 0.0723, 0.0602)'$. We further note that $\underline{P}Q = 0$ and the balance equations are also satisfied. Solving the matrix equation and using the method based on eigenvalues we get the same $\underline{P}$. Observe that $\underline{P}$ is the same as $\underline{\eta}$, as derived in Example 6.6.1. From Eq. 6.6.2, we note that limit of the marginal distribution is also given by $\underline{P}$. Thus, in the long run, the proportions of time system in states 1, 2, 3, 4 are 0.1687, 0.6988, 0.0723, 0.0602 respectively. Note that proportion of time system is in state 2 is maximum and the rate of leaving the state 2 is minimum. □

The next section presents the `R` codes.

6.7 R Codes

Following is a code for computation of the transition probability matrix P of the embedded Markov chain from Q.

Code 6.7.1 Computation of the transition probability matrix P: Following are two different approaches to compute P. These are illustrated for Q matrix in Example 6.4.4.

```
# Part I: Input Q matrix
state=c(0,1,2,3,4);ns=length(state); r1=c(-7.6,2.1,1.1,1.2,3.2)
r2=c(1.1,-7.7,3.2,2.1,1.3); r3=c(2.3,1.1,-5.7,1.2,1.1)
r4=c(1.3,1.2,1.1,-6.8,3.2); r5=c(2.3,1.1,2.1,3.2,-8.7)
Q=matrix(c(r1,r2,r3,r4,r5),nrow=ns,ncol=ns,byrow=TRUE); Q
a=rowSums(Q);a; lambda=-diag(Q); lambda
# Part II: Method 1 to compute P
P=matrix(0,nrow=ns,ncol=ns)
for(i in 1:ns)
{
for(j in 1:ns)
{
if(i!=j)
{
P[i,j]=Q[i,j]/lambda[i]
}
}
}
P=round(P,4); P; b=rowSums(P); b
# Part III: Method 2 to compute P
P1=Q-diag(diag(Q)); P=P1/rowSums(P1); P=round(P,4); P
b=rowSums(P);b
```

□

In Code 6.7.1, it is implicitly assumed that diagonal elements of Q are not 0. However, for an absorbing state i, $q_{ii} = 0$. Following code incorporates this feature while computing P from Q.

Code 6.7.2 Computation of P from Q when a state is absorbing: It is illustrated in Example 6.4.4.

```
# Part I: Input Q matrix
state=c(0,1,2,3); ns=length(state); r1=c(-5.2,2.3,1.4,1.5)
r2=c(0,0,0,0); r3=c(2.4,1.8,-7.4,3.2); r4=c(1.9,1.5,2.6,-6.0)
Q=matrix(c(r1,r2,r3,r4),nrow=ns,ncol=ns,byrow=TRUE); Q
a=rowSums(Q); a
```

```
# Part II: Function to find P matrix
TP=function(Q)
{
P=matrix(rep(0,ns*ns),byrow=TRUE,ncol=ns)
for(i in 1:ns)
{
for(j in 1:ns)
{
if(i!=j)
{
d=(sum(Q[i,])-Q[i,i])
if(d!=0)
{
P[i,j]=Q[i,j]/d
}else
{
P[i,j]=0
P[i,i]=1
}
}
}
}
return(P)
}
round(TP(Q),4)
b=rowSums(TP(Q)); b
```

□

Following code is for the realization of a continuous time Markov chain, when the process is observed till a fixed number n of transitions occur.

Code 6.7.3 Realization of a continuous time Markov chain: Suppose $\{X(t), t \geq 0\}$ is continuous time Markov chain in Example 6.4.3.

```
# Part I: Input Q matrix
state=c(0,1,2); ns=length(state); a=c(-1,1,0,.1,-.6,.5,0,.2,-.2)
Q=matrix(a, nrow=ns,ncol=ns,byrow=TRUE);Q
# Part II: To find P from Q
lambda=-diag(Q); lambda; P1=Q-diag(diag(Q)); P=P1/rowSums(P1); P
# Part III: Realization for fixed number of transitions
inistate=2; n=11; x=y=c(); x[1]=inistate; set.seed(111)
for(i in 1:(n-1))
{
y[i]=rexp(1,rate=lambda[x[i]+1])
```

```
x[i+1]=sample(state,1,P[x[i]+1,],replace=T)
}
x; table(x); round(y,4)
# Part IV: Graph of realization
w=cumsum(y); round(w,4); u=rep(w,each=2)
w1=c(0,u); w2=c(0,w); length(w2); length(x)
x1=rep(x,each=2); x2=x1[1:length(x1)-1]
plot(w1,x2,"l",ylab="States",xlab="Time",yaxt="n",xaxt="n",
main="Realization of CTMC for a Fixed Number of Transitions",
col="dark blue",lty=4)
axis(2,at=sort(unique(x)),labels=sort(unique(x)))
axis(1,at=round(w,2),las=2,cex.axis=0.8)
points(w2,x,pch=20,col="dark blue")
```

□

Remark 6.7.1 In the above code, in two lines

```
y[i]=rexp(1,rate=lambda[x[i]+1])
x[i+1] = sample(state,1,P[x[i]+1,],replace=T)
```

we have to take argument for lambda and P as x[i]+1, since the state space is $S = \{0, 1, 2\}$. If it is $S = \{1, 2, 3\}$, these two lines will change to

```
y[i]=rexp(1,rate=lambda[x[i]])
x[i+1] = sample(state,1,P[x[i],],replace=T)
```

Following code is for the realization of the process when it is observed for a fixed time period $[0, T]$.

Code 6.7.4 Realization of a continuous time Markov chain: Suppose $\{X(t), t \geq 0\}$ is continuous time Markov chain in Example 6.4.3.

```
# Part I: Input Q matrix
state=c(0,1,2); ns=length(state); a=c(-1,1,0,.1,-.6,.5,0,.2,-.2)
Q=matrix(a, nrow=ns,ncol=ns,byrow=TRUE);Q
# Part II: Find P from Q
lambda=-diag(Q);lambda; P1=Q-diag(diag(Q)); P=P1/rowSums(P1)
# Part III: Realization for the fixed period
inistate=2; x=y=c(); set.seed(110); sumy=0; T=40
x[1]=inistate; i=1
while(sumy<T)
{
y[i]=rexp(1,rate=lambda[x[i]+1])
x[i+1]=sample(state,1,replace=T,P[x[i]+1,])
sumy=sum(y)
i=i+1
}
```

```
l=length(y); l; y=y[-1]; length(y); w=cumsum(y); m=length(x); m
x=x[-m]; x; table(x)
# Part IV: Graph of realization
round(w,2); round(y,2); u=rep(w,each=2)
w1=c(0,u,T); length(w1); x1=rep(x,each=2); x1; length(x1)
w2=c(0,w); length(w2); length(x)
plot(w1,x1,"l",main="Realization of CTMC for a Fixed Time Period",
ylab="States",xlab="Time",yaxt="n",xaxt="n",col="dark blue",lty=4)
axis(2,at=sort(unique(x1)),labels=sort(unique(x1)))
axis(1,at=round(c(w,T),2),las=2,cex.axis=0.7)
points(w2,x,pch=20,,col="dark blue")
abline(v=T,col="dark blue")
```

□

Same remark as in Code 6.7.3 applies in this code also.

Following is a code for computation of $P(t)$ using 5 methods as listed in Sect. 6.4.

Code 6.7.5 Computation of $P(t)$: It is illustrated in Example 6.5.5.

```
# Part I: Input Q matrix
state=1:4; ns=length(state)
a=c(-5,3,1,1,1,-1,0,0,2,1,-4,1,0,2,2,-4)
Q=matrix(a,nrow=ns,ncol=ns,byrow=TRUE); Q
# Part II: Method 1 based on spectral decomposition
E=eigen(Q); V=E$vectors; d=E$values; D=diag(d); t=c(1,2,5)
for(i in 1:length(t))
{
Dt=diag(exp(d *t[i]))
Pt=Re(V%*%Dt%*%solve(V))
Pt=round(Pt,3)
print(Pt)
}
# Part III: Approximation Methods 1,2,3
n=60; I=diag(c(1,1,1,1)); Pt1=Pt2=Pt3=list()
require(expm)
for (i in 1:3)
{
Pt1[[i]]=round(expm(Q*t[i]),3) # Method 1
Pt2[[i]]=round((I+(Q*t[i]/n))%^%n,3) # Method 2
Pt3[[i]]=round((solve(I-(Q*t[i]/n)))%^%n,3) # Method 3
}
list("Method 1"= Pt1,"Method 2"= Pt2,"Method 3"= Pt3)
# Part IV: Approximation Method 4, known as uniformization
lambda=-diag(Q); m=max(lambda);m
P=matrix(0,nrow=ns,ncol=ns)
for(i in 1:ns)
{
for(j in 1:ns)
{
P[i,j]=Q[i,j]/m
P[i,i]=1-lambda[i]/m
}
}
P; e=exp(1); Pt4=list()
for(i in 1:length(t))
```

```
{
N=max(m*t[i]+5*sqrt(m*t[i]),20)
N=round(N,0)
sum=I*e^(-m*t[i])
for(n in 1:N)
{
sum=sum+(e^(-m*t[i])*(m*t[i])^n/factorial(n))*(P)%^%n
}
Pt4[[i]]=round(sum,3)
}
list("Method 4"=Pt4)
```

□

Code 6.7.6 Computation of the probability of trouble free flight: This code computes the probability of trouble free flight, as discussed in Example 6.5.6, for three values of t and three values of λ.

```
# Part I: Input Q matrix
a=.01 # value of lambda
r1=c(0,0,0,0,0,0,0,0,0); r2=c(a,-a,0,0,0,0,0,0,0)
r3=c(0,2*a,-2*a,0,0,0,0,0,0); r4=c(a,0,0,-a,0,0,0,0,0)
r5=c(0,a,0,a,2*a,0,0,0,0); r6=c(0,0,a,0,2*a,-3*a,0,0,0)
r7=c(0,0,0,2*a,0,0,-2*a,0,0); r8=c(0,0,0,0,2*a,0,a,-3*a,0)
r9=c(0,0,0,0,0,2*a,0,2*a,-4*a)
Q=matrix(c(r1,r2,r3,r4,r5,r6,r7,r8,r9),nrow=9,ncol=9,byrow=TRUE)
Q
# Part II: Computation of probability
I=diag(rep(1,9)); library(expm); t=c(4,9,15); n=50
pr=matrix(0,nrow=3,ncol=3); b=c(1,.5,.25)
# To find Q for 3 values of lambda;1 corresponds to lambda=0.01,
# .5 corresponds to lambda=0.005 and .25 to lambda=0.0025
for(i in 1:length(t))
{
for(j in 1:length(b))
{
P=(I+b[j]*Q*t[i]/n)%^%n
pr[i,j]=P[9,5]+P[9,6]+P[9,8]+P[9,9]
}
}
pr=round(pr,4);pr
```

□

Following code computes the stationary distribution using three methods.

Code 6.7.7 Computation of the stationary distribution: We illustrate the code for the continuous time Markov chain in Example 6.6.1.

```
# Part I: Input Q matrix
state=1:4; ns=length(state)
a=c(-5,3,1,1,1,-1,0,0,2,1,-4,1,0,2,2,-4)
Q=matrix(a, nrow=ns,ncol=ns,byrow=TRUE); Q
# Part II: Stationary distribution by Method 1
A=t(Q); e=rep(1,ns); B=rbind(A,e); b=c(rep(0,ns),1)
library(MASS); P3=round(ginv(B)%*%b,4); P3
```

```
# Part III: Stationary distribution by Method 2
A=t(Q); eigen(A); e1=eigen(A)$values; e1; e2=which(round(e1,4)==0)
v=eigen(A)$vectors
b=v[,e2] # eigen vector corresponding to eigen value 0
eta=b/sum(b);eta=round(Re(eta),4); eta
# Part IV: Stationary distribution by Method 3
P1=Q-diag(diag(Q)); P=P1/rowSums(P1); P # tpm
B=t(P); eigen(B); e3=eigen(B)$values; e3; e4=which(round(e3,4)==1)
v1=eigen(B)$vectors; d=v1[,e4]; pi=d/sum(d); pi
d2=-diag(Q);d2 # Vector of intensity rates
d3=pi/d2; d4=d3/sum(d3) # eta=d4
MeC=1/(eta*d2); MeD=1/pi
d5=round(rbind(eta,pi,d4,MeC,MeD),4); d5
```

□

Code 6.7.8 Computation of the long run distribution: This code computes the long run distribution. It is illustrated for the continuous time Markov chain in Example 6.6.1.

```
# Part I: Input Q matrix
state=1:4; ns=length(state)
a=c(-5,3,1,1,1,-1,0,0,2,1,-4,1,0,2,2,-4)
Q=matrix(a,nrow=ns,ncol=ns,byrow=TRUE); Q
# Part II: To find P(t) till rows are identical
library(expm)
g=function(t,Q)
{
Pt=expm(Q*t)
P=Pt[1,]
P1=round(Pt[1,],4)
return(P1)
}
t=2:6; G=matrix(nrow=length(t),ncol=ns)
for(i in 1:length(t))
{
G[i,]=g(t[i],Q)
}
G
P=G[length(t),];P
# Part III: To verify that PQ = 0, a null vector
P2=round(P%*%Q,4); P2
# Part IV: To verify that P satisfies balance equations
lam=as.vector(-diag(Q)); x1=x2=c()
for(j in 1:ns)
```

```
{
x1[j]=P[j]*lam[j]
x2[j]=sum(P*Q[,j])-P[j]*Q[j,j]
}
x1; x2; diff=round(x1-x2,6); diff
# Part V: To solve matrix equation
A=t(Q); e=rep(1,ns); B=rbind(A,e); b=c(rep(0,ns),1)
library(MASS); P3=round(ginv(B)%*%b,4); P3
# Part VI: To find limiting distribution using eigen vectors
e=eigen(A); e1=e$values; e1; e2=which(round(e1,4)==0)
v=e$vectors[,e2]; P4=v/sum(v); P4=round(Re(P4),4); P4
```

□

A quick recap of the results discussed in the present chapter is given below.

Summary

1. Suppose $\{S_n, n \geq 0\}$ is a sequence of positive random variables such that (i) $S_0 = 0$, (ii) $S_n < S_{n+1}\ \forall\ n$ and (iii) $S_n \to \infty$ as $n \to \infty$. Suppose $\{X_n, n \geq 0\}$ is a sequence of random variables with countable state space S. A stochastic process $\{X(t), t \geq 0\}$ defined as
$$X(t) = X_n, \text{ if } S_n \leq t < S_{n+1}, \forall\ n \geq 0$$
is known as a pure jump process.
2. A pure jump process is said to be a Markov pure jump process, if (i) $\{X_n, n \geq 0\}$ is a homogeneous Markov chain with countable state space S and transition probability matrix $P = [p_{ij}]$, $i, j \in S$ such that, for each $i \in S$, $p_{ii} = 0$ or 1, (ii) $\{T_n, n \geq 1\}$, where $T_n = S_n - S_{n-1}$, is a sequence of independent random variables and T_n has exponential distribution with parameter $\lambda_{X_{n-1}}$ for $n = 1, 2, \ldots$ and (iii) $\hat{\lambda} = \sup_{i \in S} \lambda_i < \infty$ and $\lambda_i = 0$ if and only if $p_{ii} = 1$. The process $\{X_n, n \geq 0\}$ is known as the embedded Markov chain of the process $\{X(t), t \geq 0\}$. The process $\{X(t), t \geq 0\}$ satisfies the Markov property.
3. Alternatively, $\{X(t), t \geq 0\}$ is a continuous time Markov chain or Markov process with discrete state space if
$$P[X(t) = j | X(u), 0 \leq u \leq s] = P[X(t) = j | X(s)] \ \ a.s.$$
It is time homogeneous if
$$P[X(t+s) = j | X(s) = i] = P[X(t) = j | X(0) = i] = P_{ij}(t), \ \ s, t \geq 0, \ i, j \in S.$$

4. In a continuous time Markov chain, sojourn time random variables are independent and the distribution of sojourn time random variable T_i in state i is exponential.
5. For any $n \geq 1, i, \; j \in S$ and $u > 0$,

$$P[X_n = j, H_n > u | X_0, X_1, \ldots, X_{n-1} = i; H_1, \ldots, H_{n-1}] = p_{ij} e^{-\lambda_i u}.$$

6. Chapman-Kolmogorov equations: For any $i, \; j \in S$ and for any $s, t \geq 0$,

$$P_{ij}(t+s) = \sum_{k \in S} P_{ik}(t) P_{kj}(s) \iff P(t+s) = P(t)P(s).$$

7. The probability transition function $P_{ij}(t)$ is continuous $\forall \; t \geq 0$. If the continuous time Markov chain is a regular process, then $P_{ij}(t)$ for $t \geq 0$ is differentiable and the derivative is continuous.
8. $P_{ij}(t)$ satisfies the integral equation

$$P_{ij}(t) = e^{-\lambda_i t} \delta_{ij} + \int_0^t \lambda_i e^{-\lambda_i u} \sum_{k \neq i} p_{ik} P_{kj}(t-u) \, du.$$

9. With $P_{ii}(0) = 1$ and $P_{ij}(0) = 0$,

$$q_{ij} = \lim_{h \to 0} \frac{P_{ij}(h) - P_{ij}(0)}{h} = P'_{ij}(0) = \lambda_i p_{ij}$$

and $-\lambda_i = \lim_{h \to 0} \frac{P_{ii}(h) - P_{ii}(0)}{h} = P'_{ii}(0) = q_{ii}$. In matrix notation, $\lim_{h \to 0}(P(h) - I)/h = Q$.
10. A matrix $Q = [q_{ij}]$ with diagonal elements as $q_{ii} = -\lambda_i$ and off-diagonal elements as the transition rates q_{ij} is the infinitesimal generator or an intensity matrix of the Markov process. The infinitesimal generator Q determines the Markov pure jump process
11. Kolmogorov's backward differential equations: $P'_{ij}(t) = \sum_{k \in S} q_{ik} P_{kj}(t)$. In matrix notation, it is $P'(t) = QP(t)$.
12. Kolmogorov's forward differential equations: $P'_{ij}(t) = \sum_{k \in S} P_{ik}(t) q_{kj}$. In matrix notation, it is $P'(t) = P(t)Q$.
13. $P(t) = e^{tQ} = \sum_{n=0}^{\infty} (tQ)^n / n!$
14. Suppose $Q = VDV^{-1}$, where D is the diagonal matrix of eigenvalues α_i's of Q, V is the matrix of right eigenvectors. Then $P(t) = e^{tQ} = VD_tV^{-1}$, where D_t denotes the diagonal matrix with ith diagonal element as $e^{\alpha_i t}$.
15. $P(t) \approx (I + Qt/n)^n$ for sufficiently large n.
16. $P(t) = e^{Qt} \approx (I - Qt/n)^{-n} = \left[(I - Qt/n)^{-1}\right]^n$ for large n.
17. A row vector $\underline{\eta} = \{\eta_i, \; i \in S\}$ with $\eta_i \geq 0$ for all i and $\sum_{i \in S} \eta_i = 1$, is said to be a stationary distribution if

$$\underline{\eta} = \underline{\eta}P(t) \iff \eta_j = \sum_{i \in S} P_{ij}(t)\eta_i \;\; \forall \; j \in S \;\; \& \quad \text{for all } \; t \geq 0.$$

η_i is interpreted as the long run expected proportion of time spent in state i.

18. $\underline{\eta} = \underline{\eta}P(t) \;\; \forall \;\; t \geq 0 \iff \underline{\eta}Q = \underline{0}$. Thus, $\underline{\eta}$ is a left eigenvector of Q corresponding to eigenvalue 0.
19. If the distribution of $X(0)$ is the stationary distribution $\underline{\eta}$, then the distribution of $X(t)$ will also be $\underline{\eta}$ for all $t > 0$.
20. An irreducible non-null persistent process has a unique stationary distribution $\underline{\eta}$. Further, $\eta_i = 1/(\lambda_i E(\tau_i)), i \in S$.
21. Suppose $\underline{\eta}$ and $\underline{\pi}$ are stationary distributions of $\{X(t), t \geq 0\}$ and of the corresponding embedded Markov chain $\{X_n, n \geq 0\}$ respectively. If $\lambda_i \neq 0 \; \forall \; i \in S$, then
$$\pi_i = c\eta_i\lambda_i \;\; \forall \;\; i \in S, \;\; \text{where } \; c = (\sum_{i \in S} \eta_i\lambda_i)^{-1}.$$
If $\lambda_i = 0$ for $i = k$ say, then the vector $\underline{\eta}$, with kth component 1 and all other components 0, is a stationary distribution of $\{X(t), t \geq 0\}$. $\underline{\pi}$ is the same as $\underline{\eta}$.
22. If a continuous time Markov chain is irreducible and non-null persistent, then $P_j = \lim_{t\to\infty} P_{ij}(t)$ exists and independent of i with $\sum_{j \in S} P_j = 1$. The limit P_j is interpreted as a long run proportion of time spent in state j, $j \in S$.
23. The solution of $\underline{P}Q = \underline{0}$ gives the limiting distribution $\underline{P}$. The equation $\underline{P}Q = \underline{0}$ indicates that $\underline{P}$ is a left eigenvector corresponding to eigenvalue 0 of Q.
24. Balance equations: $\lambda_j P_j = \sum_{i \neq j} q_{ij} P_i$. These are interpreted as "long run rate out of state j is the same as the long run rate into state j".

6.8 Conceptual Exercises

6.8.1 Suppose $\{X(t), t \geq 0\}$ is a continuous time Markov chain with state space $S = \{1, 2, 3, 4\}$ and intensity matrix Q as given below.

$$Q = \begin{array}{c} \\ 1 \\ 2 \\ 3 \\ 4 \end{array} \begin{array}{c} \begin{array}{cccc} 1 & 2 & 3 & 4 \end{array} \\ \begin{pmatrix} -3 & 2 & 0 & 1 \\ 0 & -2 & 1/2 & 3/2 \\ 1 & 1 & -4 & 2 \\ 1 & 0 & 0 & -1 \end{pmatrix} \end{array}.$$

(i) Find the parameters of the sojourn time random variables. What are the expected sojourn times in the 4 states? (ii) Find the transition probability matrix of the embedded Markov chain. (iii) Examine if the continuous time Markov chain is irreducible. (iv) Examine whether the states are transient or persistent. (v) Write the system of balance equations and solve it to get the long run distribution. (vi) Find the stationary distribution. Is it the same as the long run distribution? (vii) Find the long run mean fraction of time system is in states $1, 2, 3, 4$.

6.8.2 Suppose $\{X(t), t \geq 0\}$ is a continuous time Markov chain with state space $S = \{1, 2\}$ and the intensity rates $q_{12} = 2$ and $q_{21} = 3$. Find the matrix $P(t)$ of transition probability functions, by solving Kolmogorov's forward and backward differential equations.

6.8.3 In a workshop there are two machines, operating simultaneously and independently, where both machines have an exponentially distributed time to failure with mean $1/\mu$. There is a single repair facility, and the repair times are exponentially distributed with rate λ. In the long run, what is the probability that no machine is operating?

6.8.4 A factory has five machines. The operating time until failure of a machine has exponential distribution with rate parameter 0.20 per hour. The repair time of a failed machine also has exponential distribution with parameter 0.50 per hour. The failures of the machines are independent. Further we assume that all the failed machines can be repaired simultaneously. Suppose $X(t)$ denotes the number of machines working at time t. (i) Can we model $\{X(t), t \geq 0\}$ as a continuous time Markov chain? Justify your answer. (ii) If yes, write down the intensity matrix and the transition probability matrix of the corresponding embedded Markov chain. (iii) Is the continuous time Markov chain irreducible? (iv) Classify the states as transient or persistent.

6.8.5 A system consists of two machines. The amount of time that an operating machine works before breaking down is exponentially distributed with mean 5 h. The amount of time that it takes a repairman to fix a machine is exponentially distributed with mean 4 h. Suppose $X(t)$ is the number of machines in operating condition at time t. (i) Find the long run distribution of $\{X(t), t \geq 0\}$. (ii) If an operating machine produces 100 units of output per hour, what is the long run average output per hour of the system?

6.8.6 Suppose a data scientist at a business analytics company can be a trainee, a junior data scientist or a senior data scientist. Suppose the three levels are denoted by $1, 2, 3$ respectively. If $X(t)$ denotes the level of the person at time t, we assume that X(t) evolves as a Markov chain in continuous time. Suppose the mean sojourn times in the three states 1, 2 and 3 are $0.1, 0.2, 2$ years respectively. It is given that a trainee is promoted to a junior data scientist with probability $2/3$ and to a senior data scientist with probability $1/3$. A junior data scientist leaves and is replaced by a trainee with probability $2/5$ and is promoted to a senior data scientist with probability $3/5$. A senior data scientist leaves and is replaced by a trainee with probability $1/4$ and by a junior data scientist with probability $3/4$. Find the long run average proportion of time a data scientist is a senior data scientist.

6.8.7 There are two photo copying machines in the office, one is operating and the other is a standby machine. The operating machine fails after an exponentially distributed duration having rate μ and is replaced by the standby. Lifetime of standby machine is also exponentially distributed with rate μ. It is given that at a time only one machine will be repaired and repair times are exponentially distributed with rate λ. Suppose $X(t)$ is the number of machines in operating condition at time t. (i) Can $\{X(t), t \geq 0\}$ be modeled as a continuous time Markov chain? Justify your answer. (ii) Write down the generator matrix. (iii) Find the long run proportion of time one machine is working. (iv) How will the generator matrix change if both the machines can be repaired simultaneously? (v) How will it further change if both machines are operating simultaneously?

6.9 Computational Exercises

6.9.1 Suppose $\{X(t), t \geq 0\}$ is a continuous time Markov chain with state space $S = \{1, 2, 3, 4\}$ and intensity matrix Q as given below.

$$Q = \begin{array}{c} \\ 1 \\ 2 \\ 3 \\ 4 \end{array} \begin{array}{c} \begin{array}{cccc} 1 & 2 & 3 & 4 \end{array} \\ \begin{pmatrix} -3 & 2 & 0 & 1 \\ 0 & -2 & 1/2 & 3/2 \\ 1 & 1 & -4 & 2 \\ 1 & 0 & 0 & -1 \end{pmatrix} \end{array}.$$

Find a realization of the continuous time Markov chain, when it is observed (i) for a fixed number of transitions and (ii) for a fixed period of time. Draw the plot of realization in both the cases and comment on your findings.

6.9.2 For the Q matrix as in Exercise 6.9.1, obtain approximately the matrix $P(t)$ of transition probability functions, for any three values of t. Use all the five methods. Comment on the findings.

6.9.3 For the Q matrix as in Exercise 6.9.1, obtain the long run distribution, assuming it exists, using following three methods: (i)Find $P(t)$ for sufficiently large t till the rows are identical. (ii) Solve the system of equations $\underline{P}Q = 0$ under the condition that sum of the components of $\underline{P}$ is 1.
(iii) Use eigenvalue and eigenvector approach. Verify that $\underline{P}$ satisfies the balance equations.

6.9.4 For the Q matrix as in Exercise 6.9.1, obtain the stationary distribution by solving $\eta = \eta P(t)$ for three values of t. Verify it remains the same for any 3 values of t. Also solve $\eta Q = 0$ and examine whether you get the same answer. Examine whether it is the same as the long run distribution.

6.9.5 Suppose $\underline{\eta}$ is a stationary distribution of the continuous time Markov chain with Q matrix as in Exercise 6.9.1 and $\underline{\pi}$ is a stationary distribution of the corresponding embedded discrete time Markov chain. Examine whether $\eta_j = c\pi_j/\lambda_j \;\; \forall \;\; j \in S$, where $c^{-1} = \sum_{j \in S} \pi_j/\lambda_j$ and λ_j is the transition rate of the sojourn time random variable T_j. Find $E(\tau_j)$ and μ_j for $j \in S$. Comment on the results.

6.9.6 A factory has five machines and five repairmen. The operating time until failure of a machine is an exponentially distributed random variable with rate parameter 0.20 per hour. The repair time of a failed machine is an exponentially distributed random variable with rate parameter 0.50 per hour. Up to five machines may be operating at any given time, their failures being independent of one another. We assume that all the failed machines can be repaired simultaneously. Suppose $X(t)$ denotes the number of machines working at time t and $\{X(t), t \geq 0\}$ is modeled as a continuous time Markov chain. In the long run, what fraction of time are all the repairmen idle?

6.9.7 Suppose $\{X(t), t \geq 0\}$ is a continuous time Markov chain with state space $S = \{1, 2, 3, 4\}$ and intensity matrix Q as given below.

$$Q = \begin{array}{c} \\ 1 \\ 2 \\ 3 \\ 4 \end{array} \begin{array}{c} \begin{array}{cccc} 1 & 2 & 3 & 4 \end{array} \\ \begin{pmatrix} -3 & 2 & 0 & 1 \\ 0 & -2 & 1/2 & 3/2 \\ 0 & 0 & 0 & 0 \\ 1 & 0 & 0 & -1 \end{pmatrix} \end{array}.$$

Find the stationary and long run distributions. Also find the stationary distribution of the embedded Markov chain.

6.10 Multiple Choice Questions

Note: In each of the questions, multiple options may be correct.

Note: In each of the following questions, $\{X(t), t \geq 0\}$ is a time homogeneous continuous time Markov chain with transition probability function $P = [P_{ij}(t)]$ and intensity rates $Q = [q_{ij}]$.

6.10.1 Suppose T_i is the sojourn time random variable in state i. Following are three statements. The distribution of T_i is (I) gamma with scale parameter λ_i and shape parameter 1, (II) exponential with scale parameter λ_i and (III) the same as that of $X/2\lambda_i$, where $X \sim \chi^2_2$.
Which of the following options is correct?

(a) Only (II) is true
(b) Only (I) and (II) are true
(c) Only (I) and (III) are true
(d) All three are true

6.10.2 Suppose T_i is a sojourn time random variable in state i. Following are three statements. The distribution of $T_2 + T_3$ is (I) gamma with scale parameter $\lambda_2 + \lambda_3$ and shape parameter 2, (II) exponential with scale parameter $\lambda_2 + \lambda_3$, (III) a convolution of distributions of two independent exponential random variables with scale parameters λ_2 and λ_3. (It is known as a hypo-

exponential distribution.)
Which of the following options is correct?

(a) Only (II) is true
(b) Both (I) and (III) are true
(c) Only (III) is true
(d) Only (I) is true

6.10.3 A unique solution of the functional equation $g(s+t) = g(t)g(s)\ s, t \geq 0$, where g is decreasing and $g(0) = 1$ is

(a) $g(s) = e^{cs}$, where $c > 0$ is any constant
(b) $g(s) = \log(cs)$, where $c > 0$ is any constant
(c) $g(s) = e^{-cs}$, where $c > 0$ is any constant
(d) $g(s) = e^{c+s}$, where $c > 0$ is any constant

6.10.4 If $X(0) = i$, then for $\lambda_i > 0$, which of the following options is/are correct?

(a) probability of no transition in $(0, h)$ is $\lambda_i h + o(h)$
(b) probability of exactly one transition in $(0, h)$ is $1 - \lambda_i h + o(h)$
(c) probability of at most one transition in $(0, h)$ is $1 - o(h)$
(d) probability of 2 transitions in $(0, h)$ is $2\lambda_i h + o(h)$

6.10.5 Which of the following options is/are correct? As $h \to 0$,

(a) $\lim(P_{ii}(h) - P_{ii}(0))/h = q_{ii}$
(b) $\lim P_{ii}(h)/h = q_{ii}$
(c) $\lim(P_{ij}(h) - P_{ij}(0))/h = q_{ij}$
(d) $\lim P_{ij}(h)/h = q_{ij}$

6.10.6 Suppose $\{\lambda_i, i \in S\}$ are the rate parameters. If a state i is absorbing then which of the following options is/are correct?

(a) $q_{ij} = 1\ \forall\ j \neq i \in S$ and $\lambda_i = 0$
(b) $q_{ij} = 0\ \forall\ j \neq i \in S$ and $\lambda_i > 0$
(c) $q_{ij} = 0\ \forall\ j \neq i \in S$ and $\lambda_i = 0$
(d) $q_{ij} = 0\ \forall\ j \neq i \in S$ and $\lambda_i = 1$

6.10.7 Following are two statements. (I) Kolmogorov's forward differential equation in a matrix form is $P'(t) = P(t)Q$. (II) Kolmogorov's backward differential equation in a matrix form is $P'(t) = QP(t)$.
Which of the following options is correct?

(a) Both (I) and (II) are false
(b) Both (I) and (II) are true
(c) (I) is true but (II) is false
(d) (I) is false but (II) is true

6.10.8 Following are three statements. (I) $P(5) = e^{5Q}$.(II) $P(5) = 5e^{Q}$.
(III) $P(5) = (P(1))^5$.
Which of the following options is correct?

(a) Only (I) is true
(b) Only (III) is true
(c) Both (I) and (III) are true
(d) Only (II) is true

6.10.9 For large n, which of the following options is/are correct?

(a) $P(t) \approx \left(I + Q\frac{t}{n}\right)^n$
(b) $P(t) \approx \left(I - Q\frac{t}{n}\right)^n$
(c) $P(t) \approx \left(I - Q\frac{t}{n}\right)^{-n}$
(d) $P(t) \approx \left(I + Q\frac{t}{n}\right)^{-n}$

6.10.10 Suppose $\lim_{t\to\infty} P_{ij}(t) = P_j$ exists and $\underline{P} = (P_0, P_1, P_2, \ldots,)$. Following are four statements. (I)$\underline{P}Q = \underline{0}$. (II) $\underline{P}Q = \underline{e}$, where $\underline{e} = (1, 1, 1, \ldots)'$. (III)$\underline{P}$ is a left eigenvector corresponding to the eigenvalue 0 of Q.
(IV) $\underline{P}$ is a right eigenvector corresponding to the eigenvalue 0 of Q.
Which of the following options is/are correct?

(a) Both (I) and (IV) are true.
(b) Both (II) and (IV) are true.
(c) Both (II) and (III) are true.
(d) Both (I) and (III) are true.

6.10.11 A row vector $\underline{\eta}$ is a stationary distribution associated with the continuous time Markov chain. Following are two statements. (I) $\underline{\eta} = \underline{\eta}P(15)$.
(II) $\underline{\eta}Q = \underline{0}$.
Which of the following options is correct?

(a) Both (I) and (II) are false
(b) Both (I) and (II) are true
(c) (I) is true but (II) is false
(d) (I) is false but (II) is true

6.10.12 Suppose $\{X(t), t \geq 0\}$ is an irreducible and non-null persistent continuous time Markov chain. A row vector $\underline{\eta}$ is a stationary distribution associated with the continuous time Markov chain. Following are two statements. (I) $\underline{\eta} = \underline{\eta}P(t) \ \forall \ t > 0$. (II) $\underline{\eta}Q = \underline{\eta}$.
Which of the following options is correct?

(a) Both (I) and (II) are false
(b) Both (I) and (II) are true
(c) (I) is true but (II) is false
(d) (I) is false but (II) is true

6.10.13 Suppose $\{X(t), t \geq 0\}$ is a continuous time irreducible and positive recurrent Markov chain with finite state space. Following are three statements. (I) As $t \to \infty$, $\lim P'_{ij}(t) = 0$. (II) As $h \to 0$, $\lim P_{ij}(h)/h = q_{ij}$.

(III) As $h \to 0$, $\lim P_{ii}(h)/h = -q_{ii}$.
Which of the following options is correct?

(a) Only (II) is true
(b) Both (I) and (III) are true
(c) Both (I) and (II) are true
(d) Only (III) is true

6.10.14 Following are two statements. (I) A continuous time stochastic process with stationary and independent increments is a time homogeneous Markov process. (II) A time homogeneous continuous time Markov chain is always a process with stationary and independent increments.
Which of the following options is correct?

(a) Both (I) and (II) are false
(b) Both (I) and (II) are true
(c) (I) is true but (II) is false
(d) (I) is false but (II) is true

References

1. Cinlar, E. (1975). *Introduction to stochastic processes*. New Jersey: Prentice Hall.
2. Hoel, P. G., Port, S. C., & Stone, C. J. (1972). *Introduction to stochastic processes*. Wiley Eastern.
3. Karlin, S., & Taylor, H. M. (1975). *A first course in stochastic processes*. New York: Academic Press.
4. Kulkarni, V. G. (2011). *Introduction to modeling and analysis of stochastic systems*. New York: Springer.
5. Miranda Holmes-Cerfon. (2019). *Applied stochastic analysis, Spring 2019*, Lecture notes on web.
6. Norris, J. R. (1997). *Markov chains*. UK: Cambridge University Press.
7. Ross, S. M. (2014). *Introduction to probability models* (11th ed.). New York: Academic Press.

Chapter 7
Poisson Process

7.1 Introduction

A Poisson process is a simple but the most widely used stochastic process for modeling the time epochs of arrivals into a system. It is named after the celebrated French mathematician Simon Denis Poisson (1781–1840). Poisson is a french word for fish. The book by Karlin and Taylor [4] presents an illustrative example of a Poisson process for fishing.

We frequently come across systems where transitions among states are generated by a series of events occurring over time, for example, arrivals at a supermarket, restaurant, bank, ATMs, failures of components of a system, admissions to a care center, occurrence of accidents in a given city, claims to an insurance company, etc. We observe the epochs of occurrence of events and also the total number of events in an interval $(0, t]$, $t > 0$. Suppose S_n denotes the epoch of occurrence of nth event, where $0 = S_0 < S_1 < \cdots < S_n$. Then the sequence $\{S_n, n \geq 1\}$ of random variables is a point process, as defined below.

Definition 7.1.1 Point Process: A sequence $\{S_n, n \geq 1\}$ of non-negative random variables such that $0 < S_1 < S_2 < \cdots$ is said to be a point process provided $S_n \to \infty$ as $n \to \infty$.

Since $S_n < S_{n+1}$ almost surely for every n, the sample paths of $\{S_n, n \geq 1\}$ are increasing almost surely. Further, the condition $S_n \to \infty$ as $n \to \infty$ implies that only finitely many events can occur in a finite interval. The sequence $\{S_n, n \geq 1\}$ is similar to the sequence defined in the pure jump process in Sect. 6.1.

We now define a counting process corresponding to the point process.

Definition 7.1.2 Counting Process: Suppose $\{S_n, n \geq 1\}$ is a point process and $X(t)$ is defined as

$$X(0) = 0 \;\&\; X(t) = \max\{n \geq 0 | S_n \leq t\}, \;\; t > 0$$

or equivalently for $n = 0, 1, 2, \ldots,$

S. Madhira and S. Deshmukh, *Introduction to Stochastic Processes Using R*,
https://doi.org/10.1007/978-981-99-5601-2_7

$$X(0) = 0 \;\&\; X(t) = n \text{ if } S_n \leq t < S_{n+1},\; t > 0.$$

Then $\{X(t), t \geq 0\}$ is known as a counting process corresponding to the point process $\{S_n, n \geq 1\}$.

Note that $X(t)$ represents the number of "events" that occur during the interval $(0, t]$. For example, $X(t)$ may be the number of customers entering a mall in the time interval $(0, t]$ or the number of transactions in a bank in the time interval $(0, t]$ or the number of admissions to a hospital in the time interval $(0, t]$. From its definition, it is clear that a counting process $\{X(t), t \geq 0\}$ satisfies the following properties:

(i) $X(t) \geq 0$ with $X(0) = 0$.
(ii) The state space of $\{X(t), t \geq 0\}$ is the set of whole numbers.
(iii) If $s < t$, then $X(s) \leq X(t)$.
(iv) For $s < t$, $X(t) - X(s)$ equals the number of events that occur in the interval $(s, t]$.
(v) The sample paths of $\{X(t), t \geq 0\}$ are right continuous.
(vi) $X(t) \to \infty$ as $t \to \infty$ almost surely. Otherwise, $X(t) = m$ for all $t \geq t_0$, for some t_0, implies that $S_n \leq t_0$ for $n \geq m$. This contradicts the assumption that $S_n \to \infty$ as $n \to \infty$.
(vii) $X(t) \geq n \iff S_n \leq t$. This is a connecting link between a point process and the corresponding counting process.

We have defined a process with stationary and independent increments in Sect. 1.3. Using the definition, we say that a counting process is a process with independent increments if the numbers of events that occur in disjoint time intervals are independent random variables. Further, it is a process of stationary increments if the distribution of the number of events that occur in any interval of time depends only on the length of the time interval, that is, if the number of events in the interval $(s, s + t]$ has the same distribution for all s.

Suppose $T_n = S_n - S_{n-1}$ with $S_0 = 0$, then T_n denotes the time interval between occurrence of nth and $(n-1)$th events. The random pattern of the sequence $\{T_n, n \geq 1\}$ decides the different types of stochastic processes. If $\{T_n, n \geq 1\}$ is a sequence of independent and identically distributed non-negative random variables, then the process $\{S_n, n \geq 0\}$ or the corresponding counting process $\{X(t), t \geq 0\}$ is known as a renewal process. In this setup, S_n is the epoch of nth renewal and T_n is the random duration between nth and $(n-1)$th renewals. In particular, if the common distribution of T_n is exponential, then the renewal process reduces to a Poisson process. In this case, it can be proved that for each fixed t, $X(t)$ has a Poisson distribution and hence the process is termed as a Poisson process. The present chapter is devoted to a detailed study of a Poisson process, while Chap. 10 is concerned with general renewal processes. The Poisson process and its generalizations have been extensively studied in the literature. Interested readers may refer to Cinlar [1], Karlin and Taylor [4], Kulkarni [5], Ross [6] and Taylor and Karlin [7] for various illustrations of a Poisson process.

There are several approaches to define a Poisson process. Each definition in its own way, gives some insight into the structure and properties of a Poisson process. We give three definitions and establish their equivalence. The first two definitions in Sect. 7.2 are in terms of stationary and independent increments property of a process. The third definition in Sect. 7.3 is in terms of a point process or the corresponding counting process. In the present chapter, we concentrate on time-homogeneous Poisson processes. A brief introduction to a non-homogeneous Poisson process is given in Sect. 7.4. With a suitable time translation, a non-homogeneous Poisson process can be transformed into a homogeneous Poisson process. Hence, it is enough to concentrate on a homogeneous Poisson process. Section 7.5 is concerned with two operations on a Poisson process, known as decomposition and superposition. Section 7.6 is devoted to a compound Poisson process. It is a generalization of a Poisson process and has many applications in different areas. Section 7.7 presents R codes used in solving examples.

7.2 Poisson Process as a Process with Stationary and Independent Increments

We begin with two definitions of a homogeneous Poisson process, in terms of a stochastic process with stationary and independent increments.

Definition 7.2.1 Poisson Process: A continuous time stochastic process $\{X(t), t \geq 0\}$ with state space $S = \{0, 1, \ldots, \}$ is said to be a homogeneous Poisson process with rate $\lambda \in (0, \infty)$ if the following axioms are satisfied.

(i) $X(0) = 0$, (ii) $\{X(t), t \geq 0\}$ is a process with stationary and independent increments and (iii) $P_{ii}(h) = P[X(t+h) = i | X(t) = i] = 1 - \lambda h + o(h)$, $P_{i,i+1}(h) = P[X(t+h) = i+1 | X(t) = i] = \lambda h + o(h)$ and $P_{ij}(h) = P[X(t+h) = j | X(t) = i] = o(h), \ \forall \ j \neq i, \ i+1$.

The third condition in the definition conveys that in an interval of length h, the probability of occurrence of exactly one event is $\lambda h + o(h)$ and that of more than one occurrence is $o(h)$. This condition is referred to as a regularity condition. It states that the occurrences of events are regular or orderly.

The next theorem establishes the Markov property of a Poisson process.

Theorem 7.2.1 *(i) A Poisson process with rate λ is a continuous time Markov chain with state space S. (ii) The sojourn time random variables are independent and identically distributed random variables, each having exponential distribution with rate parameter λ. (iii) In the embedded Markov chain, the transition probabilities are given by $p_{i,i+1} = 1 \ \forall \ i \in S$.*

Proof (i) In Theorem 1.3.1, it is proved that a stochastic process with stationary and independent increments is a homogeneous Markov process. Hence, Definition 7.2.1 implies that a Poisson process is a continuous time Markov chain.

(ii) In Chap. 6, it has been proved that for a continuous time Markov chain, the sojourn time random variables are independent and have exponential distribution, where the parameter depends on the state from which transition occurs. From Definition 7.2.1, we note that the infinitesimal transition probabilities are given by,

$$P_{ii}(h) = 1 - \lambda h + o(h), \;\; P_{i,i+1}(h) = \lambda h + o(h) \quad \& \quad P_{ij}(h) = o(h) \;\forall\; j \neq i+1.$$

Hence, the intensity rates q_{ij} and the transition probabilities of the corresponding embedded Markov chain are given by

$$q_{ij} = \lambda \quad \& \;\; p_{ij} = 1 \;\text{ for }\; j = i+1, \;\; q_{ii} = -\lambda, \;\; q_{ij} = p_{ij} = 0 \;\text{ for }\; j \neq i+1.$$

Thus, for a Poisson process, the intensity rates $q_{ii} = -\lambda$ do not depend on i. Hence, sojourn time random variables are independent and identically distributed, each having exponential distribution with scale parameter λ.

(iii) The generator matrix Q and the transition probability matrix P of the corresponding embedded Markov chain for a Poisson process with rate λ are given by

$$Q = \begin{array}{c} \\ 0 \\ 1 \\ 2 \\ \vdots \end{array} \begin{array}{c} \begin{array}{ccccc} 0 & 1 & 2 & \cdots \end{array} \\ \begin{pmatrix} -\lambda & \lambda & 0 & \cdots \\ 0 & -\lambda & \lambda & \cdots \\ 0 & 0 & -\lambda & \cdots \\ \vdots & \vdots & \vdots & \vdots \end{pmatrix} \end{array} \qquad P = \begin{array}{c} \\ 0 \\ 1 \\ 2 \\ \vdots \end{array} \begin{array}{c} \begin{array}{cccc} 0 & 1 & 2 & \cdots \end{array} \\ \begin{pmatrix} 0 & 1 & 0 & \cdots \\ 0 & 0 & 1 & \cdots \\ 0 & 0 & 0 & \cdots \\ \vdots & \vdots & \vdots & \vdots \end{pmatrix} \end{array}.$$

From the matrix P, we note that $p_{i,i+1} = 1 \;\forall\; i \in S$. □

From the matrix P, we note that starting from 0, the next state will be 1, next to that will be 2 and so on. Further, we observe that with $X(0) = 0$, the process remains in state 0 for a random time T_1, which has exponential distribution with scale parameter λ, then it jumps to 1 with probability 1, remains in state 1 for random time T_2, having exponential distribution with scale parameter λ and then jumps to 2 with probability 1 and so on. From the definition of the Markov pure jump process in Sect. 6.2, we note that the Poisson process $\{X(t), t \geq 0\}$ is a Markov pure jump process, where

$$X(t) = n \;\text{ if }\; S_n \leq t < S_{n+1}, \;\; n \geq 0, \;\text{ with }\; S_n = \sum_{i=1}^{n} T_i.$$

Observe that each jump is of size 1. The process $\{X_n = n, n \geq 0\}$ is the corresponding embedded Markov chain with transition probability matrix P as specified above. Further, $\{T_n = S_n - S_{n-1}, n \geq 1\}$ is a sequence of independent and identically distributed random variables, each having an exponential distribution with scale parameter $\lambda \in (0, \infty)$.

In a Poisson process, a change of state is usually labeled as an "arrival" or "occurrence of event". We use the term "arrival" with the understanding that its real meaning will vary, depending on the application involved. As a consequence, the distribution of T_n is labeled as the inter-arrival or interval distribution. The random variable S_n is the epoch of occurrence of nth event.

Remark 7.2.1 (i) From the transition probability matrix P of the embedded Markov chain, we note that

$$0 \to 1 \to 2 \to 3 \to 4 \cdots \quad \Rightarrow \quad i \to i+k, \quad i = 0, 1, \ldots, \quad k = 1, 2, \ldots.$$

Thus, any state $i \in S$ leads to $i + k$ for $k = 1, 2, \ldots$, however, $i + k$ does not lead to i for any $i \in S$. Thus, states in S do not communicate. Further, for any $i \in S, i \to i + 1$ but, $i + 1 \nrightarrow i$, which implies that each state in S is an inessential state and hence is a transient state. Therefore, by Theorem 3.3.6, a stationary distribution of the embedded Markov chain does not exist. Since all the states in the embedded Markov chain are transient, all states in the Poisson process are also transient. Consequently, its long run distribution also does not exist.

(ii) Note that all states are of the same type, "transient". However, these do not communicate with each other. In Chap. 2, we have proved that in a finite state space Markov chain, all the states cannot be transient. Observe that the state space of the embedded Markov chain of a Poisson process is countably infinite and all the states are transient. Some more stochastic processes where the set of transient states is countably infinite are (a) unrestricted simple random walk, when $p \neq 1/2$, studied in Chap. 4 and (b) BGW branching process, discussed in Chap. 5.

(iii) Observe that for any state i in the embedded Markov chain of a Poisson process, $p_{ii}^{(n)} = 0 \ \forall \ n \geq 1$, implying that period of each state in the embedded Markov chain is 0. All states have the same period although the states do not communicate with each other.

From Definition 7.2.1, it is not clear why such a process is labeled as a Poisson process. The second definition of the Poisson process given below answers this issue.

Definition 7.2.2 Poisson Process: A continuous time stochastic process $\{X(t), t \geq 0\}$ with state space $S = \{0, 1, \ldots,\}$ is a homogeneous Poisson Process with rate $\lambda \in (0, \infty)$ if the following axioms are satisfied. (i) $X(0) = 0$, (ii) $\{X(t), t \geq 0\}$ is a process with stationary and independent increments and (iii) $P[X(t) = k] = e^{-\lambda t}(\lambda t)^k/k!, \quad k = 0, 1, \ldots$ or equivalently, (iv) $\{X(t), t \geq 0\}$ is a process with independent increments and (v) for $s < t$, $P[X(t) - X(s) = k] = e^{-\lambda(t-s)}\left(\lambda(t-s)\right)^k/k!, \quad k = 0, 1, \ldots$.

Remark 7.2.2 In Definition 7.2.2, conditions (i), (ii) and (iii) imply that,

$$X(t) - X(s) \stackrel{d}{=} X(t-s) - X(0) \stackrel{d}{=} X(t-s) \quad \sim Poi(\lambda(t-s))$$

and condition (v) follows. Conversely, the condition (v) implies that the process $\{X(t)\}$ has stationary increments. Consequently, the condition (ii) follows. Further, with $s = 0$ and $X(0) = 0$,

$$P[X(t) - X(s) = k] = P[X(t) = k] = e^{-\lambda t}(\lambda t)^k/k!, \quad k = 0, 1, \ldots,$$

and condition (iii) follows.

In Definition 7.2.2, the third condition states that $X(t) \sim Poi(\lambda t)$ for each fixed t and that is the reason for labeling $\{X(t), t \geq 0\}$ as a Poisson process with rate λ.

In the following theorem, we establish the equivalence of the two definitions.

Theorem 7.2.2 *Definitions 7.2.1 and 7.2.2 of a Poisson process are equivalent.*

Proof We assume that $\{X(t), t \geq 0\}$ is a Poisson process according to Definition 7.2.2. Using the stationarity and independence of increments and the third condition, we have for $h > 0$

$$\begin{aligned} P_{ii}(h) &= P[X(t+h) = i|X(t) = i] = P[X(h) = 0] \\ &= e^{-\lambda h} = 1 - \lambda h + (\lambda h)^2/2 - \cdots = 1 - \lambda h + o(h). \\ P_{i,i+1}(h) &= P[X(t+h) = i+1|X(t) = i] = P[X(h) = 1] \\ &= e^{-\lambda h}\lambda h = \lambda h(1 - \lambda h + o(h)) = \lambda h + o(h). \\ P_{ij}(h) &= P[X(t+h) = j|X(t) = i] = P[X(h) = j - i] \\ &= e^{-\lambda h}(\lambda h)^{j-i}/(j-i)! = o(h) \ \text{ for } \ j = i+2, i+3, \ldots. \end{aligned}$$

Thus, Definition 7.2.1 follows from Definition 7.2.2. To prove that Definition 7.2.2 follows from Definition 7.2.1, we have to show that

$$\begin{aligned} & P_{ii}(h) = 1 - \lambda h + o(h), \ P_{i,i+1}(h) = \lambda h + o(h) \\ \& \ & P_{ij}(h) = o(h) \ \forall \ j \neq i+1 \ \Rightarrow \ P[X(t) = k] = e^{-\lambda t}(\lambda t)^k/k!, \quad k = 0, 1, \ldots. \end{aligned}$$

In view of the fact that $X(0) = 0$ with probability 1, note that

$$\begin{aligned} P_k(t) = P[X(t) = k] &= P[X(t) = k, X(0) = 0] \\ &= P[X(t) = k|X(0) = 0]P[X(0) = 0] \\ &= P[X(t) = k|X(0) = 0] = P_{0k}(t), \end{aligned}$$

which is the transition probability function of a Poisson process for transition from 0 to k. Hence, to find the expression for $P_k(t) = P[X(t) = k]$, we derive a system of differential equations as in the case of Kolmogorov's forward differential equations and solve it. An event can occur k times in $(0, t+h)$ for $k > 0$, in three mutually exclusive ways as follows:

(i) $[X(t) = k, X(t+h) - X(t) = 0]$,
(ii) $[X(t) = k-1, X(t+h) - X(t) = 1]$ and
(iii) $[X(t) = k-l, X(t+h) - X(t) = l], \ 2 \leq l \leq k$.

The third axiom from Definition 7.2.1 implies that probability of the third possibility is $o(h)$. Hence, using the stationary and independence of increments property, we have, for $k = 1, 2, \ldots$,

$$\begin{aligned}
P_k(t+h) &= P[X(t+h)=k] = P[X(t)=k,\ X(t+h)-X(t)=0] \\
&+ P[X(t)=k-1,\ X(t+h)-X(t)=1] + o(h) \\
&= P[X(t)=k]P[X(t+h)-X(t)=0] \\
&+ P[X(t)=k-1]P[X(t+h)-X(t)=1] + o(h) \\
&= (1-\lambda h+o(h))P_k(t) + (\lambda h+o(h))P_{k-1}(t) + o(h) \\
&\Rightarrow \frac{P_k(t+h)-P_k(t)}{h} = -\lambda P_k(t) + \lambda P_{k-1}(t) + o(h)/h \\
&\Rightarrow \lim_{h\to 0}(P_k(t+h)-P_k(t))/h = -\lambda P_k(t) + \lambda P_{k-1}(t) \\
&\Rightarrow P_k'(t) = -\lambda P_k(t) + \lambda P_{k-1}(t).
\end{aligned}$$

The other approach to derive these differential equations is as follows:

$$\begin{aligned}
P_k(t+h) &= P[X(t+h)=k] = P[X(t+h)=k|X(t)=k]P[X(t)=k] \\
&+ P[X(t+h)=k|X(t)=k-1]P[X(t)=k-1] + o(h) \\
&= P_{kk}(h)P_k(t) + P_{k-1,k}(h)P_{k-1}(t) + o(h) \\
&= (1-\lambda h+o(h))P_k(t) + (\lambda h+o(h))P_{k-1}(t) + o(h) \\
&\Rightarrow P_k'(t) = -\lambda P_k(t) + \lambda P_{k-1}(t).
\end{aligned}$$

Suppose $k = 0$. An event occurs 0 times in $(0, t+h]$, if it does not occur in $(0, t]$ and it does not occur in $(t, t+h]$. Thus,

$$\begin{aligned}
P_0(t+h) &= P[X(t+h)=0] = P[X(t)=0, X(t+h)-X(t)=0] \\
&= P[X(t)=0]P[X(t+h)-X(t)=0] \\
&= (1-\lambda h+o(h))P_0(t) \\
&\Rightarrow (P_0(t+h)-P_0(t))/h = -\lambda P_0(t) + o(h)/h \\
&\Rightarrow \lim_{h\to 0}(P_0(t+h)-P_0(t))/h = -\lambda P_0(t) \\
&\Rightarrow P_0'(t) = -\lambda P_0(t).
\end{aligned}$$

Thus, to find $P_k(t)$, we solve the system of differential equations given by

$$P_0'(t) = -\lambda P_0(t) \ \ \& \ \ P_k'(t) = -\lambda P_k(t) + \lambda P_{k-1}(t), \ \ k \ge 1\ ,$$

subject to the conditions

$$P[X(0)=0] = P_0(0) = 1 \ \ \& \ \ P[X(0)=k] = P_k(0) = 0 \ \forall \ k \ge 1.$$

Now,

$$P_0'(t) = -\lambda P_0(t) \ \Rightarrow \ P_0(t) = ce^{-\lambda t} = e^{-\lambda t} \text{ since } P_0(0) = 1 \ \Rightarrow \ c = 1.$$

Thus, $P_0(t) = P[X(t) = 0] = e^{-\lambda t}$. To obtain $P_k(t)$ for $k \geq 1$, note that

$$\begin{aligned}
P_k'(t) &= -\lambda P_k(t) + \lambda P_{k-1}(t) \\
\Rightarrow \; e^{\lambda t} P_k'(t) + e^{\lambda t}\lambda P_k(t) &= e^{\lambda t}\lambda P_{k-1}(t) \\
\Rightarrow \; \frac{d}{dt}\left(e^{\lambda t} P_k(t)\right) &= \lambda e^{\lambda t} P_{k-1}(t) \\
\Rightarrow \; \frac{d}{dt} Q_k(t) &= \lambda Q_{k-1}(t), \quad \text{where} \quad Q_k(t) = e^{\lambda t} P_k(t)
\end{aligned}$$

We now verify by induction that $Q_k(t) = (\lambda t)^k/k!$. Observe that $P_0(t) = e^{-\lambda t} \Rightarrow Q_0(t) = 1 = (\lambda t)^0/0!$. Thus, $Q_k(t) = (\lambda t)^k/k!$ is true for $k = 0$. Suppose for some integer $m \geq 1$, $Q_m(t) = (\lambda t)^m/m!$. Now,

$$\begin{aligned}
\frac{d}{dt} Q_{m+1}(t) &= \lambda Q_m(t) = \lambda(\lambda t)^m/m! = \lambda^{m+1} t^m/m! \\
\Rightarrow \; \int_0^s \frac{d}{dt} Q_{m+1}(t)\, dt &= (\lambda^{m+1}/m!) \int_0^s t^m \, dt \\
\Rightarrow \; Q_{m+1}(s) &= (\lambda s)^{m+1}/(m+1)!.
\end{aligned}$$

Thus, by induction, $Q_k(t) = (\lambda t)^k/k! \Rightarrow P_k(t) = e^{-\lambda t}(\lambda t)^k/k!$. Hence, $\forall\ k \geq 0$, $P_k(t) = P[X(t) = k] = e^{-\lambda t}(\lambda t)^k/k!$ and Definition 7.2.1 implies Definition 7.2.2. □

Using the result that $X(t) \sim Poi(\lambda t)$ and independence and stationarity of increments, we derive the mean function, variance function and the covariance function of the process in the following theorem.

Theorem 7.2.3 *Suppose $\{X(t), t \geq 0\}$ is a Poisson process with rate λ. Then*

$$M(t) = E(X(t)) = \lambda t, \quad Var(X(t)) = \lambda t \quad \& \quad Cov(X(s), X(t)) = \lambda \min\{s, t\}.$$

Proof Since for each fixed t, $X(t) \sim Poi(\lambda t)$, we immediately get

$$M(t) = E(X(t)) = \lambda t = E(X(1)t \quad \& \quad Var(X(t)) = \lambda t = Var(X(1))t.$$

To find the covariance function $C(s, t) = Cov(X(s), X(t))$, observe that for $s < t$

$$\begin{aligned}
C(s, t) &= E(X(s)X(t)) - E(X(s))\,E(X(t)) = E(X(s)X(t)) - \lambda^2 st \\
&= E(X(s) - X(0))(X(t) - X(s) + X(s)) - \lambda^2 st \\
&= E(X(s) - X(0))E((X(t) - X(s)) + E((X(s))^2) - \lambda^2 st \\
&= \lambda s \lambda(t - s) + \lambda s + \lambda^2 s^2 - \lambda^2 st = \lambda s.
\end{aligned}$$

Similarly, for $t < s$, we get $Cov(X(s), X(t)) = \lambda t$. Thus,

$$Cov(X(s), X(t)) = \lambda \min\{s, t\} = Var(X(1)) \min\{s, t\}.$$

□

Remark 7.2.3 (i) Note that $M(t)$, the expected number of events up to t, is a linear function of t. It increases as λ increases. This justifies calling λ the rate of a Poisson process. If λ is high, average number of events in $(0, t]$ will be high, at the same time variability also increases.

(ii) In Theorem 1.3.3 in Chap. 1, it is proved that for a stochastic process with stationary and independent increments,

$$\begin{aligned} E(X(t)) &= E(X(1)t, \quad Var(X(t)) = Var(X(1))t \ \text{ and} \\ Cov(X(s), X(t)) &= Var(X(1)) \min\{s, t\}. \end{aligned}$$

Observe that forms of the mean function, the variance function and the covariance function of a Poisson process, are consistent with these results.

(iii) In Theorem 1.3.4 in Chap. 1, it is proved that for a process $\{X(t), t \geq 0\}$ with stationary and independent increments, the distribution of $X(1)$ determines the distribution of $X(t)$ for all t, since $\phi_t = \{\phi_1\}^t$ where ϕ_t denotes the characteristic function of $X(t)$. For a Poisson distribution with mean λt,

$$\phi_t(u) = e^{-\lambda t(1-\exp(iu))} = (e^{-\lambda(1-\exp(iu))})^t = \{\phi_1(u)\}^t, \quad \forall\ u \in \mathbb{R}.$$

Thus, the distribution of $X(1)$ determines the distribution of $X(t)$ for all t. From the distribution of $X(1)$, we get λ and that determines the distribution of $X(t)$ for all t. It is also conveyed by Definition 7.2.2.

Using the expression for the mean function of a Poisson process, in the next theorem, we prove that a Poisson process is an evolutionary stochastic process.

Theorem 7.2.4 *A Poisson process is an evolutionary stochastic process.*

Proof If a stochastic process $\{X(t), t \geq 0\}$ is a stationary process, then it is known that all the marginal distributions are the same. As a consequence, $E(X(t))$ is the same for all t, which is possible only if $E(X(t))$ is a constant. For a Poisson process with rate λ, the mean function is given by $E(X(t)) = \lambda t$ which depends on t. Hence, we conclude that a Poisson process is not a stationary process, but it is an evolutionary process. □

In Theorem 7.2.1, we proved that inter-arrival random variables are independent and identically distributed each having exponential distribution with scale parameter λ. The random variable $S_n = \sum_{i=1}^{n} T_i$ denotes the epoch of occurrence of nth event. It is a sum of n independent and identically distributed random variables, each having exponential distribution with scale parameter λ. Hence, the distribution of S_n is gamma $G(\lambda, n)$ with scale parameter λ and shape parameter n. Since shape parameter is an integer, it is also known as an Erlang distribution. Its probability density function is

$$f(t) = \lambda^n e^{-\lambda t} t^{n-1} / \Gamma(n), \quad t > 0.$$

We define $S_0 = 0$. Both these results can be proved in a different way, as shown in the following theorem. In this theorem, using the independence and stationarity of increments, we first obtain the joint probability density function of $S_1, S_2, \ldots, S_n$ and from that obtain the joint probability density function of $T_1, T_2, \ldots, T_n$ which shows that $T_1, T_2, \ldots, T_n$ are independent and identically distributed random variables, each having exponential distribution with scale parameter λ. We also obtain two conditional distributions.

Theorem 7.2.5 *Suppose $\{X(t), t \geq 0\}$ is a Poisson process with rate λ, $\{S_n, n \geq 0\}$ is a sequence of the epochs of occurrence of the events in the Poisson process and $\{T_n, n \geq 1\}$ is a sequence of the inter-occurrence times. Suppose $\underline{s}_n = (s_1, s_2, \ldots, s_n)$. Then,*

(i) the joint probability density function $f(\underline{s}_n) = f_{S_1,S_2,\ldots,S_n}(s_1, s_2, \ldots, s_n)$ of $S_1, S_2, \ldots, S_n$ is given by

$$f(\underline{s}_n) = \begin{cases} \lambda^n \exp\{-\lambda s_n\} & \text{if } 0 < s_1 < s_2 < \cdots < s_n < \infty \\ 0, & \text{otherwise.} \end{cases}$$

(ii) the inter-occurrence times $T_n, n \geq 1$ are independent and identically distributed random variables, each having exponential distribution with scale parameter λ.

(iii) the conditional probability density function $f(\underline{s}_{n-1} | S_n = t)$ of $S_1, S_2, \ldots, S_{n-1}$ given $S_n = t$ is given by

$$f(\underline{s}_{n-1} | S_n = t) = \begin{cases} (n-1)!/t^{n-1} & \text{if } 0 < s_1 < s_2 < \cdots < s_{n-1} < t \\ 0, & \text{otherwise.} \end{cases}$$

(iv) the conditional probability density function $f(\underline{s}_n | X(t) = n)$ of $S_1, S_2, \ldots, S_n$ given $X(t) = n$ is given by

$$f(\underline{s}_n | X(t) = n) = \begin{cases} n!/t^n & \text{if } 0 < s_1 < s_2 < \cdots < s_n < t \\ 0, & \text{otherwise.} \end{cases}$$

Proof (i) To obtain the joint probability density function $f(\underline{s}_n)$ of $S_1, S_2, \ldots, S_n$, note that for $s_0 = 0, h_0 = 0, s_1 < s_2 < \cdots < s_n$ and sufficiently small $h_1, h_2, \ldots, h_n$,

$$
\begin{aligned}
& P\big[s_k < S_k < s_k + h_k, k = 1, 2, \ldots n\big] \\
&= P\big[X(s_k + h_k) - X(s_k) = 1, X(s_k) - X(s_{k-1} + h_{k-1}) = 0, 1 \leq k \leq n\big] \\
&= \prod_{k=1}^{n} P\big[X(s_k + h_k) - X(s_k) = 1\big] \prod_{k=1}^{n} P\big[X(s_k) - X(s_{k-1} + h_{k-1}) = 0\big] \\
&= \prod_{k=1}^{n} P\big[X(h_k) = 1\big] \prod_{k=1}^{n} P\big[X(s_k - s_{k-1} - h_{k-1}) = 0\big] \\
&= \lambda^n \Big(\prod_{k=1}^{n} h_k\Big) \exp\{-\lambda \sum_{k=1}^{n} h_k\} \times \exp\{-\lambda \sum_{k=1}^{n} (s_k - s_{k-1} - h_{k-1})\}
\end{aligned}
$$

Now, dividing both the sides by $\prod_{k=1}^{n} h_k$ and taking limits as $h_1, h_2, \ldots, h_n \to 0$, we get,

$$
f(\underline{s}_n) = \begin{cases} \lambda^n \exp\{-\lambda \sum_{k=1}^{n} (s_k - s_{k-1})\}, & \text{if } 0 < s_1 < s_2 < \cdots < s_n < \infty \\ 0, & \text{otherwise.} \end{cases},
$$

which simplifies to

$$
f(\underline{s}_n) = \begin{cases} \lambda^n \exp\{-\lambda s_n\} & \text{if } 0 < s_1 < s_2 < \cdots < s_n < \infty \\ 0, & \text{otherwise.} \end{cases}
$$

(ii) Since, $T_k = S_k - S_{k-1}, k = 1, 2, \ldots, n$ where $S_0 = 0$, the Jacobian of the transformation is 1. By substituting S_k in terms of T_k, we get

$$
\begin{aligned}
f(\underline{t}_n) &= f_{T_1, T_2, \ldots, T_n}(t_1, t_2, \ldots, t_n) = f_{S_1, S_2, \ldots, S_n}(t_1, t_1 + t_2, \ldots, t_1 + t_2 + \cdots + t_n) \\
&= \lambda^n \exp\{-\lambda \sum_{k=1}^{n} t_k\}, \quad t_1, t_2, \ldots, t_n \geq 0.
\end{aligned}
$$

Hence, the inter-occurrence times $T_k, k = 1, 2, \ldots, n$ are independent and identically distributed random variables, each having exponential distribution with scale parameter λ.

(iii) Note that $S_n = \sum_{i=1}^{n} T_i$ follows a gamma distribution with scale parameter λ and shape parameter n. Hence, result (iii) follows from (i).

(iv) For sufficiently small $h_1, h_2, \ldots, h_n$ and $s_0 = 0, h_0 = 0, s_1 < s_2 < \cdots < s_n < s_n + h_n < t$, we have

$$
\begin{aligned}
& P\big[s_k < S_k < s_k + h_k, k = 1, 2, \ldots n, X(t) = n\big] \\
&= P\big[s_k < S_k < s_k + h_k, k = 1, 2, \ldots n, S_n \leq t, S_{n+1} > t\big] \\
&= P\big[s_k < S_k < s_k + h_k, k = 1, 2, \ldots n, S_{n+1} > t\big] \\
&= P\big[X(s_k + h_k) - X(s_k) = 1, X(s_k) - X(s_{k-1} + h_{k-1}) = 0, 1 \leq k \leq n, \\
&\qquad X(t) - X(s_n + h_n) = 0\big] \\
&= \prod_{k=1}^{n} P\big[X(s_k + h_k) - X(s_k) = 1\big] \prod_{k=1}^{n} P\big[X(s_k) - X(s_{k-1} + h_{k-1}) = 0\big] \\
&\quad \times P[X(t) - X(s_n + h_n) = 0]
\end{aligned}
$$

$$= \prod_{k=1}^{n} P\big[X(h_k) = 1\big] \prod_{k=1}^{n} P\big[X(s_k - s_{k-1} - h_{k-1}) = 0\big] P[X(t - s_n - h_n] = 0$$
$$= \lambda^n \left(\prod_{k=1}^{n} h_k\right) \exp\{-\lambda \sum_{k=1}^{n} h_k\} \times \exp\{-\lambda \sum_{k=1}^{n} (s_k - s_{k-1} - h_{k-1})\}$$
$$\times \exp\{-\lambda(t - s_n - h_n)\}$$

Now, dividing both the sides by $\prod_{k=1}^{n} h_k$ and taking limits as $h_1, h_2, \ldots, h_n \to 0$, we get for $0 < s_1 < s_2 < \cdots < s_n < t$,

$$f_{S_1,S_2,\ldots,S_n,X(t)}(s_1, s_2, \ldots, s_n, n) = \lambda^n \exp\{-\lambda \sum_{k=1}^{n} (s_k - s_{k-1})\} \exp\{-\lambda(t - s_n)\}$$

and it is 0 otherwise. Thus,

$$f_{S_1,S_2,\ldots,S_n,X(t)}(s_1, s_2, \ldots, s_n, n) = \begin{cases} \lambda^n \exp\{-\lambda t\} & \text{if } 0 < s_1 < s_2 < \cdots < s_n < t \\ 0, & \text{otherwise.} \end{cases}$$

Dividing this density by $P[X(t) = n]$, the conditional probability density function of $S_1, S_2, \ldots, S_n$ given $X(t) = n$ is given by

$$f(\underline{s}_n | X(t) = n) = \begin{cases} n!/t^n & \text{if } 0 < s_1 < s_2 < \cdots < s_n < t \\ 0, & \text{otherwise.} \end{cases}$$

□

Remark 7.2.4 It is to be noted that the conditional probability density function of $S_1, S_2, \ldots, S_n$ given $X(t) = n$ is the same as the joint probability density function of the order statistics $\{U_{(1)}, U_{(2)}, \ldots, U_{(n)}\}$ corresponding to a random sample of size n from uniform $U(0, t)$ distribution. Thus, given $X(t)$, each arrival is uniformly distributed over $(0, t)$, irrespective of values of λ. Result (iii) of Theorem 7.2.5 states that given S_n, the times of occurrence of first $n - 1$ events are distributed as the ordered values of a set of $n - 1$ random variables which are uniformly distributed on $(0, S_n)$. Thus, a Poisson process distributes arrival epochs at random, just as the uniform distribution distributes points at random over an interval. In view of the result that occurrences are equally likely to happen anywhere in $(0, t)$, given that n events have occurred in $(0, t)$, the occurrences in a Poisson process are described as purely random events. Hence, a Poisson process is called as a purely random or completely random process. In spatial processes, if the objects, such as trees, are distributed according to a Poisson process on a plane, then the pattern is described as a completely random forest, Cressie [2].

Remark 7.2.5 In the result (iv) of Theorem 7.2.5, suppose we take $n = 1$. Then it follows that the conditional distribution of S_1, given that only 1 event has occurred in $(0, t]$, is uniform $U(0, t)$. We know that a Poisson process possesses stationary and independent increments and hence it seems reasonable that each interval in [0, t] of equal length should have the same probability of containing the event.

Further, it also conveys that the probability that the event occurs in the interval $(s_1, s_2) \subset [0, t]$ is $(s_2 - s_1)/t$, given that only 1 event has occurred in $(0, t]$. Result (iv) of Theorem 7.2.5, with $n = 1$ can also be proved in a different way as shown in the following theorem. This approach uses the property that a Poisson process has stationary and independent increments.

Theorem 7.2.6 *Suppose $\{X(t), t \geq 0\}$ is a Poisson process with rate λ. The conditional distribution of time of occurrence of an event in $(0, t]$, given that 1 event has occurred has occurred in $[0, t]$ is uniform $U(0, t)$.*

Proof For $s \leq t$, observe that the event $[T_1 \leq s, X(t) = 1]$ is equivalent to the event that one event occurs in $(0, s]$ and 0 events occur in $(s, t]$.

Hence,

$$\begin{aligned} P[T_1 \leq s | X(t) = 1] &= \frac{P[T_1 \leq s, X(t) = 1]}{P[X(t) = 1]} = \frac{P[X(s) = 1, X(t) - X(s) = 0]}{P[X(t) = 1]} \\ &= \frac{P[X(s) = 1]P[X(t) - X(s) = 0]}{P[X(t) = 1]} = \frac{\lambda s e^{-\lambda s} e^{-\lambda(t-s)}}{\lambda t e^{-\lambda t}} \\ &= s/t\ , \end{aligned}$$

which is the distribution function at s of a random variable having uniform $U(0, t)$ distribution. Thus, given that 1 event has occurred in $(0, t)$, the conditional distribution of the time of occurrence of the event in $(0, t)$ is uniform $U(0, t)$. Note that it does not depend on λ. □

The distribution of S_n can also be obtained from the distribution of $X(t)$ using two approaches, as shown in the following theorem. In one, we use the link $X(t) \geq n \iff S_n \leq t$, and in the other, the property of independence of increments.

Theorem 7.2.7 *Suppose $\{X(t), t \geq 0\}$ is a Poisson process with rate λ and the random variable S_n denotes the epoch of occurrence of nth event. Then $S_n \sim G(\lambda, n)$ distribution.*

Proof Using the link $X(t) \geq n \iff S_n \leq t$, we have

$$F_n(t) = P[S_n \leq t] = P[X(t) \geq n] = \sum_{i=n}^{\infty} \frac{e^{-\lambda t}(\lambda t)^i}{i!}$$

$$\begin{aligned} \Rightarrow\ f_n(t) &= \frac{d}{dt} F_n(t) = -\sum_{i=n}^{\infty} \frac{\lambda e^{-\lambda t}(\lambda t)^i}{i!} + \sum_{i=n}^{\infty} \frac{\lambda e^{-\lambda t}(\lambda t)^{i-1}}{(i-1)!} \\ &= \frac{\lambda e^{-\lambda t}(\lambda t)^{n-1}}{(n-1)!} + \sum_{i=n+1}^{\infty} \frac{\lambda e^{-\lambda t}(\lambda t)^{i-1}}{(i-1)!} - \sum_{i=n}^{\infty} \frac{\lambda e^{-\lambda t}(\lambda t)^i}{i!} \\ &= \frac{\lambda e^{-\lambda t}(\lambda t)^{n-1}}{(n-1)!} + \sum_{i=n}^{\infty} \frac{\lambda e^{-\lambda t}(\lambda t)^i}{i!} - \sum_{i=n}^{\infty} \frac{\lambda e^{-\lambda t}(\lambda t)^i}{i!} = \frac{\lambda e^{-\lambda t}(\lambda t)^{n-1}}{(n-1)!} \\ &= \lambda^n e^{-\lambda t} t^{n-1} / \Gamma(n)\ , \end{aligned}$$

which is the probability density function of a gamma $G(\lambda, n)$ distribution. In the second approach, observe that

$$\begin{aligned} P[t < S_n \leq t+h] &= P[X(t) = n-1, 1 \text{ event in } (t, t+h]] + o(h) \\ &= P[X(t) = n-1]P[1 \text{ event in } (t, t+h]] + o(h) \\ &= \frac{e^{-\lambda t}(\lambda t)^{n-1}}{(n-1)!}(\lambda h + o(h)) + o(h) = \frac{\lambda e^{-\lambda t}(\lambda t)^{n-1}}{(n-1)!}h + o(h) \\ \Rightarrow \quad f_n(t) &= \lambda^n e^{-\lambda t} t^{n-1} / \Gamma(n) , \end{aligned}$$

the second step follows due to the independence of increments. □

Following examples illustrate some applications of the properties of a Poisson process.

Example 7.2.1 Messages arrive at a mobile according to a Poisson process with rate of 15 messages per hour.

(i) The probability that no message arrives during 10:00 a.m. to 10:20 a.m. is the probability that no event occurs in 20 min, that is, in $(0, 1/3]$, in the unit of hours. It is the same as the probability that waiting time for the first event to occur is larger than $1/3$ h. Thus, it is given by

$$P[X(1/3) = 0] = P[T_1 > 1/3] = e^{-5} = 0.0067 \quad \text{as} \quad X(1/3) \sim Poi(15 \times 1/3).$$

(ii) The distribution of the time at which the first afternoon message arrives is the same as the distribution of the waiting time of the first event to occur after 12:00 noon and it is exponential with rate 15 messages per hour.

(iii) The probability that the first message in the afternoon arrives by 12:30 p.m. is

$$P[X(1/2) = 1] = e^{-15\times 1/2}(15/2) = 0.0041 \text{ since } X(1/2) \sim Poi(15 \times 1/2). \quad \square$$

Example 7.2.2 Suppose defects occur along a telephone cable according to a Poisson process of rate $\lambda = 0.1$ per kilometer.

(i) The probability that no defects appear in the first five kilometers of the cable is given by $P[X(5) = 0] = e^{-0.5} = 0.6065$.

(ii) Suppose it is given that there are no defects in the first five kilometers of the cable, the conditional probability of no defects between seven kilometers and nine kilometers is given by

$$\begin{aligned} P[X(9) - X(7) = 0|X(5) = 0] &= P[X(9) - X(7) = 0|X(5) - X(0) = 0] \\ &= P[X(9) - X(7) = 0, X(5) - X(0) = 0] \\ &\times (P[X(5) - X(0) = 0])^{-1} \\ &= P[X(9) - X(7) = 0] = P[X(2) = 0] \\ &= e^{-0.2} = 0.8187 , \end{aligned}$$

where we use the property that a Poisson process is a process with stationary and independent increments. □

Example 7.2.3 Customers arrive at a mall according to a Poisson process with rate $\lambda = 10$ per hour. The store opens at 10:00 a.m.

(i) The probability that five customers arrived between 11:15 and 11:45 is

$$P[X(1/2) = 5] = e^{-5}(5)^5/5! = 0.1755.$$

(ii) The probability that exactly one customer has arrived by 10:15 and a total of 10 have arrived by 11 a.m. is given by

$$\begin{aligned} P[X(1/4) = 1, X(1) = 10] &= P[X(1/4) = 1, X(1) - X(1/4) = 9] \\ &= P[X(1/4) = 1]P[X(1) - X(1/4) = 9] \\ &= P[X(1/4) = 1]P[X(3/4) = 9] \\ &= 2.5e^{-2.5}e^{-7.5}(7.5)^9/9! = 0.0235. \quad \square \end{aligned}$$

Following example illustrates Theorem 7.2.5.

Example 7.2.4 Customers enter a store according to a Poisson process of rate $\lambda = 6$ per hour.

(i) Suppose it is known that a single customer entered during the first hour. Then the conditional distribution of time to arrival given $X(1) = 1$ has $U(0, 1)$ distribution. Hence, the conditional probability that this person entered during the first fifteen minutes is $1/4$.

(ii) Given that 3 customers arrived during the first hour, the conditional probability that the first customer arrived in first 15 min, the second customer arrived during the first 20–30 min and the third customer arrived during the first 30–45 min can be computed using the conditional joint distribution of $\{S_1, S_2, S_3\}$. From Theorem 7.2.5, it is the same as the joint distribution of the order statistics $\{U_{(1)}, U_{(2)}, U_{(3)}\}$ from uniform $U(0, 1)$ distribution. Hence, the required probability p_3 is given by

$$\begin{aligned} p_3 &= P[0 < S_1 < 1/4, 1/3 < S_2 < 1/2, 1/2 < S_3 < 3/4 | X(1) = 3] \\ &= P[0 < U_{(1)} < 1/4, 1/3 < U_{(2)} < 1/2, 1/2 < U_{(3)} < 3/4] \\ &= \int_0^{1/4} \int_{1/3}^{1/2} \int_{1/2}^{3/4} 3! du_1 du_2 du_3 = 6(1/4)(1/6)(1/4) = 1/16. \end{aligned}$$

This probability can also be computed as follows using independence of increments property of the Poisson process. Suppose an event A is defined as $A = [X(1/4) = 1, X(1/3) - X(1/4) = 0, X(1/2) - X(1/3) = 1, X(3/4) - X(1/2) = 1, X(1) - X(3/4) = 0]$. Then

$$
\begin{aligned}
p_3 &= P(A)/P[X(1) = 3]] \\
&= P[X(1/4) = 1]P[X(1/3) - X(1/4) = 0]P[X(1/2) - X(1/3) = 1] \\
&\times P[X(3/4) - X(1/2) = 1]P[X(1) - X(3/4) = 0](P[X(1) = 3])^{-1} \\
&= \frac{e^{-6/4}(6/4)e^{-6/12}e^{-6/6}(6/6)e^{-6/4}(6/4)e^{-6/4}}{e^{-6}6^3/3!} = \frac{3!(6/4)^2}{6^3} = \frac{1}{16}.
\end{aligned}
$$

If it is given that 4 customers arrived during the first hour, then the probability of the same event is given by

$$
\begin{aligned}
p_4 &= P[0 < S_1 < 1/4, 1/3 < S_2 < 1/2, 1/2 < S_3 < 3/4 | X(1) = 4] \\
&= P[0 < U_{(1)} < 1/4, 1/3 < U_{(2)} < 1/2, 1/2 < U_{(3)} < 3/4, 3/4 < U_{(4)} < 1] \\
&= \int_{3/4}^{1}\int_{1/2}^{3/4}\int_{1/3}^{1/2}\int_{0}^{1/4} 4! du_1 du_2 du_3 du_4 \\
&= 24(1/4)(1/6)(1/4)(1/4) = 1/16.
\end{aligned}
$$

This probability can also be computed as follows using independence of increments property of the Poisson process. Suppose an event B is defined as $B = [X(1/4) = 1, X(1/3) - X(1/4) = 0, X(1/2) - X(1/3) = 1, X(3/4) - X(1/2) = 1, X(1) - X(3/4) = 1]$. Then

$$
\begin{aligned}
p_4 &= P(B)/P[X(1) = 4] \\
&= P[X(1/4) = 1]P[X(1/3) - X(1/4) = 0]P[X(1/2) - X(1/3) = 1] \\
&\times P[X(3/4) - X(1/2) = 1]P[X(1) - X(3/4) = 1](P[X(1) = 4])^{-1} \\
&= \frac{e^{-6/4}(6/4)e^{-6/12}e^{-6/6}(6/6)e^{-6/4}(6/4)e^{-6/4}(6/4)}{e^{-6}6^4/4!} = \frac{4!(6/4)^3}{6^4} = \frac{1}{16}.
\end{aligned}
$$

□

In Chap. 6, in Example 6.4.3 we have obtained a realization of a Markov process given the generator matrix Q. Using a similar approach, we can find a realization of a Poisson process with rate λ, when it is observed for a fixed time period $[0, T]$. We have already noted that for a Poisson process $p_{ij} = 1$ for $j = i + 1$ and $p_{ij} = 0$ for $j \neq i + 1$. Thus, at every epoch of occurrence of an event, the state of the process increases by a jump of size 1. Further, inter-occurrence random variables are independent and identically distributed, each having an exponential distribution with rate λ. Thus, to find a realization of a Poisson process, it is enough to generate a random samples of size 1 from the exponential distribution with rate λ, till the sum of the realized values is bigger than T. For example, if $\sum_{i=1}^{5} T_i \leq T$ and $\sum_{i=1}^{6} T_i > T$, then we conclude that 5 events have occurred in $(0, T]$ and $X(T) = 5$. The generation of exponential random variables stops when $\sum_{i=1}^{6} T_i > T$. Hence, in a vector which stores the values of T_i, we delete the last observation. Following example uses Code 7.7.1 to obtain a realization of a Poisson process with rate λ using such an approach.

Table 7.1 Arrival epochs in a Poisson Process

Arrival epochs	λ			
	0.5	1.1	1.6	2.9
S_1	1.51	1.70	1.08	0.06
S_2	3.87	2.06	1,47	1.54
S_3	4.17	2.20	2.24	1.84
S_4	4.44	3.77	2.86	2.12
S_5		3.85	2.99	2.34
S_6		4.46	3.12	2.59
S_7			4.55	2.74
S_8			4.56	2.76
S_9			4.60	2.94
S_{10}			4.67	3.23
S_{11}			4.72	4.15
S_{12}			4.97	4.25
S_{13}				4.39
S_{14}				4.54
S_{15}				4.65
$X(T)$	4	6	12	15

Example 7.2.5 Suppose $\{X(t), t \geq 0\}$ is a Poisson process with rate λ. We consider four values of λ, $\lambda = 0.5, 1.1, 1.6$ and $\lambda = 2.9$, to compare the performance of the process. We simulate the process for $T = 5$ time units. The output in terms of arrival epochs is organized in Table 7.1. Once we know the arrival epochs, we know how many events have occurred in $(0, T]$ and the realized values of inter-occurrence random variables.

From Table 7.1, we observe that as λ increases, the number $X(T)$ of events in $(0, T]$ also increases. The length of the interval between consecutive occurrences decreases, since the mean inter-occurrence time decreases. Realization of the process $\{X(t), t \geq 0\}$ for a fixed time interval, is presented below for $\lambda = 1.1$.

$$X(t) = \begin{cases} 0, \text{ if } \ 0 \leq t < 1.70 \\ 1, \text{ if } 1.70 \leq t < 2.06 \\ 2, \text{ if } 2.06 \leq t < 2.20 \\ \vdots \\ 6, \text{ if } \ 4.46 \leq t \leq 5. \end{cases}$$

Figure 7.1 shows sample paths of a Poisson process corresponding to four values of rates. The four graphs clearly display a difference among realizations of the process corresponding to four values of λ.

□

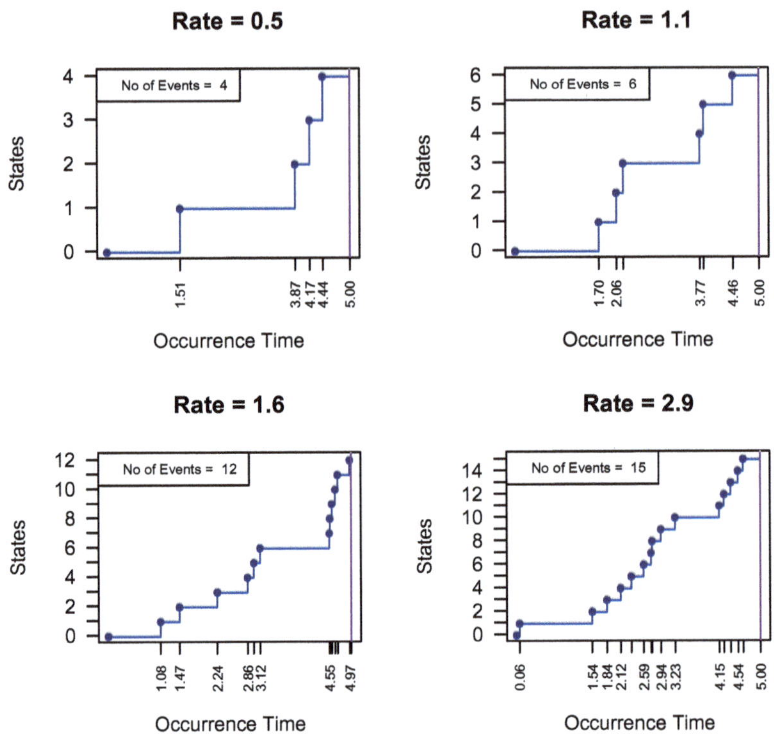

Fig. 7.1 Realization of a Poisson Process for a Time Interval [0, 5]

Remark 7.2.6 From Example 7.2.5, we note that the sample path of a Poisson process is a non-decreasing and right continuous step function. Further, $\{X(t), t \geq 0\}$ is also a non-decreasing process and increases with jumps of size 1 only. It counts the number of occurrences of events and hence is known as a corresponding counting process. We also note that to find a realization of the Poisson process with rate λ, it is sufficient to know that the inter-arrival random variables are independent and identically distributed and have exponential distribution with mean $1/\lambda$.

The definition of a Poisson process with rate λ states that $X(t) \sim Poi(\lambda t)$ for each fixed t. In the following example, we simulate a Poisson process multiple times and verify this result. We use Code 7.7.2.

Example 7.2.6 Suppose $\{X(t), t \geq 0\}$ is a Poisson process with rate $\lambda = 0.2$ and is observed for $T = 20$ time units. We simulate the process 100 times and based on 100 values of $X(T)$, examine whether a Poisson distribution $Poi(4)$ is a good model, using the chi-square goodness of fit test. Visually, we compare the graphs of observed and expected probability distributions, expected under Poisson $Poi(4)$ model. These distributions are presented in Table 7.2.

Table 7.2 Observed and expected frequency distributions

x	Observed frequency	Expected frequency
0	1	1.83
1	9	7.33
2	13	14.65
3	17	19.54
4	21	19.54
5	16	15.63
6	10	10.42
7	10	5.95
8	3	2.98
≥ 9	0	2.14

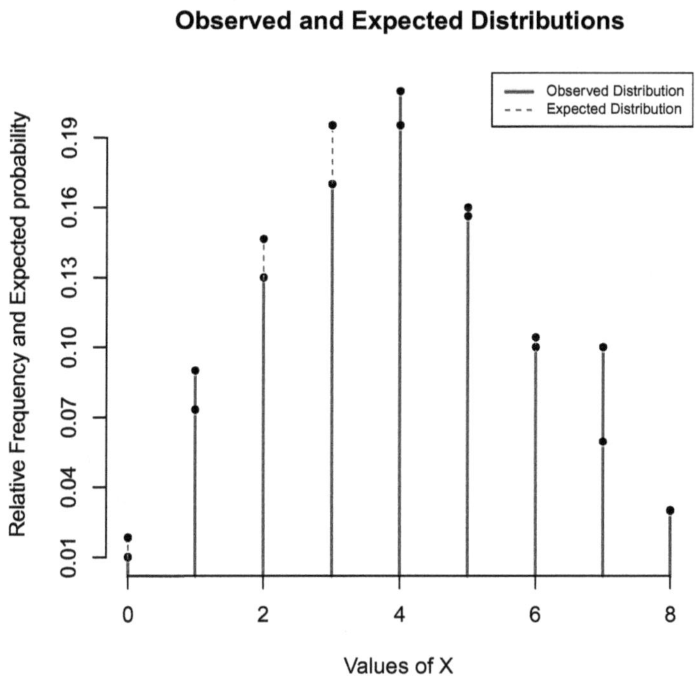

Fig. 7.2 Verification: $X(T) \sim Poi(\lambda T)$

From Table 7.2, we note that the observed and the expected frequencies are close to each other, which suggests that Poisson $Poi(4)$ may be a good model for the observed data. Figure 7.2 also displays the close agreement.

We confirm the visual impression by Karl Pearson's test procedure for goodness of fit. We pool the frequencies for the first two and the last three classes. The value of the test statistic $T_{100} = \sum_{i=1}^{7}(o_i - e_i)^2/e_i = 1.0660$. Under H_0, $T_{100} \sim \chi^2_6$ distribution. From the output, we note that $T_{100} < \chi^2_{0.95,6} = 12.5916$, and hence H_0 is not rejected at 5% level of significance. The corresponding p-value is 0.9830. Thus, data do have strong support to the theoretical result that $X(T) \sim Poi(\lambda T)$ distribution. Without pooling the frequencies which are less than 5, the corresponding values are $T_{100} = \sum_{i=1}^{10}(o_i - e_i)^2/e_i = 6.2968$, $\chi^2_{0.95,9} = 16.9190$ and p-value $= 0.7099$. A built-in function `chisq.test` produces the following results.

$$\text{X-squared} = 6.2968, \quad \text{df} = 9, \quad \text{p-value} = 0.7099.$$

These are exactly the same as those for the chi-square test procedure without pooling the frequencies which are less than 5. □

Suppose in Example 7.2.6, the values of λ are 1, 1.5 and 2.2. Using Code 7.7.2, we can test whether distribution of $X(t)$ is Poisson with appropriate mean. For these values of λ, $\lambda * 20$ is 20, 30, 44, respectively. A Poisson distribution with large values of mean m, can always be approximated by a normal $N(m, m)$ distribution. Hence, one can use Shapiro–Wilk test of normality to examine goodness of fit. In the following example, we adopt this approach for these three values of λ, using Code 7.7.3.

Example 7.2.7 Suppose $\{X(t), t \geq 0\}$ is a Poisson process with rate λ as 1, 1.5 and 2.2. Suppose the process is observed for $T = 20$ time units. We simulate the process 100 times and based on 100 values of $X(T)$, examine whether a normal $N(\lambda T, \lambda T)$ distribution is a good model using Shapiro–Wilk test. From the output, the p-values of the Shapiro–Wilk test procedure are $0.4319, 0.5023, 0.7934$ corresponding to $\lambda = 1, 1.5, 2.2$, respectively. Hence, we conclude that for fixed T, a Poisson $Poi(\lambda T)$ is a good model for $X(T)$, supporting the theoretical result. □

The next section is concerned with the third definition of a Poisson process. It is in terms of a point process and the corresponding counting process as defined in Sect. 7.1.

7.3 Poisson Process as a Point Process

Suppose $\{S_n, n \geq 1\}$ is a point process and $\{X(t), t \geq 0\}$ is the corresponding counting process.

Definition 7.3.1 Poisson Process: A point process $\{S_n, n \geq 1\}$ is said to be a Poisson process with rate $\lambda > 0$, if $\{T_n = S_n - S_{n-1}, n \geq 1\}$ is a sequence of independent and identically distributed random variables each having exponential distribution with scale parameter λ. The corresponding counting process $\{X(t), t \geq 0\}$ is also known as a Poisson process, where $X(t)$ denotes the number of events occurring in $(0, t], t \geq 0$.

The random variable $S_n = \sum_{i=1}^{n} T_i$ denotes the epoch of occurrence of nth event. As discussed earlier, it is a sum of n independent and identically distributed random variables, each having exponential distribution with scale parameter λ. Hence, $S_n \sim G(\lambda, n)$ distribution. The distribution of $X(t)$ can be obtained using the link $X(t) \geq n \iff S_n \leq t$. We use this approach in the following theorem to obtain the distribution of $X(t)$ using the distribution of S_n.

Theorem 7.3.1 *Suppose $\{T_n, n \geq 1\}$ is a sequence of independent and identically distributed random variables each having exponential distribution with scale parameter λ, $S_n = \sum_{i=1}^{n} T_i$ and $\{X(t), t \geq 0\}$ is the corresponding counting process. Then for fixed t, $X(t) \sim Poi(\lambda t)$ distribution.*

Proof With the link $X(t) \geq n \iff S_n \leq t$, we have

$$\begin{aligned} P[X(t) = n] &= P[X(t) \geq n] - P[X(t) \geq n+1] = P[S_n \leq t] - P[S_{n+1} \leq t] \\ &= F_n(t) - F_{n+1}(t), \end{aligned}$$

where F_n is a distribution function of S_n and $S_n \sim G(\lambda, n)$ distribution. Using integration by parts, we obtain a relation between $F_n(t)$ and $F_{n+1}(t)$ as follows:

$$\begin{aligned} F_n(t) &= \frac{\lambda^n}{\Gamma(n)} \int_0^t x^{n-1} e^{-\lambda x}\, dx \;=\; \frac{\lambda^n}{\Gamma(n)} \int_0^t \frac{1}{n} \left(\frac{d}{dx} x^n \right) e^{-\lambda x}\, dx \\ &= \frac{\lambda^n}{\Gamma(n)} \frac{e^{-\lambda t} t^n}{n} + \frac{\lambda^n}{\Gamma(n)} \frac{\lambda}{n} \int_0^t x^n e^{-\lambda x} dx \\ &= \frac{e^{-\lambda t} (\lambda t)^n}{n!} + \frac{\lambda^{n+1}}{\Gamma(n+1)} \int_0^t x^n e^{-\lambda x} dx \\ &= \frac{e^{-\lambda t} (\lambda t)^n}{n!} + F_{n+1}(t) \\ \Rightarrow \quad P[X(t) = n] &= F_n(t) - F_{n+1}(t) = e^{-\lambda t} (\lambda t)^n / n!, \quad n = 0, 1, \ldots. \end{aligned}$$

Thus, for fixed t, $X(t) \sim Poi(\lambda t)$ distribution. □

Remark 7.3.1 With the link $X(t) \geq n \iff S_n \leq t$, we have

$$\begin{aligned} P[X(t) \geq n+1] = P[S_{n+1} \leq t] &\Rightarrow P[X(t) < n+1] = P[S_{n+1} > t] \\ &\Rightarrow P[X(t) \leq n] = P[S_{n+1} > t]. \end{aligned}$$

Thus, the value of the distribution function of $X(t)$ at n is the same as the value of the survival function of S_{n+1} at t. We have already obtained the mean function $E(X(t))$ of a Poisson process with rate λ. It can also be obtained using the above link. We have

$$E(X(t)) = \sum_{n\geq 1} P[X(t) \geq n] = \sum_{n\geq 1} P[S_n \leq t] = \sum_{n\geq 1} \int_0^t \frac{\lambda^n}{\Gamma(n)} x^{n-1} e^{-\lambda x}\, dx$$
$$= \int_0^t \lambda e^{-\lambda x} \Big(\sum_{n\geq 1} \frac{(\lambda x)^{n-1}}{(n-1)!} \Big)\, dx = \int_0^t \lambda e^{-\lambda x} e^{\lambda x}\, dx = \int_0^t \lambda\, dx = \lambda t.$$

In Theorem 7.2.5, we have obtained the joint distribution of $S_1, S_2, \ldots, S_n$ and the conditional distribution of $S_1, S_2, \ldots, S_n$ given $X(t) = n$, using the result that $\{X(t), t \geq 0\}$ is a process with stationary and independent increments, according to Definition 7.2.2 of a Poisson process. In the following theorem, we obtain both these distributions using the result that $T_i, i = 1, 2, \ldots, n$ are independent and identically distributed random variables each having exponential distribution with scale parameter λ.

Theorem 7.3.2 *Suppose $\{T_n, n \geq 1\}$ is a sequence of independent and identically distributed random variables each having exponential distribution with scale parameter λ and $S_n = \sum_{i=1}^n T_i$. Then (i) the joint probability density function $f(\underline{s}_n)$ of $S_1, S_2, \ldots, S_n$, where $\underline{s}_n = (s_1, s_2, \ldots, s_n)$, is given by $f(\underline{s}_n) = \begin{cases} \lambda^n \exp\{-\lambda s_n\} & \text{if } 0 < s_1 < s_2 < \cdots < s_n < \infty \\ 0, & \text{otherwise.} \end{cases}$*
(ii) the conditional probability density function $f(\underline{s}_n | X(t) = n)$ of $S_1, S_2, \ldots, S_n$ given $X(t) = n$ is given by,

$$f(\underline{s}_n | X(t) = n) = \begin{cases} n!/t^n & \text{if } 0 < s_1 < s_2 < \cdots < s_n < t \\ 0, & \text{otherwise.} \end{cases}$$

Proof Since $\{T_n, n \geq 1\}$ is a sequence of independent and identically distributed random variables each having exponential distribution with scale parameter λ, the joint probability density function $f(\underline{t}_n)$ of $T_1, T_2, \ldots, T_n$ is given by

$$f(\underline{t}_n) = \prod_{i=1}^n \lambda \exp(-\lambda t_i) = \lambda^n \exp\Big(-\lambda \sum_{i=1}^n t_i\Big), \quad t_i > 0,\ i = 1, 2, \ldots, n.$$

To obtain the joint probability density function of $S_1, S_2, \ldots, S_n$, we consider a linear transformation A defined by

$$A(u_1, u_2, \ldots, u_n) = (u_1, u_1 + u_2, \ldots, u_1 + u_2 + \cdots + u_n).$$

The random variables $(S_1, S_2, \ldots, S_n)$ can be obtained by applying the linear transformation A to the variables $(T_1, T_2, \ldots, T_n)$. Thus,

$$(S_1, S_2, \ldots, S_n) = A(T_1, T_2, \ldots, T_n).$$

The jacobian of the transformation A is 1 since the matrix of this transformation is triangular with 1's on the diagonal. By the density transformation theorem, the joint probability density function of $S_1, S_2, \ldots, S_n$ is given by

$$f(\underline{s}_n) = \prod_{i=1}^{n} \lambda \exp(-\lambda(s_i - s_{i-1})) = \lambda^n \exp(-\lambda s_n), \quad \text{if } s_1 < s_2 < \cdots < s_n$$

and 0 otherwise. Note that $f(\underline{s}_n)$ depends only on s_n and does not depend on $(s_1, s_2, \ldots, s_{n-1})$.

(ii) Suppose $s_1 < s_2 < \cdots < s_n < t$. Note that the conditional probability density function $f(\underline{s}_n | X(t) = n)$ is given by

$$\begin{aligned}
f(\underline{s}_n | X(t) = n) & \\
&= \lim_{\epsilon \to 0} \frac{P[s_1 < S_1 < s_1 + \epsilon, s_2 < S_2 < s_2 + \epsilon, \ldots, s_n < S_n < s_n + \epsilon | X(t) = n]}{\epsilon^n} \\
&= \lim_{\epsilon \to 0} \frac{P[s_1 < S_1 < s_1 + \epsilon, s_2 < S_2 < s_2 + \epsilon, \ldots, s_n < S_n < s_n + \epsilon, X(t) = n]}{\epsilon^n P[X(t) = n]} \\
&= \lim_{\epsilon \to 0} \frac{P[s_1 < S_1 < s_1 + \epsilon, s_2 < S_2 < s_2 + \epsilon, \ldots, s_n < S_n < s_n + \epsilon, S_{n+1} > t]}{\epsilon^n P[X(t) = n]} \\
&= \frac{\int_t^\infty \lambda^{n+1} \exp\{-\lambda s_{n+1}\} ds_{n+1}}{e^{-\lambda t} (\lambda t)^n / n!} \\
&= \frac{\lambda^n e^{-\lambda t}}{e^{-\lambda t} (\lambda t)^n / n!} = n!/t^n \ \text{ if } 0 < s_1 < s_2 < \ldots < s_n < t
\end{aligned}$$

and 0 otherwise. In step 4, we used the probability density function of $S_1, S_2, \ldots, S_{n+1}$ as derived in (i) and the fact that it depends only on s_{n+1}. We also used $P[X(t) = n]$ as derived in Theorem 7.3.1 from the marginal distribution of S_n. Note that the conditional probability density function $f(\underline{s}_n | X(t) = n)$ is the same as derived in (iv) of Theorem 7.2.5. □

In the following theorem, using Theorems 7.3.1 and 7.3.2, we prove that Definition 7.3.1 is equivalent to Definition 7.2.2, thus all the three definitions of a Poisson process are equivalent.

Theorem 7.3.3 *Definitions 7.2.2 and 7.3.1 of a Poisson process are equivalent.*

Proof Suppose that $\{X(t), t \geq 0\}$ is a Poisson process according to Definition 7.2.2. It is proved in Theorem 7.2.1 that it is a continuous time Markov chain such that the sequence $\{T_n, n \geq 1\}$ of sojourn time random variables is a sequence of independent and identically distributed random variables, each having exponential distribution with scale parameter λ. Thus, Definition 7.2.2 implies Definition 7.3.1. We now assume that $\{X(t), t \geq 0\}$ is a Poisson process according to Definition 7.3.1. Thus, using the result that $\{T_n, n \geq 1\}$ is a sequence of independent and identically distributed random variables each having exponential distribution, we have to prove that $\{X(t), t \geq 0\}$ is a process with stationary and independent increments and for fixed t, $X(t) \sim Poi(\lambda t)$. In Theorem 7.3.1, we have proved that $X(t) \sim Poi(\lambda t)$, where the proof is based on the distribution of S_n. To examine whether $\{X(t), t \geq 0\}$ is a process with stationary and independent increments, note that for $n \geq 1$ and $0 = t_0 < t_1 < t_2 < \cdots < t_n$, the event

$$A = [X(t_r) - X(t_{r-1}) = x_r, r = 1, 2, \ldots, n]$$

is equivalent to the event that x_r epochs S_k are in $(t_{r-1}, t_r]$, $r = 1, 2, \ldots, n$. Suppose $y_0 = 0$, $y_r = \sum_{j=1}^{r} x_j$, $r = 1, 2, \ldots, n$ and $x = y_n$. Then the event

$$B = [t_{r-1} < S_{y_{r-1}+1} < S_{y_{r-1}+2} < \cdots < S_{y_{r-1}+x_r} \leq t_r, r = 1, 2 \ldots, n] = A.$$

$$\& P(A) = P[B|X(t_n) = x]P[X(t_n) = x].$$

Using result (ii) of Theorem 7.3.2, $P[B|X(t_n) = x]$ can be derived using the following result. Observe that for $j = 0, 1, \ldots,$ and $l \geq 1$

$$I_j(a, b) = \int_a^b \int_{s_{j+1}}^b \int_{s_{j+2}}^b \cdots \int_{s_{j+l-1}}^b ds_{j+1} ds_{j+2} \cdots ds_{j+l} = \frac{(b-a)^l}{l!}.$$

Then,

$$\begin{aligned} & P\left[X(t_r) - X(t_{r-1}) = x_r, r = 1, 2, \ldots, n \middle| X(t_n) = x)\right] \\ &= P\left[t_{r-1} < S_{y_{r-1}+1} < S_{y_{r-1}+2} < \cdots < S_{y_{r-1}+x_r} \leq t_r, r = 1, \ldots, n \middle| X(t_n) = x\right] \\ &= (x!/t_n^x) \prod_{r=1}^{n} I_{y_{r-1}}(t_{r-1}, t_r) = (x!/t_n^x) \prod_{r=1}^{n} (t_r - t_{r-1})^{x_r} / x_r!. \end{aligned}$$

Hence,

$$\begin{aligned} P\left[X(t_r) - X(t_{r-1}) = x_r, r = 1, 2, \ldots, n\right] &= \frac{x!}{t_n^x} \prod_{r=1}^{n} \frac{(t_r - t_{r-1})^{x_r}}{x_r!} \times e^{-\lambda t_n} \frac{(\lambda t_n)^x}{x!} \\ &= \prod_{r=1}^{n} e^{-\lambda(t_r - t_{r-1})} \frac{(\lambda(t_r - t_{r-1}))^{x_r}}{x_r!}, \end{aligned}$$

which gives independence and stationarity of increments. This also proves that $X(t_r) - X(t_{r-1}) \sim Poi(\lambda(t_r - t_{r-1}))$ distribution, for $r = 1, 2, \ldots, n$. □

Remark 7.3.2 Theorem 7.3.3 conveys that $\{X(t), t \geq 0\}$ is a Poisson process, if and only if the inter-occurrence random variables are independent and identically distributed random variables, each having exponential distribution. Thus, it is a characterization of a Poisson process.

Following examples illustrate various properties of a Poisson process.

Example 7.3.1 Suppose men and women enter a supermarket according to independent Poisson processes having respective rates of two and four per minute. Suppose $\{X_m(t), t \geq 0\}$ and $\{X_w(t), t \geq 0\}$ are independent Poisson processes with rates $\lambda_m = 2$ and $\lambda_w = 4$ respectively, where $X_m(t)$ and $X_w(t)$ denote the number of men and women respectively, entering the supermarket in $(0, t]$.

(i) Starting at an arbitrary time, the probability that at least two men arrive before the first woman arrives is given by $P[X_m(T_{w1}) \geq 2]$, where T_{w1} denotes the waiting time for the first women to enter the supermarket. Since T_{w1} has exponential distribution with parameter $\lambda_w = 2$, we find $P[X_m(T_{w1}) < 2]$ as follows:

$$\begin{aligned}
P[X_m(T_{w1}) < 2] &= P[X_m(T_{w1}) = 0] + P[X_m(T_{w1}) = 1] \\
&= E(P[X_m(T_{w1}) = 0]|T_{w1}) + E(P[X_m(T_{w1}) = 1]|T_{w1}) \\
&= E(e^{-\lambda_m T_{w1}}) + E(e^{-\lambda_m T_{w1}}\lambda_m T_{w1}) \\
&= \int_0^\infty e^{-\lambda_m y}\lambda_w e^{-\lambda_w y}\,dy + \int_0^\infty e^{-\lambda_m y}\lambda_m y \lambda_w e^{-\lambda_w y}\,dy \\
&= \frac{\lambda_w}{\lambda_w + \lambda_m} + \frac{\lambda_w \lambda_m}{\lambda_w + \lambda_m}(\lambda_w + \lambda_m)\int_0^\infty y e^{-(\lambda_w + \lambda_m)}\,dy \\
&= \frac{\lambda_w}{\lambda_w + \lambda_m} + \frac{\lambda_w \lambda_m}{(\lambda_w + \lambda_m)^2} = \frac{4}{6} + \frac{4 \times 2}{6^2} = \frac{32}{36} = \frac{8}{9}.
\end{aligned}$$

Hence, $P[X_m(T_{w1}) \geq 2] = 1/9$.

(ii) On similar lines, we find the probability that at least two men arrive before the third woman arrives. The time for third women to arrive, denoted by S_{w3} follows $G(\lambda_w, 3)$ distribution. Thus, the probability that at least two men arrive before the third woman arrives is $P[X_m(S_{w3}) \geq 2] = 1 - P[X_m(S_{w3}) < 2]$. Now,

$$\begin{aligned}
P[X_m(S_{w3}) < 2] &= P[X_m(S_{w3}) = 0] + P[X_m(S_{w3}) = 1] \\
&= E(P[X_m(S_{w3}) = 0]|S_{w3}) + E(P[X_m(S_{w3}) = 1]|S_{w3}) \\
&= \int_0^\infty \frac{\lambda_w^3}{\Gamma(3)} e^{-\lambda_w y} y^{3-1} e^{-\lambda_m y}\,dy \\
&+ \int_0^\infty \frac{\lambda_w^3}{\Gamma(3)} e^{-\lambda_w y} y^{3-1} e^{-\lambda_m y}\lambda_m y\,dy \\
&= \frac{\lambda_w^3}{2}\frac{\Gamma(3)}{(\lambda_w + \lambda_m)^3} + \lambda_m \int_0^\infty \frac{\lambda_w^3}{\Gamma(3)} e^{-(\lambda_w + \lambda_m)y} y^{4-1}\,dy \\
&= \frac{\lambda_w^3}{2}\frac{\Gamma(3)}{(\lambda_w + \lambda_m)^3} + \frac{\lambda_w^3 \lambda_m}{\Gamma(3)}\frac{\Gamma(4)}{(\lambda_w + \lambda_m)^4} \\
&= \frac{\lambda_w^3}{(\lambda_w + \lambda_m)^3} + \frac{\lambda_w^3 \lambda_m}{(\lambda_w + \lambda_m)^3}\frac{3}{\lambda_w + \lambda_m} \\
&= \frac{\lambda_w^3}{(\lambda_w + \lambda_m)^3}\left(1 + \frac{3\lambda_m}{\lambda_w + \lambda_m}\right) = \frac{16}{27}.
\end{aligned}$$

Hence, $P[X_m(T_{w3}) \geq 2] = 11/27$. □

Example 7.3.2 The life time of a component of a machine is modeled by the exponential distribution with mean 1 per week. Failed component is immediately replaced by a new one. Suppose $X(t)$ denotes the number of failures in $(0, t]$. Since the inter-failure distribution is exponential, it follows that $\{X(t), t \geq 0\}$ is a Poisson process with rate 1 per week. (i) The probability that two weeks have elapsed since the last failure is $P[X(2) = 0] = e^{-2} = 0.135$.

(ii) Suppose there are 5 new components in the inventory and the next supply is not due in 10 weeks. The machine will not be out of order in the next 10 weeks if the number of failures in 10 weeks is not more than 5. Thus, the probability of such an event is given by $P[X(10) \leq 5] = \sum_{i=0}^{5} e^{-10}10^i/i! = 0.068$. □

Example 7.3.3 A certain scientific theory supposes that mistakes in cell division occur according to a Poisson process with rate 2.5 per year and that an individual dies when 200 such mistakes have occurred.

(i) Assuming this theory, we find the mean lifetime of an individual and also the variance of the lifetime of an individual as follows. If T denotes the life time, then $T = \sum_{i=1}^{200} T_i$, where T_i denotes the random interval between ith and $(i-1)$th mistakes in cell division. Since mistakes in cell division occur according to a Poisson process with rate 2.5 per year, it follows that $\{T_i, i \geq 1\}$ are independent and identically distributed random variables, each having exponential distribution with $E(T_i) = 1/2.5$ and $Var(T_i) = (1/2.5)^2$. Hence, $E(T) = 200/2.5 = 80$ and $Var(T) = 200/6.25 = 32$.

(ii) We now find the probability that an individual dies before age 70. An individual dies before age 70 if more than 200 mistakes occur before 70 years. Hence, it is given by $P[X(70) \geq 200]$. Further, $X(70)$ has Poisson distribution with mean $70 \times 2.5 = 175$, which is large. Hence distribution of $X(70)$ can be approximated by the normal $N(175, 175)$ distribution. Thus,

$$P[X(70) \geq 200] \approx P\left[Z \geq \frac{200 - 175}{\sqrt{175}}\right] = P[Z \geq 1.8898] = 0.0294,$$

where $Z \sim N(0, 1)$.

(iii) On similar lines, we find the probability that an individual survives to age 85 as follows. An individual survives to age 85 if number of mistakes in cell division is less than 200 in 85 years. Hence, it is given by $P[X(85) < 200]$, where $X(85)$ has Poisson distribution with mean $85 \times 2.5 = 212.5$, which is large. Hence, distribution of $X(85)$ can be approximated by the normal $N(212.5, 212.5)$ distribution. Thus,

$$P[X(85) < 200] \approx P\left[Z < \frac{200 - 212.5}{\sqrt{212.5}}\right] = P[Z < -0.8575] = 0.1956.$$

□

A Poisson process discussed so far is a homogeneous Poisson process. In the next section, we briefly discuss a generalization of a homogeneous Poisson process, which is a non-homogeneous Poisson process. In this case, the arrival rate at time t is a function of t.

7.4 Non-homogeneous Poisson Process

In many situations, the arrival rate is a function of time. For example, the arrival rate to a restaurant varies with the time of day and increases during the lunch and dinner times. Number of vehicles arriving at a traffic signal varies during the day, the rate is high when offices and schools open and close. The rate of arrivals at a mall also changes during the day, it being high in the morning and evening. A non-homogeneous Poisson process is a suitable model in such situations. It is defined below.

Definition 7.4.1 A stochastic process $\{X(t), t \geq 0\}$ is known as a non-homogeneous Poisson process with intensity function $\lambda(t)$, if (i) X(0) = 0, (ii) $\{X(t), t \geq 0\}$ has independent increments, (iii) $P[X(t+h) - X(t) = 1]$ is $\lambda(t)h + o(h)$ and (iv) $P[X(t+h) - X(t) \geq 2] = o(h)$.

We state below some results related to a non-homogeneous Poisson process.

(i) A function $m(t)$ defined by $m(t) = \int_0^t \lambda(y)\, dy$ is known as the mean function of the non-homogeneous Poisson process.
(ii) $X(t+s) - X(s)$ follows a Poisson distribution with mean $\int_s^{t+s} \lambda(y)\, dy$.
(iii) Increments over disjoint intervals are independent random variables.

Following example illustrates these results.

Example 7.4.1 A bank opens at 9 a.m. and close at 4 p.m. During the first two hours, the rate of arrivals increases linearly, in the next two hours also, it increases but at a lower rate, and in the last three hours, it decreases linearly. Suppose arrivals of customers at a bank are modeled by a non-homogeneous Poisson process with rate function $\lambda(t)$ as follows:

$$\lambda(t) = \begin{cases} 2 + 4(t-9), & \text{if } 9 \leq t \leq 11 \\ 10 + (t-11), & \text{if } 11 \leq t \leq 13 \\ 12 - 4(t-13), & \text{if } 13 \leq t \leq 16. \end{cases}$$

We take 9 a.m. as the origin to compute the following probabilities:

(i) The probability that 5 customers arrive during 9 a.m. and 10 a.m. is $P[X(1) - X(0) = 5]$. Now $X(1) - X(0) \sim Poi(\mu)$, where $\mu = \int_0^1 (2+4t)\, dt = 4$. Hence, $P[X(1) - X(0) = 5] = e^{-4}4^5/5! = 0.1563$.

(ii) The probability that 5 customers arrive during 3 p.m. and 4 p.m. is $P[X(7) - X(6) = 5]$. Note that $X(7) - X(6) \sim Poi(\mu)$, where $\mu = \int_6^7 (12 - 4(t - 4))\, dt = 2$. Hence, $P[X(7) - X(6) = 5] = e^{-2}2^5/5! = 0.0361$. Note that mean 2 of number of arrivals in period 3 p.m. and 4 p.m. is smaller than that during 9 a.m.

and 10 a.m. It is in view of the fact that $\lambda(t)$ is an increasing function for the period 9 a.m. and 10 a.m. and it is a decreasing function for the period 3 p.m. and 4 p.m.

(iii) The probability that 5 customers arrive during 9 a.m. and 10 a.m. and 5 customers arrive during 3 p.m. and 4 p.m. is $0.1563 \times 0.0361 = 0.0056$ as a consequence of independence of increments.

(iv) If we want to compute the mean number of arrivals between 10 a.m. and 12 noon, we split the interval as 10 a.m. to 11 a.m. and 11 a.m. and 12 noon as the intensity rates are different for these two time periods. It is given by,

$$\mu = \int_1^2 (2 + 4t)\, dt + \int_2^3 (10 + (t - 2))\, dt = 10.5. \qquad \square$$

In the following section, we discuss some operations on a Poisson process, such as superposition of Poisson processes and decomposition or thinning of a Poisson process.

7.5 Superposition and Decomposition

Superposition of stochastic processes is equivalent to addition of random variables. In many real life situations, we come across such a phenomenon. For example, suppose $X_R(t)$ is the number of cars arriving at a certain location from right and $X_L(t)$ is the number of cars arriving at that location from left in $(0, t]$. Then $X(t) = X_R(t) + X_L(t)$ is the total number of cars arriving at the location up to time t. The process $\{X(t), t \geq 0\}$ is called the superposition of $\{X_R(t), t \geq 0\}$ and $\{X_L(t), t \geq 0\}$. The next theorem gives the probabilistic structure of the superposed process.

Theorem 7.5.1 *Suppose $\{X(t), t \geq 0\}$ and $\{Y(t), t \geq 0\}$ are two independent Poisson processes with rates λ_1 and λ_2, respectively. Then, the process $\{Z(t) = X(t) + Y(t), t \geq 0\}$ is also a Poisson process with rate $\lambda_1 + \lambda_2$.*

Proof We show that the process $\{Z(t), t \geq 0\}$ satisfies the requirements of Definition 7.2.2. Clearly $Z(0) = 0$.

(i) To prove that the process $\{Z(t), t \geq 0\}$ is a process with independent increments, we consider the random variables $Z(t_1), Z(t_2) - Z(t_1), Z(t_3) - Z(t_2)$ for $0 < t_1 < t_2 < t_3$. Since the $\{X(t), t \geq 0\}$ and $\{Y(t), t \geq 0\}$ are processes with independent increments, and since the two processes themselves are independent, the six random variables $X_1(t), X(t_2) - X(t_1), X(t_3) - X(t_2), Y(t_1)$, $Y(t_2) - Y(t_1), Y(t_3) - Y(t_2)$ are all independent. This implies that, the three random variables $Z(t_1) = X(t_1) + Y(t_1)$,

$$Z(t_2) - Z(t_1) = X(t_2) - X(t_1) + Y(t_2) - Y(t_1),$$
$$Z(t_3) - Z(t_2) = X(t_3) - X(t_2) + Y(t_3) - Y(t_2)$$

are independent. Extending similar argument for n time points, we claim that the process $\{Z(t), t \geq 0\}$ has independent increments.

(ii) For $s < t$, the increment $Z(t) - Z(s) = X(t) - X(s) + Y(t) - Y(s)$. We know that $X(t) - X(s) \stackrel{d}{=} X(t-s)$ and $Y(t) - Y(s) \stackrel{d}{=} Y(t-s)$ and since $X(t-s) + Y(t-s) \sim Poi(\lambda_1 + \lambda_2)(t-s)$, $Z(t) - Z(s) \sim Poi(\lambda_1 + \lambda_2)(t-s)$. This implies that, the distribution of the increments is Poisson and depends only on $(t-s)$. From Definition 7.2.2, it follows that the process $\{Z(t), t \geq 0\}$ is a Poisson process with rate $\lambda_1 + \lambda_2$. □

Remark 7.5.1 Another approach to prove Theorem 7.5.1 is based on the definition of a Poisson process in terms of a point process. Suppose $\{U_n, n \geq 1\}$ and $\{V_n, n \geq 1\}$ are sequences of inter-arrival random variables corresponding to independent Poisson processes $\{X(t), t \geq 0\}$ and $\{Y(t), t \geq 0\}$, respectively. Then $\{U_n, n \geq 1\}$ and $\{V_n, n \geq 1\}$ are independent sequences of independent and identically distributed random variables with common distribution to be exponential with parameter λ_1 and λ_2, respectively. We have $Z(t) = X(t) + Y(t)$. Thus, an event in $\{Z(t)\}$ process occurs when an event occurs in $\{X(t)\}$ or $\{Y(t)\}$, whichever is earlier. Thus, the distribution of inter-occurrence random variables in $\{Z(t)\}$ process is the distribution of $T_n = \min\{U_n, V_n\}, n \geq 1$, which is exponential with parameter $\lambda_1 + \lambda_2$. It follows that $\{T_n, n \geq 1\}$ is a sequence of independent and identically distributed random variables. Hence, $\{Z(t), t \geq 0\}$ is a Poisson process.

Theorem 7.5.1 can be extended to any finite number of independent Poisson processes. Superposition of finite number of independent Poisson processes is also known as an additive property of a Poisson process. Following examples illustrate Theorem 7.5.1.

Example 7.5.1 Suppose an auto arrives at a stand from north at a rate of 1 per minute according to a Poisson process $\{X_N(t), t \geq 0\}$ and from south at a rate of 2 per minute according to a Poisson process $\{X_S(t), t \geq 0\}$. Suppose the two arrival processes are independent. Using superposition $\{X(t) = X_N(t) + X_S(t), t \geq 0\}$ is a Poisson process with rate 3 per minute. Hence, the probability that a customer has to wait at a stand for more than two minutes is the same as the probability that no auto arrived at the stand during a period of two minutes. It is $P[X(2) = 0] = e^{-6} = 0.0025$. □

Example 7.5.2 A machine is subject to shocks arriving from two independent sources. The shocks from source A arrive according to a Poisson process $\{X_A(t), t \geq 0\}$ with rate 2 per day and those from source B arrive according to a Poisson process $\{X_B(t), t \geq 0\}$ with rate 4 per day. Thus, for a fixed t, $X_A(t) \sim Poi(2t)$, $X_B(t) \sim Poi(4t)$ and the two are independent random variables. As a consequence, the average and variance of the number of shocks in $(0, t]$ are

$$E(X_A(t) + X_B(t)) = 6t \ \& \ Var(X_A(t) + X_B(t)) = 6t.$$

Hence, the mean and variance of the total number of shocks from both the sources over an 8-hour shift will be $6 \times 1/3 = 2$ each. Suppose 4 shocks occurred in an 8-hour shift, then the probability that 1 shock is from source A is given by

$$\begin{aligned} P[X_A(1/3) = 1 | X_A(1/3) + X_B(1/3) = 4] &= \frac{P[X_A(1/3) = 1, X_B(1/3) = 3]}{P[X_A(1/3) + X_B(1/3) = 4]} \\ &= \frac{P[X_A(1/3) = 1]P[X_B(1/3) = 3]}{P[X_A(1/3) + X_B(1/3) = 4]} \\ &= \frac{e^{-2/3}(2/3)^1 e^{-4/3}(4/3)^3 4!}{3! e^{-2}(2)^4} \\ &= 32/81. \qquad \square \end{aligned}$$

Remark 7.5.2 We have a result from the distribution theory that if X and Y are independent random variables and if $X \sim Poi(\lambda_1)$ distribution and $Y \sim Poi(\lambda_2)$ distribution, then the conditional distribution of X given $X + Y = n$ is binomial $B(n, p)$ where $p = \lambda_1/(\lambda_1 + \lambda_2)$. In the above example, $X_A(1/3) \sim Poi(2/3)$ distribution and $X_A(1/3) + X_B(1/3) \sim Poi(2)$ distribution. Hence, the conditional distribution of $X_A(1/3)$ given $X_A(1/3) + X_B(1/3) = 4$ is binomial $B(4, p)$ where $p = 1/3$. Thus, the conditional probability is $\binom{4}{1}(1/3)(2/3)^3$ which is $32/81$.

We now discuss one more operation on a Poisson process, known as thinning or the decomposition of a Poisson process. In superposition, we essentially add two or more Poisson processes. In decomposition, as the term indicates, we generate two or more processes from the given process. Following theorem explains the operation and the corresponding result for a Poisson process.

Theorem 7.5.2 *Suppose $\{X(t), t \geq 0\}$ is a Poisson process with rate λ. Suppose each event that occurs is of type-1 with probability p and is of type-2 with probability $q = 1 - p$, independently of other events. Suppose $X_1(t)$ and $X_2(t)$ denote the number of type-1 and type-2 events respectively, that occur in $(0, t]$. Then the processes $\{X_1(t), t \geq 0\}$ and $\{X_2(t), t \geq 0\}$ are independent Poisson processes with rates $\lambda_1 = \lambda p$ and $\lambda_2 = \lambda q$, respectively.*

Proof As in Theorem 7.5.1, we show that the processes $\{X_i(t), t \geq 0\}$, $i = 1, 2$ satisfy the requirements of Definition 7.2.2.

(i) Observe that $X(0) = 0 \;\Rightarrow\; X_1(0) = 0, \;\&\; X_2(0) = 0$.

(ii) For any interval, $(s, t]$, $X_1(t) - X_1(s)$ is the number of type-1 events that occur in the interval. It depends only on the number of events that occur in the process $\{X(t), t \geq 0\}$ in the interval $(s, t]$, that is on $X(t) - X(s)$ only. In other words, $X_1(t) - X_1(s)$ is a function of $X(t) - X(s)$ only, say $g(X(t) - X(s))$. Since the distribution of $X(t) - X(s)$ depends only on the length of the interval $(t - s)$, the distribution of $g(X(t) - X(s))$ also depends on $(t - s)$ only. Hence, the process $\{X_1(t), t \geq 0\}$ has stationary increments. Similarly, it follows that the process $\{X_2(t), t \geq 0\}$ has stationary increments.

(iii) Suppose $0 < t_1 < t_2 < \cdots < t_k$ are k time points. By the argument similar to those in (ii), for the process $\{X_1(t), t \geq 0\}$, the ith increment $X_1(t_i) - X_1(t_{i-1}) = g(X(t_i) - X(t_{i-1}))$, $i = 1, 2, \ldots, k$ for some function g. Since the process $\{X(t), t \geq 0\}$ is of independent increments, it follows that process $\{X_1(t), t \geq 0\}$ is also of independent increments. Similarly, the process $\{X_2(t), t \geq 0\}$ is a process with independent increments.

(iv) Now for $l, m = 0, 1, \ldots,$

$$\begin{aligned}
P[X_1(t) = l, X_2(t) = m] &= P[X_1(t) = l, X_2(t) = m \big| X(t) = l + m] \\
&\quad \times P[X(t) = l + m] \\
&= \binom{l+m}{l} p^l q^m \frac{e^{-\lambda t}(\lambda t)^{l+m}}{(l+m)!} \\
&= \frac{e^{-\lambda pt}(\lambda pt)^l}{l!} \times \frac{e^{-\lambda qt}(\lambda qt)^m}{m!}.
\end{aligned}$$

Thus, for a fixed t, $X_1(t) \sim Poi(\lambda pt)$ and $X_2(t) \sim Poi(\lambda qt)$ and $X_1(t)$ and $X_2(t)$ are independent. Hence, it follows that $\{X_1(t), t \geq 0\}$ and $\{X_2(t), t \geq 0\}$ are independent Poisson processes with rates λp and λq, respectively. □

The type of decomposition discussed in Theorem 7.5.2, is known as Bernoulli thinning. It can be extended to multinomial thinning in which a given process is decomposed into more than two components. In this case also, the component processes are independent Poisson processes.

Following examples illustrate the application of Bernoulli thinning of a Poisson process.

Example 7.5.3 Customers enter a store according to a Poisson process of rate $\lambda = 10$ per hour. Independently, each customer buys something with probability $p = 0.3$ and leaves without making a purchase with probability $q = 1 - p = 0.7$. Suppose $X(t)$ denotes the number of customers entering the store in $(0, t]$, $X_1(t)$ denotes the number of customers who buy something and $X_2(t)$ denotes the number of customers leaving the store without making a purchase. Since $\{X(t), t \geq 0\}$ is a Poisson process with rate λ, $\{X_1(t), t \geq 0\}$ is a Poisson process with rate λp and $\{X_2(t), t \geq 0\}$ is a Poisson process with rate λq. Further, $\{X_1(t), t \geq 0\}$ is independent of $\{X_2(t), t \geq 0\}$. The probability that out of 9 people who entered the store during the first hour, 3 make a purchase and 6 do not, can be computed as follows:

$$\begin{aligned}
P[X_1(1) = 3, X_2(1) = 6] &= P[X_1(1) = 3]P[X_2(1) = 6] \\
&= \frac{e^{-3}3^3}{3!} \frac{e^{-7}7^6}{6!} = 0.2240 \times 0.1490 = 0.0334.
\end{aligned}$$

Further, given that 9 people entered the store during the first hour, the conditional probability that 3 of these people make a purchase is given by

$$\begin{aligned}
P[X_1(1) = 3 | X(1) = 9] &= P[X_1(1) = 3, X_2(1) = 6] \big/ P[X(1) = 9] \\
&= \frac{e^{-3}3^3}{3!} \frac{e^{-7}7^6}{6!} \Big/ \frac{e^{-10}10^9}{9!} = 0.2668. \quad \square
\end{aligned}$$

Example 7.5.4 A radioactive source emits particles at a rate of 5 per minute according to a Poisson process. Each particle emitted has a probability 0.6 of being recorded. If $X_R(t)$ denotes the number of recorded particles, then $X_R(t) \sim Poi(0.6 \times 5t)$.

Hence, the probability that in a 4 minute interval, 10 particles are recorded is $P[X_R(4) = 10] = e^{-12}12^{10}/10! = 0.104$. □

Example 7.5.5 In good years, storms occur according to a Poisson process with rate 3 per year, while in other years, they occur according to a Poisson process with rate 5 per year. Suppose $X(t)$, $X_g(t)$ and $X_b(t)$ denote the number of storms during the first t time units of year, of a good year and a bad year respectively. It is given that $\{X_g(t), t \geq 0\}$ and $\{X_b(t), t \geq 0\}$ are Poisson processes with rates 3 and 5 per year, respectively. Suppose G denotes the event that a year is a good year, while B denotes the event that a year is a bad year. It is given that $P(G) = 0.3$ and $P(B) = 0.7$.

(i) From the given information, we can obtain the probability mass function of $X(t)$ for fixed t as follows. For $n = 0, 1, 2, \ldots,$

$$\begin{aligned} P[X(t) = n] &= P[X(t) = n|G]P(G) + P[X(t) = n|B]P(B) \\ &= P[X_g(t) = n](0.3) + P[X_b(t) = n](0.7) \\ &= 0.3\frac{e^{-3t}(3t)^n}{n!} + 0.7\frac{e^{-5t}(5t)^n}{n!}. \end{aligned}$$

Hence, distribution of $X(t)$ for each fixed t is a mixture of Poisson distributions. Thus, $\{X(t), t \geq 0\}$ is not a Poisson process. Further, $\{X(t), t \geq 0\}$ is not a superposition of $\{X_g(t), t \geq 0\}$ and $\{X_b(t), t \geq 0\}$.

(ii) Suppose the next year starts off with three storms by the end of the first two months. It is of interest to find out the probability of the year to be a good year or a bad year. Hence, we find the conditional probability that it is a good year using Bayes' theorem as follows. We have

$$\begin{aligned} P[G|X(2/12) = 3] &= \frac{P[X(2/12) = 3|G]P(G)}{P[X(2/12) = 3|G]P(G) + P[X(2/12) = 3|B]P(B)} \\ &= \frac{P[X_g(2/12) = 3]P(G)}{P[X_g(2/12) = 3]P(G) + P[X_b(2/12) = 3]P(B)} \\ &= \frac{0.3 \times e^{-1/2}(1/2)^3/3!}{0.3 \times e^{-1/2}(1/2)^3/3! + 0.7 \times e^{-5/6}(5/6)^3/3!} \\ &= \left\{1 + \frac{7e^{-1/3}(5/3)^3}{3}\right\}^{-1} = 0.1144\,, \end{aligned}$$

as $X_g(2/12) \sim Poi(3 * 2/12)$ distribution and $X_b(2/12) \sim Poi(5 * 2/12)$ distribution. It is to be noted that given that there are three storms by the end of first two months, the probability that it is a good year is very small. □

In the next section, we study a generalization of a Poisson process, known as a compound Poisson process. It has applications in a variety of areas.

7.6 Compound Poisson Process

We elaborate on the concept of a compound Poisson process by a commonly used model in risk theory in non-life insurance. In this setup, it is assumed that claims arrive at an insurance company in accordance with some point process. Suppose $S_1, S_2, \ldots,$ are epochs of claim arrivals and $X_1, X_2, \ldots$ are corresponding claim sizes. Suppose $S(t)$ denotes the total claim amount up to t. Then $S(t) = X_1 + X_2 + \cdots + X_{N(t)}$ where $N(t)$ is the total number of claims up to t. To find a suitable distribution of $S(t)$ for fixed t, we have to find a suitable model for the frequency distribution, that is, the distribution of $N(t)$ and a severity distribution, that is, the distribution of X_i $i \geq 1$. If $\{N(t), t \geq 0\}$ is a Poisson process, then the model for $S(t)$ is known as a collective risk model. In insurance literature, it is termed as a Cramer–Lundberg model. As a stochastic process, $\{S(t), t \geq 0\}$ is referred to as a compound Poisson process. In this case, for fixed t, $S(t)$ is a sum of random number of random variables, where the distribution of random summand is Poisson.

A model for $S(t)$ is useful to decide the annual premium. Suppose $S(1)$ denotes risk corresponding to a portfolio for one year. Then $E(S(1))$ given by $E(S(1)) = E(X_1) * E(N(1))$ is known as a pure premium, since it does not take into account the variability in the frequency and amount of claims. Allowance for variability of $S(1)$ is incorporated in the net premium defined as $(1 + \theta)E(S(1))$ where θ is known as a loading factor. It is determined so that with very high chance, premium income exceeds claim expenditure. Further allowance for administrative costs gives a gross premium.

A compound Poisson process also arises in bulk queues, where customers arrive in a groups of random size and arrival epochs form a Poisson process. Following is the definition of a compound Poisson process.

Definition 7.6.1 Compound Poisson Process: Suppose $X(t)$ is defined as $X(t) = \sum_{i=1}^{N(t)} Y_i$, where (i) $\{N(t), t \geq 0\}$ is a Poisson process, (ii) $\{Y_i, i \geq 1\}$ is a sequence of independent and identically distributed random variables and (iii) $\{N(t), t \geq 0\}$ and $\{Y_i, i \geq 1\}$ are independent processes. Then the stochastic process $\{X(t), t \geq 0\}$ is known as a compound Poisson process.

It is to be noted that $N(0) = 0$ implies $X(0) = 0$. If Y_i is a degenerate random variable, degenerate at 1 $\forall\ i$, then $X(t) = N(t)\ \ \forall\ \ t$ and a compound Poisson process reduces to a Poisson process. Compound Poisson process is also known as a Poisson cluster process or cumulative Poisson process. The number $N(t)$ of clusters constitutes a Poisson process, while a random variable Y_i denotes the size of the ith cluster.

Following are some illustrations of a compound Poisson process.

(i) Suppose customers arrive at an ATM center according to a Poisson process $\{N(t), t \geq 0\}$ and suppose Y_i denotes the cash withdrawn by the ith customer. Then $X(t) = \sum_{i=1}^{N(t)} Y_i$ represents the cash withdrawn in $(0, t]$ and $\{X(t), t \geq 0\}$ is a compound Poisson process.

(ii) Suppose $N(t)$ denotes the number of cars arriving at a restaurant and Y_i denotes the number of individuals in the ith car. If we assume that $\{N(t), t \geq 0\}$ is a Poisson process and Y_i is a random variable with some discrete probability distribution then $X(t) = \sum_{i=1}^{N(t)} Y_i$ represents the number of individuals arriving at the restaurant in $(0, t]$ and $\{X(t), t \geq 0\}$ is a compound Poisson process.

(iii) Suppose $X(t)$ is the number of shocks to a system up to time t and Y_i is the damage or wear incurred by the ith shock. We assume that damage is positive and that the damage accumulates additively, so that $X(t) = \sum_{i=1}^{N(t)} Y_i$ represents the total damage sustained up to time t. Suppose that the system continues to operate as long as this total damage is less than some critical value "A" and fails in the contrary circumstance. If U denotes the time of system failure. Then $[U > t]$ if and only if $[X(t) < A]$. Hence,

$$P[U > t] = P[X(t) < A] = E(P[X(t) < A|N(t)]) = \sum_{i=1}^{\infty} \frac{e^{-\lambda t}(\lambda t)^n}{n!} G^{(n)}(A),$$

where $G^{(n)}(A) = P[Y_1 + Y_2 + \cdots + Y_n \leq A]$ is the distribution function of n-fold convolution at A.

Study of $\{X(t), t \geq 0\}$ in all such cases is useful for taking appropriate decisions. For example, how much cash to be kept at ATM in the first example, how many tables and chairs to be arranged in a restaurant and to decide the quantity of food to be made. We have already noted its application in insurance sector. A variety of practical problems can be reduced to a compound Poisson process. One may refer to Feller [3], Kulkarni [5] and Ross [6] for some more illustrations.

The compound Poisson process $\{X(t), t \geq 0\}$, when observed for a fixed time interval $(0, T]$, is presented as follows. Suppose $N(T) = k$ events occurred in $(0, T]$ and $T_i, i = 1, 2, \ldots, k$ denote the interval random variables in the Poisson process

$$X(t) = \begin{cases} 0, & \text{if} \quad 0 \leq t < T_1 \\ Y_1, & \text{if} \quad T_1 \leq t < T_1 + T_2 \\ Y_1 + Y_2, & \text{if } T_1 + T_2 \leq t < T_1 + T_2 + T_3 \\ & \vdots \\ Y_1 + Y_2 + \cdots + Y_k, & \text{if } T_1 + Y_2 + \cdots + T_k \leq t \leq T. \end{cases}$$

In the following example, we obtain a realization of a compound Poisson process, when it is observed for a fixed interval. We use Code 7.7.4. It is similar to that for a Poisson process.

Example 7.6.1 Suppose $N(t)$ denotes the number of cars arriving at a restaurant and Y_i denotes the number of individuals in the ith car. We assume that $\{N(t), t \geq 0\}$ is a Poisson process with rate 0.7 per 5 min and Y_i is a discrete random variable with support $\{1, 2, 3, 4, 5\}$ and respective probabilities as $\{0.2, 0.2, 0.3, 0.2, 0.1\}$. Then $X(t) = \sum_{i=1}^{N(t)} Y_i$ represents the number of individuals arriving at the restaurant in $(0, t]$ and $\{X(t), t \geq 0\}$ is a compound Poisson process. Suppose we observe the

Table 7.3 Realization of a compound Poisson Process

$N(t)$	T_i	S_i	X(t)
0	0.0000	0.0000	0
1	1.1459	1.1459	3
2	0.0303	1.1762	6
3	3.0207	4.1969	11
4	3.5191	7.7160	14
5	0.1220	7.8380	15
6	0.7268	8.5648	19
7	1.1463	9.7112	20

process for 50 min, that is, 10 time units. Using Code 7.7.4, we obtain a realization of $\{X(t), t \geq 0\}$. The output is displayed in Table 7.3. From Table 7.3, we note that the first car arrived at $T_1 = 1.1459$ with 3 individuals, the second car came at $S_2 = 1.1762$ with again 3 individuals. Thus, $X(S_2) = 6$. By time $T = 10$ time units, 7 cars and 20 individuals arrived. Thus, $N(10) = 7, \ X(10) = 20$. The realization of $\{X(t), t \geq 0\}$, when observed for a fixed time interval (0, 10], is presented as follows:

$$X(t) = \begin{cases} 0, \text{ if } & 0 \leq t < 1.1459 \\ 3, \text{ if } & 1.1459 \leq t < 1.1762 \\ 6, \text{ if } & 1.1762 \leq t < 4.1969 \\ & \vdots \\ 20, \text{ if } & 9.7112 \leq t \leq 10. \end{cases}$$

□

It is easy to find the mean function and the variance function of $X(t)$. Suppose λ is a rate of the Poisson process, then $E(N(t)) = Var(N(t)) = \lambda t$. Suppose $E(Y_i) = \mu$ and $Var(Y_i) = \sigma^2$. Then

$$\begin{aligned} E(X(t)) &= E(E(X(t))|N(t)) = E(N(t)\mu) = \lambda\mu t \\ Var(X(t)) &= E((Var(X(t)))|N(t)) + Var(E(X(t))|N(t)) \\ &= E(\sigma^2 N(t)) + Var(\mu N(t)) = \sigma^2\lambda t + \mu^2\lambda t \\ &= \lambda t(\sigma^2 + \mu^2) = \lambda t E(Y_1^2). \end{aligned}$$

Following examples illustrate some applications of a compound Poisson process in various areas.

Example 7.6.2 An insurance company pays out claims on its non-life insurance policies in accordance with a Poisson process having rate $\lambda = 5$ per week. The amount of money paid on each policy is exponentially distributed with mean Rs.2000. Then the amount of money paid by the insurance company in $(0, t]$ is modeled as a compound Poisson process $X(t) = \sum_{i=1}^{N(t)} Y_i$, where Y_i denotes the amount paid for ith claim and $N(t)$ denotes the number of claims in $(0, t]$. The mean and variance of the amount of money paid by the insurance company in a four-week span is then given by

$$\begin{aligned} E(X(4)) &= 5 \times 2000 \times 4 = 40000 \\ \& \ Var(X(4)) &= 5 \times 4 \times (2000^2 + 2000^2) = 16 \times 10^7. \end{aligned} \qquad \square$$

Example 7.6.3 The number of cars visiting a national park in $(0, t]$ is modeled by a Poisson process with rate 15 per hour. Each car has k occupants with probability p_k as given below.

$$p_1 = 0.2,\ p_2 = 0.3,\ p_3 = 0.3,\ p_4 = 0.1,\ p_5 = 0.05,\ p_6 = 0.05.$$

If a random variable Y denotes the number of occupants in a car, then from the given probability distribution of Y, we have

$$\mu = E(Y) = 2.65, \quad E(Y^2) = 8.75 \quad \& \quad \sigma^2 = Var(Y) = 1.7275.$$

Further, $X(t) = \sum_{i=1}^{N(t)} Y_i$ represents the number of visitors to the park in $(0, t]$. Then $\{X(t), t \geq 0\}$ is a compound Poisson process. Hence, the mean and the variance of the number of visitors to the park during a 10-hour period are given by,

$$E(X(10)) = 15 \times 10 \times 2.65 = 397.5 \quad \& \quad Var(X(10)) = 15 \times 10 \times 8.75 = 1312.5.$$

Suppose the national park charges Rs 50 per car plus Rs 20 per occupant as the entry fee. Thus, the total fee collected during a 10-hour period is $Z(10) = 50N(10) + 20X(10)$. The mean and the variance of the total fee collected during a 10-hour period are

$$\begin{aligned} E(Z(10)) &= 50E(N(10)) + 20E(X(10)) = 50 \times 150 + 20 \times 397.5 = 15450 \\ Var(Z(10)) &= Var(50N(10)) + Var(20X(10)) + 2\,Cov(50N(10), 20X(10)) \\ &= 2500 \times Var(N(10)) + 400 \times Var(X(10)) \\ &\quad + 2 \times 50 \times 20\,Cov(N(10), X(10)). \end{aligned}$$

Note that

$$Cov(N(t), X(t)) = E(N(t)X(t)) - E(N(t))E(X(t)) = E(N(t)X(t)) - \mu\lambda^2 t^2.$$

We find E(N(t)X(t)) by conditioning on $N(t)$. Thus,

$$
\begin{aligned}
E(N(t)X(t)) &= E(N(t)E(\sum_{i=1}^{N(t)} Y_i|N(t))) = E(N(t)\mu N(t)) \\
&= \mu(\lambda t + \lambda^2 t^2) \\
\Rightarrow \quad Cov(N(t), X(t)) &= \lambda \mu t = E(X(t)) \\
\Rightarrow \quad Cov(N(10), X(10)) &= 2.65 \times 15 \times 10 = 397.5 \\
\& \quad Var(Z(10)) &= 900000 + 795000 = 1695000,
\end{aligned}
$$

which is very large. □

Example 7.6.4 Suppose people arrive at a bus stop in accordance with a Poisson process with rate λ. The bus departs at time t. Suppose $X(t)$ denotes the total waiting time of all those who get on the bus at time t. Then $X(t)$ can be expressed as $X(t) = \sum_{i=1}^{N(t)} Y_i$, where Y_i denotes the waiting time of ith individual. It is to be noted that $Y_i = t - S_i$ where S_i is the epoch of arrival of the ith individual. Given $N(t) = n$, $S_1, S_2, \ldots, S_n$ are distributed as order statistics from uniform $U(0, t)$ distribution. For each fixed i, S_i has uniform $U(0, t)$ distribution and hence $Y_i = t - S_i$ also has uniform $U(0, t)$ distribution. Hence, for each fixed i, $E(Y_i) = t/2$ and $Var(Y_i) = t^2/12$ which implies that

$$
E(X(t)) = E(E(X(t)|N(t))) = E(N(t)t/2) = (\lambda t \times t)/2 = \lambda t^2/2.
$$

To find $Var(X(t))$, we note that

$$
\begin{aligned}
E(X(t)|N(t)) &= N(t)t/2 \ \& \ Var(X(t)|N(t)) = N(t)t^2/12 \\
\Rightarrow \quad Var(X(t)) &= E((Var(X(t)))|N(t)) + Var(E(X(t))|N(t)) \\
&= E(N(t)t^2/12) + Var(N(t)t/2) \\
&= (t^2/12) \times \lambda t + (t^2/4) \times \lambda t = \lambda t^3/3.
\end{aligned}
$$

□

For a Poisson process $\{N(t), t \geq 0\}$ for each fixed t, $N(t)$ has a Poisson distribution. The distribution of $X(t)$ for fixed t, is known as a compound Poisson distribution. In general, it is difficult to obtain its probability law. If the common distribution of Y_i is discrete with support as the set of whole numbers or its subset, then we can obtain the probability generating function of $X(t)$ for fixed t, and from it, we can compute the probabilities of certain events. Suppose the common probability generating function of Y_i is $P(s)$, $|s| \leq 1$. The probability generating function of $N(t)$ for fixed t is $G(s) = \exp\{\lambda t(s - 1)\}$. Then the probability generating function of $X(t)$ for fixed t is

$$
G(P(s)) = \exp\{\lambda t(P(s) - 1)\}.
$$

For fixed t, $P[X(t) = i]$ is the coefficient of s^i in the expansion of $G(P(s))$.

Following example illustrates the probability computation from the probability generating function of $X(t)$.

Example 7.6.5 The number $N(t)$ of two wheelers passing from a certain intersection in time $(0, t]$ follows Poisson distribution with rate $\lambda = 2$ per minute. The two wheeler has only 1 person with probability $2/3$ and two persons with probability $1/3$. Suppose we want to find the probability that 4 persons riding the two wheeler crossed the intersection in two minutes. Suppose $X(t) = \sum_{i=1}^{N(t)} Y_i$, where $\{N(t), t \geq 0\}$ is a Poisson process with rate 2 per minute, Y_i is a random variable with possible values 1 or 2 with probabilities $2/3$ and $1/3$, respectively. To find the probability that 4 persons crossed the intersection in two minutes, we find the probability generating function of $X(t)$ for fixed t as follows. The common probability generating function of Y_i is $P(s) = (2s + s^2)/3$ and of $N(t)$ for fixed t, is $G(s) = \exp\{\lambda t(s - 1)\}$. Then the probability generating function of $X(t)$ for fixed t is $G(P(s)) = \exp\{\lambda t((2s + s^2)/3 - 1)\}$. We have $\lambda = 2$ per time unit which is 1 minute. Hence, the probability generating function of $X(2)$ is $\exp\{4((2s + s^2)/3 - 1)\}$. We find the coefficient of s^4 to find the probability that 4 persons crossed the intersection in two minutes. It is given by $e^{-4}(8/9 + 128/81 + 512/243) = 0.0636$. □

We have noted above that finding the distribution of $X(t)$ for fixed t is tedious. In practice, an approximation to the distribution of $X(t)$ is always used. Under certain conditions, using random sum central limit theorem, it can be shown that as the expected number of events in a Poisson process increases, that is, as $t \to \infty$,

$$Z(t) = \frac{X(t) - E(X(t))}{\sqrt{Var(X(t))}} \xrightarrow{L} Z \sim N(0, 1).$$

Following example illustrates the application of this result.

Example 7.6.6 Customers arrive at an ATM center according to a Poisson process with rate 15 per hour. The amount of money withdrawn on each transaction is a random variable with mean Rs 5000 and standard deviation 1000. The machine is in use for 24 h. The total withdrawal in $(0, t]$ is then modeled as a compound Poisson process $X(t) = \sum_{i=1}^{N(t)} Y_i$, where Y_i denotes the amount withdrawn by the ith individual. Thus, expected total daily withdrawal and its variance are

$$\begin{aligned} E(X(24)) &= 15 \times 24 \times 5000 = 1800000 \\ \& \ Var(X(24)) &= 15 \times 24 \times (1000^2 + 5000^2) = 9.36 \times 10^9. \end{aligned}$$

The standard deviation of $X(24)$ is 96747.09. Number of daily withdrawals $N(24)$ has Poisson distribution with mean $15 \times 24 = 360$, which is large. Hence, distribution of $X(24)$ can be approximated by the normal distribution with mean 18×10^5 and standard deviation 96747.09. We compute the approximate probability that the total daily withdrawal is less than 19×10^5 as follows:

$$
\begin{aligned}
P[X(24) \leq 1900000] &= P\left[Z \leq \frac{19 \times 10^5 - 18 \times 10^5}{96747.09}\right] \\
&= P[Z \leq 1.033623] = 0.8493
\end{aligned}
$$

$$
\begin{aligned}
\text{Similarly } P[X(24) \leq 2000000] &= 0.9806 \\
\text{\& } P[X(24) \leq 2100000] &= 0.9990.
\end{aligned}
$$

Such an analysis is helpful for the management to decide on how much cash to be deposited daily at the ATM. □

An important theorem about the compound Poisson process, which has some applications in risk modeling, is stated below. It is similar to the superposition of Poisson processes.

Theorem 7.6.1 *If* $\{X_i(t), t \geq 0\}$, $i = 1, 2, \ldots, m$ *are mutually independent compound Poisson processes with Poisson parameter* λ_i *and with probability density function or probability mass function of* Y_i *as* $P_i(x)$, $i = 1, 2, \ldots, m$, *then* $\{X(t) = X_1(t) + X_1(t) + \cdots X_m(t) = \sum_{i=1}^{N(t)} U_i, t \geq 0\}$ *is a compound Poisson process with rate of Poisson process* $\{N(t), t \geq 0\}$ *as* $\lambda = \sum_{i=1}^{m} \lambda_i$ *and* $P(x) = \sum_{i=1}^{m} \lambda_i P_i(x)/\lambda$ *as a common distribution of* U_i.

Thus, the sum of independent but not necessarily identically distributed compound Poisson processes is again a compound Poisson process. This result has two important implications in building insurance models. First, if m insurance portfolios are combined, where the aggregate claims of the respective portfolios have compound Poisson distribution and are mutually independent, then the aggregate claims for the combined portfolio will again have a compound Poisson distribution. Secondly, we can consider a single insurance portfolio for a period of m years. Here we assume independence among the m annual aggregate claims and that the aggregate claims for each year has a compound Poisson distribution. It is not necessary that the annual aggregate claims distributions be identical.

Following example illustrates Theorem 7.6.1.

Example 7.6.7 Suppose $S_1(t)$, $S_2(t)$ and $S_3(t)$ denote the total risks in a collective risk model for a period $(0, t]$, corresponding to three different portfolios. It is given that $\{S_1(t), t \geq 0\}$ is a compound Poisson process with Poisson parameter $\lambda = 3$ and claim amounts are 10, 20 and 30 units with probabilities 0.3, 0.4 and 0.3, respectively. $\{S_2(t), t \geq 0\}$ is a compound Poisson process with Poisson parameter $\lambda = 4$ and claim amounts are either 20 or 40 units with probability 0.5 for each. $\{S_3(t), t \geq 0\}$ is a compound Poisson process with Poisson parameter $\lambda = 5$ and claim amounts are either 30 or 50 units with probability 0.4 and 0.6, respectively. If the three processes $\{S_1(t)\}$, $\{S_2(t)\}$ and $\{S_3(t)\}$ are independent, we find the approximate probability that the total risk $S(1)$ from three portfolios over a period of one year is less than 600 units and 700 units as follows. From Theorem 7.6.1, it follows that $S(1)$ has compound Poisson distribution with parameter of the Poisson distribution as $\lambda = 3 + 4 + 5 = 12$. The common probability distribution of U_i is as given below.

$$P[U_i = 10] = 0.0750, \quad P[U_i = 20] = 0.2666, \quad P[U_i = 30] = 0.2417$$
$$P[U_i = 40] = 0.1667 \;\; \& \;\; P[U_i = 50] = 0.2500$$
$$\Rightarrow \quad \mu = E(U_i) = 32.501, \quad E(U_i^2) = 1223.39$$
$$\Rightarrow \quad E(S(1)) = \lambda\mu = 390.012, \quad Var(S(1)) = \lambda E(U_i^2) = 14680.68$$

and the standard deviation of $S(1)$ is $Sd(S(1)) = 121.1639$. Thus, approximating distribution of $(S(1) - E(S(1)))/Sd(S(1))$ by the standard normal distribution, we have

$$P[S(1) \leq 600] = P\left[\frac{S(1) - E(S(1))}{Sd(S(1))} \leq \frac{600 - 390.012}{121.1639}\right]$$
$$\approx P[Z \leq 1.73309] = 0.9585, \quad \text{where } Z \sim N(0, 1)$$

Similarly, $P[S(1) \leq 700] = 0.9947$. □

The next section presents R codes used in solving examples.

7.7 R Codes

Following is a code to find a realization of a Poisson process with rate λ using the procedure described in Sect. 7.2. We have noted that for different values of λ, the number of occurrences of events will be different, it being a rate parameter. Hence, the vectors to store the realized values of the inter-occurrence random variables, corresponding epochs of arrivals, which is obtained by cumulative sums of the inter-occurrence random variables, and the states of the process, are of different sizes. Hence, we use the function `list` to store these values. In the following code, `int`, `x` and `arr` are the names of the lists to store realized values of the inter-occurrence random variables, states and epochs of arrivals, respectively. Other lists `u, v, w` are for plots. We illustrate the code, for Example 7.2.5.

Code 7.7.1 Realization of a Poisson process: Suppose a Poisson process $\{X(t), t \geq 0\}$, with rate $\lambda = 0.5$, 1.1, 1.6 and 2.9, is observed for $T = 5$ time units.

```
# Part I: Input values of rate parameter and T
lam=c(0.5,1.1,1.6,2.9); T=5
# Part II: Realization
int=x=arr=u=v=w=list()
N=c() # vector to store X(T) corresponding to 4 values of lambda
for(j in 1:length(lam))
{
set.seed(j); y=c(); sumy=0; i=1
while(sumy<T)
{
y[i]=rexp(1,rate=lam[j])
```

```
sumy=sum(y)
i=i+1
}
int[[j]]=y[-length(y)]
N[j]=length(int[[j]])
x[[j]]=0:N[j]
u[[j]]=rep(x[[j]],each=2)
arr[[j]]=round(cumsum(int[[j]]),2)
v[[j]]=c(0,round(rep(arr[[j]],each=2),3),T)
w[[j]]=c(0,arr[[j]])
}
N; list("States"=x,"Interval r.v."=int,"Arrival epochs"=arr)
# Part III: Graph of realization
par(mfrow=c(2,2)); rate=paste("Rate=",lam,sep=" ")
number=paste("No of Events= ",N,sep=" ")
for(j in 1:length(lam))
{
plot(v[[j]],u[[j]],"l",main=rate[j],ylab="States",
xlab="Occurrence Time",yaxt="n",xaxt="n",col="blue")
axis(2,at=sort(unique(u[[j]])),labels=sort(unique(u[[j]])),
las=2)
axis(1,at=round(c(arr[[j]],T),2),las=2,cex.axis=0.7)
points(w[[j]],x[[j]],pch=20,col="dark blue")
abline(v=T,col="purple")
legend("topleft",legend=number[j],cex=.7)
}
```

□

We use the following code to simulate the Poisson process multiple times and then to apply the chi-square goodness of fit test, to test whether the proposed model is a good model for the simulated data. We illustrate it, for Example 7.2.6.

Code 7.7.2 Multiple realizations of a Poisson process: A Poisson process $\{X(t), t \geq 0\}$ with rate $\lambda = 0.2$ is observed for $T = 20$ time units. It is simulated 100 times and based on 100 values of $X(T)$, we examine whether a Poisson distribution $Poi(\lambda T)$ is a good model, graphically and algebraically.

```
# Part I: Inpute rate parameter, T and number of simulations
lambda=.2; nsim=100; T=20
# Part II: Multiple simulations
x=c()
for(m in 1:nsim)
{
set.seed(m); y=c(); sumy=0; i=1
while(sumy<T)
{
```

```
y[i]=rexp(1,rate=lambda)
sumy=sum(y)
i=i+1
}
l=length(y); l; y=y[-l]; x[m]=length(y)
}
x; d=data.frame(table(factor(x,levels=0:max(x))));d
# Part III: Graph of observed and expected frequencies
obf=d$Freq; N=sum(obf); N; rf=obf/N; rf; mean=lambda*T
t=sort(unique(x));t; v=dpois(t,mean); v;ex=nsim*v
ex=round(ex,2);ex; da=data.frame(d,ex);da
a=min(min(rf),min(v));b=max(max(rf),max(v));a;b
u=seq(round(a,2),b,.03);u
plot(t,rf,"h",col="blue",lty=1,lwd=2,xlab="Values of X",
main="Observed and Expected Distributions",ylim=range(c(a,b)),
ylab="Relative Frequency and Expected probability",yaxt="n")
axis(2,at=u); points(t,rf,pch=20)
lines(t,v,"h",col="red",lty=2,lwd=1)
points(t,v,pch=20); abline(h=0,col="blue")
legend("topright",legend=c("Observed Distribution",
"Expected Distribution"),cex=.7,col=c("blue","red"),
lty=c(1,2),lwd=c(2,1)); max=max(t);max
# Part IV: Goodness of fit test, pooling the
# frequencies which are < 5
v1=1-ppois(max,mean); v2=c(v,v1); e=nsim*v2; e; length(e)
ob=c(obf,0) ; ob;length(ob)
o1=c(ob[1]+ob[2],ob[3:7],ob[8]+ob[9]+ob[10]); o1
e1=c(e[1]+e[2],e[3:7],e[8]+e[9]+e[10]); e1
t1=sum((o1-e1)^2/e1); t1;df=length(o1)-1;df
b1=qchisq(.95,df); b1; p1=1-pchisq(t1,df); p1
# Part V: Goodness of fit test, without pooling the
# frequencies which are less than 5
d=length(ob)-1; d
t=sum((ob-e)^2/e); t; b=qchisq(.95,d); b; p2=1-pchisq(t,d); p2
# Part VI: Built-in function for goodness of fit test
chisq.test(ob,p=v2)
```

□

The following code uses Shapiro–Wilk test of normality to examine goodness of fit of the Poisson distribution with large mean. It is illustrated, for Example 7.2.7.

Code 7.7.3 Goodness of fit test: A Poisson process $\{X(t), t \geq 0\}$ with rate $\lambda = 1, 1.5$ and 2.2 is observed for $T = 20$ time units. We simulate the process 100 times and based on 100 values of $X(T)$, examine whether a Poisson $Poi(\lambda T)$ distribution is a good model, using Shapiro–Wilk test for normality.

```
# Part I: Input rate parameter, T and number of simulations
lam=c(1,1.5,2.2); T=20; nsim=100
# Part II: Multiple simulations
x=matrix(nrow=length(lam),ncol=nsim)
for(j in 1:length(lam))
{
for(m in 1:nsim)
{
set.seed(m); y=c(); sumy=0; i=1
while(sumy<T)
{
y[i]=rexp(1,rate=lam[j])
sumy=sum(y); i=i+1
}
l=length(y); l; y=y[-1]; x[j,m]=length(y)
}
}
# Part III: Goodness of fit test
stat=pval=c()
for(j in 1:length(lam))
{
u=(x[j,]-lam[j]*T)*(lam[j]*T)^(-.5)
stat[j]=shapiro.test(u)$stat
pval[j]=shapiro.test(u)$p.value
}
d=round(data.frame(lam,stat,pval),4); d
```

□

Following code is for a realization of a compound Poisson process, when it is observed for a fixed interval. We illustrate it, for Example 7.6.1

Code 7.7.4 Realization of a compound Poisson process: Suppose $X(t) = \sum_{i=1}^{N(t)} Y_i$ is a compound Poisson process, where $\{N(t), t \geq 0\}$ is a Poisson process with rate 0.7 per 5 min and Y_i is a discrete random variable with support $\{1, 2, 3, 4, 5\}$ and with respective probabilities as $\{0.2, 0.2, 0.3, 0.2, 0.1\}$. Following code generates this compound Poisson process when observed for 50 min, that is, 10 time units.

```
# Part I: Input rate parameter, T and distribution of Y
lambda=.7; T=10; s=c(1,2,3,4,5); p=c(.2,.2,.3,.2,.1)
# Part II: Realization
x=y=u=z=c(); x[1]=z[1]=0;set.seed(11);sumy=0; i=1
while(sumy<T)
{
y[i]=rexp(1,rate=lambda)
x[i+1]=x[i]+1
sumy=sum(y)
u[i]=sample(s,1,p,replace=T)
z[i]=sum(u)
i=i+1
}
l=length(y);l; y=c(0,y[-1]);u=u[-1];z=c(0,z[-1]); m=length(x)
x=x[-m]; w=cumsum(y); d=data.frame(x,y,w,z);d1=round(d,4);d1
```

□

A quick recap of the results discussed in the present chapter is given below.

Summary

1. Definition 1: A continuous time stochastic process $\{X(t), t \geq 0\}$ with state space W is a homogeneous Poisson process with rate $\lambda > 0$ if the following axioms are satisfied. (i) $X(0) = 0$. (ii) $\{X(t), t \geq 0\}$ is a stochastic process with stationary and independent increments.
   (iii) $P_{ii}(h) = 1 - \lambda h + o(h)$, $P_{i,i+1}(h) = \lambda h + o(h)$ and $\forall\ j \neq i+1$, $P_{ij}(h) = o(h)$.
2. Definition 2: A continuous time stochastic process $\{X(t), t \geq 0\}$ with state space W is a homogeneous Poisson Process with rate $\lambda > 0$ if the following axioms are satisfied. (i) $X(0) = 0$. (ii) $\{X(t), t \geq 0\}$ is a stochastic process with stationary and independent increments.
   (iii) $P[X(t) = k] = e^{-\lambda t}(\lambda t)^k/k!$, $k = 0, 1, \ldots$ or equivalently
   (iv) $\{X(t), t \geq 0\}$ is a process with independent increments and (v) for $s < t$, $P[X(t) - X(s) = k] = e^{-\lambda(t-s)}\left(\lambda(t-s)\right)^k/k!$, $k = 0, 1, \ldots$.
3. Definition 3: A point process $\{S_n, n \geq 1\}$ is called a Poisson process with rate $\lambda > 0$ if $\{T_n = S_n - S_{n-1}, n \geq 1\}$ is a sequence of independent and identically distributed random variables each having exponential distribution with scale parameter λ. The corresponding counting process $\{X(t), t \geq 0\}$ is also known as the Poisson process, where $X(t)$ denotes the number of events occurring in $(0, t]$, $t \geq 0$.
4. For a Poisson process with rate λ, $E(X(t)) = Var(X(t)) = \lambda t$ and $Cov(X(s), X(t)) = \lambda \min\{s, t\}$.
5. The conditional distribution of a time of occurrence of an event in $[0, t]$, given that 1 event has occurred has occurred in $[0, t]$ is uniform $U(0, t)$.

6. $S_n \sim G(\lambda, n)$. The conditional distribution of n arrival times $\{S_1 < S_2 < \cdots < S_n\}$ in $(0, t]$, given that n events occurred in $(0, t]$, is the same as the distribution of the order statistics corresponding to n independent random variables each having uniform $U(0, t)$ distribution.
7. The counting process $\{X(t), t \geq 0\}$ is said to be a non-homogeneous Poisson process with intensity function $\lambda(t), t \geq 0$, if (i) X(0) = 0,
(ii) $\{X(t), t \geq 0\}$ has independent increments,
(iii) $P[X(t+h) - X(t) \geq 2] = o(h)$ and
(iv) $P[X(t+h) - X(t) = 1] = \lambda(t)h + o(h)$. The function m(t) defined by $m(t) = \int_0^t \lambda(y)\, dy$ is known as the mean function of the non-homogeneous Poisson process. For fixed t and s, $X(t+s) - X(s)$ follows a Poisson distribution with mean function $\int_s^{t+s} \lambda(y)\, dy$ and increments over disjoint intervals are independent random variables.
8. Superposition of Poisson processes: Suppose $\{X(t), t \geq 0\}$ and $\{Y(t), t \geq 0\}$ are two independent Poisson processes with rates λ_1 and λ_2 respectively and $Z(t) = X(t) + Y(t)$. Then the process $\{Z(t), t \geq 0\}$ is also a Poisson process with rate $\lambda_1 + \lambda_2$.
9. Decomposition of a Poisson process: Suppose $\{X(t), t \geq 0\}$ is a Poisson process with rate λ. Suppose each event that occurs is of type-1 with probability p and is of type-2 with probability $q = 1 - p$, independently of other events. Suppose $\{X_1(t)\}$ and $\{X_2(t)\}$ denote the number of type-1 and type-2 events, respectively, that occur in $(0, t]$. Then the processes $\{X_1(t), t \geq 0\}$ and $\{X_2(t), t \geq 0\}$ are independent Poisson processes with rates $\lambda_1 = \lambda p$ and $\lambda_2 = \lambda q$, respectively.
10. A stochastic process $\{X(t), t \geq 0\}$, where $X(t) = \sum_{i=1}^{N(t)} Y_i$, is said to be a compound Poisson process if (i) $\{N(t), t \geq 0\}$ is a Poisson process,
(ii) $\{Y_i, i \geq 1\}$ is a sequence of independent and identically distributed random variables and (iii) $\{N(t), t \geq 0\}$ and $\{Y_i, i \geq 1\}$ are independent processes.
11. For a compound Poisson process $\{X(t), t \geq 0\}$, $E(X(t)) = \lambda \mu t$ and $Var(X(t)) = \lambda t(\sigma^2 + \mu^2)$, where λ is a rate of a Poisson process and $E(Y_i) = \mu$ and $Var(Y_i) = \sigma^2$.
12. For a compound Poisson process $\{X(t), t \geq 0\}$, as $t \to \infty$,

$$Z(t) = \frac{X(t) - E(X(t))}{\sqrt{Var(X(t))}} \xrightarrow{L} Z \sim N(0, 1).$$

13. If $\{X_i(t), t \geq 0\}, i = 1, 2, \ldots, m$ are mutually independent compound Poisson processes with Poisson parameter λ_i and with probability density function or probability mass function of Y_i as $P_i(x),\ i = 1, 2, \ldots, m$, then $\{X(t) = X_1(t) + X_1(t) + \cdots X_m(t) = \sum_{i=1}^{N(t)} U_i, t \geq 0\}$ is a compound Poisson process with rate of Poisson process $\{N(t), t \geq 0\}$ as $\lambda = \sum_{i=1}^{m} \lambda_i$ and $P(x) = \sum_{i=1}^{m} \lambda_i P_i(x)/\lambda$ as a common distribution of U_i.

7.8 Conceptual Exercises

7.8.1 Suppose that customers arrive at a facility according to a Poisson process $\{X(t), t \geq 0\}$ with rate 6 per hour. (i) Find the probability that 8 customers arrive in one hour. (ii) Find the probability that 4 customers arrive in one hour and 10 customers arrive by the end of three hours.
(iii) If 10 customers arrived by the end of the first three hours, find the probability that 3 customers arrived in the first hour. (iv) If 2 customers arrived by the end of the first hour, find the probability that 6 customers will arrive in next two hours. (v) Find the probability that at most 5 customers arrive in one hour. (vi) If more than 2 customers arrived by the end of the first hour, find the probability that more than 6 customers will arrive in the next two hours. (vii) Find $E(X(2))$, $Var(X(2))$ and $Cov(X(1), X(2))$.

7.8.2 Shocks occur to a system according to a Poisson process with rate λ. Suppose that the system survives each shock with probability α, independently of other shocks, so that its probability of surviving k shocks is α^k. What is the probability that the system survives up to time t?

7.8.3 A device fails when a cumulative effect of k shocks occurs. If the shocks occur according to a Poisson process with rate λ, what is the probability density function of the life T of the device?

7.8.4 A radioactive source emits particles according to a Poisson process with rate $\lambda = 2$ particles per minute. (i) What is the probability that the first particle appears after three minutes? (ii) What is the probability that the first particle appears some time after three minutes but before six minutes? (iii) What is the probability that exactly one particle is emitted in the three to five minutes interval?

7.8.5 Customers arrive at a mall according to a Poisson process with rate $\lambda = 10$ per hour. The store opens at 10:00 a.m. (i) Find the probability that exactly one customer has arrived by 10:15 and a total of 10 have arrived by 11 a.m. (ii) Suppose it is known that a single customer entered during the first hour. What is the conditional probability that this person entered during the first thirty minutes? (iii) Given that 3 customers arrived during the first hour, what is the conditional probability that the first customer arrived in first 15 min, the second customer arrived during the first 20–30 min and the third customer arrived during the first 30–45 min interval? (iv) Given that 4 customers arrived during the first hour, what is the conditional probability that first customer arrived in first 15 min, second and the third customers arrived during the first 20–30 min and the fourth customer arrived during the first 30–45 min?

7.8.6 Suppose $\{X(t), t \geq 0\}$ is a Poisson process with rate 3 per hour. Find the conditional probability that two events occurred in the first hour, given that five events occurred in the first three hours.

7.8.7 Customers arrive at a service facility according to a Poisson process with rate $\lambda = 5$ per hour. Given that 12 customers arrived during the first two hours of service, what is the conditional probability that 5 customers arrived during the first hour?

7.8.8 Customers arrive at a bank according to a Poisson process with rate $\lambda = 10$ per hour. Find the probability that more than 5 customers arrive during a period of 15 min.

7.8.9 A critical component on a submarine has an operating lifetime that is exponentially distributed with mean 0.5 years. As soon as a component fails, it is replaced by a new one having statistically identical properties. What is the smallest number of spare components that the submarine should stock if it is leaving for a one-year tour and wishes that the probability of number failures exceeding the number of spare components to be less than 0.02?

7.8.10 For a Poisson process with rate λ, given that $X(t) = n$, find the density function for W, the time of occurrence of the rth event. What is its expectation? Assume that $r < n$.

7.8.11 Suppose $\{X(t), t \geq 0\}$ is a Poisson process with rate λ and $X(1) = n$. For $n = 1, 2, \ldots,$ determine the mean of the first arrival time W.

7.8.12 Suppose $\{X(t), t \geq 0\}$ is a Poisson process with rate λ and $W(t) = X(t+1) - X(t)$. Find $Cov(W(t), W(t+s))$.

7.8.13 Customers arrive at a bank according to a Poisson with rate λ. Suppose two customers arrived during the first hour. What is the probability that (i) both arrived during the first 20 min? (ii) at least one arrived during the first 20 min?

7.8.14 Events occur according to a Poisson process with rate $\lambda = 2$ per hour. (i) What is the probability that no event occurs between 8 p.m. and 9 p.m.? (ii) Starting at noon, what is the expected time at which the fourth event occurs? (iii) What is the probability that two or more events occur between 6 p.m. and 8 p.m.?

7.8.15 The number of industrial accidents in a factory is adequately modeled as a Poisson process with a rate of 1.3 per year. (i) What is the probability that there are no accidents in one year? (ii) What is the expected time (in days) between two consecutive accidents?

7.8.16 The interval between successive train arrivals at a station is uniformly distributed on (0, 1), time unit being one hour. Passengers arrive according to a Poisson process $\{X(t), t \geq 0\}$ with rate 70 per hour. Suppose a train has just left the station. If Y denotes the number of people who get on the next train. Find $E(Y)$ and $Var(Y)$.

7.8.17 If an individual has never had a previous automobile accident, then the probability that he or she has an accident in the next h time units, units in years, is $\beta h + o(h)$. On the other hand, if he or she has ever had a previous accident, then the probability is $\alpha h + o(h)$. Assuming that the occurrence of accidents in both the cases is modeled by independent Poisson processes, find the expected number of accidents an individual has by time t.

7.8.18 Customers demanding service at a central processing facility arrive according to a Poisson process with rate $\lambda = 8$ per unit time. Independently, each customer is classified as high priority, depending on his/her purchase history, with probability $p = 0.2$ or low priority with probability $q = 1 - p = 0.8$.

What is the probability that three high priority and five low priority customers arrive during the first unit of time?

7.8.19 Suppose 45% of the cars on the road are of non-white color and 55% are of white color. Suppose $X(t)$ denotes the number of cars crossing at a given toll booth during $(0, t]$ and $\{X(t), t \geq 0\}$ is a Poisson process with rate 40 per hour. What is the probability that four white color cars and five non-white cars cross the intersection during a 10-minute interval?

7.8.20 Shocks occur to a system according to a Poisson process with rate λ. Each shock causes some damage to the system, and these damages accumulate. Suppose $N(t)$ denotes the number of shocks up to time t and Y_i denotes the damage caused by the ith shock and $X(t)$ is the total damage up to time t. Determine the mean and variance of the total damage up to time t when the individual shock damages are exponentially distributed with parameter θ.

7.8.21 For a collective risk model, the number of claims in a year has a Poisson distribution with rate 8. Claim amounts are mutually independent with two possible values 1000 and 2000 rupees with equal chance. (i) Find the probability that total claims in one year are 0, 1000, 2000 and 3000.
(ii) Find the probability generating function of total claims in one year and hence find the probability that total claims in one year are 0, 1000, 2000 and 3000.

7.8.22 Aggregate claims process $\{X(t), t \geq 0\}$ in a collective risk model is a compound Poisson process, where $N(1)$ has a Poisson distribution with mean $\lambda = 10$. The individual claim distribution is exponential with mean $\mu = 100$ units. Find the mean and variance of $X(1)$. Using normal approximation, find $P[X(1) > 1.3E(X(1))]$.

7.8.23 The frequency of car accidents in $(0, t]$ is modeled as a Poisson process. There are 90% good drivers which on the average commit 1 accident over a period of one year, while the similar rate for bad drivers is 3. If an accident occurs, the claim amount has lognormal distribution with location parameter 3 and scale parameter 2. Calculate the mean m and variance v of total claims over a period of one year.

7.8.24 Suppose that $\{X(t), t \geq 0\}$ is modeled as a compound Poisson process with rate $\lambda = 3$ and the probability mass function of Y_i is given by $p(x) = 0.1x, x = 1, 2, 3, 4$. Calculate probabilities that aggregate claims over a period of one year equal $0, 1, 2, 3$ units. Also, find the probability that aggregate claims exceed 3 units.

7.8.25 Customers arrive at a store as a group of 1 or 2 persons with equal probability. The arrival of groups is according to a Poisson process with rate 3 per 10 min. Find the probability that 4 customers arrive in 20 min.

7.9 Computational Exercises

7.9.1 Suppose $\{X(t), t \geq 0\}$ is a Poisson process with rate λ. Obtain a realization of the process for a fixed time interval $[0, T]$. Take four values of λ and compare the performance of the process. Draw the plot of realization in all the cases. Find $X(T)$ in all the cases. Comment on your findings.

7.9.2 Suppose $\{X(t), t \geq 0\}$ is a Poisson process with rate λ and is observed for T time units. Simulate the process m times and based on m values of $X(T)$, examine whether Poisson $Poi(\lambda T)$ distribution is a good model for $X(T)$.

7.9.3 Suppose customers arrive at an ATM center according to a Poisson process $\{N(t), t \geq 0\}$ with rate θ and suppose Y_i denotes the cash withdrawn by the ith customer. Then $X(t) = \sum_{i=1}^{N(t)} Y_i$ represents the cash withdrawn in $(0, t]$ and $\{X(t), t \geq 0\}$ is a compound Poisson process. Assuming that Y_i follows gamma distribution with scale parameter α and shape parameter λ, obtain a realization of the compound Poisson process for a fixed time interval $(0, T]$. Take suitable values for the parameters.

7.10 Multiple Choice Questions

Note: In each of the questions, multiple options may be correct.

7.10.1 Suppose $\{X(t), t \geq 0\}$ is a continuous time stochastic process with state space W. Following are three statements. (I) $X(0) = 0$.
(II)$P_{ii}(h) = 1 - \lambda h + o(h)$, $P_{i,i+1}(h) = \lambda h + o(h)$ and $P_{ij}(h) = o(h)$ for all $j \neq i + 1$. (III) $\{X(t), t \geq 0\}$ is a stochastic process with stationary and independent increments.
Which of the following options is correct? It is a homogeneous Poisson process with rate $\lambda > 0$ if

(a) Only (II) is true
(b) Both (II) and (III) are true
(c) Both (I) and (II) are true
(d) All three are true

7.10.2 Suppose $\{X(t), t \geq 0\}$ is a Poisson process with rate λ. Then which of the following options is/are correct?

(a) $E(X(t)) = \lambda t$
(b) $Var(X(t)) = \lambda t$
(c) $Cov(X(s), X(t)) = \lambda \min\{s, t\}$
(d) $Cov(X(s), X(t)) = \lambda |t - s|$

7.10.3 Suppose $\{X(t), t \geq 0\}$ is a Poisson process with rate λ. Following are three statements. (I) $E(X(t)) = \lambda t$. (II) $Cov(X(s), X(t)) = \lambda |t - s|$.

(III) $Cov(X(s), X(t)) = \lambda \min\{s, t\}$.
Which of the following options is correct?

(a) Only (I) and (II) are true
(b) Only (III) is true
(c) All three are true
(d) Only (I) and (III) are true

7.10.4 Following are two statements. (I) Poisson process is a stationary process. (II) Poisson process is a process with stationary and independent increments. Which of the following options is correct?

(a) Both (I) and (II) are false
(b) Both (I) and (II) are true
(c) (I) is true but (II) is false
(d) (I) is false but (II) is true

7.10.5 Which of the following statement is NOT correct? A Poisson process is

(a) a Markov process
(b) a process with stationary and independent increments
(c) a strongly stationary process
(d) a counting process

7.10.6 Suppose $\{X(t), t \geq 0\}$ is a Poisson process with rate λ and T_1 denotes the waiting time for the first event to occur. Which of the following options is/are correct if $s < t$?

(a) $P[T_1 < s|X(t) = 1] = 1 - e^{-\lambda s}$
(b) $P[T_1 < s|X(t) = 1] = \lambda s e^{-\lambda}$
(c) $P[T_1 < s|X(t) = 1] = s/t$
(d) $P[T_1 < s|X(t) = 1] = \lambda s/t$

7.10.7 Which of the following options is correct? Messages arrive on a mobile phone according to a Poisson process with rate of 5 messages per hour. The probability that no message arrives during 10:00 a.m. to 12:00 noon is

(a) $2e^{-5}$
(b) e^{-10}
(c) $1 - e^{-10}$
(d) $e^{-2/5}$

7.10.8 Which of the following options is correct? Messages arrive on a mobile phone according to a Poisson process with rate of 5 messages per hour. The probability that the first message in the afternoon arrives by 1:20 p.m. is

(a) $(20/3)e^{-5\times 4/3}$
(b) $e^{-5\times 4/3}$
(c) $100e^{-100}$
(d) $(5/3)e^{-5\times 1/3}$

7.10.9 Which of the following options is correct? The life time of a component of a machine is modeled by an exponential distribution with mean 2 per week. Failed component is immediately replaced by a new one. Then the probability that no component is replaced in two weeks is

(a) e^{-2}
(b) e^{-4}
(c) $e^{-0.5}$
(d) e^{-1}

7.10.10 Which of the following options is correct? Customers enter a store according to a Poisson process of rate $\lambda = 6$ per hour. Suppose it is known that a single customer entered during the first hour. Then the conditional probability that this person entered during the first fifteen minutes is

(a) $6/15$
(b) $1/4$
(c) $e^{-6/4}$
(d) $1 - e^{-90}$

7.10.11 Which of the following options is correct? Suppose auto arrives at a stand from north at a rate of 1 per minute according to a Poisson process and from south at a rate of 2 per minute according to a Poisson process. Suppose the two arrival processes are independent. Then the probability that a customer has to wait at the stand for more than two minutes is

(a) e^{-3}
(b) e^{-6}
(c) e^{-4}
(d) $1 - e^{-3}$

7.10.12 Which of the following options is correct? A radioactive source emits particles at a rate of 5 per minute according to a Poisson process. Each particle emitted has a probability 0.6 of being recorded. The probability that in a 4 minute interval, 10 particles are recorded is

(a) $e^{-5}5^{10}/10!$
(b) $e^{-3}3^{10}/10!$
(c) $e^{-20}20^{10}/10!$
(d) $e^{-12}12^{10}/10!$

7.10.13 Which of the following options is correct? Suppose $\{X(t), t \geq 0\}$ is a Poisson process with rate 2. Suppose it is known that $X(1) = 5$. Then the mean of the first arrival time is

(a) $1/5$
(b) $1/6$
(c) 2
(d) $1/2$

7.10.14 Which of the following options is correct? If an individual has never had a previous automobile accident, then the probability that he or she has an accident in the next h time units, units in years, is $0.5h + o(h)$. On the other hand, if he or she has ever had a previous accident, then the probability is $0.8h + o(h)$. Assuming that the occurrence of accidents in both the cases is modeled by independent Poisson processes, the expected number of accidents an individual has in two years is

(a) 0.8
(b) 1.3
(c) 2.6
(d) 0.65

7.10.15 Customers arrive at a bank according to a Poisson process with rate 10 per hour. If two customers arrived during the first hour, the probability that both arrived during the first 20 min is

(a) $1/18$
(b) $1/36$
(c) $1/9$
(d) $1/3$

References

1. Cinlar, E. (1975). *Introduction to stochastic processes*. New Jersey: Prentice Hall.
2. Cressie, N. A. C. (1993). *Statistics for spatial data*. Interscience, New York: Wiley.
3. Feller, W. (2000). *An Introduction to Probability Theory and its Applications* (2nd ed., Vol. II). Singapore: Wiley.
4. Karlin, S., & Taylor, H. M. (1975). *A first course in stochastic processes*. New York: Academic.
5. Kulkarni, V. G. (2011). *Introduction to modeling and analysis of stochastic systems*. New York: Springer.
6. Ross, S. M. (2014). *Introduction to probability models* (11th ed.). New York: Academic.
7. Taylor, H. N., & Karlin, S. (1984). *An introduction to stochastic modeling*. New York: Academic.

Chapter 8
Birth and Death Process

8.1 Introduction

In Chap. 6, we studied the general theory of continuous time Markov chains and noted that it is completely specified by either infinitesimal transition probabilities or intensity rates. In Chap. 7, we discussed one such process, a Poisson process in detail. In this chapter, we present some more illustrations of continuous time Markov chains, which have applications in various areas. These include pure birth process, Yule Furry process, pure death process, birth and death process and its variations. We have already noted the infinitesimal transition probabilities and intensity rates of these processes in Chap. 6. In the present chapter, we derive a system of differential equations to obtain an expression for $P[X(t) = n]$ and then study its limit, when it exists.

A birth-death process is one of the most commonly used model for population growth, in which a population of individuals increases due to births and decreases due to deaths. In view of its applications in population growth models, increment is labeled as birth and decrement is labeled as death. Increase and decrease may be due to in-migration and out-migration. Occurrence of these events and time of occurrence of these events, both are governed by random mechanism. Thus, if $X(t)$ denotes the size of the population at time t, then for each $t > 0$, $X(t)$ is a random variable and we are interested in the evolution of $X(t)$ over time which motivates the study of the process $\{X(t), t \geq 0\}$.

Apart from the applications of a birth-death process in modeling population growth, it has applications in many other areas. The most frequently cited illustration of a birth and death process is in queuing theory. Suppose we have a queuing system with one service counter. Its state at any time t is represented by the number of customers in the system at that time, that is, number of people waiting in the queue and the one getting service. The terms customer, server, service time and waiting time have different meanings in different applications. For example, customers may be patients arriving at a doctor's clinic. Then the doctor is a server and the service

S. Madhira and S. Deshmukh, *Introduction to Stochastic Processes Using R*,
https://doi.org/10.1007/978-981-99-5601-2_8

times are the examination times. Customers may be aircraft arriving at an airport requiring to land, then the runways are the servers and waiting time is the time spent by an aircraft circling around the airport.

Suppose there are n customers in a system. Suppose new individuals enter the system and people in the system leave according to the following scheme. When there are n persons in the system, the random time between two arrivals is exponentially distributed with mean $1/\lambda_n$ and is independent of the time until the next departure. The random time between two departures is also exponentially distributed with mean $1/\mu_n$. If $X(t)$ denotes the number of people in the system at time t, then the stochastic process $\{X(t), t \geq 0\}$ is modeled as a birth-death process. The parameters $\lambda_n, n \geq 0$ are labeled as arrival or birth rates and the parameters $\mu_n, n \geq 1$ are labeled as departure or death rates. Different forms of these rates lead to various versions of a birth-death process and queuing models, which have different suitable labels.

In the next section, we begin with a pure birth process, a simple version of birth-death process. Sections 9.3 and 9.4 are devoted to a death process and a birth-death process, respectively. Section 9.5 is concerned with a linear birth-death process and its variations. In Sect. 9.6, we discuss the long run behavior of a birth-death process. Section 9.7 presents `R` codes for realizations of these processes.

8.2 Birth Process

A natural generalization of a Poisson process is to permit the chance of an event occurring at a given instant of time to depend upon the number of events that have occurred up to that time. An example of this phenomenon is the reproduction of living organisms and hence the process is named as a birth process. In this process, under certain conditions, such as sufficient food, no mortality, no migration, the infinitesimal probabilities of a birth at a given instant depend on the population size at that time. Following is the definition of a pure birth process.

Definition 8.2.1 Pure Birth Process: Suppose $\{X(t), t \geq 0\}$ is a time homogeneous continuous time Markov chain with state space $S = W = \{0, 1, 2, \ldots, \}$, $X(0) = a \geq 0$ and the infinitesimal transition probabilities,

$$\begin{aligned} P_{i,i+1}(h) &= P[X(t+h) = i+1 \big| X(t) = i] = \lambda_i h + o(h) \\ P_{ii}(h) &= P[X(t+h) = i \big| X(t) = i] = 1 - \lambda_i h + o(h) \\ P_{ij}(h) &= P[X(t+h) = j \big| X(t) = i] = o(h) \;\; \forall \;\; j \neq i, \; i+1 \, . \end{aligned}$$

Then $\{X(t), t \geq 0\}$ is said to be a pure birth process with birth rates $\{\lambda_i, i \geq 0\}$.

In general, $\forall \; a \geq 0$, $X(t)$ denotes the population size at t. If $X(0) = 0$, then it is also the number of births in $(0, t]$. We discuss the two cases when $X(0) = 0$ and $X(0) > 0$ separately. We note that when $X(0) = 0$ and $\lambda_i = \lambda \;\; \forall \;\; i \in S$, then pure birth process reduces to a Poisson process with rate λ.

When $X(0) = 0$, the infinitesimal generator matrix Q and the transition probability matrix P of the corresponding embedded Markov chain are given by

$$Q = \begin{array}{c} \\ 0 \\ 1 \\ 2 \\ \vdots \end{array} \begin{array}{c} \begin{array}{cccc} 0 & 1 & 2 & \cdots \end{array} \\ \begin{pmatrix} -\lambda_0 & \lambda_0 & 0 & \cdots \\ 0 & -\lambda_1 & \lambda_1 & \cdots \\ 0 & 0 & -\lambda_2 & \cdots \\ \vdots & \vdots & \vdots & \vdots \end{pmatrix} \end{array} \quad \& \quad P = \begin{array}{c} \\ 0 \\ 1 \\ 2 \\ \vdots \end{array} \begin{array}{c} \begin{array}{cccc} 0 & 1 & 2 & \cdots \end{array} \\ \begin{pmatrix} 0 & 1 & 0 & \cdots \\ 0 & 0 & 1 & \cdots \\ 0 & 0 & 0 & \cdots \\ \vdots & \vdots & \vdots & \vdots \end{pmatrix} \end{array}.$$

From the matrix P, we note that as in a Poisson process, starting from 0, the next state will be 1, next will be 2 and so on. Thus, $\{X(t), t \geq 0\}$ is an increasing process and increases by jumps of size 1 only. Further, being a Markov process, we note that with $X(0) = 0$, the process remains in state 0 for a random time T_0, which has exponential distribution with scale parameter λ_0, then it jumps to 1 with probability 1, remains in state 1 for random time T_1, having exponential distribution with scale parameter λ_1 and then jumps to 2 with probability 1 and so on.

From the transition probability matrix P of the embedded Markov chain, we note that

$$0 \to 1 \to 2 \to 3 \to 4 \cdots \quad \Rightarrow \quad i \to i + k, \quad i = 0, 1, 2, \ldots, \quad k = 1, 2, \ldots.$$

Thus, any state $i \in S$ leads to $i + k$ for $k = 1, 2, \ldots$, however, $i + k$ does not lead to i for any $i \in S$. Thus, states in S do not communicate. Further, for any $i \in S$, $i \to i + 1$, however, $i + 1 \nrightarrow i$, which implies that each state in S is an inessential state and hence is a transient state. Hence, by Theorem 3.3.6, a stationary distribution of the embedded Markov chain does not exist. Since all the states in the embedded Markov chain are transient, all states in the pure birth process are also transient. Consequently, its long run distribution does not exist. A pure birth process is thus a non-ergodic process.

Suppose $P_k(t) = P[X(t) = k]$. With the assumption $X(0) = 0$ and in view of time homogeneity,

$$P_k(t) = P[X(t) = k] = P[X(t) = k \big| X(0) = 0] = P_{0k}(t).$$

It is the probability of transition from 0 to k in t time units, it is the same as the probability of k births in $(0, t]$. In the following theorem, we find an expression for $P_k(t)$, by deriving and solving a system of differential equations.

Theorem 8.2.1 *Suppose $\{X(t), t \geq 0\}$ is a pure birth process with $X(0) = 0$ and birth rates λ_i, $i \in W$, which are assumed to be distinct. Then*

$$\begin{aligned} P_0(t) &= e^{-\lambda_0 t} \\ P_k(t) &= \lambda_0 \lambda_1 \cdots \lambda_{k-1} \sum_{j=0}^{k} e^{-\lambda_j t} \left(\prod_{i=0, i \neq j}^{k} (\lambda_i - \lambda_j) \right)^{-1}, \quad k = 1, 2, \ldots. \end{aligned}$$

Proof Using the postulates of the pure birth process, we get for $k \geq 1$,

$$\begin{aligned} P_k(t+h) &= P[X(t+h)=k] = P[X(t+h)=k \big| X(t)=k]P[X(t)=k] \\ &\quad + P[X(t+h)=k \big| X(t)=k-1]P[X(t)=k-1] + o(h) \\ &= P_{kk}(h)P_k(t) + P_{k-1,k}(h)P_{k-1}(t) + o(h) \\ &= (1-\lambda_k h + o(h))P_k(t) + (\lambda_{k-1}h + o(h))P_{k-1}(t) + o(h) \\ \Rightarrow \quad P_k'(t) &= -\lambda_k P_k(t) + \lambda_{k-1} P_{k-1}(t) \,. \end{aligned}$$

Suppose $k = 0$. An event occurs 0 times in $(0, t+h]$, if it does not occur in $(0, t]$ and it does not occur in $(t, t+h]$. Thus,

$$\begin{aligned} P_0(t+h) &= P[X(t+h)=0] = P[X(t+h)=0 \big| X(t)=0]P[X(t)=0] \\ &= (1-\lambda_0 h + o(h))P_0(t) \;\Rightarrow\; P_0'(t) = -\lambda_0 P_0(t) \,. \end{aligned}$$

Thus, to find $P_k(t) = P[X(t) = k]$, we have to solve the system of differential equations given by

$$P_0'(t) = -\lambda_0 P_0(t) \;\&\; P_k'(t) = -\lambda_k P_k(t) + \lambda_{k-1} P_{k-1}(t), \;\; k \geq 1 \,,$$

subject to the conditions

$$P[X(0)=0] = P_0(0) = 1 \;\&\; P[X(0)=k] = P_k(0) = 0 \;\forall\; k \geq 1.$$

Now,

$$P_0'(t) = -\lambda_0 P_0(t) \;\Rightarrow\; P_0(t) = ce^{-\lambda_0 t} = e^{-\lambda_0 t} \text{ as } P_0(0) = 1 \;\Rightarrow\; c = 1 \,.$$

Thus, $P_0(t) = P[X(t) = 0] = e^{-\lambda_0 t}$. For $k = 1$, we have the differential equation $P_1'(t) = -\lambda_1 P_1(t) + \lambda_0 P_0(t)$ which is solved using similar arguments as for the Poisson process. Thus,

$$\begin{aligned} P_1'(t) + \lambda_1 P_1(t) = \lambda_0 P_0(t) &\Rightarrow e^{\lambda_1 t}\left(P_1'(t) + \lambda_1 P_1(t)\right) = e^{\lambda_1 t}\lambda_0 P_0(t) \\ &\Rightarrow \frac{d}{dt} e^{\lambda_1 t} P_1(t) = e^{\lambda_1 t}\lambda_0 P_0(t) \\ &\Rightarrow e^{\lambda_1 t} P_1(t) = \int_0^t e^{\lambda_1 u}\lambda_0 P_0(u)\, du + c \\ &\Rightarrow P_1(t) = \lambda_0 e^{-\lambda_1 t} \int_0^t e^{\lambda_1 u} P_0(u)\, du + ce^{-\lambda_1 t} \,, \end{aligned}$$

where c is the constant of integration which is determined by the boundary conditions $P_0(0) = 1$ and $P_1(0) = 0$.

$$P_1(0) = 0 \;\Rightarrow\; c = 0 \;\Rightarrow\; P_1(t) = \lambda_0 e^{-\lambda_1 t}\int_0^t e^{\lambda_1 u} P_0(u)\,du.$$

Proceeding on similar lines, it can be shown that
$P_k(t) = e^{-\lambda_k t}\int_0^t \lambda_{k-1} P_{k-1}(u) e^{\lambda_k u}\,du$. Now,

$$\begin{aligned}
P_1(t) &= e^{-\lambda_1 t}\int_0^t \lambda_0 P_0(u) e^{\lambda_1 u}\,du = e^{-\lambda_1 t}\int_0^t \lambda_0 e^{-\lambda_0 u} e^{\lambda_1 u}\,du\\
&= e^{-\lambda_1 t}\int_0^t \lambda_0 e^{-(\lambda_0-\lambda_1)u}\,du = \frac{\lambda_0 e^{-\lambda_1 t}}{\lambda_0-\lambda_1}\left[1-e^{-(\lambda_0-\lambda_1)t}\right]\\
&= \frac{\lambda_0}{\lambda_0-\lambda_1}\left[e^{-\lambda_1 t}-e^{-\lambda_0 t}\right] = \lambda_0\left[\frac{e^{-\lambda_0 t}}{\lambda_1-\lambda_0}+\frac{e^{-\lambda_1 t}}{\lambda_0-\lambda_1}\right]\\
P_2(t) &= e^{-\lambda_2 t}\int_0^t \lambda_1 P_1(u) e^{\lambda_2 u}\,du\\
&= e^{-\lambda_2 t}\int_0^t \lambda_1 \frac{\lambda_0}{\lambda_1-\lambda_0}\left[e^{-\lambda_0 u}-e^{-\lambda_1 u}\right] e^{\lambda_2 u}\,du\\
&= e^{-\lambda_2 t}\lambda_1\lambda_0\int_0^t \left[\frac{e^{-(\lambda_0-\lambda_2)u}}{\lambda_1-\lambda_0}+\frac{e^{-(\lambda_1-\lambda_2)u}}{\lambda_0-\lambda_1}\right]du\\
&= e^{-\lambda_2 t}\lambda_1\lambda_0\left[\frac{1-e^{-(\lambda_0-\lambda_2)t}}{(\lambda_1-\lambda_0)(\lambda_0-\lambda_2)}+\frac{1-e^{-(\lambda_1-\lambda_2)t}}{(\lambda_0-\lambda_1)(\lambda_1-\lambda_2)}\right]\\
&= \lambda_1\lambda_0\left[\frac{e^{-\lambda_2 t}-e^{-\lambda_0 t}}{(\lambda_1-\lambda_0)(\lambda_0-\lambda_2)}+\frac{e^{-\lambda_2 t}-e^{-\lambda_1 t}}{(\lambda_0-\lambda_1)(\lambda_1-\lambda_2)}\right]\\
&= \lambda_0\lambda_1\left[\frac{e^{-\lambda_0 t}}{(\lambda_1-\lambda_0)(\lambda_2-\lambda_0)}+\frac{e^{-\lambda_1 t}}{(\lambda_0-\lambda_1)(\lambda_2-\lambda_1)}+\frac{e^{-\lambda_2 t}}{(\lambda_0-\lambda_2)(\lambda_1-\lambda_2)}\right].
\end{aligned}$$

We now assume for some $k \geq 1$,

$$P_k(t) = \lambda_0\lambda_1\cdots\lambda_{k-1}\sum_{j=0}^{k} e^{-\lambda_j t}\left(\prod_{i=0, i\neq j}^{k}(\lambda_i-\lambda_j)\right)^{-1},$$

which is true for $k = 1, 2$. Now,

$$
\begin{aligned}
P_{k+1}(t) &= e^{-\lambda_{k+1}t}\int_0^t \lambda_k P_k(u)e^{\lambda_{k+1}u}\,du\\
&= e^{-\lambda_{k+1}t}\lambda_0\lambda_1\cdots\lambda_{k-1}\lambda_k\sum_{j=0}^{k}\left(\prod_{i=0,i\neq j}^{k}(\lambda_i-\lambda_j)\right)^{-1}\int_0^t e^{-\lambda_j u}e^{\lambda_{k+1}u}\,du\\
&= e^{-\lambda_{k+1}t}\lambda_0\lambda_1\cdots\lambda_k\sum_{j=0}^{k}\left(\prod_{i=0,i\neq j}^{k}(\lambda_i-\lambda_j)\right)^{-1}\int_0^t e^{-(\lambda_j-\lambda_{k+1})u}\,du\\
&= e^{-\lambda_{k+1}t}\lambda_0\lambda_1\cdots\lambda_k\sum_{j=0}^{k}\left(\prod_{i=0,i\neq j}^{k}(\lambda_i-\lambda_j)\right)^{-1}\frac{[1-e^{-(\lambda_j-\lambda_{k+1})t}]}{(\lambda_j-\lambda_{k+1})}.
\end{aligned}
$$

$P_{k+1}(t)$ further simplifies as follows:

$$
\begin{aligned}
P_{k+1}(t) &= \lambda_0\lambda_1\cdots\lambda_k\left[\sum_{j=0}^{k}e^{-\lambda_j t}\left(\prod_{i=0,i\neq j}^{k+1}(\lambda_i-\lambda_j)\right)^{-1}\right]\\
&\quad-\lambda_0\lambda_1\cdots\lambda_k\left[\sum_{j=0}^{k}e^{-\lambda_{k+1}t}\left(\prod_{i=0,i\neq j}^{k+1}(\lambda_i-\lambda_j)\right)^{-1}\right]\\
&= \lambda_0\lambda_1\cdots\lambda_k\sum_{j=0}^{k+1}e^{-\lambda_j t}\left(\prod_{i=0,i\neq j}^{k+1}(\lambda_i-\lambda_j)\right)^{-1},
\end{aligned}
$$

where the last step follows using partial fractions. Hence, by induction, we get

$$
P_k(t) = \lambda_0\lambda_1\cdots\lambda_{k-1}\sum_{j=0}^{k}e^{-\lambda_j t}\left(\prod_{i=0,i\neq j}^{k}(\lambda_i-\lambda_j)\right)^{-1},\quad k=1,2,\ldots,
$$

$$
\text{and}\quad P_0(t) = e^{-\lambda_0 t}.
$$

□

Suppose T_k denotes the time between k-th and $(k+1)$-th birth, $k \geq 1$. We have noted above that

$$
P_0(t) = P[X(t)=0] = e^{-\lambda_0 t} = P[T_0 > t],
$$

where T_0 is the sojourn time in state 0, that is, waiting time for the first birth and it has exponential distribution with rate λ_0. From the Markov property of $\{X(t), t \geq 0\}$, it follows that T_k also has exponential distribution with rate λ_k and $\{T_k, k \geq 0\}$ are independent random variables. Thus, $S_k = \sum_{j=0}^{k-1} T_j$ is the waiting time for the k-th birth and its expectation is $E(S_k) = \sum_{j=0}^{k-1} 1/\lambda_j$. It is to be noted that $\sum_{j=0}^{\infty} 1/\lambda_j$ is the expected time for population size to become infinite. If this is finite, then the pure birth process explodes to ∞ in finite time, that is, $X(t)$ may be infinite with positive probability or $\sum_{j=0}^{\infty} P_k(t) < 1$. Thus heuristically,

$$\sum_{j=0}^{\infty} P_j(t) = 1 \iff \sum_{j=0}^{\infty} 1/\lambda_j = \infty. \tag{8.2.1}$$

For a rigorous proof refer to Feller [2].

Suppose $X(0) = a > 0$ and a random variable $Y(t)$ is defined as $Y(t) = X(t) - a$. Then the above theorem holds for the process $\{Y(t), t \geq 0\}$, so that for $k = a + 1, a + 2, \ldots$, $P_k^{(a)}(t) = P[X(t) = k|X(0) = a]$ is as follows:

$$P_k^{(a)}(t) = \lambda_a \lambda_{a+1} \cdots \lambda_{k-1} \sum_{j=a}^{k} e^{-\lambda_j t} \left(\prod_{i=a, i \neq j}^{k} (\lambda_i - \lambda_j) \right)^{-1}$$

$$\text{and} \quad P_a^{(a)}(t) = e^{-\lambda_a t}. \tag{8.2.2}$$

We now discuss one more approach to find $P_k^{(a)}(t)$ in Theorem 8.2.2. It is based on the fact that sojourn time random variables are independent exponential random variables. We assume that $\lambda_i \neq \lambda_j$ when $i \neq j$ and use the result stated in Lemma 8.2.1, Ross [4] related to the distribution of sum of independent random variables having exponential distribution with distinct rate parameters. Distribution of such a sum is known as a hypoexponential distribution.

Lemma 8.2.1 *Suppose $\{X_i, i = 1, \ldots, n\}$ are independent random variables having exponential distribution with respective rates λ_i, $i = 1, \ldots, n$. Suppose $\lambda_i \neq \lambda_j$ when $i \neq j$. Then the probability density function and survival function of $X = \sum_{i=1}^{n} X_i$ for $x \geq 0$ are given by*

$$f_X(x) = \sum_{i=1}^{n} C_{in} \lambda_i e^{-\lambda_i x}, \; P[X > x] = \sum_{i=1}^{n} C_{in} e^{-\lambda_i x}, \; \textit{where}\; C_{in} = \prod_{j=1, j \neq i}^{n} \frac{\lambda_j}{\lambda_j - \lambda_i}.$$

Theorem 8.2.2 *Suppose $\{X(t), t \geq 0\}$ is a pure birth process with birth rates λ_i, $i \in S$ and $X(0) = 0$. Then $P_0(t) = e^{-\lambda_0 t}$ and*

$$P_k(t) = \lambda_0 \lambda_1 \cdots \lambda_{k-1} \sum_{j=0}^{k} e^{-\lambda_j t} \left(\prod_{i=0, i \neq j}^{k} (\lambda_i - \lambda_j) \right)^{-1} \quad k = 1, 2, \ldots.$$

Proof Since the distribution of T_0 is exponential, $P_0(t) = P[T_0 > t] = e^{-\lambda_0 t}$. To find $P_k(t)$ for $k \geq 1$, note that $X(t) < k \iff T_0 + T_1 + \cdots + T_{k-1} > t$. Hence, by Lemma 8.2.1, we have

$$
\begin{aligned}
P_k(t) &= P[X(t) < k+1|X(0)=0] - P[X(t) < k|X(0)=0]\\
&= \sum_{i=0}^{k} e^{-\lambda_i t} \prod_{r=0,r\neq i}^{k} \frac{\lambda_r}{\lambda_r-\lambda_i} - \sum_{i=0}^{k-1} e^{-\lambda_i t} \prod_{r=0,r\neq i}^{k-1} \frac{\lambda_r}{\lambda_r-\lambda_i}\\
&= e^{-\lambda_k t} \prod_{r=0}^{k-1} \frac{\lambda_r}{\lambda_r-\lambda_k} + \sum_{i=0}^{k-1} e^{-\lambda_i t} \prod_{r=0,r\neq i}^{k} \frac{\lambda_r}{\lambda_r-\lambda_i} - \sum_{i=0}^{k-1} e^{-\lambda_i t} \prod_{r=0,r\neq i}^{k-1} \frac{\lambda_r}{\lambda_r-\lambda_i}\\
&= e^{-\lambda_k t} \prod_{r=0}^{k-1} \frac{\lambda_r}{\lambda_r-\lambda_k} + \sum_{i=0}^{k-1} e^{-\lambda_i t} \prod_{r=0,r\neq i}^{k-1} \frac{\lambda_r}{\lambda_r-\lambda_i} \left(\frac{\lambda_k}{\lambda_k-\lambda_i} - 1\right)\\
&= e^{-\lambda_k t} \prod_{r=0}^{k-1} \frac{\lambda_r}{\lambda_r-\lambda_k} + \sum_{i=0}^{k-1} e^{-\lambda_i t} \prod_{r=0,r\neq i}^{k-1} \frac{\lambda_r}{\lambda_r-\lambda_i} \left(\frac{\lambda_i}{\lambda_k-\lambda_i}\right)\\
&= \lambda_0\lambda_1\cdots\lambda_{k-1} \sum_{i=0}^{k} e^{-\lambda_i t} \prod_{r=0,r\neq i}^{k} \frac{1}{\lambda_r-\lambda_i}
\end{aligned}
$$

which is the same as the expression in Theorem 8.2.1. □

On similar lines, when $X(0) = a \geq 1$, we get $P_k^{(a)}(t)$ as in Eq. (8.2.2). The expression for $P_k^{(1)}(t)$ gets simplified with specific values of λ_i in the special case of a pure birth process. Suppose $X(0) = 1$ and $\lambda_k = k\lambda,\ k \geq 1$, then the pure birth process is known as a linear birth process or a Yule Furry process, in honor of scientists who studied this process for population growth. The Yule Furry process arises in physics and biology and describes the growth of a population in which each member has a probability $\lambda h + o(h)$ of giving birth to a new member during a time interval of length h, $\lambda > 0$. Assuming independence and no interaction among members of the population, the binomial theorem gives

$$
\begin{aligned}
P[X(t+h) - X(t) = 1\big|X(t) = k] &= \binom{k}{1}(\lambda h + o(h))(1-(\lambda h + o(h)))^{k-1}\\
&= k\lambda h + o(h)\ .
\end{aligned}
$$

Following is the definition of a Yule Furry process.

Definition 8.2.2 Yule Furry Process: Suppose $\{X(t), t \geq 0\}$ is a time homogeneous continuous time Markov chain with state space $\{1, 2, \ldots, \}$, $X(0) = 1$ and the following infinitesimal transition probabilities.

$$
\begin{aligned}
P_{i,i+1}(h) &= P[X(t+h) = i+1\big|X(t) = i] = i\lambda h + o(h)\\
P_{ii}(h) &= P[X(t+h) = i\big|X(t) = i] = 1 - i\lambda h + o(h)\\
P_{ij}(h) &= P[X(t+h) = j\big|X(t) = i] = o(h)\ \forall\ j \neq i, i+1\ .
\end{aligned}
$$

Then $\{X(t), t \geq 0\}$ is known as a Yule Furry process.

For a Yule Furry process, the transition rates are $\lambda_i = i\lambda\, i > 0$, that is, the population birth rate λ_i is directly proportional to the population size, the proportionality constant being the individual birth rate λ. As such, a Yule process forms a stochastic analogue of the deterministic population growth model given by the Malthusian law, which is represented by the differential equation

$$\frac{dX(t)}{dt} = \alpha X(t),$$

where $X(t)$ is non-random population size at time t. In the deterministic model, the rate $\frac{dX(t)}{dt}$ of population growth is directly proportional to population size $X(t)$. In the stochastic model, the infinitesimal deterministic increase $dX(t)$ is replaced by the probability of a unit increase during the infinitesimal time interval dt.

Suppose $P_k(t) = P[X(t) = k] = P[X(t) = k|X(0) = 1]$. In the following theorem we derive its expression for Yule Furry process.

Theorem 8.2.3 *Suppose* $\{X(t), t \geq 0\}$ *is a Yule Furry process with birth rate* λ *and* $X(0) = 1$. *Then for fixed* t, *the distribution of* $X(t)$ *is geometric with parameter* $p = e^{-\lambda t}$ *and support* $\{1, 2, \ldots, \}$.

Proof The differential equations derived for a pure birth process, with $\lambda_i = i\lambda$ $\forall\ i \in S$ reduce as follows. Note that $P_0(t) = 0$ for all t, hence $P_0'(t) = 0$. Further,

$$P_1'(t) = -\lambda P_1(t)\ \&\ \ P_k'(t) = -k\lambda P_k(t) + (k-1)\lambda P_{k-1}(t),\ k > 1\,. \tag{8.2.3}$$

Thus to find $P_k(t)$, we solve this system of differential equations subject to the conditions $P_0(1) = 1$ and $P_k(0) = 0\ \forall\ k > 1$. The solution of the equation $P_1'(t) = -\lambda P_1(t)$ subject to the condition $P_0(1) = 1$, is $P_1(t) = e^{-\lambda t}$. For $k = 2, 3\ldots$,

$$\begin{aligned} P_k(t) &= ce^{-k\lambda t} + e^{-k\lambda t}\int_0^t (k-1)\lambda P_{k-1}(u)e^{k\lambda u}\, du \\ &= e^{-k\lambda t}\int_0^t (k-1)\lambda P_{k-1}(u)e^{k\lambda u}\, du\,, \end{aligned} \tag{8.2.4}$$

as $P_k(0) = 0\ \forall\ k > 1$ implies $c = 0$. Now,

$$\begin{aligned} P_2(t) &= e^{-2\lambda t}\int_0^t \lambda P_1(u)e^{2\lambda u}\, du = e^{-2\lambda t}\int_0^t \lambda e^{-\lambda u}e^{2\lambda u}\, du \\ &= e^{-2\lambda t}\lambda\int_0^t e^{\lambda u}\, du = e^{-2\lambda t}(e^{\lambda t} - 1) = e^{-\lambda t}(1 - e^{-\lambda t})\,. \end{aligned}$$

We assume that $P_k(t) = e^{-\lambda t}(1 - e^{-\lambda t})^{k-1}$, which is true for $k = 1, 2$ and prove the result by induction. Now, we have the following two methods to prove the result.
Method I: In the differential equation given in (8.2.3), we substitute
$P_k(t) = e^{-\lambda t}(1 - e^{-\lambda t})^{k-1}$. Thus,

$$
\begin{aligned}
P'_{k+1}(t) &= -(k+1)\lambda P_{k+1}(t) + k\lambda e^{-\lambda t}(1-e^{-\lambda t})^{k-1} \\
\Rightarrow\ e^{(k+1)\lambda t} P'_{k+1}(t) &= e^{(k+1)\lambda t}\left(-(k+1)\lambda P_{k+1}(t)\right) \\
&\quad + e^{(k+1)\lambda t}\left(k\lambda e^{-\lambda t}(1-e^{-\lambda t})^{k-1}\right) \\
\Rightarrow\ \frac{d}{dt}\left(e^{(k+1)\lambda t} P_{k+1}(t)\right) &= k\lambda e^{k\lambda t}(1-e^{-\lambda t})^{k-1} = k\lambda e^{\lambda t}(e^{\lambda t}-1)^{k-1} \\
\Rightarrow\ \frac{d}{dt}\left(e^{(k+1)\lambda t} P_{k+1}(t)\right) &= \frac{d}{dt}\left(e^{\lambda t}-1\right)^k \\
\Rightarrow\ e^{(k+1)\lambda t} P_{k+1}(t) &= (e^{\lambda t}-1)^k \\
\Rightarrow\ P_{k+1}(t) &= e^{-\lambda t}(1-e^{-\lambda t})^k.
\end{aligned}
$$

Hence, by induction we conclude that $P_k(t) = e^{-\lambda t}(1-e^{-\lambda t})^{k-1} \quad \forall \ k \geq 1$.
Method II: In this method, we substitute $P_k(t) = e^{-\lambda t}(1-e^{-\lambda t})^{k-1}$ in the integral equation (8.2.4). Thus, we have

$$
\begin{aligned}
P_{k+1}(t) &= e^{-(k+1)\lambda t}\int_0^t k\lambda P_k(u) e^{(k+1)\lambda u}\, du \\
&= e^{-(k+1)\lambda t} k\lambda \int_0^t e^{k\lambda u}(1-e^{-\lambda u})^{k-1}\, du \\
&= e^{-(k+1)\lambda t} k\lambda \int_0^t e^{\lambda u}(e^{\lambda u}-1)^{k-1}\, du \\
&= e^{-(k+1)\lambda t} k \int_0^{e^{\lambda t}-1} y^{k-1}\, dy \ \text{ with } \ y = e^{\lambda u}-1 \\
&= e^{-(k+1)\lambda t}(e^{\lambda t}-1)^k = e^{-\lambda t}(1-e^{-\lambda t})^k .
\end{aligned}
$$

Hence, by induction, $P_k(t) = e^{-\lambda t}(1-e^{-\lambda t})^{k-1} \quad \forall \ k = 1, 2, \ldots$. Thus, given $X(0) = 1$, the distribution of $X(t)$ is geometric with parameter $p = e^{-\lambda t}$ and support $\{1, 2, \ldots, \}$. □

Remark 8.2.1 The same result can be derived using Eq. (8.2.2). Thus, with $\lambda_j = j\lambda$, we have

$$
\begin{aligned}
P_k^{(1)}(t) &= \lambda_1\lambda_2\cdots\lambda_{k-1}\sum_{j=1}^{k} e^{-j\lambda t}\left(\prod_{i=1,i\neq j}^{k}(\lambda_i-\lambda_j)\right)^{-1}\\
&= \lambda^{k-1}(k-1)!\sum_{j=1}^{k} e^{-j\lambda t}\lambda^{-(k-1)}\left(\prod_{i=1,i\neq j}^{k}(i-j)\right)^{-1}\\
&= (k-1)!\sum_{j=1}^{k} e^{-j\lambda t}\frac{(-1)^{(j-1)}}{(j-1)!(k-j)!}\\
&= \sum_{j=1}^{k}\binom{k-1}{j-1}e^{-j\lambda t}(-1)^{j-1} = e^{-\lambda t}\sum_{j=0}^{k-1}\binom{k-1}{j}e^{-j\lambda t}(-1)^{j}\\
&= e^{-\lambda t}(1-e^{-\lambda t})^{k-1}\ \ \forall\ k=1,2,\ldots
\end{aligned}
$$

Remark 8.2.2 Since the distribution of $X(t)$ is geometric with parameter $p = e^{-\lambda t}$, it is clear that $\sum_{k=1}^{\infty} P_k(t) = 1$. Further, $\sum_{k=1}^{\infty} 1/k\lambda = \infty$, thus the result of Eq. (8.2.1) is verified.

If $X(0) = a$, then the total population size $X(t)$ is the sum of a independent populations, each having the geometric distribution with parameter $p = e^{-\lambda t}$ and support $\{1, 2, \ldots, \}$. Hence, if $X(0) = a$, the distribution of $X(t)$ is negative binomial with the probability mass function, the mean function and the variance function as follows:

$$
\begin{aligned}
P[X(t) = k] &= \binom{k-1}{k-a}e^{-a\lambda t}(1-e^{-\lambda t})^{k-a}\ \ \forall\ k = a, a+1, \ldots,\\
E(X(t)) &= ae^{\lambda t}\ \ \&\ \ Var(X(t)) = ae^{\lambda t}(e^{\lambda t}-1).
\end{aligned}
$$

Remark 8.2.3 (i) Observe that $E(X(t))$ increases exponentially fast, but is finite for finite t. (ii) We have already noted above that a Yule Furry process is a stochastic analogue of the Malthusian law of population growth $\frac{dX(t)}{dt} = \lambda X(t)$. Its solution under the initial condition $X(0) = a$ is $X(t) = ae^{\lambda t}$ which is nothing but the expectation of $X(t)$ in the stochastic version.
(iii) Further, $E(X(t)) = ae^{\lambda t}$ implies that a Yule Furry process is not a stationary process, but it is an evolutionary stochastic process.

In the following example, we find a realization of a Yule Furry process, using Code 8.7.1.

Example 8.2.1 Suppose $\{X(t), t \geq 0\}$ is a Yule Furry process with birth rate λ and $X(0) = 1$. We obtain a realization of the process when it is observed for $T = 10$ time units. For comparison, we take two values of λ as 0.20 and 0.35. The output in terms of epochs of births is organized in Table 8.1. Once we know these epochs, we

Table 8.1 Epochs of Birth in a Yule Furry Process

Birth epochs		
λ	0.20	0.35
S_1	3.60	5.33
S_2	6.41	5.91
S_3	6.64	6.05
S_4	6.81	7.28
S_5	7.22	7.33
S_6	9.52	7.65
S_7		8.09
S_8		8.63
S_9		9.05
S_{10}		9.09
S_{11}		9.29
S_{12}		9.58
S_{13}		9.73
$X(T)$	7	14

know how many births have occurred in $(0, T]$ and the realized values of the inter-birth random variables. When the birth rate is 0.2, 6 births occur in a time interval $(0, 10]$ and population size at $T = 10$ is 7, while when the birth rate is 0.35, 13 births occur in a time interval $(0, 10]$ and population size at $T = 10$ is 14. Larger birth rate leads to larger number of births in the same time interval. Figure 8.1 displays a realization of the Yule Furry process for both the rates. From Table 8.1 for $\lambda = 0.35$, observe that the last few inter-birth durations are comparatively small. It is in view of the fact that birth rate $i\lambda$ increases as population size increases and hence the mean of inter-birth duration random variable decreases. Figure 8.1 also depicts this feature. □

In Theorem 8.2.3, it is proved that for a Yule Furry process, $X(t)$ follows a geometric distribution. In the next example, we verify it, using Code 8.7.2.

Example 8.2.2 Suppose $\{X(t), t \geq 0\}$ is a Yule Furry process with birth rate $\lambda = 0.12$ and $X(0) = 1$. We obtain 150 realizations of the process when it is observed for $T = 10$ time units. We want to examine whether a geometric distribution with success probability $p = e^{-\lambda T}$ and support $\{1, 2, \ldots, \}$ is a good model for these observed data. Sample mean and sample variance based on 150 realized values of $X(T)$ are 3.2867 and 6.9979, respectively. From Theorem 8.2.3, $E(X(T)) = e^{\lambda T} = 3.3201$ and $Var(X(T)) = e^{\lambda t}(e^{\lambda t} - 1) = 7.7031$. It is to be noted that the sample mean and the sample variance are close to the population mean and variance respectively. Table 8.2 displays the observed frequency (O) and the expected frequency (E), expected under the geometric distribution, of the population size $X(T)$ corresponding to 150 simulations. Figure 8.2 depicts the observed and expected frequency

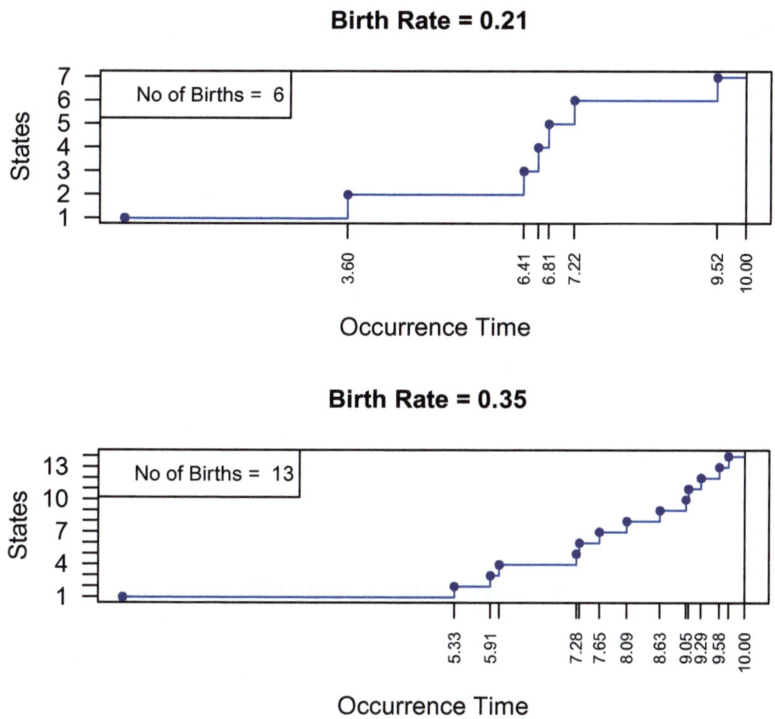

Fig. 8.1 Realization of a Yule Furry Process

distributions. From both Fig. 8.2 and Table 8.2, we observe that observed and expected frequencies are close to each other. To confirm the above observation, we carry out the goodness of fit test to test the null hypothesis H_0 that $X(T)$ follows geometric distribution. We use Karl Pearson's chi square test statistic T_n. For these data, $T_n = 9.0409$ after pooling the frequencies which are less than 5. Under H_0, $T_n \sim \chi^2_8$ distribution, the p-value comes out to be 0.3389. Without pooling the frequencies which are less than 5, $T_n = 12.4206$. Under H_0, $T_n \sim \chi^2_{12}$ distribution and the corresponding p-value is 0.4125. The built-in function `chisq.test(ob,p=v2)` give the same results. Thus, from the p-values of Karl Pearson's test procedure and from Fig. 8.2, we may conclude that geometric distribution is a good model for the population size on the basis of data simulated from the Yule Furry process. □

The next section presents the theory of a death process.

Table 8.2 Yule Furry Process: observed and expected frequency distributions

x	O	E
1	49	45.18
2	31	31.57
3	21	22.06
4	10	15.42
5	7	10.77
6	13	7.53
7	5	5.26
8	5	3.68
9	4	2.57
10	2	1.80
11	1	1.25
12	2	0.88
>12	0	2.03

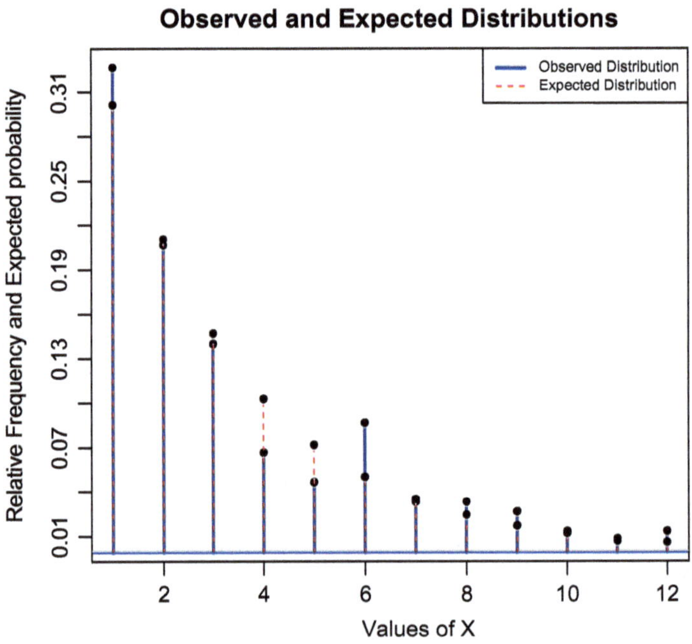

Fig. 8.2 Yule Furry Process: observed and expected distributions

8.3 Death Process

In a pure birth process, we noted that the population increases by births, more specifically increases by jumps of unit magnitude. When the population can only decrease by jumps of unit magnitude, we model the population at time t by a pure death process. We assume that initial population is a and the population decreases by deaths. A pure death process is defined on similar lines as those of a pure birth process.

Definition 8.3.1 Pure Death Process: Suppose $\{X(t), t \geq 0\}$ is a time homogeneous continuous time Markov chain with state space $S = \{0, 1, \ldots, a\}$, $X(0) = a$ and the following infinitesimal transition probabilities.

$$\begin{aligned}
P_{i,i-1}(h) &= P[X(t+h) = i-1 \big| X(t) = i] = \mu_i h + o(h), \ i \geq 1 \\
P_{ii}(h) &= P[X(t+h) = i \big| X(t) = i] = 1 - \mu_i h + o(h) \\
P_{ij}(h) &= P[X(t+h) = j \big| X(t) = i] = o(h) \ \forall \ j \neq i, \ i-1.
\end{aligned}$$

Then $\{X(t), t \geq 0\}$ is known as the pure death process and μ_i, $i = 0, 1, 2, \ldots, a$ with $\mu_0 = 0$ are known as death rates.

Note that the state space of a pure death process is finite while that of a pure birth process is countably infinite. Thus, a pure death process is a Markov process with finite state space. The infinitesimal generator matrix Q and the transition probability matrix P of the embedded Markov chain are given by

$$Q = \begin{array}{c} \\ 0 \\ 1 \\ 2 \\ \vdots \\ a \end{array}\!\!\begin{array}{c} \begin{array}{cccccc} 0 & 1 & 2 & \cdots & a & \end{array} \\ \begin{pmatrix} 0 & 0 & 0 & \cdots & 0 \\ \mu_1 & -\mu_1 & 0 & \cdots & 0 \\ 0 & \mu_2 & -\mu_2 & \cdots & 0 \\ \vdots & \vdots & \vdots & \vdots & \vdots \\ 0 & 0 & \cdots & \mu_a & -\mu_a \end{pmatrix} \end{array} \quad \& \quad P = \begin{array}{c} \\ 0 \\ 1 \\ 2 \\ \vdots \\ a \end{array}\!\!\begin{array}{c} \begin{array}{cccccc} 0 & 1 & 2 & \cdots & a & \end{array} \\ \begin{pmatrix} 1 & 0 & 0 & \cdots & 0 \\ 1 & 0 & 0 & \cdots & 0 \\ 0 & 1 & 0 & \cdots & 0 \\ \vdots & \vdots & \vdots & \vdots & \vdots \\ 0 & 0 & \cdots & 1 & 0 \end{pmatrix} \end{array}.$$

The first row of Q has all elements 0, indicating that 0 is an absorbing state. Hence, 0 is a non-null persistent and aperiodic, that is, an ergodic state. From the transition probability matrix P of the corresponding embedded Markov chain, we note that any state $i > 0$ is such that $i \to j < i$, but j does not lead to i. Hence any state $i > 0$ is an inessential state and hence a transient state.

Note that $\{0\}$ is a single closed communicating class and all other states are transient. Hence, the long run distribution and the unique stationary distribution are given by $(1, 0, 0, \ldots, 0)'$.

The process $\{X(t), t \geq 0\}$ stays in state i for a random time, which has exponential distribution with mean μ_i^{-1} and then moves to state $i-1$. This process continues until the state 0 is reached.

To find $P[X(t) = k]$, Kolmogorov's forward differential equations are derived as follows. Suppose $k = 1, 2, \ldots, a-1$. Then

$$\begin{aligned} P_k(t+h) &= P[X(t+h)=k] = P[X(t+h)=k \big| X(t)=k]P[X(t)=k] \\ &\quad + P[X(t+h)=k \big| X(t)=k+1]P[X(t)=k+1] + o(h) \\ &= P_{kk}(h)P_k(t) + P_{k+1,k}(h)P_{k+1}(t) + o(h) \\ &= (1-\mu_k h + o(h))P_k(t) + (\mu_{k+1}h + o(h))P_{k+1}(t) + o(h) \\ \Rightarrow \quad P_k'(t) &= -\mu_k P_k(t) + \mu_{k+1}P_{k+1}(t) \,. \end{aligned}$$

Further,

$$\begin{aligned} P_0(t+h) &= P[X(t+h)=0] = P[X(t+h)=0 \big| X(t)=0]P[X(t)=0] \\ &\quad + P[X(t+h)=0 \big| X(t)=1]P[X(t)=1] \\ &= 1P_0(t) + (\mu_1 h + o(h))P_1(t) \\ \Rightarrow \quad P_0'(t) &= \mu_1 P_1(t) \\ \& \quad P_a(t+h) &= P[X(t+h)=a] = P[X(t+h)=a \big| X(t)=a]P[X(t)=a] \\ &= (1-\mu_a h + o(h))P_a(t) \\ \Rightarrow \quad P_a'(t) &= -\mu_a P_a(t) \,. \end{aligned}$$

Thus, the system of differential equations for pure death process is given by

$$\begin{aligned} P_k'(t) &= -\mu_k P_k(t) + \mu_{k+1}P_{k+1}(t), \;\; k = 1, 2, \ldots, a-1 \\ P_a'(t) &= -\mu_a P_a(t) \quad \& \quad P_0'(t) = \mu_1 P_1(t) \,. \end{aligned} \tag{8.3.1}$$

These are solved iteratively, using initial conditions $P_a(0) = 1$ & $P_i(0) = 0$ $\forall\, i \neq a$, starting from $P_a'(t) = -\mu_a P_a(t)$, which has a solution $P_a(t) = e^{-\mu_a t}$.

We now study a particular case of a pure death process. Suppose the individuals act independently of each other and the probability that an individual dies in an interval $(t, t+h]$ is $\mu h + o(h)$. The probability that more than one individual dies in an interval $(t, t+h]$ is $o(h)$. Thus, if $X(t) = i$, then probability of death in an infinitesimal interval $(t, t+h]$ is given by $i\mu h + o(h)$. A pure death process in this case is known as a linear death process and μ is known as a death rate.

Definition 8.3.2 Linear Death Process: Suppose $\{X(t), t \geq 0\}$ is a time homogeneous continuous time Markov chain with $X(0) = a$, state space $S = \{0, 1, \ldots, a\}$ and the following infinitesimal transition probabilities.

$$\begin{aligned} P_{i,i-1}(h) &= P[X(t+h) = i-1 \big| X(t) = i] = i\mu h + o(h), \;\; i \geq 1 \\ P_{ii}(h) &= P[X(t+h) = i \big| X(t) = i] = 1 - i\mu h + o(h) \\ P_{ij}(h) &= P[X(t+h) = j \big| X(t) = i] = o(h) \;\; \forall \;\; j \neq i-1, \;\; j \neq i. \end{aligned}$$

Then $\{X(t), t \geq 0\}$ is known as a linear death process.

Using Code 8.7.3, we obtain a realization of a linear death process in the following example.

Table 8.3 Epochs of Death in a Linear Death Process

Death epochs		
μ	0.1	0.25
S_1	0.38	0.37
S_2	1.00	0.46
S_3	1.08	0.49
S_4	1.16	0.90
S_5	1.44	0.92
S_6	3.37	1.10
S_7	4.24	1.41
S_8	4.66	1.87
S_9	5.46	2.31
S_{10}	5.59	2.37
S_{11}	6.98	2.66
S_{12}	7.83	3.22
S_{13}	9.37	3.55
S_{14}		4.46
S_{15}		5.18
S_{16}		5.82
S_{17}		7.27
$X(T)$	7	3

Example 8.3.1 Suppose $\{X(t), t \geq 0\}$ is a linear death process with death rate μ and $X(0) = a = 20$. Suppose it is observed for $T = 10$ time units. For comparison we take two values of μ as 0.1 and 0.25. The output in terms of epochs of deaths is organized in Table 8.3. Once we know these epochs, we know how many deaths have occurred in $(0, T]$ and the realized values of inter-death random variables.

From Table 8.3 we note that, when death rate is 0.1, 13 deaths occur in a time interval $(0, 10]$ and population size at $T = 10$ is 7, while when death rate is 0.25, 17 deaths occur in a time interval $(0, 10]$ and population size at $T = 10$ is 3. For both the rates, as time increases, the realized values of inter-death duration tend to increase, since the population size decreases. Figure 8.3 displays the realization of the process for both the rates. Note that as time increases, the realized values of inter-death duration increase in Fig. 8.3. □

We solve the system of differential equations derived in Eq. (8.3.1) for the linear death process and find the distribution of $X(t)$ in the next theorem.

Theorem 8.3.1 *Suppose $\{X(t), t \geq 0\}$ is a linear death process with death rate μ and $X(0) = a > 0$. Then for fixed t, $X(t) \sim B(a, e^{-\mu t})$ distribution.*

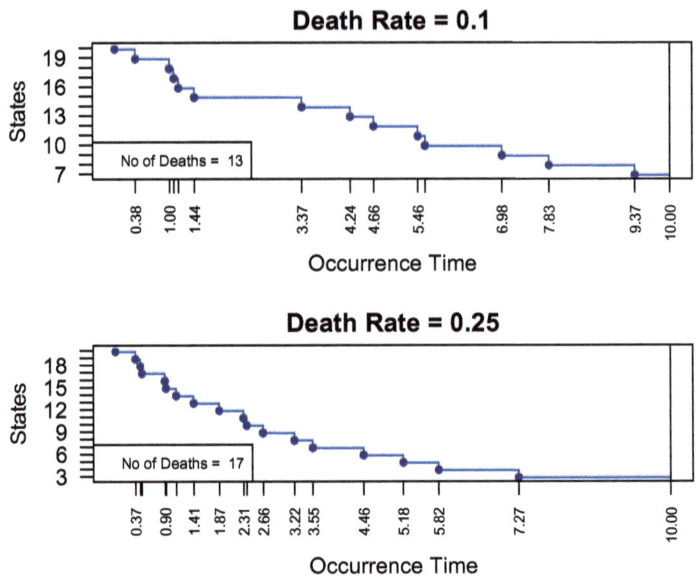

Fig. 8.3 Realization of a Linear Death Process

Proof For a linear death process, Kolmogorov's forward differential equations for a pure death process derived in Eq. (8.3.1) reduce as follows:

$$\begin{aligned} P_k'(t) &= -k\mu P_k(t) + (k+1)\mu P_{k+1}(t), \ k = 1, 2, \ldots, a-1 \\ P_a'(t) &= -a\mu P_a(t), \quad P_0'(t) = \mu P_1(t)\,. \end{aligned} \tag{8.3.2}$$

We solve these to find $P[X(t) = k]$ for $k = 0, 1, \ldots, a$, subject to the conditions that $P_a(0) = 1$ and $P_k(0) = 0 \ \forall \ k \neq a$. Suppose $k = a$. Now,

$$\begin{aligned} & P_a'(t) = -a\mu P_a(t) \ \Rightarrow \ P_a(t) = ce^{-a\mu t} = e^{-a\mu t} \ \text{ since } P_a(0) = c = 1 \\ \Rightarrow \ & P_a(t) = e^{-a\mu t} = \binom{a}{a}(e^{-\mu t})^a(1 - e^{-\mu t})^0. \end{aligned}$$

For $k = 1, 2, \ldots, a-1$, under the condition $P_k(0) = 0 \ \forall \ k \neq a$, the solution of $P_k'(t) = -k\mu P_k(t) + (k+1)\mu P_{k+1}(t)$ is

$$P_k(t) = e^{-k\mu t} \int_0^t (k+1)\mu P_{k+1}(u) e^{k\mu u} \, du \tag{8.3.3}$$

Hence,

$$P_{a-1}(t) = e^{-(a-1)\mu t} \int_0^t a\mu P_a(u) e^{(a-1)\mu u}\, du = e^{-(a-1)\mu t} \int_0^t a\mu e^{-a\mu u} e^{(a-1)\mu u}\, du$$

$$= e^{-(a-1)\mu t} \int_0^t a\mu e^{-\mu u}\, du \ = e^{-(a-1)\mu t} a(1 - e^{-\mu t})$$

$$= \binom{a}{a-1} (e^{-\mu t})^{a-1} (1 - e^{-\mu t}) \, .$$

We assume that $P_{a-j}(t) = \binom{a}{a-j}(e^{-\mu t})^{a-j}(1 - e^{-\mu t})^j$, which is true for $j = 0, 1$ and prove the result by induction.
Method I: In the differential equation given in (8.3.2), we take $k = a - j - 1$ and multiply both sides by $e^{(a-j-1)\mu t}$. Thus,

$$\begin{aligned}
P'_{a-j-1}(t) \ &= \ -(a-j-1)\mu P_{a-j-1}(t) + (a-j)\mu \binom{a}{a-j} (e^{-\mu t})^{a-j} (1 - e^{-\mu t})^j \\
\Rightarrow \frac{d}{dt}\left(e^{(a-j-1)\mu t} P_{a-j-1}(t)\right) &= \frac{d}{dt}\left(\binom{a}{a-j-1}(1 - e^{-\mu t})^{j+1}\right) \\
\Rightarrow e^{(a-j-1)\mu t} P_{a-j-1}(t) &= \binom{a}{a-j-1}(1 - e^{-\mu t})^{j+1} \\
\Rightarrow P_{a-j-1}(t) &= \binom{a}{a-j-1}(e^{-\mu t})^{a-j-1}(1 - e^{-\mu t})^{j+1}.
\end{aligned}$$

Hence by induction, $P_k(t) = \binom{a}{k}(e^{-\mu t})^k (1 - e^{-\mu t})^{a-k}$, $k = 0, 1, 2 \ldots, a$.

Method II: In this method, we substitute $P_{a-j}(t) = \binom{a}{a-j}(e^{-\mu t})^{a-j}(1 - e^{-\mu t})^j$ in the integral equation (8.3.3). Thus, we have

$$\begin{aligned}
P_{a-j-1}(t) &= e^{-(a-j-1)\mu t} \int_0^t (a-j)\mu P_{a-j}(u) e^{(a-j-1)\mu u}\, du \\
&= e^{-(a-j-1)\mu t} (a-j)\mu \binom{a}{a-j} \int_0^t e^{-\mu u}(1 - e^{-\mu u})^j\, du \\
&= e^{-(a-j-1)\mu t} (a-j) \binom{a}{a-j} \int_0^{1-e^{-\mu t}} y^j\, dy \ \text{ with } y = 1 - e^{-\mu u} \\
&= e^{-(a-j-1)\mu t} (a-j) \binom{a}{a-j} (1 - e^{-\mu t})^{j+1}/(j+1) \\
&= e^{-(a-j-1)\mu t} \binom{a}{a-j-1} (1 - e^{-\mu t})^{j+1} \\
&= \binom{a}{a-j-1} (e^{-\mu t})^{a-j-1} (1 - e^{-\mu t})^{j+1} \, .
\end{aligned}$$

Hence, by induction, we have

$$P_k(t) = \binom{a}{k}(e^{-\mu t})^k(1 - e^{-\mu t})^{a-k}, \ k = 0, 1, 2 \ldots, a \Rightarrow X(t) \sim B(a, e^{-\mu t}),$$

for fixed t. □

Since $X(t) \sim B(a, e^{-\mu t})$, the mean function and the variance function are given by $E(X(t)) = ae^{-\mu t}$ & $Var(X(t)) = ae^{-\mu t}(1 - e^{-\mu t})$. We note that $E(X(t))$ decreases exponentially fast, but is never 0 for any finite t. It is similar to the result that in the Yule Furry process, $E(X(t))$ increases exponentially fast, but is never infinite for any finite t. A linear death process is a finite state Markov process. Its long run distribution $\underline{P}$ exists and is a solution of $\underline{P}\,Q = \underline{0}$. For example, with $a = 4$, $\underline{P} = (1, 0, 0, 0, 0)$. Thus, in the long run the process becomes extinct, which is quite natural. The stationary distribution is also the same as $\underline{P}$. The time U to extinction is given by $U = \sum_{i=1}^{a} T_i$ where T_i is a sojourn time in state i and it has exponential distribution with rate parameter $\mu_i = i\mu$. Thus, the distribution of U is the distribution of sum of a independent random variables, each having exponential distribution, which is the hypoexponential distribution.

In the next example we obtain multiple realizations and verify the result that $X(t)$ has binomial distribution. We use Code 8.7.4.

Example 8.3.2 Suppose $\{X(t), t \geq 0\}$ is a linear death process with death rate $\mu = 0.12$ and $X(0) = 10$. We obtain 200 realizations of the process when it is observed for $T = 12$ time units. We want to examine whether a binomial distribution with success probability $p = e^{-\mu T}$ and support $\{0, 1, 2, \ldots, 10\}$ is a good model for the simulated data. For the simulated data, average population size is 2.55 with variance 1.8869. With $\mu = 0.12$, $X(0) = 10$ and $T = 12$, $p = e^{-\mu T} = 0.2369$, the mean $X(0)p = 2.3693$, close to 2.55 and the variance $X(0)p(1 - p)$ is 1.8079, it is close to 1.8869. Table 8.4 displays the observed frequency (O) of the population size $X(T)$ corresponding to 200 simulations and the expected frequency (E), expected under the binomial distribution. Figure 8.4 depicts the observed and expected frequency distributions.

The expected frequencies for $x = 9, 10$ are 0.0036 & 0.0001, respectively. The observed frequencies for $x = 9, 10$ are 0. These are not displayed in Table 8.4. From both Table 8.4 and Fig. 8.4, we observe that the observed frequency and the expected frequency are close to each other, except for $x = 0$. To confirm the above observation,

Table 8.4 Linear Death Process: observed and expected frequency distributions

x	0	1	2	3	4	5	6	7	8
O	9	39	52	55	30	9	5	1	0
E	13.39	41.57	58.08	48.09	26.13	9.74	2.52	0.45	0.05

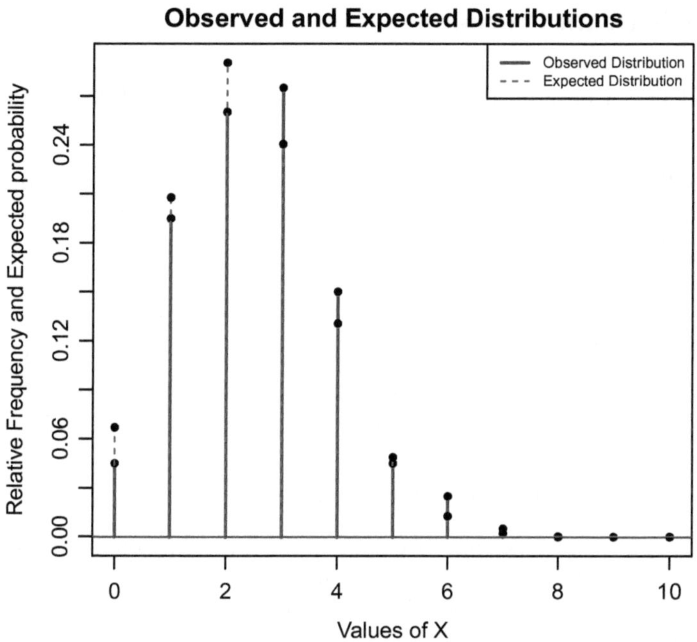

Fig. 8.4 Linear Death Process: observed and expected distributions

we carry out the goodness of fit test to test the null hypothesis H_0 that $X(T)$ follows a binomial distribution, using Karl Pearson's test procedure. For these data value of the chi square test statistic $T_n = 6.7913$, after pooling the frequencies which are less than 5. Under H_0, $T_n \sim \chi^2_6$ distribution, the p-value comes out to be 0.3406. Without pooling the frequencies which are less than 5, $T_n = 7.0394$. Under H_0, $T_n \sim \chi^2_{10}$ distribution and the corresponding p-value is 0.7217. The built-in function `chisq.test(ob,p=pr)` gives the same results. Thus, from the p-values of the chi square test procedure and from Fig. 8.4, we may conclude that a binomial distribution is a good model for the population size on the basis of data simulated from the linear death process. □

In the next section, we combine the postulates of a birth process and a death process to model a population which varies randomly due to increments and decrements. It leads to a birth-death process, which has applications in a variety of areas.

8.4 Birth-Death Process

We first define a birth-death process and discuss how the different versions are obtained from these.

Definition 8.4.1 Birth-Death Process: Suppose $\{X(t), t \geq 0\}$ is a continuous time Markov chain with state space $S = \{0, 1, ...\}$, $X(0) = a \geq 0$ and the following infinitesimal transition probabilities.

$$\begin{aligned} P_{i,i+1}(h) &= P[X(t+h) = i+1 \big| X(t) = i] = \lambda_i h + o(h) \ \forall \ i \geq 0 \\ P_{i,i-1}(h) &= P[X(t+h) = i-1 \big| X(t) = i] = \mu_i h + o(h) \ \forall \ i \geq 1 \\ P_{ii}(h) &= P[X(t+h) = i \big| X(t) = i] = 1 - (\lambda_i + \mu_i)h + o(h) \ \forall \ i \geq 0 \\ P_{ij}(h) &= P[X(t+h) = j \big| X(t) = i] = o(h) \ \forall \ j \neq i \ i+1, i-1 \ . \end{aligned}$$

Then $\{X(t), t \geq 0\}$ is known as a birth-death process.

Note that $\mu_0 = 0$, but λ_0 may or may not be 0 depending on the particular application area. For example, $\lambda_0 = 0$ for Yule Furry process, but $\lambda_0 = \lambda > 0$ in $M/M/1$ queuing model, which is a particular case of a birth-death process.

From the definition, it follows that the birth-death process is a continuous time Markov chain with state space W, in which transitions from state i are only to the neighboring states, that is either to state $i-1$ or to state $i+1$. The intensity rates for this process for $i \geq 0$ are given by

$$q_{i,i+1} = \lambda_i, \ \ q_{i,i-1} = \mu_i, \ \ q_{ii} = -(\lambda_i + \mu_i) \ \ \& \ \ q_{ij} = 0 \ \forall \ j \neq i+1, i-1, \ \ \ \mu_0 = 0.$$

Thus, the sojourn time random variable in state $i > 0$ has an exponential distribution with rate $\lambda_i + \mu_i$, for $i = 0$, the rate is λ_0. Further, if $P = [p_{ij}]$ denotes the transition probability matrix of the embedded chain, then from the transition rates we have $p_{ij} = 0$ if $j \neq i+1, i-1$, $p_{01} = 1$ or $p_{00} = 1$ and for $i \geq 1$, $p_{i,i+1}$ and $p_{i,i-1}$ are positive. Now we find the expressions of these probabilities.

$$\begin{aligned} p_{i,i+1} &= P[\text{Process enters state } i+1 \text{ starting from } i] \\ &= P[\text{A birth occurs before a death}] \\ &= P[X < Y] \ \text{ where } X \sim exp(\lambda_i) \ \& \ Y \sim exp(\mu_i) \\ &= \lambda_i/(\lambda_i + \mu_i), \qquad i \geq 0. \end{aligned}$$

Similarly,

$$\begin{aligned} p_{i,i-1} &= P[\text{Process enters state } i-1 \text{ starting from } i] \\ &= P[\text{A death occurs before a birth}] \\ &= P[X > Y] \ \text{ where } X \sim exp(\lambda_i) \ \& \ Y \sim exp(\mu_i) \\ &= \mu_i/(\lambda_i + \mu_i), \qquad i \geq 1. \end{aligned}$$

As in pure birth and pure death processes, we can obtain a system of differential equations for a birth-death process and solve these to find a marginal distribution of $X(t)$. Suppose $P_k(t) = P[X(t) = k]$ given $X(0) = a$. To find an expression for $P_k(t)$, we derive a system of differential equations as in case of Kolmogorov's forward differential equations. Using the postulates of a birth-death process we get for $k \geq 1$,

$$\begin{aligned}
P_k(t+h) &= P[X(t+h) = k] = P[X(t+h) = k \big| X(t) = k]P[X(t) = k] \\
&\quad + P[X(t+h) = k \big| X(t) = k-1]P[X(t) = k-1] \\
&\quad + P[X(t+h) = k \big| X(t) = k+1]P[X(t) = k+1] + o(h) \\
&= P_{kk}(h)P_k(t) + P_{k-1,k}(h)P_{k-1}(t) + P_{k+1,k}(h)P_{k+1}(t) + o(h) \\
&= (1 - (\lambda_k + \mu_k)h + o(h))P_k(t) + (\lambda_{k-1}h + o(h))P_{k-1}(t) \\
&\quad + (\mu_{k+1}h + o(h))P_{k+1}(t) + o(h) \\
\Rightarrow \quad P_k'(t) &= -(\lambda_k + \mu_k)P_k(t) + \lambda_{k-1}P_{k-1}(t) + \mu_{k+1}P_{k+1}(t) \,.
\end{aligned}$$

For $k = 0$,

$$\begin{aligned}
P_0(t+h) &= P[X(t+h) = 0] = P[X(t+h) = 0 \big| X(t) = 0]P[X(t) = 0] \\
&\quad + P[X(t+h) = 0 \big| X(t) = 1]P[X(t) = 1] + o(h) \\
&= (1 - \lambda_0 h + o(h))P_0(t) + (\mu_1 h + o(h))P_1(t) + o(h) \\
\Rightarrow \quad P_0'(t) &= -\lambda_0 P_0(t) + \mu_1 P_1(t) \,.
\end{aligned}$$

Thus, to find $P_k(t) = P[X(t) = k]$, we have to solve a system of differential equations given by

$$\begin{aligned}
P_0'(t) &= -\lambda_0 P_0(t) + \mu_1 P_1(t) \\
P_k'(t) &= -(\lambda_k + \mu_k)P_k(t) + \lambda_{k-1}P_{k-1}(t) + \mu_{k+1}P_{k+1}(t), \quad k \geq 1,
\end{aligned}$$

subject to the conditions $P_a(0) = 1$ and $P_k(0) = 0 \;\; \forall \;\; k \neq a$. These are infinitely many difference differential equations and in general difficult to solve. It has been proved that the system has a unique solution such that $P_k(t) \geq 0$ and $\sum_{k=0}^{\infty} P_k(t) \leq 1$ for all $t > 0$. The sum will be 1, if $\sum_{k=0}^{\infty}(\lambda_k + \mu_k)^{-1} = \infty$, Feller [2].

Observe that the embedded Markov chain corresponding to a birth-death process is a simple random walk on a set of whole numbers, where probability of transition from i to $i+1$ is $\lambda_i/(\lambda_i + \mu_i), \; i \geq 0$ and probability of transition from i to $i-1$ is $\mu_i/(\lambda_i + \mu_i), \; i \geq 1$. Note that the transition probability depends on the state from which the transition occurs. The state 0 may or may not be an absorbing state, it depends on the value of λ_0. If $\lambda_0 = 0$, then the probability of transition from 0 to 1 is zero and the process is absorbed into state 0. If $\lambda_0 > 0$, then the probability of transition from 0 to 1 is 1, since $\mu_0 = 0$. Hence, in this case, 0 is a reflecting barrier. Thus, a birth-death process is regarded as a continuous time analogue of a random walk on W, with absorbing barrier at 0, if $\lambda_0 = 0$ and with reflecting barrier at 0, if $\lambda_0 > 0$. An absorption into state 0 is not a certain event, as the population may fluctuate among the states $\{1, 2, \ldots, \}$ or possibly drift to ∞. We now compute the probability of absorption, also known as the probability of ultimate extinction when $\lambda_0 = 0$.

Suppose $X(0) = a > 0$ and q_a denotes the probability of absorption into state 0 starting from a. It is to be noted that $\lambda_0 = 0 \Rightarrow q_0 = 1$. Considering possible states after the first transition, we have the following recurrence relation to compute q_a for $a \geq 1$.

$$\begin{aligned} q_a &= (\lambda_a/(\lambda_a + \mu_a))q_{a+1} + (\mu_a/(\lambda_a + \mu_a))q_{a-1} \\ \Rightarrow (\lambda_a + \mu_a)q_a &= \lambda_a q_{a+1} + \mu_a q_{a-1} \\ \Rightarrow \lambda_a q_{a+1} - \lambda_a q_a &= \mu_a q_a - \mu_a q_{a-1} \\ \Rightarrow q_{a+1} - q_a &= (\mu_a/\lambda_a)(q_a - q_{a-1}) \\ \Rightarrow u_a &= (\mu_a/\lambda_a)u_{a-1} \text{ with } u_a = q_{a+1} - q_a \ . \end{aligned}$$

Iterating the last relation, we get

$$q_{a+1} - q_a = u_a = \prod_{i=1}^{a}(\mu_i/\lambda_i)u_0 = (q_1 - 1)\prod_{i=1}^{a}(\mu_i/\lambda_i) \ .$$

Summing these equations for $a = 1$ to $a = m \geq 1$, we get

$$\begin{aligned} q_{m+1} - q_1 &= \sum_{a=1}^{m}\left(\prod_{i=1}^{a}(\mu_i/\lambda_i)\right)(q_1 - 1) \\ \Rightarrow \lim_{m\to\infty} q_{m+1} - q_1 &= \sum_{a=1}^{\infty}\left(\prod_{i=1}^{a}(\mu_i/\lambda_i)\right)(q_1 - 1). \end{aligned} \tag{8.4.1}$$

From Eq. (8.4.1), we arrive at the following conclusions.

(i) Observe that q_m being probability, is bounded by 1 and hence if $\sum_{a=1}^{\infty}\left(\prod_{i=1}^{a}(\mu_i/\lambda_i)\right) = \infty$, then q_1 must be 1, which implies that $\lim_{m\to\infty} q_{m+1} = 1$. Thus, if $\sum_{a=1}^{\infty}\left(\prod_{i=1}^{a}(\mu_i/\lambda_i)\right) = \infty$, then the probability of ultimate absorption into state 0 is 1, whatever may be the initial size.

(ii) Suppose now $0 < q_1 < 1$, then $\sum_{a=1}^{\infty}\left(\prod_{i=1}^{a}(\mu_i/\lambda_i)\right) < \infty$. Further, q_m is a decreasing function of m and $q_m \to 0$ as $m \to \infty$, Karlin and Taylor [3]. Now solving the equation $-q_1 = \sum_{a=1}^{\infty}\left(\prod_{i=1}^{a}(\mu_i/\lambda_i)\right)(q_1 - 1)$ for q_1, we get the expression for q_1. Using this expression of q_1, in $q_{m+1} - q_1 = \sum_{a=1}^{m}\left(\prod_{i=1}^{a}(\mu_i/\lambda_i)\right)(q_1 - 1)$, we get q_{m+1}. Thus, q_1 and q_a for $a \geq 1$ are as given below.

$$q_1 = \frac{\sum_{l=1}^{\infty}\left(\prod_{i=1}^{l}(\mu_i/\lambda_i)\right)}{1+\sum_{l=1}^{\infty}\left(\prod_{i=1}^{l}(\mu_i/\lambda_i)\right)} \quad \& \quad q_a = \frac{\sum_{l=a}^{\infty}\left(\prod_{i=1}^{l}(\mu_i/\lambda_i)\right)}{1+\sum_{l=1}^{\infty}\left(\prod_{i=1}^{l}(\mu_i/\lambda_i)\right)}, \quad a \geq 1. \quad (8.4.2)$$

We can simplify the expression for q_a for a linear birth-death process. The next section is concerned with such a process.

8.5 Linear Birth-Death Process

We begin with the definition of a linear birth-death process, which is a particular case of a birth-death process.

Definition 8.5.1 Linear Birth-Death Process: Suppose $\{X(t), t \geq 0\}$ is a birth-death process with birth rates λ_i and death rates $\mu_i, i \geq 0$. If $\lambda_i = i\lambda$ and $\mu_i = i\mu$, then $\{X(t), t \geq 0\}$ is known as a linear birth-death process.

Note that $\lambda_0 = 0$ in a linear birth-death process and hence 0 is an absorbing state. We now examine whether some derivations in the previous section can be simplified in view of linearity of birth and death rates.
(i) We note that a system of difference differential equations can be solved for a linear birth-death process, using the probability generating function. Suppose for $|s| \leq 1$, $P(s,t) = \sum_{k=0}^{\infty} P_k(t)s^k$ denotes the probability generating function of $P_k(t),\ k \geq 0$, then for $X(0) = a$, it has been shown that Bhat [1]

$$P(s,t) = \begin{cases} \left[\frac{\mu(1-e^{-(\lambda-\mu)t})-(\mu-\lambda e^{-(\lambda-\mu)t})s}{\lambda-\mu e^{-(\lambda-\mu)t}-\lambda(1-e^{-(\lambda-\mu)t})s}\right]^a, & \text{if } \mu \neq \lambda \\ \left[\frac{\lambda t+s(1-\lambda t)}{1+\lambda t-\lambda t s)}\right]^a, & \text{if } \mu = \lambda\,. \end{cases}$$

Expanding $P(s,t)$ in powers of s, the coefficient of s^k gives $P_k(t)$. If $a = 1$, then $\{P_k(t), k = 0, 1, \ldots,\}$ is a probability mass function of a geometric distribution with modified probability for $k = 0$. If $a > 1$, then it is the a fold convolution of this modified geometric distribution. From the generating function, we can find the mean function and the variance function of a linear birth-death process.

(ii) The mean function $M(t) = E(X(t))$, when $X(0) = a$, is as follows:

$$M(t) = \begin{cases} ae^{(\lambda-\mu)t}, & \text{if } \mu \neq \lambda \\ a, & \text{if } \mu = \lambda\,. \end{cases}$$

Thus, the expected population size increases exponentially fast if $\lambda > \mu$ and decreases exponentially fast if $\lambda < \mu$. If $\lambda = \mu$, then $E(X(t))$ is constant at a for all $t > 0$.

(iii) When $a = 1$, the variance function $V(t) = Var(X(t))$ is given by

$$V(t) = \begin{cases} \frac{\lambda+\mu}{\lambda-\mu} e^{(\lambda-\mu)t}(e^{(\lambda-\mu)t} - 1), & \text{if } \mu \neq \lambda \\ 2\lambda t, & \text{if } \mu = \lambda\,. \end{cases}$$

(iv) When $\mu \neq \lambda$, the mean function $M(t)$ is a function of t and the variance function $V(t)$ is a function of t for all λ and μ. Thus, a linear birth-death process is not a stationary process, but an evolutionary stochastic process.

(v) In a linear birth-death process, $\lambda_0 = 0$ and hence state 0 is an absorbing state. Suppose $P_0(t)$ denotes a probability of extinction on or before time t. When $X(0) = a = 1$, from $P(s, t)$ we find a constant term, that is, a coefficient of s^0 and it gives $P_0(t)$ as,

$$P_0(t) = \begin{cases} \frac{\mu(1-e^{-(\lambda-\mu)t})}{\lambda-\mu e^{-(\lambda-\mu)t}}, & \text{if } \lambda \neq \mu \\ \frac{\lambda t}{1+\lambda t}, & \text{if } \lambda = \mu\,. \end{cases}$$

From these expressions for $P_0(t)$, it follows that as $t \to \infty$,

$$P_0(t) \to \begin{cases} 1, & \text{if } \lambda \leq \mu \\ \mu/\lambda, & \text{if } \lambda > \mu\,. \end{cases}$$

Thus, the probability of ultimate extinction is 1, if $\lambda \leq \mu$, that is, if the death rate is larger than or equal to the birth rate. It is less than 1 if $\lambda > \mu$. If $\lambda > \mu$, the population explodes to infinity with probability $1 - \mu/\lambda$. It then follows that $\lim_{t\to\infty} X(t)$ is either 0 or ∞. If $X(0) = a > 1$, then

$$P_0(t) \to \begin{cases} 1, & \text{if } \lambda \leq \mu \\ (\mu/\lambda)^a, & \text{if } \lambda > \mu\,. \end{cases}$$

In Eq. (8.4.2), we have derived the probability q_a of absorption into state 0 for a birth-death process with $X(0) = a$, when the series $\sum_{a=1}^{\infty}\left(\prod_{i=1}^{a}(\mu_i/\lambda_i)\right)$ is finite. For a linear birth-death process, when $\lambda > \mu$, the series is convergent. In this case, the expression for q_a, $a \geq 1$ reduces as follows:

$$\begin{aligned} q_a &= \frac{\sum\limits_{l=a}^{\infty}\left(\prod\limits_{i=1}^{l}(\mu_i/\lambda_i)\right)}{1+\sum\limits_{l=1}^{\infty}\left(\prod\limits_{i=1}^{l}(\mu_i/\lambda_i)\right)} = \frac{\sum\limits_{l=a}^{\infty}(\mu/\lambda)^l}{1+\sum\limits_{l=1}^{\infty}(\mu/\lambda)^l} \\ &= \frac{(\mu/\lambda)^a(1-\mu/\lambda)^{-1}}{1+\mu/\lambda(1-\mu/\lambda)^{-1}} = \left(\frac{\mu}{\lambda}\right)^a. \end{aligned}$$

When $\lambda \leq \mu$, the series is divergent and $q_a = 1$. Thus, the probability q_a of absorption into state 0 for a linear birth-death process is given by

$$q_a = \begin{cases} 1, & \text{if } \lambda \leq \mu \\ (\mu/\lambda)^a, & \text{if } \lambda > \mu\,, \end{cases}$$

which is exactly the same as the limit of $P_0(t)$.

(vi) For a linear birth-death process, when $\mu > \lambda$, the mean time T to absorption in state 0, when $X(0) = 1$ is given by, $T = -(1/\lambda)\log(1 - \lambda/\mu)$, refer to Karlin and Taylor [3].

The following example illustrates how to obtain a realization of a linear birth-death process using Code 8.7.5.

Example 8.5.1 Suppose $\{X(t), t \geq 0\}$ is a linear birth-death process with $X(0) = 10$, birth rate λ and death rate μ. We obtain its realization for the fixed time interval (0, 10]. For comparison, we take two sets of birth and death rates as $\lambda = 0.12$, $\mu = 0.1$ and $\lambda = 0.1$, $\mu = 0.4$. Figure 8.5 displays the realization of the linear birth-death process for these two sets of parameters. When birth rate is 0.12 and death rate is 0.10, 30 births and deaths occurred in the interval (0, 10] and $X(10) = 20$. When birth rate is 0.1 and death rate is 0.4, 24 births and deaths occurred in the interval (0, 10] and $X(10) = 2$. It is a consequence of higher death rate. □

We now note some more variations of a birth-death process.

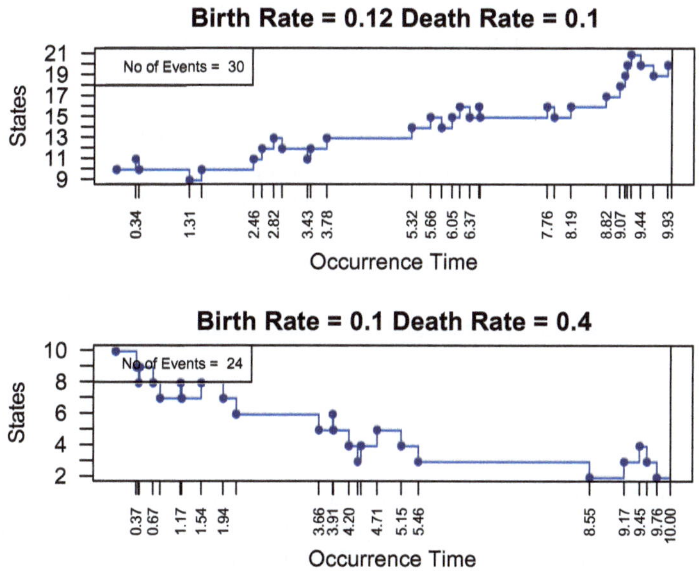

Fig. 8.5 Realization of a Linear Birth-Death Process

(i) Suppose $\lambda_k = \lambda$ & $\mu_k = k\mu$, then a birth-death process is known as a immigration-death process. Thus, μ denotes the death rate per individual and λ is the immigration rate, not depending on the population size.

(ii) Suppose $\lambda_k = k\lambda$ & $\mu_k = \mu$, then a birth-death process is known as a birth-emigration process. Thus, λ denotes the birth rate per individual and μ is the emigration rate, not depending on the population size.

(iii) Suppose $\mu_k = k\mu, \quad k \geq 1$ and $\lambda_k = k\lambda + \alpha \;\; k \geq 0$, then a birth-death process is known as a linear growth process with immigration. It has been already discussed in Sect. 6.4.

(iv) Suppose $\lambda_k = \lambda$ and $\mu_k = \mu$, then a birth-death process is known as a immigration-emigration process. Thus, λ and μ denote the immigration and the emigration rates, respectively. These do not depend on the population size. In this case, the process $\{X(t), t \geq 0\}$ increases by arrivals which is a Poisson process with rate λ and decreases by another Poisson process with rate μ, whenever $X(t) > 0$. Thus, emigration rate is also independent of the state $X(t)$ at time t, it being 0 if $X(t) = 0$. Such an immigration-emigration process is a queuing model with one server as described in the introductory section. In queuing theory such a model is referred to as $M/M/1$ model. The first M indicates that the arrival process is Poisson with rate λ, and hence is a Markov process, the second M indicates that the service times are exponentially distributed with rate μ, which is equivalent to the assumption that the departure process is also a Poisson process and hence it is also a Markov process. The letter M stands for the memoryless property of the exponential distribution. The number 1 in $M/M/1$ indicates that there is a single server. Further, these two Poisson processes are independent.

(v) Suppose we have a $M/M/s$ queuing system, that is, the arrival process is Poisson with rate λ, the service times are exponentially distributed with rate μ and there are s servers. Suppose $X(t)$ denotes the number of customers in the system at time t, that is, number of customers waiting in the queue and the customers getting service. If $X(t) \leq s$, then all the $X(t)$ are persons are being served and the departure rate is $\mu X(t)$. If $X(t) > s$, then the departure rate is $s\mu$. In view of the assumption of exponential inter-arrival time, exponential service time and their mutual independence, $\{X(t), t \geq 0\}$ is a Markov process and it is a birth-death process with

$$\lambda_k = \lambda, k = 0, 1, \ldots, \quad \mu_k = k\mu \text{ if } k = 1, 2, \ldots, s$$
$$\& \;\; \mu_k = s\mu \text{ if } k = s + 1, s + 2, \ldots.$$

In this model, it is implicitly assumed that the system has unlimited waiting capacity, which is not the case in practice. Suppose further that the waiting room for the customers has a limited capacity, say C. The customers will not join the system if the waiting room is full. Hence, $\{X(t), t \geq 0\}$ is a birth-death process with

$$\lambda_k = \lambda, k = 0, 1, \ldots, C-1, \ \lambda_k = 0 \text{ if } k \geq C,$$
$$\mu_k = k\mu \text{ if } k = 1, 2, \ldots, s \ \& \ \mu_k = s\mu, \ \text{ if } k = s+1, s+2, \ldots$$

(vi) Suppose the population size $X(t)$, which ranges between two fixed integers N_1 and N_2, $(N_1 < N_2)$, is modeled as a birth-death process with birth rate $\lambda_k = \alpha k(N_2 - k)$ and death rate $\mu_k = \beta k(k - N_1)$. Then $\{X(t), t \geq 0\}$ is known as a logistic process. Observe that if $X(t)$ is near N_2, then the birth rate will be low and the death rate will be high and then $X(t)$ will tend toward N_1. Similarly, if $X(t)$ is near N_1, then the birth rate will be high and the death rate will be low and then $X(t)$ will tend toward N_2. Thus, $X(t)$ fluctuates between N_1 and N_2.

In general it is difficult to find exact solutions of the differential equations in all these versions of birth-death processes. Hence, in the next section, we study the long run behavior of the process $\{X(t), t \geq 0\}$ as $t \to \infty$.

8.6 Long Run Behavior of a Birth-Death Process

We can study the limiting behavior of the process, if there is a stability in the system, in the sense that $X(t)$ does not diverge to infinity as $t \to \infty$. The long run distribution then provides approximation for $P_k(t)$ for large t. In most of the cases, $\lim_{t\to\infty} P_k(t) = P_k$ exists and is independent of the initial conditions. We now discuss the conditions under which there is a stability in the system and under these conditions derive the long run distribution. In order to find a long run distribution for a birth-death process, we consider the balance equations discussed in Chap. 6. These state that in the long run, the rate at which the process enters a state must match with the rate at which the process leaves that state. For a birth-death process, the balance equations are given as

State	Rate at which leave = rate at which enter
0	$\lambda_0 P_0 = \mu_1 P_1$
1	$(\lambda_1 + \mu_1)P_1 = \mu_2 P_2 + \lambda_0 P_0$
2	$(\lambda_2 + \mu_2)P_2 = \mu_3 P_3 + \lambda_1 P_1$
$n, \ n \geq 1$	$(\lambda_n + \mu_n)P_n = \mu_{n+1} P_{n+1} + \lambda_{n-1} P_{n-1}$

By adding to each equation, the equation preceding to it, we obtain

$$\lambda_0 P_0 = \mu_1 P_1, \ \lambda_1 P_1 = \mu_2 P_2, \ \lambda_2 P_2 = \mu_3 P_3, \ldots, \ \lambda_n P_n = \mu_{n+1} P_{n+1} \, .$$

From these equations, we have $P_1 = (\lambda_0/\mu_1)P_0$, $P_2 = (\lambda_1/\mu_2)P_1 = (\lambda_1\lambda_0/\mu_2\mu_1)P_0$. Continuing in this way, finally we get for $n \geq 1$,

$$P_n = \frac{\lambda_{n-1}}{\mu_n} P_{n-1} = \frac{\lambda_{n-1}\lambda_{n-2}\cdots\lambda_1\lambda_0}{\mu_n\mu_{n-1}\cdots\mu_2\mu_1} P_0 = \alpha_n P_0, \ \text{ where } \alpha_n = \prod_{i=0}^{n-1}\lambda_i \Big/ \prod_{i=1}^{n}\mu_i .$$

$$\text{Now, } 1 = \sum_{n=0}^{\infty} P_n \Rightarrow P_0 + P_0 \sum_{n=1}^{\infty} \alpha_n = 1$$

$$\Rightarrow P_0 = \left(1 + \sum_{n=1}^{\infty} \alpha_n\right)^{-1} \Rightarrow P_n = \alpha_n \Big/ \left(1 + \sum_{n=1}^{\infty} \alpha_n\right), \; n \geq 1,$$

provided $\sum_{n=1}^{\infty} \alpha_n$ is convergent. Hence, it is a necessary condition for the stability of a birth-death process. Under this condition $\{P_n, n \geq 0\}$ is the long run distribution, also known as the stable distribution, as well as the stationary distribution of a birth-death process. Recall that we arrived at the similar condition of convergence in Theorem 4.5.1 while deriving the stationary distribution of a birth-death chain.

Some particular cases are listed below.

(i) $M/M/1$ queuing model: In this model, when $\lambda_k = \lambda$, $k = 0, 1, \ldots,$ and $\mu_k = \mu$, $k = 1, 2, \ldots,$ $P_n = (1 - \lambda/\mu)(\lambda/\mu)^n, n = 0, 1, \ldots,$ provided

$$\sum_{n=0}^{\infty} (\lambda/\mu)^n < \infty \iff \rho = \lambda/\mu < 1.$$

Thus, the long run distribution is geometric with parameter $1 - \rho$. In the long run, the expected number of customers in the system is $\rho/(1 - \rho)$. The parameter $\rho = \lambda/\mu$ is known as the traffic intensity. It is the expected number of arrivals per unit of rate of service time. It is a good measure of the long-run behavior of the queue size. If $\lambda/\mu > 1$, it is clear that the server will be unable to keep up with the arrivals and then the queue size increases without limit. Hence the long run distribution does not exist and $X(t)$ will explode to infinity. The condition $\lambda/\mu < 1$ implies that $1/\mu < 1/\lambda$, that is, the mean service time is smaller than the mean arrival time and it seems reasonable that it has to be satisfied for long run distribution to exist. If $\rho < 1$, the server will be able to clear the work load presented to it, so we expect the queue to be empty again and again.

If the process has been going on for a long time, then P_n is the probability that there are n customers in the system, $n - 1$ in the queue and 1 being served. In particular, $P_0 = 1 - \rho$ is the probability that there is no customer in the system, hence a customer will be served immediately upon arrival, his waiting time will be 0. Thus, P_0 denotes the probability of being served immediately upon arrival, it is also interpreted as the probability of the server being idle. The probability that a server is busy is thus $1 - P_0$.

In this system, if the waiting room has finite capacity C, then the long run distribution always exists, since it is a finite state Markov process. It is given by $P_n(C), n = 0, 1, \ldots, C$ where

$$P_n(C) = \begin{cases} \frac{(1-\rho)\rho^n}{1-\rho^{C+1}}, \text{ if } \rho \neq 1 \\ \\ \frac{1}{C+1}, \text{ if } \rho = 1 , \end{cases}$$

Thus, for $\rho \neq 1$ the long run distribution is right truncated geometric distribution, truncated at C. The long-run fraction of the time the server is idle is given by $P_0(C)$ and the long-run fraction of the time the system is full is $P_C(C)$.

(ii) $M/M/s$ queuing model: In this model, when $\lambda_k = \lambda,\ k = 0, 1, \ldots,$ $\mu_k = k\mu$ if $k = 1, 2, \ldots, s$ and $\mu_k = s\mu$ if $k = s+1, s+2, \ldots,$

$$P_n = \begin{cases} \left(\frac{\lambda}{\mu}\right)^n \frac{1}{n!} P_0, \text{ if } \quad n = 1, 2, \ldots, s \\ \\ \left(\frac{\lambda}{s\mu}\right)^n \frac{1}{s!s^{n-s}} P_0, \text{ if } n = s+1, s+2, \ldots , \end{cases}$$

where

$$P_0 = \left[1 + \sum_{n=1}^{s} \left(\frac{\lambda}{\mu}\right)^n \frac{1}{n!} + \sum_{n=s+1}^{\infty} \left(\frac{\lambda}{s\mu}\right)^n \frac{1}{s!s^{n-s}}\right]^{-1} .$$

It is to be noted that $P_0 > 0$ if the corresponding series converges, which is true if $\lambda/s\mu < 1$. Thus, the long run distribution exists if $\lambda/s\mu < 1$. The parameter $\rho = \lambda/s\mu$ is known as a traffic intensity of the $M/M/s$ queuing system. It is the relative rate of arrival to that of the maximum departure rate. If $\rho > 1$, the queue size would increase to infinity and there is no stability to the system and the long run distribution does not exist.

(iii) In a immigration-death process when $\lambda_k = \lambda,\ k = 0, 1, \ldots,$ and $\mu_k = k\mu$, $k = 1, 2, \ldots,$

$$P_n = e^{-(\lambda/\mu)}(\lambda/\mu)^n/n!, \quad n = 0, 1, \ldots,$$

and the long run distribution is Poisson with parameter λ/μ. In this case it exists for all values of λ and μ.

In the next example, we compute the long run distribution when the birth-death process has a finite state space.

Example 8.6.1 Suppose in a workshop there are m machines and a single repair facility. Probability that a machine working at time t, fails in time $(t, t+h]$ is $\lambda h + o(h)$, independently of other machines. If the repair facility is busy at time t, probability that the repair is complete in $(t, t+h]$ is $\mu h + o(h)$ independently of state of the other machines. Suppose $X(t)$ denotes the number of failed machines at time t, including the one in repair. The possible values of $X(t)$ are $\{0, 1, \ldots, m\}$. Then $\{X(t), t \geq 0\}$ is modeled as a continuous time Markov chain with finite state space $S = \{0, 1, \ldots, m\}$ and infinitesimal transition probabilities as given below.

$$\begin{aligned} P_{k,k+1}(h) &= (m-k)\lambda h + o(h), \quad k = 0, 1, \ldots, m-1 \\ P_{k,k-1}(h) &= \mu h + o(h), \quad k = 1, 2, \ldots, m \\ P_{kk}(h) &= 1 - [(m-k)\lambda + \mu]h + o(h), \quad k = 0, 1, \ldots, m \\ P_{kj}(h) &= o(h), \quad j \neq k+1, k-1, k\,. \end{aligned}$$

From the infinitesimal probabilities, it follows that $\{X(t), t \geq 0\}$ is birth-death process with finite state space $S = \{0, 1, \ldots, m\}$ and birth rates $\lambda_k = (m-k)\lambda$, $k = 0, 1, \ldots, m-1$ and death rates $\mu_k = \mu$, $k = 1, 2, \ldots, m$ and $\mu_0 = 0$. The long run distribution exists and is obtained by solving the following balance equations.

$$m\lambda P_0 = \mu P_1, \;\; \mu P_m = \lambda P_{m-1} \;\; \& \;\; [(m-k)\lambda + \mu]P_k = (m-k+1)\lambda P_{k-1} + \mu P_{k+1},$$

for $k = 1, 2, \ldots, m-1$. With $k = m-1$, from the above equations we get,

$$\mu P_m = \lambda P_{m-1} \;\;\Rightarrow\;\; P_{m-1} = (\mu/\lambda)P_m \;\;\Rightarrow\;\; P_{m-2} = (1/2)(\mu/\lambda)^2 P_m.$$

Now, we show by induction that $P_k = (\mu/\lambda)^{m-k} P_m/(m-k)!, \quad k = 0, 1, \ldots, m.$ The result is true for $k = m, m-1$ and $k = m-2$. Now suppose that the result is true for $k = j$ and $k = j-1$. For $k = j-1$, we have,

$$[\lambda(m-j+1) + \mu]P_{j-1} = \lambda(m-j+2)P_{j-2} + \mu P_j.$$

Hence, $\quad \lambda(m-j+2)P_{j-2}$

$$\begin{aligned} &= \lambda(m-j+1) + \mu]P_{j-1} - \mu P_j \\ &= \left[\frac{(\lambda(m-j+1)+\mu)(\mu/\lambda)^{m-j+1}}{(m-j+1)!} - \frac{\mu(\mu/\lambda)^{m-j}}{(m-j)!}\right] P_m \\ &= \left[\frac{\mu^{m-j+1}(1/\lambda)^{m-j}}{(m-j)!} + \frac{\mu(\mu/\lambda)^{m-j+1}}{(m-j+1)!} - \frac{\mu^{m-j+1}(1/\lambda)^{m-j}}{(m-j)!}\right] P_m \\ &= \frac{\mu(\mu/\lambda)^{m-j+1}}{(m-j+1)!} P_m \\ \Rightarrow \;\; P_{j-2} &= \frac{(\mu/\lambda)^{m-j+2}}{(m-j+2)!} P_m. \end{aligned}$$

Thus the result is true for $k = j-2$, if it is true for $k = j$ and $k = j-1$. Hence by induction $P_k = (\mu/\lambda)^{m-k} P_m/(m-k)!$ for $k = 0, 1, 2, \ldots, m$. Now the condition

$$\sum_{k=0}^{m} P_k = 1 \;\;\Longrightarrow\;\; P_m = \left(\sum_{k=0}^{m} \frac{1}{(m-k)!}\left(\frac{\mu}{\lambda}\right)^{m-k}\right)^{-1}.$$

We now derive the long run expected number E of machines in a failed state. It is given by $E = \sum_{k=1}^{m} kP_k$. To find the expression for kP_k note that

$$
\begin{aligned}
P_{k+1}/P_k &= \frac{(m-k)!(\mu/\lambda)^{m-k-1}}{(m-k-1)!(\mu/\lambda)^{m-k}} \\
\Rightarrow\ P_{k+1} &= \frac{(m-k)\lambda P_k}{\mu} \quad \Rightarrow\ kP_k = mP_k - (\mu/\lambda)P_{k+1} \\
\Rightarrow\ E &= \sum_{k=1}^{m} kP_k = \sum_{k=1}^{m} (mP_k - (\mu/\lambda)P_{k+1}) \\
&= m(1-P_0) - (\mu/\lambda)(1-P_0-P_1) \\
&= m(1-P_0) - (\mu/\lambda)(1-P_0) + \frac{\mu}{\lambda}P_1 \\
&= m(1-P_0) - (\mu/\lambda)(1-P_0) + mP_0 \\
&= m - (\mu/\lambda)(1-P_0)\ .
\end{aligned}
$$

Suppose in the workshop there are m machines and $r < m$ repair facilities. Then proceeding on similar lines, we note that $\{X(t), t \geq 0\}$ is a birth-death process with finite state space $S = \{0, 1, \ldots, m\}$ and birth rates $\lambda_k = (m-k)\lambda$, $k = 0, 1, \ldots, m-1$ and death rates $\mu_k = k\mu$ if $k < r$ and $\mu_k = r\mu$ if $k \geq r$. □

The next section presents R codes used to find the realization of the processes studied in this chapter.

8.7 R Codes

Following is a R code to find a realization of a Yule Furry process. It is similar to that of a Poisson process. It is illustrated for the Yule Furry process in Example 8.2.1.

Code 8.7.1 Realization of a Yule Furry process: We obtain a realization of a Yule Furry process with $X(0) = 1$ and birth rate λ as 0.20 and 0.35, when it is observed for $T = 10$ time units.

```
# Part I: Input birth rates, X(0)and T
la=c(0.20,0.35); a=1; T=10
# Part II: Realizations
int=x=arr=u=v=w=list(); N=c()
for(j in 1:length(la))
{
set.seed(j)
y=s=c();sumy=0; i=1;s[1]=a
while(sumy<T)
{
```

```
y[i]=rexp(1,rate=s[i]*la[j])
s[i+1]=s[i]+1
sumy=sum(y)
i=i+1
}
int[[j]]=y[-length(y)]
N[j]=length(int[[j]])
x[[j]]=s[-length(s)]
u[[j]]=rep(x[[j]],each=2)
arr[[j]]=round(cumsum(int[[j]]),2)
v[[j]]=c(0,round(rep(arr[[j]],each=2),3),T)
w[[j]]=c(0,arr[[j]])
}
N
list("States"=x,"Interval r.v."=int,"Birth epochs"=arr)
# Part III: Graphs of realizations
par(mfrow=c(2,1))
rate=paste("Birth Rate=",la,sep=" ")
number=paste("No of Births=",N,sep=" ")
for(j in 1:length(la))
{
plot(v[[j]],u[[j]],"l",main=rate[j],ylab="States",
xlab="Occurrence Time", yaxt="n",xaxt="n",col="blue")
axis(2,at=sort(unique(u[[j]])),labels=sort(unique(u[[j]])),
las=2)
axis(1,at=round(c(arr[[j]],T),2),las=2,cex.axis=0.7)
points(w[[j]],x[[j]],pch=20,col="dark blue")
abline(v=T)
legend("topleft",legend=number[j],cex=.8)
}
```

□

In the next code, we obtain multiple realizations of a Yule Furry process and verify the result that $X(t)$ has geometric distribution. It is illustrated for the process in Example 8.2.2.

Code 8.7.2 Multiple realizations of a Yule Furry process: Suppose $\{X(t), t \geq 0\}$ is a Yule Furry process with birth rate $\lambda = 0.12$ and $X(0) = 1$. With this code, we obtain 150 realizations of the process when it is observed for $T = 10$ time units and verify whether $X(T)$ follows a geometric distribution.

```
# Part I: Input birth rate, T and number of simulations
lambda=.12; T=10; nsim=150
# Part II: Multiple simulations
x=c() # population size at T
for(m in 1:nsim)
```

```
{
set.seed(m); y=c(); sumy=0; i=1
while(sumy<T)
{
y[i]=rexp(1,rate=i*lambda)
sumy=sum(y); i=i+1
}
l=length(y); l; y=y[-l]; x[m]=length(y)+1
}
x; mean(x);var(x); e=exp(1); p=e^(-lambda*T); mu=1/p;mu
var=(1-p)/p^2; var
# Part III: Observed and expected distributions
d=data.frame(table(factor(x, levels=1:max(x))))
obf=d$Freq; N=sum(obf); rf=obf/N; t=sort(unique(x))
v=p*(1-p)^(t-1); ex=nsim*v; ex=round(ex,2);
a=min(min(rf),min(v));b=max(max(rf),max(v))
u=seq(round(a,2),b,.03)
plot(t,rf,"h",col="blue",lty=1,lwd=2,xlab="Values of X",
main="Observed and Expected Distributions",
ylim=range(c(a,b)),ylab="Relative Frequency and Expected probability",
yaxt="n")
axis(2,at=u); points(t,rf,pch=20)
lines(t,v,"h",col="red",lty=2,lwd=1)
points(t,v,pch=20)
abline(h=0, col="blue")
legend("topright",cex=.7,col=c("blue","red"),lty=c(1,2),
legend=c("Observed Distribution","Expected Distribution"),
lwd=c(2,1))
# Part IV: Goodness of fit test with pooling frequencies
v1=1-sum(v); v2=c(v,v1); e=nsim*v2; length(e);ob=c(obf,0)
length(ob); y=c(t,"> 12"); da=round(data.frame(ob,e),2)
da1=data.frame(y,da);da1
o1=c(ob[1:8],ob[9]+ob[10]+ob[11]+ob[12])
e1=c(e[1:8],e[9]+e[10]+e[11]+e[12])
t1=sum((o1-e1)^2/e1); t1;df=length(o1)-1;df
b1=qchisq(.95,df); b1; p1=1-pchisq(t1,df); p1
# Part V: Goodness of fit test without pooling frequencies
d=length(ob)-1; d
t=sum((ob-e)^2/e); t; b=qchisq(.95,d); b; p=1-pchisq(t,d); p
# Part VI: Goodness of fit test with built-in function
chisq.test(ob,p=v2)
```

□

The next R code is to obtain a realization of a linear death process. It is illustrated for the process in Example 8.3.1.

Code 8.7.3 Realization of a linear death process: We obtain a realization of a linear death process with death rate μ as 0.1 and 0.25 and $X(0) = a = 20$, when it is observed for $T = 10$ time units.

```
# Part I: Input death rates, X(0) and T
mu=c(.1,.25); a=20; T=10
# Part II: Realizations
int=x=arr=u=v=w=list(); N=c()
for(j in 1:length(mu))
{
set.seed(j); y=s=c();sumy=0; i=1;s[1]=a
while(sumy<T)
{
y[i]=rexp(1,rate=s[i]*mu[j])
s[i+1]=s[i]-1
if(s[i]==0){break}
else
sumy=sum(y); i=i+1
}
int[[j]]=y[-length(y)]
N[j]=length(int[[j]])
x[[j]]=s[-length(s)]
u[[j]]=rep(x[[j]],each=2)
arr[[j]]=round(cumsum(int[[j]]),2)
v[[j]]=c(0,round(rep(arr[[j]],each=2),3),T)
w[[j]]=c(0,arr[[j]])
}
# x;int;arr;u;v;w
N
list("States"=x,"Interval r.v."=int,"Death epochs"=arr)
# Part III: Graphs of realizations
par(mfrow=c(2,1))
rate=paste("Death Rate=",mu,sep=" ")
number=paste("No of Deaths=",N,sep=" ")
for(j in 1:length(mu))
{
plot(v[[j]],u[[j]],"l",main=rate[j],ylab="States",
xlab="Occurrence Time",yaxt="n",xaxt="n",col="blue")
axis(2,at=sort(unique(u[[j]])),labels=sort(unique(u[[j]])),
las=2)
axis(1,at=round(c(arr[[j]],T),2),las=2,cex.axis=0.7)
points(w[[j]],x[[j]],pch=20,col="dark blue")
abline(v=T)
legend("bottomleft",legend=number[j],cex=.7)
}
```

□

Using the following code, we obtain multiple realizations of a linear death process and verify the result that $X(t)$ follows a binomial distribution. It is illustrated for the linear death process in Example 8.3.2.

Code 8.7.4 Multiple realizations of a linear death process: We obtain 200 realizations of a linear death process with death rate $\mu = 0.12$ and $X(0) = 10$, when it is observed for $T = 12$ time units and verify whether $X(T)$ follows the binomial $B(10, e^{-\mu T})$ distribution.

```
# Part I: Input death rate, X(0),T and number of simulations
mu=.12; nsim=200; T=12; ini=10 # X(0)
# Part II: Multiple realizations
y=s=x=c();
for(m in 1:nsim)
{
set.seed(m); sumy=0; s[1]=ini; i=1
while(sumy<T)
{
y[i]=rexp(1,rate=s[i]*mu)
s[i+1]=s[i]-1
if(s[i]==0){break}
else
sumy=sum(y); i=i+1
}
l=length(y); y=y[-l]; x[m]=ini-length(y)
}
mean(x);var(x); e=exp(1);p=e^(-mu*T);p
mu1=ini*p;mu1; var=ini*p*(1-p);var
d=data.frame(table(factor(x, levels=0:max(x))));d
# Part III: Observed and expected distributions
obf=d$Freq; N=sum(obf); rf=obf/N;rf
ob=c(obf,0,0,0); r=0:ini; rf1=c(rf,0,0,0)
pr=dbinom(r,size=ini,prob=p);sum(pr)
ex=nsim*pr; ex=round(ex,2);ex;
da=data.frame(r,ob,ex);da
a=min(min(rf1),min(pr));b=max(max(rf1),max(pr))
u=seq(round(a,2),b,.03);u
plot(r,rf1,"h",col="blue",lty=1,lwd=2,xlab="Values of X",
main="Observed and Expected Distributions",ylim=range(c(a,b)),
ylab="Relative Frequency and Expected probability",yaxt="n")
axis(2,at=u); points(r,rf1,pch=20)
lines(r,pr,"h",col="red",lty=2,lwd=1)
points(r,pr,pch=20)
abline(h=0, col="blue")
legend("topright",cex=.7,col=c("blue","red"),lty=c(1,2),
legend=c("Observed Distribution","Expected Distribution"),
```

```
lwd=c(2,1))
# Part IV: Goodness of fit test, with pooling the frequencies
e=nsim*pr;e; o1=c(obf[1:6],obf[7]+obf[8])
e1=c(e[1:6],e[7]+e[8]+e[9]+e[10]+e[11]); e1
t1=sum((o1-e1)^2/e1); t1;df=length(o1)-1;df
b1=qchisq(.95,df); b1; p1=1-pchisq(t1,df); p1
# Part V: Goodness of fit test, without pooling the frequencies
ob=c(obf,0,0,0);df1=length(ob)-1; df1
t=sum((ob-e)^2/e); t; b2=qchisq(.95,df1); b2
p2=1-pchisq(t,df1); p2
# Part VI: Goodness of fit test, with built-in function
chisq.test(ob,p=pr)
```

□

The following code obtains a realization of a linear birth-death process using R. It is illustrated for the process in Example 8.5.1.

Code 8.7.5 Realization of a linear birth-death process: We obtain the realization of a linear birth-death process with birth rate λ, death rate μ and $X(0) = 10$, for a fixed time interval (0, 10]. For comparison we take two sets of birth and death rates as $\lambda = 0.12$, $\mu = 0.1$ and $\lambda = 0.1$, $\mu = 0.4$

```
# Part I: Input birth and death rates, X(0) and T
la=c(0.12,0.1); mu=c(0.1,0.4); a=10; T=10
# Part II: Realizations
P=matrix(c(mu[1]/(la[1]+mu[1]),la[1]/(la[1]+mu[1]),
mu[2]/(la[2]+mu[2]),la[2]/(la[2]+mu[2])), ncol=2,byrow=TRUE)
P # tpm of embedded Markov chain
S=c(-1,1); int=x=arr=u=v=w=list(); N=c()
for(j in 1:length(mu))
{
set.seed(j)
y=s=c();sumy=0; i=1;s[1]=a
while(sumy<T)
{
y[i]=rexp(1,rate=s[i]*(la[j]+mu[j]))
s[i+1]=s[i]+sample(S,1,P[,j],replace=T)
if(s[i]==0){break}
else
sumy=sum(y); i=i+1
}
int[[j]]=y[-length(y)]
N[j]=length(int[[j]])
x[[j]]=s[-length(s)]
u[[j]]=rep(x[[j]],each=2)
arr[[j]]=round(cumsum(int[[j]]),2)
```

```
v[[j]]=c(0,round(rep(arr[[j]],each=2),3),T)
w[[j]]=c(0,arr[[j]])
}
N
list("States"=x,"Interval r.v."=int,"epochs"=arr)
# Part III: Graphs of realizations
par(mfrow=c(2,1))
rate=paste("Birth Rate=",la,"Death Rate=",mu,sep="  ")
number=paste("No of Events=",N,sep="  ")
for(j in 1:length(mu))
{
plot(v[[j]],u[[j]],"l",main=rate[j],ylab="States",
xlab="Occurrence Time",yaxt="n",xaxt="n",col="blue")
axis(2,at=sort(unique(u[[j]])),labels=sort(unique(u[[j]])),
las=2)
axis(1,at=round(c(arr[[j]],T),2),las=2,cex.axis=0.7)
points(w[[j]],x[[j]],pch=20,col="dark blue")
abline(v=T)
legend("topleft",legend=number[j],cex=.7,col="red")
}
```

□

A quick recap of the results discussed in the present chapter is given below.

Summary

1. Suppose $\{X(t), t \geq 0\}$ is a time homogeneous continuous time Markov chain with state space W, $X(0) = 0$ and the following infinitesimal transition probabilities.

$$\begin{aligned} P_{i,i+1}(h) &= \lambda_i h + o(h) \;\; P_{ii}(h) = 1 - \lambda_i h + o(h) \\ P_{ij}(h) &= o(h) \;\; \forall \;\; j \neq i, \; i+1 \,. \end{aligned}$$

   Then $\{X(t), t \geq 0\}$ is known as a pure birth process and $\{\lambda_i, i \geq 0\}$ are known as birth rates. If $X(0) = a > 0$ and $\lambda_i = i\lambda, \; i \geq 1$, then it is known as a linear birth process or Yule Furry process.
2. For a pure birth process with distinct birth rates $\lambda_i, \; i \in S$ and $X(0) = a$, for $k = a+1, a+2, \ldots, P_k^{(a)}(t) = P[X(t) = k|X(0) = a]$ is given by,

$$P_k^{(a)}(t) = \lambda_a \lambda_{a+1} \cdots \lambda_{k-1} \sum_{j=a}^{k} e^{-\lambda_j t} \left(\prod_{i=a, i \neq j}^{k} (\lambda_i - \lambda_j) \right)^{-1},$$

   and $\;\; P_a^{(a)}(t) \; = e^{-\lambda_a t}.$

8.8.6 Suppose there are m welders. The probability that a welder not using an electric supply at time t starts using it in $(t, t+h]$ is $\lambda h + o(h)$. The probability that a welder using an electric supply at time t stops using it in $(t, t+h]$ is $\mu h + o(h)$. Welders are assumed to work independently of each other. Suppose $X(t)$ denotes the number of welders using electric supply at time t. Examine the long run behavior of the process.

8.8.7 A birth and death process has parameters $\lambda_k = \alpha(k+1)$ for $k = 0, 1, 2, \ldots,$ and $\mu_k = \beta(k+1)$ for $k = 1, 2, \ldots$. (i) Examine whether the long run distribution of the process exists. (ii) If it exists, find it. (iii) Find the long run distribution if $\alpha = 0.2$ and $\beta = 0.5$.

8.8.8 Suppose customers arrive at a service facility with a single service counter, according to the Poisson process with rate $\lambda = 5$ per hour. The service time random variable for each customer has exponential distribution with mean $1/\mu = 10$ min. The facility has a limited waiting capacity of 10 chairs. (i) Find the long-run fraction of the time the service facility is idle. (i) Find the long-run fraction of the time the service facility is full.
(iii) Solve(i) and (ii) if the mean service time is 20 min. (iv) Comment on the results.

8.8.9 A time-shared computer system has three terminals that are attached to a central processing unit that can simultaneously handle at most two active users. If a person logs on and requests service when two other users are active, then the request is held in a buffer until it can receive service. Suppose $X(t)$ is the total number of requests that are either active or in the buffer at time t. Assume that X(t) is a birth and death process with parameters $\lambda_k = \lambda$ for $k = 0, 1, 2$ and $\lambda_k = 0$ for all $k \geq 3$, $\mu_k = k\mu$ for $k = 0, 1, 2$ and $\mu_k = 2\mu$ for $k = 3$. (i) Determine the long run probability that the system is idle. (ii) Determine the long run probability that the system has two active users. (iii) For $\lambda = 0.3$ and $\mu = 0.4$, find the long run distribution. What is the long run mean proportion of time the system is idle?

8.8.10 Suppose in a workshop there are 3 machines and a single repair facility. Probability that a machine working at time t, fails in time $(t, t+h]$ is $\lambda h + o(h)$, independently of other machines. If the repair facility is busy at time t, probability that the repair is complete in $(t, t+h]$ is $\mu h + o(h)$ independently of state of the other machines. Suppose $X(t)$ denotes the number of machines not working at time t, then $\{X(t), t \geq 0\}$ is modeled as a continuous time Markov chain with finite state space $S = \{0, 1, 2, 3\}$. (i) Express it as a birth-death process. Identify the birth and death rates. (ii) In the long run, find the probability that all machines are in working condition when $\lambda = 0.4$ and $\mu = 0.3$. (iii) For these values of λ and μ, find the expected number of idle machines, in the long run.

8.9 Computational Exercises

8.9.1 Suppose $\{X(t), t \geq 0\}$ is a Yule Furry process with birth rate λ and $X(0) = 1$. Obtain a realization of the process when it is observed for T time units. For comparison take two values of λ.

8.9.2 Suppose $\{X(t), t \geq 0\}$ is a Yule Furry process with birth rate λ and $X(0) = 1$. Obtain multiple realizations of the process when it is observed for T time units and verify whether that $X(T)$ follows a geometric distribution, graphically and using the appropriate test procedure.

8.9.3 Suppose $\{X(t), t \geq 0\}$ is a linear death process with death rate μ and $X(0) = a$. Obtain a realization of the process when it is observed for T time units. For comparison take two values of μ.

8.9.4 Suppose $\{X(t), t \geq 0\}$ is a linear death process with death rate μ and $X(0) = a$. Obtain multiple realizations of the process when it is observed for T time units and verify whether $X(T)$ follows a binomial $B(a, e^{-\mu T})$ distribution, graphically and using the appropriate test procedure.

8.9.5 Suppose $\{X(t), t \geq 0\}$ is a linear birth-death process with $X(0) = a$, birth rate λ and death rate μ. Obtain its realization for a fixed time interval $(0, T]$. For comparison take two sets of birth and death rates.

8.9.6 Find a realization of $M/M/1$ queuing system when it is observed for a fixed time interval $(0, T]$.

8.9.7 Find a realization of $M/M/5$ queuing system when it is observed for a fixed time interval $(0, T]$.

8.10 Multiple Choice Questions

Note: In each of the questions, multiple options may be correct.

8.10.1 Which of the following options is/are correct? Suppose $\{X(t), t \geq 0\}$ is a Yule Furry process with birth rate λ and $X(0) = 1$. Then the distribution of $X(t)$ is geometric with success probability p where

(a) $p = e^{-\lambda t}$ and support $\{0, 1, 2, \ldots,\}$
(b) $p = e^{-\lambda t}$ and support $\{1, 2, \ldots,\}$
(c) $p = 1 - e^{-\lambda t}$ and support $\{1, 2, \ldots,\}$
(d) $p = 1 - e^{-\lambda t}$ and support $\{0, 1, 2, \ldots,\}$

8.10.2 Suppose $\{X(t), t \geq 0\}$ is a Yule Furry process with birth rate λ and $X(0) = a$. Which of the following options is/are correct?

(a) $E(X(t)) = ae^{-\lambda t}$ & $Var(X(t)) = ae^{\lambda t}(e^{\lambda t} - 1)$
(b) $E(X(t)) = ae^{\lambda t}$ & $Var(X(t)) = a^2e^{\lambda t}(e^{\lambda t} - 1)$
(c) $E(X(t)) = ae^{\lambda t}$ & $Var(X(t)) = ae^{\lambda t}(e^{\lambda t} - 1)$
(d) $E(X(t)) = ae^{\lambda t}$ & $Var(X(t)) = ae^{2\lambda t}(e^{\lambda t} - 1)$

8.10.3 Which of the following options is/are correct? Suppose $\{X(t), t \geq 0\}$ is a linear death process with $X(0) = a > 0$ and death rate μ. Then for fixed t, the distribution of $X(t)$ is binomial

(a) $B(a, e^{-\mu t})$
(b) $B(a, e^{-\mu}t)$
(c) $B(a, \mu e^{-\mu t})$
(d) $B(a, e^{\mu t}/(1 + e^{\mu t}))$

8.10.4 Which of the following options is/are correct? A linear death process is a continuous time Markov chain, where

(a) all states are non-null persistent
(b) all states are transient
(c) 0 is a non-null persistent state and all other states are null persistent
(d) 0 is a non-null persistent state and all other states are transient

8.10.5 Which of the following options is/are correct? A birth and death process is a continuous time Markov chain with state space $\{0, 1, \ldots, \}$ and the intensity rates for $i \geq 0$ are

(a) $q_{i,i+1} = i\mu,\ q_{i,i-1} = \lambda_i,\ q_{ii} = -(\lambda_i + i\mu)\ \&\ q_{ij} = 0\ \forall\ j \neq i+1, i-1$
(b) $q_{i,i+1} = \lambda_i,\ q_{i,i-1} = \mu_i,\ q_{ii} = -(\lambda_i + \mu_i)\ \&\ q_{ij} = 0\ \forall\ j \neq i+1, i-1$
(c) $q_{i,i+1} = i\lambda,\ q_{i,i-1} = i\mu,\ q_{ii} = -i(\lambda + \mu)\ \&\ q_{ij} = 0\ \forall\ j \neq i+1, i-1$
(d) $q_{i,i+1} = \lambda_i,\ q_{i,i-1} = \mu_i,\ q_{ii} = -(\lambda_i + \mu_i)\ \&\ q_{ij} = 1\ \forall\ j \neq i+1, i-1$

8.10.6 Which of the following options is/are correct? In the embedded Markov chain of a birth and death process

(a) $p_{i,i+1} = \lambda_i/(\lambda_i + \mu_i),\ \ i \geq 0$
(b) $p_{i,i-1} = \mu_i/(\lambda_i + \mu_i),\ \ i \geq 1$
(c) $p_{i,i+1} = \lambda_i/\mu_i,\ \ i \geq 0$
(d) $p_{i,i-1} = \mu_i/\lambda_i,\ \ i \geq 1$

8.10.7 Which of the following options is/are correct? For a linear birth-death process with birth rate λ, death rate μ and $X(0) = a$, the mean function $M(t) = E(X(t))$ is

(a) $M(t) = ae^{\lambda\mu t}$
(b) $M(t) = ae^{-(\lambda-\mu)t}$
(c) $M(t) = ae^{(\lambda-\mu)t}$
(d) $M(t) = ae^{(\lambda+\mu)t}$

8.10.8 Which of the following options is/are correct? Suppose $\{X(t), t \geq 0\}$ is a pure birth process with birth rate $\lambda_i = \lambda\ \forall\ i \in S$ and $X(0) = 0$. Then

(a) $\{X(t), t \geq 0\}$ is a process with stationary and independent increments
(b) $E(X(t)) = e^{\lambda t}$
(c) $X(t) \sim P(\lambda t)$ for fixed t
(d) $X(0)$ cannot be 0 in a pure birth process

References

1. Bhat, B. R. (2000). *Stochastic models: Analysis and applications*. New Delhi: New Age International.
2. Feller, W. (1978). *An introduction to probability theory and its applications* (Vol. I). New York: Wiley.
3. Karlin, S., & Taylor, H. M. (1975). *A first course in stochastic processes*. New York: Academic Press.
4. Ross, S. M. (2014). *Introduction to probability models* (11th ed.). New York: Academic Press.

Chapter 9
Brownian Motion Process

9.1 Introduction

Brownian motion process is the most renowned process and is the first which was thoroughly investigated. Historically, in the summer of 1827 the British botanist Robert Brown, using a microscope, observed that microscopic pollen grains suspended in a drop of water moved constantly in haphazard zigzag trajectories. Since then, not only biologists but physicists also observed the movement of charged particles to follow a similar pattern. Small particles suspended in fluid or dust particles or smoke floating in air execute ceaseless irregular motion due to collisions with the molecules in the gas or liquid in which these are suspended. In honor of Robert Brown, such motion has been labeled as Brownian motion. Over the years, it was established that finer particles move more rapidly, that the motion is stimulated by heat and that the movement becomes more active with a decrease in fluid viscosity.

A mathematical description of this phenomenon was first derived from the laws of physics by Einstein in 1905. He asserted that the Brownian motion originates in the continual bombardment of the pollen grains by the molecules of the surrounding water, with successive molecular impacts coming from different directions and contributing different impulses to the particles. Einstein argued that as a result of the continual collisions, the particles themselves had the same average kinetic energy as the molecules. Belief in molecules and atoms was not universal in 1905, and the success of Einstein's explanation of the well-documented existence of Brownian motion convinced a number of distinguished scientists that such things as atoms actually exist. Incidentally, 1905 is the same year in which Einstein set forth his theory of relativity and his quantum explanation for the photoelectric effect.

Brownian motion is complicated because the molecular bombardment of the suspended particles is itself a complicated process. It took more than a decade to get a clear picture of the Brownian motion stochastic process. In 1923, the probabilist Norbert Wiener set forth the modern mathematical foundation. The first concise mathematical formulation of the theory and the rigorous mathematical framework

S. Madhira and S. Deshmukh, *Introduction to Stochastic Processes Using R*,
https://doi.org/10.1007/978-981-99-5601-2_9

for its description were given by Wiener in his 1918 dissertation and later papers. Hence in his honor, the Brownian motion process is called a Wiener process. It is also known as the Wiener-Einstein process. After Wiener's basic formulation, the Brownian motion process has undergone extensive mathematical development. Predating Einstein by some years, in 1900 in Paris, Louis Bachelier proposed a Brownian motion model for the movement of prices in the French bond market. While Bachelier's paper was largely ignored by academics for many decades, his work now stands as the innovative first step in the mathematical theory of stock markets that has greatly altered the financial world of today.

Suppose $X(t)$ denotes the position of a particle at time t, then the possible value of $X(t)$ is a three-dimensional point. Thus, the natural state space of the stochastic process $\{X(t), t \geq 0\}$ is $\mathbb{R}^3$. However, in this book we restrict to the study of one-dimensional Brownian motion only that corresponds to one coordinate of the three-dimensional point. Thus, we consider the state space as $\mathbb{R}$ or its subset. In the framework of stochastic processes with state space as $\mathbb{R}$, the Brownian motion process is a continuous time, continuous state space Markov process having stationary and independent increments. A Brownian motion is considered as an approximation to an unrestricted random walk on the set of integers. Secondly, it also arises as a solution to a partial differential equation. We elaborate on these two approaches in the next section, after providing the definition of a Brownian motion process.

The Brownian motion process was originally developed as a model of diffusion of a particle suspended in a liquid or gas. However, nowadays the process and its many generalizations and extensions occur in numerous and diverse areas of pure and applied sciences such as finance, in particular in the analysis of the price levels of the stock market, quantum mechanics, economics, genetics, demography, communication theory, biology and management science. It is used as a model for storage systems such as dams and inventories and as a model for replenishment of reserves. In queuing theory, it is applied as a model to approximate heavy traffic.

In the next section, we give the definition of a Brownian motion process and study some of its properties. One of the peculiar properties of the Brownian motion is continuity but non-differentiability of its sample paths. Using the continuity property of sample paths and the celebrated reflection principle, we can find the distribution of the maximum and the minimum of a Brownian motion process when t varies over $[0, T]$. These results are discussed in Sect. 9.3. There are many variations and extensions of a Brownian motion process. Sections 9.4 and 9.5 present two such extensions, Brownian bridge and geometric Brownian motion, respectively. In Sect. 9.6, we briefly introduce some more variations including the Ornstein-Uhlenbeck process. `R` codes used for illustrations are given in Sect. 9.7.

9.2 Definition and Properties

Physical considerations of a Brownian motion of a particle lead to the following definition of a Brownian motion process.

Definition 9.2.1 Brownian Motion Process: A continuous time and continuous state space stochastic process $\{X(t), t \geq 0\}$ with state space $\mathbb{R}$ is said to be a Brownian motion process or the Wiener process with drift coefficient μ and diffusion coefficient σ^2, if (i) X(0) = 0, (ii) $\{X(t), t \geq 0\}$ has stationary and independent increments and (iii) $\forall\, t > 0$, $X(t) \sim N(\mu t, \sigma^2 t)$ distribution.

Remark 9.2.1 The condition $X(0) = 0$ can be relaxed. As we have proved in Sect. 1.3, we can define $Y(t)$ as $X(t) - X(0)$, so that $Y(0) = 0$ which satisfies all other properties of the process $\{X(t), t \geq 0\}$. Hence, without loss of generality we can take $X(0) = 0$. If $\mu = 0$ and $\sigma^2 = 1$, the process is said to be a standard Brownian motion process. Suppose $\{W(t), t \geq 0\}$ denotes the standard Brownian motion process with $W(0) = 0$ and $X(t)$ is defined as $X(t) = x_0 + \mu t + \sigma W(t)$, then $\{X(t), t \geq 0\}$ is a Brownian motion process with drift coefficient μ, diffusion coefficient σ^2 and $X(0) = x_0$. For further study of a Brownian motion process, we restrict to the standard Brownian motion process.

We now discuss how a Brownian motion process can be considered as an approximation to an unrestricted random walk on the set of integers, (Karlin and Taylor [5]).

Brownian motion process as an approximation to random walk: In an unrestricted simple random walk, in each time unit a transition is to the right with probability p and to the left with probability $q = 1 - p$. Suppose this process is accelerated by taking smaller and smaller steps in smaller and smaller time intervals. More precisely, suppose each step is of size Δx and time between two consecutive steps is Δt. It is assumed that Δt and Δx are small and tend to 0 at a certain rate. Suppose $X(t)$ denotes the position of a particle at time t. It can be expressed as

$$X(t) = \Delta x(X_1 + X_2 + \cdots + X_{[t/\Delta t]}),$$

where $[t/\Delta t]$ is the integer part of $t/\Delta t$ and $\{X_1, X_2, \ldots\}$ are independent and identically distributed random variables such that

$$X_i = \begin{cases} 1, & \text{if step is to the right} \\ -1, & \text{if step is to the left.} \end{cases}$$

Thus,

$$E(X_i) = \Delta x(p - q) \;\; \& \;\; Var(X_i) = (\Delta x)^2[p + q - (p - q)^2] = 4pq(\Delta x)^2 .$$

It then follows that

$$E(X(t)) = \Delta x(p-q)\,[t/\Delta t] \;\;\&\;\; Var(X(t)) = 4pq(\Delta x)^2\;[t/\Delta t]\;.$$

Suppose

$$p = 1/2 + (\mu/2\sigma^2)\Delta x \;\;\&\;\; q = 1/2 - (\mu/2\sigma^2)\Delta x,\;\; \mu \in \mathbb{R},\; \sigma^2 > 0\;.$$

Then, $E(X(t))$ and $Var(X(t))$ can be expressed as

$$\begin{aligned} E(X(t)) &= \Delta x(p-q)\,[t/\Delta t] = \mu((\Delta x)^2/\sigma^2)\,[t/\Delta t] \\ \&\;\; Var(X(t)) &= 4pq(\Delta x)^2\,[t/\Delta t] = (\Delta x)^2\Big(1-\big(\mu\Delta x/\sigma^2\big)^2\Big)\,[t/\Delta t]\;. \end{aligned}$$

Further, Δt and Δx are allowed to tend to 0 so that limit of $E(X(t))$ and $Var(X(t))$ exist. Thus, it is required that $\Delta x \to 0$ and $\Delta t \to 0$ such that $(\Delta x)^2/\Delta t \to \sigma^2$. It then follows that $E(X(t)) \to \mu t$ and $Var(X(t)) \to \sigma^2 t$. Note that the total displacement $X(t) - X(0)$ during a time interval of length t is a sum of $n \approx [t/\Delta t]$ independent and identically distributed Bernoulli type random variables. Hence, by the central limit theorem as $n \to \infty$, given $X(0) = 0$, the distribution of $X(t)$ can be approximated by normal $N(\mu t, \sigma^2 t)$ distribution. When $X(0) = x_0$, the position at t is $X(t) + x_0$. Conditional on $X(0) = x_0$, the distribution of $X(t)$ can be approximated by normal $N(x_0 + \mu t, \sigma^2 t)$ distribution. In general, given $X(s) = x_s$, the displacement $X(t) - X(s)$ during $(s, t]$ can be approximated by normal $N(x_s + \mu(t-s), \sigma^2(t-s))$ distribution. Thus, $\{X(t), t \geq 0\}$ is a Markov process with stationary and independent increments, as the underlying random walk has this property and for each fixed t, $X(t)$ has normal distribution. These properties characterize a Brownian motion process. The normal distribution plays an important role in the analysis of a Brownian motion process, analogous to the role played by a Poisson distribution in a Poisson process. For a symmetric random walk, $p = q = 1/2$ and hence $\mu = 0$. However, as $\Delta x \to 0$, $p \to 1/2$ and $q \to 1/2$ even if $\mu \neq 0$.

We now elaborate on how a Brownian motion process also arises as a solution to a partial differential equation, Karlin and Taylor [5].

Brownian motion process as a solution of a differential equation: Suppose $X(t)$ denotes the x coordinate of the position of a particle in a Brownian motion at time t and x_0 denotes the same at time t_0. Since motion of a particle is subject to perpetual collision with the molecules of the surrounding medium, it follows that for fixed t, $X(t)$ is a random variable. It seems reasonable to assume that the distribution of $X(t) - X(s)$ is the same as that of $X(t+h) - X(s+h)$ for any h, if it can be assumed that the medium is in equilibrium. Thus, it is intuitively clear that the distribution of $X(t) - X(s)$ should depend on $t - s$ and not on the time when we begin the observation. Similarly, it can be assumed that the displacement of the particle over non-overlapping intervals are independent random variables. Suppose $p(x, t|x_0)$ denotes the probability density function of $X(t + t_0)$ given that $X(t_0) =$

x_0. Since the probability law governing the transition is stationary in time, $p(x, t|x_0)$ does not depend on t_0 and hence we take $t_0 = 0$. Further, for small t, $X(t)$ is likely to be very close to $X(0)$, hence mathematically we have $\lim_{t\to 0} p(x, t|x_0) = 0 \ \forall \ x \neq x_0$. From principles of physics, Einstein in 1905 showed that $p(x, t|x_0)$ satisfies the partial differential equation,

$$\frac{\partial}{\partial t} p(x, t|x_0) = \frac{\sigma^2}{2} \frac{\partial^2}{\partial x^2} p(x, t|x_0), \ \ \sigma^2 > 0 \, . \qquad (9.2.1)$$

This equation is known as a diffusion equation or heat equation and σ^2 is known as diffusion coefficient. Small particles execute Brownian motion owing to collisions with the molecules in the gas or liquid in which those are suspended. The evaluation of σ^2 is based on the formula $\sigma^2 = RT/Nf$ where R is the gas constant, T is the temperature, N is Avogadro's number and f is the coefficient of friction. By choosing the proper scale, σ^2 can be taken as 1. We now verify that

$$p(x, t|x_0) = \frac{1}{\sqrt{2\pi t}} \exp\{-\frac{1}{2t}(x - x_0)^2\}$$

is a solution to the diffusion equation (9.2.1). Note that

$$\begin{aligned}
\frac{\partial}{\partial t} p(x, t|x_0) &= \frac{1}{\sqrt{2\pi}} \frac{-1}{2t^{3/2}} \exp\{-\frac{1}{2t}(x - x_0)^2\} \\
&+ \frac{1}{\sqrt{2\pi t}} \exp\{-\frac{1}{2t}(x - x_0)^2\} \frac{(x - x_0)^2}{2t^2} \\
&= \frac{1}{2\sqrt{2\pi t}} \exp\{-\frac{1}{2t}(x - x_0)^2\} \Big[-\frac{1}{t} + \frac{(x - x_0)^2}{t^2} \Big]
\end{aligned}$$

$$\begin{aligned}
\text{Further,} \ \frac{\partial}{\partial x} p(x, t|x_0) &= -\frac{1}{\sqrt{2\pi t}} \exp\{-\frac{1}{2t}(x - x_0)^2\} \frac{(x - x_0)}{t} \\
\Rightarrow \ \frac{\partial^2}{\partial x^2} p(x, t|x_0) &= -\frac{1}{\sqrt{2\pi t}} \exp\{-\frac{1}{2t}(x - x_0)^2\} \frac{1}{t} \\
&+ \frac{1}{\sqrt{2\pi t}} \exp\{-\frac{1}{2t}(x - x_0)^2\} \frac{(x - x_0)^2}{t^2} \\
&= \frac{1}{\sqrt{2\pi t}} \exp\{-\frac{1}{2t}(x - x_0)^2\} \Big[-\frac{1}{t} + \frac{(x - x_0)^2}{t^2} \Big] .
\end{aligned}$$

Thus, $p(x, t|x_0) = \frac{1}{\sqrt{2\pi t}} \exp\{-\frac{1}{2t}(x - x_0)^2\}$ is a solution to the diffusion equation, subject to the conditions that

$$p(x, t|x_0) \geq 0, \ \ \int_{-\infty}^{\infty} p(x, t|x_0)\, dx = 1 \ \ \& \ \ \lim_{t\to 0} p(x, t|x_0) = 0 \ \forall \ x \neq x_0 \, .$$

It can be shown that the above solution is unique. Hence, $\{X(t), t \geq 0\}$ is a process with stationary and independent increments and for fixed t, $X(t) - X(0)$ follows normal $N(x_0, t)$ distribution.

Both the approaches suggest that the marginal distribution of $X(t)$ is normal. We now derive various properties of a Brownian motion process.

The second condition in the definition of a Brownian motion process states that it is a stochastic process with stationary and independent increments. It then follows that a Brownian motion process is a Markov process. It is proved in the next theorem.

Theorem 9.2.1 *A Brownian motion process is a time homogeneous Markov process with state space* $\mathbb{R}$.

Proof In Sect. 1.3, it is proved that a process with stationary and independent increments is a time homogeneous Markov process, when the state space is a countably infinite set. Using a similar approach, the same result can be proved when the state space is continuous. It then follows that the Brownian motion process is a Markov process. Further, since the increments are stationary, the Markov process is time homogeneous. The transition distribution function for the standard Brownian motion process, for $x, x_0 \in \mathbb{R}$ and $s < t$, is as follows:

$$F(x, t, x_0, s) = P[X(t) \leq x | X(s) = x_0] = \int_{-\infty}^{x} \frac{1}{\sqrt{2\pi(t-s)}} \exp\left\{-\frac{(u - x_0)^2}{2(t-s)}\right\} du.$$

□

Theorem 9.2.2 *Suppose* $\{X(t), t \geq 0\}$ *is a Brownian motion process with drift coefficient* μ *and diffusion coefficient* σ^2*. Then*

$$E(X(t)) = \mu t, \quad Var(X(t)) = \sigma^2 t \quad \& \quad Cov(X(t), X(s)) = \sigma^2 \min\{s, t\}\,.$$

Proof Since $\{X(t), t \geq 0\}$ is a Brownian motion process, for each fixed t,

$$X(t) \sim N(\mu t, \sigma^2 t) \;\Rightarrow\; E(X(t)) = \mu t \;\&\; Var(X(t)) = \sigma^2 t.$$

To find the covariance function, we note that it is a process with stationary and independent increments. In Sect. 1.3, it is proved that for such a process, $Cov(X(t), X(s)) = Var(X(1)) \min\{s, t\}$. Since $Var(X(1)) = \sigma^2$, $Cov(X(t), X(s)) = \sigma^2 \min\{s, t\}$. □

Remark 9.2.2 From Theorem 9.2.2, we note that a Brownian motion process with drift coefficient μ and diffusion coefficient σ^2 is not a stationary process as its mean function and variance function are not constant. With $\mu = 0$, the mean function is constant, however the variance function depends on t. Thus, it is an evolutionary stochastic process.

We now prove that the standard Brownian motion process is a martingale. We first define a martingale.

Definition 9.2.2 A stochastic process $\{X(t), t \geq 0\}$ such that

$$E(X(t)|X(u), 0 \leq u \leq s) = X(s), \quad \text{almost surely}$$

is known as a martingale.

Theorem 9.2.3 *The standard Brownian motion process is a martingale.*

Proof Suppose $\{W(t), t \geq 0\}$ is the standard Brownian motion process. Using independent increments property, we find $E(W(t)|W(u), 0 \leq u \leq s)$, as follows:

$$\begin{aligned} E(W(t)|W(u), 0 \leq u \leq s) &= E((W(t) - W(s) + W(s))|W(u), 0 \leq u \leq s) \\ &= E((W(t) - W(s))|W(u), 0 \leq u \leq s) \\ &+ E(W(s)|W(u), 0 \leq u \leq s) \\ &= E((W(t) - W(s)) + W(s) \;\; \text{almost surely} \\ &= 0 + W(s) = W(s). \end{aligned}$$

Hence, $\{W(t), t \geq 0\}$ is a martingale. □

In Chap. 1, it is stated that the distribution of a stochastic process is completely determined by the associated family of finite dimensional distribution functions. We now derive a typical element of such a family for the standard Brownian motion process in the following theorem.

Theorem 9.2.4 *Suppose $\{W(t), t \geq 0\}$ is the standard Brownian motion process and $0 < t_1 < t_2 < \cdots < t_n \in (0, \infty)$ are positive real numbers. Then the joint distribution of $(W(t_1), W(t_2), \ldots, W(t_n))'$ is n-variate normal with mean vector $\underline{0}$ and dispersion matrix $\Sigma = [\sigma_{ij}]$, where $\sigma_{ii} = t_i$ and $\sigma_{ij} = \min\{t_i, t_j\}$.*

Proof Suppose we define

$$\underline{Z} = (W(t_1), W(t_2), \ldots, W(t_n))' \;\; \& \;\; Y = \underline{a}'\underline{Z} = \sum_{i=1}^{n} a_i W(t_i)$$

where $\underline{a} \neq \underline{0}$ is any n-dimensional vector of real numbers. By definition, for fixed t, $W(t) \sim N(0, t)$ distribution and by Theorem 9.2.2, $Cov(W(t), W(s)) = \min\{s, t\}$. Hence,

$$Y \sim N(0, v) \;\; \text{where} \;\; v = \sum_{i=1}^{n} a_i^2 t_i + 2\sum_{i=1}^{n} \sum_{j \neq i=1}^{n} a_i a_j \min\{t_i, t_j\} = \underline{a}'\Sigma\underline{a},$$

where $\Sigma = [\sigma_{ij}]$ and $\sigma_{ii} = t_i$, $\sigma_{ij} = \min\{t_i, t_j\}$. It then follows that $\underline{Z} \sim N(\underline{0}, \Sigma)$ by the Cramer-Wold principle. □

From Theorem 9.2.4, we have the following corollary.

Corollary 9.2.1 *Suppose $\{X(t), t \geq 0\}$ is a Brownian motion process with drift coefficient μ and diffusion coefficient σ^2. Suppose $0 < t_1 < t_2 < \cdots < t_n$ are positive real numbers. Then the distribution of $(X(t_1), X(t_2), \ldots, X(t_n))'$ is n-variate normal with mean vector $\mu(t_1, t_2, \ldots, t_n)'$ and dispersion matrix $\Sigma = \sigma^2[\sigma_{ij}]$, where $\sigma_{ii} = t_i$ and $\sigma_{ij} = \min\{t_i, t_j\}$.*

From Theorem 9.2.4, we can establish a relation between Brownian motion process and Gaussian process. We first define a Gaussian process.

Definition 9.2.3 Gaussian Process: A stochastic process $\{X(t), t \geq 0\}$ is said to be a Gaussian or a normal process if $\{X(t_1), X(t_2), \ldots, X(t_n)\}$ has a multivariate normal distribution for any finite $n \geq 1$ and for any finite set $\{t_1, t_2, \ldots, t_n\} \in [0, \infty)$.

Remark 9.2.3 (i) If $\{X(t), t \geq 0\}$ is a Brownian motion process, then from Theorem 9.2.4, the joint distribution of $(X(t_1), X(t_2), \ldots, X(t_n))'$ is a multivariate normal distribution for all $t_1, t_2, \ldots, t_n \in [0, \infty)$. It thus follows that a Brownian motion process is a Gaussian process. (ii) In general a Gaussian process is not a Brownian motion process, since it need not be a process with stationary and independent increments. (iii) However, with a typical covariance structure, a Gaussian process is a Brownian motion process. It is known that a multivariate normal distribution is completely determined by its mean vector and the dispersion matrix. Similarly, a Gaussian process is determined uniquely by its two functions, the mean function and the covariance function, which is positive definite. Conversely, given an arbitrary mean value function $M(t)$ and a positive definite covariance function, there exists a corresponding Gaussian process; refer to Karlin and Taylor [5]. Thus, if we take mean function as $M(t) = \mu t$ and covariance function as $Cov(X(t), X(s)) = \sigma^2 \min\{s, t\}$, then there exists a Gaussian process and it is also a Brownian motion process.

In view of the above remark, a Brownian motion process can also be defined in terms of a Gaussian process as follows.

Definition 9.2.4 Brownian Motion Process: A Gaussian process $\{X(t), t \geq 0\}$ with mean value function $E(X(t)) = \mu t$ and covariance function as $Cov(X(t), X(s)) = \sigma^2 \min\{s, t\}$ is a Brownian motion process with drift coefficient μ and diffusion coefficient σ^2.

Remark 9.2.4 In Chap. 1, we have noted that a stationary process possessing finite first two moments is covariance stationary; however, the converse is usually not true. It is true only for a Gaussian process. As the finite dimensional distributions of a Gaussian process are determined by their means and covariances, it follows that a covariance stationary Gaussian process is a stationary process. However, a Brownian motion process is not even covariance stationary.

The following examples illustrate various properties of a Brownian motion process.

Example 9.2.1 Suppose $\{W(t), t \geq 0\}$ is the standard Brownian motion process. Thus, for fixed t, $W(t) \sim N(0, t)$ distribution. Hence, we find $a(t)$ such that $W(t)$ lies between $-a(t)$ and $a(t)$ at time $t > 0$, with probability 0.9 as follows:

$$\begin{aligned} P[-a(t) \leq W(t) \leq a(t)] = 0.9 \Rightarrow \quad & \Phi(a(t)/\sqrt{t}) - \Phi(-a(t)/\sqrt{t}) = 0.9 \\ \Rightarrow \quad & 2\Phi(a(t)/\sqrt{t}) = 1.9 \;\Rightarrow\; a(t) = 1.65\sqrt{t}\,. \end{aligned}$$

□

Example 9.2.2 Suppose $X(t) = \mu t + \sigma W(t)$, where $\{W(t), t \geq 0\}$ is the standard Brownian motion process. We find $P_{1,2} = P[X(1) < E(X(1)), X(2) > E(X(2))]$ as follows:

$$\begin{aligned} P_{1,2} &= P[\mu + \sigma W(1) < \mu,\; 2\mu + \sigma W(2) > 2\mu] \\ &= P[W(1) < 0, W(2) > 0] \\ &= E(P[W(1) < 0, W(2) > 0 \big| W(1)]) \\ &= \int_{-\infty}^{0} P[W(2) - W(1) > -x \big| W(1) = x]\phi(x)\, dx \\ &= \int_{-\infty}^{0} P[W(2) - W(1) > -x]\phi(x)\, dx \\ &= \int_{-\infty}^{0} (1 - \Phi(-x))\phi(x)\, dx, \;\; \text{since } W(2) - W(1) \sim N(0, 1) \\ &= \int_{-\infty}^{0} \Phi(x)\phi(x)\, dx = \int_{0}^{0.5} u\, du, \;\; \text{with } \Phi(x) = u \\ &= 1/8\,, \end{aligned}$$

where the fifth step follows in view of independence of increments. □

Inventory at time t is usually modeled as a Brownian motion process. Positive values imply there is a stock of items in the inventory while negative values imply backlogs, that is, the stock of items is not sufficient to satisfy the demand. The following example illustrates the application of a Brownian motion process in inventory models. We use the result stated in the following lemma, regarding the conditional distribution of X_2 given $X_1 = x_1$ in a bivariate normal distribution.

Lemma 9.2.1 *Suppose* $(X_1, X_2)' \sim N_2(\underline{\mu}, \Sigma)$ *distribution, where* $\underline{\mu} = (\mu_1, \mu_2)'$ *and* $\Sigma = [\sigma_{ij}]$. *Suppose* ρ *denotes the correlation coefficient between* X_1 *and* X_2. *Then the conditional distribution of* X_2 *given* $X_1 = x_1$ *is normal* $N(\theta, v)$ *where* $\theta = \mu_2 + (\sigma_{21}/\sigma_{11})(x_1 - \mu_1)$ *and* $v = \sigma_{22}(1 - \rho^2)$.

Example 9.2.3 Suppose an inventory of items is modeled by a Brownian motion process with drift coefficient -2 and diffusion coefficient 4 and initial stock has 10 items. Then for fixed t, $X(t) \sim N(10 - 2t, 4t)$ distribution. The joint distribution of

$(X(s), X(t))'$ for $s < t$ is bivariate normal with mean vector $(10 - 2s, 10 - 2t)'$ and dispersion matrix $\Sigma = [\sigma_{ij}]$ where $\sigma_{11} = 4s$, $\sigma_{22} = 4t$ and $\sigma_{12} = \sigma_{21} = 4s$. Hence, using Lemma 9.2.1, the conditional distribution of $X(t)$ given $X(1) = 7$ for $t > 1$ is normal $N(\theta(t), v(t))$ where

$$\theta(t) = E(X(t)|X(1)) = 10 - 2t + (4/4)(7 - 8) = 9 - 2t \;\;\&\;\; v(t) = 4t(1 - \rho^2).$$

To find ρ, note that

$$\begin{aligned} 4 &= Cov(X(1), X(t)) = \rho(Var(X(1)) * Var(X(t))^{1/2} = \rho \times 2 \times 2 \times t^{1/2} \\ \Rightarrow \;\; \rho &= 1/\sqrt{t} \;\; \Rightarrow \;\; v(t) = 4t(1 - 1/t) = 4(t - 1). \end{aligned}$$

Hence, given the inventory level at $t = 1$ to be 7, the expected inventories at $t = 2, 3, 4, 5$ are as given below

$$\begin{aligned} &E(X(2)|X(1) = 7) = 5, \quad E(X(3)|X(1) = 7) = 3, \\ &E(X(4)|X(1) = 7) = 1 \;\;\&\;\; E(X(5)|X(1) = 7) = -1. \end{aligned}$$

Observe that the expected inventory $\theta(t)$ is a decreasing function of t, as expected since the drift coefficient is negative. It will be 0 at $t = 4.5$. However, note that $Var(X(t)|X(1) = 7) = v(t)$ is an increasing function of t. Using the information on variance and the fact that $P[X(t) \in (\theta(t) - 3\sqrt{v(t)}, \theta(t) + 3\sqrt{v(t)})|X(1) = 7] = 0.9973$, we note that

$$\begin{aligned} P[X(2) \in (-1, 11)|X(1) = 7] &= 0.9973 \\ P[X(3) \in (-5.4853, 11.4853)|X(1) = 7] &= 0.9973 \\ P[X(4) \in (-9.3923, 11.3923)|X(1) = 7] &= 0.9973 \\ P[X(5) \in (-13, 11)|X(1) = 7] &= 0.9973. \end{aligned}$$

Thus, at time points $t = 2, 3, 4, 5$, the inventory can be negative with some positive probability. We now find t at which given $X(1) = 7$, the inventory will be negative with probability 0.9 as follows:

$$\begin{aligned} P[X(t) < 0|X(1) = 7] &= 0.9 \\ &\Rightarrow P\left[\frac{X(t) - 9 + 2t}{2\sqrt{t - 1}} < \frac{-9 + 2t}{2\sqrt{t - 1}}\right] = 0.9 \\ &\Rightarrow \Phi((-9 + 2t)/2\sqrt{t - 1}) = 0.9 \\ &\Rightarrow (-9 + 2t)/2\sqrt{t - 1} = 1.2816 \\ &\Rightarrow 4t^2 - 42.57t + 87.57 = 0 \\ &\Rightarrow t = 2.7868 \;\text{ or }\; 7.8556\,. \end{aligned}$$

Note that we have to select that root for which $(-9+2t)/2\sqrt{t-1} > 0$, which implies $t = 7.8556$. Proceeding on similar lines, we find t such that $P[X(t) < 0|X(1) = 7]$ with probability 0.99. In this case, we get two roots as $t = 12.3305$ or $t = 2.0812$. We select one for which $(-9+2t)/2\sqrt{t-1} > 0$. hence, $t = 12.3305$. Thus, at $t = 12.3305$ the inventory is almost negative. □

Example 9.2.4 Suppose inventory $W(t)$ of items is modeled as the standard Brownian motion process. Initial inventory is x units and the inventory is continuously updated at a rate of 20% items per unit of time. If $X(t)$ denotes the inventory level at t, then $X(t) = (1.2)^t x + W(t),\ t \geq 0$ so that for fixed t, $X(t) \sim N((1.2)^t x, t)$ distribution. It is of interest to decide the initial stock x so that the probability of stock-out at some time t is bounded above by 0.05. Thus,

$$\begin{aligned} P[X(t) < 0] \leq 0.05 &\Rightarrow P[(X(t) - (1.2)^t x)/\sqrt{t} < -(1.2)^t x/\sqrt{t}] \leq 0.05 \\ &\Rightarrow \Phi(-((1.2)^t x)/\sqrt{t}) \leq 0.05 \\ &\Rightarrow -((1.2)^t x)/\sqrt{t} \leq -1.6449 \\ &\Rightarrow x \geq 1.6449\sqrt{t}/(1.2)^t, \end{aligned}$$

for $t = 0.5, 1.0, 1.5, 2.0, 2.5, 3.0, 3.5, 4.0, 4.5, 5.0$, values of x are 1.0618, 1.3708, 1.5325, 1.6154, 1.6488, 1.6488, 1.6257, 1.5865, 1.5361, 1.4781, respectively. It is to be noted that x values increase initially and then decrease. To find the maximum value of x, we find t for which $g(t) = 1.6449\sqrt{t}/(1.2)^t$ is maximum which is equivalent to finding t for which $\log g(t)$ is maximum. Solving $\frac{d}{dt}\log g(t) = 0$, we get $t = (2\log(1.2))^{-1} = 2.7424$. For $t = 2.7424$, $g(t) = 1.6522$. Thus, if the initial inventory is 1.6522, then the probability of stock-out is bounded above by 0.05. □

In the following theorem, we prove that a Brownian motion process remains invariant under certain operations.

Theorem 9.2.5 *Suppose $\{W(t), t \geq 0\}$ is the standard Brownian motion process. Then each of the following processes is also a standard Brownian motion process.*

1. $\{X(t) = cW(t/c^2), t \geq 0\}$. *(Scale symmetry)*
2. $\{X(t) = W(t+h) - W(h), t \geq 0\},\ h \geq 0$. *(Translation of increments)*
3. $\{X(t) = tW(1/t), t > 0\}, X(0) = 0$. *(Inversion)*
4. $\{X(t) = -W(t), t \geq 0\}$. *(Reflection)*

Proof We examine that each of these processes have stationary and independent increments and for fixed t, $X(t)$ has normal $N(0, t)$ distribution.
(i) Since $W(0) = 0$, we have $X(0) = 0$. Further,

$$\begin{aligned} X(t) = cW(t/c^2) \ &\Rightarrow\ E(X(t)) = 0 \\ Var(X(t)) &= c^2 Var(W(t/c^2)) = c^2(t/c^2) = t \\ \&\ X(t) &\sim N(0, t) \ \text{ distribution.} \end{aligned}$$

For $s < t$, the increment $X(t) - X(s) = cW(t/c^2) - cW(s/c^2) \sim N(0, v)$ distribution, where $v = t + s - 2cov(X(t), X(s))$ and

$$\begin{aligned} Cov(X(t), X(s)) &= c^2 Cov(W(t/c^2), W(s/c^2)) = c^2 min\{t/c^2, s/c^2\} \\ &= c^2(s/c^2) = s \;\Rightarrow\; v = t - s. \end{aligned}$$

Thus, it follows $X(t) - X(s) \sim N(0, t - s)$ distribution. Hence $\{X(t), t \geq 0\}$ is a process with stationary increments. We now examine whether $\{X(t), t \geq 0\}$ has independent increments. For simplicity of notation, we verify the property for $0 < t_1 < t_2 < t_3$. Since $\{W(t), t \geq 0\}$ is the standard Brownian motion process, $W(t_2) - W(t_1)$ and $W(t_3) - W(t_2)$ are independent random variables, and it then follows that $c(W(t_2/c^2) - W(t_1/c^2))$ and $c(W(t_3/c^2) - W(t_2/c^2))$ are also independent random variables. Thus, $\{X(t), t \geq 0\}$ is a process with stationary and independent increments, for each fixed t, $X(t) \sim N(0, t)$ and hence it is the standard Brownian motion process.
(ii) Note that $X(t) = W(t + h) - W(h) \;\Rightarrow\; X(0) = 0 \;\&\; X(t) \sim N(0, v)$ distribution, with $v = t + h + h - 2h = t$. Further, for $s < t$, it is $X(t) - X(s) = W(t + h) - W(s + h) \sim N(0, t - s)$ distribution. Hence $\{X(t), t \geq 0\}$ is a process with stationary increments. For $0 < t_1 < t_2 < t_3$,

$$X(t_3) - X(t_2) = W(t_3 + h) - W(t_2 + h) \;\&\; X(t_2) - X(t_1) = W(t_2 + h) - W(t_1 + h)\,.$$

$\{W(t), t \geq 0\}$ is a process with independent increments implying that $W(t_3 + h) - W(t_2 + h)$ and $W(t_2 + h) - W(t_1 + h)$ are independent random variables, and hence it follows that $\{X(t), t \geq 0\}$ is also a process with independent increments. Thus, it is the standard Brownian motion process.
(iii) It is given that $X(0) = 0$. Further,

$$\begin{aligned} X(t) &= tW(1/t) \;\Rightarrow\; E(X(t)) = 0 \\ Var(X(t)) &= t^2 Var(W(1/t)) = t^2/t = t \;\&\; X(t) \sim N(0, t) \\ \text{For } s < t,\;\; X(t) - X(s) &= tW(1/t) - sW(1/s) \\ &\sim N(0, v) \text{ where } v = t + s - 2cov(X(t), X(s)) \\ \&\; Cov(X(t), X(s)) &= tsCov(W(1/t), W(1/s)) = ts\, min\{1/t, 1/s\} \\ &= s \;\Rightarrow\; v = t - s \\ &\Rightarrow X(t) - X(s) \sim N(0, t - s). \end{aligned}$$

Hence, $\{X(t), t \geq 0\}$ is a process with stationary increments. We now examine whether $\{X(t), t \geq 0\}$ has independent increments. For $0 < t_1 < t_2 < t_3$, observe that $\{W(t), t \geq 0\}$ is the standard Brownian motion process implying that $(W(1/t_1), W(1/t_2), W(1/t_3))'$ has trivariate normal distribution, which further implies that $(t_1W(1/t_1), t_2W(1/t_2), t_3W(1/t_3))'$ has trivariate normal distribution. Hence, $(t_1W(1/t_1), t_2W(1/t_2) - t_1W(1/t_1), t_3W(1/t_3) - t_2W(1/t_2))'$ follows trivariate normal distribution. Thus, $(X(t_1), X(t_2) - X(t_1), X(t_3) - X(t_2))$ has trivariate normal distribution. In view of multivariate normality, to examine whether these three random variables are independent, it is enough to show that the covariance between any pair of these three random variables is zero. Note that

$$\begin{aligned}
Cov(X(t_1), X(t_2) - X(t_1)) &= Cov(X(t_1), X(t_2)) - Var(X(t_1)) \\
&= t_1 t_2 Cov(W(1/t_1), W(1/t_2)) \\
&- t_1^2 Var(W(1/t_1)) \\
&= t_1 t_2 (1/t_2) - t_1^2 (1/t_1) = 0 \\
Cov(X(t_1), X(t_3) - X(t_2)) &= Cov(X(t_1), X(t_3)) - Cov(X(t_1), X(t_2)) \\
&= t_1 t_3 (1/t_3) - t_1 t_2 (1/t_2) = 0 \\
Cov(X(t_2) - X(t_1), X(t_3) - X(t_2)) &= t_2 t_3 (1/t_3) - t_2^2 (1/t_2) \\
&- t_1 t_3 (1/t_3) + t_1 t_2 (1/t_2) = 0.
\end{aligned}$$

Hence, it follows that $X(t_1), X(t_2) - X(t_1), X(t_3) - X(t_2)$ are independent. Proceeding on similar lines, we claim that $\{X(t), t \geq 0\}$ has independent increments. For each fixed t, $X(t) \sim N(0, t)$, thus $\{X(t), t \geq 0\}$ is the standard Brownian motion process.

(iv) We have $X(t) = -W(t)$, hence it follows immediately that $X(0) = 0$ and $X(t)$ follows $N(0, t)$ distribution. Similarly for $s < t$,

$$X(t) - X(s) = -(W(t) - W(s)) \sim N(0, v) \;\text{ where }\; v = t + s - 2s = t - s.$$

Hence $\{X(t), t \geq 0\}$ is a process with stationary increments. For $0 < t_1 < t_2 < t_3$,

$$X(t_3) - X(t_2) = -(W(t_3) - W(t_2)) \;\&\; X(t_2) - X(t_1) = -(W(t_2) - W(t_1)) \,.$$

In general, for any $i \geq 1$ and $t_{i-1} < t_i$, $X(t_i) - X(t_{i-1}) = -(W(t_i) - W(t_{i-1}))$. Hence, $\{W(t), t \geq 0\}$ is a process with independent increments implying that $\{X(t), t \geq 0\}$ is also a process with independent increments. Thus, it is proved that $\{X(t), t \geq 0\}$ is the standard Brownian motion process. □

The next section is devoted to the realization and the properties of sample paths of a Brownian motion process. The structure and the properties of sample paths of a Brownian motion process have been studied extensively in the literature.

9.3 Realization and Properties of Sample Path

In the previous chapters, we have obtained the realization of a Markov chain, a continuous time Markov chain, including a Poisson process, Yule-Furry process, linear death process and birth-death process. In all these processes, the state space is discrete. A Brownian motion process is a continuous time and continuous state space Markov process. To obtain its realization for a fixed interval $[0, T]$, we divide the interval into a number of small intervals of length h, say. We have $X(0) = 0$. The position $X(h)$ of a particle at time point h is governed by a displacement $X(h) - X(0)$ of the particle in a length of interval h. Suppose the drift coefficient is 0 and the diffusion coefficient is σ^2. Then it is $X(h) - X(0) \sim N(0, \sigma^2 h)$ distribution. Hence, $X(h)$ will be decided by a random sample of size 1 from $N(0, \sigma^2 h)$ distribution. Proceeding on these lines for any $t \in [0, T]$, it is $X(t+h) - X(t) \sim N(0, \sigma^2 h)$ distribution. Thus, $X(t+h) = X(t) + u$ where u is the realized value of a random sample of size 1 from $N(0, \sigma^2 h)$ distribution. We adopt such a procedure to write a code to obtain a realization of a Brownian motion process for a fixed interval. It is illustrated in the next example.

Example 9.3.1 Suppose $\{X(t), t \geq 0\}$ is a Brownian motion process, with drift parameter 0 and diffusion parameter σ^2 where $\sigma = 0.5, 1, 1.7$. We use Code 9.7.1 to obtain the realization, when it is observed for the interval $[0, 10]$.

Figure 9.1 presents the realization. From the graph, we note that the sample path of a Brownian motion process is continuous, but not differentiable. Further as σ increases, variation in the values of $X(t)$ increases, as expected. □

The next example also presents a realization of the Brownian motion process, but we adopt a slightly different approach. In Example 9.3.1, the random sample from the standard normal distribution is the same for three realizations; only the values of diffusion coefficients are different. This approach is used in Code 9.7.1.

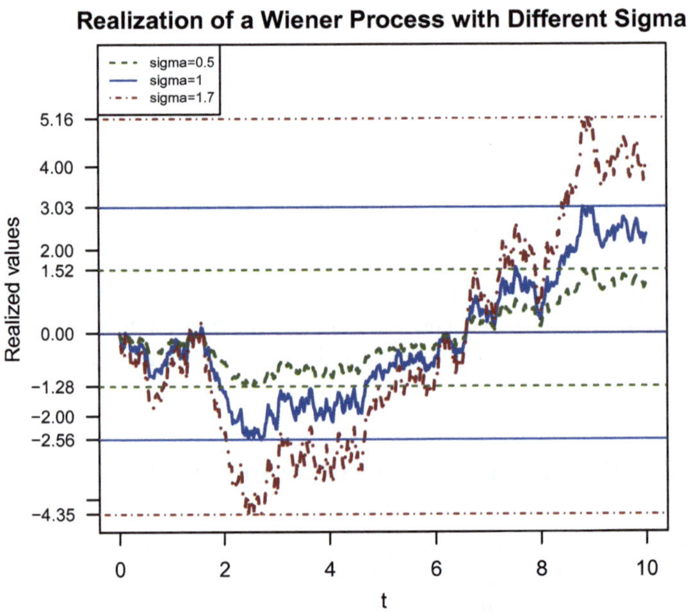

Fig. 9.1 Realization of a Brownian Motion Process with different values of diffusion coefficient

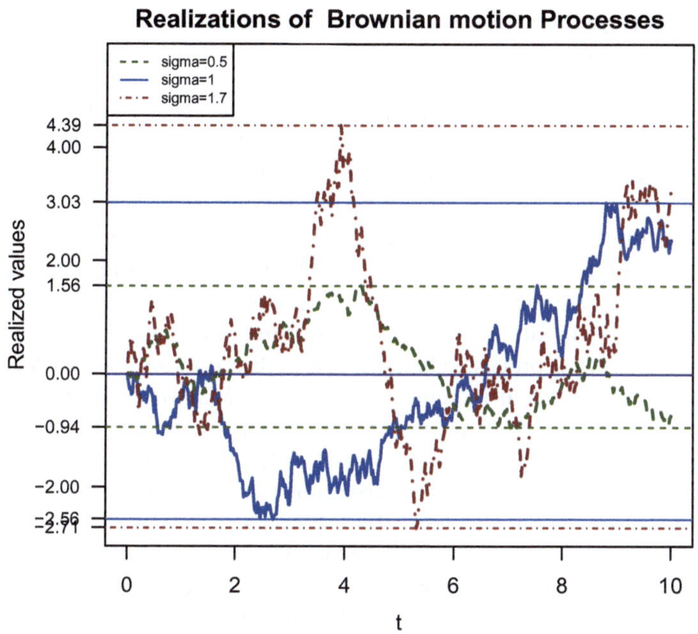

Fig. 9.2 Realizations of Brownian Motion Processes

In the next example, we draw different random samples from the standard normal distribution corresponding to different values of diffusion coefficients. In Code 9.7.2 this approach is adopted.

Example 9.3.2 Suppose $\{X(t), t \geq 0\}$ is a Brownian motion process, with drift parameter 0 and diffusion parameter σ^2 where $\sigma = 0.5, 1, 1.7$. We use Code 9.7.2 to obtain the realizations for the interval [0, 10]. Figure 9.2 presents the realization of three different Brownian motion processes. □

It we compare Figs. 9.1 and 9.2, we note that in Fig. 9.1, the shape of the three curves is similar since the random sample from the standard normal distribution is the same for three realizations. The spread is different, depending upon the values of diffusion parameters. In Fig. 9.2, the shapes of three curves are different since we draw different random samples from the standard normal distribution corresponding to different values of diffusion coefficients.

From the above two examples, we note that the realization of a Brownian motion process is the graph of the position of a particle against time. The sample path is a continuous function, however it is very wrinkled. The physical origin of a Brownian motion suggests that the particle moves randomly due to its continuous collisions in the surrounding medium. Thus, it is expected that a sample path would be a continuous function. However, its derivative does not exist anywhere. We state these results in the following theorem. For proofs refer to Billingsley [1].

Theorem 9.3.1 *Suppose $\{X(t), t \geq 0\}$ is a Brownian motion process. Its sample paths are continuous everywhere, but differentiable nowhere, with probability one.*

Heuristically, we explain these properties as follows. Since $X(t+h) - X(t)$ has $N(0, \sigma^2 h)$ distribution, it converges in law and hence in probability to a degenerate random variable, degenerate at 0 as $h \to 0$. Thus, $X(t+h)$ converges to $X(t)$ in probability. Further, $(X(t+h) - X(t))/h$ follows $N(0, \sigma^2/h)$ distribution and hence has no limit as $h \to 0$. Intuitively, this implies that the derivative of $X(t)$ at t does not exist. Figures 9.1 and 9.2 support the conclusion that the sample paths are continuous and give some idea that these are not differentiable.

We now note some similarities and some differences between a Brownian motion process and a Poisson process. Observe that both processes (i) are processes with stationary and independent increments, (ii) satisfy Markov property, (iii) are evolutionary processes and (iv) have the same covariance function. On the other hand, (i) for fixed t, in a Poisson process $X(t) \sim Poi(\lambda t)$ distribution while in a Brownian motion process $X(t) \sim N(\mu t, \sigma^2 t)$ distribution. (ii) The sample paths of a Brownian motion process are continuous, while for a Poisson process these are step functions and hence only right continuous. At jump points, there is a discontinuity.

Using the continuity property of a sample path and the celebrated reflection principle (Feller [3]), we now derive the results related to a maximum and a minimum of a Brownian motion process when t varies over $[0, T]$. The reflection principle states that there is a one-to-one correspondence between all paths from $A = (a_1, a_2)$ to $B = (b_1, b_2)$, which touch or cross the x-axis and all paths from $A_1 = (a_1, -a_2)$ to $B_1 = (b_1, -b_2)$. Using this principle, in the next two theorems we derive the distribution of the maximum and of the minimum of a Brownian motion process. From these theorems, we derive the distribution of the first passage time, T_a, to a fixed point $a \in \mathbb{R}$ for the standard Brownian motion process. It is defined as follows:

$$T_a = T_a(\omega) = \min\{t \geq 0 | X(t, \omega) = a\}, \quad a \in \mathbb{R}, \quad \omega \in \Omega.$$

The definition is analogous to the first passage random variable defined for a Markov chain in Chap. 2. Since the sample paths are continuous, there exists a time T_a at which $X(t)$ first attains the value a, or $X(t)$ hits a for the first time. Hence, T_a is also known as a hitting time random variable. It is clear that T_a is a random variable since its value changes as the sample path changes. It is a well-defined random variable, since the sample paths of a Brownian motion process are continuous and eventually visit every $a \in \mathbb{R}$ with probability one.

Theorem 9.3.2 *Suppose $\{W(t), t \geq 0\}$ is the standard Brownian motion process with $W(0) = 0$. Suppose $U(T) = \max_{0 \leq t \leq T} W(t)$. Then for any $a > 0$, the survival function and the probability density function of $U(T)$ are given by*

$$P[U(T) \geq a] = 2P[W(T) \geq a] = 2[1 - \Phi(a/\sqrt{T})]$$
$$\& \quad f_{U(T)}(a) = \sqrt{2/\pi T}\; e^{-a^2/2T}, \; a \geq 0.$$

Proof It is to be noted that $U(T) \geq 0$ almost surely, since $W(0) = 0$. We consider a collection of sample paths $\{\omega | W(t, \omega), 0 \leq t \leq T\}$ such that $W(T)(\omega) \geq a > 0$. Since $W(0) = 0$ and the sample paths are continuous, there exists a time $T_a = T_a(\omega)$ at which $W(t) = W(t, \omega)$ first attains the value a. For $t > T_a$ we reflect $W(t)$ around a line $W(t) = a$ and define

$$\tilde{W}(t) = \begin{cases} W(t), & \text{if } t \leq T_a \\ a - (W(t) - a), & \text{if } t > T_a. \end{cases}$$

Note that $W(T) \geq a \Rightarrow \tilde{W}(T) \leq a$. Further, $W(t)$ touches the line $W(t) = a$ at $t = T_a$; it may or may not cross the line for $t \in (T_a, T]$. If it does not cross, then $W(t)$ will be below the line $W(t) = a$ for $t \in (T_a, T]$ and $\tilde{W}(t)$ will be above the line $W(t) = a$ for $t \in (T_a, T]$. In such a case, $\tilde{U}(T) = \max_{0 \leq t \leq T} \tilde{W}(t) \geq a$. If it crosses the line at least once for $t \in (T_a, T]$, then for such t, $W(t) < a$ and hence $\tilde{W}(t) > a$ at least once for $t \in (T_a, T]$. It then follows that $\tilde{U}(T) = \max_{0 \leq t \leq T} \tilde{W}(t) \geq a$. Hence we have

$$U(T) = \max_{0 \leq t \leq T} W(t) \geq a \text{ as well as } \tilde{U}(T) = \max_{0 \leq t \leq T} \tilde{W}(t) \geq a .$$

Thus, corresponding to every sample path $\{W(t), 0 \leq t \leq T\}$, we have a sample path $\{\tilde{W}(t), 0 \leq t \leq T\}$ such that for both the sample paths, respective maximum values of the processes for $t \in [0, T]$ are at least a, that is $[U(T) \geq a]$ and $[\tilde{U}(T) \geq a]$. Conversely, observe that

$$\begin{aligned}[U(T) \geq a] = [U(T) \geq a, W(T) > a] \cup [U(T) \geq a, W(T) < a] \\ \cup [U(T) \geq a, W(T) = a] .\end{aligned}$$

The three events in the union are mutually exclusive, the probability of third event is 0 and the other two are mapped onto each other by reflection around a line $W(t) = a$, as is clear from the following arguments. Note that

$$[U(T) \geq a, W(T) > a] = [W(T) > a] \text{ as } [W(T) > a] \subset [U(T) \geq a].$$

Similarly,

$$[U(T) \geq a, W(T) < a] = [\tilde{U}(T) \geq a, \tilde{W}(T) > a] = [\tilde{W}(T) > a].$$

Thus, the event $[U(T) \geq a]$ corresponds to two sample paths for which $[W(T) > a]$ and $[\tilde{W}(T) > a]$. To examine whether these two events have the same probability, we proceed as follows. Suppose $T_a = s \leq T$, this event depends on $W(u), 0 \leq u \leq s$. Further, $W(T_a) = W(s) = a$ and $\tilde{W}(T_a) = a$. Using the independence of increments property of the Brownian motion process, $W(T) = W(T) - W(s) + a$ follows $N(a, T - s)$ distribution. Using symmetry of $N(a, T - s)$ distribution around a, we have for all $s \leq T$,

$$\begin{aligned} P[W(T) \geq a | T_a = s] &= P[W(T) \leq a \big| T_a = s] \\ &= P[2a - W(T) \geq 2a - a \big| T_a = s] \\ &= P[\tilde{W}(T) \geq a \big| T_a = s] \,. \end{aligned} \tag{9.3.1}$$

Using Eq. (9.3.1) we note that

$$\begin{aligned} P[W(T) > a] &= E_{T_a}(P[W(T) > a] \big| T_a) = E_{T_a}(P[\tilde{W}(T) \geq a] \big| T_a) \\ &= P[\tilde{W}(T) \geq a] \\ \Rightarrow \quad P[U(T) \geq a] &= P[[U(T) \geq a, W(T) > a] \cup [U(T) \geq a, W(T) < a]] \\ &= P[[W(T) > a] \cup [\tilde{W}(T) > a]] \\ &= P[W(T) > a] + P[\tilde{W}(T) > a] \\ &= 2P[W(T) \geq a] = \frac{2}{\sqrt{2\pi T}} \int_a^{\infty} e^{-u^2/2T} \, du \\ &= \frac{2}{\sqrt{2\pi}} \int_{a/\sqrt{T}}^{\infty} e^{-u^2/2} \, du = 2[1 - \Phi(a/\sqrt{T})] \,. \end{aligned}$$

Further, the probability density function of $U(T)$ is given by

$$\begin{aligned} f_{U(T)}(a) &= -\frac{d}{da} 2[1 - \Phi(a/\sqrt{T})] = 2\phi(a/\sqrt{T})(1/\sqrt{T}) \\ &= \sqrt{2/\pi T} \, e^{-a^2/2T} \,, \quad a \geq 0 \,. \end{aligned}$$

□

The proof given above is heuristic; rigorous proof involves the strong Markov property. In the next theorem, we obtain the distribution of minimum of a Brownian motion process; the proof is based on the proof of Theorem 9.3.2.

Theorem 9.3.3 *Suppose* $\{W(t), t \geq 0\}$ *is the standard Brownian motion process with* $W(0) = 0$. *Suppose* $L(T) = \min_{0 \leq t \leq T} W(t)$. *Then for any* $a < 0$, *the distribution function and the probability density function of* $L(T)$ *are given by*

$$P[L(T) \leq a] = 2[1 - \Phi(-a/\sqrt{T})] \quad \& \quad f_{L(T)}(a) = \sqrt{2/\pi T} \, e^{-a^2/2T}, \quad a < 0.$$

Proof Since $W(0) = 0$, $L(T) \leq 0$ almost surely. For $a < 0$,

$$\begin{aligned} P[L(T) \leq a] &= P[\min_{0 \leq t \leq T} W(t) \leq a | W(0) = 0] = P[\min_{0 \leq t \leq T} W(t) \leq 0 | W(0) = a] \\ &= P[\max_{0 \leq t \leq T} W(t) \geq 0 | W(0) = -a] \ \text{ by symmetry} \\ &= P[\max_{0 \leq t \leq T} W(t) \geq -a | W(0) = 0] \\ &= 2[1 - \Phi(-a/\sqrt{T})] = 2\Phi(a/\sqrt{T}) \,. \end{aligned}$$

Hence, the probability density function $f_{L(T)}(a)$ of $L(T)$ is given by

$$f_{L(T)}(a) = \sqrt{2/\pi T}\, e^{-a^2/2T} \ , \ \ a < 0 \, .$$

□

Figures 9.1 and 9.2 show the maximum and a minimum of $W(t)$ when $t \in [0, 10]$. If $Y \sim N(0, T)$, then note that the probability density function of $U(T)$ is the same as that of $|Y|$, while the probability density function of $L(T)$ is the same as that of $-|Y|$. Thus, $E(U(T)) = \sqrt{2T/\pi}$ and $E(L(T)) = -\sqrt{2T/\pi}$. Using the distributions of $U(T)$ and of $L(T)$, we now obtain the distribution of the first passage time T_a, $a \in \mathbb{R}$ for the standard Brownian motion process. We prove an interesting result which states that although the standard Brownian motion process reaches any level $a \neq 0$ with probability 1, it takes on an average an infinite time to do so.

Theorem 9.3.4 *Suppose $\{W(t), t \geq 0\}$ is the standard Brownian motion process with $W(0) = 0$. Suppose $T_a = \min\{t \geq 0 | W(t) = a\}$, $a \in \mathbb{R}$ is the first passage time random variable. Then for any $a \in \mathbb{R}$, the distribution function and the probability density function of T_a are given by*

$$P[T_a \leq t] = 2[1 - \Phi(|a|/\sqrt{t})] \quad \& \quad f_{T_a}(t) = (|a|/\sqrt{2\pi})t^{-3/2}e^{-a^2/2t} \ , \ \ t > 0.$$

Further, all raw moments of T_a are infinite.

Proof Suppose $a > 0$. Then $T_a \leq t \iff U(t) \geq a$. Hence, the distribution function $F(t)$ of T_a and the probability density function $f_{T_a}(t)$ are given by

$$\begin{aligned} F(t) &= P[T_a \leq t] = P[U(t) \geq a] = 2[1 - \Phi(a/\sqrt{t})] \\ \Rightarrow \ f_{T_a}(t) &= -2\frac{d}{dt}\Phi(a/\sqrt{t}) = 2\phi(a/\sqrt{t})\frac{a}{2t^{3/2}} \\ &= (a/\sqrt{2\pi})t^{-3/2}e^{-a^2/2t} \ , \ \ t > 0 \, . \end{aligned}$$

For $a < 0$, the distribution function of T_a is obtained using the link

$$T_a \leq t \iff L(t) \leq a.$$

Hence for $a < 0$, the distribution function $F(t)$ of T_a and the probability density function $f_{T_a}(t)$ are given by

$$\begin{aligned} F(t) &= P[T_a \leq t] = P[L(t) \leq a] = 2\Phi(a/\sqrt{t}) = 2[1 - \Phi(-a/\sqrt{t})] \\ \Rightarrow \ f_{T_a}(t) &= (-a/\sqrt{2\pi})t^{-3/2}e^{-a^2/2t} \ , \ \ t > 0 \, . \end{aligned}$$

Since $W(0) = 0$, $T_0 = 0$ with probability 1 and combining both the above cases, the distribution function $F(t)$ of T_a and the probability density function $f_{T_a}(t)$ of T_a for $a \neq 0$ are given by

$$F(t) = 2[1 - \Phi(|a|/\sqrt{t})] \;\; \& \;\; f_{T_a}(t) = (|a|/\sqrt{2\pi})t^{-3/2}e^{-a^2/2t}\;, \quad t > 0\,.$$

Further for $a \neq 0$,

$$\begin{aligned} E(T_a) &= \frac{|a|}{\sqrt{2\pi}} \int_0^\infty t^{-1/2} e^{-a^2/2t}\, dt \\ &= \frac{|a|}{\sqrt{2\pi}} \left[\int_0^1 t^{-1/2} e^{-a^2/2t}\, dt + \int_1^\infty t^{-1/2} e^{-a^2/2t}\, dt \right]. \end{aligned}$$

In the second integral of the above expression,

$$\begin{aligned} t > 1 \quad &\Rightarrow \quad -a^2/2t > -a^2/2 \;\Rightarrow\; t^{-1/2}e^{-a^2/2t} > t^{-1/2}e^{-a^2/2} \\ &\Rightarrow \quad \int_1^\infty t^{-1/2} e^{-a^2/2t}\, dt > \int_1^\infty t^{-1/2} e^{-a^2/2}\, dt \\ &= \quad e^{-a^2/2} \int_1^\infty t^{-1/2}\, dt. \end{aligned}$$

However, $\int_1^\infty t^{-1/2}\, dt$ is divergent. Hence, $E(T_a) = \infty$. It then follows that all higher order moments are also infinite. Thus, the standard Brownian motion process reaches any level $a \neq 0$ with probability 1; however, it takes on average an infinite time to hit a. □

The distribution of T_a is an inverse Gaussian distribution. It is interesting to see why it is labeled as the inverse Gaussian distribution. Toward this end, we find a relation between the distribution of T_a and the normal distribution. Suppose for $a \neq 0$, it follows $Y \sim N(0, a^{-2})$ distribution, that is $|a|Y \sim N(0, 1)$ distribution. Suppose a random variable U is defined as $U = Y^{-2}$. U is thus a non-negative random variable. Hence for $u \geq 0$,

$$\begin{aligned} P[U \leq u] = P[Y^{-2} \leq u] &= P[Y^2 \geq 1/u] = 1 - P[Y^2 \leq 1/u] \\ &= 1 - P[-|a|/\sqrt{u} \leq |a|Y \leq |a|/\sqrt{u}] \\ &= 1 - \left[\Phi(|a|/\sqrt{u}) - \Phi(-|a|/\sqrt{u})\right] \\ &= 2\Phi(-|a|/\sqrt{u})\,. \end{aligned}$$

Hence, the probability density function of U for $u > 0$ is given by

$$f_U(u) = \frac{d}{du} 2\Phi(-|a|/\sqrt{u}) = 2\phi(-|a|/\sqrt{u})\frac{|a|}{2u^{3/2}} = \frac{|a|}{\sqrt{2\pi}} u^{-3/2} e^{-a^2/2u}\,,$$

which is the same as the probability density function of T_a. Thus, the distributions of $U = Y^{-2}$ and T_a are the same, where $Y \sim N(0, a^{-2})$ distribution. Hence, the distribution of T_a is known as the inverse Gaussian distribution. Another reason to label it as an inverse Gaussian distribution is that there is inverse relationship between the cumulant generating function of this distribution and that of a normal distribution. Such a distribution was also obtained by Wald as the limiting form for the distribution of the sample size in a sequential probability ratio test. For properties of this distribution, one may refer to Johnson et al. [4]. Treating a as a time parameter, $\{T_a, a > 0\}$ is a stochastic process with stationary and independent increments with increments having the inverse Gaussian distribution.

Using an alternative approach (Cox and Miller [2]), which uses differential equations, the probability density function $f_{T_a}(t)$ of T_a for $a \neq 0$, for a Brownian motion process with drift coefficient μ and diffusion coefficient σ^2 is obtained as

$$f_{T_a}(t) = \frac{|a|}{\sigma\sqrt{2\pi}} t^{-3/2} e^{-(a-\mu t)^2/2\sigma^2 t} \,, \quad t > 0 \,.$$

This distribution is also the inverse Gaussian distribution. The mean and variance of T_a for $\mu \neq 0$ are given by

$$E(T_a) = a/\mu \quad \& \quad Var(T_a) = a\sigma^2/\mu^3 \,.$$

It is to be noted that these results are consistent with the result that $E(T_a) = \infty$ for the standard Brownian motion process.

In the following examples, we use the distribution of T_a to compute the probabilities of some events.

Example 9.3.3 The probability that the standard Brownian motion process reaches level 5 for the first time, by time 6, is obtained as follows:

$$P[T_5 \leq 6] = 2(1 - \Phi(5/\sqrt{6})) = 0.04123. \quad \square$$

Example 9.3.4 Suppose water level $X(t)$ in a tank at time t is modeled as $X(t) = 6 + 3W(t)$, where $\{W(t), t \geq 0\}$ is the standard Brownian motion process. To compute the probability that the tank becomes empty by time 12 units, we define $T = \min\{t | X(t) = 6 + 3W(t) = 0\} = \min\{t | W(t) = -2\} = T_{-2}$. Thus, we compute

$$P[T_{-2} \leq 12] = P[L(12) \leq -2] = 2(1 - \Phi(2/\sqrt{12})) = 0.5637. \quad \square$$

In the following example, we obtain the initial level of inventory, modeled as a Brownian motion process, using the distribution of T_a.

Example 9.3.5 Suppose the inventory at time t is modeled as $x + W(t)$, where $\{W(t), t \geq 0\}$ is the standard Brownian motion process and x is the initial inventory.

We find the smallest value of x, so that probability of stock-out in (0, 10] is less than 0.1. Now,

$$x + W(t) \geq 0 \ \text{ for } \ 0 < t \leq 10 \quad \Longleftrightarrow \quad T_{-x} = \min\{t \geq 0 | W(t) = -x\} > 10 \,.$$

Hence,

$$\begin{aligned} P[x + W(t) \geq 0 \text{ for } 0 < t \leq 10] \geq 0.9 &\Longleftrightarrow P[T_{-x} > 10] \geq 0.9 \\ &\Rightarrow P[T_{-x} \leq 10] \leq 0.1 \\ &\Rightarrow 2[1 - \Phi(x/\sqrt{10})] \leq 0.1 \\ &\Rightarrow \Phi(x/\sqrt{10}) \geq 0.95 \\ &\Rightarrow x \geq \sqrt{10}\Phi^{-1}(0.95) = 5.20. \end{aligned}$$

Thus, initial inventory should be set at 5.20 units, so that the probability of stock-out in (0, 10] is less than 0.1. □

As a consequence of the reflection principle and the probability density function of T_a, it can be shown that the probability that the standard Brownian motion $\{W(t), t \geq 0\}$ with $W(0) = 0$ will cross the time axis at least once in the time interval $(t, t+s]$ for $t, s > 0$ is $(2/\pi)\arctan(\sqrt{s/t}) = (2/\pi)\arccos(\sqrt{t/(t+s)})$. For details, one may refer to the book by Feller [3].

Suppose $\{X(t), t \geq 0\}$ is a Brownian motion process with $X(0) = 0$, drift coefficient μ and diffusion coefficient σ^2. Suppose $a < 0$ and $b > 0$ are two given numbers. Then a random variable $T(a, b)$ defined as

$$T(a, b) = \min\{t \geq 0 | X(t) \in \{a, b\}\}$$

denotes the first time epoch when $X(t)$ visits a or b. In the following theorem, we state some results related to $T(a, b)$, Kulkarni [6].

Theorem 9.3.5 *Suppose $\{X(t), t \geq 0\}$ is a Brownian motion process with $X(0) = 0$, drift coefficient μ and diffusion coefficient σ^2. Suppose $a < 0$ and $b > 0$ are two given numbers and $\theta = -2\mu/\sigma^2$. For $\mu \neq 0$,*

$$P[X(T(a, b)) = b] = \frac{e^{\theta a} - 1}{e^{\theta a} - e^{\theta b}} \ \ \& \ \ E(T(a, b)) = \frac{b(e^{\theta a} - 1) - a(e^{\theta b} - 1)}{\mu(e^{\theta a} - e^{\theta b})}.$$

$$\text{For } \mu = 0, \quad P[X(T(a, b)) = b] = |a|/(|a| + b) \ \ \& \ \ E(T(a, b)) = |a|b/\sigma^2.$$

Remark 9.3.1 The expressions for $\mu = 0$ can be obtained from those of $\mu \neq 0$, by allowing $\theta \to 0$ and using the L'Hospital's rule.

The following example illustrates Theorem 9.3.5.

Example 9.3.6 Suppose $\{X(t), t \geq 0\}$ is a Brownian motion process with $X(0) = 4$, drift coefficient $\mu = 3$ and diffusion coefficient $\sigma^2 = 2$. To compute the probability that $X(t)$ hits 9 before hitting 3, we define $Y(t) = X(t) - 4$ so that $Y(0) = 0$. Thus, we compute the probability that $Y(t)$ hits $5 = b$ say, before hitting $-1 = a$ say, as $P[Y(T(-1, 5)) = 5]$. With $\theta = -2\mu/\sigma^2 = -3$,

$$P[Y(T(-1,5)) = 5] = (e^3 - 1)/(e^3 - e^{-15}) = 0.9502 \;\; \& \;\; E(T(-1,5)) = 1.5671.$$

Thus, the probability that $X(t)$ hits 9 before hitting 3 is 0.9502, the probability that $X(t)$ hits 3 before hitting 9 is $1 - 0.9502 = 0.0498$ and expected time of visit to either 9 or 3 is 1.5671 time units. For $\mu = -2$ and $\theta = -2\mu/\sigma^2 = 1$,

$$P[Y(T(-1,5)) = 5] = (e^{-1} - 1)/(e^{-1} - e^5) = 0.0042 \;\; \& \;\; E(T(-1,5)) = 0.4872.$$

Thus, the probability that $X(t)$ hits 9 before hitting 3 is 0.0042, the probability that $X(t)$ hits 3 before hitting 9 is $1 - 0.0042 = 0.9958$ and expected time of visit to either 9 or 3 is 0.4872 time units. It is to be noted that if $\mu > 0$ then the probability that $X(t)$ hits 9 before hitting 3 is higher than that when $\mu < 0$, as expected. □

Example 9.3.7 Suppose water level $X(t)$ in a tank at time t is modeled as $X(t) = 6 + 3W(t)$, where $\{W(t), t \geq 0\}$ is the standard Brownian motion process. Suppose the tank overflows when the water level reaches 15 units. To compute the probability that the tank overflows before it becomes empty, we define $Y(t) = X(t) - 6$. Further,

$$X(t) = 15 \;\Rightarrow\; Y(t) = 9 = b, \text{ say } \& \; X(t) = 0 \;\Rightarrow\; Y(t) = -6 = a \text{ say.}$$

Since $\mu = 0$ and $\sigma^2 = 9$,

$$P[Y(T(-6,9)) = 9] = |a|/(|a| + b) = 6/15 = 0.4 \;\; \& \;\; E(T(-6,9)) = |a|b/\sigma^2 = 6.$$

□

Suppose $\{X(t), t \geq 0\}$ is a Brownian motion process with $X(0) = 0$, drift coefficient μ and diffusion coefficient σ^2. The random variables U and L are defined as follows:

$$U = \max\{X(t)|0 \leq t < \infty\} \;\; \& \;\; L = \min\{X(t)|0 \leq t < \infty\}.$$

The next theorem states results related to the maximum U and the minimum L of the Brownian motion process over $[0, \infty)$, Kulkarni [6].

Theorem 9.3.6 *Suppose $\{X(t), t \geq 0\}$ is a Brownian motion process with $X(0) = 0$, drift coefficient μ and diffusion coefficient σ^2. Suppose*

$$U = \max\{X(t)|0 \leq t < \infty\}, \;\; L = \min\{X(t)|0 \leq t < \infty\} \;\; \& \;\; \theta = -2\mu/\sigma^2 .$$

Then (i) For $\mu < 0$, the probability density function of U is given by $f_U(x) = \theta e^{-\theta x}$, $x \geq 0$ and $L = -\infty$ with probability 1. (ii) For $\mu > 0$, the probability density function of L is given by $f_L(x) = -\theta e^{-\theta x}$, $x \leq 0$ and $U = \infty$ with probability 1 and (iii) If $\mu = 0$, then $U = \infty$ and $L = -\infty$ with probability one.

The following examples illustrate Theorem 9.3.6.

Example 9.3.8 Suppose $\{X(t), t \geq 0\}$ is a Brownian motion process with $X(0) = 6$, drift coefficient $\mu = -2$ and diffusion coefficient $\sigma^2 = 4$. To compute the probability that $X(t)$ never goes above 12, we define $Y(t) = X(t) - 6$ so that $Y(0) = 0$. Thus, the probability that $X(t)$ never goes above 12, is the same as the probability that $Y(t)$ never goes above 6. With $\theta = -2\mu/\sigma^2 = 1$,

$$\begin{aligned} P[\max\{X(t)|0 \leq t < \infty\} \leq 12] &= P[\max\{Y(t)|0 \leq t < \infty\} \leq 6] \\ &= P[U \leq 6] = 1 - \exp(-6\theta) \\ &= 1 - \exp(-6) = 0.9975. \end{aligned}$$

If $\mu = 2$, then $P[U \leq 6] = 0$, since $U = \infty$ with probability 1. □

Example 9.3.9 Suppose $\{X(t), t \geq 0\}$ is a Brownian motion process with $X(0) = 3$, drift coefficient $\mu = 2$ and diffusion coefficient $\sigma^2 = 4$. To compute the probability that $X(t)$ never goes below -2, we define $Y(t) = X(t) - 3$ so that $Y(0) = 0$. Thus, the probability that $X(t)$ never goes below -2 is the same as the probability that $Y(t)$ never goes below -5. With $\theta = -2\mu/\sigma^2 = -1$,

$$\begin{aligned} P[\min\{X(t)|0 \leq t < \infty\} \geq -2] &= P[\min\{Y(t)|0 \leq t < \infty\} \geq -5] \\ &= P[L \geq -5] = \int_{-5}^{0} (-\theta)e^{-\theta x} dx \\ &= 1 - \exp(-5) = 0.9932. \end{aligned}$$

If $\mu = -2$, then $P[L \geq -5] = 0$, since $L = -\infty$ with probability 1. □

Example 9.3.10 Suppose the inventory at a store at time t is modeled as $X(t) = 5 - 3t + 4W(t)$ where $\{W(t), t \geq 0\}$ is the standard Brownian motion process. Thus, $\{X(t), t \geq 0\}$ is a Brownian motion process with $X(0) = 5$, drift coefficient $\mu = -3$ and diffusion coefficient $\sigma^2 = 4^2$. Suppose we want to decide how large the storage area A should be so that the probability that it is ever full is less than 0.01, that is, we have to find A such that $P[\max\{X(t)|0 \leq t < \infty\} \geq A] \leq 0.01$. Observe that, with $\theta = -2\mu/\sigma^2 = 3/8$ and $Y(t) = X(t) - 5$, we have

$$\begin{aligned} & P[\max\{X(t)|0 \leq t < \infty\} \geq A] \leq 0.01 \\ \Rightarrow \ & P[\max\{Y(t)|0 \leq t < \infty\} \geq A - 5] \leq 0.01 \\ & \Rightarrow \ P[U \geq A - 5] \leq 0.01 \ \Rightarrow \ \int_{A-5}^{\infty} \theta e^{-\theta x} dx \leq 0.01 \\ & \Rightarrow \ e^{-\theta(A-5)} \leq 0.01 \ \Rightarrow \ A \geq 7.9886. \end{aligned}$$

Thus, A has to be at least 7.9886 units. □

In the following three sections, we discuss some variations of Brownian motion process, such as Brownian bridge and geometric Brownian motion process.

9.4 Brownian Bridge

Suppose $\{W(t), t \geq 0\}$ is the standard Brownian motion process. We begin with a particular conditional distribution of $W(t)$ in the following example.

Example 9.4.1 Suppose $\{W(t), t \geq 0\}$ is the standard Brownian motion process. We obtain the conditional distribution of $W(t)$ given $W(t_1) = a$ and $W(t_2) = b$ for $t_1 < t < t_2$ as follows. Suppose $Y = W(t) - W(t_1)$ and $Z = W(t_2) - W(t)$. Then from the definition of a Brownian motion process, it follows that Y and Z are independent random variables having normal $N(0, t - t_1)$ and $N(0, t_2 - t)$ distributions, respectively. Hence, it follows that the joint distribution of $(Y, Y + Z)' = (W(t) - W(t_1), W(t_2) - W(t_1))'$ is bivariate normal with mean vector $\underline{0}$ and dispersion matrix D given

$$D = \begin{pmatrix} t - t_1 & t - t_1 \\ t - t_1 & t_2 - t_1 \end{pmatrix}.$$

From this bivariate distribution, we obtain the conditional distribution of $Y = W(t) - W(t_1)$ given $Y + Z = W(t_2) - W(t_1) = b - a$. It is again normal with mean μ and variance σ^2 where

$$\mu = \frac{(b-a)(t-t_1)}{t_2 - t_1} \ \ \& \ \sigma^2 = t - t_1 - \frac{(t-t_1)^2}{t_2 - t_1} = (t - t_1)\frac{t_2 - t}{t_2 - t_1}.$$

But $Y = W(t) - a$ when $W(t_1) = a$. Thus, the conditional distribution of $W(t)$ given $W(t_1) = a$ and $W(t_2) = b$ is normal with mean $\mu + a$ and variance σ^2. In particular if $t_1 = 0$ and $t_2 = 1$, with $a = 0$ and $b = 0$, the conditional distribution of $W(t)$ is normal $N(0, t(1-t))$.

From Example 9.4.1, we now define a version of a Brownian motion process, known as a Brownian bridge. The Brownian bridge is defined from the standard Brownian motion process by conditioning on the event $W(0) = W(1) = 0$. The

Brownian bridge is used to describe certain random functionals arising in nonparametric statistics, and as a model for the prices of publicly traded bonds having a specified redemption value on a fixed expiration date. Following is the definition of a Brownian bridge.

Definition 9.4.1 Brownian Bridge: Suppose $\{W(t), t \geq 0\}$ is the standard Brownian motion process. Then $\{X(t), 0 \leq t \leq 1\}$ is known as a Brownian bridge, if $X(0) = X(1) = 0$ and for $t \in (0, 1)$ the distribution of $X(t)$ is the same as the conditional distribution of $W(t)$ given $W(0) = W(1) = 0$, which is normal $N(0, t(1-t))$.

It is known as a Brownian bridge, as it is tied at both the ends 0 and 1 to take a particular value. The common value of $X(0)$ and $X(1)$ can be any arbitrary value, not necessarily 0.

From the definition of a Brownian bridge, it follows that its mean function and variance function are given by $E(X(t)) = 0$ & $Var(X(t)) = t(1-t)$, $0 < t < 1$. The variance function attains maximum at $t = 0.5$.

From Theorem 9.2.4, it is known that for the standard Brownian motion process $\{W(t), t \geq 0\}$, the joint distribution of $(W(t_1), W(t_2), \ldots, W(t_n))'$ is multivariate normal for all $t_1, t_2, \ldots, t_n \in [0, \infty)$. We now examine whether a similar result holds for a Brownian bridge.

Theorem 9.4.1 *A Brownian bridge is a Gaussian process with mean value function* 0 *and covariance function* $c(s,t) = s(1-t)$ *for* $s \leq t \in (0, 1)$.

Proof Suppose $\{W(t), t \geq 0\}$ is the standard Brownian motion process. For $0 < t_1 < t_2 < 1$, we define

$$Y_1 = W(t_1) - W(0), \quad Y_2 = W(t_2) - W(t_1) \quad \& \quad Y_3 = W(1) - W(t_2).$$

Then from the definition of a Brownian motion process, it follows that $\{Y_1, Y_2, Y_3\}$ are independent random variables each having normal distribution with means 0 and variances t_1, $t_2 - t_1$ and $1 - t_2$, respectively. Thus, the joint distribution of $\underline{Y} = (Y_1, Y_2, Y_3)'$ is $N_3(\underline{0}, D)$ where D is a diagonal matrix, with diagonal elements given by $(t_1, t_2 - t_1, 1 - t_2)'$. To obtain the conditional distribution of $(W(t_1), W(t_2))'$ given $W(0) = W(1) = 0$ for $0 < t_1 < t_2 < 1$, note that given $W(0) = W(1) = 0$, $W(t_1) = Y_1$ & $W(t_2) = -Y_3$. The condition $W(0) = W(1) = 0$ is equivalent to $Y_1 + Y_2 + Y_3 = 0$. Thus, the conditional distribution of $(W(t_1), W(t_2))'$ given $W(0) = W(1) = 0$ is the same as the conditional distribution of $(Y_1, -Y_3)'$ given $Y_1 + Y_2 + Y_3 = 0$. We first obtain the joint distribution of $\underline{Z} = (Y_1, -Y_3, Y_1 + Y_2 + Y_3)'$ as follows. Observe that $\underline{Z}$ can be expressed as $\underline{Z} = A\underline{Y}$ and $\underline{Y} \sim N_3(\underline{0}, D)$ implies $\underline{Z} \sim N_3(\underline{0}, ADA')$ where A and $\Sigma = ADA'$ are given by

$$A = \begin{pmatrix} 1 & 0 & 0 \\ 0 & 0 & -1 \\ 1 & 1 & 1 \end{pmatrix}, \quad \Sigma = \begin{pmatrix} t_1 & 0 & t_1 \\ 0 & 1-t_2 & -1+t_2 \\ t_1 & -1+t_2 & 1 \end{pmatrix} = \begin{pmatrix} \Sigma_{11} & \Sigma_{12} \\ \Sigma_{21} & \Sigma_{22} \end{pmatrix},$$

where Σ_{11} is of order 2×2. Hence, the conditional distribution of $(Y_1, -Y_3)'$ given $Y_1 + Y_2 + Y_3 = 0$ is $N_2(\underline{0}, V)$ where

$$V = \Sigma_{11} - \Sigma_{12}\Sigma_{22}^{-1}\Sigma_{21} = \begin{pmatrix} t_1 & 0 \\ 0 & 1 - t_2 \end{pmatrix} - (t_1, -1 + t_2)' \times (t_1, -1 + t_2) .$$

V is thus given by

$$V = \begin{pmatrix} t_1(1 - t_1) & t_1(1 - t_2) \\ t_1(1 - t_2) & t_2(1 - t_2) \end{pmatrix} .$$

Hence, the conditional distribution of $(W(t_1), W(t_2))'$ given $W(0) = W(1) = 0$ for $0 < t_1 < t_2 < 1$ is bivariate normal. Using the same arguments, it can be proved that the conditional distribution of $(W(t_1), W(t_2), \ldots, W(t_n))'$ given $W(0) = W(1) = 0$ for $0 < t_1 < t_2 < \cdots < t_n < 1$ is n-variate normal with mean vector $\underline{0}$ and dispersion matrix $V = [V_{ij}]$ where $V_{ii} = t_i(1 - t_i)$ and $V_{ij} = t_i(1 - t_j),\ t_i < t_j$. Thus, the Brownian bridge is a Gaussian process, with covariance function $c(s, t) = s(1 - t)$ for $s \le t \in (0, 1)$. □

Theorem 9.4.1 leads to another approach to define a Brownian bridge, as shown below.

Definition 9.4.2 Brownian Bridge: A Gaussian process on $[0, 1]$ with covariance function $c(s, t) = s(1 - t)$ for $s \le t \in (0, 1)$ is a Brownian bridge.

Using this definition, in the next example for a Brownian bridge $\{X(t), t \ge 0\}$, we obtain the distribution of $X(t + h) - X(t)$ for $h > 0$ and a joint distribution of $X(t_2) - X(t_1)$ and $X(t_3) - X(t_2)$ for $0 < t_1 < t_2 < t_3 < 1$. It helps us to investigate the nature of displacement random variables in a Brownian bridge.

Theorem 9.4.2 *A Brownian bridge is a stochastic process with stationary increments but the increments are not independent.*

Proof Suppose $\{X(t), 0 \le t \le 1\}$ is a Brownian bridge. Hence, it is a Gaussian process with mean value function 0 and covariance function $c(s, t) = s(1 - t)$ for $s \le t$. Thus, the joint distribution of $(X(t), X(t + h))'$ is bivariate normal with mean vector $(0, 0)'$ and dispersion matrix

$$\Sigma = \begin{pmatrix} t(1 - t) & t(1 - t - h) \\ t(1 - t - h) & (t + h)(1 - t - h) \end{pmatrix}.$$

Hence, the distribution of $X(t + h) - X(t)$ is normal with mean 0 and variance v given by

$$v = t(1 - t) + (t + h)(1 - t - h) - 2t(1 - t - h) = h(1 - h).$$

Thus, the distribution of $X(t+h) - X(t)$ depends only on h and not on t. Hence, the Brownian bridge is a stochastic process with stationary increments. To obtain the joint distribution of $\underline{Z}_1 = (X(t_2) - X(t_1), X(t_3) - X(t_2))'$ for $0 < t_1 < t_2 < t_3$, we note that the joint distribution of $\underline{Z}_2 = (X(t_1), X(t_2), X(t_3))'$ is trivariate normal with mean vector $(0, 0, 0)'$ and dispersion matrix

$$\Sigma = \begin{pmatrix} t_1(1-t_1) & t_1(1-t_2) & t_1(1-t_3) \\ t_1(1-t_2) & t_2(1-t_2) & t_2(1-t_3) \\ t_1(1-t_3) & t_2(1-t_3) & t_3(1-t_3) \end{pmatrix}.$$

Further, $\underline{Z}_1 = A\underline{Z}_2 \sim N_2(\underline{0}, A\Sigma A')$ where

$$A = \begin{pmatrix} -1 & 1 & 0 \\ 0 & -1 & 1 \end{pmatrix}$$

$$\& \;\; A\Sigma A' = \begin{pmatrix} (t_2-t_1)(1-(t_2-t_1)) & -(t_2-t_1)(t_3-t_2) \\ -(t_2-t_1)(t_3-t_2) & (t_3-t_2)(1-(t_3-t_2)) \end{pmatrix}.$$

Observe that off-diagonal elements of $A\Sigma A'$ are not 0. Thus, the increments $X(t_2) - X(t_1)$ and $X(t_3) - X(t_2)$ are not independent random variables. Hence, a Brownian bridge is a process which does not have independent increments. □

Definition 9.4.2 of a Brownian bridge is also useful to verify whether a given process is a Brownian bridge. It is illustrated in the following example.

Example 9.4.2 Suppose $\{W(t), t \geq 0\}$ is the standard Brownian motion process and $Z(t)$ is defined as $Z(t) = W(t) - tW(1),\ 0 \leq t \leq 1$. Note that $Z(0) = Z(1) = 0$. Since $\{W(t), t \geq 0\}$ is the standard Brownian motion process, the joint distribution of $\underline{Y} = (W(t_1), W(t_2), W(t_3))'$ for $0 < t_1 < t_2 < t_3 = 1$ is $N_3(\underline{0}, \Sigma)$ where $\Sigma = [\sigma_{ij}]$ and $\sigma_{ii} = t_i$ for $i = 1, 2, 3$ and $\sigma_{ij} = \min\{t_i, t_j\}$ for $i \neq j$. Suppose $\underline{Z} = (Z(t_1), Z(t_2))'$, then $\underline{Z} = A\underline{Y} \sim N_2(\underline{0}, A\Sigma A')$ distribution where A, Σ and $A\Sigma A'$ are given by

$$A = \begin{pmatrix} 1 & 0 & -t_1 \\ 0 & 1 & -t_2 \end{pmatrix}, \Sigma = \begin{pmatrix} t_1 & t_1 & t_1 \\ t_1 & t_2 & t_2 \\ t_1 & t_2 & 1 \end{pmatrix} \& \; A\Sigma A' = \begin{pmatrix} t_1(1-t_1) & t_1(1-t_2) \\ t_1(1-t_2) & t_2(1-t_2) \end{pmatrix}.$$

Using a similar approach, it can be proved that the distribution of $(Z(t_1), Z(t_2), \ldots, Z(t_n))'$ for $0 < t_1 < t_2 < \cdots t_n < 1$ is multivariate normal with mean vector $\underline{0}$ and dispersion matrix $V = [V_{ij}]$ where $V_{ii} = t_i(1-t_i)$ and $V_{ij} = t_i(1-t_j),\ t_i < t_j$. Thus, $\{Z(t), 0 \leq t \leq 1\}$ is a Gaussian process. Further, its mean function is 0 and covariance function is $s(1-t)$ for $s \leq t$. Hence, $\{Z(t), 0 \leq t \leq 1\}$ is a Brownian bridge. □

Theorem 9.4.2 and Example 9.4.2 are both useful to obtain a realization of a Brownian bridge. We use the following two approaches to write a code for obtaining a realization. (i) From Theorem 9.4.2, $X(t+h) - X(t) \sim N(0, h(1-h))$. Hence, as in the code of a realization of a Brownian motion process, for each t, $X(t+h) = X(t) + u$ where u is a realized value of a random sample of size 1 from $N(0, h(1-h))$ distribution. (ii) From Example 9.4.2, if $\{W(t), t \geq 0\}$ is the standard Brownian motion process, then the process $\{Z(t), 0 \leq t \leq 1\}$, where $Z(t) = W(t) - tW(1),\ 0 \leq t \leq 1$, is a Brownian bridge. Code 9.7.3 obtains a realization of a Brownian bridge, using both the approaches. It is illustrated in the next example.

Example 9.4.3 (i) Suppose $\{X(t), t \geq 0\}$ is a Brownian bridge, then $X(t+h) = X(t) + u$, where u is a realized value of a random sample of size 1 from $N(0, h(1-h))$ distribution. (ii) Suppose $Z(t)$ is defined as $Z(t) = W(t) - tW(1),\ 0 \leq t \leq 1$, where $\{W(t), t \geq 0\}$ is the standard Brownian motion process. Then $Z(t),\ 0 \leq t \leq 1$ is a Brownian bridge. Hence to obtain a realization of a Brownian bridge, we obtain a realization of $\{W(t), 0 \leq t \leq 1\}$ and from it get a realization of $\{Z(t),\ 0 \leq t \leq 1\}$. Figure 9.3 presents the realization of the Brownian bridge. The upper panel displays the realization using the first approach, and the lower panel displays the realization using the second approach. Realizations from both the approaches are similar. □

Using Theorem 9.4.1, in the next theorem we prove that the family of finite dimensional distribution functions of a properly normed and scaled empirical process, corresponding to a random sample from uniform $U(0, 1)$ distribution, can be approximated by that of a Brownian bridge.

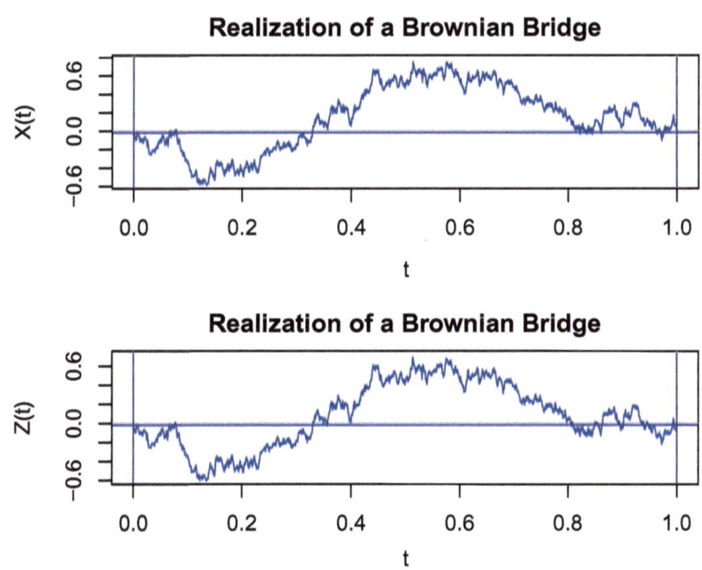

Fig. 9.3 Realization of a Brownian Bridge

Theorem 9.4.3 *Suppose $F_n(s)$ is the empirical distribution function corresponding to a random sample of size n from the uniform $U(0, 1)$ distribution. Suppose $Z_n(s) = \sqrt{n}(F_n(s) - s), 0 \le s \le 1$. Then for large n, for any $k \ge 1$ and $0 < s_1 < s_2 < \cdots < s_k < 1$, the joint distribution of $(Z_n(s_1), Z_n(s_2), \ldots, Z_n(s_k))'$ can be approximated by the joint distribution of $(X(s_1), X(s_2), \ldots, X(s_k))'$, where $\{X(s), 0 \le s \le 1\}$ is a Brownian bridge.*

Proof Suppose $\{X_i, i = 1, 2, \ldots, n\}$ are independent and identically distributed random variables each having uniform $U(0, 1)$ distribution. For $0 \le s \le 1$, a random variable $Y_i(s)$ is defined as

$$Y_i(s) = \begin{cases} 1, & \text{if } X_i \le s \\ 0, & \text{if } X_i > s. \end{cases}$$

Thus, $Y_i(s)$ is a Borel function of X_i and hence $\{Y_i, i = 1, 2, \ldots, n\}$ are also independent and identically distributed random variables each having Bernoulli distribution with

$$E(Y_i(s)) = P[X_i \le s] = F(s) = s \;\; \& \;\; Var(Y_i(s)) = s(1 - s) < \infty .$$

The function $F_n(s) = \sum_{i=1}^n Y_i(s)/n$ is the empirical distribution function. By the strong law of large numbers, for fixed s, $F_n(s) \xrightarrow{a.s} F(s) = s \;\; \forall \;\; s \in [0, 1]$. By the Glivenko-Cantelli theorem, almost sure convergence is uniform in s. Further, by the central limit theorem, for fixed s,

$$Z_n(s) = \sqrt{n}(F_n(s) - s) \xrightarrow{L} X(s) \sim N(0, s(1 - s)) \text{ distribution.}$$

Using multivariate central limit theorem, for a finite $k \ge 1$ and for fixed $0 < s_1 < s_2 < \cdots < s_k < 1$,

$$(Z_n(s_1), Z_n(s_2), \ldots, Z_n(s_k))' \xrightarrow{L} \underline{X}_k \sim N(\underline{0}, \Sigma) \text{ where } \Sigma = [\sigma_{ij}],$$

$\underline{X}_k = (X(s_1), X(s_2), \ldots, X(s_k))'$ and $\sigma_{ii} = s_i(1 - s_i)$. To find σ_{ij}, we find $Cov(Z_n(s), Z_n(t))$ for $s < t$. Suppose $N_n(s) = \sum_{i=1}^n Y_n(s)$, then $N_n(s)$ follows binomial $B(n, s)$ distribution. It can be proved that the conditional distribution of $N_n(t) - N_n(s)$ given $N_n(s)$ is binomial $B\,(n - N_n(s), (t - s)/(1 - s))$. Observe that

$$\begin{aligned} Cov(Z_n(s), Z_n(t)) &= nCov(F_n(s), F_n(t)) = Cov(N_n(s), N_n(t))/n \\ &= [E(N_n(s)N_n(t)) - n^2 st]/n . \end{aligned}$$

Now,

$$\begin{aligned}
E(N_n(s)N_n(t)) &= E(E(N_n(s)N_n(t)|N_n(s))) \\
&= E(E(N_n(s)[N_n(t) - N_n(s) + N_n(s)]|N_n(s))) \\
&= E(N_n(s)(E(N_n(t) - N_n(s))|N_n(s))) + E(N_n^2(s)) \\
&= E(N_n(s)\left(\frac{(n - N_n(s))(t - s)}{1 - s}\right) + E(N_n^2(s)) \\
&= \frac{n(t - s)}{1 - s}E(N_n(s)) - \frac{t - s}{1 - s}E(N_n^2(s)) + E(N_n^2(s)) \\
&= \frac{n(t - s)}{1 - s}ns + \frac{1 - t}{1 - s}E(N_n^2(s)) \\
&= \frac{n^2 s(t - s)}{1 - s} + \frac{1 - t}{1 - s}(ns(1 - s) + n^2 s^2) \\
&= \frac{ns}{1 - s}(1 - s)(1 - t + nt) = ns(1 - t + nt)\ .
\end{aligned}$$

Hence,

$$\begin{aligned}
Cov(Z_n(s), Z_n(t)) &= \frac{1}{n}[E(N_n(s)N_n(t)) - n^2 st] \\
&= \frac{1}{n}[ns - nst + n^2 st - n^2 st] = s(1 - t)\ ,
\end{aligned}$$

and it is the same for all n. Thus, for $i \neq j$, $\sigma_{ij} = s_i(1 - s_j)$ for $s_i < s_j$. Hence for large n, the joint distribution of $(Z_n(s_1), Z_n(s_2), \dots, Z_n(s_k))'$ can be approximated by the k-dimensional normal distribution with $\underline{0}$ mean vector and dispersion matrix Σ, where $\sigma_{st} = s(1 - t), s \leq t$. It is the same as the joint distribution of $(X(s_1), X(s_2), \dots, X(s_k))'$, where $\{X(s), 0 \leq s \leq 1\}$ is a Brownian bridge. □

Theorem 9.4.3 conveys that $F_n(s)$ can be approximated by $s + X(s)/\sqrt{n}$ for $0 \leq s \leq 1$, where $\{X(s), 0 \leq s \leq 1\}$ is a Brownian bridge. In the next example, we verify it using Code 9.7.4.

Example 9.4.4 Suppose $F_n(s)$ is the empirical distribution function, corresponding to a random sample of size $n = 500$ from uniform $U(0, 1)$ distribution. We examine whether $F_n(s)$ can be approximated by $s + X(s)/\sqrt{n}$ for $0 \leq s \leq 1$, where $\{X(s), 0 \leq s \leq 1\}$ is a Brownian bridge. In Fig. 9.4, the curves corresponding to $F_n(s)$ and $s + X(s)/\sqrt{n}$ are imposed on each other. These show close agreement and $F_n(s)$ can be approximated by $s + X(s)/\sqrt{n}$ for $0 \leq s \leq 1$. □

In the next section, we discuss a geometric Brownian motion process, which has many applications in finance.

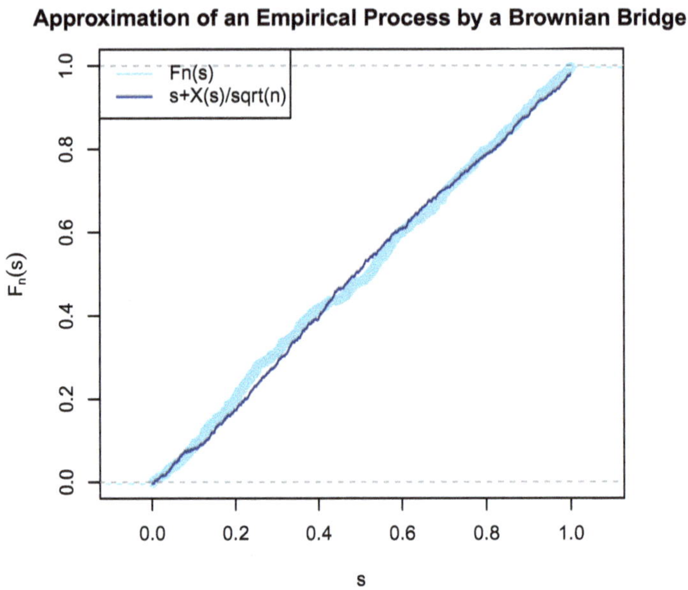

Fig. 9.4 Approximation of empirical process by Brownian Bridge

9.5 Geometric Brownian Motion Process

We begin with the definition of a geometric Brownian motion process.

Definition 9.5.1 Geometric Brownian Motion Process: Suppose $\{Z(t), t \geq 0\}$ is a Brownian motion process with drift coefficient μ and diffusion coefficient σ^2. Suppose $X(t) = e^{Z(t)}$. Then the process $\{X(t), t \geq 0\}$ is known as a geometric Brownian motion process.

Some properties of a geometric Brownian motion process are listed below.

(i) The state space of a geometric Brownian motion process is $(0, \infty)$. Hence, many economists prefer a geometric Brownian motion process as a model for market prices, in contrast to a Brownian motion process, even if its drift coefficient $\mu > 0$. Note that $X(0) = x = e^{Z(0)}$ and $x = 1$ if $Z(0) = 0$.

(ii) $X(t)$ can be expressed as $X(t) = X(0)e^{Z(t)-Z(0)}$ and $Z(t) - Z(0)$ follows $N(\mu t, \sigma^2 t)$ distribution. If $Z(0) = 0$, then $X(0) = 1$ and $\log X(t) = Z(t)$ follows normal $N(\mu t, \sigma^2 t)$ distribution. Thus for each t, $X(t)$ has a lognormal distribution.

(iii) Suppose $0 = t_0 < t_1 < t_2 < \cdots < t_n = t$. Observe that

$$X(t_i)/X(t_{i-1}) = e^{Z(t_i)-Z(t_{i-1})} = Y_i, \ i = 1, 2, \ldots, n$$

are independent random variables having a lognormal distribution. If $t_i - t_{i-1} = h$ for all $i = 1, 2, \ldots, n$, then $\{Y_1, Y_2, \ldots, Y_n\}$ are independent and identically distributed random variables, each having lognormal distribution with parameters μh and $\sigma^2 h$. Thus, $X(t)$ can be expressed as $X(t) = X(0)Y_1 Y_2 \cdots Y_n$.

(iv) The moment-generating function $M(u)$ of normal $N(\mu, \sigma^2)$ distribution is given by

$$M(u) = e^{\mu u + u^2\sigma^2/2} \Rightarrow M(1) = e^{\mu+\sigma^2/2} = e^{\delta}, \text{ where } \delta = (\mu + \sigma^2/2).$$

From this expression, we obtain below the mean function and the variance function of a geometric Brownian motion process:

$$\begin{aligned} E(X(t)|X(0) = x) &= xE(e^{Z(t)-Z(0)}) = xe^{\mu t + t\sigma^2/2} = xe^{\delta t} \\ E(X^2(t)|X(0) = x) &= x^2 E(e^{2(Z(t)-Z(0))}) = x^2 e^{2\mu t + 2t\sigma^2} \\ Var(X(t)|X(0) = x) &= x^2 e^{2\delta t}(e^{t\sigma^2} - 1)\,. \end{aligned}$$

(v) For $s < t$,

$$\begin{aligned} E(X(t)\big|X(u), 0 \le u \le s) &= E(e^{Z(t)}\big|Z(u), 0 \le u \le s) \\ &= E(e^{Z(s)+Z(t)-Z(s)}\big|Z(u), 0 \le u \le s) \\ &= e^{Z(s)}E(e^{Z(t)-Z(s)}\big|Z(u), 0 \le u \le s) \\ &= X(s)E(e^{Z(t)-Z(s)}) \\ &= X(s)\exp\{\mu(t-s) + \sigma^2(t-s)/2\} \\ &= X(s)\exp\{\delta(t-s)\} \text{ almost surely,} \end{aligned}$$

where $\delta = (\mu + \sigma^2/2)$. The fourth step is due to independence of increments and the second last step follows since $Z(t) - Z(s) \sim N(\mu(t-s), \sigma^2(t-s))$ distribution. Thus, the expected value of the process at time t, given the history of the process up to time s, depends only on s. Further, observe that

$$\begin{aligned} & E(X(t)|X(u), 0 \le u \le s) = X(s)e^{\delta(t-s)} \text{ almost surely} \\ \Rightarrow \quad & E(e^{-\delta t}X(t)|X(u), 0 \le u \le s) = X(s)e^{-\delta s} \text{ almost surely.} \end{aligned}$$

Thus, if $\{X(t), t \ge 0\}$ is a geometric process with $X(0) = 1$, then $\{e^{-\delta t}X(t) = X(t)/E(X(t)), t \ge 0\}$ is a martingale.

Just as a Brownian motion process is a Markov process, so is the geometric Brownian motion process. We prove it in the following theorem.

Theorem 9.5.1 *A geometric Brownian motion process $\{X(t), t \ge 0\}$ is a time homogeneous Markov Process.*

Proof To prove the Markov property, observe that for any Borel set A,

$$\begin{aligned} &P[X(t+h) \in A | X(u), 0 \le u \le t] \\ &= P[e^{Z(t+h)-Z(t)+Z(t)} \in A | X(u), 0 \le u \le s] \\ &= P[e^{Z(t)}e^{Z(t+h)-Z(t)} \in A | X(u), 0 \le u \le t] \\ &= P[X(t)e^{Z(t+h)-Z(t)} \in A | X(u), 0 \le u \le t] \\ &= P[X(t)e^{Z(t+h)-Z(t)} \in A | X(t)], \end{aligned}$$

since a Brownian motion process has independent increments. Thus, given the entire past $\{X(u), 0 \le u \le t\}$, the conditional distribution of $X(t+h)$ depends on only $X(t)$ and hence the Markov property is established. Further, the conditional distribution of $X(t+h)$ given $X(t)$ is a function of h, hence the Markov process is time homogeneous. □

We now discuss how to obtain a realization of a geometric Brownian motion process. From the definition of a geometric Brownian motion process, we have $X(t) = e^{Z(t)}$ where $\{Z(t), t \ge 0\}$ is a Brownian motion process with drift coefficient 0 and diffusion coefficient σ^2. If $Z(0) = 0$ then $X(0) = 1$. Further, for any $t, h > 0$

$$\begin{aligned} X(t+h) = e^{Z(t+h)} \ \& \ X(t) = e^{Z(t)} &\Rightarrow X(t+h)/X(t) = e^{Z(t+h)-Z(t)} \\ &\Rightarrow X(t+h) = X(t)e^{Z(t+h)-Z(t)}, \end{aligned}$$

where for each fixed t, it follows $Z(t+h) - Z(t) \sim N(0, \sigma^2 h)$ distribution. Hence, by adopting similar steps as in the procedure to obtain a realization of a Brownian motion process for a fixed interval, we have $X(t+h) = X(t)e^u$, where u is the realized value of a random sample of size 1 from $N(0, \sigma^2 h)$ distribution. We use this procedure in Code 9.7.5 to obtain a realization of a geometric Brownian motion process. It is illustrated in the following example.

Example 9.5.1 Suppose $\{X(t) = e^{Z(t)}, \ t \ge 0\}$ is a geometric Brownian motion process, where $\{Z(t), t \ge 0\}$ is a Brownian motion process with drift coefficient 0 and diffusion coefficient σ^2. We obtain the realizations for three different values of σ as 0.5, 1, 1.2. Figure 9.5 displays the realizations of a geometric Brownian motion process for a fixed period [0, 5] for the three values of σ. Graphs of mean functions $E(W(t))$, $E(X(t))$ and $E(Z(t))$ are imposed on the sample paths, and these are corresponding to $\sigma = 1, 0.5$ and 1.2, respectively. As in Fig. 9.1, we note that the shapes of three curves are similar as we have used the same random sample from the standard normal distribution, to generate these three curves. We observe that as σ increases, the variability in the process increases, as expected. We may take different random samples from the standard normal distribution, to generate the three curves. Table 9.1 presents the output. The first three columns display the first six realized values for three values of σ, and the next three columns display the last six realized

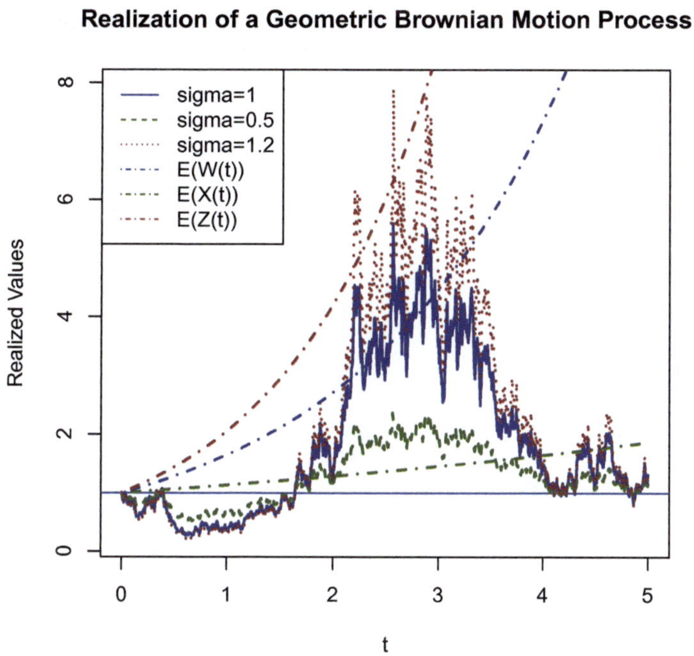

Fig. 9.5 Realization of a geometric Brownian Motion Process

Table 9.1 Realization of a geometric Brownian Motion Process: approach I

$\sigma = 0.5$	$\sigma = 1$	$\sigma = 1.2$	$\sigma = 0.5$	$\sigma = 1$	$\sigma = 1.2$
1.00	1.00	1.00	1.16	1.34	1.42
0.97	0.94	0.93	1.09	1.20	1.24
0.94	0.88	0.85	1.09	1.19	1.23
0.94	0.89	0.87	1.08	1.17	1.20
0.92	0.84	0.81	1.06	1.13	1.16
0.92	0.86	0.83	1.15	1.32	1.40

values for three values of σ. We compare these values with those obtained in the next example using a different approach to find a realization from a geometric Brownian motion process. □

Another approach to find a realization of a geometric Brownian motion process is to obtain a realization of a Brownian motion process and exponentiate it to get a realization of a geometric Brownian motion process. This procedure is adopted in Code 9.7.6. It is illustrated in the following example.

Example 9.5.2 Suppose $\{X(t) = e^{Z(t)},\ t \geq 0\}$ is a geometric Brownian motion process, as in Example 9.5.1. We take the same values of σ for comparison. Figure 9.6

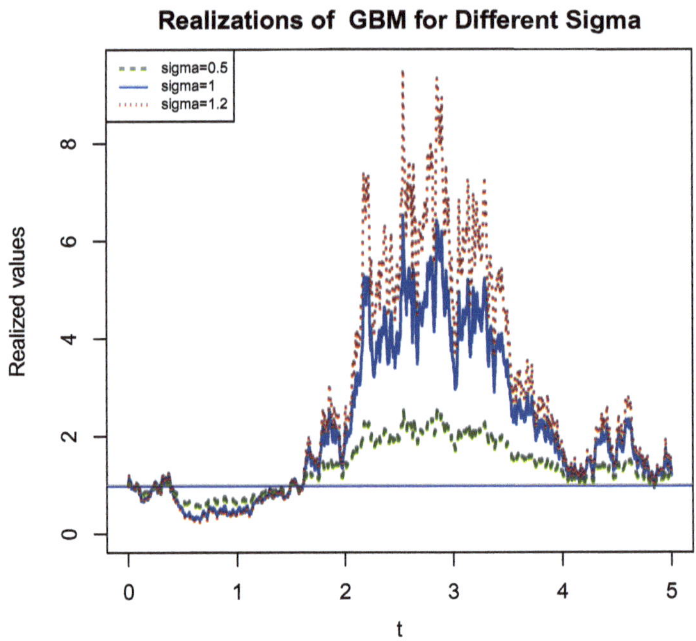

Fig. 9.6 Realization of a geometric Brownian Motion Process: second approach

Table 9.2 Realization of a geometric Brownian Motion Process: approach II

$\sigma = 0.5$	$\sigma = 1$	$\sigma = 1.2$	$\sigma = 0.5$	$\sigma = 1$	$\sigma = 1.2$
1.00	1.00	1.00	1.16	1.34	1.42
0.97	0.94	0.93	1.09	1.20	1.24
0.94	0.88	0.85	1.09	1.19	1.23
0.94	0.89	0.87	1.08	1.17	1.20
0.92	0.84	0.81	1.06	1.13	1.16
0.92	0.86	0.83	1.15	1.32	1.40

displays the realizations for a fixed period [0, 5] for three values 0.5, 1, 1.2 of σ. We note that the realization with the two approaches is exactly the same, as expected. Table 9.2 presents the output. The first three columns display the first six realized values for three values of σ, and the next three columns display the last six realized values for three values of σ. Note that the values in Tables 9.1 and 9.2 are the same up to two decimal places of accuracy, since the same random samples are generated as the seed (`set.seed(i)`) is the same in both the codes. If in Code 9.7.6 we change the seed as `set.seed(i+5)`, then we get different values. These are reported in Table 9.3. □

Table 9.3 Realization of a geometric Brownian Motion Process: different Seed

$\sigma = 0.5$	$\sigma = 1$	$\sigma = 1.2$	$\sigma = 0.5$	$\sigma = 1$	$\sigma = 1.2$
1.00	1.00	1.00	1.24	1.54	1.68
1.08	1.18	1.21	1.24	1.53	1.66
1.08	1.17	1.21	1.23	1.52	1.65
1.05	1.11	1.13	1.21	1.46	1.57
1.05	1.11	1.13	1.16	1.36	1.44
1.03	1.06	1.08	1.11	1.22	1.27

Modern mathematical economists usually prefer a geometric Brownian motion process over a Brownian motion process as a model for prices of assets, say shares of stock, that are traded in a perfect market. Such prices are non-negative and exhibit random fluctuations about a long-term exponential decay or growth curve. Both of these properties are possessed by a geometric Brownian motion, but not by a Brownian motion process itself. In particular, a geometric Brownian motion is useful in the modeling of stock prices over time when it can be assumed that the percentage changes are independent and identically distributed. For example, suppose $X(t)$ is the price of some stock at time t. Then it is reasonable to assume that $Y_t = X(t)/X(t-1), t \geq 1$ are independent and identically distributed random variables. Observe that

$$\begin{aligned} Y_t = X(t)/X(t-1) \;\; &\Rightarrow \;\; X(t) = Y_t X(t-1) = Y_t Y_{t-1} X(t-2) \\ &\Rightarrow \;\; X(t) = Y_t Y_{t-1} \cdots Y_1 X(0) \\ &\Rightarrow \;\; \log(X(t)) = \sum_{i=1}^{t} \log Y_i + \log X(0)\,. \end{aligned}$$

If $\{Y_i, i \geq 1\}$ is a sequence of independent and identically distributed random variables, by the central limit theorem, $Z(t) = \log X(t)$ can be approximated by a normal distribution. If we assume that $\{\log X(t), t \geq 0\}$ is a process with stationary and independent increments, then $\{X(t) = e^{Z(t)},\ t \geq 0\}$ is a geometric Brownian motion.

We have derived the identity $E(X(t)|X(0) = x) = xe^{\delta t}$. It is interpreted as follows. The expected price of a stock grows like a fixed-income security with continuously compounded interest rate δ. In practice, δ is usually very high than r, the real fixed-income interest rate. Hence, one invests in stocks. But unlike a fixed-income investment, there is risk involved in investment of stocks. The stock price has variability due to the randomness. Note that
$Var(X(t)|X(0) = x) = x^2 e^{2\delta t}(e^{t\sigma^2} - 1)$ and it increases as δ increases. As a consequence, the value of the stock could drop, causing one to lose the money.

Example 9.5.3 Suppose the price (in rupees) of a stock at time t (in days) is modeled as $X(t) = e^{\sigma W(t)}$, where $\sigma = 2$ and $\{W(t), t \geq 0\}$ is the standard Brownian motion process. The diffusion parameter σ^2 in this field is known as a volatility parameter of the stock. Suppose an investor owns 500 shares of the stock at time 0. He plans to sell the shares as soon as its price reaches Rs 100. We find the probability that he has to wait more than 30 days to sell the stock. Suppose

$$T = \min\{t \geq 0 | X(t) = 100\} = \min\{t \geq 0 | W(t) = 2.3026\} \;\Rightarrow\; T = T_{2.3026} \;.$$

Thus,

$$P[T > 30] = P[T_{2.3026} > 30] = 2\Phi(2.3026/\sqrt{30}) - 1 = 0.3258 \;.$$

Hence, the probability that the investor has to wait for more than 30 days is 0.3258. Suppose a dealer offers an option to buy a financial instrument that pays Rs 25 if the stock value during the next 30 days goes above Rs 100 at any time and zero rupees otherwise. One can buy this option by paying Rs 10. It is of interest to decide whether to buy such an option. It is worth buying the option if the expected payoff from the option is more than Rs 10. Suppose $U(30) = \max\{X(t), 0 \leq t \leq 30\}$. The payoff from the option is Rs 25 if $U(30) > 100$ and 0 otherwise. Hence the expected payoff is

$$25P[U(30) > 100] = 25P[T_{2.3026} \leq 30] = 25(1 - 0.3258) = 16.86 \;.$$

It is larger than Rs 10 and hence it is worth buying the option. □

Example 9.5.4 Suppose the price (in rupees) of a stock at time t (in days) is modeled as $X(t) = X(0)e^{\sigma W(t)}$, where $\sigma = 2$ and $\{W(t), t \geq 0\}$ is the standard Brownian motion process. The initial value of the stock is Rs 300. The investor plans to sell the stock when it reaches Rs 400 or falls to Rs 250. Note that

$$\begin{aligned} X(t) = 400 \;&\Rightarrow\; W(t) = 0.5\log(4/3) = 0.1438 = b, \;\text{ say} \\ \& \; X(t) = 250 \;&\Rightarrow\; W(t) = 0.5\log(5/6) = -0.0912 = a, \;\text{ say.} \end{aligned}$$

To compute the probability that the investor ends up selling at a loss, we compute the probability that $W(t)$ hits a before b, that is, we compute $P[W(T(a, b)) = a]$. Since $\mu = 0$, from Theorem 9.3.5,

$$P[W(T(a, b)) = a] = b/(|a| + b) = 0.6121.$$

The expected time of hitting either a or b is

$$E(T(a, b)) = |a|b/\sigma^2 = |a|b = 0.0131,$$

where time units as σ^2 for $W(t)$ process is 1. In finance terminology, 0.0131 is the expected time when the investor liquidates his holdings in the stock. The expected net profit in this strategy is

$$P[W(T(a,b)) = a](-50) + P[W(T(a,b)) = b](100) = 8.1873. \qquad \square$$

We now discuss some more variations of a Brownian motion process in the next section.

9.6 Variations of a Brownian Motion Process

Definition 9.6.1 Integrated Brownian Motion Process: Suppose $\{X(t), t \geq 0\}$ is a Brownian motion process and $Z(t)$ is defined as $Z(t) = \int_0^t X(s)\, ds$, then the process $\{Z(t), t \geq 0\}$ is known as an integrated Brownian motion process.

As an illustration of this process, suppose $Z(t)$ denotes the price of a commodity at time t and the rate of change of $Z(t)$ is modeled by a Brownian motion process. One may come across such a situation if the rate of change of the commodity's price is the current inflation rate and the inflation rate varies according to a Brownian motion process. Thus,

$$\frac{d}{dt}Z(t) = X(t) \quad \Longleftrightarrow \quad Z(t) = Z(0) + \int_0^t X(s)\, ds\, .$$

Remark 9.6.1 If $\{X(t), t \geq 0\}$ is a Brownian motion process with drift coefficient μ and diffusion parameter σ^2, then it can be proved that an integrated Brownian motion process $\{Z(t), t \geq 0\}$ is a Gaussian process with mean function $\mu t^2/2$ and covariance function $c(s,t) = \sigma^2 s^2 (t/2 - s/6)$ for $s \leq t$ (Ross [7]).

Definition 9.6.2 Reflected Brownian Motion Process: Suppose $\{W(t), t \geq 0\}$ is the standard Brownian motion process and $Z(t)$ is defined as $Z(t) = |W(t)|$. Then $\{Z(t), t \geq 0\}$ is known as a Brownian motion process reflected at the origin.

It is called a reflected Brownian motion process because whenever its sample path hits zero, it gets reflected back into the positive half of the real line. It is used to model the movement of a pollen grain in the vicinity of a container boundary that the grain cannot cross. The state space of the process $\{Z(t), t \geq 0\}$ is $[0, \infty)$. In view of the spatial symmetry of the Brownian motion process, the reflected Brownian motion process is also a Markov process. We obtain below the distribution function $F(x)$ and the probability density function $f(x)$ of $Z(t)$ for fixed t. For $x \geq 0$,

$$\begin{aligned} F(x) &= P[Z(t) \leq x] = P[|W(t)| \leq x] = P[-x \leq W(t) \leq x] \\ &= P[-x/\sqrt{t} \leq W(t)/\sqrt{t} \leq x/\sqrt{t}] \\ &= \Phi(x/\sqrt{t}) - \Phi(-x/\sqrt{t}) = 2\Phi(x/\sqrt{t}) - 1 \\ \Rightarrow \quad f(x) &= \sqrt{2/\pi t}\, e^{-x^2/2t}, \quad x \geq 0\, . \end{aligned}$$

Since the moments of $Z(t)$ are the same as those of $|W(t)|$, we have under the condition that $Z(0) = 0$,

$$E(Z(t)) = \sqrt{2t/\pi} \quad \& \quad Var(Z(t)) = (1 - 2/\pi)t \ ,$$

Karlin and Taylor [5].

Definition 9.6.3 Absorbed Brownian Motion Process: Suppose $\{W(t), t \geq 0\}$ is the standard Brownian motion with $W(0) = x > 0$. Suppose τ denotes the first time the process reaches 0. Suppose $Z(t)$ is defined as

$$Z(t) \;\; = \;\; \begin{cases} W(t), & \text{if } t \leq \tau \\ 0, & \text{if } t > \tau. \end{cases}$$

Then $\{Z(t), t \geq 0\}$ is known as a Brownian motion process absorbed at the origin.

It is used to model the price of a share of stock in a company that may become bankrupt at some future instant. It can be shown that absorbed Brownian motion process is also a Markov process, Karlin and Taylor [5]. One can evaluate the transition probabilities for absorbed Brownian motion process using the reflection principle.

Definition 9.6.4 Ornstein-Uhlenbeck Process: Suppose $\{W(t), t \geq 0\}$ is the standard Brownian motion process and $Z(t) = e^{-\alpha t/2}W(e^{\alpha t}),\ \alpha > 0$. Then the process $\{Z(t), t \geq 0\}$ is known as an Ornstein-Uhlenbeck process.

This process is labeled as the Ornstein-Uhlenbeck process to honor the two statisticians who first formulated and studied this process. In March 2019, Karen Uhlenbeck has been awarded the Abel prize for mathematics, which is equivalent to a Nobel prize. She is the first woman to win it.

The Ornstein-Uhlenbeck process is a time translation of a Brownian motion process, in the sense that the time is measured in an exponential scale. The Brownian motion process is a Markov process and hence Ornstein-Uhlenbeck process is also a Markov process, but does not possess independent increments. It has been proposed as a model for describing the velocity of a particle immersed in a liquid or gas, and as such is useful in statistical mechanics. This process has been applied extensively in finance, management and economics to model buffer stock control, for pricing in a large system of cash bonds, as well as short-term interest rate behavior. It is also used as a model for continuous industrial processes in chemical plants and for process control in thermal plants.

We now find its mean function and covariance function:

$$\begin{aligned} E(Z(t)) &= 0 \text{ as } E(X(t)) = 0. \\ \text{For } s < t, \ \ Cov(Z(s), Z(t)) &= e^{-\alpha t/2}e^{-\alpha s/2}Cov(X(e^{\alpha s}), X(e^{\alpha t})) \\ &= e^{-\alpha t/2}e^{-\alpha s/2}\min\{e^{\alpha s}, e^{\alpha t}\} = e^{-\alpha t/2}e^{-\alpha s/2}e^{\alpha s} \\ &= e^{-\alpha t/2}e^{\alpha s/2} \ . \end{aligned}$$

It then follows that $Cov(Z(t), Z(t+s)) = e^{\alpha t/2}e^{-\alpha(s+t)/2} = e^{-\alpha s/2}$. Thus, $\{Z(t), t \geq 0\}$ is a weakly stationary process. Since Brownian motion is a Gaussian process, $\{Z(t), t \geq 0\}$ is also a Gaussian process. As a consequence, weak stationarity of $\{Z(t), t \geq 0\}$ implies its strong stationarity.

The next section presents R codes.

9.7 R Codes

Following is a code for the realization of a Brownian motion process. We have illustrated it in Example 9.3.1.

Code 9.7.1 Realization of a Brownian motion process: Suppose $\{X(t), t \geq 0\}$ is a Brownian motion process, with drift coefficient 0 and diffusion coefficient σ^2 where $\sigma = 0.5, 1, 1.7$. We obtain its realization when it is observed for the interval [0, 10]:

```
# Part I: Input the parameters
sig1=.5; sig2=1.7
# Part II: Realization
h=.02; t=seq(0,10,h); n=length(t); n
u=y=z=x=c(); y[1]=z[1]=x[1]=0
for(i in 2:n)
{
set.seed(i)
u[i]=rnorm(1,0,1)
x[i]=x[i-1]+ u[i]*sqrt(h)
y[i]=y[i-1]+ u[i]*sqrt(h)*sig1
z[i]=z[i-1]+ u[i]*sqrt(h)*sig2
}
M=max(x);M;m=min(x);m; M1=max(y);M1;m1=min(y);m1
M2=max(z);M2;m2=min(z);m2
w=round(c(m2,-4,m,-2,m1,0,M1,2,M,4,M2),2);w
# Part III: Graph of realization
plot(t,x,"l",ylab="Realized values",xlab="t",yaxt="n",col="blue",
main="Realization of a Brownian motion Process with Different Sigma",
ylim=c(m2,6.5),lwd=2)
axis(2,at=w,las=2,cex.axis=0.8)
lines(t,y,"l",col="dark green",lty=2,lwd=2)
lines(t,z,"l",col="dark red",lty=4,lwd=2)
abline(h=0,col="dark blue"); abline(h=M,col="blue")
abline(h=m,col="blue"); abline(h=M1,lty=2,col="dark green")
abline(h=m1,lty=2,col="dark green")
abline(h=M2,lty=4,col="dark red")
abline(h=m2,lty=4,col="dark red")
legend("topleft",legend=c("sigma=0.5","sigma=1","sigma=1.7"),
cex=.7,col=c("dark green","blue","dark red"),lty=c(2,1,4))
```

□

In Code 9.7.1, the random sample from the standard normal is the same for three realizations; only the values of diffusion coefficients are different. In Code 9.7.2, we draw different random samples from the standard normal distribution corresponding to different values of diffusion coefficients. We have illustrated it in Example 9.3.2.

Code 9.7.2 Realization of a Brownian motion process: We consider the same Brownian motion process as in Example 9.3.1. In this code, we draw different random samples from the standard normal distribution corresponding to different values of diffusion coefficients as $\sigma = 0.5, 1, 1.7$, when the process is observed for the interval $[0, 10]$:

```
# Part I: Input the parameters
sig1=.5; sig2=1.7
# Part II: Realization
del=.02; t=seq(0,10,del); n=length(t); n
y=z=x=c(); y[1]=z[1]=x[1]=0;
for(i in 2:n)
{
set.seed(i)
x[i]=x[i-1]+rnorm(1,0,sqrt(del))
y[i]=y[i-1]+rnorm(1,0,sig1*sqrt(del))
z[i]=z[i-1]+rnorm(1,0,sig2*sqrt(del))
}
M=max(x);M;m=min(x); m; M1=max(y);M1;m1=min(y);m1
M2=max(z);M2;m2=min(z);m2
w=round(c(m2,m,-2,m1,0,M1,2,M,4,M2),2);w
# Part III: Graph of the realization
plot(t,x,"l",main="Realizations of Brownian motion Processes",
ylab="Realized values",xlab="t",yaxt="n",col="blue",
ylim=c(m2,5.5),lwd=2)
axis(2,at=w,las=2,cex.axis=0.8)
lines(t,y,"l",col="dark green",lty=2,lwd=2)
lines(t,z,"l",col="dark red",lty=4,lwd=2)
abline(h=0,col="dark blue")
abline(h=M,col="blue"); abline(h=m,col="blue")
abline(h=M1,lty=2,col="dark green")
abline(h=m1,lty=2,col="dark green")
abline(h=M2,lty=4,col="dark red")
abline(h=m2,lty=4,col="dark red")
legend("topleft",legend=c("sigma=0.5","sigma=1","sigma=1.7"),
cex=.7,col=c("dark green","blue","dark red"),lty=c(2,1,4))
```

□

The following code is for a realization of a Brownian bridge, using the two approaches described in Sect. 9.4. It is illustrated in Example 9.4.3.

Code 9.7.3 Realization of a Brownian bridge: Suppose $\{X(t),\ 0 \le t \le 1\}$ is a Brownian bridge. In the following code, we obtain a realization of a Brownian bridge, when observed over [0, 1]:

```
# Part I: Realization
h=.001; t=seq(0,1,h); n=length(t); n; u=z=w=x=c()
z[1]=x[1]=0; v=sqrt(h*(1-h))
for(i in 2:n)
{
set.seed(i)
u[i]=rnorm(1,0,1)
x[i]=x[i-1]+sqrt(h)*u[i]
z[i]=z[i-1]+v*u[i]
}
w=x-t*x[n]; z1=c(z[-n],0);length(w); length(z1)
# Part II: Graphs of realization
par(mfrow=c(2,1))
plot(t,z1,"l",main="Realization of a Brownian Bridge",ylab="X(t)",
xlab="t",col="blue")
abline(h=0,col="dark blue")
abline(v=1,col="dark blue")
abline(v=0,col="dark blue")
plot(t,w,"l",main="Realization of a Brownian Bridge",ylab="Z(t)",
xlab="t",col="blue")
abline(h=0,col="dark blue")
abline(v=1,col="dark blue")
abline(v=0,col="dark blue")
```

□

With the following code, we verify Theorem 9.4.3. It is illustrated in Example 9.4.4.

Code 9.7.4 Approximation of an empirical process by a Brownian bridge: Suppose $F_n(s)$ is the empirical distribution function, corresponding to a random sample of size $n = 500$ from uniform $U(0, 1)$ distribution. Using the following code, we examine whether $F_n(s)$ can be approximated by $s + X(s)/\sqrt{n}$ for $0 \le s \le 1$, where $\{X(s), 0 \le s \le 1\}$ is a Brownian bridge:

```
# Part I: Realization of a Brownian bridge
del=.001; t=seq(0,1,del); z=x=c(); x[1]=0; m=length(t); m
for(i in 2:m)
{
set.seed(i)
x[i]=x[i-1]+rnorm(1,0,sqrt(del))
}
z=x-t*x[n]; z1=t+z/sqrt(n)
```

```
# Part II: Random sample from U(0,1) distribution
n=500; set.seed(25); u=runif(n,0,1)
# Part III: Graph of empirical distribution function
# and Brownian bridge
plot(ecdf(u),xlab="s",ylab=expression(paste("F"[n](s))),
main="Approximation of an Empirical Process by a Brownian Bridge",
col="light blue")
lines(t,z1,"l",lty=1,col="dark blue")
legend("topleft",legend=c("Fn(s)","s+X(s)/sqrt(n)"),
col=c("light blue","dark blue"),lty=c(1,1))
```

□

The next code is to obtain a realization of a geometric Brownian motion process. It is illustrated in Example 9.5.1.

Code 9.7.5 Realization of a geometric Brownian motion process: Suppose $\{X(t) = e^{Z(t)},\ t \geq 0\}$ is a geometric Brownian motion process, where $\{Z(t), t \geq 0\}$ is a Brownian motion process with drift coefficient 0 and diffusion coefficient σ^2. We obtain the realizations for three different values of σ as 0.5, 1, 1.2:

```
# Part I: Input the parameters
sig1=0.5; sig2=1.2
# Part II: Realization
h=.005; t=seq(0,5,h); n=length(t); n; u=x=z=w=c()
x[1]=z[1]=w[1]=1; v=sqrt(h)
for(i in 2:n)
{
set.seed(i)
u[i]=rnorm(1,0,1)
w[i]=w[i-1]*exp(v*u[i])
x[i]=x[i-1]*exp(sig1*v*u[i])
z[i]=z[i-1]*exp(sig2*v*u[i])
}
ew=exp(t/2);ex=exp(sig1^2*t/2);ez=exp(sig2^2*t/2)
# Part III: Graph of realization
plot(t,w,"l",main="Realization of a Geometric Brownian Motion
Process",ylab="Realized Values",xlab="t",
ylim=c(min(z),max(z)),col="light blue",lwd=2)
abline(h=1,col="light blue")
lines(t,x,col="dark blue",lty=2,lwd=2)
lines(t,z,col="dark red",lty=3,lwd=2)
lines(t,ew,col="light blue",lty=4,lwd=2)
lines(t,ex,col="dark blue",lty=4,lwd=2)
lines(t,ez,col="dark red",lty=4,lwd=2)
legend("topleft",legend=c("sigma=1","sigma=0.5","sigma=1.2",
"E(W(t))","E(X(t))","E(Z(t))"),col=c("light blue","dark blue",
```

```
"dark red","light blue","dark blue","dark red"),lty=c(1,2,3,4,4,4))
# Part IV: First 6 and last 6 realized values
d=data.frame(head(x),head(w),head(z),tail(x),tail(w),tail(z))
d1=round(d,2);d1
```

□

In the following code, a different procedure is adopted to obtain a realization from a geometric Brownian motion process. It is illustrated in Example 9.5.2.

Code 9.7.6 Realization of a geometric Brownian motion process: Suppose $\{X(t) = e^{Z(t)},\ t \geq 0\}$ is a geometric Brownian motion process, as in Example 9.5.1:

```
# Part I: Input the parameters
sig1=.5; sig2=1.2
# Part II: Realization
h=.005; t=seq(0,5,h); n=length(t); n
u=y=z=x=c(); y[1]=z[1]=x[1]=0;
for(i in 2:n)
{
set.seed(i)
u[i]=rnorm(1,0,1)
x[i]=x[i-1]+ u[i]*sqrt(h)
y[i]=y[i-1]+ u[i]*sqrt(h)*sig1
z[i]=z[i-1]+ u[i]*sqrt(h)*sig2
}
x1=exp(x); y1=exp(y); z1=exp(z)
# Part III: Graph of realization
plot(t,x1,"l",main="Realizations of GBM for Different Sigma",
ylab="Realized values",xlab="t",col="blue",lwd=2,ylim=c(0,max(z1)))
lines(t,y1,"l",col="dark green",lty=2,lwd=2)
lines(t,z1,"l",col="dark red",lty=3,lwd=2)
abline(h=1,col="dark blue")
legend("topleft",legend=c("sigma=0.5","sigma=1","sigma=1.2"),cex=.7,
      col=c("dark green","blue","dark red"),lty=c(2,1,3))
# Part IV: # First 6 and last 6 realized values
d=round(data.frame(head(y1),head(x1),head(z1),tail(y1),tail(x1),
tail(z1)),2); d;
# Part V: Realization with different seed
y[1]=z[1]=x[1]=0
for(i in 2:n)
{
set.seed(i+5)
u[i]= rnorm(1,0,1)
x[i]=x[i-1]+ u[i]*sqrt(h)
y[i]=y[i-1]+ u[i]*sqrt(h)*sig1
z[i]=z[i-1]+ u[i]*sqrt(h)*sig2
```

```
}
x1=exp(x); y1=exp(y); z1=exp(z)
# Part VI: First 6 and last 6 realized values
d1=round(data.frame(head(y1),head(x1),head(z1),tail(y1),tail(x1),
tail(z1)),2); d1
```

□

A quick recap of the results discussed in the present chapter is given below.

Summary

1. A continuous time and continuous state space stochastic process $\{X(t), t \geq 0\}$ with state space $\mathbb{R}$ is said to be a Brownian motion process with drift coefficient μ and diffusion coefficient σ^2, if the following conditions are satisfied. (i) X(0) = 0, (ii) $\{X(t), t \geq 0\}$ has stationary and independent increments and (iii) for every $t > 0$, $X(t) \sim N(\mu t, \sigma^2 t)$ distribution. If $\mu = 0$ and $\sigma^2 = 1$, then it is known as the standard Brownian motion process.
2. The standard Brownian motion process is a Markov process with the transition distribution function, for $x, x_0 \in \mathbb{R}$ and $s < t$, given by
$$F(x, t, x_0, s) = P[X(t) \leq x | X(s) = x_0] = \int_{-\infty}^{x} \frac{1}{\sqrt{2\pi(t-s)}} \exp\left\{-\frac{(u - x_0)^2}{2(t-s)}\right\}.$$
3. For a Brownian motion process $\{X(t), t \geq 0\}$ with drift coefficient μ and diffusion coefficient σ^2,
$$E(X(t)) = \mu t, \quad Var(X(t)) = \sigma^2 t \quad \& \quad Cov(X(s), X(t)) = \sigma^2 \min\{s, t\}\,.$$
4. For the standard Brownian motion process $\{X(t), t \geq 0\}$, the joint distribution of $(X(t_1), X(t_2), \ldots, X(t_n))'$ is n-variate normal with mean vector $\underline{0}$ and dispersion matrix $\Sigma = [\sigma_{ij}]$, where $\sigma_{ii} = t_i$ and $\sigma_{ij} = \min\{t_i, t_j\}$, where $t_1 < t_2 < \cdots < t_n$.
5. A stochastic process $\{X(t), t \geq 0\}$ is called a Gaussian or a normal process if $\{X(t_1), X(t_2), \ldots, X(t_n)\}$ has a multivariate normal distribution for any finite $n \geq 1$ and for any finite set $\{t_1, t_2, \ldots, t_n\} \in [0, \infty)$.
6. Brownian motion process is a Gaussian process, however, in general a Gaussian process is not a Brownian motion process.
7. A Gaussian process $\{X(t), t \geq 0\}$ with mean value function $E(X(t)) = 0$ and covariance function as $Cov(X(t), X(s)) = \min\{s, t\}$ is the standard Brownian motion process.
8. Suppose $\{W(t), t \geq 0\}$ is the standard Brownian motion process. Then each of the processes $\{X(t) = cW(t/c^2), t \geq 0\}$, $\{X(t) = W(t + h) - W(h), t \geq 0\}$, $h \geq 0$, $\{X(t) = tW(1/t), t > 0\}$, $X(0) = 0$ and $\{X(t) = -W(t), t \geq 0\}$ is also the standard Brownian motion process.

9. Sample paths of a Brownian motion process are continuous everywhere, but differentiable nowhere, with probability one.
10. Suppose $U(T) = \max_{0 \leq t \leq T} X(t)$. Then for any $a > 0$, the survival function and the probability density function of $U(T)$ are given by

$$P[U(T) \geq a] = 2[1 - \Phi(a/\sqrt{T})] \ \& \ f_{U(T)}(a) = \sqrt{2/\pi T} e^{-a^2/2T}, \ a \geq 0.$$

Suppose $L(T) = \min_{0 \leq t \leq T} X(t)$. Then for any $a < 0$, the distribution function and the probability density function of $L(T)$ are given by

$$P[L(T) \leq a] = 2[1 - \Phi(-a/\sqrt{T})] \quad \& \quad f_{L(T)}(a) = \sqrt{2/\pi T} e^{-a^2/2T}, \quad a < 0.$$

11. The probability density function $f_{T_a}(t)$ of T_a, the first passage time to a fixed point $a \neq 0$, is given by

$$f_{T_a}(t) = (|a|/\sqrt{2\pi}) t^{-3/2} e^{-a^2/2t}, \quad t > 0 \ \text{ with } \ E(T_a) = \infty.$$

The distribution of T_a is an inverse Gaussian distribution.
12. Suppose $\{X(t), t \geq 0\}$ is a Brownian motion process with $X(0) = 0$, drift coefficient μ and diffusion coefficient σ^2. Suppose $a < 0$ and $b > 0$ are two given numbers and $\theta = -2\mu/\sigma^2$:

$$\text{For } \mu \neq 0, \quad P[X(T(a,b)) = b] = (e^{\theta a} - 1)/(e^{\theta a} - e^{\theta b})$$

and

$$E(T(a,b)) = (b(e^{\theta a} - 1) - a(e^{\theta b} - 1))/(\mu(e^{\theta a} - e^{\theta b})).$$

$$\text{For } \mu = 0, \quad P[X(T(a,b)) = b] = |a|/(|a| + b) \ \& \ E(T(a,b)) = |a|b/\sigma^2.$$

13. Suppose $\{X(t), t \geq 0\}$ is a Brownian motion process with $X(0) = 0$, drift coefficient μ and diffusion coefficient σ^2. Suppose $U = \max\{X(t)|0 \leq t < \infty\}$, $L = \min\{X(t)|0 \leq t < \infty\}$ and $\theta = -2\mu/\sigma^2$:
    (a) For $\mu < 0$, the probability density function of U is given by $f_U(x) = \theta e^{-\theta x}$, $x \geq 0$ and $L = -\infty$ with probability 1.
    (b) For $\mu > 0$, the probability density function of L is given by $f_L(x) = -\theta e^{-\theta x}$, $x \leq 0$ and $U = \infty$ with probability 1.
    (c) If $\mu = 0$, then $U = \infty$ and $L = -\infty$ with probability 1.
14. Suppose $\{W(t), t \geq 0\}$ is the standard Brownian motion process. Then $\{X(t), 0 \leq t \leq 1\}$ is known as a Brownian bridge, if $X(0) = X(1) = 0$ and for $t \in (0, 1)$ the distribution of $X(t)$ is the same as the conditional distribution of $W(t)$ given $W(0) = W(1) = 0$, which is normal $N(0, t(1 - t))$.

15. A Brownian bridge is a Gaussian process with mean value function 0 and covariance function $c(s,t) = s(1-t)$ for $s \leq t$.
16. Suppose $\{Z(t), t \geq 0\}$ is a Brownian motion process with drift coefficient μ and diffusion coefficient σ^2. Then the process $\{X(t), t \geq 0\}$ defined by $X(t) = e^{Z(t)}$ is known as a geometric Brownian motion process.
17. A geometric Brownian motion process is a time homogeneous Markov Process.
18. Suppose $\{W(t), t \geq 0\}$ is the standard Brownian motion process and $Z(t) = e^{-\alpha t/2}W(e^{\alpha t})$, $\alpha > 0$. Then the process $\{Z(t), t \geq 0\}$ is known as an Ornstein-Uhlenbeck process.

9.8 Conceptual Exercises

9.8.1 Suppose $\{X(t), t \geq 0\}$ is a Brownian motion process, with drift coefficient $\mu = 3$ and diffusion coefficient $\sigma^2 = 4$. Find the conditional distribution of $X(3)$ given $X(0) = 6$ and $X(5) = 7$.

9.8.2 Suppose $\{W(t), t \geq 0\}$ is the standard Brownian motion process. Find the joint distribution of $W(t_1)$ and $W(t_2) - W(t_1)$ and hence the joint distribution of $W(t_1)$ and $W(t_2)$.

9.8.3 Suppose $\{W(t), t \geq 0\}$ is the standard Brownian motion process. (i) Obtain the distribution of $W(s) - (s/t)W(t)$ when $s < t$. (ii) Obtain the joint distribution of $(W(t), W(s) - (s/t)W(t))'$. (iii) Examine whether $W(t)$ and $W(s) - (s/t)W(t)$ are independent.

9.8.4 Suppose $\{X(t), t \geq 0\}$ is a Brownian motion process, with drift coefficient μ and diffusion coefficient σ^2. Examine whether $\{X(t) - \mu t, t \geq 0\}$ is a martingale.

9.8.5 Suppose $\{W(t), t \geq 0\}$ is the standard Brownian motion process. Suppose $Y(t) = W^2(t) - t$ and $Z(t) = \exp\{cW(t) - c^2t/2\}$. Examine whether $\{Y(t), t \geq 0\}$ and $\{Z(t), t \geq 0\}$ are martingales.

9.8.6 Suppose $\{X(t), t \geq 0\}$ is a Brownian motion process, with drift coefficient $\mu = -1$, diffusion coefficient $\sigma^2 = 2$ and $X(0) = 6$. Find the probability that $X(t)$ is above its initial level at time 7.

9.8.7 Suppose $\{X(t), t \geq 0\}$ is a Brownian motion process, with drift coefficient $\mu = 2$, diffusion coefficient $\sigma^2 = 3$ and $X(0) = 2$. Find the joint distribution of $X(4) - X(1)$ and $X(9) - X(5)$.

9.8.8 Suppose $\{X(t), t \geq 0\}$ is a Brownian motion process, with drift coefficient $\mu = 2$, diffusion coefficient $\sigma^2 = 3$ and $X(0) = 2$. Find the joint distribution of $X(4) - X(1)$ and $X(3)$.

9.8.9 Suppose $\{X(t), t \geq 0\}$ is a Brownian motion process, with drift coefficient μ and diffusion coefficient σ^2. Find $r \neq 0$ so that $E(e^{rX(t)}) = 1$.

9.8.10 Suppose $\{X(t), t \geq 0\}$ is a Brownian motion process, with drift coefficient μ and diffusion coefficient σ^2. Suppose $\theta = -2\mu/\sigma^2$. Examine whether $\{e^{\theta X(t)}, t \geq 0\}$ is a martingale.

9.8.11 Suppose the amount in a bank account at time t is modeled as $X(t) = 5000 + 500W(t)$, where $\{W(t), t \geq 0\}$ is the standard Brownian motion process. Find the probability that the account is not overdrawn by time 10 units.

9.8.12 Suppose the inventory at a store at time t is modeled as $X(t) = 6 - 3t + 4W(t)$ where $\{W(t), t \geq 0\}$ is the standard Brownian motion process. Suppose the inventory storage area has a capacity of 15 units. (i) Find the probability that a stock-out occurs before the storage area overflows. (ii) Find the expected time for either stock-out or overflow of storage area.

9.8.13 Suppose $Z(t) = (t+1)X(t/(t+1))$, where $\{X(t), 0 \leq t \leq 1\}$ is a Brownian bridge. Show that $\{Z(t), t \geq 0\}$ is the standard Brownian motion process.

9.8.14 Suppose the price of a stock at time t is modeled by $X(t) = e^{\sigma W(t)}$, a geometric Brownian motion process, with volatility parameter $\sigma = 1/2$ and $\{W(t), t \geq 0\}$ is the standard Brownian motion process with $W(0) = 0$. Find the expected price of a stock at time 2 and also obtain its variance. Find the probability that the stock price is above 4 units at time 2.

9.8.15 What is the probability that a geometric Brownian motion process with parameters $\mu = -\sigma^2/2$ and σ ever rises to more than twice its original value? What is the probability if $\mu = 0$. (In financial terms, if you buy a stock or index fund whose fluctuations are described by the geometric Brownian motion, what are the chances to double your money?)

9.8.16 Suppose $\{X(t), t \geq 0\}$ is a geometric Brownian motion process with $\mu = 0.01$. If $X(0) = 100$, find $E(X(10))$, $P[X(10) > 100]$ and $P[X(10) < 110]$ for three values of σ given by $\sigma = 0.2, 0.4, 0.6$.

9.8.17 Suppose stock price $\{X(t), t \geq 0\}$ is a geometric Brownian motion process with drift coefficient $\mu = 2$ and diffusion coefficient $\sigma^2 = 7.5\%$ per annum. Assume that the current price of the stock is $X(0) = 100$. Find $E(X(3))$ and $P[X(3) > 40000]$.

9.9 Computational Exercises

9.9.1 Suppose $\{X(t), t \geq 0\}$ is a Brownian motion process, with drift coefficient $\mu = 2$ and diffusion coefficient σ^2. Generate a Brownian motion process for a fixed time period and for three values of a diffusion coefficient.
(i) Take the same random sample from standard normal distribution for three values of σ and (ii) take the different random samples from standard normal distribution for three values of σ. Plot the graphs of realization. Comment on the findings.

9.9.2 Obtain a realization of a Brownian bridge, using both the approaches adopted in Sect. 9.4.

9.9.3 Suppose $\{X(t) = e^{Z(t)}, t \geq 0\}$ is a geometric Brownian motion process, where $\{Z(t), t \geq 0\}$ is a Brownian motion process with drift coefficient 0 and diffusion coefficient σ^2. Obtain a realization of $\{X(t), t \geq 0\}$ for three different values of σ. Plot the graphs of realization with graphs of mean function imposed on it.

9.9.4 Suppose a stock price $\{X(t), t \geq 0\}$ is modeled by a geometric Brownian motion process with drift coefficient $\mu = 0.2$ and diffusion coefficient σ^2. If the initial price of the stock is $X(0) = 100$ units, find the price of the stock for the next 10 time points. Take different values of σ^2 and comment on your findings.

9.10 Multiple Choice Questions

Note: In each of the questions, multiple options may be correct.

9.10.1 A continuous time and continuous state space stochastic process $\{X(t), t \geq 0\}$ with state space $\mathbb{R}$ is a Brownian motion process with drift coefficient μ and diffusion coefficient σ^2. Which of the following options is/are correct?

(a) $X(0) = 0$
(b) $\{X(t), t \geq 0\}$ has independent increments
(c) $\{X(t), t \geq 0\}$ has stationary increments
(d) For every $t > 0$, it follows $X(t) \sim N(\mu t, \sigma^2 t)$ distribution.

9.10.2 Suppose $\{X(t), t \geq 0\}$ is a Brownian motion process with drift coefficient μ and diffusion coefficient σ^2. Which of the following options is/are correct?

(a) It is a process with stationary and independent increments
(b) It is a time homogeneous Markov process
(c) $E(X(t)) = \mu t, \quad Var(X(t)) = \sigma^2 t$
(d) $Cov(X(s), X(t)) = \sigma^2 |s - t|$.

9.10.3 Suppose $\{X(t), t \geq 0\}$ is a Brownian motion process with drift coefficient μ and diffusion coefficient σ^2. Then which of the following options is NOT true?

(a) It is a process with stationary and independent increments
(b) It is a time homogeneous Markov process
(c) $E(X(t)) = \mu t, \quad Var(X(t)) = \sigma^2 t$
(d) $Cov(X(s), X(t)) = \sigma^2 |s - t|$.

9.10.4 Suppose $\{X(t), t \geq 0\}$ is a Brownian motion process with drift coefficient μ and diffusion coefficient σ^2. Which of the following options is/are correct?

(a) It is a process with stationary and independent increments
(b) It is a time homogeneous Markov process
(c) $E(X(t)) = \mu t, \quad Var(X(t)) = \sigma^2 t$
(d) $Cov(X(s), X(t)) = \sigma^2 \min\{s, t\}$.

9.10.5 Which of the following options is/are correct? Suppose $\{W(t), t \geq 0\}$ is the standard Brownian motion process. Then the joint distribution of $(W(t_1), W(t_2) - W(t_1))$, where $t_1 < t_2$, is bivariate normal with

(a) mean vector $(0, 0)'$ and dispersion matrix $D = \text{diag}(t_1, t_2)$
(b) mean vector $(0, 0)'$ and dispersion matrix $D = \text{diag}(1, 1)$
(c) mean vector $(0, 0)'$ and dispersion matrix $D = \text{diag}(t_1, t_2 - t_1)$
(d) mean vector $(t_1, t_2 - t_1)'$ and dispersion matrix $D = \text{diag}(t_1, t_2 - t_1)$.

9.10.6 Which of the following options is/are correct? Suppose $\{W(t), t \geq 0\}$ is the standard Brownian motion process. Then the joint distribution of $(W(t_1), W(t_2))$, where $t_1 < t_2$, is bivariate normal with

(a) mean vector $(0, 0)'$ and dispersion matrix $D = \sigma_{ij}$, where $\sigma_{11} = \sigma_{12} = \sigma_{21} = t_1$ and $\sigma_{22} = t_2$
(b) mean vector $(0, 0)'$ and dispersion matrix $D = \text{diag}(1, 1)$
(c) mean vector $(0, 0)'$ and dispersion matrix $D = \text{diag}(t_1, t_2)$
(d) mean vector $(t_1, t_2)'$ and dispersion matrix $D = \text{diag}(t_1, t_2)$.

9.10.7 Following are two statements. (I) A Brownian motion process is a Gaussian process. (II) A Gaussian process is not always a Brownian motion process. Which of the following options is correct?

(a) Both (I) and (II) are false
(b) Both (I) and (II) are true
(c) (I) is true but (II) is false
(d) (I) is false but (II) is true.

9.10.8 Following are two statements. (I) A Brownian motion process is a Gaussian process. (II) A Gaussian process is always a Brownian motion process. Which of the following options is correct?

(a) Both (I) and (II) are false
(b) Both (I) and (II) are true
(c) (I) is true but (II) is false
(d) (I) is false but (II) is true.

9.10.9 Following are two statements. (I) A Brownian motion process is a Gaussian process. (II) Brownian motion process is a unique Gaussian process having continuous trajectories, zero mean and covariance function as $\min\{s, t\}$. Which of the following options is correct?

(a) Both (I) and (II) are false
(b) Both (I) and (II) are true
(c) (I) is true but (II) is false
(d) (I) is false but (II) is true.

9.10.10 Following are two statements. (I) A Brownian motion process is a Gaussian process. (II) A Gaussian process having continuous trajectories, zero mean and covariance function as $|s - t|$ is a Brownian motion process. Which of the following options is correct?

(a) Both (I) and (II) are false
(b) Both (I) and (II) are true
(c) (I) is true but (II) is false
(d) (I) is false but (II) is true.

9.10.11 Following are two statements. (I) A Brownian motion process is a Gaussian process. (II) A Gaussian process having continuous trajectories, zero mean and covariance function as $s(1 - t)$ for $s \leq t$ is a Brownian motion process. Which of the following options is correct?

(a) Both (I) and (II) are false
(b) Both (I) and (II) are true
(c) (I) is true but (II) is false
(d) (I) is false but (II) is true.

9.10.12 Following are three statements. (I) A Brownian motion process is a process with stationary and independent increments. (II) A time homogeneous Poisson process is a process with stationary and independent increments. (III) A time homogeneous continuous time Markov chain is always a process with stationary and independent increments.
Which of the following is a correct option?

(a) Only (I) and (II) are true
(b) Only (I) and (III) are true
(c) Only (II) and (III) are true
(d) All are true.

9.10.13 Following are four statements. (I) A Poisson process satisfies the Markov property. (II) A renewal process satisfies the Markov property. (III) A Brownian motion process satisfies the Markov property. (IV) Yule-Furry process satisfies the Markov property.
Which of the following is a correct option?

(a) Only (I) and (II) are true
(b) Only (I) and (III) are true
(c) Only (I), (III) and (IV) are true
(d) Only (I), (II) and (III) are true.

9.10.14 Following are three statements. (I) A Brownian bridge is a process with stationary and independent increments. (II) A time homogeneous Poisson process is a process with stationary and independent increments. (III) A Brownian motion process is a process with stationary and independent increments.
Which of the following is a correct option?

(a) Only (I) and (II) are true
(b) Only (I) and (III) are true
(c) Only (II) and (III) are true
(d) All are true.

9.10.15 Following are three statements. (I) A Poisson process satisfies the Markov property. (II) A Brownian motion process satisfies the Markov property. (III) A linear death process satisfies the Markov property.
Which of the following is a correct option?

(a) Only (I) and (II) are true
(b) Only (I) and (III) are true
(c) Only (II) and (III) true
(d) All three are true.

9.10.16 Following are three statements. (I) A time homogeneous Poisson process is a process with stationary and independent increments. (II) A linear death process is a process with stationary and independent increments. (III) A Brownian motion process is a process with stationary and independent increments.
Which of the following is a correct option?

(a) Only (I) and (II) are true
(b) Only (I) and (III) are true
(c) Only (II) and (III) are true
(d) All are true.

9.10.17 Following are three statements. (I) A birth-death process satisfies the Markov property. (II) A Brownian motion process satisfies the Markov property. (III) A renewal process satisfies the Markov property.
Which of the following is a correct option?

(a) Only (I) and (II) are true
(b) Only (I) and (III) are true
(c) Only (II) and (III) true
(d) All three are true

9.10.18 Suppose $\{W(t), t \geq 0\}$ is the standard Brownian motion process. Which of the following processes is/are also standard Brownian motion processes?

(a) $\{X(t) = cW(t/c^2), t \geq 0\}$
(b) $\{X(t) = W(t+h) - W(h), t \geq 0\}, \ h \geq 0$

(c) $\{X(t) = tW(1/t), t > 0\}$, $X(0) = 0$
(d) $\{X(t) = -W(t), t \geq 0\}$.

9.10.19 Suppose $\{W(t), t \geq 0\}$ is the standard Brownian motion process. Which of the following processes is/are also standard Brownian motion processes?

(a) $\{X(t) = cW(t/c), t \geq 0\}$
(b) $\{X(t) = W(t+h) - W(h), t \geq 0\}$, $h \geq 0$
(c) $\{X(t) = tW(1/t), t > 0\}$, $X(0) = 0$
(d) $\{X(t) = -W(t), t \geq 0\}$.

9.10.20 Suppose $\{W(t), t \geq 0\}$ is the standard Brownian motion process. Which of the following processes is/are also standard Brownian motion processes?

(a) $\{X(t) = cW(t/c), t \geq 0\}$
(b) $\{X(t) = W(t+h) - W(h), t \geq 0\}$, $h \geq 0$
(c) $\{X(t) = tW(t), t > 0\}$, $X(0) = 0$
(d) $\{X(t) = -W(t), t \geq 0\}$.

9.10.21 Suppose $\{X(t), t \geq 0\}$ is a Brownian motion process. Following are two statements: (I) Its sample paths are continuous everywhere. (II) Its sample paths are differentiable nowhere, with probability one.
Which of the following options is correct?

(a) Both (I) and (II) are false
(b) Both (I) and (II) are true
(c) (I) is true but (II) is false
(d) (I) is false but (II) is true.

9.10.22 Suppose $\{X(t), t \geq 0\}$ is a Brownian motion process. Following are two statements: (I) Its sample paths are continuous everywhere. (II) Its sample paths are differentiable everywhere.
Which of the following options is correct?

(a) Both (I) and (II) are false
(b) Both (I) and (II) are true
(c) (I) is true but (II) is false
(d) (I) is false but (II) is true.

9.10.23 Which of the following options is/are correct? Suppose $\{W(t), t \geq 0\}$ is the standard Brownian motion process with $W(0) = 0$. Suppose $U(T) = \max_{0 \leq t \leq T} W(t)$. Then for any $a > 0$, the survival function $P[U(T) \geq a]$ of $U(T)$ is

(a) $[1 - \Phi(a/\sqrt{T})]$
(b) $2[1 - \Phi(a/T)]$
(c) $2[1 - \Phi(a/\sqrt{T})]$
(d) $2[1 - \phi(a/\sqrt{T})]$.

9.10.24 Which of the following options is/are correct? Suppose $\{W(t), t \geq 0\}$ is the standard Brownian motion process with $W(0) = 0$. Suppose $L(T) = \min_{0 \leq t \leq T} W(t)$. Then for any $a < 0$, the probability density function $f_{L(T)}(a)$ of $L(T)$ is

(a) $\sqrt{\frac{2}{\pi T}} e^{-a^2/T}$
(b) $\sqrt{\frac{2}{\pi T}} e^{-a^2/2T}$
(c) $\frac{2}{\pi T} e^{-a^2/2T}$
(d) $\sqrt{\frac{2T}{\pi}} e^{-a^2/2T}$.

9.10.25 Suppose $\{W(t), t \geq 0\}$ is the standard Brownian motion process with $W(0) = 0$. Suppose $U(T) = \max_{0 \leq t \leq T} W(t)$ and $L(T) = \min_{0 \leq t \leq T} W(t)$. Suppose T_a denotes the first passage time to a fixed point $a \in \mathbb{R}$. Following are two statements. (I) For $a > 0$, $T_a \leq t \iff U(t) \geq a$. (II) For $a < 0$, $T_a \leq t \iff L(t) \leq a$. Which of the following options is correct?

(a) Both (I) and (II) are false
(b) Both (I) and (II) are true
(c) (I) is true but (II) is false
(d) (I) is false but (II) is true.

9.10.26 Suppose $\{W(t), t \geq 0\}$ is the standard Brownian motion process with $W(0) = 0$. Suppose $U(T) = \max_{0 \leq t \leq T} W(t)$ and $L(T) = \min_{0 \leq t \leq T} W(t)$. Suppose T_a denotes the first passage time to a fixed point $a \in \mathbb{R}$. Following are two statements. (I) For $a > 0$, $T_a \leq t \iff U(t) \leq a$. (II) For $a < 0$, $T_a \leq t \iff L(t) \geq a$.
Which of the following options is correct?

(a) Both (I) and (II) are false
(b) Both (I) and (II) are true
(c) (I) is true but (II) is false
(d) (I) is false but (II) is true.

9.10.27 Suppose $\{W(t), t \geq 0\}$ is the standard Brownian motion process with $W(0) = 0$. Suppose $U(T) = \max_{0 \leq t \leq T} W(t)$ and $L(T) = \min_{0 \leq t \leq T} W(t)$. Suppose T_a denotes the first passage time to a fixed point $a \in \mathbb{R}$. Following are two statements. (I) For $a > 0$, $T_a \leq t \iff U(t) \leq a$. (II) For $a < 0$, $T_a \leq t \iff L(t) \leq a$. Which of the following options is correct?

(a) Both (I) and (II) are false
(b) Both (I) and (II) are true
(c) (I) is true but (II) is false
(d) (I) is false but (II) is true.

9.10.28 Suppose $\{W(t), t \geq 0\}$ is the standard Brownian motion process with $W(0) = 0$. Suppose $U(T) = \max_{0 \leq t \leq T} W(t)$ and $L(T) = \min_{0 \leq t \leq T} W(t)$. Suppose T_a denotes the first passage time to a fixed point $a \in \mathbb{R}$. Following are two statements. (I) For $a > 0$, $T_a \leq t \iff U(t) \geq a$.
(II) For $a < 0$, $T_a \leq t \iff L(t) \geq a$. Which of the following options is correct?

(a) Both (I) and (II) are false
(b) Both (I) and (II) are true
(c) (I) is true but (II) is false
(d) (I) is false but (II) is true.

9.10.29 Which of the following options is/are correct? Suppose $\{X(t), 0 \leq t \leq 1\}$ is a Brownian bridge. Then for fixed t, the distribution of $X(t)$ is normal

(a) $N(0, t)$
(b) $N(1, t(1-t))$
(c) $N(0, 1-t)$
(d) $N(0, t(1-t))$.

9.10.30 Which of the following options is/are correct? A Brownian bridge is a Gaussian process with mean value function 0 and covariance function

(a) $c(s, t) = t(1-s)$ for $s \leq t$
(b) $c(s, t) = \min\{s, t\}$
(c) $c(s, t) = s(1-t)$ for $s \leq t$
(d) $c(s, t) = |s-t|$.

9.10.31 Following are two statements. (I) A Brownian bridge is a Gaussian process. (II) A Gaussian process is always a Brownian bridge. Which of the following options is correct?

(a) Both (I) and (II) are false
(b) Both (I) and (II) are true
(c) (I) is true but (II) is false
(d) (I) is false but (II) is true.

9.10.32 Following are two statements. (I) A Brownian bridge is a Gaussian process. (II) A Gaussian process with zero mean function and covariance function as $c(s, t) = s(1-t)$ for $s \leq t$ is a Brownian bridge. Which of the following options is correct?

(a) Both (I) and (II) are false
(b) Both (I) and (II) are true
(c) (I) is true but (II) is false
(d) (I) is false but (II) is true.

9.10.33 Following are two statements. (I) A Brownian bridge is a Gaussian process. (II) A Gaussian process with zero mean function and covariance function as $c(s,t) = \min\{s,t\}$ is a Brownian bridge. Which of the following options is correct?

(a) Both (I) and (II) are false
(b) Both (I) and (II) are true
(c) (I) is true but (II) is false
(d) (I) is false but (II) is true.

9.10.34 Suppose $\{X(t), t \geq 0\}$ is a geometric Brownian motion process. Which of the following options is correct?

(a) The state space is $(0, \infty)$
(b) $E(X(t)|X(0) = x) = xe^{\mu t + t\sigma^2/2}$
(c) $Var(X(t)|X(0) = x) = x^2 e^{2\mu t + t\sigma^2}(e^{t\sigma^2} - 1)$
(d) $Var(X(t)|X(0) = x) = x^2 e^{\mu t + t^2\sigma^2}(e^{t\sigma^2} - 1)$.

9.10.35 Which of the following options is/are correct? Suppose $\{W(t), t \geq 0\}$ is the standard Brownian motion process. Then the process $\{Z(t), t \geq 0\}$ is known as an Ornstein-Uhlenbeck process if

(a) $Z(t) = e^{-\alpha t} W(e^{\alpha t}),\ \alpha > 0$
(b) $Z(t) = e^{-\alpha t/2} W(e^{-\alpha t}),\ \alpha > 0$
(c) $Z(t) = e^{\alpha t/2} W(e^{-\alpha t}),\ \alpha > 0$
(d) $Z(t) = e^{-\alpha t/2} W(e^{\alpha t}),\ \alpha > 0$.

9.10.36 Following are two statements. (I) A covariance stationary Gaussian process is a stationary process. (II) A Gaussian process with zero mean function and covariance function as $c(s,t) = \min\{s,t\}$ is a Brownian bridge. Which of the following options is correct?

(a) Both (I) and (II) are false
(b) Both (I) and (II) are true
(c) (I) is true but (II) is false
(d) (I) is false but (II) is true.

9.10.37 Following are two statements. (I) Brownian motion process is a Gaussian process. (II) Brownian motion process is a stationary process. Which of the following options is correct?

(a) Both (I) and (II) are false
(b) Both (I) and (II) are true
(c) (I) is true but (II) is false
(d) (I) is false but (II) is true.

9.10.38 Following are two statements. (I) Brownian motion process is a process with stationary and independent increments. (II) Brownian motion process is a stationary process. Which of the following options is correct?

(a) Both (I) and (II) are false
(b) Both (I) and (II) are true

Chapter 10
Renewal Process

10.1 Introduction

In Chap. 7, we studied the Poisson process, which is the frequently used Markov process for modeling the time epochs of occurrence of events and for modeling the number of occurrences in $(0, t]$. In the present chapter, we discuss a stochastic process, which is also used to model the time epochs of occurrence of events, but is not a Markov process. As in the case of a Poisson process, suppose we observe a series of events occurring randomly over time. For example, the events may be accidents taking place at a particular traffic intersection, the failures of a repairable system, the arrivals of customers at a service center and transactions at a bank. Suppose S_n denotes the epoch of occurrence of nth event, $n \geq 1$ and $S_0 = 0$. Then $T_n = S_n - S_{n-1}$ is known as an inter-occurrence or interval random variable. In a Poisson process, interval random variables are independent and identically distributed random variables, each having exponential distribution. A renewal process $\{S_n, n \geq 0\}$ assumes that the interval random variables have any distribution, not necessarily exponential. A formal definition of a renewal process is as follows.

Definition 10.1.1 Renewal Process: A point process $\{S_n, n \geq 0\}$ with $S_0 = 0$ is said to be a renewal process, if $\{T_n = S_n - S_{n-1}, n \geq 1\}$ is a sequence of independent and identically distributed non-negative random variables. The corresponding counting process $\{X(t), t \geq 0\}$ is known as a renewal counting process.

$\{X(t), t \geq 0\}$ is also referred to as a renewal process. In the renewal process, S_n is known as the epoch of nth renewal and T_n is known as the inter-renewal time, which is the random duration between nth and $(n-1)$th renewals.

Remark 10.1.1 By definition, T_n is a non-negative random variable having distribution function F, which may be discrete, continuous or mixture of discrete and continuous distributions. However, in most of the applications, F is a continuous distribution function, such as gamma, Weibull, log-normal and Pareto.

Following are some illustrations of a renewal process.

S. Madhira and S. Deshmukh, *Introduction to Stochastic Processes Using R*,
https://doi.org/10.1007/978-981-99-5601-2_10

(i) A point process $\{S_n, n \geq 1\}$ may represent arrival times of customers at a service facility, such as a supermarket, a bank or an ATM center and a reservation counter. The inter-arrival random variables are modeled by any continuous distribution with support $\mathbb{R}^+$.

(ii) Renewal processes arise naturally in risk theory in non-life insurance where claims arrive at an insurance company in accordance with some point process $\{S_n, n \geq 1\}$, S_n being the epoch of arrival of nth claim. $X(t)$ is the total number of claims up to t. As in a compound Poisson process, if $\{Y_i, i \geq 1\}$ is a sequence of claim amounts, then $Z(t) = \sum_{i=1}^{X(t)} Y_i$ represents total claim amount in $(0, t]$, and $\{Z(t), t \geq 0\}$ is known as a compound renewal process. In non-life insurance, Pareto distribution is found to be a suitable model for the claim amounts (Boland [2]).

(iii) Suppose $\{X_n, n \geq 1\}$ is a time homogeneous Markov chain. With $X_0 = i$, suppose S_n denotes the epoch of nth visit to state $i, n \geq 1$. In view of Markov property and time homogeneity, it follows that $T_n, n \geq 1$ are independent and identically distributed random variables with distribution given by $P[T_n = k] = f_{ii}^{(k)}, \ k \geq 1$. Thus, $\{S_n, n \geq\}$ is a renewal process with inter-renewal distribution to be discrete. If the state i is non-null persistent, the distribution of T_n is proper with finite mean; if it is null persistent, then the distribution of T_n is proper but the mean is infinite. If the state i is transient, then the distribution of T_n is improper. If $X(t)$ denotes the number of visits to state i in $(0, t]$, then $\{X(t), t \geq 0\}$ is the corresponding renewal counting process.

(iv) In a simple random walk with the state space I, suppose the event of interest is returned to state 0. In Chap. 4, it is shown that the probability of return to 0 in $2n$ steps is $f_{00}^{(2n)} = (-1)^{n-1}\binom{1/2}{n}(4pq)^n$ and in odd number of steps, it is 0. Suppose S_n denotes the epoch of nth visit to state 0, then

$$\begin{aligned}
P[T_n = 2n] &= f_{00}^{(2n)} = (-1)^{n-1}\binom{1/2}{n}(4pq)^n \\
\Rightarrow \quad f_{00} &= \sum_{n=1}^{\infty} f_{00}^{(2n)} = 1 - (1 - 4pq)^{1/2} = 1 - |2p - 1| \\
&= 1 \ \text{ if } \ p = 1/2 \\
&< 1 \ \text{ if } \ p \neq 1/2.
\end{aligned}$$

Thus, for a symmetric random walk the distribution of T_n is proper and $\{S_n, n \geq 0\}$ is a renewal process, again with discrete inter-renewal distribution.

(v) As an illustration of a renewal process in industrial systems, suppose we have a system which fails if certain component stops working. We assume that a component having a lifetime T_1 is put into operation at time $t = 0$. It functions for the random time $S_1 = T_1$ and fails. Upon failure, it is replaced by an identical new component having lifetime T_2. Now, the second component fails at time $S_2 = T_1 + T_2$. Again a new identical component having lifetime T_3 is put into operation. This process continues, that is, upon failure of the nth component at time $S_n = T_1 + T_2 + \cdots + T_n$, it is replaced by an identical new component with lifetime T_{n+1}. Thus, the functioning of the system is renewed for the nth time at S_n. If $\{T_n, n \geq 1\}$ is a sequence of independent and identically distributed non-negative random variables, then $\{S_n, n \geq 0\}$ is a renewal process. The term "renewal process" comes from such applications.

In all these applications, $t \geq 0$ represents time from a chosen origin. The origin may be the time at which a reservation counter opens or a new component is put into operation or may be the time at which observation of a process starts. There onwards, "events" such as replacements of failed components or customer arrivals or state changes in a process occur over time. Since inter-renewal random variables are independent and identically distributed, after every renewal the probabilistic structure of the future process is the same as the probabilistic structure of the process that started at time 0. It may be noted that the renewal that has occurred at $t = 0$ is not counted in $X(t)$. Thus, we have $X(0) = 0$.

Remark 10.1.2 In general, a renewal process is not a continuous time Markov chain; otherwise the inter-renewal random variables must have exponential distribution. However, in a renewal process the inter-renewal random variables need not be exponential. A Poisson process is the only renewal process which is a continuous time Markov chain.

We can obtain a realization of a renewal process once we know the inter-renewal distribution F. In the following example, we find a realization of a renewal process using Code 10.6.1. The code is similar to that given in the case of a Poisson process.

Example 10.1.1 Suppose $\{X(t), t \geq 0\}$ is a renewal process where inter-renewal distribution is Gamma $G(\alpha, \lambda)$, so that the mean inter-renewal time $\mu = \lambda/\alpha$. To find a realization of the process for a fixed $T = 10$ time units, we draw a random sample of size one at a time, from $G(\alpha, \lambda)$ distribution, until their sum is > 10. For comparison, we generate realizations with parameters as $\alpha = 1.44$, $\lambda = 3.6$ with mean $\mu = 2.5$ and $\alpha = 3$, $\lambda = 4.8$ with mean $\mu = 1.6$. The output in terms of renewal epochs is given in Table 10.1. Once we know the renewal epochs, we know how many renewals have occurred in $(0, T]$ and the realized values of inter-renewal random variables.

From Table 10.1, we note that when $\mu = 2.5$, the number of renewals in $(0, 10]$ is 3 while when $\mu = 1.6$, the number of renewals is 8. Realization of the renewal process $\{X(t), t \geq 0\}$ is presented below for mean 2.5:

Table 10.1 Renewal Epochs: gamma inter-renewal distribution

Mean	S_1	S_2	S_3	S_4	S_5	S_6	S_7	S_8
2.5	1.45	5.54	9.53					
1.6	0.88	1.73	2.48	4.01	4.99	6.26	7.22	8.96

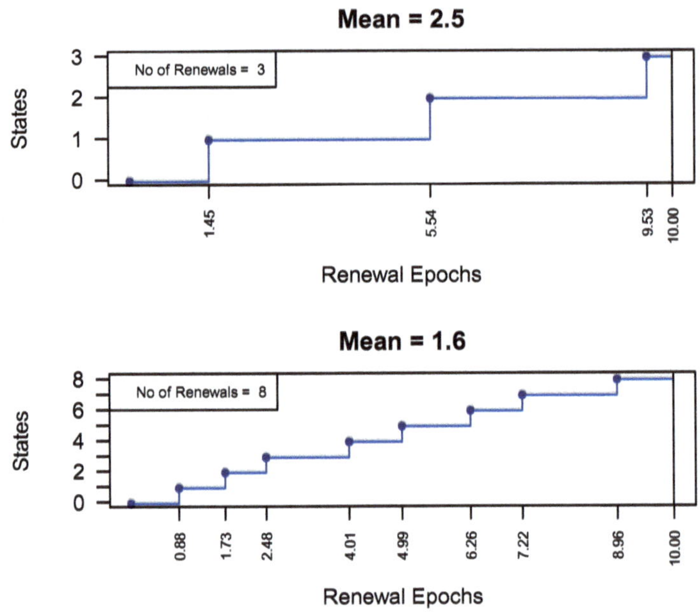

Fig. 10.1 Realization of a renewal process with gamma inter-renewal distribution

$$X(t) = \begin{cases} 0, & \text{if} \quad 0 \le t < 1.45 \\ 1, & \text{if} \quad 1.45 \le t < 5.54 \\ 2, & \text{if} \quad 5.54 \le t < 9.53 \\ 3, & \text{if} \quad 9.53 \le t \le 10. \end{cases}$$

Figure 10.1 displays the realization of the renewal process. It clearly shows changes in the realization as the mean inter-renewal time changes. Note that inter-renewal times are of longer duration when $\mu = 2.5$ as compared to $\mu = 1.6$. □

Remark 10.1.3 From Example 10.1.1, we note that as in a Poisson process, the sample path of a renewal process is a non-decreasing and right continuous step function. Further, $\{X(t), t \ge 0\}$ is also a non-decreasing process and increases with jumps of size one.

The principal objective of the renewal theory is to derive properties of some random variables associated with $\{S_n, n \ge 0\}$ or equivalently with $\{X(t), t \ge 0\}$ from the knowledge of inter-renewal distribution specified by F. We discuss some of these in the next section. From Example 10.1.1, we note that the mean μ of the inter-renewal random variable plays a major role. It is indeed true and is elaborated later. Similarly, the mean function $M(t) = E(X(t))$, known as a renewal function, also plays a key role. An important property of the renewal function is that it uniquely determines the renewal process. We discuss these properties in Sect. 10.2. In Sect. 10.3, we show that with probability 1, $\lim_{t \to \infty} X(t)/t = 1/\mu$ and give some of its applications.

Section 10.4 is devoted to some more limit theorems related to $M(t)$. The elementary renewal theorem asserts that $\lim_{t\to\infty} M(t)/t = 1/\mu$. The key renewal theorem is the refinement of the asymptotic relation $M(t) \approx t/\mu$ for large t. It states that for a fixed $h > 0$, $\lim_{t\to\infty}(M(t+h) - M(t)) = h/\mu$, if F is a continuous distribution function. In Sect. 10.4, we also state the central limit theorem for a renewal process and illustrate with an example. In Sect. 10.5, we briefly introduce some variations of a renewal process, such as delayed renewal process, stationary renewal process, renewal reward process and alternating renewal process. The last section presents R codes used to solve the examples.

10.2 Renewal Function

Suppose $\{X(t), t \geq 0\}$ is a renewal process where the inter-renewal distribution is specified by a distribution function F. We assume that $F(0) = P[T_n = 0] < 1$. Note that $P[T_n = 0] = 0$ implies that two or more renewals cannot occur simultaneously at the same epoch, that is, $P[S_n = S_{n+1}] = 0 \;\forall\, n \geq 1$. In some situations, $P[T_n = 0] = \alpha$, $0 < \alpha < 1$ and with probability $(1 - \alpha)$, F is a continuous distribution function. Thus, the inter-renewal distribution is a mixture distribution.

Suppose $\mu = E(T_n)$ is the mean inter-renewal time. From the non-negativity of T_n and the fact that T_n is not degenerate at 0, it follows that $\mu > 0$. In the theory of renewal process, the first question of interest is whether an infinite number of renewals can occur in a finite amount of time. To show that such an event cannot occur, note that $X(t) = \max\{n \geq 0 | S_n \leq t\}$, $t \geq 0$. By the strong law of large numbers,

$$S_n/n \stackrel{a.s.}{\to} \mu \text{ as } n \to \infty.$$

Further, $\mu > 0$ implies that S_n must diverge to infinity almost surely as $n \to \infty$. Thus, $S_n \leq t$ for at most a finite number of values of n and hence $X(t)$ must be finite for finite t. However, it is true that $X(t) \stackrel{a.s.}{\to} \infty$ as $t \to \infty$, since the only way in which $X(\infty)$, the total number of renewals in $(0, \infty)$, can be finite is one of the inter-renewal times to be infinite. Since F is a proper distribution function, the probability of such an event is 0.

As in the case of a Poisson process, the distribution of $X(t)$ can be obtained by the connecting link $X(t) \geq n \iff S_n \leq t$. Hence,

$$\begin{aligned} P[X(t) = n] &= P[X(t) \geq n] - P[X(t) \geq n+1] = P[S_n \leq t] - P[S_{n+1} \leq t] \\ &= F_n^*(t) - F_{n+1}^*(t) , \end{aligned}$$

where F_n^* is a distribution function of S_n; it is a n-fold convolution of F with itself. In general, it is difficult to find the explicit expression for the convolution and hence for $P[X(t) = n]$. Another approach to find $P[X(t) = n]$ by conditioning on S_n is as follows:

$$\begin{aligned}P[X(t) = n] &= \int_0^{\infty} P[X(t) = n | S_n = y] f_{S_n}(y)\, dy \\ &= \int_0^{t} P[T_{n+1} > t - y | S_n = y] f_{S_n}(y)\, dy \\ &= \int_0^{t} (1 - F(t - y)) f_{S_n}(y)\, dy\ ,\end{aligned}$$

since T_{n+1} is independent of S_n. Here $f_{S_n}(y)$ is the probability density function of S_n. However, this also does not give an explicit expression in general.

Example 10.2.1 Suppose the common distribution of T_n is a geometric distribution with probability mass function given by $P[T_n = i] = p(1-p)^{i-1}, i \geq 1$. Then the distribution of S_n is a negative binomial distribution with probability mass function given by

$$P[S_n = k] = \begin{cases} \binom{k-1}{n-1} p^n (1-p)^{k-n}, & \text{if } k \geq n \\ 0, & \text{if } k < n\ . \end{cases}$$

If the common distribution of T_n is Gamma $G(\alpha, \lambda)$, then the distribution of S_n is Gamma $G(\alpha, n\lambda)$, by the additive property of the gamma distribution. From the distribution of S_n, we can in principle obtain the distribution of $X(t)$, but it is difficult to get the exact expression. □

In the following example, we obtain $P[X(t) = n]$ for some values of t for a particular renewal counting process.

Example 10.2.2 Suppose $\{X(t), t \geq 0\}$ is a renewal process with inter-renewal distribution as $Poi(\mu)$. Then $S_n \sim Poi(n\mu)$. Hence, for $n \geq 1$

$$P[X(t) = n] = P[X(t) \geq n] - P[X(t) \geq n+1] = P[S_n \leq t] - P[S_{n+1} \leq t].$$

Further, $P[X(t) = 0] = P[T_1 > t]$ and $T_1 = S_1 \sim Poi(\mu)$. We take $\mu = 2$ and four values for t as 3, 6, 9, 12. We use Code 10.6.2 to compute the probability distribution of $X(t)$ for these values of t. It is presented in Table 10.2.

We note that $P[X(t) = n]$ for $n = 0$ to $n = 12$ add to 1.0000, 0.9998, 0.9999, 0.9983 for $t = 3, 6, 9, 12$, respectively. □

Even though it is not possible to find the distribution of $X(t)$ explicitly, it is possible to find its mean function in some cases. The mean function $E(X(t)) = M(t)$ is the expected number of renewals in $(0, t]$. However, in view of its far-reaching significance and relevance beyond its interpretation as the mean number of renewals, $M(t)$ has been given a special name as "renewal function". An important property of a renewal function is that it uniquely determines the renewal process. Specifically, there is a one-to-one correspondence between F and M. We prove this result in Theorem 10.2.4. We first derive an expression for the renewal function.

Table 10.2 $P[X(t) = n]$ for Poisson inter-renewal distribution

n	$P[X(3) = n]$	$P[X(6) = n]$	$P[X(9) = n]$	$P[X(12) = n]$
0	0.1429	0.0045	0.0000	0.0000
1	0.4237	0.1061	0.0081	0.0003
2	0.2823	0.2830	0.0758	0.0086
3	0.1088	0.2929	0.1995	0.0550
4	0.0320	0.1832	0.2587	0.1446
5	0.0080	0.0843	0.2155	0.2156
6	0.0018	0.0316	0.1330	0.2175
7	0.0004	0.0102	0.0661	0.1653
8	0.0001	0.0030	0.0279	0.1015
9	0.0000	0.0008	0.0104	0.0527
10	0.0000	0.0002	0.0035	0.0239
11	0.0000	0.0000	0.0011	0.0097
12	0.0000	0.0000	0.0003	0.0036

Theorem 10.2.1 *The renewal function $M(t)$ is given by $M(t) = \sum_{k=1}^{\infty} F_k^*(t)$ where $F_k^*(t)$ is the distribution function of S_k and it is given by*

$$F_k^*(t) = \begin{cases} F(t), & \text{if } k = 1 \\ \int_0^{\infty} F_{k-1}^*(t-y)dF(y) = \int_0^t F_{k-1}^*(t-y)dF(y), & \text{if } k > 1\,. \end{cases}$$

Thus, $F_k^*(t)$ is the k-fold convolution of F with itself.

Proof Using the link $X(t) \geq k \iff S_k \leq t$, we get

$$\begin{aligned} M(t) = E(X(t)) &= \sum_{k=1}^{\infty} kP[X(t) = k] = \sum_{k=1}^{\infty} P[X(t) \geq k] \\ &= \sum_{k=1}^{\infty} P[S_k \leq t] = \sum_{k=1}^{\infty} F_k^*(t). \end{aligned}$$

□

In the next theorem, we prove that $M(t) < \infty \ \forall \ t > 0$.

Theorem 10.2.2 *The renewal function $M(t)$ given by $M(t) = \sum_{k=1}^{\infty} F_k^*(t)$ is finite $\forall \ t > 0$.*

Proof Since the distribution function F is a non-decreasing function, $F^*_{k-1}(t-y) \leq F^*_{k-1}(t)$. Hence,

$$
\begin{aligned}
& F^*_k(t) = \int_0^\infty F^*_{k-1}(t-y)dF(y) \leq \int_0^t F^*_{k-1}(t)dF(y) = F^*_{k-1}(t)F(t) \\
\Rightarrow\; & F^*_k(t) \leq F^*_{k-2}(t)(F(t))^2 \\
\Rightarrow\; & F^*_k(t) \leq F^*_{k-(k-1)}(t)(F(t))^{k-1} = (F(t))^k \\
\Rightarrow\; & M(t) = \sum_{k=1}^\infty F^*_k(t) \leq \sum_{k=1}^\infty (F(t))^k \\
\Rightarrow\; & M(t) < \infty \text{ since } 0 < F(t) < 1 \;\forall\; t > 0.
\end{aligned}
$$

□

Example 10.2.3 Suppose $\{S_n, n \geq 1\}$ is a renewal process with inter-renewal distribution as exponential with rate λ. Then $S_n \sim G(\lambda, n)$ and hence,

$$
\begin{aligned}
M(t) &= \sum_{k=1}^\infty F^*_k(t) = \sum_{k=1}^\infty \int_0^t \frac{\lambda^k}{\Gamma(k)} e^{-\lambda x} x^{k-1}\, dx \\
&= \lambda \int_0^t e^{-\lambda x} \left(\sum_{k=1}^\infty \frac{(\lambda x)^{k-1}}{(k-1)!} \right) dx = \lambda \int_0^t e^{-\lambda x} e^{\lambda x}\, dx = \lambda t,
\end{aligned}
$$

as expected, since $\{X(t), t \geq 0\}$ in this case is a Poisson process with rate λ. □

In general, it is difficult to find the renewal function $M(t)$ using the formula derived in Theorem 10.2.1. For example, if inter-renewal distribution is uniform $U(0, 1)$, then we do not have an explicit formula for the convolution $F^*_k(t)$ and we cannot find the renewal function using convolution formula. However, an integral equation satisfied by the renewal function can be obtained by conditioning on the time of the first renewal. Establishing the result by conditioning on the time of the first renewal is known as renewal argument. We use it to derive an integral equation for the renewal function in the following theorem. We assume that the inter-renewal distribution F is continuous. We first state a lemma needed in the proof.

Lemma 10.2.1 *Suppose $\{X(t), t \geq 0\}$ is a renewal process with inter-renewal distribution F. Suppose $X^*(t)$ is defined as $X^*(t) = X(t+T^*) - X(T^*)$, where T^* is a renewal epoch. Then $\{X^*(t), t \geq 0\}$ is also a renewal process with the same inter-renewal distribution F. Further, $\{X(t), t \geq 0\}$ and $\{X^*(t), t \geq 0\}$ are independent stochastic processes.*

Theorem 10.2.3 *The renewal function $M(t)$ satisfies the equation*

$$
M(t) = F(t) + \int_0^t M(t-x)dF(x)\ . \tag{10.2.1}
$$

Proof Suppose the first renewal occurs at time x. If $x > t$, then there are no renewals in $(0, t]$ implying that $X(t) = 0$. If $x \leq t$, number of renewals in $(0, t]$ is equal to the number of renewals in $(0, x]$ plus number of renewals in $(x, t]$. By Lemma 10.2.1, it follows that the number of renewals in $(x, t]$ would have the same distribution as the number of renewals in $(0, t - x]$ since x is a renewal epoch. Thus we have

$$X(t) \overset{d}{=} \begin{cases} 0, & \text{if } x > t \\ 1 + X(t - x), & \text{if } x \leq t\,. \end{cases}$$

By conditioning on the time T_1 of the first renewal, we have

$$\begin{aligned} M(t) = E(X(t)) &= \int_0^\infty E(X(t)|T_1 = x)dF(x) \\ &= \int_0^t E(X(t)|T_1 = x)dF(x) + \int_t^\infty E(X(t)|T_1 = x)dF(x) \\ &= \int_0^t E(1 + X(t - x))dF(x) + \int_t^\infty 0\, dF(x) \\ &= F(t) + \int_0^t M(t - x)dF(x). \end{aligned}$$

□

Equation (10.2.1) is known as the renewal equation. In the following example, we verify that the mean function of a Poisson process satisfies the renewal equation.

Example 10.2.4 Suppose $\{S_n, n \geq 1\}$ is a Poisson process with rate λ. Then $F(x) = 1 - e^{-\lambda x}$ for $x > 0$ and $M(t) = \lambda t$. Observe that

$$\begin{aligned} &F(t) + \textstyle\int_0^t M(t - x)dF(x) \\ &= 1 - e^{-\lambda t} + \int_0^t \lambda(t - x)\lambda e^{-\lambda x}\, dx \\ &= (1 - e^{-\lambda t}) + \lambda t(1 - e^{-\lambda t}) - \lambda^2 \int_0^t x e^{-\lambda x}\, dx \\ &= (1 - e^{-\lambda t}) + \lambda t(1 - e^{-\lambda t}) - \lambda^2 \int_0^t x \frac{d}{dx}\left(\frac{e^{-\lambda x}}{-\lambda}\right) dx \\ &= (1 - e^{-\lambda t}) + \lambda t(1 - e^{-\lambda t}) + \lambda t e^{-\lambda t} - \lambda \int_0^t e^{-\lambda x}\, dx \\ &= (1 - e^{-\lambda t}) + \lambda t(1 - e^{-\lambda t}) + \lambda t e^{-\lambda t} - (1 - e^{-\lambda t}) \\ &= \lambda t = M(t)\,. \end{aligned}$$

□

The renewal equation can sometimes be solved to obtain the renewal function. For example, if the inter-renewal distribution is uniform $U(0, 1)$, the renewal function M is obtained in Sect. 10.4 by solving the renewal equation.

An important property of the renewal function $M(t)$ is that it uniquely determines the renewal process. To prove this result, we need some properties of a Laplace transform of a function. A Laplace transform is similar to a characteristic function and has similar properties.

Definition 10.2.1 Laplace Transform: A Laplace transform of a non-negative function $g(t), t \geq 0$, denoted by $\tilde{g}$, is defined as $\tilde{g}(s) = \int_0^\infty e^{-st} g(t)\, dt, \quad s > 0$.

We list below some properties of a Laplace transform.

(i) Uniqueness theorem of Laplace transforms: If $\tilde{g}$ and $\tilde{h}$ are Laplace transforms of the functions g and h respectively, and if $\tilde{g} = \tilde{h}$, then $g = h$.
(ii) The Laplace transform of $(\alpha g + \beta h) = \alpha\, \tilde{g} + \beta\, \tilde{h}$.
(iii) The Laplace transform of the convolution of two functions is the product of the Laplace transforms of the component functions in the convolution, that is, if $\tilde{g}$ and $\tilde{h}$ are Laplace transforms of the functions g and h respectively, and if $A(t) = \int_0^t g(t - x)h(x)\, dx$, then $\tilde{A} = \tilde{g}\tilde{h}$.

Table 10.3 displays the Laplace transforms of some functions.

From Table 10.3, note that the Laplace transform of exponential distribution with parameter λ is $\tilde{g}(s) = \lambda/(s + \lambda)$. Since Gamma $G(\lambda, 2)$ distribution is a convolution of $g(t) = \lambda e^{-\lambda t}$ with itself, it follows that the Laplace transform of the gamma distribution is given by $\tilde{g}(s) = \lambda^2/(s + \lambda)^2$.

Theorem 10.2.4 *Suppose the renewal function $M(t)$ and the distribution function F both are differentiable. Then the renewal function $M(t)$ uniquely determines F and hence the renewal process.*

Proof Suppose $F'(t) = f(t)$ & $M'(t) = m(t)$. Suppose $\tilde{m}(s)$ and $\tilde{f}(s)$ denote the Laplace transforms of $m(t)$ and $f(t)$, respectively. By the renewal equation,

Table 10.3 Laplace Transforms

$g(t)$	a	t	t^λ	$\lambda e^{-\lambda t}$	$\lambda^2 t e^{-\lambda t}$
$\tilde{g}(s)$	a/s	$1/s^2$	$\Gamma(\lambda + 1)/s^{\lambda+1}$	$\lambda/(s + \lambda)$	$\lambda^2/(s + \lambda)^2$

$$
\begin{aligned}
& M(t) = F(t) + \int_0^t M(t-x)dF(x) \\
\Rightarrow\quad & M'(t) = m(t) = f(t) + \int_0^t m(t-x) f(x)\, dx \qquad \text{by Leibnitz' rule} \\
\Rightarrow\quad & \tilde{m}(s) = \tilde{f}(s) + \text{Laplace transform of } \left(\int_0^t m(t-x) f(x)\, dx \right) \\
\Rightarrow\quad & \tilde{m}(s) = \tilde{f}(s) + \tilde{f}(s)\tilde{m}(s) \quad \text{by Property (iii)} \\
\Rightarrow\quad & \tilde{f}(s) = \tilde{m}(s)/(1+\tilde{m}(s)) \;\; \& \;\; \tilde{m}(s) = \tilde{f}(s)/(1-\tilde{f}(s)).
\end{aligned}
$$

By Property (i), the Laplace transform determines the distribution uniquely. Hence, given $\tilde{m}(s)$, $\tilde{f}(s)$ is determined and hence $F(t)$ is determined uniquely. □

The following corollary proves a characterizing property of a Poisson process.

Corollary 10.2.1 *A Poisson process is the only renewal process whose mean value function is linear.*

Proof Suppose for a renewal process, $M(t) = \lambda t$, where $\lambda > 0$. Observe that

$$
\begin{aligned}
M(t) = \lambda t \Rightarrow M'(t) = \lambda \;\; &\Rightarrow \;\; \tilde{m}(s) = \lambda/s \\
&\Rightarrow \tilde{f}(s) = \lambda/(\lambda+s) \;\; \Rightarrow \;\; f(t) = \lambda\, e^{-\lambda t}.
\end{aligned}
$$

Thus, inter-renewal time has exponential distribution with rate λ. Hence, the renewal process is a Poisson process.

□

The next example illustrates how Theorem 10.2.4 is useful to verify that a given function is a renewal function.

Example 10.2.5 Suppose $\{X(t), t \geq 0\}$ is a renewal process where the probability density function of inter-renewal distribution is given by $f(x) = \alpha^2 x e^{-\alpha x}$, $x > 0$. We examine whether $M(t) = \alpha t/2 - 1/4 + (1/4)e^{-2\alpha t}$ is a renewal function of the renewal process $\{X(t), t \geq 0\}$. It is known that $M(t)$ can be obtained from the convolution as derived in Theorem 10.2.1 or by solving the renewal equation. We adopt the approach based on the Laplace transform. Observe that $f(x)$ can be expressed as

$$
\begin{aligned}
& f(x) = \alpha^2 e^{-\alpha x} x^{2-1} \;\; \Rightarrow \;\; T_1 \sim G(\alpha, 2) \Rightarrow \;\; \tilde{f}(s) = \alpha^2/(\alpha+s)^2 \\
\Rightarrow\quad & \tilde{m}(s) = \tilde{f}(s)/(1-\tilde{f}(s)) = \alpha^2/(2\alpha s + s^2) \\
\text{Now,}\quad & M(t) = \alpha t/2 - 1/4 + (1/4)e^{-2\alpha t} \\
\Rightarrow\quad & M'(t) = m(t) = \alpha/2 - (\alpha/2)e^{-2\alpha t} \\
\Rightarrow\quad & \tilde{m}(s) = \int_0^\infty e^{-st} m(t) dt \\
& \qquad = (\alpha/2s) - (\alpha/(2(s+2\alpha)) = \alpha^2/(2\alpha s + s^2).
\end{aligned}
$$

Hence, $M(t) = \alpha t/2 - 1/4 + (1/4)e^{-2\alpha t}$ is the renewal function of the renewal process $\{X(t), t \geq 0\}$, where inter-renewal times follow gamma $G(\alpha, 2)$ distribution. □

In the next section, we discuss a limit theorem concerning the long run renewal rate for a renewal process and its applications.

10.3 Long Run Renewal Rate

It has been shown in Sect. 10.2 that $X(t) \to \infty$ as $t \to \infty$, with probability 1. However, it is of interest to know the rate at which $X(t)$ tends to infinity. It is conveyed by the almost sure limit of $X(t)/t$. To determine the rate at which $X(t)$ grows, consider a random variable $S_{X(t)}$. If $X(t) = 3$, then $S_{X(t)} = S_3$ represents the epoch of the third renewal. Since there are only three renewals that have occurred by time t, S_3 also represents the time of occurrence of the last renewal prior to or at time t. Thus, $S_{X(t)}$ represents the time of the last renewal prior to or at time t. Similar reasoning leads to the conclusion that $S_{X(t)+1}$ represents the time of the first renewal after time t. With these preliminaries, we now study the almost sure limit of $X(t)/t$. First we consider a special case when the inter-renewal random variables are degenerate at c, that is, the nth renewal takes place at time nc. Then the renewal process $\{X(t), t \geq 0\}$ is a deterministic process and $X(cu) = u$, that is, $X(t) = t/c$ which implies $X(t)/t = 1/c$. Hence, the long run rate at which the events take place is $\lim_{t\to\infty} X(t)/t = 1/c$, where c is the mean interval time. In the next theorem, we prove that such a result holds for all renewal processes.

Theorem 10.3.1 *Suppose $\{S_n, n \geq 1\}$ is a renewal process with $\{X(t), t \geq 0\}$ as the corresponding counting process. Suppose μ is the mean inter-renewal time. Then $\lim_{t\to\infty} X(t)/t = 1/\mu$ with probability one.*

Proof In a renewal process $\{S_n, n \geq 1\}$, $S_n = \sum_{i=1}^{n} T_i$, where $\{T_n, n \geq 1\}$ is a sequence of independent and identically distributed random variables with finite mean μ. Hence, by the Kolmogorov strong law of large numbers, $S_n/n = \sum_{i=1}^{n} T_i/n \stackrel{a.s.}{\to} \mu$. Since $X(t) \to \infty$ almost surely when $t \to \infty$, we have

$$\begin{aligned} & S_{X(t)}/X(t) \stackrel{a.s.}{\to} \mu, \quad S_{X(t)+1}/(X(t)+1) \stackrel{a.s.}{\to} \mu \\ \Rightarrow \quad & (S_{X(t)+1})/X(t) = (S_{X(t)+1})/(X(t)+1) \times ((X(t)+1)/X(t)) \stackrel{a.s.}{\to} \mu. \end{aligned}$$

Further, $S_{X(t)}$ represents the time of the last renewal prior to or at time t and $S_{X(t)+1}$ represents the time of the first renewal after time t. Hence we have

$$
\begin{aligned}
S_{X(t)} \leq t < S_{X(t)+1} &\Rightarrow \frac{S_{X(t)}}{X(t)} \leq \frac{t}{X(t)} < \frac{S_{X(t)+1}}{X(t)} \\
&\Rightarrow \lim_{t\to\infty} \frac{S_{X(t)}}{X(t)} \leq \lim_{t\to\infty} \frac{t}{X(t)} \leq \lim_{t\to\infty} \frac{S_{X(t)+1}}{X(t)} \quad a.s. \\
&\Rightarrow \mu \leq \lim_{t\to\infty} t/X(t) \leq \mu \ a.s. \\
&\Rightarrow \lim_{t\to\infty} X(t)/t = 1/\mu \ a.s.
\end{aligned}
$$

and the proof is complete. □

Remark 10.3.1 Since $X(t)$ is the number of renewals in $(0, t]$, $X(t)/t$ represents the rate of renewals per unit of time. Then $1/\mu$ is interpreted as long run renewal rate per unit of time. Since the average time between renewals is μ, it is quite appealing that the average rate at which renewals occur is 1 per μ time units. Thus, the long run renewal rate $1/\mu$ equals the reciprocal of the mean inter-renewal time, as intuitively expected.

This result proves to be very powerful in the long run analysis of renewal processes. We illustrate the result with several examples.

Example 10.3.1 A mobile works on a single battery. As soon as the battery in use fails, suppose it is immediately replaced with a new battery. If the lifetime of a battery, in years, is distributed uniformly over the interval $(1, 3)$, then the mean lifetime is $\mu = 2$ years. Using Theorem 10.3.1, $\lim_{t\to\infty} X(t)/t$ is $1/\mu = 1/2$. Thus, the long run rate of changing batteries is $1/2$, that is, in the long run, the battery will be replaced every 2 years. Here, we assume that the battery is immediately replaced with a new one, which may not be the case. Suppose for example the amount of time in days to procure a new battery has uniform $U(0, 1)$ distribution, that is, it has $U(0, 1/365)$ distribution when units are in years. Consequently, the mean time between renewals is $\mu_1 = \mu + 1/730 = 1461/730$ and hence the long run rate of changing batteries is $730/1461 = 0.4996578 \approx 0.5$. Thus, in the long run, the battery will be replaced approximately every 2 years. It is to be noted that compared to mean life 2 years of a battery, mean time $1/730$ years to procure a battery is very small and hence long run rate does not change much. □

The next example is similar to Example 10.3.1, but introduces the concept of an age-replacement policy.

Example 10.3.2 Mr. Bapat replaces the battery in his car as soon as it gets discharged. The time required to replace the battery can be ignored since it is small compared to the lifetime of the battery. Suppose $X(t)$ is the number of batteries replaced during the first t years of the life of the car, not counting the one that was installed at the purchase of the car. Assuming that the lifetime T_n (in years) of the nth battery for $n \geq 1$ are independent and identically distributed random variables each having uniform $U(1, 4)$ distribution, $\{X(t), t \geq 0\}$ is a renewal process with $\mu = 2.5$ years. By Theorem 10.3.1, $\lim_{t\to\infty} X(t)/t = 1/\mu = 2/5 = 0.4$. Thus in the long run, the battery will be replaced after 2.5 years. In order to avoid the inconvenience when the battery gets discharged, Mr. Bapat adopts the policy to replace the battery once it becomes 3 years old, even if it has not failed yet. Of course, if the battery fails before 3 year, he has to replace it anyway. Suppose $X_1(t)$ is the number of batteries replaced up to time t, planned or unplanned. We examine whether $\{X_1(t), t \geq 0\}$ can be modeled as a renewal process. Suppose Y_n is the inter-replacement time between $(n-1)$th and nth replacement. Then Y_n is distributed as $\min\{T_n, 3\}$. Since $\{T_n, n \geq 1\}$ are independent and identically distributed random variables, it follows that $\{Y_n, n \geq 1\}$ are also independent and identically distributed random variables and hence $\{X_1(t), t \geq 0\}$ can be modeled as a renewal process. Further,

$$E(Y_n) = \int_1^3 x f_{T_n}(x)\, dx + \int_3^4 3 f_{T_n}(x)\, dx = 4/3 + 3(1/3) = 7/3\,.$$

Another approach to evaluate the same is as follows:

$$\begin{aligned} E(Y_n) &= E(E(Y_n|T_n)) = E(Y_n|T_n \leq 3)P[T_n \leq 3] + E(Y_n|T_n > 3)P[T_n > 3] \\ &= E(T_n|T_n \leq 3)(2/3) + E(3|T_n > 3)(1/3) = (2/3)E(T_n|T_n \leq 3) + 1\,. \end{aligned}$$

Now to find $E(T_n|T_n \leq 3)$, observe that for $1 \leq x \leq 3$

$$\begin{aligned} P[T_n \leq x|T_n \leq 3] &= P[T_n \leq x, T_n \leq 3]/P[T_n \leq 3] \\ &= P[T_n \leq x]/P[T_n \leq 3] = (x-1)/2 \\ \Rightarrow\ E(T_n|T_n \leq 3) &= \int_1^3 x(1/2) = 2 \\ \Rightarrow\ E(Y_n) &= 7/3 = 2.333. \end{aligned}$$

By Theorem 10.3.1, $\lim_{t\to\infty} X_1(t)/t = 3/7$. Thus, in the long run, the battery will be replaced after 2.333 years. Thus, the long run rate is larger than the rate of replacing the battery when it gets discharged. □

Remark 10.3.2 The policy to replace the battery once it becomes 3 years old in the above example is in general known as an age-replacement policy. In this policy, an item is replaced upon failure or attaining an age A, whichever occurs first. The probability that an item fails before A is $F(A)$; it can be interpreted as the long run fraction of failure replacements for the items which fail before A. Similarly, the long run fraction of planned replacements for the items which do not fail before A is $1 - F(A)$. The distribution function F_A of a renewal interval for this age-replacement policy is given by

$$F_A(x) = \begin{cases} F(x), & \text{if } x < A \\ 1, & \text{if } x > A\,. \end{cases}$$

Hence, the mean renewal duration μ_A is given by

$$\mu_A = \int_0^\infty (1 - F_A(x))\,dx = \int_0^A (1 - F(x))\,dx < \int_0^\infty (1 - F(x))\,dx = \mu.$$

It is to be noted that $\mu_A < \mu \iff (1/\mu_A) > (1/\mu)$. It implies that the long run rate of renewals in the age-replacement policy will be always larger than that in a renewal process where the items are replaced upon failure.

Example 10.3.3 Suppose a machine works for a random amount U of time having an exponential distribution with rate λ_1. Once it fails, it gets repaired. The repair time V has an exponential distribution with rate $\lambda_2 \neq \lambda_1$. We assume that the machine is as good as new after the repair is complete. Thus, if $Y(t)$ denotes the state of the machine at time t, it is 1 if it is working and is 0 if it is down. We assume that the repair time and the working time of a machine are independent random variables. Suppose the machine is in a working condition 1 at time 0 and $X(t)$ denotes the number of visits to state 1 in $(0, t]$. The machine is in a working condition at time 0; it remains in that state 1 for a random duration U and then fails; after the repair of random duration V, it is again in state 1. Thus, the random interval T_i between $(i-1)$th and ith visits to state 1 is distributed as $U + V$, and it has an hypoexponential distribution (Ross [6]), with probability density function given by

$$f(t) = \frac{\lambda_2}{\lambda_2 - \lambda_1}\lambda_1 e^{-\lambda_1 t} + \frac{\lambda_1}{\lambda_1 - \lambda_2}\lambda_2 e^{-\lambda_2 t}, \quad t > 0.$$

Further in view of independence of U_i's and V_i's, it follows that T_i's are also independent. Thus, $\{X(t), t \geq 0\}$ is a renewal process with inter-renewal times $\{T_i, i \geq 1\}$ with common mean $\mu = 1/\lambda_1 + 1/\lambda_2 = (\lambda_1 + \lambda_2)/\lambda_1\lambda_2$. By Theorem 10.3.1, $\lim_{t\to\infty} X(t)/t$ is $\lambda_1\lambda_2/(\lambda_1 + \lambda_2)$ a.s. Thus, the long run rate of visits to state 1 is $\lambda_1\lambda_2/(\lambda_1 + \lambda_2)$. If $1/\lambda_1 = 50$ h and $1/\lambda_2 = 2$ h then $\mu = 52$. Thus, the long run rate of visits to state 1 of the machine is $1/52$. If visit to state 1 is interpreted as repair completion, then the long run rate of repair completions is $1/52$ h. It is to be noted that the long run rate of visits to state 0 is also the same. □

Example 10.3.4 Suppose customers arrive at an ATM center in accordance with a Poisson process having rate 10 per hour. If there is no one at the ATM, then the customer enters in. However, if there is already any customer at the ATM, then the customer leaves. Suppose on the average the amount of time required at the ATM is 6 min. Thus, the mean time between entering the ATM center is $\mu = 1/10 + 6/60 = 1/5$ h. Hence, the long run rate at which customers enter the ATM center is 5. It is given that the rate of arrivals of customers to the ATM center is 10. Thus, the proportion of customers who actually enter the ATM center is $5/10 = 0.5$, thus 50% customers do not use the facility. □

In the next section, we discuss some more limit theorems.

10.4 Limit Theorems

Suppose $\{X(t), t \geq 0\}$ is a Poisson process with rate λ. Then

$$M(t) = E(X(t)) = \lambda t \Rightarrow M(t)/t = \lambda \text{ for fixed } t$$
$$\Rightarrow \lim_{t\to\infty} M(t)/t = \lambda = 1/\mu\,.$$

Thus, the expected number of events per unit time in a Poisson process with rate λ is also $\lambda = 1/\mu$. In the following elementary renewal theorem, we prove that for a renewal process with mean inter-renewal time μ, $\lim_{t\to\infty} M(t)/t$ is also $1/\mu$. It is proved that $\lim_{t\to\infty} X(t)/t = 1/\mu$ with probability 1. However, we cannot conclude that $\lim_{t\to\infty} M(t)/t = \lim_{t\to\infty} E(X(t))/t = 1/\mu$, since almost sure convergence does not imply convergence in mean.

Using renewal argument as in the derivation of the renewal equation, we derive the expression for $E(S_{X(t)+1})$. In this derivation, we use the following lemma as given in Karlin and Taylor [3].

Lemma 10.4.1 *Suppose a is a bounded function. There exists one and only one function A bounded on a finite interval that satisfies the integral equation $A(t) = a(t) + \int_0^t A(t-y)dF(y)$ and it is given by $A(t) = a(t) + \int_0^t a(t-y)dM(y)$, where M is the renewal function corresponding to F.*

Theorem 10.4.1 *Suppose $\{X(t), t \geq 0\}$ is a renewal process with the renewal function M. Then*

$$E(S_{X(t)+1}) = E\Big(\sum_{i=1}^{X(t)+1} T_i\Big) = E(T_1) \times (M(t) + 1)\,. \tag{10.4.1}$$

Proof Suppose $A(t) = E(S_{X(t)+1})$. As in Theorem 10.2.1, suppose the first renewal occurs at time x. If $x > t$, then there are no renewals in $(0, t]$, so that $X(t) = 0$ and $S_{X(t)+1} = x$. If $x \leq t$, the number of renewals in $(0, t]$ is equal to the number of renewals in $(0, x]$ plus number of renewals in $(x, t]$ which is distributed as $1 + X(t-x)$ since at x there is a renewal. Hence,

$$E(S_{X(t)+1} | T_1 = x) = \begin{cases} x, & \text{if } x > t \\ x + A(t-x), & \text{if } x \leq t\,. \end{cases}$$

Hence,

$$
\begin{aligned}
A(t) = E(S_{X(t)+1}) &= \int_0^\infty E(S_{X(t)+1} | T_1 = x) dF(x) \\
&= \int_0^t (x + A(t-x)) dF(x) + \int_t^\infty x dF(x) \\
&= \int_0^\infty x dF(x) + \int_0^t A(t-x) dF(x) \\
&= E(T_1) + \int_0^t A(t-x) dF(x).
\end{aligned}
$$

Thus, $A(t)$ satisfies the equation $A(t) = a(t) + \int_0^t A(t-y)dF(y)$ with $a(t) = E(T_1)\ \forall\ t > 0$. Hence by Lemma 10.4.1,

$$
\begin{aligned}
E(S_{X(t)+1}) = A(t) = a(t) + \int_0^t a(t-y)dM(y) &= E(T_1) + \int_0^t E(T_1)dM(y) \\
= E(T_1) + E(T_1)M(t) = E(T_1)(1 + M(t)).
\end{aligned}
$$

□

Equation (10.4.1) resembles the identity $E(\sum_{i=1}^N X_i) = E(X_1)E(N)$, where $\{X_1, X_2, \ldots\}$ are independent and identically distributed integrable random variables and N is an integer-valued integrable random variable, which is independent of X_i's. The crucial difference between this identity and the identity in Eq. (10.4.1) is that the number of summands $X(t) + 1$ is not independent of $T_i's$. Equation (10.4.1) is a special case of the well-known Wald's identity. We use Theorem 10.4.1 to prove the elementary renewal theorem.

Theorem 10.4.2 *Elementary Renewal Theorem: Suppose $M(t)$ is the renewal function of the renewal process $\{X(t), t \geq 0\}$. Then $\lim_{t\to\infty} M(t)/t = 1/\mu$.*

Proof To prove the result, we prove that

$$
1/\mu \ \leq \ \liminf_{t\to\infty} M(t)/t \ \leq \ \limsup_{t\to\infty} M(t)/t \ \leq \ 1/\mu.
$$

Observe that

$$
\begin{aligned}
t < S_{X(t)+1} &\Rightarrow t \ < \ E(S_{X(t)+1}) = \mu \times (M(t) + 1) \ \text{by Equation (10.4.1)} \\
&\Rightarrow t/t\mu \ < \ \mu \times (M(t)+1)/t\mu \ \Rightarrow \ 1/\mu - 1/t \ < \ M(t)/t \\
&\Rightarrow 1/\mu \ \leq \ \liminf_{t\to\infty} M(t)/t.
\end{aligned}
$$

In order to prove $\limsup_{t\to\infty} M(t)/t \ \leq \ 1/\mu$, we use the truncation technique and define a random variable $T_k^{(c)}$ for $c > 0$ and $k \geq 1$ as follows:

$$
T_k^{(c)} = \begin{cases} T_k, & \text{if } T_k \leq c \\ c, & \text{if } T_k > c\,. \end{cases}
$$

Note that $T_k^{(c)} = g(T_k)$, where g is a Borel function. Hence, $\{T_k^{(c)}, k \geq 1\}$ is a sequence of independent and identically distributed random variables. Suppose $\{X^{(c)}(t), t \geq 0\}$ denotes a renewal process corresponding to the inter-renewal times $\{T_k^{(c)}\}$ and $M^{(c)}(t)$ is the corresponding renewal function. Also, suppose $F^{(c)}$ is the distribution function of $T_k^{(c)}$ and $\mu^{(c)}$ is its expectation. Observe that

$$\mu^c = E(T_1^{(c)}) = \int_0^c (1 - F^{(c)}(x))dx \to \int_0^\infty (1 - F(x))\, dx = \mu \text{ as } c \to \infty$$

$$\& \ T_k^{(c)} \leq T_k \ \Rightarrow \ X^{(c)}(t) \geq X(t) \ \& \ M^{(c)}(t) \geq M(t)\,.$$

Further,

$$\begin{aligned} S^c_{X^{(c)}(t)+1} - t \leq T_k^{(c)} \leq c &\Rightarrow E[S^{(c)}_{X^{(c)}(t)+1} - t] \leq E[T_k^{(c)}] \leq c \\ &\Rightarrow E[S^{(c)}_{X^{(c)}(t)+1}] \leq t + c \ \Rightarrow \ \mu^{(c)} \times [M^{(c)}(t) + 1] \leq t + c \\ &\Rightarrow \mu^{(c)} \times [M(t) + 1] \leq t + c. \end{aligned}$$

Dividing both sides of the inequality $\mu^{(c)} \times [M(t) + 1] \leq t + c$ by $t\mu^{(c)}$, we get

$$\begin{aligned} \frac{M(t)}{t} + \frac{1}{t} \leq \frac{1}{\mu^{(c)}} + \frac{c}{t\mu^{(c)}} \Rightarrow \frac{M(t)}{t} &\leq \frac{1}{\mu^{(c)}} + \frac{1}{t}\left(\frac{c}{\mu^{(c)}} - 1\right) \\ \Rightarrow \limsup_{t\to\infty} \frac{M(t)}{t} &\leq \frac{1}{\mu^{(c)}}, \ \ \forall \ c > 0 \\ \Rightarrow \ \lim_{c\to\infty} \limsup_{t\to\infty} \frac{M(t)}{t} &\leq \lim_{c\to\infty} \frac{1}{\mu^{(c)}} \\ \Rightarrow \ \limsup_{t\to\infty} \frac{M(t)}{t} &\leq \frac{1}{\mu}. \end{aligned}$$

Thus,

$$\begin{aligned} 1/\mu \leq \liminf_{t\to\infty} M(t)/t \ \leq \ \limsup_{t\to\infty} M(t)/t \ &\leq \ 1/\mu \\ \Rightarrow \ \lim_{t\to\infty} M(t)/t &= 1/\mu. \end{aligned}$$

□

Remark 10.4.1 In a Poisson process with rate λ, $M(t) = \lambda t = t/\mu$ and the renewal function is a linear function. The elementary renewal theorem conveys that for a renewal process, the renewal function is asymptotically linear.

The key renewal theorem, also known as Blackwell's theorem (Bhat [1]), is a refinement of the asymptotic relation $M(t) \approx t/\mu$ for large t. It is stated below.

Theorem 10.4.3 *Key Renewal Theorem: Suppose F is a continuous distribution function of a positive random variable with mean μ and $M(t)$ is the renewal function associated with F. Then for a fixed $h > 0$,*

$$\lim_{t\to\infty}(M(t+h) - M(t)) = h/\mu.$$

The key renewal theorem states that the expected number of renewals in an interval of length h is approximately h/μ, provided the process has been in operation for a long duration. If a renewal process is a Poisson process with rate λ, then

$$M(t+h) - M(t) = \lambda(t+h) - \lambda t = \lambda h = h/\mu, \quad \text{since} \quad \mu = 1/\lambda.$$

Thus, for a Poisson process, for every t, $M(t+h) - M(t) = h/\mu$ and hence in limit as $t \to \infty$.

Using the key renewal theorem, it is proved that (Bhat [1]), if F is a continuous distribution function of a positive random variable with mean μ and variance σ^2, then

$$\lim_{t\to\infty}(M(t) - t/\mu) = (\sigma^2 - \mu^2)/2\mu^2.$$

In the next example, we verify the elementary renewal theorem and the key renewal theorem.

Example 10.4.1 Suppose $\{X(t), t \geq 0\}$ is a renewal process where the probability density function of inter-renewal distribution is $f(x) = \alpha^2 x e^{-\alpha x}, \ x > 0$. Thus, the mean inter-renewal time is $\mu = 2/\alpha$. In Example 10.2.5, we have shown that $M(t) = (\alpha t)/2 - 1/4 + (1/4)e^{-2\alpha t}$ is a renewal function of the renewal process $\{X(t), t \geq 0\}$. Observe that

$$\begin{aligned}
\lim_{t\to\infty} M(t)/t &= \lim_{t\to\infty}(\alpha t/2 - 1/4 + (1/4)e^{-2\alpha t})/t = \alpha/2 = 1/\mu \\
\lim_{t\to\infty}(M(t+h) - M(t)) &= (\alpha h)/2 + \lim_{t\to\infty}(1/4)(e^{-2\alpha(t+h)} - e^{-2\alpha t}) \\
&= (\alpha h)/2 = h/\mu.
\end{aligned}$$

Thus, the renewal process $\{X(t), t \geq 0\}$ satisfies the elementary renewal theorem and the key renewal theorem. □

In the next example, using renewal equation and the elementary renewal theorem, we find $M(t)$ when the inter-renewal distribution is uniform $U(0, 1)$.

Example 10.4.2 Suppose $\{S_n, n \geq 0\}$ is the renewal process with inter-renewal distribution to be uniform $U(0, 1)$. Using renewal equation we find $M(t)$. Suppose $t \leq 1$. By Eq. (10.2.1), we have

$$
\begin{aligned}
& M(t) = t + \int_0^t M(t-x)dx = t + \int_0^t M(y)dy, \text{ by } y = t - x \\
\Rightarrow\ & M'(t) = 1 + M(t), \quad \text{by Leibnitz' rule} \\
\Rightarrow\ & h'(t) = h(t) \text{ where } h(t) = 1 + M(t) \\
\Rightarrow\ & h(t) = ce^t \ \Rightarrow\ M(t) = ce^t - 1 \\
\Rightarrow\ & M(t) = e^t - 1 \text{ as } M(0) = 0 \ \Rightarrow\ c = 1.
\end{aligned}
$$

Thus, $M(t) = e^t - 1 \ \ \forall\ t \le 1$. Suppose $t > 1$. Then

$$
\begin{aligned}
M(t) &= 1 + \int_0^t M(t-x)dx = 1 + \int_0^1 M(t-x)dx + 0 \\
&= 1 + \int_{t-1}^t M(y)dy, \text{ by } y = t - x \\
\Rightarrow\ M'(t) &= M(t) - M(t-1), \text{ by Leibnitz' rule}\,.
\end{aligned}
$$

Suppose $M(t) = at + b$. Hence, $M(t) - M(t-1) = at + b - a(t-1) - b = a$ and $M'(t) = a$. Thus, $M(t) = at + b$ satisfies the differential equation $M'(t) = M(t) - M(t-1)$. To decide a and b, note that

$$
\lim_{t\to\infty} M(t)/t = a = 2 \ \& \ M(1) = a + b = e - 1 \ \Rightarrow\ b = e - 3 = -0.2817.
$$

Thus,

$$
M(t) = \begin{cases} e^t - 1, & \text{if } t \le 1 \\ 2t - 0.2817, & \text{if } t > 1\,. \end{cases}
$$

□

Almost sure convergence implies convergence in probability which further implies convergence in distribution. Thus,

$$
X(t)/t \stackrel{a.s}{\to} 1/\mu \ \Rightarrow\ X(t)/t \stackrel{L}{\to} 1/\mu \text{ as } t \to \infty.
$$

However, the degenerate limit law is not useful. It is known that (Bhat [1]), with suitable normalizing factor, the asymptotic distribution of $X(t)/t$ is a non-degenerate distribution. Such an important limit theorem is the central limit theorem for a renewal process. It is stated below.

Theorem 10.4.4 *Central Limit Theorem: Suppose $\{X(t), t \ge 0\}$ is a renewal process with μ and σ^2 as the mean and variance respectively of the inter-renewal distribution. Then $\forall\ x \in \mathbb{R}$,*

$$
\lim_{t\to\infty} P\left[\frac{X(t) - t/\mu}{\sqrt{t\sigma^2/\mu^3}} \le x\right] = \frac{1}{\sqrt{2\pi}} \int_{-\infty}^x e^{-u^2/2}\, du\,.
$$

In particular, if $\{X(t), t \geq 0\}$ is a Poisson process with rate λ then $\mu = 1/\lambda$ and $\sigma^2 = 1/\lambda^2$. Hence for large t, distribution of $(X(t) - \lambda t)/\sqrt{\lambda t}$ can be approximated by the standard normal distribution.

The following example illustrates Theorem 10.4.4.

Example 10.4.3 Two machines continuously process a number of jobs. The random time to process a job on machine 1 follows the gamma $G(2, 4)$ distribution, whereas the time to process a job on machine 2 is uniformly distributed over $(0, 4)$. Suppose $X_i(t)$ denotes the number of jobs that machine i can process by time t, then $\{X_i(t), t \geq 0\}$, $i = 1, 2$ are independent renewal processes. The inter-renewal distribution of $\{X_1(t), t \geq 0\}$ is gamma with mean 2 and variance 1 while the inter-renewal distribution of $\{X_2(t), t \geq 0\}$ is uniform with mean 2 and variance 16/12. By Theorem 10.4.4, the approximate distributions of $X_1(100)$ and $X_2(100)$ are normal $N(50, 100/8)$ and normal $N(50, 100/6)$, respectively. Hence, the approximate distribution of $X_1(100) + X_2(100)$ is normal $N(100, 175/6)$. Thus, the approximate probability that the two machines together can process at least 90 jobs by $t = 100$ time units is given by

$$\begin{aligned} P[X_1(100) + X_2(100) \geq 90] &= P\left[\frac{X_1(100) + X_2(100) - 100}{\sqrt{175/6}} > \frac{90 - 100}{\sqrt{175/6}}\right] \\ &= \Phi(1.8516) = 0.9680. \end{aligned}$$

□

In the following example, we verify the limit theorems using R.

(i) Almost sure convergence implies convergence in probability. We verify that $X(t)/t \xrightarrow{P} 1/\mu$ as $t \to \infty$. By definition of the convergence in probability, $X(t)/t \xrightarrow{P} 1/\mu$, if

$$\forall\ \epsilon > 0\ \&\ \forall\ \mu > 0,\ \ P[|X(t)/t - 1/\mu| < \epsilon] \to 1 \text{ as } t \to \infty.$$

We simulate the renewal process m times for a time period $(0, t]$ for $t = t_0$ to $t = t_N$ with an increment of h. For given ϵ, we compute the estimate of the coverage probability as a relative frequency $R_t = \sum_{i=1}^{m}[\text{No of } |X_i(t)/t - 1/\mu| < \epsilon]/m$. We examine whether R_t approaches 1 as t increases.

(ii) To examine whether $\lim_{t\to\infty} M(t)/t = 1/\mu$, we estimate $M(t)$ by $\hat{M}(t) = \sum_{i=1}^{m} X_i(t)/m$ and examine whether $\hat{M}(t)/t$ approaches $1/\mu$ as t increases.

(iii) To verify the key renewal theorem, we compute $\hat{M}(t+h) - \hat{M}(t)$ and examine whether it is close to h/μ for large t.

(iv) For the verification of the central limit theorem, we compute m values of $(X(t) - t/\mu)/\sqrt{t\sigma^2/\mu^3}$ corresponding to m simulations and find the p-value of the Shapiro-Wilk test.

Code 10.6.3 generates a renewal process m times and verifies these theorems using the procedure outlined above in the next example.

Table 10.4 Verification of Limit Theorems

t	R_t	$\hat{M}(t)/t$	$\hat{M}(t+50) - \hat{M}(t)$	p-value
50	0.712	0.2955	14.988	0.0001
100	0.816	0.2976	14.880	0.0088
150	0.892	0.2976	15.040	0.0702
200	0.960	0.2984	14.876	0.1246
250	0.980	0.2982	15.008	0.2192
300	0.988	0.2986	15.048	0.1914
350	1.000	0.2989	15.056	0.2354
400	1.000	0.2992	15.080	0.0742
450	1.000	0.2994	15.064	0.1527
500	0.996	0.2996		0.1648

Example 10.4.4 Suppose $\{X(t), t \geq 0\}$ is a renewal process where inter-renewal distribution is Gamma(α, λ) with $\alpha = 1.2$ and $\lambda = 4$. Thus, the mean inter-renewal time is $\mu = \lambda/\alpha = 3.3333$ and variance $\sigma^2 = 2.7778$. Further, $1/\mu = 0.3$. We generate $m = 250$ realizations of $X(t)$ up to time t for $t = 50, 100, \ldots, 500$ in steps of $h = 50$. Then we compute R_t for $\epsilon = 0.04$. The output is organized in Table 10.4.

From Table 10.4, we note that as t increases (i) R_t approaches 1 implying that $X(t)/t \xrightarrow{P} 1/\mu = 0.3$, (ii) $\hat{M}(t)/t$ approaches $1/\mu = 0.3$, (iii) $\hat{M}(T+50) - \hat{M}(T) \approx h/\mu = 15$ and (iv) for $t \geq 150$, p-values of Shapiro-Wilk test indicate that the distribution of $(X(t) - t/\mu)/\sqrt{t\sigma^2/\mu^3}$ can be approximated by the standard normal distribution. □

In the next section, we briefly introduce some variations of the renewal processes.

10.5 Generalizations and Variations of Renewal Processes

Suppose $\{X(t), t \geq 0\}$ is a renewal process with $\{T_n, n \geq 1\}$ as the sequence of inter-renewal random variables having mean μ. We now define three more random variables related to the renewal process as follows:

$$U_t = S_{X(t)+1} - t, \quad V_t = t - S_{X(t)} \quad \& \quad W_t = U_t + V_t .$$

U_t is known as the excess or residual lifetime random variable, V_t is known as the spent time or current life or age random variable and W_t is known as the total life random variable. Using renewal argument, it can be shown that (Karlin and Taylor [3])

$$\lim_{t\to\infty} P[U_t \le x] = \lim_{t\to\infty} P[V_t \le x] = \mu^{-1}\int_0^x (1 - F(y))dy$$
$$\& \quad \lim_{t\to\infty} P[W_t \le x] = \mu^{-1}\int_0^x y dF(y).$$

These results are useful in defining some variations of the renewal processes. We begin with the concept of a delayed renewal process.

Delayed renewal process: A delayed renewal process is a counting process $\{X(t), t \ge 0\}$ in which the first inter-renewal time has possibly a different distribution than the remaining inter-renewal random variables. We define it as follows.

Definition 10.5.1 Delayed Renewal Process: Suppose $\{T_n, n \ge 1\}$ is a sequence of independent random variables with support $\mathbb{R}^+$. Suppose the distribution function of T_1 is G and $\{T_n, n \ge 2\}$ are identically distributed random variables with distribution function F. Suppose $S_n = \sum_{i=1}^n T_i,\ n \ge 1$ and $X_D(t)$ is the number of renewals up to t. Then $\{X_D(t), t \ge 0\}$ is said to be a delayed or modified or general renewal process.

We now refer to the renewal process $\{X(t), t \ge 0\}$, discussed in the previous sections, as an ordinary renewal process. The two processes are similar except that the time to the first renewal has different distribution.

We come across the delayed renewal process when the component in operation at time $t = 0$ is not new, but all subsequent replacements are new. For example, if the time origin is taken as x time units after the start of the ordinary renewal process, then the time to the first renewal after the origin in the delayed process will have the distribution of the excess life at time x of an ordinary renewal process.

No new tools are needed in the analysis of the delayed renewal process. Suppose $M_D(t) = E(X_D(t))$ is a renewal function of the delayed renewal process. Then we have the following results. For proofs, one may refer to Karlin and Taylor [3] or Medhi [5]:

(i) $M_D(t) = \sum_{n=0}^{\infty} \left(G * F_n^*(t)\right)$.
(ii) $M_D(t) = G(t) + \int_0^t M(t-x)dF(x)$, where $M(\cdot)$ is the renewal function of the ordinary renewal process.
(iii) $M_D(t)/t \to 1/\mu$ as $t \to \infty$.
(iv) $(M_D(t) - M_D(t-h)) \to h/\mu$ as $t \to \infty$, provided G and F are continuous distribution functions.

Stationary renewal process: A delayed renewal process in which the distribution function G of the first renewal time random variable is given by

$$G(x) = \mu^{-1}\int_0^x (1 - F(y))dy$$

is said to be a stationary renewal process. Such a renewal process arises when we model a renewal process that has begun a long time ago. In this case, the residual life of the item in service has the limiting distribution of the excess life in an ordinary renewal process, which is given by $\mu^{-1}\int_0^x (1 - F(y))dy$, as noted above. Thus, G denotes this limiting distribution.

A stationary renewal process satisfies properties such as (i) $M_D(t) = t/\mu \ \forall\ t$ and (ii) $P[U_t^D \leq x] = G(x) \ \forall\ t$. For proofs, one may refer to Karlin and Taylor [3].

Renewal reward process: In Chap. 7, we have discussed a compound Poisson process. A renewal reward process, also known as a cumulative renewal process, is similar to the compound Poisson process. It is a generalization of an ordinary renewal process. Suppose that at each renewal epoch S_n, there is a random quantity R_n assigned with the nth renewal, which is termed as reward. For example, R_n may be the cost incurred for replacement of the failed item at S_n, R_n may be the number of units produced by a machine when in working condition at S_n or the revenue produced by a machine when in working condition at S_n and R_n may be size of the claim at S_n in an insurance company. Note that R_n can be positive, negative or zero. We assume that $\{R_n, n \geq 1\}$ is a sequence of independent and identically distributed random variables. However, we do not assume that R_n is necessarily independent of S_n, $n \geq 1$. It is the main difference between a compound Poisson process and a renewal reward process, as defined below.

Definition 10.5.2 Renewal Reward Process: Suppose $\{(T_n, R_n)', n \geq 1\}$ is a sequence of independent and identically distributed random vectors and $\{N(t), t \geq 0\}$ is a counting process with inter-renewal times $\{T_n, n \geq 0\}$. Suppose $X(t)$ for $t \geq 0$ is defined as

$$X(t) = \begin{cases} 0, & \text{if } N(t) = 0 \\ \sum_{i=1}^{N(t)} R_i, & \text{if } N(t) > 0\,. \end{cases}$$

Then $\{X(t), t \geq 0\}$ is said to be a renewal reward process.

Note that $X(t)$ gives the total accumulated reward or cost up to time t. Thus, for an insurance company with claims of sizes R_n, coming in at times S_n, $X(t)$ is the total amount of claims against the company at time t. Suppose T_n denotes the inter-arrival times of the customers in a queue and R_n denotes the size of the arrival batch. In this case, $X(t)$ represents the total number of customers in a queue.

In Theorem 10.3.1, it is proved that for an ordinary renewal process $\{X(t), t \geq 0\}$ with μ as a mean inter-renewal time, $\lim_{t\to\infty} X(t)/t = 1/\mu$ with probability 1. We prove similar result for a renewal reward process in the following theorem.

Theorem 10.5.1 *Suppose $\{X(t), t \geq 0\}$ is a renewal reward process with $\{(T_n, R_n)', n \geq 1\}$ being the corresponding sequence of independent and identically distributed random vectors. If $E(T_1) = \mu$ and $E(R_1)$ are finite, then*

$$\lim_{t\to\infty} X(t)/t = E(R_1)/\mu, \quad \textit{with probability } 1 \ \& \ \lim_{t\to\infty} E(X(t))/t = E(R_1)/\mu.$$

Proof (i) By definition when $N(t) > 0$,

$$\begin{aligned} X(t)/t &= \sum_{i=1}^{N(t)} R_i/t = \Big(\sum_{i=1}^{N(t)} R_i/N(t)\Big) \times (N(t)/t) \\ &\to E(R_1)/\mu, \text{ with probability } 1 \text{ as } t \to \infty, \end{aligned}$$

where the convergence of the first factor is due to the strong law of large numbers and the convergence of the second factor is due to Theorem 10.3.1.
(ii) It can be shown that $E(X(t)) = E(R_1)M(t)$, where $M(t) = E(N(t))$ (Karlin and Taylor [3]). Hence,

$$E(X(t))/t = E(R_1) \times M(t)/t \to E(R_1)/\mu,$$

by the elementary renewal theorem.

□

Theorem 10.5.1 conveys that $E(R_1)/\mu$ can be interpreted as the long run mean reward per unit time. We illustrate the theorem in the following example. For more applications, one may refer to Ross [6] and Kulkarni [4].

Example 10.5.1 We consider the ordinary renewal process $\{X_1(t), t \geq 0\}$ discussed in Example 10.3.2, where $X_1(t)$ is the number of batteries replaced up to time t, planned or unplanned. Suppose Y_n is the inter-replacement time between $(n-1)$th and nth replacement. Then Y_n is distributed as $\min\{T_n, 3\}$, where $T_n \sim U(1, 4)$ distribution. It is shown that $\{Y_n, n \geq 1\}$ is a sequence of independent and identically distributed random variables with $E(Y_1) = 7/3$. Suppose the cost of replacing the battery is Rs 5000/- if it is a planned replacement and Rs 7000/- if it is an unplanned replacement. Thus, $R_n = 7000$ if $T_n < 3$ and $R_n = 5000$ if $T_n \geq 3$. Note that R_n depends on T_n and hence on Y_n. Hence, $X(t) = \sum_{i=1}^{X_1(t)} R_i$ when $X_1(t) > 0$ is a renewal reward process. Now,

$$E(R_n) = 7000P[T_n < 3] + 5000P[T_n \geq 3] = 7000 \times 2/3 + 5000 \times 1/3 = 19000/3.$$

Hence, the long run mean cost per unit time is $(19000/3)(3/7) = 2714.29$. □

Alternating renewal process: In the renewal processes discussed so far, we consider the occurrence of only one type of event. For example, when we consider a sequence $\{T_n, n \geq 1\}$ of lifetimes of identical components and $X(t)$ as the number of replacements in $(0, t]$, failure of the component is the only event we observe. In this setup, we assume that the detection of a failure and the replacement of the failed component are instantaneous. In practice, such an assumption may not be valid. Hence, suppose time taken for the detection of a failure and the replacement time of the failed component are random variables. Thus, the system has two states: working state, denoted by 1, and the repair or replacement state, denoted by 0. Suppose $\{U_n, n \geq 1\}$ is a sequence of random duration in the working state. We assume that $\{U_n, n \geq 1\}$ is a sequence of independent and identically distributed random variables with common distribution function F_1. Similarly, suppose $\{V_n, n \geq 1\}$ is a sequence of random duration in the state 0. We assume that $\{V_n, n \geq 1\}$ is a sequence of independent and identically distributed random variables with common distribution function F_0. Thus, the system alternates between the two states 0 and 1. An alternating renewal process defined below is useful to model such systems.

Definition 10.5.3 Alternating Renewal Process on $\{1, 2, \ldots, k\}$: Suppose that $\{X(t), t \geq 0\}$ is a continuous time process with state space $\{1, 2, \ldots, k\}$ and that $\{(Y_{n1}, Y_{n2}, \ldots, Y_{nk})', n \geq 1\}$ is a sequence of independent and identically distributed random vectors such that for $r = 1, 2, \ldots, k$, $\{Y_{nr}, n \geq 1\}$ is the sequence of successive sojourn times of the process $\{X(t), t \geq 0\}$ in state r having distribution F_r. Then, the sequence $\{(Y_{n1}, Y_{n2}, \ldots, Y_{nk})', n \geq 1\}$ is said be an alternating renewal process on $\{1, 2, \ldots, k\}$, if $X(t)$ successively visits states $1, 2, \ldots, k$ and then returns to state 1.

Remark 10.5.1 (i) For $n \geq 1$, suppose $T_n = \sum_{r=1}^{k} Y_{nr}$ and $S_n = \sum_{j=1}^{n} T_j$. If $X(0) = 1$, then S_n is the epoch of nth return to state 1 and the point process $\{S_n, n \geq 1\}$ is a renewal process with inter-renewal times $\{T_n, n \geq 1\}$. If $X(0) \neq 1$, then $\{S_n, n \geq 1\}$ is a delayed renewal process.
(ii) If F_r is exponential with parameter λ_r, for $r \geq 1$, then $\{X(t), t \geq 0\}$ is a continuous time Markov chain, where the embedded Markov chain has period k.

The following example illustrates the alternating renewal process when $k = 2$ and justifies the name alternating renewal process.

Example 10.5.2 Suppose at any time $t \geq 0$ a machine can be in one of the states—"up" state (denoted by 1) or "down" state (denoted by 0). Suppose the successive "up" times U_n are independent and identically distributed with distribution F_1, and the successive "down" times V_n are independent and identically distributed with distribution F_0. It is obvious that the state of the machine alternates between 0 and 1. If $X(0) = 1$, then for $n \geq 1$, $S_n = \sum_{j=1}^{n} T_j = \sum_{j=1}^{n}(U_j + V_j)$ is the epoch of nth renewal to "up" state of the machine and $\{X(t), t \geq 0\}$ is a renewal process with inter-renewal times T_n. If $X(0) = 0$, then $\{X(t), t \geq 0\}$ is a delayed renewal process in which $T_1 = V_1^*$ and $T_j = U_j + V_{j+1},\ j \geq 2$, where V_1^* is the remaining down time at 0. □

In this example, the questions of interest are (i) what is the probability that the machine is in "up" state at some specific time t_0, that is, what is $P[X(t_0) = 1]$? and (ii) what proportion of time in $(0, t]$ is the machine in "up" state? Both questions can be answered using renewal theory.

We state below some results about the alternating renewal process with two states 0 and 1. For proofs one may refer to Medhi [5].

(i) Suppose the system modeled by an alternating renewal process starts with state 1 at $t = 0$, and $p_i(t),\ i = 1, 2$ denote the probability that the system will be in state i at time t. If $E(U_i) < \infty$ and $E(V_i) < \infty$, then

$$\lim_{t\to\infty} p_1(t) = \frac{E(U_i)}{E(U_i)+E(V_i)} \quad \& \quad \lim_{t\to\infty} p_0(t) = \frac{E(V_i)}{E(U_i)+E(V_i)}.$$

(ii) Suppose $M_i(t), i = 0, 1$ denote the renewal functions. We assume that F_0 and F_1 are absolutely continuous distribution functions with probability density functions f_0 and f_1, respectively. Suppose $f_i^*(s)$ is the Laplace transform of f_i and $M_i^*(s)$ is the Laplace transforms of $M_i'(t), i = 0, 1$. Then

$$M_1^*(s) = \frac{f_1^*(s)}{s(1 - f_1^*(s)f_0^*(s))} \quad \& \quad M_0^*(s) = \frac{f_1^*(s)f_0^*(s)}{s(1 - f_1^*(s)f_0^*(s))}.$$

The next section presents R codes used to solve the examples.

10.6 R Codes

The following code is for the realization of a renewal process, corresponding to the given inter-renewal distribution. It is illustrated for the renewal process in Example 10.1.1.

Code 10.6.1 Realization of a renewal process: Suppose $\{X(t), t \geq 0\}$ is a renewal process where inter-renewal distribution is Gamma $G(\alpha, \lambda)$ with the mean inter-renewal time $\mu = \lambda/\alpha$. To find a realization of the process for a fixed $T = 10$ time units, we draw a random sample of size 1 each time from Gamma $G(\alpha, \lambda)$, till sum of these observations is > 10. For comparison, we take two sets of parameters as $\alpha = 1.44, \lambda = 3.6$ with mean $\mu = 2.5$ and $\alpha = 3, \lambda = 4.8$ with mean $\mu = 1.6$. In R, the probability density function of a Gamma distribution is expressed as $f(x) = (1/\alpha^\lambda\Gamma(\lambda))e^{-x/\alpha}x^{\lambda-1},\ x > 0$. Hence, in the following code, in the function $rgamma(1, shape = lambda[j], scale = 1/alpha[j])$, scale parameter is specified as $scale = 1/alpha[j]$):

```
# Part I: Input the parameters of inter-renewal distribution and T
alpha=c(1.44,3); lambda=c(3.6,4.8); mean=lambda/alpha; mean; T=10
# Part II: Realization
int=x=arr=u=v=w=list(); N=c()
for(j in 1:length(lambda))
{
set.seed(j); y=c(); sumy=0; i=1
while(sumy<T)
{
y[i]=rgamma(1,shape=lambda[j],scale=1/alpha[j])
sumy=sum(y)
i=i+1
}
int[[j]]=y[-length(y)]
```

```
N[j]=length(int[[j]])
x[[j]]=0:N[j]
u[[j]]=rep(x[[j]],each=2)
arr[[j]]=round(cumsum(int[[j]]),2)
v[[j]]=c(0,round(rep(arr[[j]],each=2),3),T)
w[[j]]=c(0,arr[[j]])
}
N
list("States"=x,"Interval r.v."=int,"Arrival epochs"=arr)
# Part III: Graph of realization
par(mfrow=c(2,1))
rate=paste("Mean =",mean,sep=" ")
number=paste("No of Renewals = ",N,sep=" ")
for(j in 1:length(lambda))
{
plot(v[[j]],u[[j]],"l",main=rate[j],ylab="States",
xlab="Renewal Epochs",yaxt="n",xaxt="n",col="blue")
axis(2,at=sort(unique(u[[j]])),labels=sort(unique(u[[j]])),
las=2)
axis(1,at=round(c(arr[[j]],T),2),las=2,cex.axis=0.7)
points(w[[j]],x[[j]],pch=20,col="dark blue")
abline(v=T)
legend("topleft",legend=number[j],cex=.7)
}
```

□

The following code computes the probability distribution of $X(t)$ where the common distribution of T_n is a Poisson with mean μ. It is illustrated for Example 10.2.2.

Code 10.6.2 Computation of the probability distribution of $X(t)$: Suppose $\{X(t), t \geq 0\}$ is a renewal process with inter-renewal distribution to be $Poi(\mu)$. Then $S_n \sim Poi(n\mu)$. Hence, for $n \geq 1$

$$P[X(t) = n] = P[X(t) \geq n] - P[X(t) \geq n+1] = P[S_n \leq t] - P[S_{n+1} \leq t].$$

Further, $P[X(t) = 0] = P[T_1 > t] = 1 - P[T_1 \leq t]$ and $T_1 = S_1 \sim P(\mu)$. We take $\mu = 2$ and four values of t as 3, 6, 9, 12 to find the probability distribution of $X(t)$:

```
# Part I: Input value of mu and values of t
mu=2; t=c(3,6,9,12)
# Part II: Probability distribution of X(t)
n=1:12; m=c(0,n); p0=c()
P=matrix(0,nrow=length(n),ncol=length(t))
for(j in 1:length(t))
{
```

```
p0[j]=1-ppois(t[j],mu)
for(i in 1:length(n))
{
P[i,j]=ppois(t[j],n[i]*mu)-ppois(t[j],(n[i]+1)*mu)
}
}
P1=round(rbind(p0,P),4); s=colSums(P1);s; P2=cbind(m,P1); P2
```

□

The following code verifies results of limit theorems discussed in Sects. 10.3 and 10.4. It is illustrated in Example 10.4.4.

Code 10.6.3 Verification of limit theorems: Suppose $\{X(t), t \geq 0\}$ is a renewal process where inter-renewal distribution is Gamma(α, λ). We simulate the renewal process $m = 250$ times for a time period $(0, T]$ for $T = 50$ time units to $T = 500$ time units with an increment of $h = 50$ time units. The following code computes R_t, $\hat{M}(t) = \sum_{i=1}^{m} X_i(t)/m$, m values of $(X(t) - t/\mu)/\sqrt{t\sigma^2/\mu^3}$ corresponding to m simulations and the p-values of the Shapiro-Wilk test:

```
# Part I: Input the parameters of inter-renewal distribution
# and values of T
alpha=1.2; lambda=4; mu=lambda/alpha; mu; var=lambda/alpha^2; var
nsim=250; Tinit=50; Tincr=50; Tmax=500
N=seq(Tinit,Tmax,Tincr);N
# Part II: Realizations
x=w=z=matrix(nrow=nsim,ncol=length(N))
for(j in 1:length(N))
{
T=N[j]
for(m in 1:nsim)
{
set.seed(m); y=c(); sumy=0; i=1
while(sumy<T)
{
y[i]=rgamma(1,shape=lambda,scale=1/alpha)
sumy=sum(y)
i=i+1
}
l=length(y); l; y=y[-l]; x[m,j]=length(y)
z[m,j]=length(y)/T
}
T=T+Tincr
}
head(x)
# Part III: Verification of limit theorems
for(j in 1:length(N))
```

```
for(m in 1:nsim)
{
{
w[m,j]=(x[m,j]-N[j]/mu)*(N[j]*var/mu^3)^(-.5)
}
}
eps=0.04; v=p=c()
for(j in 1:length(N))
{
v[j]=length(which(abs(z[,j]-1/mu)<eps))/nsim
p[j]=shapiro.test(w[,j])$p.value
}
v=round(v,4);p=round(p,4); mest=apply(x,2,mean)
m1=round(mest/N,4); d=data.frame(N,v,m1,p);d
m2=mest[-1]; m2; m3=mest[-length(N)]; m3; krt=m2-m3;krt
```

□

A quick recap of the results discussed in the present chapter is given below.

Summary

1. A point process $\{S_n, n \geq 0\}$ with $S_0 = 0$ or the corresponding counting process $\{X(t), t \geq 0\}$ is a renewal process if and only if $\{T_n = S_n - S_{n-1}, n \geq 1\}$ is a sequence of independent and identically distributed non-negative random variables.
2. The renewal function $M(t) = E(X(t))$ is given by $M(t) = \sum_{k=1}^{\infty} F_k^*(t)$ where $F_k^*(t)$ is the k-fold convolution of F with itself and F is the common distribution function of the inter-renewal random variables.
3. $M(t)$ satisfies the renewal equation $M(t) = F(t) + \int_0^t M(t-x)dF(x)$.
4. $M(t)$ uniquely determines the distribution F of the inter-renewal random variable and hence the renewal process.
5. Poisson process is the only renewal process whose mean value function is linear.
6. For a renewal process $\{X(t), t \geq 0\}$, suppose μ is a mean inter-renewal time. Then (i) $\lim_{t\to\infty} X(t)/t = 1/\mu$ with probability 1. (ii) $\lim_{t\to\infty} M(t)/t = 1/\mu$ (Elementary renewal theorem). (iii) $\lim_{t\to\infty}(M(t+h) - M(t))$ is h/μ, for fixed $h > 0$, provided F is a continuous distribution function (Key renewal theorem).
7. Suppose $\{X(t), t \geq 0\}$ is a renewal process with μ and σ^2 as the mean and variance respectively of the inter-renewal distribution. Then,

$$\lim_{t\to\infty} P\left[\frac{X(t) - t/\mu}{\sqrt{t\sigma^2/\mu^3}} \leq x\right] = \frac{1}{\sqrt{2\pi}}\int_{-\infty}^{x} e^{-u^2/2}\, du\ .$$

8. Some variations of a renewal process are delayed renewal process, stationary renewal process, renewal reward process and alternating renewal process.

10.7 Conceptual Exercises

10.7.1 Suppose $\{X_1(t), t \geq 0\}$ and $\{X_2(t), t \geq 0\}$ are two independent renewal processes and $X(t) = X_1(t) + X_2(t)$. Examine whether $\{X(t), t \geq 0\}$ is a renewal process. Find the inter-renewal distribution.

10.7.2 Suppose $\{X(t), t \geq 0\}$ is a renewal process with renewal function $M(t) = 1.5t$. Identify the inter-renewal distribution. Find its Laplace transform. Can you label the renewal process?

10.7.3 Suppose a system has two types of components, and the life of each is exponentially distributed with failure rates 2 and 3, respectively. The system fails with probability $1/4$ if type 1 component fails, and it fails with probability $3/4$ if type 2 component fails. If the epochs of failures form a renewal process, find the long run failure rate.

10.7.4 Mr. Sunil changes jobs frequently. On average, he works with one company for 2 years. The average duration of time he does not have a job is 2 months. In the long run, what proportion of time is Sunil working?

10.7.5 (i) Mr. Anil replaces the battery in his hearing aid machine as soon as it gets discharged. Suppose $X(t)$ is the number of batteries replaced during the first t hours of the life of the machine, not counting the one that was installed at the purchase. Assume that the lifetime T_n of the nth battery for $n \geq 1$ are independent and identically distributed random variables each having average life of 5 h. Find the long run rate at which the batteries will be replaced. (ii) In order to avoid the inconvenience when the battery gets discharged, Mr. Anil adopts the policy to replace the battery once it is used for 4 h, even if it has not failed yet. Suppose $X_1(t)$ is the number of batteries replaced up to time t, planned or unplanned. Show that $\{X_1(t), t \geq 0\}$ can be modeled as a renewal process. Find the long run rate the battery will be replaced in such a strategy. Compare this long run rate with the rate of replacing the battery when it gets discharged.

10.7.6 Suppose $\{X(t), t \geq 0\}$ is a renewal process where the probability density function of inter-renewal distribution is given by $f(x) = (x/100)e^{-x/10}$ for $x > 0$. (i) Find the long run renewal rate. (ii) Examine whether the renewal function is given by $M(t) = t/20 - 1/4 + (1/4)e^{-t/5}$. (iii) Verify the elementary renewal theorem and the key renewal theorem. (iv) Find a suitable normalization, so that the distribution of normalized $X(t)$ can be approximated by the standard normal distribution for large t.

10.7.7 Suppose $X(t)$ denotes the number of vehicles passing through a certain intersection. Suppose $\{X(t), t \geq 0\}$ is modeled as a renewal process where the inter-renewal distribution is uniform $U(0, 2)$, time unit being minutes. Find the approximate probability that the number of vehicles passing through that intersection is (i) larger than 560 and (ii) smaller than 620 in the time period 9 a.m. to 7 p.m.

10.8 Computational Exercises

10.8.1 Suppose $\{X(t), t \geq 0\}$ is a renewal process where inter-renewal distribution is geometric with success probability p. Obtain a realization of the renewal process for two different values of p. Draw the plots of realization and comment on the results.

10.8.2 Suppose $\{X(t), t \geq 0\}$ is a renewal process where inter-renewal distribution is lognormal with location parameter $\theta = 2$ and scale parameter $\sigma^2 = 4$. Simulate the renewal process m times for a time period $(0, T]$. Verify elementary renewal theorem, key renewal theorem and central limit theorem. Also verify that $X(t)/t$ converges to $1/\mu$ in probability.

10.8.3 Repeat Exercise 10.8.2 if inter-renewal distribution is Weibull with probability density function

$$f(x, \theta) = \theta x^{\theta-1} e^{-x^\theta}, x > 0, \ \ \theta > 0.$$

Comment on the results. (Hint: The mean and the variance of the Weibull distribution are $\Gamma(1/\theta + 1)$ and $\Gamma(2/\theta + 1) - (\Gamma(1/\theta + 1))^2$, respectively.

10.8.4 Suppose $\{X(t), t \geq 0\}$ is a renewal process where the common distribution of T_n is geometric with success probability p. Find probability distribution of $X(t)$ for four different values of t.

10.8.5 Suppose $\{X(t), t \geq 0\}$ is a renewal process where the common distribution of T_n is χ^2_3. Find probability distribution of $X(t)$ for four different values of t.

10.9 Multiple Choice Questions

Note: In each of the questions, multiple options may be correct.

10.9.1 Following are two statements. (I) A Poisson process is a Markov process. (II) A renewal process is a Markov process.
Which of the following options is correct?

(a) Both (I) and (II) are false
(b) Both (I) and (II) are true
(c) (I) is true but (II) is false
(d) (I) is false but (II) is true.

10.9.2 Which of the following options is/are correct? A point process $\{S_n, n \geq 0\}$ with $S_0 = 0$ is a renewal process if and only if

(a) $\{S_n - S_{n-1}, n \geq 1\}$ is a sequence of independent and identically distributed non-negative random variables

(b) $\{S_n, n \geq 1\}$ is a sequence of independent and identically distributed non-negative random variables
(c) $\{S_n - S_{n-1}, n \geq 1\}$ is a sequence of independent random variables
(d) $\{S_n, n \geq 1\}$ is a sequence of independent random variables.

10.9.3 Suppose $\{X(t), t \geq 0\}$ is a renewal process with inter-renewal distribution function F. Suppose $F_n^*(\cdot)$ denotes a n-fold convolution of F with itself. Following are three statements. (I) $P[X(t) = n] = F_{n+1}^*(t) - F_n^*(t)$. (II) $P[X(t) \leq n] = F_n^*(t)$. (III)$P[X(t) \geq n] = F_n^*(t)$.
Which of the following options is/are correct?

(a) Only (I) is true
(b) Only (II) is true
(c) Only (III) is true
(d) Both (I) and (III) are true.

10.9.4 Suppose $\{X(t), t \geq 0\}$ is a renewal process with inter-renewal distribution function F. Suppose S_n denotes an epoch of nth renewal. Following are two statements. (I) $P[X(t) \leq n] = P[S_{n+1} > t]$.
(II)$P[X(t) \geq n] = P[S_n \leq t]$.
Which of the following options is correct?

(a) Both (I) and (II) are false
(b) Both (I) and (II) are true
(c) (I) is true but (II) is false
(d) (I) is false but (II) is true.

10.9.5 Following are two statements. (I) A Poisson process is a renewal process. (II) Poisson process is a unique renewal process with a linear mean value function.
Which of the following options is correct?

(a) Both (I) and (II) are false
(b) Both (I) and (II) are true
(c) (I) is true but (II) is false
(d) (I) is false but (II) is true.

10.9.6 Following are two statements. (I) A renewal function uniquely determines the renewal process. (II) The distribution of inter-renewal random variables uniquely determines the renewal process.
Which of the following options is correct?

(a) Both (I) and (II) are false
(b) Both (I) and (II) are true
(c) (I) is true but (II) is false
(d) (I) is false but (II) is true.

10.9.7 Suppose $\{X(t), t \geq 0\}$ is a renewal process and μ is the mean inter-renewal time. Following are three statements.

(I) $\lim_{t\to\infty} X(t)/t = 1/\mu$ a.s. (II) $\lim_{t\to\infty} X(t)/t = \mu$ a.s.
(III) $\lim_{t\to\infty} X(t) = \infty$ a.s.
Which of the following options is correct?

(a) Only (I) is true
(b) Both (II) and (III) are true
(c) Both (I) and (III) are true
(d) Only (III) is true.

10.9.8 Suppose $\{X(t), t \geq 0\}$ is a renewal process and μ is the mean inter-renewal time. Following are two statements. (I) $\lim_{t\to\infty} X(t)/t = \mu$ a.s.
(II) $\lim_{t\to\infty} X(t) = \infty$ a.s.
Which of the following options is correct?

(a) Both (I) and (II) are false
(b) Both (I) and (II) are true
(c) (I) is true but (II) is false
(d) (I) is false but (II) is true.

10.9.9 Suppose $\{X(t), t \geq 0\}$ is a renewal process with renewal function $M(t)$ and μ is the mean inter-renewal time. Following are three statements. (I) $\lim_{t\to\infty} M(t)/t = 1/\mu$. (II) $\lim_{t\to\infty} M(t)/t = \mu$.
(III) $\lim_{t\to\infty} M(t) = \infty$ a.s.
Which of the following is a correct option?

(a) Only (I) is true
(b) Both (II) and (III) are true
(c) Both (I) and (III) are true
(d) Only (III) is true.

10.9.10 Suppose $M(t)$ is the renewal function of the renewal process $\{X(t), t \geq 0\}$ and μ is the mean inter-renewal time. Which of the following options is/are correct? For $h > 0$, as $t \to \infty$

(a) $\lim M(t)/t = \mu$
(b) $\lim(M(t+h) - M(t))/t = 0$
(c) $\lim(M(t+h) - M(t)) = h/\mu$
(d) $\lim M(t+h)/t = h/\mu$.

10.9.11 Suppose $\{X(t), t \geq 0\}$ is a renewal process with μ and σ^2 as the mean and variance respectively of the inter-renewal distribution. Which of the following options is correct?
For large t, $X(t)$ is approximately normally distributed with

(a) mean μt and variance $t\sigma^2/\mu^3$
(b) mean t/μ and variance $t\sigma^2/\mu^2$
(c) mean t/μ and variance $t\sigma^2/\mu^3$
(d) mean μt and variance $t\sigma^2\mu^2$.

10.9.12 Suppose $\{X(t), t \geq 0\}$ is a renewal process and S_n denotes the epoch of nth renewal, $n \geq 1$. Which of the following options is/are correct?

(a) $X(t) \geq n \iff S_n \leq t$
(b) $X(t) > n \iff S_n \leq t$
(c) $X(t) < n \iff S_n \geq t$
(d) $X(t) < n \iff S_n > t$.

References

1. Bhat, B. R. (2000). *Stochastic models: Analysis and applications*. New Delhi: New Age International.
2. Boland, P. J. (2007). *Statistical and probabilistic methods in actuarial science*. London: Chapman and Hall.
3. Karlin, S., & Taylor, H. M. (1975). *A first course in stochastic processes*. New York: Academic Press.
4. Kulkarni, V. G. (2011). *Introduction to modeling and analysis of stochastic systems*. New York: Springer.
5. Medhi, J. (1994). *Stochastic processes*. New Delhi: Wiley Eastern.
6. Ross, S. M. (2014). *Introduction to probability models* (11th ed.). New York: Academic Press.

Appendix A
Solutions to Conceptual Exercises

A.1 Chapter 2

2.9.1 Suppose $\{X_n, n \geq 0\}$ is a sequence of independent and identically distributed random variables with $P[X_n = i] = a_i,\ a_i > 0$ and $\sum_{i \in S} a_i = 1$, where $S = \{1, 2, \ldots, \}$. (i) Examine whether $\{X_n, n \geq 0\}$ is a Markov chain. (ii) Classify the states.

Solution: (i) Since $\{X_n, n \geq 0\}$ is a sequence of independent random variables, the conditional distribution of X_n given $X_{n-1}, X_{n-2}, \ldots, X_0$ is the same as the marginal distribution of X_n, which is the same as the conditional distribution of X_n given X_{n-1}. Hence, $\{X_n, n \geq 0\}$ is a Markov chain. Further, it is time homogeneous in view of the fact that $\{X_n, n \geq 0\}$ is a sequence of identically distributed random variables. The transition probability matrix P has all identical rows given by $(a_1, a_2, \ldots,)$.

(ii) Since all the elements of P matrix are positive, all states communicate with each other, so the nature of all the states is the same. Since diagonal elements are positive, all states are aperiodic. Further,

$$P^n = P \ \forall \ n \geq 1 \Rightarrow \lim_{n \to \infty} P^n = P \ \Rightarrow \ \sum_{n=1}^{\infty} p_{ii}^{(n)} = \sum_{n=1}^{\infty} a_i = \infty.$$

Hence, state i is a persistent state. Further, $\lim_{n \to \infty} p_{ii}^{(n)} = a_i > 0$ which implies that i is a non-null persistent state. Thus, all the states are non-null persistent states, which further implies that all the states are ergodic.

2.9.2 Suppose $\{X_n, n \geq 0\}$ is a sequence of independent and identically distributed random variables with $S = I$ as a set of possible values. Examine whether a sequence $\{Y_n, n \geq 0\}$ defined as $Y_n = X_0 - X_1 + X_2 - \cdots + (-1)^n X_n$ is a Markov chain.

S. Madhira and S. Deshmukh, *Introduction to Stochastic Processes Using R*,
https://doi.org/10.1007/978-981-99-5601-2

Solution: Note that $Y_n = X_0 - X_1 + X_2 - \cdots + (-1)^{n-1}X_{n-1} + (-1)^n X_n = Y_{n-1} + (-1)^n X_n$, which implies that $\{Y_n, n \geq 0\}$ is a Markov chain.

2.9.3 Suppose $\{Y_n, n \geq 0\}$ is a sequence of independent and identically distributed random variables with state space $\{0, 1, 2, 3\}$ and with respective probabilities $\{0.1, 0.3, 0.2, 0.4\}$. (i) Suppose $X_n = \min\{Y_0, Y_1, \ldots, Y_n\}$. Examine whether $\{X_n, n \geq 0\}$ is a Markov chain. If yes, determine its state space and the transition probability matrix. Determine the nature of the states in all respects. (ii) It has been shown in Section 2.1 that if $X_0 = 0$ and $X_n = \max\{Y_1, Y_2, \ldots, Y_n\}$, then $\{X_n, n \geq 0\}$ is a Markov chain. Determine the nature of the states in all respects.

Solution: Observe that by the definition of X_n,

$$\begin{aligned} X_n &= \min\{Y_0, Y_1, \ldots, Y_n\} = \min\{\min\{Y_0, Y_1, \ldots, Y_{n-1}\}, Y_n\} \\ &= \min\{X_{n-1}, Y_n\}. \end{aligned}$$

Hence, $\{X_n, n \geq 0\}$ is a Markov chain. More precisely, note that for any $i, j, x_{n-1}, \ldots, x_0$ and for any $n \geq 1$,

$$\begin{aligned} & P[X_n = j | X_{n-1} = i, \ldots, X_0 = x_0] \\ &= P[\min\{X_{n-1}, Y_n\} = j | X_{n-1} = i, \ldots, X_0 = x_0] \\ &= P[\min\{X_{n-1}, Y_n\} = j | X_{n-1} = i] \\ &= P[\min\{i, Y_n\} = j] = P[X_n = j | X_{n-1} = i]. \end{aligned}$$

Thus, $\{X_n, n \geq 0\}$ is a Markov chain with state space $\{0, 1, 2, 3\}$. Since $\{Y_n, n \geq 0\}$ is a sequence of identically distributed random variables, the Markov chain is time homogeneous. To determine the transition probabilities, note that if $X_{n-1} = 0$, then X_n is 0, whatever may be the value of Y_n. Thus, if $Y_n = 0, 1, 2, 3$ then $X_n = 0$. If $X_{n-1} = 1$, then X_n cannot be larger than 1. X_n is 0, if $Y_n = 0$. X_n is 1, if $Y_n = 1, 2, 3$. If $X_{n-1} = 2$, then X_n cannot be 3; X_n is 0, if $Y_n = 0$; X_n is 1, if $Y_n = 1$; X_n is 2 if $Y_n = 2, 3$. If $X_{n-1} = 3$, then X_n is 0, 1, 2, if $Y_n = 0, 1, 2$, respectively. $X_n = 3$ if $Y_n = 3$. Hence, the one step transition probability matrix P is given by

$$P = \begin{array}{c} \\ 0 \\ 1 \\ 2 \\ 3 \end{array} \begin{array}{c} \begin{array}{cccc} 0 & 1 & 2 & 3 \end{array} \\ \begin{pmatrix} 1 & 0 & 0 & 0 \\ 0.1 & 0.9 & 0 & 0 \\ 0.1 & 0.3 & 0.6 & 0 \\ 0.1 & 0.3 & 0.2 & 0.4 \end{pmatrix} \end{array}.$$

Observe that 0 is an absorbing state and hence non-null persistent. States $1, 2, 3 \to 0$ but $0 \nrightarrow 1, 2, 3$. Therefore, states $1, 2, 3$ are inessential and hence transient. Since all the diagonal elements are positive, all the states are aperiodic.

(ii) It is shown in Section 2.1 that $\{X_n, n \geq 0\}$ is a Markov chain with state space $\{0, 1, 2, 3\}$ and the one step transition probability matrix P given by

$$P = \begin{array}{c} \\ 0 \\ 1 \\ 2 \\ 3 \end{array} \begin{array}{c} \begin{array}{cccc} 0 & 1 & 2 & 3 \end{array} \\ \begin{pmatrix} 0.1 & 0.3 & 0.2 & 0.4 \\ 0 & 0.4 & 0.2 & 0.4 \\ 0 & 0 & 0.6 & 0.4 \\ 0 & 0 & 0 & 1 \end{pmatrix} \end{array}.$$

From the transition probability matrix P, we note that state 3 is an absorbing state and hence non-null persistent and aperiodic. Observe that for $i = 0, 1, 2$, state $i \to 3$ but $3 \nrightarrow i$. Hence 0, 1, 2 are inessential and hence transient states. All the diagonal elements are positive, which implies that all the states are aperiodic.

2.9.4 Prove or disprove: Product of two stochastic matrices is a stochastic matrix.
Solution: Suppose $P = [p_{ij}]$ and $Q = [q_{ij}]$ are two stochastic matrices. Hence,

$$p_{ij} \geq 0, \ \ q_{ij} \geq 0, \ \ \sum_{j \in S} p_{ij} = 1 \ \forall \ i \in S \ \& \ \sum_{j \in S} q_{ij} = 1 \ \forall \ i \in S.$$

Suppose $A = PQ = [a_{ij}]$. Then $a_{ij} = \sum_{r \in S} p_{ir} q_{rj}$. Hence, $a_{ij} \geq 0$ for all i and $j \in S$. Further, for any $i \in S$,

$$\sum_{j \in S} a_{ij} = \sum_{j \in S} \sum_{r \in S} p_{ir} q_{rj} = \sum_{r \in S} p_{ir} \sum_{j \in S} q_{rj} = \sum_{r \in S} p_{ir} = 1.$$

The sums in the second step can be interchanged as both are convergent series of non-negative terms. Thus, the product of two stochastic matrices is a stochastic matrix.

2.9.5 Suppose a student takes admission to a course to be completed in four semesters. At the end of the ith semester, depending on the performance during the semester, a student either proceeds to the next semester with probability p_i or quits the course with probability q_i and remains in the same semester with probability $1 - p_i - q_i$, $i = 1, 2, 3, 4$. Assuming that the movement among the semesters can be modeled by a Markov chain, find the one step transition probability matrix.
Solution: We define state 0 as leaving the course and state 5 as completing the course. Then transitions among these states are modeled by a Markov chain with transition probability matrix P as given below

$$P = \begin{array}{c} \\ 0 \\ 1 \\ 2 \\ 3 \\ 4 \\ 5 \end{array} \begin{array}{c} \begin{array}{cccccc} 0 & 1 & 2 & 3 & 4 & 5 \end{array} \\ \begin{pmatrix} 1 & 0 & 0 & 0 & 0 & 0 \\ q_1 & 1-p_1-q_1 & p_1 & 0 & 0 & 0 \\ q_2 & 0 & 1-p_2-q_2 & p_2 & 0 & 0 \\ q_3 & 0 & 0 & 1-p_3-q_3 & p_3 & 0 \\ q_4 & 0 & 0 & 0 & 1-p_4-q_4 & p_4 \\ 0 & 0 & 0 & 0 & 0 & 1 \end{pmatrix} \end{array}.$$

2.9.6 Suppose the transitions among states in a care center are governed by a homogeneous Markov chain with state space $S = \{1, 2, 3\}$, where 1 denotes healthy state, 2 denotes critically ill state and 3 stands for death. Suppose the transition probability matrix P is as given below, where time unit is taken as a day:

$$P = \begin{array}{c} \\ 1 \\ 2 \\ 3 \end{array} \begin{array}{c} \begin{array}{ccc} 1 & 2 & 3 \end{array} \\ \begin{pmatrix} 0.92 & 0.05 & 0.03 \\ 0.00 & 0.76 & 0.24 \\ 0 & 0 & 1 \end{pmatrix} \end{array}.$$

Compute the probability that an individual healthy on day 1 (i) remains healthy for the next 6 days, (ii) is critically ill for the first time on day 5, (iii) is critically ill on day 5 and (iv) is critically ill for the first time on day 6 and dies on day 9.

Solution: Suppose X_n denotes the state of the individual at 6 : 00 a.m. on day n. It is given that the individual is in the healthy state on day 1, so that $P[X_1 = 1] = 1$.

(i) The probability that the individual who is healthy on day 1 remains healthy for the next 6 days is obtained using Markov property as follows:

$$\begin{aligned} P[X_r = 1, r = 2, 3, 4, 5, 6, 7 | X_1 = 1] &= \prod_{r=1}^{6} P[X_{r+1} = 1 | X_r = 1] \\ &= (0.92)^6 = 0.6064. \end{aligned}$$

(ii) The probability that the individual who is healthy on day 1 is critically ill for the first time on day 5 is obtained as follows:

$$\begin{aligned} P[X_r = 1, r = 2, 3, 4, X_5 = 2 | X_1 = 1] &= \prod_{r=1}^{3} P[X_{r+1} = 1 | X_r = 1] \\ &\times P[X_5 = 2 | X_4 = 1] \\ &= (0.92)^3 \times 0.05 = 0.0389. \end{aligned}$$

(iii) The probability that a healthy individual admitted to the center on day 1 is critically ill on day 5 is

$$P[X_5 = 2|X_1 = 1] = p_{12}^{(4)} = 0.1196.$$

(iv) The probability that an individual who is healthy on day 1 is critically ill for the first time on day 6 and dies on day 9 is given by

$$P[X_r \neq 2, r = 2, 3, 4, 5, X_6 = 2, X_9 = 3|X_1 = 1]$$

$$= P[X_r = 1, r = 2, 3, 4, 5, X_l = 2, l = 6, 7, 8, X_9 = 3|X_1 = 1]$$

$$= \prod_{r=1}^{4} P[X_{r+1} = 1|X_r = 1]P[X_6 = 2|X_5 = 1]$$

$$\times \prod_{l=6}^{7} P[X_{l+1} = 2|X_l = 2] \times P[X_9 = 3|X_8 = 2]$$

$$= (0.92)^4 \times 0.05 \times (0.76)^2 \times 0.24 = 0.004965.$$

2.9.7 Weather in a city is classified as sunny, cloudy and rainy, and the weather condition is modeled as a Markov chain $\{X_n, n \geq 0\}$, where X_n is defined as follows:

$$X_n = \begin{cases} 1, & \text{if nth day is sunny} \\ 2, & \text{if nth day is cloudy} \\ 3, & \text{if nth day is rainy.} \end{cases}$$

Further, the one step transition probability matrix P is given by

$$P = \begin{array}{c} \\ 1 \\ 2 \\ 3 \end{array} \begin{array}{c} \begin{array}{ccc} 1 & 2 & 3 \end{array} \\ \begin{pmatrix} 0.4 & 0.4 & 0.2 \\ 0.6 & 0.2 & 0.2 \\ 0.5 & 0.4 & 0.1 \end{pmatrix} \end{array}.$$

(i) Find the probability that weather is cloudy for second, third and fourth days, given that the initial day is sunny. (ii) Find the probability that day 2 is sunny, day 3 is cloudy and day 4 is rainy given that day 1 is sunny.

Solution: With the given transition probability matrix, using Markov property repeatedly, we obtain the required probabilities as follows:

$$(i) P[X_4 = 2, X_3 = 2, X_2 = 2|X_1 = 1] = p_{22}^2 p_{12} = (0.2)^2 \times (0.4) = 0.016.$$

$$(ii) P[X_4 = 3, X_3 = 2, X_2 = 1|X_1 = 1] = p_{23}p_{12}p_{11} = 0.032.$$

2.9.8 Operating condition of a machine at any time is classified as follows: State 1: Good; State 2: Deteriorated but operating; State 3: In repair. We observe the condition of the machine at 6 : 00 pm every day. Suppose X_n denotes the state of the machine on the nth day for $n = 1, 2, \ldots$. We assume that the sequence of machine conditions is a Markov chain with transition probability matrix P as given below

$$P = \begin{array}{c} \\ 1 \\ 2 \\ 3 \end{array} \begin{array}{c} \begin{array}{ccc} 1 & 2 & 3 \end{array} \\ \begin{pmatrix} 0.9 & 0.1 & 0 \\ 0 & 0.9 & 0.1 \\ 1 & 0 & 0 \end{pmatrix} \end{array}.$$

Find the probability that the machine is in good condition on day 5 given that it is in good condition on day 1.

Solution: To find $P[X_5 = 1|X_1 = 1] = P_{11}^{(4)}$, we find P^4. Thus, $P_{11}^{(4)} = 0.6831$ is the required probability. Alternatively, there are four paths to reach to 1 from 1 in four steps:

$$1 \to 1 \to 1 \to 1 \to 1, \quad 1 \to 1 \to 2 \to 3 \to 1, \quad 1 \to 2 \to 2 \to 3 \to 1,$$

$$1 \to 2 \to 3 \to 1 \to 1.$$

The probabilities of these four paths are $0.9^4 = 0.6531, 0.009, 0.009, 0.009$, respectively. Hence, $P[X_4 = 1|X_0 = 1] = 0.6531 + 0.027 = 0.6831$.

2.9.9 Suppose $\{X_n, n \geq 0\}$ is a Markov chain as defined in Exercise 2.9.7, with the initial distribution $\underline{p}^{(0)} = (1/3, 1/3, 1/3)'$. Suppose 0.44, 0.81, 0.34, 0.56, 0.18, 0.62 is a random sample of size 6 from uniform $U(0, 1)$ distribution. Using this random sample, in the given order, find X_n for $n = 0, 1, \ldots, 5$.

Solution: As discussed in Section 2.3, to obtain a realization from the given Markov chain, at each time point, we draw a sample of size 1 from the state space, using the probability distribution specified in a row corresponding to the previous state of the transition probability matrix. To obtain an initial state, we draw a random sample of size 1 from the initial distribution. In general suppose, the support of a discrete random variable X is $S = \{1, 2, \ldots, M\}$, with probabilities $p_i, i \in S$. Suppose y denotes a random observation from uniform distribution over (0, 1). Then the procedure to obtain a random sample from the distribution of X is as follows:

$$\text{If } \sum_{i=1}^{r-1} p_i < y \leq \sum_{i=1}^{r} p_i, \quad \text{then } X = r, \quad r = 1, 2, \ldots, M.$$

We define $\sum_{i=1}^{0} p_i = 0$. To obtain a random sample of size 1 from the initial distribution, we write cumulative sums of the probability vector

$\underline{p}^{(0)} = (1/3, 1/3, 1/3)'$, which is $(1/3, 2/3, 1)'$. Now the random observation $y = 0.44$ is between $1/3$ and $2/3$. Hence, $X_0 = 2$. The transition probability matrix is given by

$$P = \begin{array}{c} \\ 1 \\ 2 \\ 3 \end{array} \begin{array}{c} \begin{array}{ccc} 1 & 2 & 3 \end{array} \\ \begin{pmatrix} 0.4 & 0.4 & 0.2 \\ 0.6 & 0.2 & 0.2 \\ 0.5 & 0.4 & 0.1 \end{pmatrix} \end{array}.$$

To find X_1, we consider the row of P that corresponds to state 2 and cumulative sums of the probabilities in this row, which are $(0.6, 0.8, 1)'$. Now the second observation from the uniform distribution is 0.81, hence $X_1 = 3$. To find X_2, we observe the cumulative sums of the probabilities in the third row, which are $(0.5, 0.9, 1)'$. With $y = 0.34$, $X_2 = 1$. Proceeding on these lines, we get $X_3 = 2$, $X_4 = 1$ and $X_5 = 2$. Thus, the realization of the Markov chain is $\{2, 3, 1, 2, 1, 2\}$.

2.9.10 Suppose $\{X_n, n \geq 0\}$ is a time homogeneous Markov chain. Find the probability that the initial state is i given that $X_n = j$.
Solution: We find $P[X_0 = i|X_n = j]$ as follows:

$$\begin{aligned} P[X_0 = i|X_n = j] &= \frac{P[X_0 = i, X_n = j]}{P[X_n = j]} = \frac{P[X_n = j|X_0 = i]P[X_0 = i]}{P[X_n = j]} \\ &= \frac{p_{ij}^{(n)} p_i^{(0)}}{\sum_{k \in S} P[X_n = j|X_0 = k]P[X_0 = k]} = \frac{p_{ij}^{(n)} p_i^{(0)}}{\sum_{k \in S} p_{kj}^{(n)} p_k^{(0)}}. \end{aligned}$$

2.9.11 Show that for any two states i and j, $\sum_{n \geq 1} p_{ij}^{(n)} \geq f_{ij}$. Show that for a persistent state i such that $i \leftrightarrow j$, $\sum_{n \geq 1} p_{ij}^{(n)} \geq 1$.
Solution: We have the recurrence relation

$$\begin{aligned} p_{ij}^{(n)} &= f_{ij}^{(n)} + \sum_{r=1}^{n-1} p_{jj}^{(n-r)} f_{ij}^{(r)} \\ \Rightarrow \quad p_{ij}^{(n)} &\geq f_{ij}^{(n)} \;\Rightarrow\; \sum_{n \geq 1} p_{ij}^{(n)} \geq \sum_{n \geq 1} f_{ij}^{(n)} = f_{ij}. \end{aligned}$$

If the state i is persistent and $i \leftrightarrow j$, then $f_{ij} = 1$. Hence, it follows that $\sum_{n \geq 1} p_{ij}^{(n)} \geq 1$.

2.9.12 If state j is transient, prove that for any state i, $\sum_{n=1}^{\infty} p_{ij}^{(n)} < 1/(1 - f_{jj})$.
Solution: By the ratio theorem, for any two states i and j in S,

$$\lim_{N \to \infty} \left[\sum_{n=1}^{N} p_{ij}^{(n)} \Big/ \left(1 + \sum_{n=1}^{N} p_{jj}^{(n)}\right) \right] = f_{ij}.$$

If state j is transient, we have proved that for any state $i \to j$, $f_{ij} < 1$. Hence,

$$\frac{\sum_{n=1}^{\infty} p_{ij}^{(n)}}{1+\sum_{n=1}^{\infty} p_{jj}^{(n)}} = f_{ij} < 1 \Rightarrow \sum_{n=1}^{\infty} p_{ij}^{(n)} < 1 + \sum_{n=1}^{\infty} p_{jj}^{(n)}$$

$$\Rightarrow \sum_{n=1}^{\infty} p_{ij}^{(n)} < 1 + f_{jj}/(1-f_{jj}) = 1/(1-f_{jj}).$$

2.9.13 Suppose $\{X_n, n \geq 0\}$ is a Markov chain with state space $S = \{1, 2, 3, 4\}$ and transition probability matrix P given by

$$P = \begin{array}{c} \\ 1 \\ 2 \\ 3 \\ 4 \end{array} \begin{array}{c} \begin{array}{cccc} 1 & 2 & 3 & 4 \end{array} \\ \begin{pmatrix} 1/3 & 2/3 & 0 & 0 \\ 1/4 & 3/4 & 0 & 0 \\ 1/3 & 1/3 & 1/6 & 1/6 \\ 1/4 & 1/4 & 1/4 & 1/4 \end{pmatrix} \end{array}.$$

(i) Examine which states communicate with each other. (ii) What are the communicating classes of states? (iii) Which are the closed classes?
(iv) Does a minimal closed class exist? If yes, specify. (v) Is the Markov chain reducible or irreducible? Justify. (vi) Decide the nature of the states.
(vii) Find f_{11} and μ_1.
Solution: (i) Observe that $1 \to 2$ and $2 \to 1$, thus $1 \leftrightarrow 2$. Further, $3 \to 1$ but $1 \nrightarrow 3$, similarly, $3 \to 2$ but $2 \nrightarrow 3$. Further, $3 \to 4$ and $4 \to 3$, thus, $3 \leftrightarrow 4$.
(ii) $\{1, 2\}$ and $\{3, 4\}$ are two communicating classes. Thus, the state space S is partitioned into two disjoint communicating classes $\{1, 2\}$ and $\{3, 4\}$.
(iii) $\{1, 2\}$ is a closed class.
(iv) $\{1, 2\}$ is a minimal closed class.
(v) Markov chain is reducible.
(vi) $3 \to 1$ but $1 \nrightarrow 3$, similarly, $4 \to 1$ but $1 \nrightarrow 4$. Thus, 3 and 4 are inessential states and hence transient states. In a finite state space Markov chain, all states cannot be transient, so either 1 or 2 is persistent, but both communicate with each other and hence must be of the same nature and hence are persistent. Again in a finite Markov chain null persistent states do not exist, and hence both are non-null persistent states. We can also use the result that in a closed communicating class, all states are non-null persistent.
(vii) Observe that

$$f_{11}^{(1)} = p_{11} = 1/3, \quad f_{11}^{(2)} = p_{12}p_{21} = (2/3)(1/4)$$
$$f_{11}^{(3)} = p_{12}p_{22}p_{21} = (2/3)(3/4)(1/4).$$

In general for $n \geq 2$,

$$\begin{aligned} f_{11}^{(n)} &= p_{12}p_{22}^{(n-2)}p_{21} = (2/3)(3/4)^{n-2}(1/4) \\ \Rightarrow \quad f_{11} &= 1/3 + \sum_{n\geq 2}(2/3)(3/4)^{n-2}(1/4) \\ &= 1/3 + (1/6)(1-3/4)^{-1} = 1/3 + 2/3 = 1 \\ \text{Hence,} \quad \mu_1 &= 1/3 + \sum_{n\geq 2} n(2/3)(3/4)^{n-2}(1/4) \\ &= 1/3 + (1/6)(4/3)\sum_{n\geq 2} n(3/4)^{n-1} = 11/3. \end{aligned}$$

2.9.14 Suppose $\{X_n, n \geq 0\}$ is a Markov chain with state space $S = \{1, 2, 3, 4\}$ and transition probability matrix P given by

$$P = \begin{array}{c} \\ 1 \\ 2 \\ 3 \\ 3 \end{array} \begin{array}{c} \begin{array}{cccc} 1 & 2 & 3 & 4 \end{array} \\ \begin{pmatrix} 1/3 & 2/3 & 0 & 0 \\ 1/4 & 3/4 & 0 & 0 \\ 1/3 & 1/3 & 1/3 & 0 \\ 1/4 & 1/4 & 1/4 & 1/4 \end{pmatrix} \end{array}.$$

(i) Examine which states communicate with each other. (ii) What are the communicating classes of states? (iii) Which are the closed classes?
(iv) Does a minimal closed class exist? If yes, specify. (v) Is the Markov chain reducible or irreducible? Justify. (vi) Decide the nature of the states.
Solution: (i) Observe that $1 \rightarrow 2$ and $2 \rightarrow 1$, thus $1 \leftrightarrow 2$. Further, $3 \rightarrow 1$ but $1 \nrightarrow 3$, similarly, $3 \rightarrow 2$ but $2 \nrightarrow 3$. Further, $3 \rightarrow 4$ but $4 \nrightarrow 3$, thus, 3 and 4 do not communicate with each other. Note that $4 \rightarrow 1$ but $1 \nrightarrow 4$, similarly, $4 \rightarrow 2$ but $2 \nrightarrow 4$.
(ii) $\{1, 2\}$, $\{3\}$ and $\{4\}$ are three communicating classes. Thus, the state space S is partitioned into three disjoint communicating classes $\{1, 2\}$, $\{3\}$ and $\{4\}$.
(iii) $\{1, 2\}$ is a closed class.
(iv) $\{1, 2\}$ is a minimal closed class.
(v) Markov chain is reducible.
(vi) $3 \rightarrow 1$ but $1 \nrightarrow 3$, $3 \rightarrow 4$ but $4 \nrightarrow 3$, similarly, $4 \rightarrow 1$ but $1 \nrightarrow 4$. Thus, 3 and 4 are inessential states and hence transient states. In a finite state space Markov chain, all states cannot be transient, so either 1 or 2 is persistent, but both communicate with each other and hence must be of the same nature and hence are persistent. Again in a finite Markov chain, null persistent states do not exist and hence both are non-null persistent states. Alternatively, $\{1, 2\}$ being a closed class, both states are non-null persistent.

2.9.15 For a Markov chain $\{X_n, n \geq 0\}$ with state space $S = \{1, 2, 3\}$, the transition probability matrix P is given by

$$P = \begin{array}{c} \\ 1 \\ 2 \\ 3 \end{array} \begin{array}{c} \begin{array}{ccc} 1 & 2 & 3 \end{array} \\ \begin{pmatrix} 0 & 1 & 0 \\ 1 & 0 & 0 \\ 1/3 & 1/3 & 1/3 \end{pmatrix} \end{array}.$$

(i) Find the period of each state. (ii) Classify the states. (iii) Find f_{33}. (iv) Find the probability of absorption from a transient state. Verify Theorem 2.6.10.

Solution: (i) From the transition probability matrix P, we note that

$$D_1 = \{n | p_{11}^{(n)} > 0\} = \{2, 4, 6, \ldots, \} \Rightarrow d_1 = 2.$$

1 and 2 communicate with each other, hence $d_2 = 2$. Further, $p_{33} > 0$, hence $d_3 = 1$. Thus, states 1 and 2 are periodic with period 2 and state 3 is aperiodic.
(ii) The class $\{1, 2\}$ is a closed class, hence both the states 1 and 2 are non-null persistent. Observe that $3 \to 1$, however $1 \nrightarrow 3$, hence the state 3 is inessential and hence transient.
(iii) To compute f_{33}, note that $f_{33}^{(1)} = p_{33} = 1/3$. From 3 transition to 3 in two or more steps, without intermediate visit to 3 is not possible, since $3 \to 1$ and $3 \to 2$, but $\{1, 2\}$ is a closed class. Hence, $f_{33} = 1/3$.
(iv) Note that $\{1, 2\}$ is a single closed communicating class and hence the probability of absorption from transient state 3 is 1. In the notation of Theorem 2.6.10, it is given by $(I_1 - Q)^{-1}\underline{d}$. Merging the closed class $\{1, 2\}$, $\underline{d} = 2/3$ and $Q = [1/3]$. Hence, $(I_1 - Q)^{-1}\underline{d} = 1$.

2.9.16 Suppose $\{X_n, n \geq 1\}$ is a Markov chain with state space $S = \{1, 2, 3, 4\}$ and transition probability matrix P given by

$$P = \begin{array}{c} \\ 1 \\ 2 \\ 3 \\ 4 \end{array} \begin{array}{c} \begin{array}{cccc} 1 & 2 & 3 & 4 \end{array} \\ \begin{pmatrix} 0 & 1 & 0 & 0 \\ 1 & 0 & 0 & 0 \\ 1/8 & 1/2 & 1/8 & 1/4 \\ 1/3 & 1/6 & 1/6 & 1/3 \end{pmatrix} \end{array}.$$

(i) Find the period of each state. (ii) Classify the states. (iii) Find f_{33}, f_{44}. (iv) Find the probability of absorption from transient states.

Solution: (i) From the transition probability matrix P, we note that

$$D_1 = \{n | p_{11}^{(n)} > 0\} = \{2, 4, 6, \ldots, \} \Rightarrow d_1 = 2.$$

1 and 2 communicate with each other, hence $d_2 = 2$. Further, $p_{33} > 0$, hence $d_3 = 1$. Similarly, $p_{44} > 0$, hence $d_4 = 1$. Thus, states 1 and 2 are periodic with period 2 and states 3 and 4 are aperiodic.
(ii) The class $\{1, 2\}$ is a closed class, hence both the states 1 and 2 are non-null persistent. Observe that $3 \to 1$, however $1 \nrightarrow 3$, hence the state 3 is

inessential and hence transient. Similarly, $4 \to 1$, however $1 \nrightarrow 4$, hence the state 4 is inessential and hence transient.
(iii) To compute f_{33}, note that $f_{33}^{(1)} = p_{33} = 1/8$. Further,

$$3 \to 4 \to 3 \Rightarrow f_{33}^{(2)} = (1/4)(1/6), 3 \to 4 \to 4 \to 3 \Rightarrow f_{33}^{(3)} = (1/4)(1/3)(1/6).$$

Continuing in this manner, $f_{33}^{(n)} = (1/4)(1/3)^{n-2}(1/6), n \geq 2$. Hence,

$$f_{33} = 1/8 + \sum_{n \geq 2} (1/4)(1/3)^{n-2}(1/6) = 3/16.$$

Similarly, $f_{44}^{(1)} = p_{44} = 1/3$ and $f_{44}^{(n)} = (1/6)(1/8)^{n-2}(1/4), n \geq 2$. Hence,

$$f_{44} = 1/3 + \sum_{n \geq 2} (1/6)(1/8)^{n-2}(1/4) = 8/21.$$

(iv) Note that $\{1, 2\}$ is a single closed communicating class and hence the probability of absorption from both the transient states 3 and 4 is 1.

2.9.17 Suppose $\{X_n, n \geq 0\}$ is a Markov chain with state space $S = \{1, 2, 3, 4, 5, 6\}$ and transition probability matrix P given by

$$P = \begin{array}{c} \\ 1 \\ 2 \\ 3 \\ 4 \\ 5 \\ 6 \end{array} \begin{array}{c} \begin{array}{cccccc} 1 & 2 & 3 & 4 & 5 & 6 \end{array} \\ \begin{pmatrix} 1/4 & 0 & 3/4 & 0 & 0 & 0 \\ 0 & 1/3 & 1/3 & 0 & 1/3 & 0 \\ 2/7 & 0 & 5/7 & 0 & 0 & 0 \\ 0 & 1/4 & 1/4 & 1/4 & 0 & 1/4 \\ 0 & 0 & 0 & 0 & 4/9 & 5/9 \\ 0 & 0 & 0 & 0 & 1/3 & 2/3 \end{pmatrix} \end{array}.$$

(i) Identify the communicating classes and closed classes. (ii) Is the Markov chain reducible or irreducible? Justify your answer. (iii) Classify the states as transient, null persistent and non-null persistent.
(iv) Find the period of each state.
Solution: (i) From the given transition probability matrix P, we note that $\{1, 3\}, \{2\}, \{4\}, \{5, 6\}$ are communicating classes, out of these $\{1, 3\}$, $\{5, 6\}$ are closed classes.
(ii) The Markov chain is reducible as the state space is not a minimal closed class; there are two proper subsets of S which are closed classes.
(iii) Since $\{1, 3\}, \{5, 6\}$ are closed classes, states 1 and 3 are non-null persistent, as well as 5 and 6 are non-null persistent. Further, states 2 and 4 are inessential and hence transient.
(iv) Since all the diagonal elements are positive, the period of each state is 1.

2.9.18 Suppose the transition probability matrix P of a Markov chain is as given below

$$P = \begin{array}{c} \\ 1 \\ 2 \\ 3 \\ 4 \\ 5 \\ 6 \end{array} \begin{array}{c} \begin{array}{cccccc} 1 & 2 & 3 & 4 & 5 & 6 \end{array} \\ \begin{pmatrix} 2/3 & 0 & 1/3 & 0 & 0 & 0 \\ 1/4 & 1/4 & 1/2 & 0 & 0 & 0 \\ 3/5 & 0 & 2/5 & 0 & 0 & 0 \\ 1/8 & 1/2 & 1/4 & 1/8 & 0 & 0 \\ 0 & 0 & 0 & 0 & 1/3 & 2/3 \\ 0 & 0 & 0 & 0 & 1/2 & 1/2 \end{pmatrix} \end{array}.$$

It is given that $f_{22} = 0.25$, $f_{42} = 0.5714$ and $f_{44} = 0.125$. Identify the remaining elements in $F = [f_{ij}]$ without computing. Use all the relevant results. Justify your answers. Find the probability of absorption from transient states to a class of persistent states.
Solution: The solution is similar to that of Example 2.2.2.

Answers to the multiple choice questions, based on Chap. 2, are given in Table A.1.

Table A.1 Answer key to MCQs in Chap. 2

Q. No.	1	2	3	4	5	6	7	8	9	10
Ans	b, c	b	b	c	c	d	a	b	b	a, b, c, d
Q. No.	11	12	13	14	15	16	17	18	19	20
Ans	d	b	a	b	d	b	a, b	d	c, d	c, d
Q. No.	21	22	23	24	25	26	27	28	29	30
Ans	d	b	a	d	a, d	d	c	b, c	a, c	c
Q. No.	31	32	33	34	35	36	37	38	39	40
Ans	b, d	c	c, d	c	c	a, c	a, c	c	c	d
Q. No.	41	42	43	44	45	46	47	48	49	50
Ans	b	a	c	a, c, d	a, b, c	a, c, d	a, d	d	a, d	c
Q. No.	51	52	53	54	55	56	57	58	59	60
Ans	b	a, c	c	d	b	d	a	a, b, c, d	a	a
Q. No.	61	62	63	64	65	66	67	68	69	70
Ans	b	b	d	b	b	d	c	d	d	a, b, c, d
Q. No.	71	72	73	74	75	76	77	78		
Ans	d	c	c	d	b	a	c	d		

A.2 Chapter 3

3.8.1 Suppose $\{X_n, n \geq 0\}$ is a sequence of independent and identically distributed random variables with $P[X_n = i] = a_i,\ a_i > 0$ and $\sum_{i \in S} a_i = 1$, where $S = \{1, 2, \ldots, \}$. It is known to be a Markov chain. Find the long run and stationary distributions, if they exist.

Solution: Since $\{X_n, n \geq 0\}$ is a sequence of independent and identically distributed random variables, it is a time homogeneous Markov chain. The transition probability matrix P has all identical rows given by $(a_1, a_2, \ldots,)$. Further, $P^n = P\ \forall\ n \geq 1$ implies $\lim_{n\to\infty} P^n = P$. Thus, the long run distribution and hence the stationary distribution exist; the two are the same and are given by $(a_1, a_2, \ldots,)$. Since $a_i > 0\ \forall\ i \in S$, the Markov chain is irreducible and hence the stationary distribution is unique.

3.8.2 Suppose $\{X_n, n \geq 1\}$ is a Markov chain with state space $S = \{1, 2, 3, 4, 5\}$ and transition probability matrix P given by

$$P = \begin{array}{c} \\ 1 \\ 2 \\ 3 \\ 4 \\ 5 \end{array}\begin{array}{c} \begin{array}{ccccc} 1 & 2 & 3 & 4 & 5 \end{array} \\ \begin{pmatrix} 1/3 & 2/3 & 0 & 0 & 0 \\ 3/4 & 1/4 & 0 & 0 & 0 \\ 0 & 0 & 1/8 & 1/4 & 5/8 \\ 0 & 0 & 0 & 1/2 & 1/2 \\ 0 & 0 & 1/3 & 0 & 2/3 \end{pmatrix} \end{array}.$$

(i) Classify the states. (ii) Examine whether the long run distribution exists. (iii) If yes, find it. (iv) Examine whether the stationary distribution exists. (v) If yes, find it. (vi) Comment on the link between the long run distribution and the stationary distribution. (vii) Find the matrix of $F = f_{ij}$. (viii) Verify the relation between f_{ij} and $\lim_{n\to\infty} p_{ij}^{(n)}$.

Solution: (i) There are two closed classes $\{1, 2\}$ and $\{3, 4, 5\}$, thus the Markov chain is reducible. The two Markov chains with these two as state spaces can be studied separately. Each is a finite state space and irreducible Markov chain and hence all the states in both the chains are non-null persistent. Further, all the diagonal elements are positive, which implies that all the states are aperiodic and hence ergodic.

(ii) Since there are two closed communicating classes, in view of Theorem 3.2.6, the long run distribution does not exist. However, for $i \in \{1, 2\}$,

$$p_{ij}^{(n)} \to \begin{cases} 1/\mu_j, & \text{if } \ j \in \{1, 2\} \\ 0, & \text{if } j \in \{3, 4, 5\}, \end{cases}$$

and for $i \in \{3, 4, 5\}$,

$$p_{ij}^{(n)} \to \begin{cases} 1/\mu_j, & \text{if } j \in \{3, 4, 5\} \\ 0, & \text{if } j \in \{1, 2\}. \end{cases}$$

Thus, $\lim_{n\to\infty} p_{ij}^{(n)}$ exists but it depends on i. Hence, the long run distribution does not exist.
(iv) and (v) Although the long run distribution does not exist, a stationary distribution may exist. We obtain stationary distributions concentrated on the two closed classes. The stationary distribution corresponding to the Markov chain with state space $\{1, 2\}$ is $(9/17, 8/17)$, and the stationary distribution corresponding to the Markov chain with state space $\{3, 4, 5\}$ is $(8/33, 4/33, 21/33)$. Thus, the stationary distributions concentrated on respective classes are $\underline{\pi}_1 = (9/17, 8/17, 0, 0, 0)$ and $\underline{\pi}_2 = (0, 0, 8/33, 4/33, 21/33)$. Any convex combination of these two is also a stationary distribution of the Markov chain. Since there are two closed classes, the matrix P has two eigenvalues equal to 1. $\underline{\pi}_1$ and $\underline{\pi}_2$ are stationary distributions corresponding to these two eigenvalues.
(vi) The long run distribution does not exist but an uncountable family of the stationary distributions exists. Further in $\lim_{n\to\infty} P^n$, the first two rows are the same as $\underline{\pi}_1$ and the last three rows are the same as $\underline{\pi}_2$.
(vii) In Markov chains with $\{1, 2\}$ and $\{3, 4, 5\}$ as state spaces, the states communicate with each other and are persistent, hence corresponding f_{ij} are 1. States from one chain do not communicate with the states in the other chain, hence corresponding f_{ij} are 0. Thus, the matrix $F = [f_{ij}]$ is given by

$$F = \begin{array}{c} \\ 1 \\ 2 \\ 3 \\ 4 \\ 5 \end{array} \begin{array}{c} \begin{array}{ccccc} 1 & 2 & 3 & 4 & 5 \end{array} \\ \begin{pmatrix} 1 & 1 & 0 & 0 & 0 \\ 1 & 1 & 0 & 0 & 0 \\ 0 & 0 & 1 & 1 & 1 \\ 0 & 0 & 1 & 1 & 1 \\ 0 & 0 & 1 & 1 & 1 \end{pmatrix} \end{array}.$$

(viii) Suppose L denotes the matrix $[f_{ij}/\mu_j]$. We find the values of μ_j from the stationary distributions concentrated on the two closed classes. Hence, the mean recurrence times of the 5 states are given by a vector $\underline{\mu} = (17/9, 17/8, 33/8, 33/4, 33/21)$. Then L is given by

$$L = \begin{array}{c} \\ 1 \\ 2 \\ 3 \\ 4 \\ 5 \end{array} \begin{array}{c} \begin{array}{ccccc} 1 & 2 & 3 & 4 & 5 \end{array} \\ \begin{pmatrix} 9/17 & 8/17 & 0 & 0 & 0 \\ 9/17 & 8/17 & 0 & 0 & 0 \\ 0 & 0 & 8/33 & 4/33 & 21/33 \\ 0 & 0 & 8/33 & 4/33 & 21/33 \\ 0 & 0 & 8/33 & 4/33 & 21/33 \end{pmatrix} \end{array}.$$

It is to be noted that L is exactly the same as the matrix of $\lim_{n\to\infty} p_{ij}^{(n)}$. Thus, we have verified the result that for an aperiodic chain $\lim_{n\to\infty} p_{ij}^{(n)} = f_{ij}/\mu_j$.

3.8.3 Suppose $\{X_n, n \geq 0\}$ is a Markov chain with state space $S = \{1, 2, 3, 4, 5\}$ and P is given by

$$P = \begin{array}{c} \\ 1 \\ 2 \\ 3 \\ 4 \\ 5 \end{array} \begin{array}{c} \begin{array}{ccccc} 1 & 2 & 3 & 4 & 5 \end{array} \\ \begin{pmatrix} 0.7 & 0.3 & 0 & 0 & 0 \\ 0.3 & 0.7 & 0 & 0 & 0 \\ 0 & 0 & 0.3 & 0.3 & 0.4 \\ 0 & 0 & 0.4 & 0.5 & 0.1 \\ 0 & 0 & 0.3 & 0.2 & 0.5 \end{pmatrix} \end{array}.$$

(i) Obtain the long run distribution if it exists.
(ii) Examine if stationary distributions exist. If yes, find all the stationary distributions.

Solution: (i) It is clear that the Markov chain is reducible with two closed communicating classes $C_1 = \{1, 2\}$ and $C_2 = \{3, 4, 5\}$. Hence, the long run distribution does not exist. Since transition probability matrices corresponding to both C_1 and C_2 are doubly stochastic, the unique stationary distributions concentrated on C_1 and C_2 are given by $\underline{a} = (1/2, 1/2)'$ and $\underline{b} = (1/3, 1/3, 1/3)'$, respectively. Thus,

$$\underline{\pi}_1 = (1/2, 1/2, 0, 0, 0)' \;\;\&\;\; \underline{\pi}_2 = (0, 0, 1/3, 1/3, 1/3)'$$

are two stationary distributions of the Markov chain. Hence, the family of stationary distributions is given by $\alpha\underline{\pi}_1 + (1 - \alpha)\underline{\pi}_2, 0 \leq \alpha \leq 1$.

3.8.4 On a Southern Pacific island, a sunny day is followed by another sunny day with probability 0.9, whereas a rainy day is followed by another rainy day with probability 0.2. Suppose that there are only sunny or rainy days. In the long run what fraction of days is sunny? Find an expression for corresponding $f_{ii}^{(n)}$ and the mean recurrence time. Examine whether the fraction of sunny days is the reciprocal of the corresponding mean recurrence time.

Solution: To find the long run fraction of days it is sunny, we find the stationary distribution associated with this Markov chain. From the given information, the one step transition probability matrix P is given by

$$P = \begin{array}{c} \\ 1 \\ 2 \end{array} \begin{array}{c} \begin{array}{cc} 1 & 2 \end{array} \\ \begin{pmatrix} 0.9 & 0.1 \\ 0.8 & 0.2 \end{pmatrix} \end{array},$$

where 1 and 2 indicate the sunny and the rainy day, respectively. Observe that

$$\begin{aligned}
\underline{\pi} = \underline{\pi}P &\Rightarrow \pi_1 = 0.9\pi_1 + 0.8\pi_2 \ \ \& \ \ \pi_2 = 0.1\pi_1 + 0.2\pi_2 \\
&\Rightarrow \pi_1 = 8\pi_2 \\
\pi_1 + \pi_2 = 1 &\Rightarrow \pi_1 = 8/9 \ \ \& \ \ \pi_2 = 1/9 \\
&\Rightarrow \underline{\pi} = (8/9, 1/9)'.
\end{aligned}$$

Thus, in the long run 8/9 fraction of days is sunny. Further, to examine whether π_1 is the reciprocal of mean recurrence time, we obtain μ_1 as follows:

$$f_{11}^{(1)} = 0.9,\ f_{11}^{(2)} = 0.1 \times 0.8,\ f_{11}^{(3)} = 0.1 \times 0.2 \times 0.8 \ \Rightarrow \ f_{11}^{(n)} = 0.1(0.2)^{n-2}0.8,$$

for $n \geq 3$. Hence,

$$\begin{aligned}
\mu_1 = \sum_{n\geq 1} n f_{11}^{(n)} &= 0.9 + \sum_{n\geq 2} n(0.1) \times (0.2)^{n-2} \times 0.8 \\
&= 9/10 + 9/40 = 45/40 = 9/8 = 1/\pi_1.
\end{aligned}$$

3.8.5 Examine whether a Markov chain with state space $S = \{1, 2, 3\}$ and

$$P = \begin{array}{c} \\ 1 \\ 2 \\ 3 \end{array}\begin{array}{c} \begin{array}{ccc} 1 & 2 & 3 \end{array} \\ \begin{pmatrix} 1/3 & 2/3 & 0 \\ 1/4 & 1/2 & 1/4 \\ 1 & 0 & 0 \end{pmatrix} \end{array}$$

is ergodic. Examine whether the long run distribution and stationary distributions exist. If yes, find the distributions and the mean recurrence times.
Solution: It is to be noted that all states communicate with each other. Thus, the Markov chain is irreducible. It being a finite state space Markov chain, all states are non-null persistent. Observe that $p_{11} > 0$, $p_{22} > 0$ and hence both 1 and 2 are aperiodic states. State 3 communicates with 1 and 2 and hence it is also aperiodic. Further, $3 \to 1 \to 2 \to 3$ and $3 \to 1 \to 2 \to 2 \to 3$ and hence $D_3 = \{3, 4, \ldots\}$ and its g.c.d. is 1. Hence, state 3 is also aperiodic. Thus, the Markov chain is ergodic and hence the long run and stationary distributions exist and are the same. Suppose $\underline{\pi}$ denotes the stationary distribution. Solving the system of equations $\underline{\pi} = \underline{\pi}P$, subject to the condition that sum of the elements of $\underline{\pi}$ is 1, we get $\underline{\pi} = (3/8, 1/2, 1/8)'$. Further, the mean recurrence times in three states are reciprocals of the components of $\underline{\pi}$. Thus, $\mu_1 = 8/3$, $\mu_2 = 2$ and $\mu_3 = 8$.

3.8.6 For a Markov chain $\{X_n, n \geq 0\}$ with state space $S = \{1, 2, 3\}$, the transition probability matrix P is given by

$$P = \begin{matrix} & \begin{matrix} 1 & 2 & 3 \end{matrix} \\ \begin{matrix} 1 \\ 2 \\ 3 \end{matrix} & \begin{pmatrix} 0.3 & 0.2 & 0.5 \\ 0.5 & 0.1 & 0.4 \\ 0.5 & 0.2 & 0.3 \end{pmatrix} \end{matrix}.$$

Each visit that the process makes to states 1, 2, 3 incurs cost of Rs.200, 500, 300, respectively. What is the long run cost per visit associated with this Markov chain?

Solution: Suppose $\underline{\pi}$ denotes the stationary distribution. Solving the system of equations $\underline{\pi} = \underline{\pi}P$, subject to the condition that sum of the elements of $\underline{\pi}$ is 1, we get $\underline{\pi} = (0.4167, 0.1818, 0.4015)'$. Hence, the long run cost per period associated with this Markov chain is

$$200 \times 0.4167 + 500 \times 0.1818 + 300 \times 0.4015 = 296.69.$$

3.8.7 Operating condition of a machine at any time is classified as follows. State 1: Good; State 2: Deteriorated but operating; State 3: In repair. Suppose for $n \geq 1$, X_n denotes the condition of the machine at the end of period n. We assume that the sequence of machine conditions is a Markov chain with the transition probability matrix P as given below

$$P = \begin{matrix} & \begin{matrix} 1 & 2 & 3 \end{matrix} \\ \begin{matrix} 1 \\ 2 \\ 3 \end{matrix} & \begin{pmatrix} 0.9 & 0.1 & 0 \\ 0 & 0.9 & 0.1 \\ 1 & 0 & 0 \end{pmatrix} \end{matrix}.$$

What is the long run rate of repairs per unit time?

Solution: Suppose $\underline{\pi}$ denotes the stationary distribution. From the system of equations $\underline{\pi} = \underline{\pi}P$, we get $\underline{\pi} = (0.4762, 0.4762, 0.0476)'$. Hence, the long run rate of repairs per unit time is 0.0476.

3.8.8 Suppose there are two groups of drivers: a group of 10,000 relatively good drivers and 10,000 relatively bad drivers. Suppose discount levels of insurance premiums are 0 (no discount), 1 (20% discount) and 2 (40% discount). The full premium is Rs. 5000. The discount level of a driver changes according to the rule "reduce by one discount level if one claim is made, and move to no discount level if more than one claim is made". Assume that a Markov chain is a suitable model for transitions from one class to another. Probability distributions for number of claims N in a year for the two groups are given below

$$\text{Good Drivers: } P[N = 0] = 0.7, \quad P[N = 1] = 0.2, \quad P[N \geq 2] = 0.1$$

$$\text{Bad Drivers: } P[N = 0] = 0.4, \quad P[N = 1] = 0.4, \quad P[N \geq 2] = 0.2.$$

(i) Obtain the one step transition probability matrices for both the groups. (ii) Assuming that all drivers start in class 0 at the beginning of the year, compute the expected premium income from the two groups for years 1, 2, 3, 4, 8, 16, 32. (iii) Compute the long run expected premium income from the two groups and comment.

Solution: From the given discount rules and probability distributions for number of claims, one step transition probability matrices for the two groups are

$$P_G = \begin{array}{c} \\ 0 \\ 1 \\ 2 \end{array} \begin{array}{c} \begin{array}{ccc} 0 & 1 & 2 \end{array} \\ \begin{pmatrix} 0.3 & 0.7 & 0 \\ 0.3 & 0 & 0.7 \\ 0.1 & 0.2 & 0.7 \end{pmatrix} \end{array} \quad P_B = \begin{array}{c} \\ 0 \\ 1 \\ 2 \end{array} \begin{array}{c} \begin{array}{ccc} 0 & 1 & 2 \end{array} \\ \begin{pmatrix} 0.6 & 0.4 & 0 \\ 0.6 & 0 & 0.4 \\ 0.2 & 0.4 & 0.4 \end{pmatrix} \end{array}.$$

Assuming that all drivers start in level 0 at the beginning of the year, we compute the expected income from the premiums. Table A.2 displays expected premium income in rupees from the two groups of drivers, computed using similar arguments as in Section 3.6. Last row of Table A.2 corresponding to year ∞ corresponds to the stationary distributions for the two groups given by

$$\underline{\pi}_G = (0.1860, 0.2442, 0.5698)' \quad \& \quad \underline{\pi}_B = (0.5238, 0.2857, 0.1905)'.$$

It is to be noted that after 8 years, for both the groups, there is relatively little difference in expected premium incomes. However, the group of bad drivers pay approximately 20% more than the group of good drivers in the long run.

Table A.2 Expected premiums from two groups

Year n	Good drivers	Bad drivers
0	5,000,0000	5,00,0000
1	4,300,0000	4,600,0000
2	3,810,0000	4,440,0000
3	3,712,0000	4,376,0000
4	3,643,4000	4,350,4000
8	3,616,8110	4,333,7700
16	3,616,2790	4,333,3340
32	3,616,2790	4,333,3340
∞	3,616,2790	4,333,3330

Table A.3 Answer key to MCQs in Chap. 3

Q. No.	1	2	3	4	5	6	7	8	9	10
Ans	d	a, b, d	a, d	c	a, b, c, d	c	b	a	a, b	a, b
Q. No.	11	12	13	14	15	16	17	18	19	20
Ans	c	c	c	c	b, c, d	c	b, c	d	c	

3.8.9 Suppose $\{X_n, n \geq 0\}$ is a Markov chain with P given by

$$P = \begin{array}{c} \\ 1 \\ 2 \\ 3 \\ 4 \\ 5 \\ 6 \\ 7 \end{array} \begin{array}{c} \begin{array}{ccccccc} 1 & 2 & 3 & 4 & 5 & 6 & 7 \end{array} \\ \begin{pmatrix} 0 & 0 & 1/2 & 1/4 & 1/4 & 0 & 0 \\ 0 & 0 & 1/3 & 0 & 2/3 & 0 & 0 \\ 0 & 0 & 0 & 0 & 0 & 1/3 & 2/3 \\ 0 & 0 & 0 & 0 & 0 & 1/2 & 1/2 \\ 0 & 0 & 0 & 0 & 0 & 3/4 & 1/4 \\ 1/2 & 1/2 & 0 & 0 & 0 & 0 & 0 \\ 1/4 & 3/4 & 0 & 0 & 0 & 0 & 0 \end{pmatrix} \end{array}.$$

(i) Examine whether the Markov chain is irreducible. (ii) Find the period of each state. (iii) Find the cyclically moving classes.
Solution: The solution is similar to that of Example 3.4.11.

Answers to the multiple choice questions, based on Chap. 3, are given in Table A.3.

A.3 Chapter 4

4.6.1 Suppose $\{X_n, n \geq 0\}$ is an unrestricted symmetric random walk, with state space I. Find (i) $\liminf_{n\to\infty} p_{66}^{(n)}$, (ii) $\limsup_{n\to\infty} p_{88}^{(n)}$, (iii) $p_{00}^{(8)}$ and $f_{00}^{(8)}$, (iv) $p_{00}^{(9)}$, $f_{00}^{(9)}$ and (v) approximate value of $p_{00}^{(40)}$ and $p_{00}^{(49)}$.
Solution: Since $\{X_n, n \geq 0\}$ is an unrestricted symmetric random walk, with state space I, all the states are null persistent. Hence $\forall\ i \in S$,

$$\lim_{n\to\infty} p_{ii}^{(n)} = 0 \Rightarrow (i) \liminf_{n\to\infty} p_{66}^{(n)} = 0 \ \&\ (ii) \limsup_{n\to\infty} p_{88}^{(n)} = 0.$$

$$(iii) p_{00}^{(8)} = \binom{8}{4}(1/2)^8 = 0.2734 \ \&\ (iv) f_{00}^{(8)} = (1/7)p_{00}^{(8)} = 0.03906.$$

(iv) For this random walk, $p_{00}^{(n)}$ and $f_{00}^{(n)}$ are 0 if n is odd. Hence, $p_{00}^{(9)} = 0$ and $f_{00}^{(9)} = 0$.

(v) Approximate value of $p_{00}^{(40)} = 1/\sqrt{20\pi} = 0.1262$. Further, $p_{00}^{(49)} = 0$, since the probability of transition from 0 to 0 in an odd number of steps is 0.

4.6.2 Suppose $\{X_n, n \geq 0\}$ is an unrestricted random walk, with state space I and $p = 1/3$. Find (i) $\liminf_{n\to\infty} p_{66}^{(n)}$, (ii) $\limsup_{n\to\infty} p_{88}^{(n)}$, (iii) $p_{00}^{(8)}$ and $f_{00}^{(8)}$, (iv) $p_{00}^{(9)}$, $f_{00}^{(9)}$ and (v) approximate value of $p_{00}^{(40)}$ and $p_{00}^{(49)}$.

Solution: With $p = 1/3$, the unrestricted random walk is not symmetric. Therefore, all the states are transient. Hence $\forall\ i \in S$,

$$\lim_{n\to\infty} p_{ii}^{(n)} = 0 \ \Rightarrow \ (i) \liminf_{n\to\infty} p_{66}^{(n)} = 0 \ \& \ (ii) \limsup_{n\to\infty} p_{88}^{(n)} = 0.$$

$$(iii) p_{00}^{(8)} = \binom{8}{4}(1/3)^4(2/3)^4 = 0.1707 \ \& \ (iv) f_{00}^{(8)} = (1/7) p_{00}^{(8)} = 0.02439.$$

(iv) For this random walk, $p_{00}^{(n)}$ and $f_{00}^{(n)}$ are 0 if n is odd. Hence, $p_{00}^{(9)} = 0$ and $f_{00}^{(9)} = 0$.

(v) Approximate value of $p_{00}^{(40)} = (1/3)^4(2/3)^4/\sqrt{20\pi} = 0.012$. Further, $p_{00}^{(49)} = 0$, since the probability of transition from 0 to 0 in an odd number of steps is 0.

4.6.3 Suppose $\{X_n, n \geq 0\}$ is a random walk with state space W and absorbing barrier at 0. Find as $n \to \infty$ (i) $\liminf p_{55}^{(n)}$, (ii) $\limsup p_{66}^{(n)}$, (iii) $\limsup p_{00}^{(n)}$ and (iv) $\limsup f_{00}^{(n)}$ and $\liminf f_{00}^{(n)}$.

Solution: For a random walk with state space W and with absorbing barrier at 0, all the states $i > 0$ are transient. Hence $\forall\ i \in S - \{0\}$,

$$\lim_{n\to\infty} p_{ii}^{(n)} = 0 \ \Rightarrow \ (i) \liminf_{n\to\infty} p_{55}^{(n)} = 0 \ \& \ (ii) \limsup_{n\to\infty} p_{66}^{(n)} = 0.$$

(iii) Further, the state 0 is absorbing and hence non-null persistent. Hence, $p_{00}^{(n)} = 1$ for all $n \geq 1$, which implies $\limsup_{n\to\infty} p_{00}^{(n)} = 1$.

(iv) Since the state 0 is absorbing, $f_{00}^{(1)} = 1$ and $f_{00}^{(n)} = 0$ for all $n \geq 2$. Hence, $\limsup_{n\to\infty} f_{00}^{(n)} = 0$ and $\liminf_{n\to\infty} f_{00}^{(n)} = 0$.

4.6.4 Suppose $\{X_n, n \geq 0\}$ is a random walk with state space W and with reflecting barrier at 0. If $p = 2/3$, find (i) $\lim_{n\to\infty} p_{00}^{(n)}$ and (ii) $\limsup_{n\to\infty} p_{77}^{(n)}$.

Solution: In a random walk $\{X_n, n \geq 0\}$, with state space W and with reflecting barrier at 0, if $p = 2/3 > 1/2$, all states are either transient or null persistent. Hence, (i) $\lim_{n\to\infty} p_{00}^{(n)} = 0$ and (ii) $\limsup_{n\to\infty} p_{77}^{(n)} = 0$.

4.6.5 Suppose $\{X_n, n \geq 0\}$ is a random walk with state space W and with reflecting barrier at 0. If $p = 1/3$, find the stationary distribution.

Solution: In a random walk $\{X_n, n \geq 0\}$, with state space W and with reflecting barrier at 0, if $p = 1/3 < 1/2$, then the stationary distribution $\underline{\pi}$ exists and is given by

$$\pi_0 = \frac{1}{2}\left(1 - \frac{p}{q}\right) = 1/4$$

$$\begin{aligned}\pi_j &= \frac{1}{2q}\left(1 - \frac{p}{q}\right)\left(\frac{p}{q}\right)^{j-1} = (3/8)(1/2)^{j-1}\\ &= (3/4)(1/2)(1/2)^{j-1}, \quad j \geq 1.\end{aligned}$$

4.6.6 Suppose $\{X_n, n \geq 0\}$ is a random walk with state space W and with partially reflecting barrier at 0. If $p = 2/3$, find (i) $\lim_{n\to\infty} p_{00}^{(n)}$ and (ii) $\limsup_{n\to\infty} p_{77}^{(n)}$.
Solution: In a random walk $\{X_n, n \geq 0\}$, with state space W and with partially reflecting barrier at 0, if $p = 2/3 > 1/2$, all states are either transient or null persistent. Hence,

$$(i) \lim_{n\to\infty} p_{00}^{(n)} = 0 \;\; \& \;\; (ii) \limsup_{n\to\infty} p_{77}^{(n)} = 0.$$

4.6.7 Suppose $\{X_n, n \geq 0\}$ is a random walk with state space W and with partially reflecting barrier at 0. If $p = \delta = 1/3$, find the stationary distribution.
Solution: In a random walk $\{X_n, n \geq 0\}$, with state space W and with partially reflecting barrier at 0, if $p = \delta = 1/3 < 1/2$, then the stationary distribution $\underline{\pi}$ exists and it is a geometric distribution given by

$$\pi_j = \left(1 - \frac{p}{1-p}\right)\left(\frac{p}{1-p}\right)^j = (1/2)^{j+1}, \quad j \geq 0.$$

4.6.8 Suppose $\{X_n, n \geq 0\}$ is a random walk with state space $S = \{0, 1, 2, 3\}$ and with absorbing barriers at 0 and 3. The transition probability matrix P is as given below

$$P = \begin{array}{c} \\ 0 \\ 1 \\ 2 \\ 3 \end{array}\begin{array}{c} \begin{array}{cccc} 0 & 1 & 2 & 3 \end{array} \\ \begin{pmatrix} 1 & 0 & 0 & 0 \\ 1/6 & 0 & 5/6 & 0 \\ 0 & 2/7 & 0 & 5/7 \\ 0 & 0 & 0 & 1 \end{pmatrix} \end{array}.$$

(i) Decide the nature of the states. (ii) Find the period of each state.
(iii) Find the probability of absorption into $\{0\}$ and $\{3\}$ from states 1, 2.
Solution: (i) The states 0 and 3 are absorbing and hence non-null persistent states. Further, states 1 and 2 communicate with each other. $1 \to 0$ but $0 \nrightarrow 1$, similarly, $2 \to 3$ but $3 \nrightarrow 2$. Hence, 1 and 2 are inessential states and hence transient.

(ii) The states 0 and 3 have period 1. Note that

$$1 \to 2 \to 1 \to 2 \to 1 \;\Rightarrow\; D_1 = \{2, 4, 6, \ldots\} \;\Rightarrow\; d_1 = 2.$$

Further, states $1 \leftrightarrow 2$ and hence the period of state 2 is also 2.
(iii) We find the probability of absorption into $C_1 = \{0\}$ and $C_2 = \{3\}$ from states in $T = \{1, 2\}$, using the formula $G = (I - Q)^{-1}D$ derived in Theorem 2.6.9. With rearrangement of rows and columns of P, the matrices D, Q and G are as given below

$$D = \begin{pmatrix} 1/6 & 0 \\ 0 & 5/7 \end{pmatrix}, \quad Q \quad \begin{pmatrix} 0 & 5/6 \\ 2/7 & 0 \end{pmatrix} \quad \& \quad G = \begin{pmatrix} 0.2188 & 0.7812 \\ 0.0625 & 0.9375 \end{pmatrix}.$$

It is to be noted that the probability of absorption into C_2 is higher for both the transient states, since probability of transition to the right is $> 1/2$. Observe that the sum of each row in G is 1.

4.6.9 Suppose $\{X_n, n \geq 0\}$ is a random walk with state space $\{0, 1, 2, 3, 4\}$ and with partially reflecting barrier at 0 and 4. If $p = 1/4$, what is the long run mean fraction of time the random walk is in states 0 and 4?
Solution: A random walk $\{X_n, n \geq 0\}$ with partially reflecting barrier at 0 and 4 is the Markov chain with state space $\{0, 1, 2, 3, 4\}$ and one step transition probability matrix P as given below

$$P = \begin{matrix} & \begin{matrix} 0 & 1 & 2 & 3 & 4 \end{matrix} \\ \begin{matrix} 0 \\ 1 \\ 2 \\ 3 \\ 4 \end{matrix} & \begin{pmatrix} 3/4 & 1/4 & 0 & 0 & 0 \\ 3/4 & 0 & 1/4 & 0 & 0 \\ 0 & 3/4 & 0 & 1/4 & 0 \\ 0 & 0 & 3/4 & 0 & 1/4 \\ 0 & 0 & 0 & 3/4 & 1/4 \end{pmatrix} \end{matrix}.$$

To find the long run fraction of time the random walk is in states 0 and 4, we find the stationary distribution $\underline{\pi}$ associated with the Markov chain by solving the matrix equation $\underline{\pi} = \underline{\pi}P$. These are given by

$$\begin{aligned} \pi_0 &= (3/4)\pi_0 + (3/4)\pi_1, \quad \pi_1 = (1/4)\pi_0 + (3/4)\pi_2 \\ \pi_2 &= (1/4)\pi_1 + (3/4)\pi_3, \quad \pi_3 = (1/4)\pi_2 + (3/4)\pi_4 \\ \pi_4 &= (1/4)\pi_3 + (1/4)\pi_4 \\ \Rightarrow\; \pi_1 &= (1/3)\pi_0, \quad \pi_2 = (1/9)\pi_0, \quad \pi_3 = (1/27)\pi_0, \quad \pi_4 = (1/81)\pi_0. \end{aligned}$$

The condition

$$\pi_0 + \pi_1 + \cdots + \pi_4 = 1 \Rightarrow \; \underline{\pi} = (81/121, 27/121, 9/121, 3/121, 1/121)'.$$

Hence, the long run mean fraction of time the random walk is in states 0 and 4 is $81/121 = 0.6694$ and $1/121 = 0.0083$, respectively. Observe that with $p = 1/4$, the probability of transition to the right is small, which is reflected in π_0, being the largest and π_4 being the smallest.

4.6.10 Suppose $\{X_n, n \geq 0\}$ is a random walk with partially reflecting barriers at 0 and M, that is, it is an irreducible Markov chain with finite state space $S = \{0, 1, 2, \ldots, M\}$ and a transition probability matrix $P = [p_{ij}]$ given by

$$p_{00} = 1 - p = q, \quad p_{01} = p, p_{ij} = q \text{ if } j = i - 1 \;\&\; p_{ij} = p \text{ if } j = i + 1,$$

$i = 1, 2, 3, \ldots, M - 1$ and $p_{Mj} = q$ if $j = M - 1$, $p_{MM} = p$. Find a stationary distribution associated with this Markov chain.

Solution: To find a stationary distribution, we solve the following system of equations:

$$q\pi_0 + q\pi_1 = \pi_0, \quad p\pi_0 + q\pi_2 = \pi_1$$
$$\pi_j = p\pi_{j-1} + q\pi_{j+1}, \quad j = 2, \ldots, M - 1 \;\&\; \pi_M = p\pi_{M-1} + p\pi_M .$$

From these equations, we have $\pi_1 = (p/q)\pi_0, \;\; \pi_2 = (p/q)\pi_1 = (p/q)^2\pi_0$. Proceeding on these lines, we have $\pi_j = (p/q)^j\pi_0$ for $j = 0, 1, 2, \ldots, M$. Using the condition $\sum_{j=0}^{M} \pi_j = 1$, we get

$$\pi_0 = \left(\sum_{j=0}^{M}(p/q)^j\right)^{-1} = \frac{1 - p/q}{1 - (p/q)^{M+1}}$$
$$\Rightarrow \quad \pi_j = (p/q)^j \frac{1 - p/q}{1 - (p/q)^{M+1}}, \quad j = 0, 1, 2, \ldots, M.$$

Since the stationary distribution exists for all $p \in (0, 1)$, the Markov chain is non-null persistent and the mean recurrence times are given by $\mu_i = 1/\pi_i, i \in S$.

4.6.11 Suppose Ajay and Vijay play a game of gambling. Ajay has probability 0.4 of winning at each flip of the coin. Suppose Ajay's initial capital is 2000 rupees while Vijay's initial capital is 3000 rupees. It is decided that the one who wins at the trial has to give 200 rupees to the other.

(i) Compute the probability that Ajay will win the game. Find his expected gain. (ii) Compute the probability that Vijay will win the game. Find his expected gain. (iii) Find the expected duration of the game.

Solution: We take Rs 200 as 1 unit, so that Ajay and Vijay have initial capital to be 10 and 15 units respectively and total capital is $M = 25$ units.

(i) To compute the probability that Ajay will win the game, we compute the probability that Vijay with initial capital of 15 units is ruined. Thus, we compute P_{15} when $p = 0.6$ and $M = 25$ units. It is given by

$$P_{15}^{(V)} = \frac{(4/6)^{15} - (4/6)^{25}}{1 - (4/6)^{25}} = 0.002244 .$$

Thus, Ajay will win the game with probability 0.002244, which is very low. His expected gain $E(G(10))$ is given by

$$E(G(10)) = (25 - 10)P_{15}^{(V)} - 10(1 - P_{15}^{(V)}) = -9.9439 = -1988.78 \text{ Rupees.}$$

(ii) To compute the probability that Vijay will win the game, we compute the probability that Ajay with initial capital of 10 units is ruined. Thus, we compute P_{10} when $p = 0.4$ and $M = 25$ units. It is given by

$$P_{10}^{(A)} = \frac{(6/4)^{10} - (6/4)^{25}}{1 - (6/4)^{25}} = 0.9978 .$$

Thus, Vijay will win the game with probability 0.9978, which is the same as $1 - 0.002244$. His expected gain $E(G(15))$ is given by

$$E(G(15)) = (25 - 10)P_{10}^{(A)} - 15(1 - P_{10}^{(A)}) = 14.934 = 2986.8 \text{ Rupees.}$$

It is quite high as the probability that Vijay will win is high.
(iii) The expected duration of the game, when the gambler has initial capital of a units, is given by $D_a = \frac{a}{q-p} - \frac{M}{q-p}\frac{1-(q/p)^a}{1-(q/p)^M}$. Labeling Ajay as a gambler, we have $a = 10$, $M = 25$ and $p = 0.4$. Hence, $D_{10} = 49.7195$. Thus, on average, the game will continue for 50 trials. If we label Vijay as a gambler, we have to take $p = 0.6$ and again we get $D_{15} = 49.7195$, as expected.

Answers to the multiple choice questions, based on Chap. 4, are given in Table A.4.

Table A.4 Answer key to MCQs in Chap. 4

Q. No.	1	2	3	4	5	6	7	8
Ans	a, b, c, d	a, b, c, d	a, c, d	b	a, b, d	c	a, b	a, d
Q. No.	9	10	11	12	13	14	15	16
Ans	c	d	b	d	d	b	a	a
Q. No.	17	18	19	20	21	22	23	24
Ans	d	d	a, b, c, d	a, b, c	d	a, b	a, b, c, d	a, b, c, d
Q. No.	25	26	27	28	29	30	31	32
Ans	a, b	a, c, d	a, b, c, d	a, b	a, c, d	c	d	d
Q. No.	33	34	35	36	37	38	39	40
Ans	d	a	d	b	b	b, c, d	a, c	b, d

A.4 Chapter 5

5.7.1 Suppose $\{Z_n, n \geq 0\}$ is a branching process with $Z_0 = 1$ and offspring distribution given by $p_0 = 0.5,\ p_1 = 0.1,\ p_3 = 0.4$. What is the probability that the population becomes extinct in the second generation, given that it is not extinct in the first generation?
Solution: We have to find out the probability

$$P[Z_2 = 0|Z_1 > 0, Z_0 = 1] = \frac{P[Z_2 = 0, Z_1 > 0, Z_0 = 1]}{P[Z_1 > 0, Z_0 = 1]}.$$

Observe that

$$\begin{aligned}
&P[Z_2 = 0, Z_1 > 0, Z_0 = 1] \\
&= P[Z_2 = 0, Z_1 = 1, Z_0 = 1] \\
&+ P[Z_2 = 0, Z_1 = 3, Z_0 = 1] \\
&= P[Z_2 = 0|Z_1 = 1]P[Z_1 = 1|Z_0 = 1]P[Z_0 = 1] \\
&+ P[Z_2 = 0|Z_1 = 3]P[Z_1 = 3|Z_0 = 1]P[Z_0 = 1] \\
&= (0.5)(0.1)(1) + (0.5)^3(0.4)(1) = 0.1.
\end{aligned}$$

$$\begin{aligned}
\text{Similarly, } P[Z_1 > 0, Z_0 = 1] &= P[Z_1 = 1|Z_0 = 1]P[Z_0 = 1] \\
&+ P[Z_1 = 3|Z_0 = 1]P[Z_0 = 1] \\
&= (0.1)(1) + (0.4)(1) = 0.5
\end{aligned}$$

Hence, $P[Z_2 = 0|Z_1 > 0, Z_0 = 1] = 1/5$.

5.7.2 Suppose $\{Z_n, n \geq 0\}$ is a branching process with $Z_0 = 1$. (i) Show that the probability generating function of the conditional distribution of Z_n given that $Z_n > 0$ is given by $(P_n(s) - P_n(0))/(1 - P_n(0)),\ |s| \leq 1$, where $P_n(s)$ is the probability generating function of Z_n given $Z_0 = 1$. Find
(ii) $P[Z_n = i|Z_n > 0],\ \ i \geq 1$ and (iii) $E(Z_n|Z_n > 0)$, when the offspring distribution is geometric with parameter $1/2$.
Solution: Observe that for $i > 0$,

$$P[Z_n = i|Z_n > 0] = \frac{P[Z_n = i, Z_n > 0]}{P[Z_n > 0]} = \frac{P[Z_n = i, Z_n > 0]}{1 - P_n(0)}.$$

(i) Hence, the probability generating function of the conditional distribution of Z_n given that $Z_n > 0$ is given by

$$\sum_{i=1}^{\infty} P[Z_n = i|Z_n > 0]s^i = \sum_{i=1}^{\infty} P[Z_n = i, Z_n > 0]s^i \Big/ (1 - P_n(0))$$
$$= \frac{\left(\sum_{i=0}^{\infty} P[Z_n = i]s^i\right) - P[Z_n = 0]}{1 - P_n(0)}$$
$$= (P_n(s) - P_n(0)) \Big/ (1 - P_n(0)).$$

(ii) Suppose $P[Z_1 = i|Z_0 = 1] = (1/2)^{i+1}, \ i = 0, 1, \ldots$. In Example 5.3.1, it is shown that for this offspring distribution $P_n(s) = (n - (n-1)s)/(n + 1 - ns)$. Hence, the probability generating function of the conditional distribution of Z_n given that $Z_n > 0$ is given by

$$\frac{P_n(s) - P_n(0)}{1 - P_n(0)} = \frac{1}{1 - (n/(n+1))}\left(\frac{(n - (n-1)s)}{n + 1 - ns} - \frac{n}{n+1}\right)$$
$$= (n+1)\frac{s}{(n + 1 - ns)(n+1)}$$
$$= s(n + 1 - ns)^{-1} = \frac{s}{n+1}(1 - (n/(n+1))s)^{-1}$$
$$= \frac{s}{n+1}\sum_{i=0}^{\infty}(n/(n+1))^i s^i$$
$$= \sum_{i=0}^{\infty}(n^i/(n+1)^{i+1})s^{i+1} = \sum_{i=1}^{\infty}(n^{i-1}/(n+1)^i)s^i$$
$$\Rightarrow \quad P[Z_n = i|Z_n > 0] = n^{i-1}/(n+1)^i, \quad i = 1, 2, \ldots.$$

Observe that

$$\sum_{i=1}^{\infty}(n^{i-1}/(n+1)^i) = (1/(n+1))\sum_{i=0}^{\infty}(n^i/(n+1)^i)$$
$$= (1/(n+1))(1 - n/(n+1))^{-1} = 1.$$

Further,

$$E(Z_n|Z_n > 0) = \sum_{i=1}^{\infty} i(n^{i-1}/(n+1)^i) = \frac{1}{n+1}\sum_{i=1}^{\infty} i(n^{i-1}/(n+1)^{i-1})$$
$$= \frac{1}{n+1}(1 - n/(n+1))^{-2} = n + 1.$$

5.7.3 In every generation of a population, each individual in the population dies with probability 1/2 or doubles with probability 1/2. Suppose Z_n denotes

the number of individuals in the population in the nth generation. Find the mean and variance of Z_n.
Solution: We are given that $Z_0 = 1$ and from the given information, we have the offspring distribution to be $p_0 = p_2 = 1/2$, with mean $\mu = 1$ and variance $\sigma^2 = 1$. Assuming $\{Z_n, n \geq 0\}$ to be a BGW branching process, we have $E(Z_n) = \mu^n = 1$. Since $\sigma^2 = 1$, $Var(Z_n) = n\sigma^2 = n$.

5.7.4 The number of offspring of an individual in a population is 0, 1 or 2 with respective probabilities $a > 0$, $b > 0$ and $c > 0$, where $a + b + c = 1$. Express the mean and variance of the offspring distribution in terms of b and c. Find the mean and variance of Z_5 given that $Z_0 = 1$.
Solution: For the given offspring distribution, mean μ is $\mu = b + 2c$ and variance σ^2 is $\sigma^2 = b + 4c - (b + 2c)^2 = b - b^2 - 4bc - 4c^2 + 4c$. Hence, $E(Z_5) = (b + 2c)^5$ and $Var(Z_5)$ is given by

$$Var(Z_5) = \sigma^2(b + 2c)^4(1 - (b + 2c)^5)/(1 - b - 2c).$$

5.7.5 Suppose a parent has no offspring with probability 1/2 and has two offspring with probability 1/2. If a population of such individuals begins with a single parent and evolves as a branching process, find the probability that the population is extinct by the nth generation, for $n = 1, 2, 3, 4, 5$.
Solution: We are given that $Z_0 = 1$ and from the given information, the offspring distribution is $p_0 = p_2 = 1/2$. We find $P[Z_n = 0]$ for $n = 1, 2, 3, 4, 5$ by finding $P_n(0) = P[Z_n = 0]$ using the recurrence relation $P_{n+1}(s) = P(P_n(s))$. For the given offspring distribution, the probability generating function $P(s)$ is given by $P(s) = P_1(s) = (1 + s^2)/2$, hence $P_1(0) = P[Z_1 = 0] = 1/2$. Now,

$$\begin{aligned}
P_2(0) &= P(P_1(0)) = (1 + P_1^2(0))/2 = 5/8 = 0.625 \\
P_3(0) &= P(P_2(0)) = (1 + P_2^2(0))/2 = 89/128 = 0.6953 \\
P_4(0) &= P(P_3(0)) = (1 + P_3^2(0))/2 = 0.7417 \\
P_5(0) &= P(P_4(0)) = (1 + P_4^2(0))/2 = 0.7751.
\end{aligned}$$

Observe that as n increases, $P_n(0) = P[Z_n = 0]$ also increases, as expected. In this case, the offspring mean is 1 and hence the probability of ultimate extinction is 1.

5.7.6 At each stage of an electron multiplier, each electron upon striking the plate generates a number of electrons for the next stage, which has Poisson distribution with mean λ. Determine the mean and variance for the number of electrons at the nth stage.
Solution: Suppose Z_n denotes the number of electrons generated at stage n. We assume that $\{Z_n, n \geq 0\}$ is a BGW branching process with $Z_0 = 1$. It is given that the offspring distribution is Poisson $P(\lambda)$. Thus, offspring mean and offspring variance both are λ. Hence, $E(Z_n) = \lambda^n$ and $Var(Z_n)$ is given by

$$Var(Z_n) = \begin{cases} \lambda^n \left(\frac{1-\lambda^n}{1-\lambda}\right), & \text{if } \lambda \neq 1 \\ n\lambda, & \text{if } \lambda = 1 . \end{cases}$$

5.7.7 At time 0, a blood culture starts with one red cell. At the end of one minute, the red cell dies and is replaced by one of the following combinations: 2 red cells with probability 1/4, 1 red and 1 white cells with probability 2/3 and 2 white cells with probability 1/12. Each red cell lives for one minute and gives birth to offspring in the same way as the parent cell. Each white cell lives for one minute and dies without reproducing. Assume that individual cells behave independently. (i) At time $n + 1$ minutes after the culture begins, what is the probability that no white cells have yet appeared? (ii) What is the probability that the entire culture eventually dies out?

Solution: Suppose Z_n denotes the number of red cells at time n. It is given that $Z_0 = 1$ and the offspring distribution is given by $p_0 = 1/12$, $p_1 = 2/3$, $p_2 = 1/4$.

(i) We want to find the probability that no white cells have appeared till time $n + 1$. For such an event to occur, at every generation the red cells reproduce only red cells. With $Z_0 = 1$, the red blood cell produces two red cells with probability 1/4. Thus, $Z_1 = 2 = 2^1$, and both the red cells produce two red cells each, with probability $(1/4)^2$. Thus $Z_2 = 4 = 2^2$ with probability $(1/4)^{Z_1}$. These 4 red cells produce 2 red cells each with probability $(1/4)^4 = (1/4)^{Z_2}$. Hence, $Z_3 = 8 = 2^3$. Thus, proceeding on these lines we have $Z_n = 2^n$ with probability $(1/4)^{Z_{n-1}}$. Consequently, at time $n + 1$ minutes after the culture begins, the probability that no white cells have yet appeared is $(1/4)^{Z_n} = (1/4)^{2^n}$.

(ii) The probability that the entire culture eventually dies out is the probability q of eventual extinction. Observe that the offspring mean $\mu = 7/6 > 1$. Hence, q is the smallest positive root of the equation $P(s) = s$. Now

$$P(s) = 1/12 + 2s/3 + s^2/4 = s \;\Rightarrow\; (s-1)(3s-1) = 0 \;\Rightarrow\; q = 1/3.$$

5.7.8 Suppose $\{Z_n, n \geq 1\}$ is a BGW branching process with $P(s) = as^2 + bs + c$, where a, b, c are positive and $P(1) = 1$. Assume that the probability of extinction $q \in (0, 1)$. Prove that (i) $c < a$ and (ii) $q = c/a$.

Solution: It is given that $0 < q < 1$, thus offspring mean μ must be > 1. Now, $P(1) = 1 \;\Rightarrow\; a + b + c = 1$. Hence,

$$\mu = 2a + b = 2a + 1 - a - c = 1 + a - c > 1 \;\Rightarrow\; a > c.$$

To find q we solve the equation $P(s) = s$.

$$\begin{aligned} P(s) = s &\Rightarrow as^2 + (b-1)s + c = 0 \\ &\Rightarrow as^2 - (a+c)s + c = 0 \;\Rightarrow\; (as - c)(s - 1) = 0 \\ &\Rightarrow q = c/a \text{ as } 0 < q < 1. \end{aligned}$$

5.7.9 The offspring distributions of some branching processes with $Z_0 = 1$ are as follows:

(a) $p_0 = 0.3,\ p_1 = 0.6,\ p_2 = 0.05,\ p_3 = 0.05$
(b) $p_0 = 0.2,\ p_1 = 0.2,\ p_2 = 0.3,\ p_3 = 0.3$
(c) $p_0 = 0.25,\ p_1 = 0.50,\ p_2 = 0.25$
(d) $p_0 = 0.25,\ p_1 = 0.40,\ p_2 = 0.35$
(e) $p_0 = 0.5,\ p_1 = 0.1,\ p_2 = 0.4.$

Find the offspring mean and determine the probability of eventual extinction in each case. Comment on the relation between the offspring mean and the extinction probability.
Solution: (i) For the offspring distribution in (a), the offspring mean is $0.85 < 1$. Hence, the extinction probability q must be 1. We verify it by solving the equation $P(s) = s$. Now,

$$\begin{aligned}
P(s) = s &\Rightarrow 0.3 - 0.4s + 0.05s^2 + 0.05s^3 = 0\\
&\Rightarrow (1 - s)(0.05s^2 + 0.01s - 0.3) = 0\\
&\Rightarrow s = 1, s = 1.65, s = -3.65.
\end{aligned}$$

Thus, being probability, the only acceptable root is 1. Hence, $q = 1$.
(ii) For the offspring distribution in (b), the offspring mean is $1.7 > 1$. Hence, the extinction probability q is the smallest positive root of the equation $P(s) = s$. Now,

$$\begin{aligned}
P(s) = s &\Rightarrow 0.2 - 0.8s + 0.3s^2 + 0.3s^3 = 0\\
&\Rightarrow (1 - s)(0.3s^2 + 0.6s - 0.2) = 0\\
&\Rightarrow s = 1, s = 0.291, s = -2.291 \quad \Rightarrow \quad q = 0.291.
\end{aligned}$$

(iii) For the offspring distribution in (c), the offspring mean is 1. Hence, the extinction probability q must be 1. The equation $P(s) = s$ is a quadratic equation $s^2 - 2s + 1 = 0$ and both the roots are 1.
(iv) For the offspring distribution in (d), the offspring mean is 1.1. Hence, the extinction probability q is the smallest positive root of the equation $P(s) = s$, which is a quadratic equation $0.35s^2 - .6s + 0.25 = 0$ and the roots are 1 and 0.7143. Hence, $q = 0.7143$.
(v) For the offspring distribution in (e), the offspring mean is 0.9. Hence, the extinction probability q must be 1. The equation $P(s) = s$ is a quadratic equation $4s^2 - 9s + 5 = 0$ and the roots are 1 and 1.25. Hence, $q = 1$.

5.7.10 One-fourth of the married couples in a society have no children. The other three-fourths of families continue to have children until they have a girl and then cease childbearing. Assume that each child is equally likely to be a boy or a girl. (i) What is the probability that a particular husband will have k male offspring, $k = 0, 1, 2, \ldots,$? (ii) What is the probability that the husband's male line will cease to exist by the 5th generation?

Table A.5 Answer key to MCQs in Chap. 5

Q. No.	1	2	3	4	5	6	7	8	9	10
Ans	a, b	c, d	a	a	d	c	b	c	d	a, b, c, d
Q. No.	11	12	13	14	15	16	17	18	19	20
Ans	b, c, d	a	a, c	c	a, b, c	d	d	d	a	

Solution: It is given that a family continues to have children until the first girl and then cease childbearing. Further, each child is equally likely to be a boy or a girl. Hence (i) the probability that a particular husband will have k male offspring is $p_k = (1/2)^{k+1}$, $k = 0, 1, 2, \ldots$. Thus, the offspring distribution is the geometric distribution with parameter (1/2). Suppose the society is divided in two groups C_1 and C_2, say, where C_1 denotes the group of families who have no children and C_2 denotes the group of families who produce according to a geometric distribution. Suppose Z_n denotes the number of male children in the nth generation corresponding to a particular family in group C_2. Thus, we have $Z_0 = 1$. Hence, as shown in Example 5.3.1, $P[Z_5 = 0] = P_5(0) = 5/6$. If the family is from group C_1, then $P[Z_5 = 0] = 1$. Hence, the required probability is $P(C_1) \times 1 + P(C_2) \times 5/6 = 1/4 + (3/4)(5/6) = 7/8$.

Answers to the multiple choice questions, based on Chap. 5, are given in Table A.5.

A.5 Chapter 6

6.8.1 Suppose $\{X(t), t \geq 0\}$ is a Markov process with state space $S = \{1, 2, 3, 4\}$ and intensity matrix Q as given below

$$Q = \begin{matrix} & \begin{matrix} 1 & 2 & 3 & 4 \end{matrix} \\ \begin{matrix} 1 \\ 2 \\ 3 \\ 4 \end{matrix} & \begin{pmatrix} -3 & 2 & 0 & 1 \\ 0 & -2 & 1/2 & 3/2 \\ 1 & 1 & -4 & 2 \\ 1 & 0 & 0 & -1 \end{pmatrix} \end{matrix}.$$

(i) Find the parameters of the sojourn time random variables. What are the expected sojourn times in the 4 states? (ii) Find the transition probability matrix of the embedded Markov chain. (iii) Examine if the Markov process is irreducible. (iv) Examine whether the states are transient or persistent. (v) Write the system of balance equations and solve it to get the long run distribution. (vi) Find the stationary distribution. Is it the same as the long run distribution? (vii) Find the log run mean fraction of a time system in states 1, 2, 3, 4.

Solution: (i) The sojourn time random variables in each of the four states have exponential distribution, with rate parameters $\lambda_1 = 3, \lambda_2 = 2, \lambda_3 = 4, \lambda_4 = 1$.

The expected sojourn times in the 4 states are $1/\lambda_i$ for $i = 1, 2, 3, 4$. (ii) The transition probability matrix P of embedded Markov chain is given by

$$P = \begin{array}{c} \\ 1 \\ 2 \\ 3 \\ 4 \end{array} \begin{array}{c} \begin{array}{cccc} 1 & 2 & 3 & 4 \end{array} \\ \begin{pmatrix} 0 & 2/3 & 0 & 1/3 \\ 0 & 0 & 1/4 & 3/4 \\ 1/4 & 1/4 & 0 & 2/4 \\ 1 & 0 & 0 & 0 \end{pmatrix} \end{array}.$$

(iii) From P, we note that $1 \to 2 \to 3 \to 4 \to 1$. Thus, all states communicate with each other. Hence, the embedded Markov chain is irreducible which implies that the Markov process is irreducible.
(iv) It is a finite state space irreducible Markov chain, hence all states are persistent.
(v) Suppose $\underline{P} = (P_1, P_2, P_3, P_4)$. Then the system of balance equations is obtained from $\underline{P}Q = \underline{0}$ or by equating the rate of transition into the state to the rate of transition out of state. These equations are as follows:

$$\begin{aligned} 3P_1 &= P_3 + P_4, \quad 2P_2 = 2P_1 + P_3 \\ 4P_3 &= (1/2)P_2, \quad P_4 = P_1 + (3/2)P_2 + 2P_3. \end{aligned}$$

We solve these equations subject to the condition that $P_1 + P_2 + P_3 + P_4 = 1$ and get $P_1 = 15/76$, $P_2 = 16/76$, $P_3 = 2/76$, $P_4 = 43/76$. Thus, the long run distribution is given by $\underline{P}$.
(vi) The stationary distribution η is a solution of $\underline{\eta}Q = \underline{0}$, subject to the condition that components of η add up to 1. Hence, η is the same as $\underline{P}$ and is given by $\underline{P} = \underline{\eta} = (15/76, 16/76, 2/76, 43/76)'$.
(vii) The log run mean fraction of time system in states 1, 2, 3, 4 is given by η_i, $i = 1, 2, 3, 4$.

6.8.2 Suppose $\{X(t), t \geq 0\}$ is a Markov process with state space $S = \{0, 1\}$ and the intensity rates $q_{01} = 2$ and $q_{10} = 3$. Find the matrix $P(t)$ of transition probability functions, by solving Kolmogorov's forward and backward differential equations.
Solution: Since the intensity rates are $q_{01} = 2$ and $q_{10} = 3$, the generator matrix Q is given by

$$Q = \begin{array}{c} \\ 0 \\ 1 \end{array} \begin{array}{c} \begin{array}{cc} 0 & 1 \end{array} \\ \begin{pmatrix} -2 & 2 \\ 3 & -3 \end{pmatrix} \end{array}.$$

Since the state space is finite, Kolmogorov's forward differential equations given by $P'(t) = P(t)Q$ exist. We solve these with initial condition $P(0) = I$. Thus,

$$P'_{00}(t) = -2P_{00}(t) + 3P_{01}(t) = -2P_{00}(t) + 3(1 - P_{00}(t)) = 3 - 5P_{00}(t)\,.$$

Suppose $h(t) = P_{00}(t) - 3/5$. Then

$$h'(t) = P'_{00}(t) = 3 - 5\,(h(t) + 3/5) = -5h(t)\,.$$

Hence, $\log h(t) = -5t + c$, that is, $h(t) = ke^{-5t}$ and hence $P_{00}(t) = ke^{-5t} + 3/5$. Now, $P_{00}(0) = 1$ gives $k = 2/5$. Thus,

$$P_{00}(t) = 3/5 + (2/5)e^{-5t} \;\;\&\;\; P_{01}(t) = 1 - P_{00}(t) = 2/5 - (2/5)e^{-5t}\,.$$

Similarly from the equation $P'(t) = P(t)Q$, we have

$$P'_{10}(t) = -2P_{10}(t) + 3P_{11}(t) = -2P_{10}(t) + 3(1 - P_{10}(t)) = 3 - 5P_{10}(t)\,.$$

Suppose $h(t) = P_{10}(t) - 3/5$. Then

$$h'(t) = P'_{10}(t) = 3 - 5\,(h(t) + 3/5) = -5h(t)\,.$$

Hence, $\log h(t) = -5t + c$, that is, $h(t) = ke^{-5t}$ and hence $P_{10}(t) = ke^{-5t} + 3/5$. Now, $P_{10}(0) = 0$ gives $k = -3/5$. Hence,

$$P_{10}(t) = 3/5 - (3/5)e^{-5t} \;\;\&\;\; P_{11}(t) = 1 - P_{10}(t) = 2/5 + (3/5)e^{-5t}\,.$$

We now discuss how to solve Kolmogorov's backward differential equations $P'(t) = QP(t)$, with initial condition $P(0) = I$. The matrix equation $P'(t) = QP(t)$ gives

$$\begin{aligned}
P'_{00}(t) &= -2P_{00}(t) + 2P_{10}(t) \;\;\&\;\; P'_{10}(t) = 3P_{00}(t) - 3P_{10}(t)\\
&\Rightarrow 3P'_{00}(t) + 2P'_{10}(t) = 0\\
&\Rightarrow 3P_{00}(t) + 2P_{10}(t) = c, \quad \text{a constant free from t}\,.
\end{aligned}$$

At $t = 0$, $P_{00}(t) = 1$ & $P_{10}(t) = 0$ implies that $c = 3$. Thus, $3P_{00}(t) + 2P_{10}(t) = 3$. Using this relation, we get

$$P'_{00}(t) = -2P_{00}(t) + 2P_{10}(t) = -2P_{00}(t) + 3 - 3P_{00}(t) = 3 - 5P_{00}(t)\,,$$

which is the same as in Kolmogorov's forward differential equations. Hence, we get

$$P_{00}(t) = 3/5 + (2/5)e^{-5t} \;\;\&\;\; P_{01}(t) = 1 - P_{00}(t) = 2/5 - (2/5)e^{-5t}\,.$$

Further,

$$P_{10}(t) = 3/2 - (3/2)P_{00}(t) = 3/2 - (3/2)\left(3/5 + (2/5)e^{-5t}\right) = 3/5 - (3/5)e^{-5t}.$$

From $P_{10}(t)$, we get

$$P_{11}(t) = 1 - P_{10}(t) = 2/5 + (3/5)e^{-5t}.$$

Thus, both the forward and the backward differential equations lead to the same solution. Hence, the matrix $P(t)$ of transition probability functions is given by

$$P(t) = \begin{array}{c} \\ 0 \\ 1 \end{array}\begin{array}{c} \begin{array}{cc} \quad 0 & \qquad\qquad 1 \end{array} \\ \begin{pmatrix} 3/5 + (2/5)e^{-5t} & 2/5 - (2/5)e^{-5t} \\ 3/5 - (3/5)e^{-5t} & 2/5 + (3/5)e^{-5t} \end{pmatrix} \end{array}.$$

6.8.3 In a workshop there are two machines, operating simultaneously and independently, where both the machines have an exponentially distributed time to failure with mean $1/\mu$. There is a single repair facility, and the repair times are exponentially distributed with rate λ. In the long run, what is the probability that no machine is operating?
Solution: Suppose $X(t)$ denotes the number of machines working at time t. Then $\{X(t), t \geq 0\}$ is a continuous time Markov chain with state space $S = \{0, 1, 2\}$. From the given information, the intensity matrix Q can be written as follows:

$$Q = \begin{array}{c} \\ 0 \\ 1 \\ 2 \end{array}\begin{array}{c} \begin{array}{ccc} 0 & \qquad 1 & \qquad 2 \end{array} \\ \begin{pmatrix} -\lambda & \lambda & 0 \\ \mu & -(\lambda + \mu) & \lambda \\ 0 & 2\mu & -2\mu \end{pmatrix} \end{array}.$$

To find the probability that no machines are operating in the long run, we find the stationary distribution which is the same as the long run distribution. Suppose $\underline{P} = (P_0, P_1, P_2)$, then $\underline{P}Q = \underline{0}$ gives the long run distribution. We thus have the following system of equations:

$$-\lambda P_0 + \mu P_1 = 0, \quad \lambda P_0 - (\mu + \lambda)P_1 + 2\mu P_2 = 0, \quad \lambda P_1 - 2\mu P_2 = 0.$$

Solving these equations subject to the condition that $P_0 + P_1 + P_2 = 1$, we get $P_0 = 1/C$, $P_1 = \lambda/\mu C$ and $P_2 = \lambda^2/2\mu^2 C$ where $C = (1 + \lambda/\mu + \lambda^2/2\mu^2)$. Hence, in the long run, the probability that no machine is operating is P_0.

6.8.4 A factory has five machines. The operating time until failure of a machine has exponential distribution with rate parameter 0.20 per hour. The repair time of a failed machine also has exponential distribution with rate parameter 0.50 per hour. The failures of the machines are independent. Further we assume that all the failed machines can be repaired simultaneously. Suppose $X(t)$ denotes the number of machines working at time t.
(i) Can we model $\{X(t), t \geq 0\}$ as a continuous time Markov chain? Justify your answer. (ii) If yes, write down the intensity matrix and the transition probability matrix of the corresponding embedded Markov chain. (iii) Is the continuous time Markov chain irreducible? (iv) Classify the states as transient or persistent.
Solution: Suppose $X(t)$ denotes the number of machines working at time t. Then the state space of the process $\{X(t), t \geq 0\}$ is $S = \{0, 1, 2, 3, 4, 5\}$. If $X(t) = 0$, that is, if all the 5 machines have failed, 5 undergo repair simultaneously. Suppose W_i denotes the repair time of ith machine, $i = 1, 2, \ldots, 5$. Then W_i follows exponential distribution with rate parameter 0.5. Thus, sojourn time in state 0 is $T_0 = \min\{W_1, W_2, \ldots, W_5\}$ and T_0 has exponential distribution with rate $5 \times 0.5 = 2.5$. If $X(t) = 1$, one machine is working and 4 machines have failed and all 4 undergo repair simultaneously. The possible transitions from 1 are to 0 if the working machine fails and 2 if one of the four machines gets repaired. Suppose Y_1 denotes the time to failure of a working machine, then Y_1 follows exponential distribution with rate 0.2. Thus, sojourn time in state 1 is $T_1 = \min\{\min\{W_1, W_2, \ldots, W_4\}, Y_1\}$. Thus, T_1 has exponential distribution with rate $4 \times 0.5 + 0.2 = 2 + 0.2 = 2.2$. Similarly, the sojourn times in states 2, 3, 4, 5 have exponential distribution with rate parameters 1.9, 1.6, 1.3, 1, respectively. Hence, we can model $\{X(t), t \geq 0\}$ as a continuous time Markov chain.
(ii) The intensity matrix Q is given by

$$Q = \begin{array}{c} \\ 0 \\ 1 \\ 2 \\ 3 \\ 4 \\ 5 \end{array} \begin{array}{c} \begin{array}{cccccc} 0 & 1 & 2 & 3 & 4 & 5 \end{array} \\ \begin{pmatrix} -2.5 & 2.5 & 0.0 & 0.0 & 0.0 & 0.0 \\ 0.2 & -2.2 & 2.0 & 0.0 & 0.0 & 0.0 \\ 0.0 & 0.4 & -1.9 & 1.5 & 0.0 & 0.0 \\ 0.0 & 0.0 & 0.6 & -1.6 & 1.0 & 0.0 \\ 0.0 & 0.0 & 0.0 & 0.8 & -1.3 & 0.5 \\ 0.0 & 0.0 & 0.0 & 0.0 & 1 & -1 \end{pmatrix} \end{array}.$$

The transition probability matrix P of the corresponding embedded Markov chain is as given below

$$P = \begin{array}{c} \\ 0 \\ 1 \\ 2 \\ 3 \\ 4 \\ 5 \end{array} \begin{array}{c} \begin{array}{cccccc} 0 & 1 & 2 & 3 & 4 & 5 \end{array} \\ \begin{pmatrix} 0 & 1 & 0.0 & 0.0 & 0.0 & 0.0 \\ 0.0909 & 0 & 0.9091 & 0.0 & 0.0 & 0.0 \\ 0.0 & 0.2106 & 0 & 0.7894 & 0.0 & 0.0 \\ 0.0 & 0.0 & 0.3750 & 0 & 0.6250 & 0.0 \\ 0.0 & 0.0 & 0.0 & 0.6154 & 0 & 0.3846 \\ 0.0 & 0.0 & 0.0 & 0.0 & 1 & 0 \end{pmatrix} \end{array}.$$

(iii) From the transition probability matrix P, we note that all states communicate with each other and hence, the continuous time Markov chain is irreducible.

(iv) Since state space is finite and all states communicate with each other, all the states are persistent. It may be noted that the embedded Markov chain is a random walk with two reflecting barriers.

6.8.5 A system consists of two machines. The amount of time that an operating machine works before breaking down is exponentially distributed with mean 5 hours. The amount of time that it takes a repairman to fix a machine is exponentially distributed with mean 4 hours. Suppose $X(t)$ is the number of machines in operating condition at time t. (i) Find the long run distribution of $\{X(t), t \geq 0\}$. (ii) If an operating machine produces 100 units of output per hour, what is the long run average output per hour of the system?

Solution: Suppose $X(t)$ is the number of machines in operating condition at time t. Then its possible values are 0, 1, 2. If $X(t) = 0$, two machines are under repair. Since there is a single repairman, $q_{01} = 1/4 = 0.25$. Using similar arguments as in the previous example, the other elements in the intensity matrix Q can be obtained and it is given by

$$Q = \begin{array}{c} \\ 0 \\ 1 \\ 2 \end{array} \begin{array}{c} \begin{array}{ccc} 0 & 1 & 2 \end{array} \\ \begin{pmatrix} -0.25 & 0.25 & 0.0 \\ 0.20 & -0.45 & 0.25 \\ 0.0 & 0.40 & -0.40 \end{pmatrix} \end{array}.$$

(i) Suppose $\underline{P} = (P_0, P_1, P_2)$, then $\underline{P}Q = \underline{0}$ gives the long run distribution. We thus have the following system of equations:

$$-0.25P_0 + 0.20P_1 = 0, 0.25P_0 - 0.45P_1 + 0.40P_2 = 0, 0.25P_1 - 0.40P_2 = 0.$$

Solving these equations subject to the condition that $P_0 + P_1 + P_2 = 1$, we get $P_0 = 32/97$, $P_1 = 40/97$ & $P_2 = 25/97$. (ii) An operating machine produces 100 units of output per hour, hence the long run average output per hour of the system is given by $100P_1 + 200P_2 = 9000/97 = 92.7835$.

6.8.6 Suppose a data scientist at a business analytics company can be a trainee, a junior data scientist or a senior data scientist. Suppose the three levels are denoted by $1, 2, 3$, respectively. If $X(t)$ denotes the level of the person at time t, we assume that X(t) evolves as a Markov chain in continuous time. Suppose the mean sojourn times in the three states 1, 2 and 3 are $0.1, 0.2, 2$ years, respectively. It is given that a trainee is promoted to a junior data scientist with probability $2/3$ and to a senior data scientist with probability $1/3$. A junior data scientist leaves and is replaced by a trainee with probability $2/5$ and is promoted to a senior data scientist with probability $3/5$. A senior data scientist leaves and is replaced by a trainee with probability $1/4$ and by a junior data scientist with probability $3/4$. Find the long run average proportion of time a data scientist is a senior data scientist.

Solution: From the given information, the generator matrix Q can be written as follows:

$$Q = \begin{matrix} \\ 1 \\ 2 \\ 3 \end{matrix} \begin{matrix} \begin{matrix} 1 & \quad 2 & \quad 3 \end{matrix} \\ \begin{pmatrix} -10 & 20/3 & 10/3 \\ 2 & -5 & 3 \\ 1/8 & 3/8 & -4/8 \end{pmatrix} \end{matrix}.$$

Solving the matrix equation $\underline{P}Q = 0$, we get the limiting distribution, which is the same as the stationary distribution. It is given by $(0.0323, 0.1075, 0.8602)'$. Hence, the long run average proportion of time a data scientist is a senior data scientist is 0.8602.

6.8.7 There are two photo copying machines in the office; one is operating and the other is a standby machine. The operating machine fails after an exponentially distributed duration having rate μ and is replaced by the standby. Lifetime of standby machine is also exponentially distributed with rate μ. It is given that at a time only one machine will be repaired and repair times are exponentially distributed with rate λ. Suppose $X(t)$ is the number of machines in operating condition at time t. (i) Can $\{X(t), t \geq 0\}$ be modeled as a continuous time Markov chain? Justify your answer. (ii) Write down the generator matrix. (iii) Find the long run average proportion of time one machine is working. (iv) How will the generator matrix change if both the machines can be repaired simultaneously? (v) How will it further change if both machines are operating simultaneously?

Solution: Suppose $X(t)$ is the number of machines in operating condition at time t. Hence, the state space is $\{0, 1, 2\}$. From the given information, the sojourn time in each state has an exponential distribution. Further, we assume that sojourn times are independent of each other and hence the stochastic process $\{X(t), t \geq 0\}$ can be modeled as a continuous time Markov chain. Suppose Q, Q_1 and Q_2 denote the intensity matrices when (i) only one machine is repaired, (ii) if both the machines can be repaired simultaneously and (iii) if both the machines are operating simultaneously and both can be repaired simultaneously, respectively. These are given by

Table A.6 Answer key to MCQs in Chap. 6

Q. No.	1	2	3	4	5	6	7	8	9	10
Ans	d	c	c	c	a, c, d	c	b	c	a, c	d
Q. No.	11	12	13	14	15	16	17	18	19	20
Ans	b	c	c	c						

$$Q = \begin{array}{c} \\ 0 \\ 1 \\ 2 \end{array} \begin{array}{c} \begin{array}{ccc} 0 & 1 & 2 \end{array} \\ \begin{pmatrix} -\lambda & \lambda & 0 \\ \mu & -(\lambda+\mu) & \lambda \\ 0 & \mu & -\mu \end{pmatrix} \end{array} \quad Q_1 = \begin{array}{c} \\ 0 \\ 1 \\ 2 \end{array} \begin{array}{c} \begin{array}{ccc} 0 & 1 & 2 \end{array} \\ \begin{pmatrix} -2\lambda & 2\lambda & 0 \\ \mu & -(\lambda+\mu) & \lambda \\ 0 & \mu & -\mu \end{pmatrix} \end{array}.$$

$$Q_2 = \begin{array}{c} \\ 0 \\ 1 \\ 2 \end{array} \begin{array}{c} \begin{array}{ccc} 0 & 1 & 2 \end{array} \\ \begin{pmatrix} -2\lambda & 2\lambda & 0 \\ \mu & -(\lambda+\mu) & \lambda \\ 0 & 2\mu & -2\mu \end{pmatrix} \end{array}.$$

Corresponding to Q, the long run distribution is given by

$$\pi_0 = \frac{1}{1+\lambda/\mu+(\lambda/\mu)^2},\ \pi_1 = \frac{\lambda/\mu}{1+\lambda/\mu+(\lambda/\mu)^2},\ \pi_2 = \frac{(\lambda/\mu)^2}{1+\lambda/\mu+(\lambda/\mu)^2}.$$

Hence, the long run average proportion of time one machine is working is $\pi_1 = (\lambda/\mu)/(1+\lambda/\mu+(\lambda/\mu)^2)$.

Answers to the multiple choice questions, based on Chap. 6, are given in Table A.6.

A.6 Chapter 7

7.7.1 Suppose that customers arrive at a facility according to a Poisson process $\{X(t), t \geq 0\}$ with rate 6 per hour. (i) Find the probability that 8 customers arrive in one hour. (ii) Find the probability that 4 customers arrive in one hour and 10 customers arrive by the end of three hours. (iii) If 10 customers arrived by the end of the first three hours, find the probability that 3 customers arrived in the first hour. (iv) If 2 customers arrived by the end of the first hour, find the probability that 6 customers will arrive in the next two hours. (v) Find the probability that at most 5 customers arrive in one hour. (vi) If more than 2 customers arrived by the end of the first hour, find the probability that more than 6 customers will arrive in the next two hours. (vii) Find $E(X(2))$, $Var(X(2))$ and $Cov(X(1), X(2))$.

Solution: (i) $P[X(1) = 8] = e^{-6}6^8/8! = 0.1033$.
Using stationary and independence of increments property of a Poisson process, we find the required probabilities as follows:

$$\begin{aligned}(ii)P[X(1) = 4, X(3) = 10] &= P[X(1) - X(0) = 4, X(3) - X(1) = 6]\\ &= P[X(1) - X(0) = 4]P[X(3) - X(1) = 6]\\ &= e^{-6}6^4/4! \times e^{-12}12^6/6! = 0.0034.\end{aligned}$$

$$\begin{aligned}(iii)P[X(1) = 3|X(3) = 10] &= P[X(1) = 3, X(3) = 10]/P[X(3) = 10]\\ &= \frac{P[X(1) - X(0) = 3, X(3) - X(1) = 7]}{P[X(3) = 10]}\\ &= (e^{-6}6^3/3!)(e^{-12}12^7/7!)/(e^{-18}18^{10}/10!)\\ &= 0.2601.\end{aligned}$$

$$\begin{aligned}(iv)P[X(3) = 8|X(1) = 2] &= P[X(3) = 8, X(1) = 2]/P[X(1) = 2]\\ &= \frac{P[X(3) - X(1) = 6, X(1) - X(0) = 2]}{P[X(1) = 2]}\\ &= (e^{-6}6^2/2!)(e^{-12}12^6/6!)/(e^{-6}6^2/2!)\\ &= e^{-12}12^6/6! = 0.0255.\end{aligned}$$

(v) $P[X(1) \leq 5] = \sum_{r=0}^{5} e^{-6}6^r/r! = 0.4457$.

$$\begin{aligned}(vi)P[X(3) - X(1) \geq 6|X(1) \geq 2] &= P[X(3) - X(1) \geq 6|X(1) - X(0) \geq 2]\\ &= P[X(3) - X(1) \geq 6]\\ &= 1 - \sum_{r=0}^{5} e^{-12}12^r/r! = 0.9797.\end{aligned}$$

(vii) $E(X(2)) = 12$, $Var(X(2)) = 12$. It is shown that $Cov(X(s), X(t)) = Var(X(1)) \min\{s, t\}$. Hence, $Cov(X(1), X(2)) = 6$. Note that it is positive.

7.7.2 Shocks occur to a system according to a Poisson process with rate λ. Suppose that the system survives each shock with probability α, independently of other shocks, so that its probability of surviving k shocks is α^k. What is the probability that the system survives up to time t?
Solution: It is given that shocks occur to a system according to a Poisson process of rate λ. Suppose $X(t)$ denotes the number of shocks in $(0, t]$. Then $P[X(t) = r] = e^{-\lambda t}(\lambda t)^r/r!, r = 0, 1, 2, \ldots$. Hence, the probability that the system survives up to time t is given by

$$\sum_{r\geq 0} \alpha^r \, e^{-\lambda t}(\lambda t)^r/r! = E(\alpha^{X(t)}) = \exp\{\lambda t(\alpha - 1)\},$$

which is the value of the probability generating function at α of a Poisson distribution with mean λt.

7.7.3 A device fails when a cumulative effect of k shocks occurs. If the shocks occur according to a Poisson process with rate λ, what is the probability density function of the life T of the device?
Solution: Since a device fails when a cumulative effect of k shocks occurs, the life T of the device is the same as the epoch of kth occurrence. Hence, T follows a Gamma $G(\lambda, k)$ distribution with the probability density function,

$$f(x) = \lambda^k x^{k-1} e^{-\lambda x} / \Gamma(k), \quad x > 0.$$

7.7.4 A radioactive source emits particles according to a Poisson process with rate $\lambda = 2$ particles per minute. (i) What is the probability that the first particle appears after three minutes? (ii) What is the probability that the first particle appears some time after three minutes but before six minutes? (iii) What is the probability that exactly one particle is emitted in the three to five minutes?
Solution: (i) $P[T_1 > 3] = e^{-6} = 0.002479$,
(ii) $P[3 < T_1 < 6] = e^{-6} - e^{-12} = 0.002473$ and
(iii) $P[X(5) - X(3) = 1] = P[X(2) = 1] = e^{-4} * 4 = 0.07326$.

7.7.5 Customers arrive at a mall according to a Poisson process with rate $\lambda = 10$ per hour. The store opens at 10:00 a.m. (i) Find the probability that exactly one customer has arrived by 10:15 and a total of 10 have arrived by 11 a.m. (ii) Suppose it is known that a single customer entered during the first hour. What is the conditional probability that this person entered during the first thirty minutes? (iii) Given that 3 customers arrived during the first hour, what is the conditional probability that first customer arrived in the first 15 minutes, second customer arrived during the first 20 to 30 minutes and the third customer arrived during the first 30 to 45 minutes? (iv) Given that 4 customers arrived during the first hour, what is the conditional probability that first customer arrived in the first 15 minutes, second and the third customers arrived during the first 20 to 30 minutes and the fourth customer arrived during the first 30 to 45 minutes?
Solution: Using stationary and independence of increments property of a Poisson process, we find the required probabilities as follows:

$$\begin{aligned}(i) P[X(1/4) = 1, X(1) = 10] &= P[X(1/4) = 1, X(1) - X(1/4) = 9] \\ &= P[X(1/4) = 1]P[X(1) - X(1/4) = 9] \\ &= e^{-5/2}(5/2)e^{-15/2}(15/2)^9/9! = 0.02348.\end{aligned}$$

(ii) $P[T_1 < 1/2 | X(1) = 1] = \frac{1}{2}/1 = 1/2$ or

$$\begin{aligned}P[X(1/2) = 1 | X(1) = 1] &= P[X(1/2) = 1, X(1) - X(1/2) = 0] \\ &\quad \times (P[X(1) = 1])^{-1} \\ &= e^{-5}5e^{-5}/e^{-10}10 = 1/2.\end{aligned}$$

(iii) We compute the required probability
$P[0 < S_1 < 1/4, 1/3 < S_2 < 1/2, 1/2 < S_3 < 3/4|X(1) = 3]$ as
$\int_0^{1/4}\int_{1/3}^{1/2}\int_{1/2}^{3/4} 3!du_1du_2du_3 = 1/16.$
(iv) We compute the required probability
$P[0 < S_1 < 1/4, 1/3 < S_2 < S_3 < 1/2, 1/2 < S_4 < 3/4|X(1) = 4]$ as
$\int_0^{1/4}\int_{1/3}^{1/2}\int_{u_2}^{1/2}\int_{1/2}^{3/4} 4!du_1du_2du_3du_4 = 1/8.$

7.7.6 Suppose $\{X(t), t \geq 0\}$ is a Poisson process with rate 3 per hour. Find the conditional probability that there were two events in the first hour, given that there were five events in the first three hours.
Solution: Using stationary and independence of increments property of a Poisson process, we find the probability as follows:

$$\begin{aligned} P[X(1) = 2|X(3) = 5] &= P[X(1) = 2, X(3) = 5]/P[X(3) = 5] \\ &= P[X(1) = 2, X(3) - X(1) = 3]/P[X(3) = 5] \\ &= P[X(1) = 2]P[X(2) = 3]/P[X(3) = 5] \\ &= (e^{-3}3^2/2!)(e^{-6}6^3/3!)\Big/(e^{-9}9^5/5!) \\ &= 80/243 = 0.3292. \end{aligned}$$

7.7.7 Customers arrive at a service facility according to a Poisson process with rate $\lambda = 5$ per hour. Given that 12 customers arrived during the first two hours of service, what is the conditional probability that 5 customers arrived during the first hour?
Solution: Using stationary and independence of increments property of a Poisson process, we find the probability as follows:

$$\begin{aligned} P[X(1) = 5|X(2) = 12] &= P[X(1) = 5, X(2) - X(1) = 7]/P[X(2) = 12] \\ &= (e^{-5}5^5/5!)(e^{-5}5^7/7!)\Big/(e^{-10}10^{12}/12!) \\ &= 0.1934. \end{aligned}$$

7.7.8 Customers arrive at a bank according to a Poisson process with rate $\lambda = 10$ per hour. Find the probability that more than 5 customers arrive during a period of 15 minutes.
Solution: The required probability is given by

$$P[X(1/4) > 5] = 1 - P[X(1/4) \leq 5] = 1 - \sum_{i=0}^{5} e^{-2.5}(2.5)^i/i! = 0.0420.$$

7.7.9 A critical component on a submarine has an operating lifetime that is exponentially distributed with mean 0.5 years. As soon as a component fails, it is replaced by a new one having statistically identical properties. What is the smallest number of spare components that the submarine should stock if it is leaving for a one-year tour and wishes that the probability of number of failures exceeding the number of spare components to be less than 0.02?
Solution: Suppose k is the number of spare parts. If $k+1$ failures occur in a year, then the unit will be inoperable. Thus, we want to find k such that $P[X(1) = k+1] \leq 0.02$. We solve $P[X(1) = k+1] = 0.02$ to find the smallest value of k. Now $X(1) \sim Poi(2)$. Hence, $P[X(1) = 5] = 0.03609$ and $P[X(1) = 6] = 0.0120 < 0.02$. Thus, the smallest value of k is 5.

7.7.10 For a Poisson process with rate λ, given that $X(t) = n$, find the density function for W, the time of occurrence of the rth event. What is its expectation? Assume that $r < n$.
Solution: The probability density function g of W is the same as that of rth order statistics from uniform $U(0, t)$. It is given by

$$\begin{aligned} g(x) &= \frac{n!}{(r-1)!(n-r)!}(x/t)^{r-1}(1-x/t)^{n-r}(1/t) \\ &= \frac{n!}{(r-1)!(n-r)!}\frac{x^{r-1}(t-x)^{n-r}}{t^n}, \quad 0 < x < t. \end{aligned}$$

Further, $E(X_{(r)}) = rt/(n+1)$.

7.7.11 Suppose $\{X(t), t \geq 0\}$ is a Poisson process with rate λ and $X(1) = n$. For $n = 1, 2, \ldots$, determine the mean of the first arrival time W.
Solution: Given $X(1) = n$, the distribution of the first arrival time W is the same as that of the first-order statistics from uniform $U(0, 1)$ distribution. Hence its mean is $1/(n+1)$.

7.7.12 Suppose $\{X(t), t \geq 0\}$ is a Poisson process with rate λ and $W(t) = X(t+1) - X(t)$. Find $Cov(W(t), W(t+s))$.
Solution: Since $\{X(t), t \geq 0\}$ is a Poisson process with rate λ, $Cov(X(t), X(s)) = \lambda s$ for $s < t$. Suppose $s < 1$, hence $t+s < t+1$. Observe that

$$\begin{aligned} Cov(W(t), W(t+s)) &= Cov(X(t+1) - X(t), X(t+s+1) - X(t+s)) \\ &= Cov(X(t+1), X(t+s+1)) \\ &- Cov(X(t+s), X(t+1)) - Cov(X(t+s+1), X(t)) \\ &+ Cov(X(t+s), X(t)) \\ &= \lambda(t+1-t-s-t+t) = \lambda(1-s). \end{aligned}$$

Suppose $s > 1$, hence $t+s > t+1$. In this case,

$$Cov(W(t), W(t+s)) = \lambda(t+1-t-1-t+t) = 0.$$

Observe that for $s > 1$, $t+s > t+1$. Hence, $X(t+1) - X(t)$ and $X(t+s+1) - X(t+s)$ are independent random variables, in view of independence of increments over disjoint intervals.

7.7.13 Customers arrive at a bank according to a Poisson process with rate λ. Suppose two customers arrived during the first hour. What is the probability that (i) both arrived during the first 20 minutes? (ii) at least one arrived during the first 20 minutes?

Solution: It is given that $X(1) = 2$, where time unit is taken as 1 hour. (i) Thus, the probability that both arrived during the first 20 minutes is given by $P[X(1/3) = 2|X(1) = 2]$. Using stationary and independence of increments property of a Poisson process, we find the probability as follows:

$$\begin{aligned}
P[X(1/3) = 2|X(1) = 2] &= \frac{P[X(1/3) = 2, X(1) = 2]}{P[X(1) = 2]} \\
&= \frac{P[X(1/3) - X(0) = 2, X(1) - X(1/3) = 0]}{P[X(1) = 2]} \\
&= \frac{P[X(1/3) - X(0) = 2]P[X(1) - X(1/3) = 0]}{P[X(1) = 2]} \\
&= \frac{P[X(1/3) = 2]P[X(2/3) = 0]}{P[X(1) = 2]} = \frac{1}{9}\,.
\end{aligned}$$

In the third step, we use the independence of increments and in the fourth step we use stationarity of increments.

(ii) The probability that at least one arrived during the first 20 minutes is given by $1 - P[X(1/3) = 0|X(1) = 2]$. Using similar arguments, it can be computed as

$$1 - P[X(1/3) = 0]P[X(2/3) = 2]/P[X(1) = 2] = 1 - 4/9 = 5/9.$$

Note that in both the cases, the probabilities do not depend on λ.

7.7.14 Events occur according to a Poisson process with rate $\lambda = 2$ per hour. (i) What is the probability that no event occurs between 8 p.m. and 9 p.m.? (ii) Starting at noon, what is the expected time at which the fourth event occurs? (iii) What is the probability that two or more events occur between 6 p.m. and 8 p.m.?

Solution: (i)$P[X(1) = 0] = e^{-2} = 0.1353$, (ii) $E(S_4) = 4/2 = 2$, that is, 2 p.m. and
(iii) $1 - P[X(2) = 0] - P[X(2) = 1] = 1 - e^{-4} - 4e^{-4} = 0.9084$.

7.7.15 The number of industrial accidents in a factory is adequately modeled as a Poisson process with a rate of 1.3 per year. (i) What is the probability that there are no accidents in one year? (ii) What is the expected time (in days) between two consecutive accidents?

Solution: (i) $P[X(1){=}0] = e^{-1.3} = 0.2725$. (ii) $365 \times (1/1.3) = 280.7692$ days.

7.7.16 The interval between successive train arrivals at a station is uniformly distributed on (0, 1), time unit being one hour. Passengers arrive according to a Poisson process $\{X(t), t \geq 0\}$ with rate 70 per hour. Suppose a train has just left the station. If Y denotes the number of people who get on the next train, find $E(Y)$ and $Var(Y)$.
Solution: If U denotes the time between successive train arrivals, then it follows $U \sim U(0, 1)$ distribution. The train has just left and hence the number of people who get on the next train is the same as the number of people who arrive during random time U. Thus, $Y = X(U)$ and hence

$$E(Y) = E(E(X(U))|U) = E(70U) = 35 \quad \text{and}$$

$$\begin{aligned} Var(Y) &= E(Var(X(U))|U) + Var(E(X(U))|U) \\ &= E(70U) + Var(70U) = 35 + 4900/12 = 443.33\,. \end{aligned}$$

7.7.17 If an individual has never had a previous automobile accident, then the probability that he or she has an accident in the next h time units is $\beta h + o(h)$. On the other hand, if he or she has ever had a previous accident, then the probability is $\alpha h + o(h)$. Assuming that the occurrence of accidents in both the cases is modeled by independent Poisson processes, find the expected number of accidents an individual has by time t.
Solution: Suppose $X_1(t)$ denotes the number of accidents in $(0, t]$ by an individual who never had a previous automobile accident. Then $\{X_1(t), t \geq 0\}$ is a Poisson process with rate β. Similarly, suppose $X_2(t)$ denotes the number of accidents in $(0, t]$ by an individual who had a previous automobile accident. Then $\{X_2(t), t \geq 0\}$ is a Poisson process with rate α. Then $\{X(t) = X_1(t) + X_2(t), t \geq 0\}$ is a Poisson process with rate $\alpha + \beta$. Hence $E(X(t)) = (\alpha + \beta)t$.

7.7.18 Customers demanding service at a central processing facility arrive according to a Poisson process with rate $\lambda = 8$ per unit time. Independently, each customer is classified as high priority, depending on his/her purchase history, with probability $p = 0.2$ or low priority with probability $q = 1 - p = 0.8$. What is the probability that three high priority and five low priority customers arrive during the first unit of time?
Solution: Suppose $X_h(t)$ denotes the number of high priority customers in $(0, t]$. Then $\{X_h(t), t \geq 0\}$ is a Poisson process with rate $8 * 0.2 = 1.6$. Suppose $X_l(t)$ denotes the number of low priority customers in $(0, t]$. Then $\{X_l(t), t \geq 0\}$ is a Poisson process with rate $8 * 0.8 = 6.4$. Thus, for fixed t, $X_h(t) \sim Poi(1.6t)$ and $X_l(t) \sim Poi(6.4t)$. Hence,

$$\begin{aligned} P[X_h(1) = 3, X_l(1) = 5] &= P[X_h(1) = 3]P[X_l(1) = 5] \\ &= e^{-1.6}(1.6)^3 e^{-6.4}(6.4)^5/3!5! \\ &= e^{-8}(1.6)^3(6.4)^5/720 = 0.02049. \end{aligned}$$

7.7.19 Suppose 45% of the cars on the road are of non-white color and 55% are of white color. Suppose $X(t)$ denotes the number of cars crossing at a given toll booth during $(0, t]$ and $\{X(t), t \geq 0\}$ is a Poisson process with rate 40 per hour. What is the probability that four white color cars and five non-white cars cross the intersection during a 10-minute interval?
Solution: Suppose $X_1(t)$ and $X_2(t)$ denote number of non-white and white color cars respectively, crossing at a given toll booth during $(0, t]$. Then $\{X_1(t), t \geq 0\}$ and $\{X_2(t), t \geq 0\}$ are Poisson processes with rates 40×0.45 and 40×0.55, respectively. Thus, for fixed t, $X_1(t) \sim Poi(18t)$ and $X_2(t) \sim Poi(22t)$ and the two are independent. Hence,

$$\begin{aligned} P[X_2(1/6) = 4, X_1(1/6) = 5] &= P[X_2(1/6) = 4]P[X_1(1/6) = 5] \\ &= \frac{e^{-3}3^5 e^{-22/6}(22/6)^4}{4!5!} \\ &= \frac{e^{-20/3}3^5(11/3)^4}{24 \times 120} = 0.01941\,. \end{aligned}$$

7.7.20 Shocks occur to a system according to a Poisson process with rate λ. Each shock causes some damage to the system, and these damages accumulate. Suppose $N(t)$ denotes the number of shocks up to time t and Y_i denotes the damage caused by the ith shock and $X(t)$ is the total damage up to time t. Determine the mean and variance of the total damage up to time t when the individual shock damages are exponentially distributed with parameter θ.
Solution: The total damage $X(t) = \sum_{i=1}^{N(t)} Y_i$ up to time t is modeled as a compound Poisson process. Hence, $E(X(t)) = \lambda t/\theta$ and $Var(X(t)) = \lambda t(2/\theta^2)$.

7.7.21 For a collective risk model, the number of claims in a year has a Poisson distribution with rate 8. Claim amounts are mutually independent with two possible values 1000 and 2000 rupees with equal chance. (i) Find the probability that total claims in one year are 0, 1000, 2000 and 3000.
(ii) Find the probability generating function of total claims in one year and hence find the probability that total claims in one year are 0, 1000, 2000 and 3000. Comment on the results.
Solution: Suppose we take the possible values 1000 and 2000 as 1 and 2 units, that is, 1 unit is 1000 rupees. Suppose $X(t) = \sum_{i=1}^{N(t)} Y_i$, where $\{N(t), t \geq 0\}$ is the Poisson process with rate 8 and Y_i is a random variable with possible values as 1 and 2 with probability 1/2 each. Then $\{X(t), t \geq 0\}$ is a compound Poisson process. (i) We compute the probabilities as follows:

$$\begin{aligned}
P[X(1) = 0] &= P[N(1) = 0] = e^{-8} = 0.000335 \\
P[X(1) = 1] &= P[Y_1 = 1|N(1) = 1]P[N(1) = 1] = P[Y_1 = 1]P[N(1) = 1] \\
&= 0.5e^{-8}8^1 = 4 \times e^{-8} = 0.001342 \\
P[X(1) = 2] &= P[Y_1 = 2|N(1) = 1]P[N(1) = 1] \\
&+ P[Y_1 + Y_2 = 2|N(1) = 2]P[N(1) = 2] \\
&= 0.5e^{-8}8^1 + P[Y_1 = 1]P[Y_2 = 1]e^{-8}8^2/2! \\
&= 0.5e^{-8}8^1 + (1/2)^2e^{-8}8^2/2! = 0.004026.
\end{aligned}$$

$$\begin{aligned}
P[X(1) = 3] &= P[Y_1 = 2, Y_2 = 1|N(1) = 2]P[N(1) = 2] \\
&+ P[Y_1 = 1, Y_2 = 2|N(1) = 2]P[N(1) = 2] \\
&+ P[Y_1 = 1, Y_2 = 1, Y_3 = 1|N(1) = 3]P[N(1) = 3] \\
&= 2 \times (1/2)^2e^{-8}8^2/2! + (1/2)^3e^{-8}8^3/3! \\
&= 16 \times e^{-8} + (32/3)e^{-8} = 0.008945.
\end{aligned}$$

(ii) The probability generating function of $X(t)$ for fixed t is $H(s) = \exp\{\lambda t(P(s) - 1)\}$, where $P(s) = 0.5(s^1 + s^2)$. With $t = 1$, the probability generating function of $X(1)$ can be expressed as follows:

$$\begin{aligned}
H(s) &= \exp\{8(0.5(s + s^2) - 1)\} = e^{-8}e^{4(s+s^2)} \\
&= e^{-8}\{1 + 4(s + s^2) + (4(s + s^2))^2/2! + (4(s + s^2))^3/3! + \cdots\} \\
&= e^{-8}\{1 + 4s + 12s^2 + 16s^3 + 8s^4 + (32/3)s^3 + \cdots\} \\
&= e^{-8}\{1 + 4s + 12s^2 + (80/3)s^3 + \cdots\}.
\end{aligned}$$

Hence,

$$\begin{aligned}
P[X(1) = 0] &= e^{-8} = 0.000335, \quad P[X(1) = 1] = 4 \times e^{-8} = 0.001342 \\
P[X(1) = 2] &= 12 \times e^{-8} = 0.004026 \\
\& \ P[X(1) = 3] &= (80/3) \times e^{-8} = 0.008945.
\end{aligned}$$

With the two approaches, we get the same probabilities, as these should be. Observe that all these probabilities are very small, as expected, since the rate of occurrence of claims is 8 per year.

7.7.22 Aggregate claims process $\{X(t), t \geq 0\}$ in a collective risk model is a compound Poisson process, where $N(1)$ has a Poisson distribution with mean $\lambda = 10$. The individual claim distribution is exponential with mean $\mu = 100$ units. Find the mean and variance of $X(1)$. Using normal approximation find $P[X(1) > 1.3E(X(1))]$.

Solution: Since $\{X(t), t \geq 0\}$ in a collective risk model is a compound Poisson process,

$$E(X(1)) = \lambda\mu = 1000 \;\&\; Var(X(1)) = \lambda(100^2 + 100^2) = 200000.$$

With normal approximation, $X(1) \sim N(1000, \sigma^2)$, where $\sigma = \sqrt{200000} = 447.2136$. Hence,

$$\begin{aligned} P[X(1) > 1.3E(X(1))] &= P[(X(1) - 1000)/\sigma > (1300 - 1000)/\sigma] \\ &= P[Z > 0.6708] = 0.2512, \end{aligned}$$

where $Z \sim N(0, 1)$.

7.7.23 The frequency of car accidents in $(0, t]$ is modeled as a Poisson process. There are 90% good drivers which on the average commit 1 accident over a period of one year while a similar rate for bad drivers is 3. If an accident occurs, the claim amount has lognormal distribution with location parameter 3 and scale parameter 2. Calculate the mean m and variance v of total claims over a period of one year.

Solution: Suppose $N_g(t)$ and $N_b(t)$ denote the frequency of car accidents in $(0, t]$ by good and bad drivers, respectively. Then $\{N_g(t), t \geq 0\}$ and $\{N_b(t), t \geq 0\}$ are Poisson processes with rates 1 and 3 per year, respectively. Suppose $\{N(t), t \geq 0\}$ denote the frequency of car accidents in $(0, t]$, then $\{N(t), t \geq 0\}$ is a Poisson process with rate $\lambda = 0.9 \times 1 + 0.1 \times 3 = 1.2$. Hence $E(N(t)) = 1.2t$. The claim amount random variable Y has lognormal distribution with location parameter $\mu = 3$ and scale parameter $\sigma^2 = 2$. Hence,

$$E(Y) = e^{\mu+\sigma^2/2} = e^4 = 54.59815 \;\&\; E(Y^2) = e^{2\mu+2\sigma^2} = e^{10} = 22026.47.$$

Hence,

$$m = 1.2 \times 54.59815 = 65.5178 \;\&\; v = 1.2 \times 22026.47 = 26431.76.$$

7.7.24 Suppose $\{X(t), t \geq 0\}$ is modeled as a compound Poisson process with rate $\lambda = 3$, and the probability mass function of Y_i is given by $p(x) = 0.1x, x = 1, 2, 3, 4$. Calculate the probabilities that aggregate claims over a period of one year equal 0, 1, 2, 3 units. Also find the probability that aggregate claims exceed 3 units.

Solution: It is given that $X(t) = \sum_{i=1}^{N(t)} Y_i$, where $\{N(t), t \geq 0\}$ is a Poisson process with rate $\lambda = 3$. The probability mass function of Y_i is given by $p(x) = 0.1x, x = 1, 2, 3, 4$. Hence,

$$
\begin{aligned}
P[X(1) = 0] &= P[N(1) = 0] = e^{-3} = 0.04979 \\
P[X(1) = 1] &= Pa[Y_1 = 1|N(1) = 1]P[N(1) = 1] = P[Y_1 = 1]P[N(1) = 1] \\
&= 0.1e^{-3}3^1 = 0.01494 \\
P[X(1) = 2] &= P[Y_1 = 2|N(1) = 1]P[N(1) = 1] \\
&+ P[Y_1 + Y_2 = 2|N(1) = 2]P[N(1) = 2] \\
&= 0.2e^{-3}3^1 + P[Y_1 = 1]P[Y_2 = 1]e^{-3}3^2/2! \\
&= 0.2e^{-3}3^1 + (0.1)^2e^{-3}3^2/2! = 0.03211 \\
P[X(1) = 3] &= P[Y_1 = 3|N(1) = 1]P[N(1) = 1] \\
&+ P[Y_1 = 2, Y_2 = 1|N(1) = 2]P[N(1) = 2] \\
&+ P[Y_1 = 1, Y_2 = 2|N(1) = 2]P[N(1) = 2] \\
&+ P[Y_1 = 1, Y_2 = 1, Y_3 = 1|N(1) = 3]P[N(1) = 3] \\
&= (0.3)e^{-3}3 + 2 \times (0.2)(0.1)e^{-3}3^2/2! + (0.1)^3e^{-3}3^3/3! \\
&= 0.05399 \\
P[X(1) > 3] &= 0.84917.
\end{aligned}
$$

7.7.25 Customers arrive at a store as a group of 1 or 2 persons with equal probability. The arrival of groups is according to a Poisson process with rate 3 per 10 minutes. Find the probability that 4 customers arrive in 20 minutes.
Solution: Suppose $X(t) = \sum_{i=1}^{N(t)} Y_i$, where $\{N(t), t \geq 0\}$ is a Poisson process with rate 3 per 10 minutes, and Y_i is a random variable with possible values 1 or 2 with equal probability 1/2. Then $\{X(t), t \geq 0\}$ is a compound Poisson process. To find the probability that 4 customers arrive in 20 minutes, we find the probability generating function of $X(t)$ for fixed t as follows. The common probability generating function of Y_i is given by $P(s) = (s + s^2)/2$. The probability generating function of $N(t)$ for fixed t is $G(s) = \exp\{\lambda t(s - 1)\}$. Then the probability generating function of $X(t)$ for fixed t is $G(P(s)) = \exp\{\lambda t((s + s^2)/2 - 1)\}$. We have $\lambda = 3$ per time unit which is 10 minutes. Thus 20 minutes is 2 time units. Hence, the probability generating function of $X(2)$ is $\exp\{6((s + s^2)/2 - 1)\}$. We find the coefficient of s^4, to find the probability that 4 customers arrive in 20 minutes. It is given by $e^{-6}(9/2! + 27/2! + 81/4!) = 0.053$.

Answers to the multiple choice questions, based on Chap. 7, are given in Table A.7.

Table A.7 Answer key to MCQs in Chap. 7

Q. No.	1	2	3	4	5	6	7	8	9	10
Ans	d	a, b, c	d	d	c	c	b	a	d	b
Q. No.	11	12	13	14	15	16	17	18	19	20
Ans	b	d	b	c	b					

A.7 Chapter 8

8.8.1 Suppose a population of organisms evolves according to a Yule-Furry process $\{X(t), t \geq 0\}$ with birth rate λ and $X(0) = 1$. For $\lambda = 0.1, 0.2, 0.3, 0.4$, find the mean population size and variance of the population size at $t = 10$. Find the probability that the population size at $t = 10$ is larger than its expected value. Comment on the findings.
Solution: For a Yule-Furry process with birth rate λ and $X(0) = 1$, the distribution of $X(t)$ is geometric with parameter $p = e^{-\lambda t}$ and support $\{1, 2, \ldots, \}$. Thus, for $i = 1, 2, \ldots$, $P[X(t) = i] = e^{-\lambda t}(1 - e^{-\lambda t})^{i-1}$. Hence, $P[X(t) \geq i] = (1 - e^{-\lambda t})^{i-1}$. Thus,

$$E(X(t)) = e^{\lambda t}, \ Var(X(t)) = e^{\lambda t}(e^{\lambda t} - 1) \ \& \ P[X(t) \geq e^{\lambda t}] = (1 - e^{-\lambda t})^{e^{\lambda t}-1}.$$

The values of $E(X(10))$, $Var(X(10))$ and $P[X(10) \geq E(X(10))]$ for $\lambda = 0.1, 0.2, 0.3, 0.4$ are presented in Table A.8.
From the table, we observe that as λ increases, $E(X(10))$ and $Var(X(10))$ increase, the rate of increase for variance is very high, as expected.

8.8.2 Suppose in Exercise 8.8, $\{X(t), t \geq 0\}$ is a Yule-Furry process with birth rate λ and $X(0) = a = 5$. (i) Find its mean function and variance function. (ii) Find $E(X(7))$, $Var(X(7))$ and $P[X(7) = 20]$ for $\lambda = 0.2, 0.3$.
Solution: (i) For a Yule-Furry process with birth rate λ and $X(0) = a$, the distribution of $X(t)$ is negative binomial with probability mass function as given below

$$P[X(t) = k] = \binom{k-1}{k-a} e^{-a\lambda t}(1 - e^{-\lambda t})^{k-a} \ \forall \ k = a, a+1, \ldots,$$

$$\text{Hence, } E(X(t)) = ae^{\lambda t} \ \& \ Var(X(t)) = ae^{\lambda t}(e^{\lambda t} - 1).$$

(ii) For $\lambda = 0.2$,

$$E(X(7)) = 20.27, \ \ Var(X(7)) = 61.95 \ \& \ P[X(7) = 20] = 0.0506.$$

For $\lambda = 0.3$,

$$E(X(7)) = 40.83, \ \ Var(X(7)) = 292.60 \ \& \ P[X(7) = 20] = 0.0150.$$

8.8.3 Suppose a population consists of a individuals at time $t = 0$ and the lifetime of each individual is a random variable with exponential distribution with parameter μ. Suppose $X(t)$ is the number of survivors in this population at time t and $\{X(t), t \geq 0\}$ is modeled as a linear death process with death rate μ and $X(0) = a = 6$. (i) Find its mean function and variance function. (ii) For $\mu = 0.3, 0.4$, find the expected population size and the variance of the

Table A.8 Yule-Fury Process: $E(X(10))$, $Var(X(10))$

λ	$E(X(10)$	$Var(X(10)$	$P[X(10) \geq E(X(10))]$
0.1	2.72	4.67	0.4547
0.2	7.39	47.21	0.3949
0.3	20.09	383.34	0.3773
0.4	54.60	2926.36	0.3713

population size at $t = 5$. Find the probability that at $t = 5$, the population size is less than 4. (iii) Find the long run distribution of the process.
Solution: (i) For a linear death process with $X(0) = a > 0$ and death rate μ, it follows $X(t) \sim B(a, e^{-\mu t})$ distribution. Hence,

$$E(X(t)) = ae^{-\mu t} \quad \& \quad Var(X(t)) = ae^{-\mu t}(1 - e^{-\mu t}).$$

(ii) For $\mu = 0.3$,

$$E(X(5)) = 1.3388, \quad Var(X(5)) = 1.0401 \;\; \& \;\; P[X(5) < 4] = 0.9898.$$

For $\mu = 0.4$,

$$E(X(5)) = 0.8120, \quad Var(X(5)) = 0.7021 \;\; \& \;\; P[X(5) < 4] = 0.9985.$$

Thus, with $\mu = 0.4$, population size less than 4 in (0, 5] is almost 1.
(iii) The long run distribution is given by $(1, 0, 0, 0, 0, 0, 0)'$.

8.8.4 For the linear birth-death process with birth rate $\lambda = 1.8$, death rate $\mu = 0.7$ and $X(0) = 1$, (i) find the mean and variance function at $t = 5$.
(ii) Find the probability of absorption into state 0. (iii) Find the probability of extinction on or before time $t = 5$.
Solution: In a linear birth-death process when $X(0) = 1$, the mean function $M(t) = E(X(t))$ is

$$M(t) = \begin{cases} e^{(\lambda-\mu)t}, & \text{if } \mu \neq \lambda \\ 1, & \text{if } \mu = \lambda . \end{cases}$$

The variance function $V(t) = Var(X(t))$ is given by

$$V(t) = \begin{cases} \frac{\lambda+\mu}{\lambda-\mu} e^{(\lambda-\mu)t}(e^{(\lambda-\mu)t} - 1), & \text{if } \mu \neq \lambda \\ 2\lambda t, & \text{if } \mu = \lambda . \end{cases}$$

(i) With $\lambda = 1.8$, $\mu = 0.7$ and $X(0) = 1$, $M(5) = e^{5.5} = 244.69$ and $V(5) = 135521.5$, which is quite high.

(ii) The probability q_1 of absorption into state 0 is $\mu/\lambda = 7/18 = 0.3889$.
(iii) The probability of extinction on or before time t is

$$P_0(t) = (\mu(1 - e^{-(\lambda-\mu)t}))/(\lambda - \mu e^{-(\lambda-\mu)t}).$$

For $t = 5, 7, 10$, $P_0(t) = 0.3879, 0.3888, 0.3889$, respectively. Observe that $P_0(10)$ and the probability of absorption into state 0 is almost the same, which supports the result that $\lim_{t\to\infty} P_0(t) = q_1$.

8.8.5 Find the mean function for a linear growth process with immigration.
Solution: Suppose $X(t)$ denotes the population size at time t with $X(0) = a$ and $M(t) = E(X(t))$. We determine $M(t)$ by deriving and solving a differential equation. We derive an equation for $M(t+h)$ by conditioning on $X(t)$ as $M(t+h) = E(X(t+h)) = E(E(X(t+h)|X(t)))$. Since h is assumed to be a small positive real number, the population at time $t+h$ will either increase in size by 1 if a birth or an immigration occurs in $(t, t+h]$, or decrease by 1 if a death occurs in this interval, or remain the same if neither of these two events occurs. Thus, given $X(t)$,

$$X(t+h) = \begin{cases} X(t)+1, & \text{with probability} \quad (\alpha + X(t)\lambda)h + o(h) \\ X(t)-1, & \text{with probability} \quad X(t)\mu h + o(h) \\ X(t), & \text{with probability } 1 - (\alpha + X(t)\lambda + X(t)\mu)h + o(h)\,. \end{cases}$$

Hence,

$$E(X(t+h)|X(t)) = X(t) + (\alpha + X(t)\lambda - X(t)\mu)h + o(h)$$

Taking expectations, we get

$$M(t+h) = M(t) + (\lambda-\mu)M(t)h + \alpha h + o(h)$$
$$\Rightarrow \quad \frac{M(t+h) - M(t)}{h} = (\lambda-\mu)M(t) + \alpha + o(h)/h\,.$$

Taking the limit as $h \to 0$, we get a differential equation $M'(t) = (\lambda-\mu)M(t) + \alpha$. Suppose $h(t) = (\lambda-\mu)M(t) + \alpha$. Hence, $h'(t) = (\lambda-\mu)M'(t)$. Thus, the differential equation can be rewritten as

$$\frac{h'(t)}{h(t)} = (\lambda-\mu) \Rightarrow h(t) = ke^{(\lambda-\mu)t}$$
$$\Rightarrow \alpha + (\lambda-\mu)M(t) = ke^{(\lambda-\mu)t}\,.$$

To determine the value of the constant k, we use the fact that $M(0) = a$ and get $k = \alpha + (\lambda-\mu)a$. Thus,

$$M(t) = (\alpha/(\lambda-\mu))[e^{(\lambda-\mu)t} - 1] + ae^{(\lambda-\mu)t}.$$

It is to be noted that we have implicitly assumed that $\lambda \neq \mu$ in the above derivation. If $\lambda = \mu$ then differential equation reduces to $M'(t) = \alpha$. Integrating and using $M(0) = a$ gives the solution $M(t) = \alpha t + a$.

8.8.6 Suppose there are m welders. The probability that a welder not using an electric supply at time t starts using it in $(t, t+h]$ is $\lambda h + o(h)$. The probability that a welder using an electric supply at time t stops using it in $(t, t+h]$ is $\mu h + o(h)$. Welders are assumed to work independently of each other. Suppose $X(t)$ denotes the number of welders using electric supply at time t. Find the long run distribution of $X(t)$.

Solution: Since $X(t)$ denotes the number of welders using electric supply at time t, the possible values of $X(t)$ are $\{0, 1, \ldots, m\}$ and $\{X(t), t \geq 0\}$ is modeled as a continuous time Markov chain with finite state space $S = \{0, 1, \ldots, m\}$ and infinitesimal probabilities as given by

$$\begin{aligned}
P_{k\,k+1}(h) &= (m-k)\lambda h + o(h), \quad k = 0, 1, \ldots, m-1 \\
P_{k\,k-1}(h) &= k\mu h + o(h), \quad k = 1, 2, \ldots, m \\
P_{k\,k}(h) &= 1 - [(m-k)\lambda + k\mu]h + o(h), \quad k = 0, 1, \ldots, m \\
P_{k\,j}(h) &= o(h), \quad j \neq k+1, k-1, k\,.
\end{aligned}$$

From the infinitesimal probabilities, it follows that $\{X(t), t \geq 0\}$ is birth-death process with finite state space $S = \{0, 1, \ldots, m\}$ and birth rates $\lambda_k = (m-k)\lambda,\ k = 0, 1, \ldots, m-1$ and death rates $\mu_k = k\mu,\ k = 1, 2, \ldots, m$ and $\mu_0 = 0$. The long run distribution exists and is obtained by solving the following balance equations:

$$\begin{aligned}
m\lambda P_0 &= \mu P_1, \quad m\mu P_m = \lambda P_{m-1} \\
\{(m-k)\lambda + k\mu\}P_k &= (m-k+1)\lambda P_{k-1} + (k+1)\mu P_{k+1},
\end{aligned}$$

for $k = 1, 2, \ldots, m-1$. From the equation $m\mu P_m = \lambda P_{m-1}$, we get $P_{m-1} = m(\mu/\lambda)P_m$. In the last equation taking $k = m-1$, we have

$$P_{m-2} = \frac{m(m-1)}{2}\left(\frac{\mu}{\lambda}\right)^2 P_m = \binom{m}{2}\left(\frac{\mu}{\lambda}\right)^2 P_m.$$

We assume $P_{m-n} = \binom{m}{n}\left(\frac{\mu}{\lambda}\right)^n P_m$ for all $n \leq k$. From the balance equations, replacing k by $m-k$ we have

$$(k\lambda + (m-k)\mu)P_{m-k} = (k+1)\lambda P_{m-k-1} + (m-k+1)\mu P_{m-k+1}.$$

Hence,

$(k+1)\lambda P_{m-k-1}$

$$= [k\lambda + (m-k)\mu]\binom{m}{m-k}\left(\frac{\mu}{\lambda}\right)^k P_m - (m-k+1)\mu\binom{m}{m-k+1}\left(\frac{\mu}{\lambda}\right)^{k-1} P_m$$
$$= [k\lambda + (m-k)\mu]\frac{m!}{k!(m-k)!}\left(\frac{\mu}{\lambda}\right)^k P_m - \frac{(m-k+1)\mu\, m!}{(m-k+1)!(k-1)!}\left(\frac{\mu}{\lambda}\right)^{k-1} P_m$$
$$= \frac{m!}{(k-1)!(m-k)!}\frac{\mu^k P_m}{\lambda^{k-1}} + \frac{m!}{k!(m-k-1)!}\frac{\mu^{k+1} P_m}{\lambda^k} - \frac{m!}{(m-k)!(k-1)!}\frac{\mu^k P_m}{\lambda^{k-1}}$$
$$= \frac{m!}{k!(m-k-1)!}\frac{\mu^{k+1}}{\lambda^k}P_m.$$

Hence,

$$P_{m-k-1} = \frac{1}{(k+1)}\frac{m!}{k!(m-k-1)!}\frac{\mu^{k+1}}{\lambda^{k+1}}P_m = \binom{m}{k+1}\left(\frac{\mu}{\lambda}\right)^{k+1} P_m.$$

Thus by induction,

$$P_{m-k} = \binom{m}{k}\left(\frac{\mu}{\lambda}\right)^k P_m, \quad \Longleftrightarrow \quad P_k = \binom{m}{m-k}\left(\frac{\mu}{\lambda}\right)^{m-k} P_m,$$

$k = 0, 1, \ldots, m$. The condition

$$\sum_{k=0}^{m} P_k = 1 \Rightarrow P_m = \left(\sum_{k=0}^{m}\binom{m}{m-k}\left(\frac{\mu}{\lambda}\right)^{m-k}\right)^{-1} = \left(\frac{\lambda}{\lambda+\mu}\right)^m.$$

Thus,

$$P_k = \binom{m}{k}\left(\frac{\lambda}{\lambda+\mu}\right)^k\left(\frac{\mu}{\lambda+\mu}\right)^{m-k}, \quad k = 0, 1, \ldots, m$$

and the long run distribution is a binomial $B(m, \lambda/(\lambda+\mu))$. In particular, if $\lambda = \mu$, $P_k = \binom{m}{k}(1/2)^m$, $k = 0, 1, \ldots, m$ and it is independent of the common value of λ and μ.

8.8.7 Suppose customers arrive at a service facility with a single service counter, according to the Poisson process with rate $\lambda = 5$ per hour. The service time random variable for each customer has exponential distribution with mean $1/\mu = 10$ minutes. The facility has a limited waiting capacity of 10 chairs. (i) Find the long run fraction of the time the service facility is idle. (i) Find the long run fraction of the time the service facility is full.
(iii) Solve (i) and (ii) if the mean service time is 20 minutes. (iv) Comment on the results.
Solution: Suppose $X(t)$ denotes the number of customers at time t at the service facility, that is, the one getting the service and waiting for the service. Since the customers arrive according to the Poisson process and the service times are exponentially distributed, $\{X(t), t \geq 0\}$ is a immigration-emigration process, that is, $M/M/1$ queuing model with finite capacity $C = 10$ of the

waiting room, $\lambda_k = \lambda$, $k = 0, 1, \ldots, C-1$ and $\mu_k = \mu$, $k = 1, 2, \ldots, C$. Since it is a finite state Markov process, the long run distribution exists and is given by $P_n(C), n = 0, 1, \ldots, C$ where $\rho = \lambda/\mu$:

$$P_n(C) = \begin{cases} \frac{(1-\rho)\rho^n}{1-\rho^{C+1}}, & \text{if } \rho \neq 1 \\ \\ \frac{1}{C+1}, & \text{if } \rho = 1\,. \end{cases}$$

Now $1/\mu = 10$ minutes is the same as $1/6$ hours. Hence, with $\lambda = 5$ and $\mu = 6$, $\rho = 5/6$. Hence,

$$P_n(10) = \frac{(1/6)(5/6)^n}{1-(5/6)^{11}} = 0.1926(5/6)^n, \quad n = 0, 1, \ldots, 10.$$

(i) The long run fraction of the time the service facility is idle is $P_0(10) = 0.1926$.
(ii) The long run fraction of the time the service facility is full is $P_{10}(10) = 1.592626e^{-08} \approx 0$.
(iii) If the mean service time is 20 minutes, $\mu = 3$ and $\rho = 5/3$. Thus, we have $P_0(10) = 0.002428$ and $P_{10}(10) = 0.4015$.
(iv) When the traffic intensity $\rho = 5/6 < 1$, the long run fraction of the time the service facility is full is almost 0, as expected. When the traffic intensity $\rho = 5/3 > 1$, it is higher and is given by 0.4015. On the other hand, when the traffic intensity $\rho = 5/6 < 1$, the long run fraction of the time the service facility is idle is 0.1926. When the traffic intensity $\rho = 5/3 > 1$, it is small and is given by 0.002428.

8.8.8 A birth and death process has parameters $\lambda_k = \alpha(k+1)$ for $k = 0, 1, 2, \ldots,$ and $\mu_k = \beta(k+1)$ for $k = 1, 2, \ldots$. (i) Examine whether the long run distribution of the process exists. (ii) If it exists, find it. (iii) Find the long run distribution if $\alpha = 0.2$ and $\beta = 0.5$.

Solution: (i) For a birth-death process, the long run distribution $P_n = \lim_{t\to\infty} P_n(t)$ exists if the series $E = \sum_{n=1}^{\infty} \frac{\lambda_{n-1}\lambda_{n-2}\ldots\lambda_1\lambda_0}{\mu_n\mu_{n-1}\ldots\mu_2\mu_1}$ is convergent. With $\lambda_k = \alpha(k+1)$ for $k = 0, 1, 2, \ldots,$ and $\mu_k = \beta(k+1)$ for $k = 1, 2, \ldots,$ we examine when this series is convergent. With these values of birth and death rates, observe that

$$E = \sum_{n=1}^{\infty} \frac{n!\alpha^n}{(n+1)!\beta^n} = \sum_{n=1}^{\infty} \frac{(\alpha/\beta)^n}{n+1} = \sum_{n=1}^{\infty} a_n, \text{ say}$$

$$\text{For } \alpha < \beta, \quad \lim_{n\to\infty} \frac{a_{n+1}}{a_n} = \lim_{n\to\infty} \frac{\alpha}{\beta}\frac{n+1}{n+2} = \frac{\alpha}{\beta} < 1$$

and by the ratio test the series E is convergent. Thus, the long run distribution of the process exists if $\alpha < \beta$. For $\alpha \geq \beta$, the series E is divergent and the

long run distribution does not exist.
(ii) Under the condition $\alpha < \beta$, the long run distribution is given by

$$P_0 = (1+E)^{-1} \;\&\; P_n = \left\{\prod_{i=0}^{n-1} \lambda_i\right\}\left\{\prod_{i=1}^{n} \mu_i (1+E)\right\}^{-1}, \; n \geq 1.$$

Now the series E can be simplified as follows:

$$\begin{aligned} E &= \sum_{n=1}^{\infty} \frac{(\alpha/\beta)^n}{n+1} = \frac{\beta}{\alpha}\sum_{n=1}^{\infty} \frac{(\alpha/\beta)^{n+1}}{n+1} \\ &= \frac{\beta}{\alpha}\left(\sum_{n=1}^{\infty} \frac{(\alpha/\beta)^n}{n+1} + (\alpha/\beta) - (\alpha/\beta)\right) \\ &= (\beta/\alpha)\left(-\log(1-(\alpha/\beta)) - (\alpha/\beta)\right) \\ &= -(\beta/\alpha)(\log(1-(\alpha/\beta)) - 1 \\ \Rightarrow \; P_0 &= (1+E)^{-1} = \frac{(\alpha/\beta)}{(-\log(1-(\alpha/\beta)))} \\ \Rightarrow \; P_n &= \frac{(\alpha/\beta)^n}{n+1} P_0, \; n \geq 1 \\ \Rightarrow \; P_n &= \frac{(\alpha/\beta)^{n+1}}{(n+1)(-\log(1-(\alpha/\beta)))}, n \geq 0. \end{aligned}$$

Observe that $\sum_{n=0}^{\infty} P_n = 1$.
(iii) For $\alpha = 0.2$ and $\beta = 0.5$, the long run distribution is presented in Table A.9 $P_n \approx 0 \; \forall \; n \geq 12$.

8.8.9 A time-shared computer system has three terminals that are attached to a central processing unit that can simultaneously handle at most two active users. If a person logs on and requests service when two other users are active, then the request is held in a buffer until it can receive service. Suppose $X(t)$ is the total number of requests that are either active or in the buffer at time t. Assume that X(t) is a birth and death process with parameters $\lambda_k = \lambda$ for $k = 0, 1, 2$ and $\lambda_k = 0$ for all $k \geq 3$, $\mu_k = k\mu$ for $k = 0, 1, 2$ and $\mu_k = 2\mu$ for $k = 3$. (i) Determine the long run probability that the system is idle. (ii) Determine the long run probability that the system has two active users.

Table A.9 Birth-Death Process: long run distribution

n	0	1	2	3	4	5
P_n	0.7830	0.1566	0.0418	0.0125	0.0040	0.0013
n	6	7	8	9	10	11
P_n	0.0005	0.0002	0.0001	0	0	0

(iii) For $\lambda = 0.3$ and $\mu = 0.4$, find the long run distribution. What is the long run mean proportion of time the system is idle?

Solution: To find the long run distribution $\underline{P} = (P_0, P_1, P_2, P_3)'$, we write balance equations as follows:

$$\lambda P_0 = \mu P_1, \quad (\lambda + \mu)P_1 = 2\mu P_2 + \lambda P_0$$
$$(\lambda + 2\mu)P_2 = 2\mu P_3 + \lambda P_1, \quad 2\mu P_3 = \lambda P_2$$
$$\Rightarrow \; P_1 = \frac{\lambda}{\mu}P_0, \quad P_2 = \frac{\lambda^2}{2\mu^2}P_0 \;\&\; P_3 = \frac{\lambda^3}{4\mu^3}P_0.$$

The condition

$$P_0 + P_1 + P_2 + P_3 = 1 \;\Rightarrow\; P_0 = \left(1 + \frac{\lambda}{\mu} + \frac{\lambda^2}{2\mu^2} + \frac{\lambda^3}{4\mu^3}\right)^{-1},$$

which is the long run probability that the system is idle.

(ii) The long run probability that the system has two active users is $P_2 = (\lambda^2/2\mu^2)P_0$.

(iii) For $\lambda = 0.3$ and $\mu = 0.4$, the long run distribution is given by $\underline{P} = (0.4680, 0.3510, 0.1316, 0.0494)'$. Thus, on average 47% of time the system will be idle in the long run.

8.8.10 Suppose in a workshop there are 3 machines and a single repair facility. The probability that a machine working at time t fails in time $(t, t+h]$ is $\lambda h + o(h)$, independently of other machines. If the repair facility is busy at time t, the probability that the repair is complete in $(t, t+h]$ is $\mu h + o(h)$ independent of state of the other machines. Suppose $X(t)$ denotes the number of machines not working at time t, then $\{X(t), t \geq 0\}$ is modeled as a continuous time Markov chain with finite state space $S = \{0, 1, 2, 3\}$. (i) Express it as a birth-death process. Identify the birth and death rates. (ii) In the long run, find the probability that all machines are in working condition when $\lambda = 0.4$ and $\mu = 0.3$. (iii) For these values of λ and μ, find the expected number of idle machines, in the long run.

Solution: The solution is similar to the derivation given in Section 8.6 for a similar model.

Answers to the multiple choice questions, based on Chap. 8, are given in Table A.10.

Table A.10 Answer key to MCQs in Chap. 8

Q. No.	1	2	3	4	5	6	7	8
Ans	b	c	a	d	b	a, b	c	a, c

A.8 Chapter 9

9.7.1 Suppose $\{X(t), t \geq 0\}$ is a Brownian motion process, with drift coefficient $\mu = 3$ and diffusion coefficient $\sigma^2 = 4$. Find the conditional distribution of $X(3)$ given $X(0) = 6$ and $X(5) = 7$.
Solution: In Example 9.4.1, we have obtained the conditional distribution of $W(t)$ given $W(t_1) = a$ and $W(t_2) = b$ for $t_1 < t < t_2$, when $\{W(t), t \geq 0\}$ is the standard Brownian motion process. The conditional distribution is normal with mean $\theta + a$ and variance v, where

$$\theta = \frac{(b-a)(t-t_1)}{t_2 - t_1} \quad \& \quad v = t - t_1 - \frac{(t-t_1)^2}{t_2 - t_1} = (t - t_1)\frac{t_2 - t}{t_2 - t_1}.$$

Now $X(t)$ can be expressed as $X(t) = X(0) + \mu t + \sigma W(t)$ and the distribution of $X(t)$ given $X(0) = 6$ is normal $N(6 + 3t, 4t)$. Further,

$$X(0) = 6 \ \Rightarrow \ W(0) = 0 \ \& \ X(5) = 7 \ \Rightarrow \ W(5) = (7 - 6 - 15)/2 = -7.$$

Suppose $t_1 = 0, t_2 = 5, t = 3, a = 0$ and $b = -7$. Then the conditional distribution of $W(3)$ given $W(t_1) = 0$ and $W(t_2) = -7$ is normal with mean $\theta + a = \theta$ and variance v, where $\theta = (-7)(3)/5 = -4.2$ and $v = 3 - 9/5 = 1.2$. Hence, the conditional distribution of $X(3) = 6 + 9 + 2W(3)$ given $X(0) = 6$ and $X(5) = 7$ is normal with mean 6.6 and variance 4.8.

9.7.2 Suppose $\{W(t), t \geq 0\}$ is the standard Brownian motion process. Find the joint distribution of $W(t_1)$ and $W(t_2) - W(t_1)$ and hence the joint distribution of $W(t_1)$ and $W(t_2)$.
Solution: Suppose $\underline{Z}_1 = (W(t_1), W(t_2))$ for $t_1 < t_2 \in (0, \infty)$. From the definition of a Wiener process, it follows

$$Y_1 = W(t_1) = W(t_1) - W(0) \sim N(0, t_1) \ \& \ Y_2 = W(t_2) - W(t_1) \sim N(0, t_2 - t_1)$$

distribution. Further due to independence of increments, Y_1 and Y_2 are independent random variables. Hence, the distribution of $\underline{Z}_2 = (W(t_1), W(t_2) - W(t_1))'$ is bivariate normal with mean vector $(0, 0)'$ and dispersion matrix D given by $D = \text{diag}(t_1, t_2 - t_1)$. Now,

$$\begin{aligned} \underline{Z}_1 &= (W(t_1), W(t_2))' = A((W(t_1), W(t_2) - W(t_1))' \\ &= A\underline{Z}_2 \ \text{ where } \ \underline{Z}_2 \sim N_2(\underline{0}, D) \\ &\Rightarrow \underline{Z}_1 \sim N_2(\underline{0}, ADA'), \end{aligned}$$

where A and ADA' are given by

$$A = \begin{pmatrix} 1 & 0 \\ 1 & 1 \end{pmatrix} \quad ADA' = \begin{pmatrix} t_1 & t_1 \\ t_1 & t_2 \end{pmatrix}.$$

9.7.3 Suppose $\{W(t), t \geq 0\}$ is the standard Brownian motion process.
(i) Obtain the distribution of $W(s) - (s/t)W(t)$ when $s < t$. (ii) Obtain the joint distribution of $(W(t), W(s) - (s/t)W(t))'$. (iii) Examine whether $W(t)$ and $W(s) - (s/t)W(t)$ are independent.
Solution: (i) Since $\{W(t), t \geq 0\}$ is the standard Brownian motion process, for $s < t$, it follows $\underline{Z} = (W(t), W(s))' \sim N_2(\underline{0}, \Sigma)$ distribution where

$$\Sigma = \begin{pmatrix} t & s \\ s & s \end{pmatrix}.$$

Hence, the distribution of $W(s) - (s/t)W(t)$ is normal $N(0, v)$ where $v = s + (s/t)^2 t - 2(s/t)s = s - s^2/t = s(1 - s/t)$.
(ii) Observe that $\underline{U} = (W(t), W(s) - (s/t)W(t))' = A\underline{Z}$ and hence $\underline{U} \sim N_2(\underline{0}, A\Sigma A')$ distribution where

$$A = \begin{pmatrix} 1 & 0 \\ -s/t & 1 \end{pmatrix} \ \& \ A\Sigma A' = \begin{pmatrix} t & 0 \\ 0 & s(1 - s/t) \end{pmatrix}.$$

(iii) Since $Cov(W(t), W(s) - (s/t)W(t)) = 0$, $W(t)$ and $W(s) - (s/t)W(t)$ are independent.

9.7.4 Suppose $\{X(t), t \geq 0\}$ is a Brownian motion process with drift coefficient μ and diffusion coefficient σ^2. Examine whether $\{X(t) - \mu t, t \geq 0\}$ is a martingale.
Solution: To examine whether $\{X(t) - \mu t, t \geq 0\}$ is a martingale, observe that

$$\begin{aligned} &E((X(t) - \mu t)|X(u), 0 \leq u \leq s) \\ &= E((X(t) - X(s) + X(s) - \mu t - \mu s + \mu s)|X(u), 0 \leq u \leq s) \\ &= E((X(t) - X(s)) - \mu(t - s))|X(u), 0 \leq u \leq s) \\ &+ E((X(s) - \mu s)|X(u), 0 \leq u \leq s) \\ &= E((X(t) - X(s)) - \mu(t - s)) + (X(s) - \mu s) \ \text{ almost surely} \\ &= 0 + (X(s) - \mu s) = X(s) - \mu s \ \text{ almost surely.} \end{aligned}$$

Hence, $\{X(t) - \mu t, t \geq 0\}$ is a martingale. Alternatively, $X(t)$ can be expressed as $X(t) = \mu t + \sigma W(t)$ and the standard Wiener process

$\{W(t), t \geq 0\}$ is a martingale, that is, $E(W(t)|W(u), 0 \leq u \leq s) = W(s)$ almost surely. Hence,

$$\begin{aligned} E(W(t)|W(u), 0 \leq u \leq s) &= W(s) \\ &\Rightarrow E\left(\frac{X(t) - \mu t}{\sigma}\Big|X(u), 0 \leq u \leq s\right) = \frac{X(s) - \mu s}{\sigma} \\ &\Rightarrow E((X(t) - \mu t)|X(u), 0 \leq u \leq s) = X(s) - \mu s. \end{aligned}$$

9.7.5 Suppose $\{W(t), t \geq 0\}$ is the standard Brownian motion process. Suppose $Y(t) = W^2(t) - t$ and $Z(t) = \exp\{cW(t) - c^2t/2\}$. Examine whether $\{Y(t), t \geq 0\}$ and $\{Z(t), t \geq 0\}$ are martingales.
Solution: Since $\{W(t), t \geq 0\}$ is the standard Brownian motion process, for $s < t$, $W(t) - W(s)$ is independent of $W(s) - W(0)$ and follows normal $N(0, t - s)$ distribution. Further, $\{W(t), t \geq 0\}$ is a martingale. Observe that

$$\begin{aligned} E(Y(t)|Y(u), 0 \leq u \leq s) &= E(W^2(t) - t|W(u), 0 \leq u \leq s) \\ &= E(W^2(t) + W^2(s) - W^2(s) - t - s + s \\ &+ 2W(t)W(s) - 2W(t)W(s))|W(u), 0 \leq u \leq s) \\ &= E(W(t) - W(s))^2 - (t - s) \\ &+ 2W(s)E(W(t)|W(u), 0 \leq u \leq s) - W^2(s) - s \\ &= W^2(s) - s = Y(s) \text{ almost surely.} \end{aligned}$$

Hence, $\{Y(t), t \geq 0\}$ is a martingale. To examine whether $\{Z(t), t \geq 0\}$ is a martingale, note that

$$\begin{aligned} &E(Z(t)|Z(u), 0 \leq u \leq s) \\ &= E(\exp\{cW(t) - c^2t/2\}|W(u), 0 \leq u \leq s) \\ &= E(\exp\{cW(t) - cW(s) + cW(s) - c^2t/2\}|W(u), 0 \leq u \leq s) \\ &= E(\exp\{cW(t) - cW(s)\}) \times \exp\{cW(s) - c^2t/2\} \\ &= \exp\{c^2(t - s)/2 + cW(s) - c^2t/2\} = \exp\{cW(s) - c^2s/2\} \\ &= Z(s) \text{ almost surely.} \end{aligned}$$

Hence, $\{Z(t), t \geq 0\}$ is a martingale.

9.7.6 Suppose $\{X(t), t \geq 0\}$ is a Brownian motion process with drift coefficient $\mu = -1$, diffusion coefficient $\sigma^2 = 2$ and $X(0) = 6$. Find the probability that $X(t)$ is above its initial level at time 7.
Solution: For fixed t, $X(t) \sim N(6 - t, 2t)$ distribution, hence $X(7) \sim N(-1, 14)$ distribution. We compute $P[X(7) > 6]$ as follows:

$$\begin{aligned} P[X(7) > 6] &= P[(X(7) + 1)/\sqrt{14} > (6 + 1)\sqrt{14}] = 1 - \Phi(1.8708) \\ &= 0.9307. \end{aligned}$$

9.7.7 Suppose $\{X(t), t \geq 0\}$ is a Brownian motion process with drift coefficient $\mu = 2$, diffusion coefficient $\sigma^2 = 3$ and $X(0) = 2$. Find the joint distribution of $X(4) - X(1)$ and $X(9) - X(5)$.
Solution: A Brownian motion process is a process with stationary and independent increments, with increment of length h having $N(\mu h, \sigma^2 h)$ distribution. Hence, $X(4) - X(1)$ and $X(9) - X(5)$ are independent random variables, having $N(6, 9)$ and $N(8, 12)$ distribution, respectively. Thus, the joint distribution of $X(4) - X(1)$ and $X(9) - X(5)$ is bivariate normal with mean vector $(6, 8)'$ and dispersion matrix to be diagonal with diagonal elements 9 and 12.

9.7.8 Suppose $\{X(t), t \geq 0\}$ is a Brownian motion process, with drift coefficient $\mu = 2$, diffusion coefficient $\sigma^2 = 3$ and $X(0) = 2$. Find the joint distribution of $X(4) - X(1)$ and $X(3)$.
Solution: A Brownian motion process is a Gaussian process with covariance function as $C(s, t) = \min\{s, t\}$. Hence, the joint distribution of $\underline{Z} = (X(1), X(3), X(4))'$ is trivariate normal $N_3(\underline{\mu}, \Sigma)$, where $\underline{\mu} = (4, 8, 10)'$ and Σ is given by

$$\Sigma = 3\begin{pmatrix} 1 & 1 & 1 \\ 1 & 3 & 3 \\ 1 & 3 & 4 \end{pmatrix}.$$

Now, it follows $\underline{U} = (X(4) - X(1), X(3))' = A\underline{Z} \sim N_2(A\underline{\mu}, A\Sigma A')$ distribution, where

$$A = \begin{pmatrix} -1 & 0 & 1 \\ 0 & 1 & 0 \end{pmatrix}, \quad A\underline{\mu} = (6, 8)' \;\;\&\;\; A\Sigma A' = 3\begin{pmatrix} 3 & 2 \\ 2 & 3 \end{pmatrix}.$$

9.7.9 Suppose $\{X(t), t \geq 0\}$ is a Brownian motion process, with drift coefficient μ and diffusion coefficient σ^2. Find $r \neq 0$ so that $E(e^{rX(t)}) = 1$.
Solution: If $\{X(t), t \geq 0\}$ is a Brownian motion process, with drift coefficient μ and diffusion coefficient σ^2, then

$$E(e^{rX(t)}) = e^{\mu rt + \sigma^2 r^2 t/2} = 1 \Rightarrow \mu rt + \sigma^2 r^2 t/2 = 0$$
$$\Rightarrow r = 0 \text{ or } r = -2\mu/\sigma^2.$$

9.7.10 Suppose $\{X(t), t \geq 0\}$ is a Brownian motion process, with drift coefficient μ and diffusion coefficient σ^2. Suppose $\theta = -2\mu/\sigma^2$. Examine whether $\{e^{\theta X(t)}, t \geq 0\}$ is a martingale.
Solution: To examine whether $\{e^{\theta X(t)}, t \geq 0\}$ is a martingale, observe that

$$\begin{aligned}E(e^{\theta X(t)}|X(u), 0 \le u \le s) &= E(e^{\theta X(t)-\theta X(s)+\theta X(s)}|X(u), 0 \le u \le s)\\ &= E(e^{\theta(X(t)-X(s))}|X(u), 0 \le u \le s)\\ &\times E(e^{\theta X(s)}|X(u), 0 \le u \le s)\\ &= E(e^{\theta(X(t)-X(s))}) \times e^{\theta X(s)} \text{ almost surely}\\ &= e^{\theta(t-s)\mu+\theta^2(t-s)\sigma^2/2} \times e^{\theta X(s)}\\ &= e^{\theta(t-s)(\mu+\theta\sigma^2/2)} \times e^{\theta X(s)}\\ &= e^{\theta X(s)}, \text{ as } \theta = -2\mu/\sigma^2.\end{aligned}$$

Hence, $\{e^{\theta X(t)}, t \ge 0\}$ is a martingale.

9.7.11 Suppose the amount in a bank account at time t is modeled as $X(t) = 5000 + 500W(t)$, where $\{W(t), t \ge 0\}$ is the standard Brownian motion process. Find the probability that the account is not overdrawn by time 10.

Solution: We find the probability that the balance is not 0 by time 10. Now, $X(t) = 5000 + 500W(t) = 0 \Rightarrow W(t) = -10$. Hence, the probability that the account is not overdrawn by time 10 is given by

$$\begin{aligned}P[\min_{0\le t\le 10} X(t) \ge 0] &= P[\min_{0\le t\le 10} W(t) \ge -10]\\ &= P[L(10) > -10] = 1 - P[L(10) \le -10]\\ &= 1 - 2(1 - \Phi(10/\sqrt{10})) = 0.2482.\end{aligned}$$

9.7.12 Suppose the inventory at a store at time t is modeled as $X(t) = 6 - 3t + 4W(t)$ where $\{W(t), t \ge 0\}$ is the standard Brownian motion process. Suppose the inventory storage area has capacity of 15 units. (i) Find the probability that a stock-out occurs before the storage area overflows. (ii) Find the expected time for either stock-out or overflow of storage area.

Solution: Suppose $Y(t)$ is defined as $Y(t) = X(t) - 6$, so that $Y(0) = 0$. (i) Further, $X(t) = 0 \Rightarrow Y(t) = -6 = a$, say, and $X(t) = 15 \Rightarrow Y(t) = 9 = b$, say. Using Theorem 9.3.5, we compute $P[Y(T(a, b)) = a]$ as follows. For given μ and σ, $\theta = -2\mu/\sigma^2 = 3/8$. Now

$$P[Y(T(a, b)) = a] = (1 - e^{\theta b})/(e^{\theta a} - e^{\theta b}) = 0.9693.$$

(ii) The expected time for either stock-out or overflow of storage area is given by

$$E(T(a, b)) = \frac{b(e^{\theta a} - 1) - a(e^{\theta b} - 1)}{\mu(e^{\theta a} - e^{\theta b})} = 1.8464 \text{ time units.}$$

9.7.13 Suppose $Z(t) = (t + 1)X(t/(t + 1))$, where $\{X(t), 0 \le t \le 1\}$ is a Brownian bridge. Show that $\{Z(t), t \ge 0\}$ is the standard Brownian motion process.

Solution: A Brownian bridge is a Gaussian process with covariance function as $c(s,t) = s(1-t)$ for $s < t$. Hence, $\{Z(t), t \geq 0\}$ is also a Gaussian process. We find its covariance function as follows. It is to be noted that

$$s < t \;\Rightarrow\; st + s < st + t \;\Rightarrow\; s(t+1) < t(s+1) \;\Rightarrow\; s/(s+1) < t/(t+1).$$

Observe that for $s < t$,

$$\begin{aligned} Cov(Z(t), Z(s)) &= Cov((t+1)X(t/(t+1)), (s+1)X(s/(s+1))) \\ &= (t+1)(s+1)Cov(X(t/(t+1)), X(s/(s+1))) \\ &= (t+1)(s+1)(s/(s+1))(1 - t/(t+1)) \\ &= s = \min\{s,t\}. \end{aligned}$$

Thus, $\{Z(t), t \geq 0\}$ is a Gaussian process with covariance function $c(s,t) = \min\{s,t\}$. Hence, $\{Z(t), t \geq 0\}$ is the standard Brownian motion process.

9.7.14 Suppose the price of a stock at time t is modeled by $X(t) = e^{\sigma W(t)}$, a geometric Brownian motion process, with volatility parameter $\sigma = 1/2$ and $\{W(t), t \geq 0\}$ is the standard Brownian motion process with $W(0) = 0$. Find the expected price of a stock at time 2 and also obtain its variance. Find the probability that the stock price is above 4 units at time 2.
Solution: Since $\{X(t) = e^{\sigma W(t)}, t \geq 0\}$ is a geometric Brownian motion process,

$$E(X(t)) = e^{\mu t + \sigma^2 t/2} \;\&\; Var(X(t)) = e^{2\mu t + \sigma^2 t}(e^{t\sigma^2} - 1).$$

With $\mu = 0 \;\&\; \sigma = 1/2, \quad E(X(2)) = e^{1/4} = 1.2840$ and $Var(X(2)) = e^{1/2}(e^{1/2} - 1) = 1.0696$. Further, the probability that the stock price is above 4 units at time 2 is given by

$$P[X(2) > 4] = P[W(2) > \log(4)/\sigma] = 1 - \Phi(log(4)/(\sqrt{2}\sigma)) = 0.02497.$$

9.7.15 What is the probability that a geometric Brownian motion process with parameters $\mu = -\sigma^2/2$ and σ ever rises to more than twice its original value? What is the probability if $\mu = 0$. (In financial terms, if you buy a stock or index fund whose fluctuations are described by the geometric Brownian motion, what are the chances to double your money?)
Solution: Suppose $\{X(t) = X(0)e^{\mu t + \sigma W(t)}, t \geq 0\}$ is a geometric Brownian motion process, with $X(0) = 1$. To compute the probability $P[X(t) > 2, 0 \leq t < \infty]$, observe that $\theta = -2\mu/\sigma^2 = 1$. Further,

$$\begin{aligned} P[X(t) > 2, 0 \le t < \infty] &= P[\mu t + \sigma W(t) > \log 2, 0 \le t < \infty] \\ &= 1 - P[\mu t + \sigma W(t) < \log 2, 0 \le t < \infty] \\ &= 1 - P[\max_{0 \le t < \infty} (\mu t + \sigma W(t)) < \log 2] \\ &= 1 - P[U < \log 2] = e^{-\theta \log 2} = 1/2. \end{aligned}$$

If $\mu = 0$, then $1 = P[U = \infty] \le P[U > \log 2]$ implies $P[U > \log 2] = 1$.

9.7.16 Suppose $\{X(t), t \ge 0\}$ is a geometric Brownian motion process with $\mu = 0.01$. If $X(0) = 100$, find $E(X(10))$, $P[X(10) > 100]$ and $P[X(10) < 110]$ for three values of σ given by $\sigma = 0.2, 0.4, 0.6$.

Solution: If $\{X(t), t \ge 0\}$ is a geometric Brownian motion process with drift coefficient μ and diffusion coefficient σ^2, then $X(t) = X(0)e^{\mu t + \sigma W(t)}$ where $\{W(t), t \ge 0\}$ is the standard Brownian motion process. Further,

$$\begin{aligned} & E(X(t)) = X(0)e^{(\mu + \sigma^2/2)t} \\ \Rightarrow \quad & E(X(10)) = 100e^{10(.01 + 0.5\sigma^2)} = 134.9859, 245.9603, 668.5894 \end{aligned}$$

for $\sigma = 0.2, 0.4, 0.6$, respectively. Now,

$$\begin{aligned} P[X(10) > 100] &= P[10\mu + \sigma W(10) > 0] = P[W(10) > -0.1/\sigma] \\ &= P[Z > -0.1/(\sqrt{10}\sigma)] \text{ where } Z \sim N(0, 1) \\ &= 0.5628, 0.5315, 0.5210 \text{ for } \sigma = 0.2, 0.4, 0.6, \end{aligned}$$

respectively. Similarly,

$$\begin{aligned} P[X(10) < 110] &= P[10\mu + \sigma W(10) < \log(110) - \log(100)] \\ &= P[W(10) < (\log(1.1) - 0.1)/\sigma] \\ &= P[Z < -0.004689/(\sqrt{10}\sigma)] \text{ where } Z \sim N(0, 1) \\ &= 0.4970, 0.4985, 0.4990 \text{ for } \sigma = 0.2, 0.4, 0.6, \end{aligned}$$

respectively.

9.7.17 Suppose stock price $\{X(t), t \ge 0\}$ is a geometric Brownian motion process with drift coefficient $\mu = 2$ and diffusion coefficient $\sigma^2 = 7.5\%$ per annum. Assume that the current price of the stock is $X(0) = 100$. Find $E(X(3))$, $P[X(3) > 40000]$.

Solution: If $\{X(t), t \ge 0\}$ is a geometric Brownian motion process with drift coefficient $\mu = 2$ and diffusion coefficient $\sigma^2 = 0.075$, then $X(t) = X(0)e^{\mu t + \sigma W(t)}$ where $\{W(t), t \ge 0\}$ is the standard Brownian motion process. Hence,

$$E(X(t)) = X(0)e^{(\mu + \sigma^2/2)t} \quad \Rightarrow \quad E(X(3)) = 100e^{6 + 0.1125} = 45146.6.$$

Table A.11 Answer key to MCQs in Chap. 9

Q. No.	1	2	3	4	5	6	7	8	9	10
Ans	a, b, c, d	a, b, c	d	a, b, c, d	c	a	b	c	b	c
Q. No.	11	12	13	14	15	16	17	18	19	20
Ans	c	a	c	c	d	b	a	a, b, c, d	b, c, d	b, d
Q. No.	21	22	23	24	25	26	27	28	29	30
Ans	b	c	c	b	b	a	d	c	d	c
Q. No.	31	32	33	34	35	36	37	38	39	40
Ans	c	b	c	a, b, c	d	c	c	c	c	b
Q. No.	41	42	43	44	45	66	47	48	49	50
Ans	b	d								

Further,

$$
\begin{aligned}
P[X(3) > 40000] &= P[\sqrt{.075}W(3) > \log(400) - 6] \\
&= P[Z > -0.01799] = 0.5072 \text{ where } Z \sim N(0, 1).
\end{aligned}
$$

Answers to the multiple choice questions, based on Chap. 9, are given in Table A.11.

A.9 Chapter 10

10.7.1 Suppose $\{X_1(t), t \geq 0\}$ and $\{X_2(t), t \geq 0\}$ are two independent renewal processes. Suppose $X(t) = X_1(t) + X_2(t)$. Examine whether $\{X(t), t \geq 0\}$ is a renewal process. Find the inter-renewal distribution.

Solution: Suppose $\{U_n, n \geq 1\}$ and $\{V_n, n \geq 1\}$ are sequences of inter-renewal random variables corresponding to the renewal processes $\{X_1(t), t \geq 0\}$ and $\{X_2(t), t \geq 0\}$, respectively. Then $\{U_n, n \geq 1\}$ is a sequence of independent and identically distributed random variables with common distribution F_1 say. Similarly, $\{V_n, n \geq 1\}$ is a sequence of independent and identically distributed random variables with common distribution F_2 say. Further, these two sequences are independent. Suppose $T_n = \min\{U_n, V_n\}, n \geq 1$. Being a Borel function of U_n and V_n, it follows that $\{T_n, n \geq 1\}$ is a sequence of independent and identically distributed random variables with common distribution F. We obtain F as follows. For $t > 0$,

$$\begin{aligned} 1 - F(t) &= P[T_n \geq t] = P[\min\{U_n, V_n\} \geq t] \\ &= P[U_n \geq t]P[V_n \geq t] = (1 - F_1(t))(1 - F_2(t)) \\ \Rightarrow \quad F(t) &= F_1(t) + F_2(t) - F_1(t)F_2(t). \end{aligned}$$

We have $X(t) = X_1(t) + X_2(t)$. Thus, count in $X(t)$ process increases when a renewal occurs in $X_1(t)$ or $X_2(t)$, whichever is earlier. Thus, the inter-renewal distribution of $X(t)$ process is the distribution of $T_n = \min\{U_n, V_n\}, n \geq 1$. We have shown that $\{T_n, n \geq 1\}$ is a sequence of independent and identically distributed random variables with common distribution $F = F_1 + F_2 - F_1 F_2$. Hence, $\{X(t), t \geq 0\}$ is a renewal process. In particular, if F_1 and F_2 have exponential distribution with rates λ_1 and λ_2, then $\{X(t), t \geq 0\}$ is a Poisson process, which is a superposition of two independent Poisson processes.

10.7.2 Suppose $\{X(t), t \geq 0\}$ is a renewal process with renewal function $M(t) = 1.5t$. Identify the inter-renewal distribution. Find its Laplace transform. Can you label the renewal process?

Solution: In Corollary 10.2.1, it is proved that a Poisson process is the unique renewal process whose mean value function is linear, that is $M(t) = \lambda t$. Hence, the given renewal process is a Poisson process with rate $\lambda = 1.5$. Consequently, the inter-renewal distribution is exponential with rate 1.5, that is, mean 2/3. The Laplace transform of the exponential distribution is $\tilde{f}(s) = (1 + 1.5/s)^{-1}$.

10.7.3 Suppose a system has two types of components; the life of each is exponentially distributed with failure rates 2 and 3, respectively. The system fails with probability 1/4 if type 1 component fails, and it fails with probability 3/4 if type 2 component fails. If the epochs of failures form a renewal process, find the long run failure rate.

Solution: The system has two types of components, life of each is exponentially distributed with failure rates $\lambda_1 = 2$ and $\lambda_2 = 3$, respectively. The system fails with probability $p = 1/4$, if type 1 component fails and it fails with probability $1 - p = 3/4$, if type 2 component fails. Thus, the time to failure of the system has a mixture distribution with probability density function given by $f(x) = p\lambda_1 e^{-\lambda_1 x} + (1 - p)\lambda_2 e^{-\lambda_2 x}, x > 0$. The mean failure time is $p/\lambda_1 + (1 - p)/\lambda_2$. Thus, the long run failure rate is $\lambda_1\lambda_2/(\lambda_2 p + (1 - p)\lambda_1) = 8/3$.

10.7.4 Mr. Sunil changes jobs frequently. On average, he works with one company for 2 years. The average duration of time he does not have a job is 2 months. In the long run, what proportion of time is Sunil working?

Solution: The solution is similar to Example 10.3.3. Suppose Sunil works for a random amount U of time having an average of 2 years. Once he quits, he does not have a job for random time V, with average 2/12 years. Thus, if $Y(t)$ denotes the state of Sunil at time t, it is 1 if he is working and is 0 if he does not have a job. We assume that U and V are independent random variables. Suppose Sunil is in state 1 at time 0 and $X(t)$ denotes the number of visits to state 1 in $(0, t]$. Thus, the random interval T_i between $(i-1)$th and ith visits to state 1 is distributed as $U + V$, with average $2 + 1/6 = 13/6$. Further, in view of independence of U_i's and V_i's it follows that T_i's are also independent. Thus, $\{X(t), t \geq 0\}$ is a renewal process with inter-renewal times $\{T_i, i \geq 1\}$ with common mean $\mu = 13/6$. By Theorem 10.3.1, $\lim_{t\to\infty} X(t)/t = 6/13$ a.s. Thus, the long run rate at which Sunil is in job is $6/13 = 0.4615$ years.

10.7.5 (i) Mr. Anil replaces the battery in his hearing aid as soon as it gets discharged. Suppose $X(t)$ is the number of batteries replaced during the first t hours of the life of the machine, not counting the one that was installed at the purchase. Assume that the lifetime T_n of the nth battery for $n \geq 1$ are independent and identically distributed random variables each having average life of 5 hours. Find the long run rate at which the batteries will be replaced.
(ii) In order to avoid the inconvenience when the battery gets discharged, Mr. Anil adopts the policy to replace the battery once it is used for 4 hours, even if it has not failed yet and also upon failure. Suppose $X_1(t)$ is the number of batteries replaced up to time t, planned or unplanned. Show that $\{X_1(t), t \geq 0\}$ can be modeled as a renewal process. Find the long run rate the battery will be replaced in such a strategy. Compare this long run rate with the rate of replacing the battery when it gets discharged.
Solution: The solution is similar to Example 10.3.2.

10.7.6 Suppose $\{X(t), t \geq 0\}$ is a renewal process where the probability density function of inter-renewal distribution is given by $f(x) = (x/100)e^{-x/10}$ for $x > 0$. (i) Find the long run renewal rate. (ii) Examine whether the renewal function is given by $M(t) = t/20 - 1/4 + (1/4)e^{-t/5}$.
(iii) Verify the elementary renewal theorem and the key renewal theorem. (iv) Find a suitable normalization, so that the distribution of normalized $X(t)$ can be approximated by the standard normal distribution for large t.
Solution: Observe that the probability density function of T_1 can be expressed as $f(x) = (1/10)^2(\Gamma(2))^{-1}e^{-x/10}x^{2-1}$. Thus, $T_1 \sim G(\alpha, 2)$ with $\alpha = 1/10$. (i) Thus, the mean renewal time is $\mu = 20$. Hence, the long run renewal rate is 1/20. (ii) The solution is similar to that given in Example 10.2.5.
(iii) Note that

$$\lim_{t\to\infty} M(t)/t = \lim_{t\to\infty} 1/20 - 1/4t + 1/(4te^{t/5}) = 1/20.$$

By the elementary renewal theorem, $\lim_{t\to\infty} M(t)/t = 1/\mu = 1/20$ and the two limits are the same. Observe that

$$\begin{aligned} M(t+10) - M(t) &= ((t+10)/20 - 1/4 + (1/4)e^{-(t+10)/5}) \\ &- (t/20 - 1/4 + (1/4)e^{-t/5}) \\ &= 1/2 + (1/4)e^{-t/5}(e^{-2} - 1) \\ &\to 1/2 \text{ as } t \to \infty. \end{aligned}$$

By the key renewal theorem,

$$\lim_{t\to\infty} M(t+10) - M(t) = 10/\mu = 1/2$$

and thus the key renewal theorem is verified.
(iv) By the central limit theorem, if $\{X(t), t \geq 0\}$ is a renewal process with μ and σ^2 as the mean and variance respectively of the inter-renewal distribution, then for large t, distribution of $(X(t) - t/\mu)/\sqrt{t\sigma^2/\mu^3}$ is approximated by the standard normal distribution. Here $\mu = 20$ and $\sigma^2 = 200$. Thus, for large t, distribution of $(X(t) - t/20)/\sqrt{t/40}$ is approximated by the standard normal distribution.

10.7.7 Suppose $X(t)$ denotes the number of vehicles passing through a certain intersection. Suppose $\{X(t), t \geq 0\}$ is modeled as a renewal process where the inter-renewal distribution is uniform $U(0, 2)$, time unit being minutes. Find the approximate probability that the number of vehicles passing through that intersection is (i) larger than 560, (ii) smaller than 620 in the time period 9 a.m. to 7 p.m.
Solution: The inter-renewal distribution is uniform $U(0, 2)$, hence its mean is $\mu = 1$ and variance $\sigma^2 = 1/3$. The time period 9 a.m. to 7 p.m. is 10 hours, that is, 600 minutes. If we take 9 a.m. as the origin, that is, $t = 0$, then we have to find the probability that $P[X(600) > 560]$ and $P[X(600) < 620]$. Since $t = 600$ is large, we use the central limit theorem to compute the probabilities approximately. Thus, the distribution of $X(600)$ is approximated by normal $N(\theta, v)$, where mean $\theta = t/\mu = 600$ and variance $v = t\sigma^2/\mu^3 = 200$. Suppose $Z \sim N(0, 1)$. Then,

$$P[X(600) > 560] = P[Z > -2.8284] = 0.9976$$
$$P[X(600) < 620] = P[Z < 1.4142] = 0.9213.$$

Answers to the multiple choice questions, based on Chap. 10, are given in Table A.12.

Table A.12 Answer key to MCQs in Chap. 10

Q. No.	1	2	3	4	5	6	7	8	9	10	11	12
Ans	c	a	c	b	b	b	c	d	c	b, c	c	a

Index

S. Madhira and S. Deshmukh, *Introduction to Stochastic Processes Using R*,
https://doi.org/10.1007/978-981-99-5601-2

MIX
Papier aus verantwortungsvollen Quellen
Paper from responsible sources
FSC® C105338

If you have any concerns about our products,
you can contact us on
ProductSafety@springernature.com

In case Publisher is established outside the EU,
the EU authorized representative is:
Springer Nature Customer Service Center GmbH
Europaplatz 3, 69115 Heidelberg, Germany

Printed by Libri Plureos GmbH
in Hamburg, Germany